WITHDRAWN

Advances in
# THE STUDY OF BEHAVIOR
VOLUME 25

# Advances in
# THE STUDY OF BEHAVIOR

*Edited by*

PETER J. B. SLATER

JAY S. ROSENBLATT

CHARLES T. SNOWDON

MANFRED MILINSKI

# Parental Care: Evolution, Mechanisms, and Adaptive Significance

*A Volume in*
Advances in
THE STUDY OF BEHAVIOR

VOLUME 25

*Edited by*

JAY S. ROSENBLATT
*Institute of Animal Behavior
Behavioral and Neural Sciences Program,
and Psychology Department
Rutgers University
Newark, New Jersey*

CHARLES T. SNOWDON
*Department of Psychology
University of Wisconsin
Madison, Wisconsin*

ACADEMIC PRESS
San Diego  London  Boston  New York  Sydney  Tokyo  Toronto

This book is printed on acid-free paper. ∞

Copyright © 1996 by ACADEMIC PRESS

All Rights Reserved.
No part of this publication may be reproduced or transmitted in any form or by any means, electronic or mechanical, including photocopy, recording, or any information storage and retrieval system, without permission in writing from the publisher.

Academic Press, Inc.
525 B Street, Suite 1900, San Diego, California 92101-4495, USA
http://www.apnet.com

Academic Press Limited
24-28 Oval Road, London NW1 7DX, UK
http://www.hbuk.co.uk/ap/

International Standard Serial Number: 0065-3454

International Standard Book Number: 0-12-004525-7

PRINTED IN THE UNITED STATES OF AMERICA
96  97  98  99  00  01  QW  9  8  7  6  5  4  3  2  1

# Contents

| | |
|---|---|
| Contributors | xiii |
| Preface | xv |
| Introduction | xvii |

## PART I

# PARENTAL CARE AMONG INVERTEBRATES AND EARLY VERTEBRATES

### Parental Care in Invertebrates
#### STEPHEN T. TRUMBO

| | | |
|---|---|---|
| I. | Introduction | 3 |
| II. | Revisiting the Prime Movers of Care | 4 |
| III. | Physiology of Care | 6 |
| IV. | Parental Care Theory and Invertebrates | 19 |
| V. | Paternal Care: Uniparental Male versus Biparental Care | 33 |
| VI. | Summary | 39 |
| | References | 40 |

### Cause and Effect of Parental Care in Fishes: An Epigenetic Perspective
#### STEPHEN S. CRAWFORD AND EUGENE K. BALON

| | | |
|---|---|---|
| I. | Introduction | 53 |
| II. | The Diversity of Parental Care in Fishes | 55 |
| III. | A General Model of Life History in Fishes | 68 |
| IV. | Energy Requirements for Reproduction | 72 |
| V. | Definitions of Parental Care | 77 |
| VI. | A Review of Recent Work on Parental Care in Fishes | 81 |
| VII. | The Epigenesis of Piscine Parental Care | 90 |
| VIII. | Conclusions | 98 |
| | References | 100 |

### Parental Care among the Amphibia
#### MARTHA L. CRUMP

| | |
|---|---:|
| I. Introduction | 109 |
| II. Phylogenetic Distribution of Parental Care | 111 |
| III. Modes of Parental Care: Occurrence and Function | 113 |
| IV. Geographic Distribution of Parental Care | 122 |
| V. Distribution of Parental Care between the Sexes | 127 |
| VI. Correlates of Parental Care | 128 |
| VII. Evolution of Parental Care | 130 |
| VIII. Flexibility in Parental Care | 136 |
| IX. Summary | 138 |
| References | 140 |

### An Overview of Parental Care among the Reptilia
#### CARL GANS

| | |
|---|---:|
| I. Introduction | 145 |
| II. The Reptilian Diversity | 146 |
| III. Structural Types of Parental Investment | 148 |
| IV. Physiological Types of Parental Investment | 150 |
| V. Behavioral Types of Parental Investment | 152 |
| VI. Opportunities for Study | 153 |
| VII. Summary | 154 |
| References | 155 |

## PART II

# ENDOCRINE, NEURAL, AND BEHAVIORAL FACTORS GOVERNING PARENTAL CARE AMONG MAMMALS AND BIRDS

### Neural and Hormonal Control of Parental Behavior in Birds
#### JOHN D. BUNTIN

| | |
|---|---:|
| I. Introduction | 161 |
| II. Incubation Behavior | 162 |

| III. Defense of the Nest and Young | 179 |
|---|---|
| IV. Parental Responses toward Young | 182 |
| V. Summary | 199 |
| References | 202 |

## Biochemical Basis of Parental Behavior in the Rat
### ROBERT S. BRIDGES

| I. Behavioral Responses of Parental Rats | 215 |
|---|---|
| II. Parental Behavior—A Developmental Perspective | 215 |
| III. Physiological Underpinnings of Parental Behavior | 217 |
| IV. Neurochemical Elements and Experiential Aspects of Parental Care | 233 |
| V. Overview: A Model for the Biochemical Regulation of Parental Care | 235 |
| References | 237 |

## Somatosensation and Maternal Care in Norway Rats
### JUDITH M. STERN

| I. Introduction | 243 |
|---|---|
| II. Somatosensation and Maternal Behavior | 244 |
| III. Trigeminal Somatosensation and Maternal Nurturance | 252 |
| IV. Ventral Trunk Somatosensation and Nursing Behavior | 267 |
| V. Somatosensation and Maternal Aggression | 275 |
| VI. Neurobiological Consequences and Implications | 277 |
| VII. General Conclusions and Summary | 286 |
| References | 288 |

## Experiential Factors in Postpartum Regulation of Maternal Care
### ALISON S. FLEMING, HYWEL D. MORGAN, AND CAROLYN WALSH

| I. Introduction | 295 |
|---|---|
| II. Types of Experience | 296 |
| III. Functional Adaptiveness of Experience in the Service of "Good" Mothering | 297 |
| IV. The Physiology of Maternal Behavior | 299 |
| V. Conclusion | 323 |
| References | 326 |

Maternal Behavior in Rabbits: A Historical and
Multidisciplinary Perspective

GABRIELA GONZÁLEZ-MARISCAL
AND JAY S. ROSENBLATT

| | |
|---|---|
| I. Introduction: Why Are Rabbits Interesting to Study? | 333 |
| II. Studying Rabbit Maternal Behavior in the Laboratory: What Can We Measure? | 336 |
| III. Participation of Estradiol, Progesterone, and Prolactin in the Initiation of Maternal Behavior | 341 |
| IV. Maintaining Maternal Behavior after Parturition: The Importance of the Interaction between Mother and Young | 348 |
| V. Recapitulation and Discussion | 351 |
| VI. Summary | 355 |
| References | 356 |

Parental Behavior in Voles

ZUOXIN WANG AND THOMAS R. INSEL

| | |
|---|---|
| I. Introduction | 361 |
| II. Parental Behavior and Social Organization | 361 |
| III. Patterns of Parental Behavior | 363 |
| IV. Species Differences in the Rate of Litter Development | 366 |
| V. Environmental Factors Regulating Parental Behavior | 369 |
| VI. Neurobiological Basis of Parental Behavior | 371 |
| VII. Summary | 380 |
| References | 381 |

Physiological, Sensory, and Experiential Factors
of Parental Care in Sheep

F. LÉVY, K. M. KENDRICK, E. B. KEVERNE, R. H. PORTER,
AND A. ROMEYER

| | |
|---|---|
| I. Introduction | 385 |
| II. The Control of Maternal Reponsiveness to the Neonate | 387 |
| III. The Control of Maternal Selectivity to the Neonate | 397 |

IV. Conclusion.......................................... 411
V. Summary............................................ 415
References............................................ 416

Socialization, Hormones, and the Regulation of Maternal Behavior in Nonhuman Simian Primates

CHRISTOPHER R. PRYCE

I. Introduction......................................... 423
II. The "Context" of Maternal Behavior Regulation....... 426
III. Preadult Socialization and Maternal Behavior......... 432
IV. Hormones and Maternal Behavior.................... 441
V. A Socialization–Neuroendocrine Model............... 460
VI. Conclusions......................................... 462
VII. Summary............................................ 463
References............................................ 465

PART III

# FUNCTIONAL, ECOLOGICAL, AND ADAPTIVE ASPECTS OF PARENTAL CARE

Field Studies of Parental Care in Birds: New Data Focus Questions on Variation among Females

PATRICIA ADAIR GOWATY

I. Introduction......................................... 477
II. What Do Males Do? What Do Females Do?........... 479
III. Male Parental Care and Female Fitness: The Adaptive Significance of Social Monogamy..................... 487
IV. Sexual Conflict and Differential Parental Allocation.... 494
V. Extra-Pair Paternity................................. 496
VI. Genetic Paternity and Paternal Care.................. 507
VII. Maternal Care....................................... 516
VIII. Adaptive Significance of Parental Care in Birds........ 517
IX. The Future.......................................... 521
X. Summary............................................ 522
References............................................ 523

## Parental Investment in Pinnipeds
### FRITZ TRILLMICH

| | |
|---|---|
| I. Introduction | 533 |
| II. Background Information on Pinniped Systematics and Phylogeny | 535 |
| III. Parental Care Patterns in Pinnipeds | 536 |
| IV. Evidence for a Cost of Reproduction | 546 |
| V. Evidence for Differential Investment in Male and Female Offspring | 555 |
| VI. Postweaning Investment? | 566 |
| VII. Conclusion | 567 |
| References | 570 |

## Individual Differences in Maternal Style: Causes and Consequences for Mothers and Offspring
### LYNN A. FAIRBANKS

| | |
|---|---|
| I. Introduction | 579 |
| II. Individual Differences in Maternal Style | 580 |
| III. Consequences of Variation in Maternal Style for the Mother | 591 |
| IV. Consequences of Variation in Maternal Style for the Offspring | 593 |
| V. Proximate Mechanisms | 602 |
| VI. Conclusions | 604 |
| References | 606 |

## Mother–Infant Communication in Primates
### DARIO MAESTRIPIERI AND JOSEP CALL

| | |
|---|---|
| I. Introduction | 613 |
| II. Vocal Communication | 614 |
| III. Visual and Tactile Communication | 626 |
| IV. Parent–Infant Communication in Evolutionary Perspective | 633 |
| References | 635 |

## Infant Care in Cooperatively Breeding Species
### CHARLES T. SNOWDON

| | |
|---|---:|
| I. Introduction | 643 |
| II. How Do Helpers Help? | 644 |
| III. What Are the Factors That Influence Helping Behavior? | 661 |
| IV. What Benefits Do Helpers Receive? | 667 |
| V. What Mechanisms Inhibit Reproduction by Helpers? | 675 |
| VI. Future Research Directions | 680 |
| VII. Summary | 681 |
| References | 683 |
| | |
| *Index* | 691 |
| *Contents of Previous Volumes* | 713 |

# Contributors

Numbers in parentheses indicate the pages on which the authors' contributions begin.

EUGENE K. BALON (53), *Institute of Ichthyology and Department of Zoology, University of Guelph, Guelph, Ontario, Canada N1G 2W1*

ROBERT S. BRIDGES (215), *Department of Comparative Medicine, Tufts University School of Veterinary Medicine, North Grafton, Massachusetts 01536*

JOHN D. BUNTIN (161), *Department of Biological Sciences, University of Wisconsin-Milwaukee, Milwaukee, Wisconsin 53201*

JOSEP CALL (613), *Department of Psychology and Yerkes Regional Primate Research Center, Emory University, Atlanta, Georgia 30322*

STEPHEN S. CRAWFORD (53), *Institute of Ichthyology and Department of Zoology, University of Guelph, Guelph, Ontario, Canada N1G 2W1*

MARTHA L. CRUMP (109), *Department of Biological Sciences, Northern Arizona University, Flagstaff, Arizona 86011*

LYNN A. FAIRBANKS (579), *Department of Psychiatry & Biobehavioral Sciences, University of California, Los Angeles, California 90024*

ALISON S. FLEMING (295), *Department of Psychology, Erindale College, University of Toronto, Mississauga, Ontario, Canada N1G 2W1*

CARL GANS (145), *Department of Biology, University of Michigan, Ann Arbor, Michigan 48103*

GABRIELA GONZÁLEZ-MARISCAL (333), *Centro de Investigación en Reproducción Animal, CINVESTAV-Universidad Autónoma de Tlaxcala, Tlaxcala, Tlax. 90,000, Mexico*

PATRICIA ADAIR GOWATY (477), *Institute of Ecology, University of Georgia, Athens, Georgia 30602*

THOMAS R. INSEL (361), *Department of Psychiatry and Behavioral Sciences, Emory University, Atlanta, Georgia 30322*

K. M. KENDRICK (385), *The Babraham Institute, Babraham, Cambridge, England*

E. B. KEVERNE (385), *Sub Department of Animal Behaviour, Madingley, Cambridge, England*

F. LÉVY (385), *Laboratoire de Comportement Animal, INRA/CNRS, URA 1291, Nouzilly, France*

DARIO MAESTRIPIERI (613), *Department of Psychology and Yerkes Regional Primate Research Center, Emory University, Atlanta, Georgia 30322*

HYWEL D. MORGAN (295), *Department of Psychology, Erindale College, University of Toronto, Mississauga, Ontario, Canada N1G 2W1*

R. H. PORTER (385), *Laboratoire de Comportement Animal, INRA/CNRS, URA 1291, Nouzilly, France*

CHRISTOPHER R. PRYCE (423), *Department of Anthropology, University of Auckland, Auckland, New Zealand*

A. ROMEYER (385), *Centro de Investigación en Reproducción Animal, CINVESTAV-Universidad Autónoma de Tlaxcala, Tlaxcala, Tlax. 90,000, Mexico*

JAY S. ROSENBLATT (333), *Institute of Animal Behavior, Rutgers University, Newark, New Jersey 07102*

CHARLES T. SNOWDON (643), *Department of Psychology, University of Wisconsin, Madison, Wisconsin 53706*

JUDITH M. STERN (243), *Department of Psychology, Rutgers University, New Brunswick, New Jersey 08903*

FRITZ TRILLMICH (533), *Lehrstuhl für Verhaltensforschung, Universität Bielefeld, D-33501 Bielefeld, Germany*

STEPHEN T. TRUMBO (3), *Department of Ecology and Evolutionary Biology, University of Connecticut, Waterbury, Connecticut 06710*

CAROLYN WALSH (295), *Department of Psychology, Erindale College, University of Toronto, Mississauga, Ontario, Canada N1G 2W1*

ZUOXIN WANG (361), *Department of Psychiatry and Behavioral Sciences, Emory University, Atlanta, Georgia 30322*

# Preface

Since its first volume in 1965, *Advances in the Study of Behavior* has had the policy of publishing articles on varied topics in each volume. Volumes were eclectic and often had articles ranging in subject matter from insect behavior to human mother–infant interactions. The series became known for the high quality of its contributions and timeliness of its coverage of new developments in the field.

The present twenty-fifth volume of this series represents a departure from that specific policy, but one which, we believe, does not violate the principles underlying this series. This is the first volume in which all of the articles are on a specific theme, namely, parental care. We chose this specific area of research because important advances are occurring across a broad taxonomic range, where researchers are using a wide variety of approaches to study this important aspect of behavior: ecological, functional, neuroendocrine, neurophysiological, psychological, developmental, social, and evolutionary. We invited as contributors leading researchers in the study of parental care, representing the broad taxonomic range and variety of approaches.

This volume, therefore, presents coverage of this area of research by those engaged in the most advanced research. To enable us to focus periodically on a specific area of behavior study, *Advances in the Study of Behavior* will periodically publish thematic volumes such as this one, and will continue to publish eclectic volumes such as the previous ones. We believe that this policy will enable us to add depth to our coverage of advances in the study of behavior without sacrificing the breadth for which this series is known.

Jay S. Rosenblatt
Charles T. Snowdon

# Introduction

Advances are occurring in the study of parental care across a wide range of taxonomic groups and using a variety of approaches, in recognition of the important role parental behavior plays in ontogeny, development, reproduction, social behavior, and evolution. This volume presents the overall contour of this research: its coverage is broader and more varied than previous volumes on this topic despite the obvious omission of human parental care, which would require a volume of its own, and the still limited number of animal species that have been studied experimentally. For the first time, we have included adequate coverage of parental care in invertebrates and early vertebrates, fishes, amphibia, and reptiles, in the same volume with parental care in mammals and birds, including several chapters on nonhuman primates. There is, we believe, an inherent fascination in studies of parental care in a wide variety of taxonomic groups. Equally important, however, is our belief that parental care needs to be studied in an evolutionary framework and this requires broad coverage of parental care from many points of view.

The early chapters of this volume provide descriptions and classifications of types of parental care in invertebrates (Trumbo), fishes (Crawford and Balon), amphibians (Crump), and reptiles (Gans). The variety shown is bewildering but in each case parental care is anchored, on the one hand, to the ecology of the species, and on the other, to the reproductive physiology and embryonic development of the species. The ingenuity of these adaptations of parental care to these niches is impressive. It is clear, as proposed by Crawford and Balon, that the evolution of parental care in these taxa is driven by the requirement of inclusive fitness to provide optimal conditions for the embryonic development of the offspring in the various environments. One theme that runs through these chapters concerns the physiological and behavioral adaptations required for the transition in reproduction from aquatic to terrestrial–arboreal ecological niches. In this evolution, we see how physiological and behavioral mechanisms of parental care leapfrog over one another: behavioral mechanisms supplement physiological mechanisms and are then replaced by more advanced physiological mechanisms that, in turn, require and give rise to more advanced behavioral mechanisms. It is difficult to separate the two, nor is there need to. A second theme is the increase in the duration of parental care, with parents attending the young for longer periods of their embryonic and posthatching development. One consequence of this development is the increasingly complex physiological and behavioral mechanisms required of the parent(s) during advanced stages of offspring development. In the separate chapters these and additional themes are traced in these four taxonomic groups.

Studies of the mechanisms of parental care have focused on the general question of what makes parents responsive to their young. How is parental care initiated? Which stimuli do parents respond to and how do these change over the course of infant development? How is parental care maintained and eventually terminated? Experimental studies of these questions have mainly focused on mammals and birds, where parental care is universally established. The chapters on mechanisms of parental care describe elegant laboratory work with a variety of birds, with rats, voles, rabbits, and sheep, and, more recently, with the common marmoset, a nonhuman primate. Different degrees of progress in understanding the interplay of behavioral, hormonal, and neural mechanisms have been achieved. Research on mechanisms of parental care has been most clearly delineated in studies on rats and we are fortunate to have three leading researchers (Bridges on the biochemical mechanisms, Stern on sensory mechanisms, and Fleming *et al.* on experiential factors) contribute to this volume.

Chapters on voles and rabbits describe specialized aspects of parental care in these species that differ in interesting ways from the rat. In the vole Wang and Insel describe the role of experience in maternal behavior as well as the neuroendocrine differences between males and females in monogamous species with biparental care and in polygynous species where maternal care is predominant. In the rabbit, González-Mariscal and Rosenblatt show that postpartum parental care is minimal, consisting of a brief nursing once a day, whereas prepartum preparatory nest building is highly developed. Lévy *et al.* have described a rich, multidisciplinary body of research on mechanisms of maternal care of highly precocious young in sheep, with emphasis on the roles of both mothers and infants in successful care. Despite considerable research on maternal–infant relationships in nonhuman primates, there has been surprisingly little work on the hormonal and motivational aspects of maternal care. Pryce has carried out the first descriptive and experimental studies of hormonal and behavioral induction of maternal care in nonhuman primates. In birds we find the greatest variety of patterns of parental care: Buntin provides a detailed review of research seeking the hormonal basis for egg incubation, posthatch brooding, and parental feeding in both altricial and precocial species.

Questions about the adaptive significance of parental care, the variation that exists among species in which parents provide care and where there are nonreproductive helpers to assist them with infants, as well as individual differences in parental care are dealt with in the third section of this volume. Understanding the importance of variation in infant care often requires field studies or studies in socially complex captive groups. Gowaty describes field studies of socially monogamous birds in which the distribution of parental care among individuals varies with the high number of extrapair fertilizations. She introduces a novel concept that focuses on the constraints on the female's reproductive success, rather than the usual focus on the male's, to understand variation in parental care. Trillmich uses pinnipeds, in which paternal care is generally absent and maternal care consists mainly of providing high energy milk, as an extreme system for testing hypotheses derived from parental investment theory. Mother–infant interactions among nonhuman primates have been studied by several groups represented in this volume. The work of Fairbanks is noteworthy for its attention to stable individual differences in maternal style: she has studied how these arise and their consequences for mothers and their infants. In particular she documents the consequences of different maternal styles on reproductive outcomes of both mothers and infants. Maestripieri and Call describe their studies on the importance of multimodal communication between mothers and their infants. Finally, Snowdon reviews his research on cooperatively breeding nonhuman primates, and he raises questions concerning helpers in these species and in cooperatively breeding birds and other mammals. Among the questions are: what are the behavioral and physiological mechanisms that influence helpers, do helpers contribute to infant survival, and what benefits do they receive by deferring their own reproduction?

By presenting in a single volume topics such as the evolution of parental care throughout the animal kingdom, the physiological and behavioral mechanisms that motivate and guide this behavior throughout the parental care cycle, and the adaptive significance of parental care, we hope we have provided a springboard for further research and theory. Theories about the evolution of parent care within taxa, particularly among early vertebrates, have been presented, but the evolution of parental care viewed across taxa has received little attention. Further research on the physiological and behavioral mechanisms underlying parental care needs to expand the database to include a wider variety of mammals and birds from different taxa before any generalizations can be formulated. Especially among birds and mammals (though not exclusively), issues of adaptive significance of parental care are most complex but are very significant for understanding the evolution of parental care. The problem of collecting valid data on these issues vies with the problem of collecting precise data, suggesting that both field and laboratory studies of animal groups must be employed in strategic ways. The chapters in this section of the volume testify to the continuing progress which is being made in collecting the data needed to test theories about adaptive significance and the increasing use of multidisciplinary research to accomplish this.

Jay S. Rosenblatt<br>
Charles T. Snowdon

# PART I
# PARENTAL CARE AMONG INVERTEBRATES AND EARLY VERTEBRATES

# Parental Care in Invertebrates

Stephen T. Trumbo

DEPARTMENT OF ECOLOGY AND EVOLUTIONARY BIOLOGY
UNIVERSITY OF CONNECTICUT
WATERBURY, CONNECTICUT 06710

## I. Introduction

Social behavior among invertebrates ranges from the asocial through subsocial (parent–offspring association) to the most complex of eusocial societies (overlap of generations, reproductive division of labor, cooperative brood care) (E. O. Wilson, 1971). The taxonomic, physiological, and ecological diversity provide ample material for comparative studies of behavior. Because parental behavior is widely dispersed taxonomically, the invertebrates represent numerous independent experiments in the evolution of the parental lifestyle. Comparative analyses should thus permit invertebrate biologists to address several questions that are inaccessible to vertebrate biologists. Unfortunately, our present understanding of parental care in invertebrates is limited. This has occurred both because of the overwhelming diversity of invertebrates and because subsocial behavior often was studied as a pretext for understanding advanced sociality.

The natural history and ecological correlates of invertebrate parental behavior have been adequately covered by previous reviewers. Excellent compilations of the natural history of parental care (with an emphasis on the insects) appear in E. O. Wilson (1971), Hinton (1981), Eickwort (1981), and in Preston-Mafham and Preston-Mafham (1993). Since E. O. Wilson's (1975) interest in the evolutionary prime movers of parental behavior, the ecological and behavioral correlates of parental care among invertebrates have also been reviewed extensively (R. L. Smith, 1980; Thornhill and Alcock, 1983; Tallamy, 1984; Zeh and Smith, 1985; Tallamy and Wood, 1986; Tallamy, 1994). After revisiting these prime movers, I review two additional areas: the physiological and behavioral mechanisms that control the onset, intensity, and termination of parental care; and the use of invertebrates to address parental care theory. This approach is taken to demonstrate that mechanistic studies of invertebrate parental behavior will provide insight as well as experimental tools for ecologists, to fill gaps in the coverage of invertebrates, and to reveal the potential for the use of invertebrate models in tests of parental care theory. Taxonomically, less emphasis will be given to the Hymenoptera and Isoptera because these groups have been reviewed extensively in works addressing eusociality, and to the marine taxa because of the lack of information beyond their natural history.

## II. REVISITING THE PRIME MOVERS OF CARE

Seminal works on the social insects and social behavior by E. O. Wilson (1971, 1975) stimulated a burst of experimental studies of parental care. Wilson detailed four ecological pressures (prime movers) that select for parental care: stable and structured environments, harsh environments, scarce and specialized food resources, and predation. A consequence of the promulgation of the prime movers was that empirical studies were quickly undertaken to identify prime functions. Search for prime functions also was encouraged by the notion that the causes of parental care in noneusocial species would be straightforward. In most parental invertebrates, particularly those that nest or carry their offspring, intensive study has revealed that caregiving has multiple functions. This finding suggests that parental care involves a suite of evolutionary changes as well as the loss of adaptations suited to the nonparental lifestyle (some of these secondary changes occurring long after the origin of parental care). The identification of the prime cause of parental care, therefore, will be difficult in many cases.

Three examples will demonstrate the complex functions of care in many invertebrates. The salt-marsh beetle, *Bledius spectabalis,* often is given as an example of an organism in which parental care evolved to cope with a harsh environment. Indeed, Wyatt (1986) elegantly demonstrated that maintenance of a wine-bottle-shaped burrow prevented rapid flooding and anoxia in the intertidal environment. Subsequent investigation has revealed that parental care serves the additional functions of provisioning young with algae, keeping the burrow mold-free, and defending against carabid predators and ichneumonid wasp parasitoids (Wyatt, 1986; Wyatt and Foster, 1989a,b).

Rudolf Diesel has enumerated multiple functions of care in the Jamaican land crab, *Metopaulias depressus,* for which it is suspected that the evolutionary prime mover is an inhospitable environment. *Metopaulias* lives in water that collects in the axils of epiphytic bromeliads. A water-filled axil serves as a nursery during a 9-week developmental period (Diesel, 1989). Untended bromeliads are unsuitable for crab larvae because leaf debris both reduces dissolved $O_2$ and lowers pH in the nursery. The maternal *Metopaulias* regularly removes debris from leaf axils, raising dissolved $O_2$.

The larval environment is further modified by the addition of snail shells that raises pH and increases $Ca^{2+}$ availability (Diesel, 1992; Diesel and Schuh, 1993). The mother also captures prey and brings it to the nursery, and protects young from spider and nymphal damselfly predators.

The prime mover of parental care in burying beetles (*Nicrophorus* spp.) is thought to be the exploitation of a scarce and valuable resource, small vertebrate carrion. In burying beetles, male–female pairs secure the carcass underground and defend it from intraspecific and interspecific competitors. Parents control the decomposition by removing hair or feathers, rounding the carcass into a ball, and applying antimicrobial secretions (Pukowski, 1933; Halffter et al., 1983). Shortly before hatching of young the female opens a small hole in the uppermost part of the carrion ball. This provides access to the carcass interior for feeding by first instar larvae. Parents supplement feeding by regurgitating liquified carrion. If the food supply is not sufficient to support the entire brood, parents will cannibalize a subset of the brood so that adequate resources remain for surviving young (Bartlett, 1987). Both parents will defend the brood against predators such as carabid beetles, and against infanticidal intruders (Scott, 1990; Trumbo, 1990a).

Each of these examples of complex care is more remarkable because each occurs in a genus with closely related asocial species. *Bledius* and *Metopaulias,* in fact, coexist with nonparental congeners. The multiple functions of parental care suggest that the initial evolution of parental care was followed by secondary parental adaptations that replaced adaptations with similar functions in the nonparental ancestor. For example, parental care in many groups has an egg care function; several authors have noted the greater vulnerability of untended eggs of parental as compared to nonparental species (Eberhard, 1975; Hinton, 1981; R. L. Smith, 1980; Tallamy, in press). This suggests that as caregiving came to include the function of facilitating hatching, resources previously devoted to egg viability in nonparental ancestors may have been employed for other purposes. The existence of a suite of adaptations, whether parental or nonparental, suggests that evolution in either direction may be impeded by having to cross a fitness valley. This is because a single environmental change is unlikely to favor every component of a parental or nonparental lifestyle. Clearly, however, natural selection has surmounted the barrier in many cases.

The complex function of care can make it more difficult to identify a single environmental prime mover. Experiments in which a parent is removed may disrupt many aspects of an organism's life history, and thus contribute little to our understanding of the origin of parental care. How then might one test the hypothesis that a prime mover such as a harsh environment has been important for the evolution of parental care? The numerous examples of the independent lineages that have evolved care provide one opportunity.

One of the earliest interspecific comparisons found that there was a relative increase in the number of marine species possessing nonpelagic larval development (associated with greater parental investment) in harsher environments (toward the poles and at greater depths) (Thorson, 1950; Mileikovsky, 1971). Environmental harshness, however, is a relative concept and will clearly depend on the taxa under consideration. Beetles utilizing marine tidal zones and crabs in terrestrial habitats may both be considered an invasion of a harsh environment. Further interspecific tests of this hypothesis will require *a priori* establishment of criteria for environmental challenge. Substantiation of the importance of this and other prime movers can perhaps be made more convincingly with invertebrates because of the large number of independent lineages thought to have evolved parental care, and the tremendous diversity in the level of sociality among closely related taxa.

Phylogenetic comparisons also have limits. The evolution of parental care, like the evolution of flight, is an adaptation that may permit subsequent radiation (but see Tallamy, in press). The cause and effect of associations such as that between parental care and environmental harshness may be difficult to resolve. Once parental care evolves, due to whichever prime mover, the protection afforded immature stages may permit invasion of an environment that was formerly not suited for juvenile development. In some cases it will be difficult to determine whether parental care evolved concurrently with invasion of the harsh environment, or whether parental care was a preadaptation. A similar caveat applies to each of the other three prime movers.

## III. Physiology of Care

Behavioral ecologists are concerned primarily with functional explanations of variation in behavior. A thorough ecological examination of behavioral plasticity includes information on stimuli eliciting behavioral changes, fitness consequences of adopting alternative behaviors, and a phylogenetic analysis of differences in social behavior. Knowledge of the physiology of behavioral plasticity is helpful for exploring how plasticity is related to ecology. For instance, the ability to manipulate the physiological state of an individual provides a naturalistic way to alter behavior and thereby examine correlations and trade-offs among behavioral states (Wingfield *et al.*, 1987); fitness consequences of alternative behaviors (Ketterson and Nolan, 1992; Ketterson *et al.*, 1992); and resulting changes in the behavior, hormonal state, and fitness of social partners (Wingfield *et al.*, 1990). In addition, physiological responses can be sensitive probes to determine which environmental stimuli organisms pay selective attention to when making critical life-history decisions (Dusenberry, 1992). Despite the fact that in many ways insects are the paradigmatic group for examining social plasticity (E. O. Wilson, 1971; Thornhill and Alcock, 1983), there has been little work on the physiological basis of intraspecific and interspecific diversity of this behavior.

Students of insect parental behavior can draw upon a solid understanding of the physiology of reproduction. Endocrinologists have outlined the control of reproductive maturation and behavior in several models (Koeppe *et al.,* 1985; Hagedorn, 1985). In the protocerebrum of the insect brain, the pars intercerebralis has neural input into the corpora allata (CA), the synthesis and release site of juvenile hormone (JH) (Feyereisen, 1985) (see Fig. 1). JH is the primary gonadotrophic hormone in insects (Koeppe *et al.,* 1985), and also regulates numerous adult behaviors (Robinson, 1987; Cusson and McNeil, 1989). The brain-CA-ovary axis of insects is structurally and functionally analogous to the neuroendocrine regulation of the adenohypophysis of the pituitary by the hypothalamus in vertebrates (Scharrer, 1987).

Invertebrate parents provide extended parental care in three principal ways: carrying young internally or externally, providing resources for young within nests, and tending young that are clustered near food sources. Al-

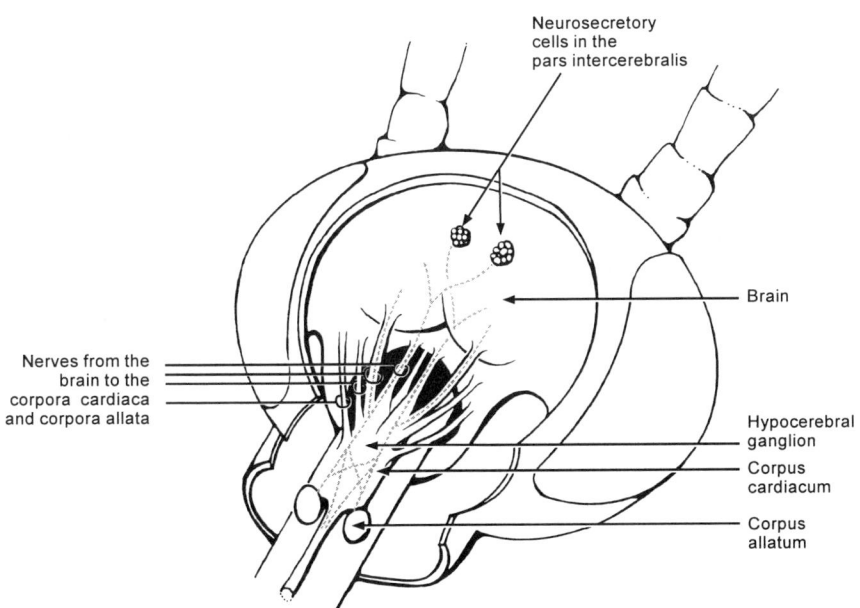

FIG. 1. Semidiagrammatic representation of the brain and endocrine tissues of insects. (Modified from Engelmann, 1970.)

though careful behavioral and ecological studies have been conducted on representatives of each of these three modes of parental care, physiological studies are limited to those that carry young and to nest builders. Here, I review the physiological correlates of reproduction and parental care in the three most intensively studied invertebrate models: the cockroach, *Diploptera punctata;* the earwigs; and the burying beetle, *Nicrophorus orbicollis.*

A. PREGNANCY IN THE VIVIPAROUS COCKROACH, *DIPLOPTERA PUNCTATA*

Cockroaches exhibit a variety of reproductive patterns including oviparity (deposition of eggs), viviparity (birth of live young), and ovoviviparity (hatching of eggs internally) (reviewed in Roth and Willis, 1960). Oviparous species with limited parental investment (e.g., *Periplaneta americana*) produce small batches of eggs continually after reaching reproductive maturity. Gonadotrophic cycles are overlapping and there is a small peak in JH synthesis prior to the onset of vitellogenesis in the first oocytes, followed by a succession of JH peaks corresponding to late vitellogenesis in basal oocytes and early vitellogenesis in penultimate oocytes (Weaver *et al.,* 1975; Feyereisen, 1985). Viviparous and ovoviviparous cockroaches, in contrast, carry young internally and exhibit more defined parental and hormonal cycles, analogous to those in vertebrates (Scharrer, 1987).

The viviparous cockroach, *Diploptera punctata,* is the best studied example of "pregnancy" in an invertebrate. *Diploptera punctata* begins gestation about 8 days after the imaginal molt. It has a 60-day pregnancy period during which a highly nutritive milk containing 45% protein and 16–22% lipid is secreted from the walls of the brood sac (Ingram *et al.,* 1977). After the imaginal molt, neural input from the brain is thought to initially inhibit the synthesis of JH in the CA (denervation results in enhanced CA activity) (Tobe and Stay, 1980, 1985). The CA of adult *Diploptera* females are thought to be released from inhibition by mating and feeding, as the CA are in other cockroaches (Gadot *et al.,* 1989a; Aclé *et al.,* 1990). A rise in JH biosynthesis is followed by vitellogenesis (Fig. 2). Initially, developing oocytes feedback positively on the CA, but once the ovary is mature, JH synthesis is inhibited. After peaking, JH synthesis declines prior to oviposition (Rankin and Stay, 1985). Ovarian ecdysteroids also increase and then decrease prior to oviposition, the peak occurring slightly after the JH peak (Stay *et al.,* 1984). The importance of changing ecdysteroid titers for regulating the reproductive cycle is not clear.

During most of gestation, synthesis of JH remains inhibited. Basal oocytes (which give rise to the second brood) are nonvitellogenic until a few days prior to parturition of the first brood (Stay and Lin, 1981). JH treatment (implantation of active CA or topical application of JH analogs) during

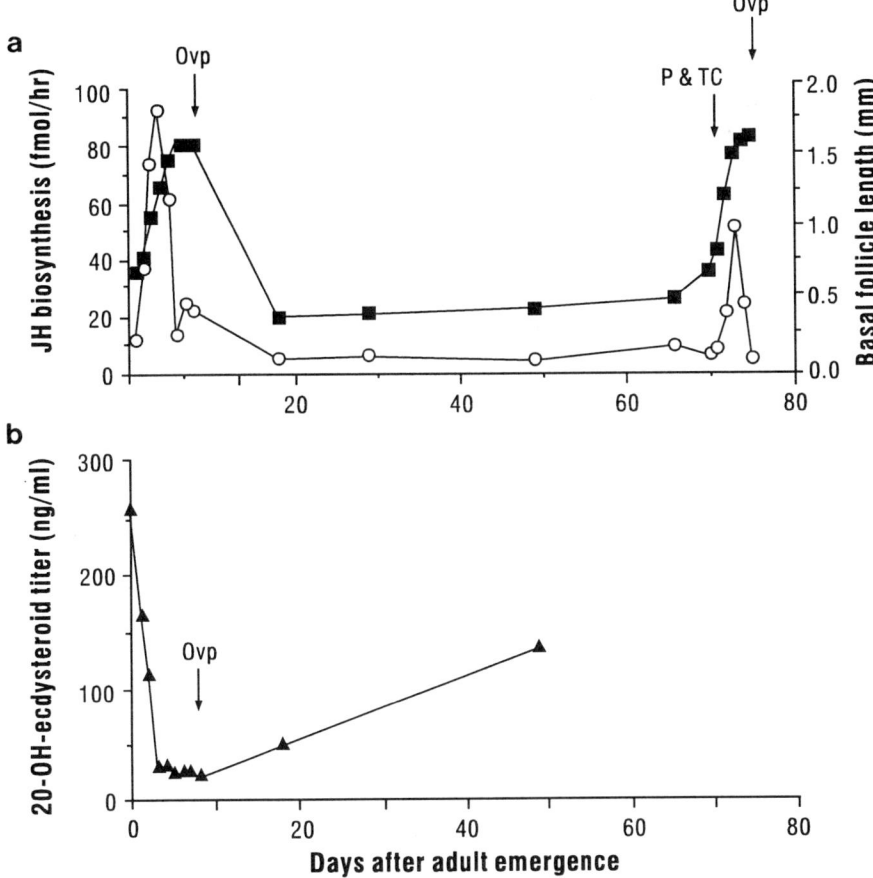

FIG. 2. (a) Changes in biosynthetic rate of juvenile hormone (JH) by the corpora allata (open circles), and length of the basal follicle (closed squares) during the pregnancy cycle of *Diploptera punctata*. Ovp = oviposition; P & TC = parturition and termination of care. (b) Blood titers of 20-OH-ecdysone during the pregnancy cycle. (Figure modified from Rankin and Stay (1985) and Stay *et al.* (1984).)

gestation causes a premature decline in protein and milk content of the brood sac (Stay and Lin, 1981), demonstrating a negative link between this hormone and parental investment. According to Rankin and Stay (1985), a low rate of JH synthesis during pregnancy is maintained by the small ovary, which cannot stimulate the CA, and by inhibition from the brain. The presence of embryos within the brood sac is apparently detected by sensory hairs and this information is relayed to the brain via the ventral nerve cord (Engelmann, 1957). Removal of embryos during late pregnancy caused CA activity to increase and accelerated the second bout of vitellogenesis and oviposition. Rankin and Stay (1985) postulated both neural and humoral inhibition of the CA, because denervation resulted in only partial release from inhibition. Neuropeptides that inhibit CA activity of *Diploptera punctata in vitro* have been isolated from both the brain and blood (Rankin and Stay, 1987; Woodhead *et al.*, 1993).

There are many variations on the mother–offspring relationship among cockroaches, only some of which have been explored at the level of physiology. The oviparous German cockroach, *Blatella germanica,* carries an egg sac externally and maintains low levels of JH synthesis during egg care in a manner similar to pregnancy in viviparous cockroaches (Gadot *et al.,* 1989b). Young of the ovoviviparous cockroach, *Leucophaea maderae,* are reported to accompany their mother on nocturnal foraging trips following parturition (Liechti and Bell, 1975). The Australian burrowing roach, *Macropanesthia rhinoceros,* and the woodroach, *Cryptocercus,* have extremely long-term familial associations (Seelinger and Seelinger, 1983; Nalepa, 1988; Ruegg and Rose, 1991; Matsumoto, 1992). Unfortunately, very little is known about the physiological correlates of these postparturition parent–offspring relationships.

B. Nest Building in Earwigs

Our most detailed account of endocrine correlates of a subsocial nest builder is of the earwigs, *Labidura riparia* and *Euborellia annulipes* (Dermaptera). Much of the research on parental behavior in this group is reported in the French literature. The account here is largely drawn from summaries and original observations contained in Lamb (1976), Baehr *et al.* (1982), Vancassel *et al.* (1984), and Rankin *et al.* (1995a,b).

There are approximately 1000 species of Dermaptera. Of the 13 species studied, all show maternal care (Lamb, 1976). After emergence as adults, females undergo a brief period of reproductive maturation and then cycle through alternating periods of ovarian development and brood care. During the sexual phase (ovaries maturing), earwigs feed (including cannibalization of young), mate, and form burrows. Ovarian development is suppressed in the parental phase, during which time the mother fasts while caring for her eggs. After hatching, the female stays with nymphs for 3–4 days, then opens the burrow, captures prey, and provisions the nest. The parental period ends when young disperse.

After emerging as an adult, the start of vitellogenesis correlates with an increase in JH titer in the blood. Neither feeding nor mating is required for this increase (Vancassel, 1973; Lamb, 1976; Vancassel *et al.*, 1984). Topical application of JH causes an earlier age of courtship but does not result in earlier mating in the ring-legged earwig, *Euborellia annulipes*

(Rankin *et al.*, 1995a). The removal of the CA prevents oviposition; CA removal accompanied by application of JH-mimics restores oviposition, demonstrating the typical role of JH in reproductive maturation (Baehr *et al.*, 1982). As in *Diploptera*, there is both a JH and ecdysteroid peak prior to oviposition. In contrast to *Diploptera*, however, *Euborellia* maintains elevated synthesis of JH at the beginning of the oviposition period, possibly because oviposition will occur over a 1- to 2-day span (Fig. 3). This suggests

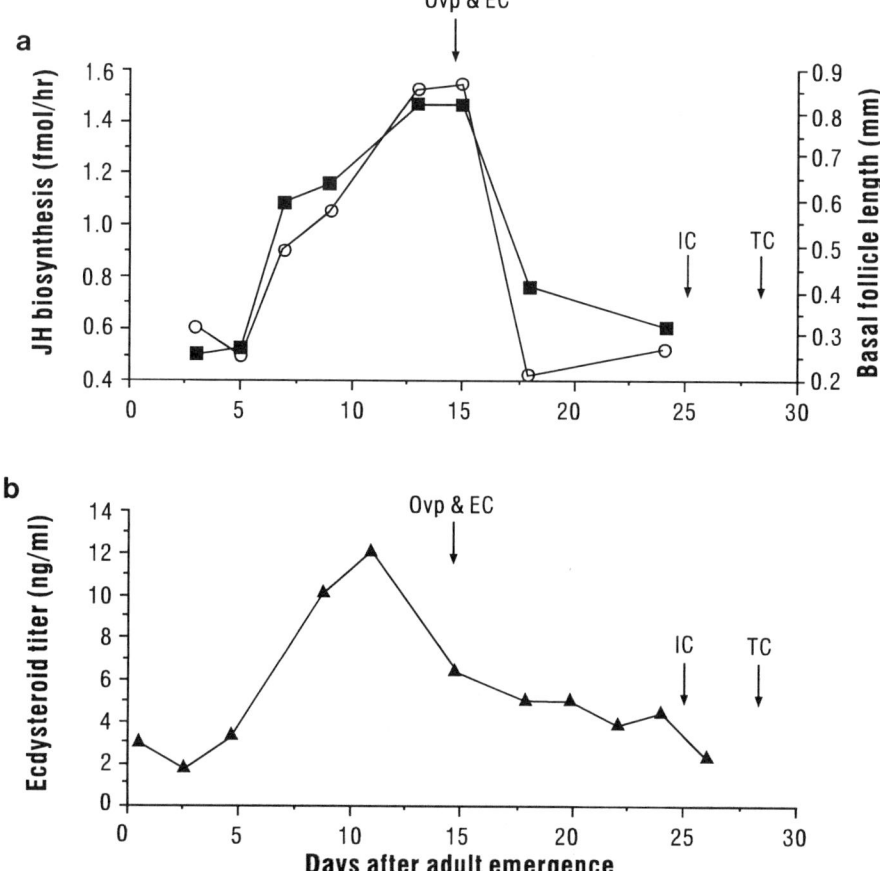

FIG. 3. (a) Changes in biosynthetic rate of juvenile hormone (JH) by the corpora allata (open circles), and length of the basal follicle (closed squares) during the parental cycle of earwigs. Ovp & EC = oviposition and egg care; IC = initiate care of nymphs; TC = terminate care. (b) Blood titers of ecdysteroids during the parental cycle. (Figure modified from Rankin *et al.* (1995) and Baehr *et al.* (1982).)

that mature ovaries do not have an immediate inhibitory effect on CA activity, as does occur in *Diploptera* (Rankin *et al.*, 1995). Topical application of JH at adult emergence results in an earlier age of first oviposition, but at the expense of a smaller clutch (Rankin *et al.*, 1995).

During parental care, blood levels of both JH and ecdysteroids decline and remain low. The ecdysone/20–hydroxyecdysone ratio changes markedly between the sexual and parental phase, but the significance of this is unclear (Vancassel *et al.*, 1991). If eggs are removed from mothers on the first day of the parental cycle, there is a short-term increase in both JH titer and oocyte size, apparently in preparation for a new sexual phase. The majority (64%) of such mothers do not accept eggs presented after a 48-h hiatus from care, and all refuse eggs after 72 h. When mothers have provided 5 days of care, however, egg removal has less dramatic effects on parental responsivity; only 11% of such mothers refused eggs after 48 h without contact (Vancassel *et al.*, 1984). It is not clear whether this difference in maternal response is caused by changes in the characteristics of maturing eggs or by the cumulative time spent providing care. Mothers normally do not hunt or eat during egg care. If food is continually provided late in the parental cycle, however, some will feed. These mothers terminate care early, have high blood levels of JH and ecdysteroids, and have premature oocyte development, all characteristic of the beginning of the next sexual phase. Conversely, enforced fasting and regular contact with young can prolong parental care (Vancassel *et al.*, 1987). Substitution of younger offspring for old also prolongs parental care (Caussanel, 1970), as occurs in parental dung beetles (Klemperer, 1983a), suggesting that offspring provide important cues that influence the endocrine state of caregivers.

In *Labidura riparia,* the parental period also can be extended by removal of the CA, presumably because a lower JH titer delays the termination of parental behavior, the reinitiation of vitellogenesis, and the onset of the next sexual phase (Pierre, 1978). In support of the role of JH in terminating parental care in *Euborellia annulipes,* Rankin *et al.* (1995a) found that topical application of JH III on the day of oviposition shortened the period of egg care, increased the probability of cannibalism, and stimulated premature ovarian maturation. Despite these findings, Baehr *et al.* (1982) do not think there is a direct connection between low levels of JH and parental care. They found that ovariectomized females had hyperactive CA and yet would accept and care for experimentally provided eggs for extended periods. Thus, they argue that parental behavior is controlled more directly by the brain. Presumably, JH treatment causes an earlier termination of care through its indirect effects on the brain via the ovaries (see Fig. 4).

Caussanel *et al.* (1978) reports neurosecretory correlates of care. Neurosecretory products of the pars lateralis and A cells in the pars intercerebralis

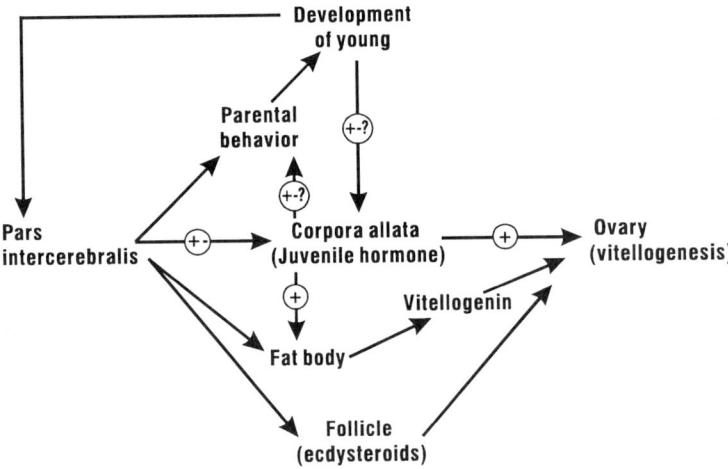

FIG. 4. Model of the regulation of reproduction and parental behavior in insects. Positive and negative influences on and from the corpora allata are shown.

decrease during the vitellogenic cycle, but are elevated during egg care. These site-specific increases in secretion occur despite an overall decrease in secretion from the pars intercerebralis during maternal care. Maternal care is also correlated with storage of neurosecretory material in the aortic walls. If the pars intercerebralis is cauterized, care is interrupted and young are cannibalized (Baehr et al., 1982).

In summary, there are several similarities between pregnancy in cockroaches and brood care by earwigs. In both models, there are peaks in both JH and ecdysteroids prior to oviposition followed by inhibited CA activity and low blood levels of JH during parental care. JH synthesis and ovarian development are then stimulated during the latter part of the parental period, presumably because of changing cues from offspring. Demonstrations of direct links between changing hormonal levels and parental behavior, however, have been few. Fortunately, endocrinological techniques are available for addressing this issue. The development of fast, sensitive, and reliable radioimmunoassays for JH (Strambi et al., 1984; Goodman et al., 1993; Huang et al., 1994) make it possible to work with the large number of individuals required for behavioral study. Classical techniques of gland removal and hormone replacement are feasible; in many cases dissected CA also can be employed *in vitro* to measure biosynthetic rates of JH. A decided advantage in behavioral experiments is that JH and JH analogs can be applied topically, allowing noninvasive manipulations with a minimum of disruption. Although JH is quickly broken down in the blood, analogs such as methoprene can be employed when a JH-like effect must be maintained for a longer period of time. It also is possible to raise JH titers in the blood by applying inhibitors of JH esterase. A number of anti-CA compounds (precocenes) have been developed to eliminate the source of JH nonsurgically; these compounds, however, do not perform equally well in all groups of insects (Bowers *et al.*, 1976). There also has been considerable recent progress, especially employing the cockroach models, in identifying allatostatins from the brain that regulate the synthetic activity of the CA (Woodhead *et al.*, 1993; Stay *et al.*, 1994). In combination, the application of these techniques should allow behavioral endocrinologists to determine whether JH has direct or indirect effects on parental care.

C. EXTENDED BIPARENTAL CARE IN THE BURYING BEETLE, *NICROPHORUS ORBICOLLIS*

I initiated endocrinological studies of biparental burying beetles, which reveal several differences from the two models previously discussed. After burying beetles emerge as adults, they feed on carrion and fly larvae for 2–3 weeks until reproductive competence is attained. Females become sexually receptive shortly after emergence and will store sperm for considerable periods (Eggert, 1992). JH titers and ovarian mass increase over the first 20 days and eventually reach a resting-state plateau (see Fig. 5; Trumbo *et al.*, 1995).

Burying beetles reproduce only when they locate a small carcass, a temporally and spatially ephemeral resource. This pattern is similar to mosquitoes in which ovaries remain in a resting state until a blood meal is obtained (Edman and Lynn, 1975). After burying beetles inter a carcass, they remove hair and deposit antibiotic anal secretions to control the decomposition (Pukowski, 1933). Competition for fresh vertebrate carcasses is keen, with competitors possessing behavioral and physiological adaptations to exploit carrion quickly. Sarcophagid flies, for example, decrease the time it takes for their young to exploit carrion by larvipositioning directly on the resource (Denno and Cothran, 1976). When a female burying beetle locates a carcass, she begins a period of assessment behavior, apparently determining whether the resource is of suitable size and condition. Within 10 minutes of the discovery, JH levels have doubled. Within 18–24 h, ovarian mass increases two to three times and oviposition in the surrounding soil begins (D. S. Wilson and Knollenberg, 1984; Scott and Traniello, 1987; Trumbo *et al.*, 1995). The rapid and substantial ovarian increase is accompanied by some of the highest measured titers of JH among insects; these titers have been confirmed by qualitative analyses of JH (Fig. 5).

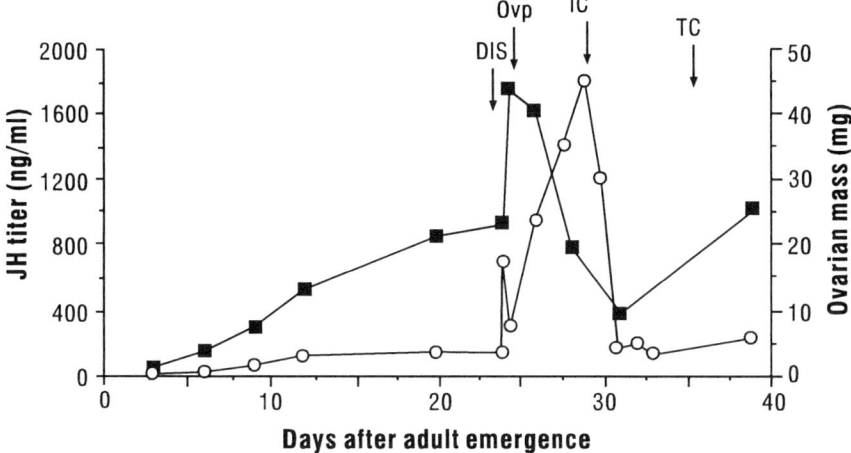

FIG. 5. Blood titers of juvenile hormone (JH) (open circles) and ovarian mass (closed squares) during the parental cycle of the burying beetle, *Nicrophorus orbicollis*. DIS = discovery of a carcass; Ovp = oviposition; IC = initiate care of larvae; TC = terminate care. (Figure modified from Trumbo *et al.* (1995) and S. T. Trumbo, unpublished data.)

Females do not feed during the initial 10 minutes following discovery, suggesting that feeding is not necessary for the initial JH surge. Substantial feeding prior to oviposition does occur, however, and may have quantitative effects on the number of eggs produced. Mating after resource discovery also does not appear to be necessary for the rapid JH increase or ovarian maturation. While a male will often aid the female and mate regularly during nest preparation, a lone female will complete the entire nesting cycle on her own if a male fails to discover the resource. Cues derived from behavioral assessment, and not from mating or feeding, are thus responsible for increases in JH and maturation of ovaries following carcass discovery. Reproduction, therefore, appears to be regulated by a complex integration of environmental, behavioral, and endocrine factors in a manner similar to many vertebrates (Lehrman, 1965; Crews, 1975).

Following oviposition, JH titers decline but still remain at levels higher than experienced prior to resource discovery. Interestingly, a second peak in JH titer occurs in both the male and female parent, coinciding with the arrival of young on the carcass (S. T. Trumbo, unpublished results; Fig. 5). The arrival of young on the carcass and the second peak in JH occur approximately 5 days after carcass discovery at a time when ovarian development is suppressed. When young arrive on the carcass, parents become extremely active, regurgitating liquified carrion and increasing nest maintenance. As young grow and become independent, parental activity subsides and care consists mainly on defense against predators (Fetherston et al., 1990, 1994). Titers of JH decline during the latter part of the parental cycle, before increasing again at the time parents desert the nest (S. T. Trumbo, unpublished results). Aggression against congeneric competitors remains high during the entire parental period and does not parallel changes in blood levels of JH.

As in *Diploptera* and the earwigs, female burying beetles experience an increase in JH prior to oviposition, low levels of JH during part of the parental cycle, and an increase in JH and ovarian growth as parental care is being terminated. In burying beetles, however, there is an intriguing additional surge of JH during the early and most active phase of the parental cycle. There are at least two possible explanations for high levels of JH during the first few days of parental care. The first possibility (parental care) is that the JH peak is connected to very active caregiving (parents rarely rest during the first 2–3 days larvae are on the carcass). A second possibility (clutch replacement) is that the maintenance of elevated JH during the initial period of care coincides with the time frame in which a female will attempt to reuse the resource should a brood failure occur. Brood failure can occur because of poor hatching success or because a conspecific male takes over the resource, kills all offspring, and mates with the resident female (Trumbo, 1990a; Scott, 1990). Late in the parental cycle, the deterioration of the resource is such that females will not produce a replacement clutch in response to brood failure. Elevated titers of JH would be maintained then, as long as the female was willing to reinitiate reproduction on a carcass. Under this hypothesis, decreasing blood levels of JH late in the parental cycle would be associated with unwillingness to oviposit near a resource that can no longer support a brood. It should be noted that in dung beetles, regression of ovarian follicles during parental care is gradual, and may be associated with a gradual decline in the probability of reusing the dung resource in the event of a brood failure (Martinez and Caussanel, 1984; Sato and Immamori, 1987; Edwards and Aschenborn, 1989). Neither the parental care nor the clutch replacement hypothesis for the second JH peak in burying beetles can be supported empirically at this time.

D. EUSOCIAL INSECTS

The study of the physiological regulation of simple family groups may provide insight into the evolution of sociality. This may depend, in part, on whether advanced sociality evolved through the subsocial (mother–offspring) or parasocial (sister–sister) route (Michener, 1969). In primitively eusocial insects such as bumblebees (*Bombus*) and paper wasps (*Polistes*), the queen actively maintains her dominant position as primary egg layer by display, aggression, and egg eating. In both *Bombus* and *Polistes,* dominance is correlated with large CA, higher blood levels of JH, greater oocyte development, aggression, and more time spent egg laying (Röseler *et al.,* 1984; Larrere and Couillaud, 1993). Individuals with low JH spend more time foraging, one of the principal forms of alloparental care. In bumblebees, the dominant individual may suppress JH levels in subordinates, in part, by producing a pheromone that decreases CA activity in nest mates (van Doorn, 1987; Larrere and Couillaud, 1993). Treatment with JH analogs tends to have little effect on social hierarchies in established association in *Polistes,* and in queenright bumblebee colonies. On the other hand, JH treatment is likely to confer dominance on individuals if administered before a hierarchy is established (Röseler *et al.,* 1985), and in queenless bumblebee colonies (van Doorn, 1987; Larrere and Couillaud, 1993). These findings parallel work on dominance hierarchies in vertebrates in which the stability of the social group and timing of hormonal manipulations is critically important (Wingfield *et al.,* 1987).

In the highly eusocial honeybee (*Apis mellifera*) and in the wasp *Polybia,* JH plays the quite different role of mediating age-based division of labor among nonreproductive workers. Low levels of JH are associated with nest activities such as caring for brood, while higher levels of JH are typically found in other forms of alloparental care such as foraging (O'Donnell and Jeanne, 1993; Robinson *et al.,* 1989).

E. The Importance of Mechanistic Studies

Unfortunately, few studies of the physiological basis of reproductive behavior of invertebrates also address areas of primary concern to behavioral ecologists. A thorough understanding of the physiology of invertebrate parental behavior promises to make three contributions to behavioral ecology: (1) permit the quantification of responses to key stimuli affecting behavior; (2) provide tools for exploring the costs–benefits of life-history trade-offs; and (3) provide insight into the role of phylogeny in adaptive evolution.

It is important to identify the exact informational cues that organisms monitor to address questions about optimality, constraints and evolutionary direction (Stephens, 1989; Ryan, 1990; Stamps, 1991; Endler, 1992). Studies of sexual selection, for example, have taken important new directions subsequent to analyses of the arbitrary location of secondary sexual characteristics (Burley, 1977), pre-existing sensory biases (Basolo, 1990), and mate copying (Dugatkin, 1992). Likewise, theoretical study of decision making by parents will be stimulated when we better understand the cues to which caregivers pay selective attention. In the biparental burying beetle, a female will spend more time performing energetically demanding caregiving activities such as feeding young and maintaining the nest when forced to provide care alone (Fetherston et al., 1994). Is her altered behavior dependent on stimuli associated with the condition of young, with demands by the young, with the presence–absence of her mate, or with parental effort of her mate? It is unlikely that we will understand parental care decisions and limits to adaptive responses of parents until we know what information parents selectively monitor. Physiological analyses may permit easily quantifiable, sensitive probes in experiments manipulating critical stimuli.

Hormones play a principal role in allocating organisms' internal resources to competing reproductive and nonreproductive needs (Finch and Rose, 1995). Manipulations of testosterone in free-living vertebrates have allowed the examination of the trade-offs among territory aquisition, mate attraction, caregiving, and survival (Marler and Moore, 1988; Ketterson et al., 1992). Similar studies soon may be possible with invertebrates. In the dung beetle, *Onthophagus bimodis,* for instance, males that complete pupal development above a critical size pursue a parental strategy, locating dung pats and assisting females with brood care. Smaller-bodied males never provide care, however, and attempt to sneak copulations as a means of achieving reproductive success (Cook, 1990). Such size-dependent switches in life-history strategy are thought to be selected in response to strong intrasexual competition along with a highly variable feeding environment for larvae (Eberhard, 1982). Unfortunately, very little is known of the hormonal control of adult male behavior in *Onthophagus.* Manipulations of larvae that alter the strict body size–reproductive strategy relationship may provide powerful tests of adaptive explanations of alternative life histories.

Mechanistic studies also may contribute to comparative studies of invertebrate parental behavior. The multiple origins of parental and social systems among so many distinct phylogenetic groups of insects, regulated by similar endocrine systems, begs for interspecific comparisons. In particular, the evolutionary potential of organisms' reproductive physiology might be addressed. Comparative approaches to the evolution of parental care physiology among vertebrates is limited. Little is known about the physiological changes that occurred as the ancestors of birds and mammals evolved parental care. Because key components of parental care likely evolved but once (e.g., lactation) we cannot know whether alternative physiological routes might have evolved to control the same trait. The multiple origins of parental behavior within the insects will allow for two types of tests. First, broad phylogenetic comparisons can be made to examine whether convergent behavior is necessarily based on convergent physiology. Second, comparisons of closely related species within groups in which both parental and nonparental lifestyles are represented will permit the testing of specific hypotheses of how evolution proceeds from one adaptive peak to another. Even with the limited knowledge presently available, it is clear that parental and social behavior is not regulated in exactly the same manner among independent lineages. Whether anatomy and physiology are extremely malleable in the hands of selection, or whether they impose formidable constraints on the evolution of behavior, will not be understood until functional and mechanistic studies are integrated.

## IV. Parental Care Theory and Invertebrates

### A. The Regulation of Brood Number

The number and size of offspring that will be raised are primary life-history variables. The trade-off between number and size of offspring is predicted to result in an optimal brood size that maximizes total fitness of the parent (C. C. Smith and Fretwell, 1974; Godfray, 1987; K. Wilson and Lessels, 1994). The optimal brood number is generally less than the maximum brood number that could be raised because body size of offspring affects their subsequent reproductive success. Small-bodied individuals experience greater mortality (J. Mappes and T. Mappes, personal communication), are less competitive for resources (Otronen, 1988), and produce fewer offspring (Tyndale-Biscoe, 1984; Cook, 1988; Bartlett and Ashworth, 1988). The production of undersized offspring can be especially disadvantageous when offspring must compete for limited resources (Brockelman, 1975; Lloyd, 1987). Females of the solitary wasp, *Euodynerus foraminatus,* for example, provide more food to offspring that are likely to engage in frequent contests for mates or nest sites (clumped nests) than when competition among offspring is expected to be less severe (Cowan, 1981). The ability to partition resources among offspring efficiently is especially critical when fitness of offspring is related to body size in a logistic fashion, in which young without adequate resources are not competitive, and young receive diminishing returns for superabundant resources (e.g., see Tyndale-Biscoe, 1984).

Theoretical and empirical tests have demonstrated that optimal clutch size can be affected by larval competition, costs of reproduction, brood parasitism, and search time for suitable locations to oviposit (K. Wilson and Lessels, 1994). Less attention has been given to constraints imposed by parents' ability to gather accurate information concerning resource availability and number of young. These constraints will not come to light until the cues that parents employ to monitor their environment are known. Brood size adjustments also may be limited by parents' ability to provide care. A number of folivorous insects, for example, shield young under their body. Eggs on the outside of an egg cluster are highly vulnerable to predation and parasitism (Odhiambo, 1959). Using the parent bug, *Elasmucha grisea*, Mappes and Kaitala (1994) demonstrated that females of different body size produce a clutch that is appropriate to the area that a female can cover and therefore defend. Small females that were experimentally provided an enlarged clutch lost all additional eggs. Interestingly, *Elasmucha* lays the largest eggs in the central, safest part of the egg mass, confirming that unequal care within a brood is expected when prospects for offspring vary (Mappes *et al.,* in press). This study nicely demonstrates that components of life history cannot be considered in isolation (Hinde, 1975). Obviously, an adequate understanding of clutch size in *Elasmucha* would require detailed examination of all ecological factors that have had a significant effect on the evolution of female body size.

The evolution of parental care allowed species to adjust brood size in ways not possible in nonparental species. Continued contact with young permits greater investment per offspring; repeated assessment of resource availability; and, when young fail, recouping of investment and redirection of resources. Parental care may thus permit a finer degree of control over brood number and hence body size of offspring. This prediction is supported by less variation in body size among wood-feeding species with parental care as compared with those without care (Haack and Slansky, 1987). Because parents often act as a buffer between the environment and their offspring, resource availability to young is often less variable than resource availability to the parent. Among a number of taxa, females of small size produce offspring of equivalent size to large-bodied females by producing fewer young per reproductive attempt (Tyndale-Biscoe, 1984; Schmidt and Smith, 1987; McLay and Hayward, 1987; but see Lee and Peng, 1981).

An understanding of the stimuli that parents monitor to assess both resource availability and number of brood will permit a more complete accounting of clutch–brood size decisions. The simplest mechanism to adjust brood number to available resources is to gather food until a single offspring can be supported, oviposit an egg into the provision, and repeat until resources are exhausted. This strategy is employed by many caregiving dung beetles, which exploit an easily divided resource (e.g., Sato and Immamori, 1987).

When food is less malleable, there is less flexibility in securing the appropriate amount of food per offspring. Many solitary wasps capture discrete prey items for their developing young. Size of offspring at adult emergence is related to the quantity of stored food. The digger wasp, *Ammophila sabulosa,* adjusts provisioning according to the size of prey items (caterpillars) captured. If the initial prey item placed in an egg cell has a mass >200 mg, the cell is usually permanently closed. If the initial caterpillar is less than this critical size, additional small prey items (<200 mg) are brought to the cell (Field, 1992). Field suggests that when a cell is multiply provisioned, second and subsequent prey items are generally small because the additional burdens of capturing and transporting large caterpillars will not be compensated for by commensurate benefits for offspring receiving a superabundant provision.

When large and indivisible resources are exploited, brood size adjustments must be made solely by altering the number of young. The parasitoid wasp, *Trichogramma,* lays a clutch that is proportional to the value of its host (insect eggs). Volume is estimated by the time interval required to traverse the host (Schmidt and Smith, 1987). Burying beetles, which exploit a wide size range of vertebrate carcasses (2–75 g for *Nicrophorus vespilloides,* Müller *et al.,* 1990a), likewise have elaborate mechanisms to adjust brood number. On very small carcasses, females oviposit a reduced clutch (Müller *et al.,* 1990a). Clutch size adjustments are crude, however, and alone cannot account for the fine ability to match brood number to food supply. The male or female working together, or either parent working alone, can regulate brood size by cannibalizing day-old larvae that make their way to the prepared brood ball (Bartlett, 1987; Trumbo and Fernandez, 1995). On carcasses that cannot support the clutch that has been oviposited, parents reduce brood number such that surviving larvae on a wide range of carcass sizes will disperse from the resource at an equivalent mass (D. S. Wilson and Fudge, 1984; Trumbo, 1990b; Scott and Traniello, 1990). The critical stimulus that parents assess appears to be volume of the brood ball during the postoviposition, prehatching period (Trumbo and Fernandez, 1995; C. Creighton, personal communication). Experimental removal of parents after oviposition results in scramble competition for resources among undersized young (Trumbo, 1990b).

When caregiving is extended over a long period of time, parents can alter brood size in response to changing conditions. Parents of many species cannibalize young if disturbed or if prospects otherwise deteriorate (e.g., Lamb, 1976; Rollo, 1984). Recycling of investment may be especially critical in species utilizing low-nitrogen food sources such as wood (Nalepa and Mullins, 1992). Burying beetles raise fewer young on carcasses that are infested with carrion fly larvae, but the mechanism for this adjustment is unknown (Trumbo, 1992). When juvenile mortality is high, but the environment remains otherwise favorable, females may lay replacement clutches (Edwards and Aschenborn, 1989). Females of the burying beetle, *Nicropho-*

*rus vespilloides,* will sometimes lay a replacement clutch in response to partial loss of brood, if the surviving number is well below the capacity that the carrion resource can support (Müller, 1987). When mortality occurs among larval dung beetles, the female will reallocate dung to brood balls of healthy young (Montieth and Storey, 1981; Tyndale-Biscoe, 1984), and form additional balls for oviposition (Klemperer, 1983a).

The cues parents use to assess brood number are poorly understood. Complete removal of the brood stimulates oviposition in many taxa (Tallamy and Denno, 1981; Vancassel *et al.,* 1987), but quantitative inhibition of increasing brood number and the importance of critical stimuli have rarely been demonstrated. Females of the dung beetle, *Onticellus cinctus,* appear to assess a correlate of the number of brood balls. Higher concentrations of a possible brood pheromone presumably inhibit further oviposition (Klemperer, 1983a,b). An understanding of the cues parents employ to assess brood number will likely provide important tools for investigating conflicts between parents and offspring over the optimal number of young, and conflicts over the appropriate investment in each offspring. Excluding the eusocial taxa, invertebrates have rarely been employed to examine such issues. Parent–offspring conflict is expected to be more pronounced when the number of brood is few, as occurs in many dung beetles. The consequences for parents and offspring when parents are manipulated to raise varying number of young can be revealed by experimentally varying signals from offspring (e.g., brood pheromone in *Onticellus* and *Copris* (Klemperer, 1983a), or auditory signals from *Cephalodesmius* (Montieth and Storey, 1981)). The investigation of the cues involved in parent–offspring conflict also may be examined by presenting parents with congeneric young with different developmental needs than parents' own young. Such manipulations are possible because many parental insects are poor discriminators of young that are experimentally presented under the appropriate conditions (Melber and Schmidt, 1975; Klemperer, 1982; Kudô, 1990; Radl and Linsenmair, 1991; Trumbo and Wilson, 1993; Kight, 1995). For example, burying beetles will raise any congeneric young that are placed in the nest at the time the parents own young are to hatch (Müller and Eggert, 1990). If burying beetle parents monitor regurgitation demands, then parents might be expected to raise broods of different size when provided young of species with different nutritional needs.

Parental investment is an attractive resource for nonparental individuals to exploit. Brood parasitism can have clear costs for parents (Tallamy and Horton, 1990; Müller *et al.,* 1990b; Trumbo, 1994). Parental care in the generally nonparental parasitoid wasps is thought to have evolved to prevent additional parasitoids from ovipositing in the same host (Hardy and Blackburn, 1991). When brood parasitism is common, reduced clutch sizes are predicted because fewer resources will be available per young, resulting in undersized offspring of both parent and the brood parasite (Andersson and Eriksson, 1982; also see K. Wilson and Lessels, 1994). When brood size is regulated by parental infanticide, however, a different outcome might be expected. To dilute the contribution of a parasite before indiscriminate brood reduction occurs, one might expect an increase in clutch size when parasitism is likely. This hypothesis is yet to be tested in a species exhibiting filial brood size regulation.

B. FACILITATING FEEDING OF OFFSPRING

Many caregiving invertebrates provide offspring with nutritive resources beyond those stored in the egg. The continued presence of a parent allows the food-gathering advantages of adults (faster location of food sources, processing of food, less vulnerability while foraging, and storage of larger reserves) to benefit young. Invertebrate parents provide extended care by: (1) carrying young internally or externally; (2) tending eggs–nymphs that are clustered near food sources; or (3) providing resources for young within nests. For each parental lifestyle there is a continuum of care from transient protection with no provisioning, to long-term care persisting until the completion of immature development.

Investment by carrying young can range from short-term protection of offspring to the provision of nutritive milk to incubating young (Ingram *et al.*, 1977) to the extreme exhibited by mites *Sitergites* and *Acarophenax*, which harbor young until they burst out of the maternal corpse (Hamilton, 1967). The tsetse fly, *Glossina*, makes perhaps the greatest relative investment in an embryo, giving birth to a single large offspring (Buxton, 1955).

Invertebrates that lay eggs in exposed environments can offer protection to the brood by remaining after oviposition. Parental care ranges from transient hovering over a newly laid clutch to long-term protection of immature stages. In the leaf-feeding tortoise beetle, *Acromis sparsa*, the mother protects eggs, nymphs, and pupae against parasitoids and predators, and terminates care only when her adult offspring complete ecolysis (Windsor, 1987). Other plant feeders take an active role in guarding the brood as they move through the environment. The lace bug, *Coryuca hewitti*, communicates to young by rapidly vibrating her abdomen and possibly providing an aggregating pheromone (Faeth, 1989) as the mother "herds" the brood toward new leaves. The membracid bug, *Umbonia crassicornis*, facilitates feeding more directly, cutting a series of spiral slits in the bark with her ovipositor prior to hatching. The mother then remains with nymphs until they reach adulthood, maintaining feeding aggregations by stroking nymphs on their backs with her forelegs (Wood, 1974). In leaf feeders, facilitation of feeding is thought to be a secondary parental adaptation, subsequent to selection for protection of exposed young.

Among nest and web builders, the simplest form of provisioning is to tolerate young at a site where the adult normally stores food for its own use (Buskirk, 1981). Tolerance of young and inhibition of long-distance movements by adults also facilitates transfer of digestive symbionts to offspring, especially in taxa specialized to feed on resources of low quality. Many wood-feeding insects have evolved extended familial associations, including complex parent–offspring communication (Schuster and Schuster, 1985; Nalepa, 1994). In other species, special food resources are provided to young in the form of trophic eggs (West and Alexander, 1963), prepared dung balls (Halffter, 1977), regurgitated carrion (Pukowski, 1933), and paralyzed prey (many subsocial Hymenopteria; Pratte and Jeanne, 1984). In earwigs, parents both bring food to the nest and regurgitate to young (Shepard *et al.,* 1973; Lamb, 1976), although they normally do not ingest food themselves.

Complex provisioning behavior in nest builders is thought to have evolved by the addition of steps onto an established behavioral sequence. Dung beetle nesting habits, for example, have been categorized into five types (Halffter, 1977). Types I–III are variations on burying food within previously prepared galleries at the site of the dung source. More "advanced" ball-rollers (IV and V) move food away from the source prior to burial and nest construction, either making single balls and nests and providing limited care (IV), or making multiple balls and providing highly developed care (V). Such classifications can be overly rigid when applied to individual species. For example, the Australian dung beetle, *Cephalodesmius,* has been classified as Type III because it forms multiple balls in a previously excavated nest. *Cephalodesmius,* however, synthesizes dung from readily available leaf litter and is not dependent on the temporally and spatially ephemeral dung resource. Thus, Montieth and Storey (1981) argue that there is no ecological necessity to delay nest building until a food source is located, and therefore it is inappropriate to include this group among the more primitive nesting dung beetles.

In a similar way, nesting patterns of subsocial wasps have been categorized according to apparent behavioral complexity without full regard for foraging ecology. The simplest nesting sequence is to provide a single large prey item for the larva, while complex nesting can entail multiple nests, monitoring of larval food requirements, sophisticated orientation abilities, and progressive provisioning (a *Microstigmus* wasp is recorded to have brought 171 captured thrips to a brood cell (Matthews, 1970)). Behaviors associated with multiple provisioning are thought to be prerequisites of social behavior in the Hymenoptera. Attempts to develop a hierarchy of nesting complexity among the Hymenoptera (Evans, 1958; Atkins, 1980) clearly do not fit other taxonomies. Many anomalous nesting habits within a phylogenetic taxon likely are due to ecological differences in the distribution, availability, and size of food resources. Phylogenies of behavior may provide insight only when comparing closely related species. The tendency to view complex nesting behavior across taxa within a single classification system may be borrowed from theories of chained responses (see Hinde, 1970) in which complex behavior was explained as programmed sequences of linked mechanical steps. The classification of nesting patterns is one area in which ecological insights need to be more fully integrated with established mechanistic interpretations of behavior.

C. Protecting Young

Defending young against predators and parasites is thought to be one of the prime movers in the evolution of parental care (E. O. Wilson, 1975), especially for nonnesting species in which young are exposed in the environment (Tallamy and Wood, 1986). The significance of defense for the maintenance of the parental life history is easily demonstrated. In many folivores, mortality approaches 100% if the parent is removed (Tallamy and Denno, 1981; Windsor, 1987; Edgerly, 1987; Choe, 1989). Although parental care in many species typically has multiple functions, avoiding detection by predators (Mappes and Kaitala, 1995) and active defense of young against predators often appear to be at least secondary adaptations (Rosenheim, 1987; Wyatt and Foster, 1989a; Diesel, 1992).

Nonparental species have evolved numerous adaptations to reduce predation and parasitism of eggs. Eggs are generally dispersed singly or in small clumps in protected places (but see Stamp, 1980). Eggs may be further protected by incorporation of toxins, possession of hard shells (Eberhard, 1975), covering with waxlike compounds (Wood, 1974), or by hiding in shell-like secretions (Miller, 1971). A completely different suite of adaptations are present in species with parental care. Once parental behavior takes over the function of egg protection, resources previously devoted to counter predation and parasitism may be better invested elsewhere. Unguarded eggs of parental species tend to be more vulnerable to predation and parasitism than eggs of nonparental species. Likewise, young of species with care may be less sclerotized than those of closely related nonparental groups (Anderson, 1982). Other coadaptations may follow the evolution of parental care. To facilitate care, eggs of parental species are often laid in one tightly clustered clump (Eberhard, 1975; Tallamy, 1984). When young become mobile, they tend to remain aggregated even in the absence of a parent (Eberhard, 1975; Sites and McPherson, 1982; Nafus and Schreiner,

1988; Windsor and Choe, 1994), although in a few species, parents actively maintain family groups (Kearns and Yamamoto, 1981). In *Elasmucha grisea,* larvae release trail pheromones that both siblings and mothers can follow (Maschwitz and Gutmann, 1979). Injured young may produce an alarm pheromone that stimulates maternal aggression (Wood, 1976; Kudô, 1990; Maschwitz and Gutmann, 1979). Among folivorous species with care, it is not surprising that removal of a parent results in nearly complete mortality. A conspicuously large grouping of eggs or immatures, absent the protective mechanisms of nonparental species, make unguarded young extremely vulnerable.

The evolution of novel defense mechanisms can entail new ecological risks. Plants that employ secondary compounds to deter generalist predators may be exploited by specialists that take their cue from these same compounds. Likewise, parental adaptations may deter some predators or parasitoids but allow specialists to cue in on an abundant and easily located resource. A number of parasitoid wasps selectively exploit young of parental species (Eberhard, 1975; Nafus and Schreiner, 1988; Wyatt and Foster, 1989b; Kudô, 1996; Edgerly, in press). In a revealing set of experiments, Eberhard (1975) demonstrated that in the pentatomid bug, *Antiteuchus tripterus,* the presence of a guarding mother increased the chance that the specialist wasp, *Phanuropsis semiflaviventris,* would parasitize eggs of *Antiteuchus.* This occurred despite antiparasitoid defensive behaviors that mothers specifically employed against the wasp (shielding the egg mass, leaning toward the wasp, antennating, shaking the body, and kicking). Eberhard suggested that *Phanuropsis* uses the parent to orient to its host, something that the parasitoid finds difficult to do when the parent is removed experimentally. The net effect of care in *Antiteuchus* is still positive, largely due to reduction in losses to generalist predators. Eberhard suggested that this species may be in an evolutionary trap, in which selection for thin egg shells and clumped oviposition, which accompanied the evolution of care, may prevent movement toward a nonparental adaptive peak. It is conceivable that the success of a specialist predator or parasitoid might be sufficient to move a parental species toward a new adaptive peak not involving care. The likelihood of this scenario would depend on the cues specialists employ (cues from parent, young, or habitat?), as well as whether care involves multiple functions.

Douglas Tallamy (in press) has pointed out that the comparative effectiveness of alternative nonparental life histories (e.g., hiding small numbers of eggs in scattered locations) cannot be tested by simple removal of parents. Obviously, nonparental ancestors of extant species with care did better than the 100% juvenile mortality that is often the outcome of parental removal experiments. Tallamy argues that parental care is pleisiomorphic within certain clades of Homoptera and the true bugs (Heteroptera), is often less successful than nonparental options, and that the subsocial "advancement" has been lost in numerous taxa because of the substantial costs of care. Under this scenario, the pentatomid hemipterans are derived from a parental ground-nesting ancestor with a lifestyle similar to that exhibited by present-day cydnid bugs (Sites and McPherson, 1982; Filippi-Tsukamoto *et al.,* 1995). Parental care has been retained (or occasionally re-evolved) when host plant seasonality allows but one reproductive attempt, reducing the costs of high parental investment (Tashikawa and Schaefer, 1985), or when care has taken on additional functions such as facilitating feeding.

The intensity of parental defense has been used to measure changes in parental effort, especially as it relates to changing reproductive value of the parent or of offspring. In general, parental effort regarding the brood on hand is predicted to be greater with increasing reproductive value of the brood, and with decreasing reproductive value of the parent. Brood defense makes an especially effective assay of parental intensity. Because many young of parental species are defenseless against predators throughout development, the need for protection will remain constant over sufficient time to conduct experimental tests. This may not be the case with provisioning, where needs of immatures change quickly. Tallamy (1982) found that younger *Gargaphia solani* mothers (greater reproductive value of parent) were less likely to defend offspring aggressively than older mothers. Females with high reproductive potential also attempted to reduce the costs of care by dumping eggs more frequently in egg masses of other females (Tallamy, 1986). Caregiving females also increased defense intensity (more wing fanning, ramming, and chasing of a simulated predator) as the reproductive value of maturing nymphs increased (Tallamy, 1982). Crespi (1990) found that the thrips *Elaphrothrips tuberculatus* were more aggressive when defending larger clutches. Unfortunately, the number of empirical studies of reproductive value theory employing parental defense is rather limited, despite the ease with which defense can be quantified, the short life span of invertebrates that causes reproductive value of parents and offspring to change quickly, and the ability to alter brood size experimentally.

D. Terminating Care

Because parental care is costly, parents are expected to re-evaluate the decision to provide care during the parental period. Parental care is terminated when young mature and become less dependent on parental attention, or when prospects for successfully rearing the brood diminish to the point that parents abandon their young prematurely. Although parent–offspring conflict theory predicts that parents generally end the familial association, in some cases care simply ends when young disperse (Vancassel *et al.,* 1987;

Ruttan, 1990). More often among invertebrates, the parent abandons the clutch or brood. The timing of desertion may not be distributed continuously, but may occur at specific stages in the development of young. Thus a maternal mosquito, *Trichoprosopon digitatum,* leaves at the time eggs hatch (Lounibos and Machado-Allison, 1986); the pentatomid bug, *Parastrachia japonensis,* stays with second instars, but deserts third instars (Tachikawa and Schaefer, 1985); and the tortoise beetle, *Acromis sparsa,* leaves when its adult offspring emerge (Windsor, 1987). Manipulation experiments, in which young of one stage are switched with those of a different stage, demonstrate that parents are monitoring the development of offspring. The duration of care can be extended by substituting young offspring for old in earwigs, dung beetles, and cydnid bugs (Caussanel, 1970; Klemperer, 1983a; S. L. Kight, personal communication). Klemperer (1982) has demonstrated that care is mediated by a chemical signal from brood; mothers typically abandon brood balls without young, but will maintain, repair, and keep empty brood balls upright if they are treated with dichloromethane extracts from viable brood. Communication between offspring and parent may also maintain care in the dung beetle, *Cephalodesmius,* in which larvae stridulate by scraping their "chin" along their "tail" (Montieth and Storey, 1981). The importance of communication of brood to caregivers has been nicely demonstrated in the honeybee, *Apis mellifera.* Workers can differentiate among brood cells of well-fed and hungry larvae (Huang and Otis, 1991), and among young and old brood (LeConte *et al.,* 1994). Changing proportions of fatty acid esters on the cuticle of maturing larvae apparently induce adult workers to terminate feeding and to begin capping brood cells (LeConte *et al.,* 1994).

The termination of care may occur at discrete points in the circadian cycle. The tsetse fly, *Glossina,* generally gives birth to her single offspring in the midafternoon (Buxton, 1955), and male burying beetles abandon their brood in the first few hours after sunset, corresponding to the activity period during nonparental phases of adult life (Trumbo, 1991).

Different species regulate the duration of caregiving by monitoring various combinations of internal and external cues. The mosquito, *Trichoprosopon digitatum,* tends eggs for approximately 30 h (at which time eggs hatch) on floating rafts within water-holding fruit husks. This parental behavior, highly unusual among the Diptera, may keep eggs from being washed from the container habitat during heavy rainfall (Lounibos and Machado-Allison, 1986). When eggs were taken from brooding females and replaced with eggs that were 12 h older, females nevertheless guarded rafts until 12 h after hatching. Conversely, brooding females that were given eggs 12 h younger, inappropriately abandoned rafts 12 h prior to hatching. Thus, the mechanism for terminating care in this species seems to be entirely internal, and is not affected by the state of the young (Lounibos and Machado-Allison, 1986).

In an elegant set of experiments, Kight (in submission) demonstrated that the duration of care in the ground-nesting *Sehirus* is controlled by both internal and external cues. *Sehirus* parents guard eggs until they hatch (approximately 10 days), and then guard and provision young for 1–3 days posthatching. Kight assayed parental responsiveness by measuring reaction to tactile disturbance. Females experimentally given older eggs in place of younger ones showed a decrease in responsiveness shortly after hatching, suggesting that mothers were affected by the experience of hatching, and that the condition of offspring was monitored. The reduction in response occurred at a time when aggression would have been high had no experimental manipulation been made. Conversely, females given younger eggs in place of older ones extended brooding for up to 6 additional days, but no longer. Thus, when hatching fails to occur, an internal process eventually terminates care, independent of the condition of the young. Kight suggests that bracketing maternal responsivity with both experiential and endogenous factors may ensure maternal care of appropriate duration.

The maintenance of care sometimes depends on mutual feedback between parent and offspring. Crayfish mothers produce a brood pheromone that is attractive to early-stage larvae (Little, 1976). Continued production of the brood pheromone is dependent on contact with larvae. When larvae reach the fourth larval stage, they are less responsive to brood pheromone. The decrease in mother–offspring contact, in turn, leads to decreased pheromone production and the cessation of maternal responsiveness (females become cannibalistic toward larval crayfish) (Little, 1976).

Prospects for both the parent and offspring can change following oviposition, prior to normal termination of care. Among iteroparous species (more than one breeding attempt per lifetime), parents often monitor environmental conditions when deciding whether to continue care or to abandon the brood. In both a staphylinid beetle that utilizes decaying mushrooms as a resource and in carrion-feeding burying beetles, parents will abandon young if the resource deteriorates prematurely (Ashe, 1987; S. T. Trumbo, personal observation). In both cases, the young are known to have rapid development in comparison to closely related nonparental genera, presumably to minimize the possibility that the resource deteriorates before young mature.

The time invested in parental care, the risk of predation, and the decreased ability to forage are often fixed costs that must be incurred regardless of the number of young being reared. Thus, it is not unexpected that parents sometimes desert or cannibalize small broods (Eberhard, 1986; Müller, 1987; Olmstead and Wood, 1990). Abandoning small broods is expected when the diminished benefits derived from caring are not matched by diminished costs. Kight and Kruse (1992) demonstrated that both clutch size and time invested affect termination of care in the back-brooding water bug, *Belostoma*. Experimentally reduced clutches were discarded by males more often than normal clutches, but the probability of desertion also depended on the length of prior investment. Males that have less time left until hatching of eggs were more likely to continue brooding than males that had just begun the parental period. In this case, past investment provides a reliable measure of the rate of return for future investment (expected benefits ÷ time remaining until hatching).

Parental care theory predicts that in semelparous (one breeding attempt per lifetime) species, parents are expected to invest maximally in their one brood. In a comparative study, A. Kaitala and J. Mappes (personal communication) demonstrated that iteroparous shield bugs (*Elasmucha* spp.) will abandon clutches if disturbed. Semelparous congeners, however, never abandon their clutches. If forcefully dislodged, mothers will spray distasteful chemicals over their young as a last resort. One of the more bizarre forms of terminal investment is found in the tick, *Ixodes kopsteini*, which dies shortly after formation of eggs. This species does not have the ability to oviposit externally. Instead, the young hatch inside the dead body of their mother, the corpse acting as a protective sanctuary.

Parental care theory predicts that parents and offspring may disagree when caregiving should terminate (Trivers, 1974). Examination of the cues that parents use to make decisions should provide insight into how parent–offspring conflict evolves. It will be of particular interest to know how cues that parents employ to terminate care may affect offspring's ability to manipulate care. Evolutionarily, offspring may have enhanced possibilities of manipulating the level of care when parents have been selected to monitor cues derived directly from young, rather than selected to monitor indirect cues that may signal the appropriate time to desert.

E. The Costs of Parental Care

Organisms have limited physiological resources that they can devote to growth, survival, production of gametes, and care of offspring (Calow, 1979). If organisms devote more to one life-history component, such as care for the brood on hand, then fewer resources are available for competing needs. Thus, parental care is assumed to have costs (Trivers, 1972).

A misleading accounting of the costs of care may be obtained by comparing the subsequent reproductive performance of formerly parental and nonparental individuals. Individuals of greater vigor may be able to devote more resources to all components of fitness, obscuring the trade-offs that are assumed to exist among parental care, survival, and future reproductive ability (van Noordwijk and de Jong, 1986). Tallamy and Denno (1982), for example, found that females of *Gargaphia solani* with larger clutches and greater parental investment early in life were the same individuals that produced larger clutches late in life, despite the expected negative relationship if all things (genetics, juvenile environment, adult feeding) were equal. To circumvent the confounding effects of differential vigor on allocation of resources among life-history components, experimental manipulations are often employed. Removal of brood from a caregiver can shorten the time to the next oviposition (Vancassel, 1977; Müller, 1987; Nalepa, 1988) and increase lifetime egg production (Tallamy and Denno, 1982; Fink, 1986).

Observations of caregivers suggest that the principal costs of care are reduced foraging and increased risk of predation. Parents often spend less time feeding themselves than nonparental individuals (Edgerly, 1987), and may experience a decrease in fat reserves (Eberhard, 1975). The reduced mobility of caregivers is especially costly for predators that must pursue prey (Odhiambo, 1959; R. L. Smith, 1976, in press; Crowl and Alexander, 1989; Kight *et al.*, 1995). Decreased mobility, either because of attendance of sedentary juvenile stages or because of the encumbrance of carrying young, can also increase predation risk (Smith, 1976; Tallamy and Horton, 1990). In many species that shield young on exposed vegetation, the tendency to run, fly, or drop to the ground when disturbed is suppressed during care (Eberhard, 1975; Tallamy, in press). Among insects in which care is provided while guarding a bonanza resource such as dung or carrion, individuals may gain substantial weight during care, and the costs of care may be primarily in the form of predation risk and time that cannot be spent searching for additional resources. Because hormones mediate numerous critical life-history decisions (Finch and Rose, 1995), endocrinological manipulations are likely to become increasingly important in revealing trade-offs among components of fitness.

The substantial costs of providing care have undoubtedly contributed to the evolution of mechanisms to reduce such costs. Young of the treehopper, *Publilia reticulata*, receive protection either from their mother or from ant mutualists. The presence of ant caretakers apparently acts as a cue for the mother to desert the brood, a behavioral response that transfers the costs of care from the mother to the ants (Bristow, 1983). Because subsocial insects generally have weak mechanisms to discriminate among conspecific young, egg dumping also is effective for minimizing the costs of care. Egg dumpers of *Polyglypta* are faster to oviposit a second clutch (Eberhard, 1986), and *Gargaphia* egg dumpers have greater lifetime fecundity than conventional caregivers (Tallamy and Horton, 1990). Tallamy (in press) further suggests that viviparity (which frees the mother from providing care in one place) and paternal behavior may have evolved to reduce the costs of maternal care.

In the woodroach, *Cryptocercus,* the costs of long-term care (3+ years) while utilizing nitrogen-poor wood, are sufficient to induce semelparity in the field (Nalepa, 1988). Pairs typically can produce a replacement brood only if the original brood is lost early in its development. Nalepa (1994) suggests that in the ancestors of termites, the transfer of some of the costs of care to latter-stage instars (which are nutritionally independent of the parents in *Cryptocercus*) may have been a critical point in the evolution of eusociality in termites. The transfer of caregiving costs from mother to older offspring sets the stage for satisfying all three conditions for eusociality (overlap of generations, brood care by workers, and nonreproductive castes). When latter-stage instars provide care, the mother can reserve her own physiological resources for survival and further egg production. The availability of younger siblings for older siblings to care for may prolong the mother–offspring association. While alloparenting provides an avenue for pre-reproductive individuals to increase their inclusive fitness, it also diminishes the possibility that the physiological resources necessary for independent reproduction will ever be accumulated. In several cockroaches, feeding and a positive energy balance are necessary to stimulate ovarian maturation (Engelmann, 1957; Stay and Coop, 1973). Clearly, hormonal studies of feeding, reproduction, and caregiving in subsocial cockroaches, and of the subsocial stage of colony founding in termites, is likely to provide insight into the evolution of eusociality in this group.

The commitment to provide parental care often entails staying in one place, greater exposure to predators, reduced feeding opportunities, and lost time searching for additional mates or valuable resources. Many of these costs are incurred whether the parent provides care for one offspring or for many. One might predict that larger clutches would evolve when there are positive economies of scale, in order to reduce the per offspring costs of care. Nonparental species are predicted to have greater lifetime fecundity, but accomplish this by dispersing many clutches of fewer eggs over a longer period of oviposition. Interspecific comparisons support the prediction of fewer but larger clutches in species providing the greatest investment (Schreiner and Nafus, 1991; Tallamy, in press; J. Mappes, personal communication).

Providing care for a larger clutch may diminish the possibility of further reproduction by more thoroughly depleting fat reserves, as well as by interfering with feeding. Selection for production of a larger clutch may establish a positive feedback whereby increased clutch size reduces subsequent reproduction, and thus selects for an increased tendency toward semelparity and a large one-time investment in young. This selective reinforcement of the parental lifestyle may keep species in a parental "bind" (Tallamy, in press). After the primary evolutionary event that initiates the parental lifestyle, subsequent secondary adaptations to reduce the costs of care are expected. The ecological avenues available to reduce the costs of care will, in large part, determine whether further selection will diminish or enhance the parental tendency. The potential for egg dumping, transfer of care to mutualists, or uniparental male care will select for reduced maternal care, while oviparity or production of a large clutch will reinforce a maternal life history of caregiving.

## V. Paternal Care: Uniparental Male versus Biparental Care

Postzygotic care by males is rare among invertebrates. Its rarity has stimulated study of the selective forces that push a species toward a paternal lifestyle. Our understanding of the physiological and behavioral mechanisms that regulate paternal behavior is extremely limited. Therefore, I confine the discussion to a behavioral and ecological comparison of uniparental male versus biparental care.

Exclusive male care apparently evolves under a very different set of circumstances than biparental care. Uniparental male care is associated with all three basic patterns of postzygotic care: carrying eggs (King and Jarvis, 1970; R. L. Smith, 1976), tending young exposed in the environment (Odhiambo, 1959; Ralston, 1977; Ichikawa, 1988; R. L. Smith and Larsen, 1993), and nesting (Kaestner, 1968; Mora, 1990). Nest building generally is not elaborate and often occurs out in the open. Complexity and exclusivity of nests utilized by uniparental males may be constrained by the need to permit easy female visitation. In contrast, biparental care is usually associated with nests. The nest may represent a substantial investment and be occupied for long periods of time. These differences are related to differences in courtship and the nature of resources exploited.

### A. Courtship

Both uniparental male care and biparental care are associated with high certainty of paternity. In uniparental care, the mechanism of paternity assurance is usually repeated copulation just prior to oviposition (King and Jarvis, 1970; R. L. Smith, 1979; Mora, 1990). A female may court a male to induce him to accept her eggs, presumably because male care is a valuable resource that increases her lifetime fecundity (Mora, 1990; R. L.

Smith and Larsen, 1993; Tallamy, in press). When males can guard eggs of several females simultaneously, there is the potential for polygyny (Odhiambo, 1959; King and Jarvis, 1970; Ralston, 1977; Mora, 1990). Males may be able to increase their access to females via caregiving when females actively discriminate against males without eggs or a nest, as occurs in the harvestman, *Zygopachylus albomarginus* (Mora, 1990). A preference for males that are guarding eggs is the equivalent of females copying the mate choice of other females. If searching for a male entails costs, a male already caring for eggs might be attractive because he has demonstrated both his ability to stimulate a previous female to oviposit and his willingness to provide care. (If adding her clutch to the nest of a caring male overextends his parental ability, however, a female may first destroy eggs before ovipositing her own (Ichikawa, 1995).) In the hemipteran *Rhinocoris*, hatching success of eggs increases with the number of eggs tended (up to a point), suggesting that locating males with egg batches may be a good female strategy (L. Thomas, personal communication). In vertebrate mating systems, mate copying in species without male care may reduce the costs of female choice and result in the selection of higher-quality mates (Dugatkin, 1992). It is possible, but not yet tested, that mate copying may promote polygyny in invertebrates with exclusive male care.

It would also be of interest to search for a connection between egg dumping by females and the evolution of uniparental male care. Among egg dumpers, females prefer to leave their eggs with females already tending a clutch. The evolution of a paternal tendency in such species might immediately be associated with females leaving eggs with caring males, female preference for males already tending eggs, and thus the potential for polygyny. Groups that exhibit both uniparental male and uniparental female care, such as *Rhinocoris* (Hemiptera) would be ideal for investigating a relationship between egg dumping and uniparental male care.

Patterns of copulation vary among biparental invertebrates. The frequency of copulation may be related to the exclusivity of the pairing. When pairs mate within an excavated nest or gallery, such as occurs in termites, the bark beetle, *Ips*, and the dung beetle *Phaneaus*, copulation may occur infrequently (Schmitz, 1972; Halffter and Lopez, 1977; Nalepa and Jones, 1991). In burying beetles, on the other hand, where visitations by rivals are common during burial of a carcass, copulations are frequent and ensure that the resident male sires most of the brood (Müller and Eggert, 1989).

Elaborate courtship might be expected in monogamous species with biparental care. Courtship can coordinate reproduction and nesting behavior of the two sexes, and can ensure that substantial investment is not wasted on a partner that is inferior or one encumbered with another mate. Although this reasoning may apply to birds (Lehrman, 1965; Wittenberger and Tilson, 1980), courtship is surprisingly cursory among most biparental invertebrates (Pukowski, 1933; Kirkendall, 1983; Nalepa and Jones, 1991; Ruegg and Rose, 1991; but see Linsenmair, 1987). Biparental care within invertebrates has generally evolved in groups that exploit a critical, contested resource such as food or shelter. Among many biparental invertebrates, each sex engages in intrasexual fights (Pukowski, 1933; Schuster and Schuster, 1985; Sato and Hiramatsu, 1993; Shellman-Reeve, 1990). Mate choice thus may be largely passive, and pairing the indirect result of both sexes excluding rivals. In fact, in many cases the adults search for the critical resource independently and will begin to nest immediately while the resource is usable, even before a partner arrives. After pairing is established, individuals may become aggressive toward intruders of either sex (Pukowski, 1933; Kirkendall, 1983; Schuster and Schuster, 1985; Linsenmair, 1987; Nalepa and Jones, 1991).

Burying beetles demonstrate a revealing change in aggressiveness as nesting ensues. Initially, only same-sex conspecifics and heterospecifics are excluded. After the nest is formed and oviposition begins, residents become aggressive toward opposite-sex intruders because successful intruders will cannibalize the entire brood (Trumbo, 1990a). Even so, residents with brood attack same-sex intruders more vigorously than opposite-sex intruders (S. T. Trumbo, unpublished results). Differential aggressiveness likely occurs because individuals expelled by a same-sex rival will receive no reproductive benefit from the carrion resource. On the other hand, expulsion of one's mate, followed by infanticide and pairing with the intruder, often results in a replacement brood (Trumbo, 1990b). The cost of replacement by an opposite-sex intruder, therefore, can be considerably less than the cost of replacement by a same-sex intruder. Less aggression toward opposite-sex intruders might therefore be expected, despite the certainty of infanticide following a takeover.

Restrictions on male promiscuity, either through the withholding of copulations or via female synchrony of sexual receptivity, have been proposed as mechanisms promoting paternal investment in biparental species (Thornhill and Alcock, 1983; Zeh and Smith, 1985). Among invertebrates, however, the association of biparental care with a bonanza, contested resource suggests that restrictions on male promiscuity may play a less important role in the evolution of biparental care than resource ecology.

The potential for polygyny is generally not as great among species with biparental care as with exclusive male care. Extended paternal care can severely limit male access to additional females (Eggert, 1992; Reid and Roitberg, 1994) and, in some cases, males may reject supernumerary females. Polygyny and biparental care, however, are compatible in some species. Where resource availability varies, males may assess resource quality and then reject or accept additional females accordingly. In the bark beetle, *Ips confusus,* males usually accept up to three females but often reject the fourth, presumably because of the negative effects of crowding within galleries (Borden, 1967). In the burying beetle, *Nicrophorus defodiens,* males that locate a suitable carcass release a sex pheromone if a female is not present. After a female discovers the resource, the male normally terminates advertisement (Pukowski, 1933). When exploiting a large carcass (>40 g) that can support the brood of more than one female, however, males will often continue advertising after the initial pairing. Release of sex pheromone is inhibited once additional females are attracted to the carcass (Trumbo and Eggert, 1995). Males, then, take into account both the quality of the resource as well as the number of females present, when deciding whether to continue releasing pheromone. Polygyny enhances the male's but not the females' reproductive success. Paired females may attempt to interfere with male advertisement by pursuing, pushing, and climbing on top of a male that continues to advertise (Eggert and Sakaluk, 1995; Eggert and Müller, in press).

B. RESOURCES

Uniparental male care among invertebrates rarely occurs in proximity to a discrete food resource. Tallamy (in press) points out that of seven arthropod taxa that evolved exclusive male care, six consist entirely of predators that hunt, and one feeds on scattered detritus. Both of these foraging strategies require mobility, and predation further necessitates agility. Thus parental care, whether involving carrying young or occurring at one site, is in direct conflict with the needs of mobile, agile foragers. The costs associated with reduced foraging may not be borne equally by females and males because poor nutrition diminishes egg production more than sperm production (Tallamy, 1994; R. L. Smith, in press). This was demonstrated using the giant water bug, *Belostoma flumineum;* fecundity of females, but not number of offspring sired by males, was reduced by a poor diet during the adult stage (R. L. Smith and D. W. Tallamy, unpublished results). When there is selection for parental care in active foragers, males may be able to bear the costs of care more easily than females ("enhanced fecundity" of female hypothesis, Tallamy, 1994, in press).

The function of male care in uniparental species has been addressed by removing the male from his brood. In each case in which it has been investigated, uniparental male care reduces predation and/or parasitism (Odhiambo, 1959; Ralston, 1977; Mora, 1990). In aquatic species in which males carry eggs, paternal care serves the additional function of facilitating hatching (King and Jarvis, 1970; R. L. Smith, 1976).

Biparental care occurs, for the most part, in species nesting in or in proximity to a critical, discrete resource that has the potential to support the nutritional or shelter needs of a large number of brood (the major exception being found among the biparental Hymenoptera). The pattern of resources utilized by species with uniparental male versus biparental care is thus quite different. Emlen and Oring (1977) suggested that uniparental male care may evolve from the ancestral state of biparental care by female desertion of caring males. Among invertebrates, however, uniparental male and biparental care exhibit little overlap in either phylogeny or the type of resource exploited, suggesting that uniparental male care rarely evolved from biparental care.

Among invertebrates with biparental care, removal of the male parent always results in reduced brood production in experimental manipulations (Peckham, 1977; Cook, 1990; Trumbo, 1991; Amman and Bartos, 1991; Nalepa, 1994; Tallamy, 1994). Where biparental care occurs among the Hymenoptera, male nest guarding is thought to have evolved from territorial behavior (maintaining access to females returning to the nest) (Alcock, 1975). The presence of the male near the nest while the female is foraging reduces attacks by predators and parasites (Hook and Matthews, 1980; Coville and Coville, 1980). In some *Polistes,* the additional task of brood care has evolved (Hunt and Noonan, 1979; Cameron, 1986). Among other biparental species, male care often serves many functions and may involve resource procurement, nest preparation, and provisioning of offspring (see Zeh and Smith, 1985; Tallamy and Wood, 1986). The function of male care varies with the type of resource. Where species exploit resources with a high nitrogen content such as carrion or dung, males help secure the resource from intraspecific and interspecific competitors (Halffter et al., 1974; Trumbo, 1994; Scott, 1994). Where paternal care is extended, males aid in provisioning young (Halffter and Edmonds, 1982; Fetherston et al., 1994). In species that exploit low-quality resources such as wood or leaves, preparing the nest and enhancing the nutritional value of the resource may be just as important as defense (Montieth and Storey, 1981; Kirkendall, 1983; Schuster and Schuster, 1985; Nalepa and Jones, 1991; Matsumoto, 1992; Tallamy, 1994).

The most complex forms of biparental care are sometimes thought to involve strict division of labor by the sexes (Zeh and Smith, 1985). In most examples of sexual division of labor, however, male care is a simple extension of nest-mate guarding, and male participation is usually limited to the oviposition and nest initiation stages, as is found in Hymenoptera and in many dung beetles (Peckham, 1977; Halffter, 1977; Sato and Immamori, 1988). Zeh and Smith cite the example of biparental *Hemilepistus* as having complex care and division of labor. Task specialization in this species is not pronounced, however, since males and females have similar parental repertoires and simply take turns foraging and guarding (Linsenmair, 1987). Similar high levels of paternal care are found in other biparental species with little division of labor, including those with the longest familial associations. In such species, each sex can take the duties of the other; specialization may be limited to quantitative differences in behavioral tendencies (Fetherston et al., 1990). This pattern of biparental care is found in *Cryptocercus* (Seelinger and Seelinger, 1983), termites (Nalepa and Jones, 1991), passalid beetles (Valenzuela-González, 1993; Schuster and Schuster, 1985), and burying beetles (Fetherston et al., 1990, 1994). Notably, in these species males and females often search for resources independently and the male may initiate nesting without the female in order to quickly exploit the resource (Pukowski, 1933; Schuster and Schuster, 1985; Nalepa and Jones, 1991).

The effectiveness of flexible parental repertoires versus strict sexual division of labor is demonstrated by mate compensation in species with complex care. When the dampwood termite, *Zootermopsis nevadensis*, is fed a nitrogen-poor diet, females restrict their activity, presumably to retain energy for egg production. Males, however, sustain high activity during colony initiation. This shifts the costs of nest initiation from the female to the male at a time when poor nutrition has a significant impact on female fecundity (Shellman-Reeve, 1990). Mate compensation has been demonstrated most convincingly in the burying beetle, *Nicrophorus orbicollis*. When his mate is absent, the male parent adjusts by staying 4 days longer with the brood (ensuring that one parent is present until the brood disperses), and also by increasing active forms of parental care such as feeding the larvae and maintaining the nest (Trumbo, 1991; Fetherston et al., 1994). When care is limited to two individuals, rigid task specialization results in little room for error. Even in biparental species such as the leaf-gathering dung beetles, *Lethrus*, and *Cephalodesmius*, in which males forage while females manufacture brood balls, females are fully competent to forage on their own if partners no longer return with provisions.

### C. Future Ecological Studies of Paternal Care

Patterns of copulation suggest that when there is exclusive male care in invertebrates, females often are quite willing to mate with males that demonstrate evidence of caregiving. Studies that examine the cues that females use to assess males (presence of eggs, quality of nest, quality of male) will provide insight into the evolution of uniparental male care. In particular, it will be interesting to establish whether mate copying is related to the potential for polygyny, and thus has been important in the evolution of uniparental male care. Physiological studies will provide needed information on the extent to which uniparental male care shifts the costs of reproduction from the female to the male.

There has been considerable debate whether females withhold copulation to elicit male assistance in biparental species. In many cases, copulations occur frequently and prior to procurement and preparation of the resource. Physiological assays may allow for sensitive analyses of whether cues from the female, the resource, or from the brood stimulate males to initiate care. Such studies will enlighten the discussion of whether female coercion, the potential for polygyny, or resource ecology is most significant for shaping patterns of paternal investment.

## VI. Summary

Ecological and physiological analyses of invertebrate parental care need to be integrated. Consideration of phylogeny provides one starting point. R. L. Smith (in press) and Tallamy (in press) have provided detailed phylogenetic and ecological comparisons of the evolution of parental care among closely related taxa. There are no formidable barriers to extending the analysis in these and other groups to the level of physiology. Burying beetles (*Nicrophorus* spp.), for example, are thought to be derived from the nonparental *Ptomascopus* (tribe Nicrophorini). *Ptomascopus* utilizes a small carcass as a breeding resource as does *Nicrophorus*, but does not build a nest or stay with the offspring following oviposition (Peck, 1982). Endocrine analyses of both genera will provide information on how the novel adaptation of parental care was inserted within the reproductive cycle of a nonparental ancestor. In some cases, the endocrinology necessary for comparative studies with closely related nonparental species has already been done. The hormonal regulation of the reproductive behavior and physiology of mosquitoes has been worked out in detail, largely because of the importance of this group as vectors for disease. The maternal mosquito, *Trichoprosopon digitatum*, (Lounibos and Machado-Allison, 1986) would provide an excellent subject for an investigation of the comparative physiology of care.

Broad phylogenetic comparisons have the potential to address how physiology constrains the expression of parental care. Parental care in numerous independent phylogenetic lines of insects, for example, is regulated by the same neural and endocrine structures. Are the importance of the various neural and endocrine factors similar in all groups, or has phylogeny and/or ecology affected the evolution of physiological regulation? Do the types of sensory cues employed to stimulate care in a particular group affect the direction that evolution has taken, the possibility of parent–offspring conflict, or mate compensation in biparental species?

For many reasons, invertebrates will be used to address these questions, not the least being the ability to perform ecological and physiological manipulations on large numbers of individuals in a short period of time. Some of the advantages of invertebrates as experimental subjects have not been exploited in studies of parental care. Breeding experiments to select for high and low lines of caregiving will prove useful for uncovering the genetic basis of care. Genetic effects on parental care among invertebrates have rarely been established (but see Tallamy and Dingle, 1986; Robinson et al., 1989). Individuals from selected genetic lines can be employed to examine the physiological differences between individuals with varying tendencies to express parental behavior; genetic lines also will be useful in field experiments investigating the ecological trade-offs of adopting alternative patterns of investment. The effects of development on the expression of care have been neglected as well. The ease with which the developmental environment of many immature invertebrates can be manipulated suggests that the lack of understanding is caused by neglect and not by experimental barriers.

The study of parental care among invertebrates will be renewed by the discovery of new species, the discovery of caregiving among previously described species, and the discovery of new functions of care in those known to be parental. The challenge of explaining the known diversity of care is formidable in its own right. The inclusion of further empirical studies as well as theoretical insights promise to make the investigation of parental care among invertebrates a paradigmatic study of how parallel social adaptations evolve among diverse taxa.

### Acknowledgments

Scott Kight, Joanna Mappes, Christine Nalepa, Susan Rankin, Gene Robinson, Michelle Scott, Robert Smith, Charles Snowdon, Douglas Tallamy, Sue Trumbo, and Kentwood Wells provided instructive comments on earlier drafts of this paper. Manuscripts and discussion of unpublished results were kindly provided by Curtis Creighton, Scott Kight, Joanna Mappes, Susan Rankin, Robert Smith, Douglas Tallamy, and L. Thomas. Mary Jane Spring contributed the figures. This work was supported by NSF grants IBN-9203261 and IBN-9420985 to the author.

### References

Aclé, D., Brookes, V. J., Pratt, G. E., and Feyereisen, R. (1990). Activity of the corpora allata of adult female *Leucophaea maderae:* Effects of mating and feeding. *Arch. Insect Biochem. Physiol.* **14,** 121–129.

Alcock, J. (1975). Territorial behavior by males of *Philanthus multimaculatus* (Hymenoptera: Sphecidae) with a review of male territoriality in male sphecids. *Anim. Behav.* **23,** 889–895.
Amman, G. D., and Bartos, D. L. (1991). Mountain pine beetle offspring characteristics associated with females producing first and second broods, male presence, and egg gallery length. *Environ. Entomol.* **20,** 1562–1567.
Anderson, R. S. (1982). Burying beetle larvae: Nearctic *Nicrophorus* and Oriental *Ptomascopus. Syst. Entomol.* **7,** 249–264.
Andersson, M., and Eriksson, O. G. (1982). Nest parasitism in goldeneyes *Bucephala clangula:* Some evolutionary aspects. *Am. Nat.* **120,** 1–16.
Ashe, J. S. (1987). Egg chamber production, egg protection and clutch size among fungivorus beetles of the genus *Eumicrota* (Coleoptera: Staphylinidae) and their evolutionary implications. *Zool. J. Linn. Soc.* **90,** 255–273.
Atkins, M. D. (1980). "Introduction to Insect Behavior." Macmillan, New York.
Baehr, J. Cl., Cassier, P., Caussanel, Cl., and Porcheron, P. (1982). Activity of corpora allata, endocrine balance and reproduction in female *Labidura riparia* (Dermaptera). *Cell and Tissue Research* **225,** 267–282.
Bartlett, J. (1987). Filial cannibalism in burying beetles. *Behav. Ecol. Sociobiol.* **21,** 179–183.
Bartlett, J., and Ashworth, C. M. (1988). Brood size and fitness in *Necrophorus vespilloides. Behav. Ecol. Sociobiol.* **22,** 429–434.
Basolo, A. L. (1990). Female preference predates the evolution of the sword in swordtail fish. *Science (Washington, D. C.)* **250,** 808–810.
Borden, J. H. (1967). Factors influencing the response of *Ips confusus* (Coleoptera: Scolytidae) to male attractant. *Can. Entomol.* **99,** 1164–1183.
Bowers, W. S., Ohta, T., Cleere, J. S., and Marsella, P. A. (1976). Discovery of insect anti-juvenile hormones in plants. *Science (Washington, D. C.)* **193,** 542–547.
Bristow, C. M. (1983). Treehoppers transfer parental care to ants: a new benefit of mutualism. *Science (Washington, D. C.)* **220,** 532–533.
Brockelman, W. Y. (1975). Competition, the fitness of offspring, and optimal clutch size. *Am. Nat.* **109,** 677–699.
Burley, N. (1977). Parental investment, mate choice and mate quality. *Proc. Natl. Acad. Sci. USA* **74,** 3476–3479.
Buskirk, R. E. (1981). Sociality in the Arachnida. *In* "Social Insects, Volume II" (H. Herrman, ed.), pp. 282–367. Academic Press, New York.
Buxton, P. A. (1955). "The Natural History of Tsetse Flies." Lewis & Co., London.
Calow, P. (1979). The cost of reproduction: A physiological approach. *Biol. Rev.* **54,** 23–40.
Cameron, S. A. (1986). Brood care by males of *Polistes major* (Hymenoptera: Vespidae). *J. Kans. Entomol. Soc.* **59,** 183–185.
Caussanel, Cl. (1970). Principles exigences écophysiologiques du forficule des sables, *Labidura riparia* (Derm. Labiduridae). *Ann. Soc. Entomol. Fr.* **6,** 589–612.
Caussanel, Cl., Breuzet, M., and Karlinsky, A. (1978). Intervention du systeme neuroendocrine cérébral dans le comportement parental de la femelle de *Labidura riparia* (Insecte, Dermaptère). *C. R. Acad. Sci. (Paris) Ser. D* **286,** 1699–1702.
Choe, J. C. (1989). Maternal care in *Labidomera suturella* Chevrolat (Coleoptera: Chyromelidae: Chyromelinae) from Costa Rica. *Psyche* **96,** 63–67.
Clutton-Brock, T. H. (1991). "The Evolution of Parental Care." Princeton Univ. Press, Princeton, NJ.
Cook, D. F. (1988). Sexual selection in dung beetles. II. Female fecundity as an estimate of male reproductive success in relation to horn size, and alternative behavioural strategies in *Onthophagus binodis* (Coleoptera: Scarabaeidae). *Aust. Zool.* **36,** 521–532.
Cook, D. F. (1990). Differences in courtship, mating and postcopulatory behavior between male morphs of the dung beetle *Onthophagus binodis* Thunberg (Coleoptera: Scarabaeidae). *Anim. Behav.* **40,** 428–436.

Coville, R. E., and Coville, P. L. (1980). Nesting biology and male behavior of *Trypoxylon* (*Trypargilum*) *tenoctitlan* in Costa Rica (Hymenoptera: Sphecidae). *Ann. Entomol. Soc. Am.* **73**, 110–119.

Cowan, D. P. (1981). Parental investment in two solitary wasps *Ancistrocerus adiabatus* and *Euodynerus foraminatus* (Eumenidae: Hymenoptera). *Behav. Ecol. Sociobiol.* **9**, 95–102.

Crespi, B. J. (1990). Subsociality and female reproductive success in a mycophagous thrips: An observational and experimental analysis. *J. Insect Behav.* **3**, 61–74.

Crews, D. (1975). Psychobiology of reptilian reproduction. *Science (Washington, D. C.)* **189**, 1059–1065.

Crowl, T. A., and Alexander, J. E., Jr. (1989). Parental care and foraging ability in male water bugs (*Belostoma Flumineum*). *Can. J. Zool.* **67**, 513–515.

Cusson, M., and McNeil, J. N. (1989). Involvement of juvenile hormone in the regulation of pheromone release activities in a moth. *Science (Washington, D. C.)* **243**, 210–212.

Denno R. F., and Cothran, W. R. (1976). Competitive interactions and ecological strategies in sarcophagid and calliphorid flies inhabiting rabbit carrion. *Ann. Entomol. Soc. Am.* **69**, 103–113.

Diesel, R. (1989). Parental care in an unusual environment: *Metopaulias depressus* (Decapoda: Grapsidae), a crab that lives in epiphytic bromeliads. *Anim. Behav.* **38**, 561–575.

Diesel, R. (1992). Managing the offspring environment: Brood care in the bromeliad crab, *Metopaulias depressus. Behav. Ecol. Sociobiol.* **30**, 125–134.

Diesel, R., and Schuh, M. (1993). Maternal care in the bromeliad crab, *Metopaulias depressus* (Decapoda): Maintaining oxygen, pH and calcium levels optimal for the larvae. *Behav. Ecol. Sociobiol.* **32**, 11–15.

Dugatkin, L. A. (1992). Sexual selection and imitation: females copy the mate choice of others. *Am. Nat.* **139**, 1384–1389.

Dusenbery, D. B. (1992). "Sensory Ecology: How Organisms Acquire and Respond to Information." W. H. Freeman, New York.

Eberhard, W. G. (1975). The ecology and behavior of a subsocial pentatomid bug and two scelionid wasps: Strategy and counterstrategy in a host and its parasitoids. *Smithson. Contrib. Zool.* **205**, 1–39.

Eberhard, W. G. (1982). Beetle horn dimorphism: Making the best of a bad lot. *Am. Nat.* **119**, 420–426.

Eberhard, W. G. (1986). Possible mutualism between females of the subsocial membracid *Polyglypta dispar* (Homoptera). *Behav. Ecol. Sociobiol.* **19**, 447–453.

Edgerly, J. S. (1987). Maternal behavior of a webspinner (Order Embiidina). *Ecol. Entomol.* **12**, 1–11.

Edgerly, J. S. (in press). Life beneath silk walls: a review of the primitively social embiidina. *In* "Social Competition and Cooperation in Insects and Arachnids, Volume II, Evolution of Sociality" (B. J. Crespi and J. C. Choe, eds.), Princeton Univ. Press, Princeton, NJ.

Edman, J. D., and Lynn, H. C. (1975). Relationship between blood meal volume and ovarian development in *Culex nigripalpus* (Diptera: Culicidae). *Entomol. Exp. Appl.* **18**, 492–496.

Edwards, P. B., and Aschenborn, H. H. (1989). Maternal care of a single offspring in the dung beetle *Kheper nigroaeneus:* The consequences of extreme parental investment. *J. Nat. Hist.* **23**, 17–27.

Eggert, A.-K. (1992). Alternative male mate-finding tactics in burying beetles. *Behav. Ecol.* **3**, 243–254.

Eggert, A.-K., and Müller, J. K. (in press). Biparental care and social evolution in burying beetles: Lessons from the larder. *In* "Social Competition and Cooperation in Insects and Arachnids, Volume II, Evolution of Sociality" (B. J. Crespi and J. C. Choe, eds.), Princeton Univ. Press, Princeton, NJ.

Eggert, A.-K., and Sakaluk, S. K. (1995). Female-coerced monogamy in burying beetles. *Behav. Ecol. Sociobiol.* **37,** 147-153.
Eickwort, G. C. (1981). Presocial insects. *In* "Social Insects, Volume II" (H. Hermann, ed.), pp. 199-280. Academic Press, New York.
Emlen, S., and Oring, L. (1977). Ecology, sexual selection and the evolution of mating systems. *Science (Washington, D. C.)* **197,** 215-223.
Endler, J. A. (1992). Signals, signal conditions, and the direction of evolution. *Am. Nat.* **139,** S125-S153.
Engelmann, F. (1957). Die Steuerung der Ovarfunktion bei der ovoviviparen Schabe *Leucophaea maderae* (Fabr.). *J. Insect Physiol* **1,** 257-278.
Engelmann, F. (1970). "The Physiology of Insect Reproduction." Pergamon Press, Oxford.
Evans, H. E. (1958). The evolution of social life in wasps. *Proc. Int. Congr. Entomol., 10th,* **1958,** 449-456.
Faeth, S. H. (1989). Maternal care in a lace bug, *Corythuca hewitti* (Hemiptera: Tingidae). *Psyche* **96,** 101-110.
Fetherston, I. A., Scott, M. P., and Traniello, J. F. A. (1990). Parental care in burying beetles: The organization of male and female brood care behavior. *Ethology* **85,** 177-190.
Fetherston, I. A., Scott, M. P., and Traniello, J. F. A. (1994). Behavioral compensation for mate loss in the burying beetle, *Nicrophorus orbicollis. Anim. Behav.* **47,** 777-785.
Feyereisen, R. (1985). Regulation of juvenile hormone titer: synthesis. *In* "Comprehensive Insect Physiology, Biochemistry and Pharmacology, Volume 7, Endocrinology II" (G. A. Kerkut and L. I. Gilbert, eds.), pp. 391-429. Pergamon Press, Oxford.
Field, J. (1992). Patterns of nest provisioning and parental investment in the solitary digger wasp, *Ammophila sabulosa. Ecol. Entomol.* **17,** 43-51.
Filippi-Tsukamoto, L., Nomakuchi, S., Kuki, K., and Tojo, S. (1995). Adaptiveness of parental care in *Parastrachia japonesis* (Hemiptera: Cydnidae). *Ann. Entomol. Soc. Am.* **88,** 374-383.
Finch, C. E., and Rose, M. R. (1995). Hormones and the physiological architecture of life history evolution. *Q. Rev. Biol.* **70,** 1-52.
Fink, L. S. (1986). Costs and benefits of maternal behavior in the green lynx spider (Oxyopidae, *Peucetia viridans*). *Anim. Behav.* **34,** 1051-1060.
Gadot, M., Burns, E., and Schal, C. (1989a). Juvenile hormone biosynthesis and oocyte development in adult female *Blatella germanica:* Effects of grouping and mating. *Arch. Insect Biochem. Physiol.* **11,** 189-200.
Gadot, M., Chiang, A.-S., and Schal, C. (1989b). Farnesoic acid-stimulated rates of juvenile hormone biosynthesis during the gonotrophic cycle in *Blatella germanica. J. Insect Physiol* **35,** 537-542.
Godfray, H. C. J. (1987). The evolution of clutch size in invertebrates. *In* "Oxford Surveys of Evolutionary Biology, Volume 4" (P. H. Harvey and L. Partridge, eds.), pp. 117-154. Oxford Press, Oxford.
Goodman, W. G., Huang, Z.-Y., Robinson, G. E., Stambi, A., and Strambi, C. (1993). A comparison of two juvenile hormone radioimmunoassays. *Arch. Insect Biochem. Physiol.* **23,** 147-152.
Haack, R. A., and Slansky, F., Jr. (1987). Nutritional ecology of wood-feeding Coleoptera, Lepidoptera, and Hymenoptera. *In* "Nutritional Ecology of Insects, Mites, Spiders and Related Invertebrates" (F. Slansky, Jr. and J. G. Rodriguez, eds.), pp. 449-486. Wiley, New York.
Hagedorn, H. H. (1985). The role of ecdysteroids in reproduction. *In* "Comprehensive Insect Physiology, Biochemistry and Pharmacology, Volume 8, Endocrinology II" (G. A. Kerkut and L. I. Gilbert, eds.), pp. 205-262. Pergamon Press, Oxford.

Halffter, G. (1977). Evolution of nidification in the Scarabaeinae (Coleoptera, Scarabaeidae). *Quaestiones Entomologicae* **13**, 231–253.

Halffter, G., and Edmonds, W. D. (1982). The nesting behavior of dung beetles (Scarabaeinae): An ecological and evolutive approach. *Publ. Inst. Ecol. (Mexico)* **10**, 1–176.

Halffter, G., and Lopez, Y. (1977). Development of the ovary and mating behavior in *Phanaeus*. *Ann. Entomol. Soc. Am.* **70**, 203–213.

Halffter, G., Halffter, V., and Lopez, I. (1974). *Phanaeus* behavior: Food transportation and bisexual cooperation. *Environ. Entomol.* **3**, 341–345.

Halffter, G., Anduaga, S., and Huerta, C. (1983). Nidification des *Nicrophorus*. *Bull. Soc. Entomol. Fr.* **88**, 648–666.

Hamilton, W. D. (1967). Extraordinary sex ratios. *Science (Washington, D.C.)* **156**, 477–488.

Hardy, I. C. W., and Blackburn, T. M. (1991). Brood guarding in a bethylid wasp. *Ecol. Entomol.* **16**, 55–62.

Hinde, R. A. (1970). "Animal Behavior: A Synthesis of Ethology and Comparative Psychology." McGraw-Hill, New York.

Hinde, R. A. (1975). The concept of function. In "Function and Evolution in Behavior" (G. P. Baerends, C. Beer, and A. Manning, eds.), p. 415. Clarendon Press, Oxford.

Hinton, H. E. (1981). "Biology of Insect Eggs, Volume I." Pergamon Press, Elmsford, NY.

Hook, A. W., and Matthews, R. W. (1980). Nesting biology of *Oxybelus sericeus* with a discussion of nest guarding by male sphecid wasps (Hymenoptera). *Psyche* **87**, 21–37.

Huang, Z.-Y., and Otis, G. W. (1991). Inspection and feeding of larvae by worker honey bees (Hymenoptera: Apidae): Effect of starvation and food quantity. *J. Insect Behav.* **4**, 305–317.

Huang, Z.-Y., Robinson, G. E., and Borst, D. W. (1994). Physiological correlates of division of labor among similarly aged honey bees. *J. Comp. Physiol. A* **174**, 731–739.

Hunt, J. H., and Noonan, K. C. (1979). Larval feeding by male *Polistes fuscatus* and *Polistes metricus* (Hymenoptera: Vespidae). *Insect Soc.* **26**, 247–251.

Ichikawa, N. (1988). Male brooding behavior of the giant water bug *Lethocerus deyrollei* Vuillefroy (Heteroptera: Belostomatidae). *J. Ethol.* **6**, 121–127.

Ichikawa, N. (1995). Male counterstrategy against infanticide of the female giant water bug *Lethocerus deyrollei* (Hemiptera: Belostomatidae). *J. Insect Behav.* **8**, 181–188.

Ingram, M. J., Stay, B., and Cain, G. D. (1977). Composition of milk from the viviparous cockroach, *Diploptera punctata*. *Insect Biochem.* **7**, 257–267.

Kaestner, A. (1968). "Invertebrate Zoology, Volume 2, Arthropod Relatives, Chelicerata, Myriapoda" (translated from the German by H. W. Levi and L. R. Levi). Interscience Publishers, New York.

Kearns, R. S., and Yamamoto, R. T. (1981). Maternal behavior and alarm response in the eggplant lace bug, *Gargaphia solani* Heidemann (Tingidae: Heteroptera). *Psyche* **88**, 215–230.

Ketterson, E. D., and Nolan, V., Jr. (1992). Hormones and life histories: an integrative approach. *Am. Nat.* **140**, 533–562.

Ketterson, E. D., Nolan, V., Jr., Wolf, L., and Ziegenfus, C. (1992). Testosterone and avian life histories: Effects of experimentally elevated testosterone on behavior and correlates of fitness in the dark-eyed junco (*Junco hyemalis*). *Am. Nat.* **140**, 980–999.

Kight, S. L. (1995). Do maternal burrower bugs *Sehirus cinctus* Palisot (Heteroptera: Cydnidae) use spatial and chemical cues for egg discrimination? *Can. J. Zool.* **73**, 815–817.

Kight, S. L., and Kruse, K. C. (1992). Factors affecting the allocation of paternal care in waterbugs (*Belostoma flumineum* Say). *Behav. Ecol. Sociobiol.* **30**, 409–414.

Kight, S. L., Sprague, J., Kruse, K. C., and Johnson, L. (1995). Are egg-bearing male water bugs, *Belostoma flumineum* Say (Hemiptera: Belostomatidae), impaired swimmers? *J. Kans. Entomol. Soc.* **68**, 468–470,

King, P. E., and Jarvis, J. H. (1970). Egg development in a littoral pyncnogonid *Nymphon gracile. Mar. Biol.* **7,** 294–304.

Kirkendall, L. R. (1983). The evolution of mating systems in bark and ambrosia beetles (Coleoptera: Scolytidae and Platypodidae). *Zool. J. Linn. Soc.* **77,** 293–352.

Klemperer, H. G. (1982). Parental behavior in *Copris lunaris* (Coleoptera, Scarabaedidae): Care and defence of brood balls and nest. *Ecol. Entomol.* **7,** 155–167.

Klemperer, H. G. (1983a). Subsocial behavior in *Oniticellus cinctus* (Coleoptera, Scarabaeidae): Effect of the brood on parental care and oviposition. *Physiol. Entomol.* **8,** 393–402.

Klemperer, H. G. (1983b). The evolution of parental behavior in Scarabaeinae (Coleoptera, Scarabaeidae): An experimental approach. *Ecol. Entomol.* **8,** 49–59.

Koeppe, J. K., Fuchs, M., Chen, T. T., Hunt, L. M., Kovalick, G. E., and Briers, T. (1985). The role of juvenile hormone in reproduction. *In* "Comprehensive Insect Physiology, Biochemistry and Pharmacology, Volume 8" (G. A. Kerkut and L. I. Gilbert, eds.), pp. 165–204. Pergamon Press, Oxford.

Kudô, S. (1990). Brooding behavior in *Elasmucha putoni* (Heteroptera: Acanthosomatidae), and a possible nymphal alarm substance triggering guarding responses. *Appl. Entomol. Zool.* **25,** 431–437.

Kudô, S. (1996). Ineffective maternal care of a subsocial bug against a nymphal parasitoid: A possible consequence of specialization to predators. *Ethology* **102,** 227–235.

Lamb, R. J. (1976). Parental behavior in the Dermaptera with special reference to *Forficula auricularia* (Dermaptera: Forficulidae). *Can. Entomol.* **108,** 609–619.

Larrere, M., and Couillaud, F. (1993). Role of juvenile hormone biosynthesis in dominance status and reproduction of the bumblebee, *Bombus terrestris. Behav. Ecol. Sociobiol.* **33,** 335–338.

LeConte, Y., Sreng, L., and Trouiller, J. (1994). The recognition of larvae by worker honeybees. *Naturwissenschaften* **81,** 462–465.

Lee, J. M., and Peng, Y. S. (1981). Influence of adult size of *Onthophagus gazella* on manure pat degradation, nest construction, and progeny size. *Environ. Entomol.* **10,** 626–630.

Lehrman, D. S. (1965). Interaction between internal and external environments in the regulation of the reproductive cycle of the ring dove. *In* "Sex and Behavior" (F. A. Beach, ed.), pp. 335–380. Wiley, New York.

Liechti, P. M., and Bell, W. J. (1975). Brooding behavior of the Cuban burrowing cockroach *Byrsotria fumigata* (Blaberidae, Blattaria). *Insect Soc.* **22,** 35–46.

Linsenmair, K. E. (1987). Kin recognition in subsocial arthropods, in particular in the desert isopod *Hemilepistus reamuri. In* "Kin Recognition in Animals" (D. J. C. Fletcher and C. D. Michener, eds.), pp. 121–208. Wiley, New York.

Little, E. E. (1976). Ontogeny of maternal behavior and brood pheromone in crayfish. *J. Comp. Physiol.* **112,** 133–142.

Lloyd, D. G. (1987). Selection of offspring size at independence and other size-versus-number strategies. *Am. Nat.* **129,** 800–817.

Lounibos, L. P., and Machado-Allison, C. E. (1986). Mosquito maternity: Egg brooding in the life cycle of *Trichoprosopon digitatum. In* "The Evolution of Insect Life Cycles" (Taylor and Karban, eds.), pp. 173–184. Springer-Verlag, Berlin.

Mappes, J., and Kaitala, A. (1994). Experiments with *Elasmucha grisea* L. (Heteroptera: Acanthosomatidae): Does a female parent bug lay as many eggs as she can defend? *Behav. Ecol.* **5,** 314–317.

Mappes, J., and Kaitala, A. (1995). Host-plant selection and predation risk for offspring of the parent bug. *Ecology* **76,** 2668–2670.

Mappes, J., Mappes, T., and Lappalainen, T. (in press). Unequal maternal investment in offspring quality in relation to predation risk. *Evol. Ecol.*

Marler, C. A., and Moore, M. C. (1988). Evolutionary costs of aggression revealed by testosterone manipulations in free-living male lizards. *Behav. Ecol. Sociobiol.* **23,** 21–26.

Martinez, I., and Caussanel, C. (1984). Modification de la pars intercerebralis, des corpora allata, des gonades et comportment reproducteur chez *Canthon cyanellus*. *C. R. Acad. Sci.* (*Paris*) III **14,** 597–602.

Maschwitz, U., and Gutmann, C. (1979). Trail and alarm pheromones in *Elasmucha grisea* (Heteroptera: Acanthosomidae). *Insect Soc.* **26,** 101–111.

Matsumoto, T. (1992). Familial association, nymphal development and population density in the Australian giant burrowing cockroach, *Macropanesthia rhinoceros* (Blattaria: Blaberidae). *Zool. Sci.* **9,** 835–842.

Matthews, R. W. (1970). A new thrips-hunting *Microstigmus* from Costa Rica (Hymenoptera: Sphecidae, Pemphredoninae). *Psyche* **77,** 120–126.

McLay, C. L., and Hayward, T. L. (1987). Reproductive biology of the intertidal spider *Desis marina* (Araneae: Desidae) on a New Zealand rocky shore. *J. Zool.* (*London*) **211,** 357–372.

Melber, A., and Schmidt, G. (1975). Sozialverhalten zweir *Elasmucha*-Arten (Heteroptera: Insecta). *Z. Tierpsychol.* **39,** 403–414.

Michener, C. D. (1969). Comparative social behavior of bees. *Annu. Rev. Entomol.* **14,** 299–342.

Mileikovsky, S. A. (1971). Types of larval development in marine bottom invertebrates, their distribution and ecological significance: a re-evaluation. *Mar. Biol.* **10,** 193–213.

Miller, N. C. E. (1971). "The Biology of the Heteroptera" Biddles Ltd., Guildford, England.

Montieth, G. B., and Storey, R. I. (1981). The biology of *Cephalodesmius,* a genus of dung beetle which synthesizes "dung" from plant material (Coleopetera: Scarabaeidae: Scarabaeinae). *Mem. Queensl. Mus.* **20,** 253–277.

Mora, G. (1990). Paternal care in a neotropical harvestman, *Zygopachylus albomarginis* (Arachhnida, Opiliones: Gonyleptidae). *Anim. Behav.* **39,** 582–593.

Müller, J. K. (1987). Replacement of a lost clutch: A strategy for optimal resource utilization in *Necrophorus vespilloides* (Coleoptera: Silphidae). *Ethology* **76,** 74–80.

Müller, J. K., and Eggert, A.-K. (1989). Paternity assurance by 'helpful' males: Adaptations to sperm competition in burying beetles. *Behav. Ecol. Sociobiol.* **24,** 245–249.

Müller, J. K., and Eggert, A.-K. (1990). Time-dependent shifts in infanticidal and parental behavior in female burying beetles: A mechanism of indirect parent-offspring recognition. *Behav. Ecol. Sociobiol.* **27,** 11–16.

Müller, J. K., Eggert, A.-K., and Furlkröger, E. (1990a). Clutch size regulation in the burying beetle *Necrophorus vespilloides* Herbst (Coleoptera: Silphidae). *J. Insect Behav.* **3,** 265–270.

Müller, J. K., Eggert, A.-K., and Dressel, J. (1990b). Intraspecific brood parasitism in the burying beetle, *Necrophorus vespilloides* (Coleoptera: Silphidae). *Anim. Behav.* **40,** 491–499.

Nafus, D. M., and Schreiner, I. H. (1988). Parental care in a tropical nymphalid butterfly *Hypolimnas anomala*. *Anim. Behav.* **36,** 1425–1431.

Nalepa, C. A. (1988). Cost of parental care in the woodroach *Cryptocercus punctulatus* Scudder (Dictyoptera: Cryptocercidae). *Behav. Ecol. Sociobiol.* **23,** 135–140.

Nalepa, C. A. (1994). Nourishment and the origin of termite sociality. *In* "Nourishment and Evolution in Insect Societies" (J. H. Hunt and C. A. Nalepa, eds.), pp. 57–104. Westview Press, Boulder, CO.

Nalepa, C. A., and Jones, S. C. (1991). Evolution of monogamy in termites. *Biol. Rev.* **66,** 83–97.

Nalepa, C. A., and Mullins, D. E. (1992). Initial reproductive investment and parental body size in *Cryptocercus punctulatus* (Dictyoptera: Cryptocercidae). *Physiol. Entomol.* **17,** 255–259.

Odhiambo, T. R. (1959). An account of parental care in *Rhinocoris albopilosus* Signoret (Hemiptera: Reduviidae), with notes on its life history. *Proc. R. Entomol. Soc. London A* **34,** 175–185.

O'Donnell S., and Jeanne, R. L. (1993). Methoprene accelerates age polyethism in workers of a social wasp (*Polybia occidentalis*). *Physiol. Entomol.* **18,** 189–194.

Olmstead, K. L., and Wood, T. K. (1990). The effect of clutch size and ant attendance on egg guarding by *Entylia bactriana* (Homoptera: Membracidae). *Psyche* **97,** 111–120.

Otronen, M. (1988). The effect of body size on the outcome of fights in burying beetles (*Nicrophorus*). *Ann. Zool. Fenn.* **25,** 191–201.

Peck, S. B. (1982). The life history of the Japanese carrion beetle *Ptomascopus morio* and the origins of parental care in *Nicrophorus* (Coleoptera: Silphidae, Nicrophorini). *Psyche* **89,** 107–111.

Peckham, D. J. (1977). Reduction of miltogrammine cleptoparasitism by male *Oxybelus subulatus* (Hymenoptera: Sphecidae). *Ann. Entomol. Soc. Am.* **70,** 823–828.

Pierre, J.-S. (1978). Effect de l'allatectomie sue la fin du cycle parental chez *Labidura riparia* Pallas (Dermapteres-Labiduridae). *Biol. Rev.* **4,** 219–226.

Pratte, M., and Jeanne, R. L. (1984). Antennal drumming behavior in *Polistes* wasps (Hymenoptera: Vespidae). *Z. Tierpsychol.* **66,** 177–188.

Preston-Mafham, R., and Preston-Mafham, K. (1993). "The Encyclopedia of Land Invertebrate Behavior." MIT Press, Cambridge, MA.

Pukowski, E. (1933). Ökologische untersuchungen an *Necrophorus* F. *Z. Morphol. Öekol. Tiere* **27,** 518–586.

Radl, R. C., and Linsenmair, K. E. (1991). Maternal behavior and nest recognition in the subsocial earwig *Labidura riparia* Pallas (Dermaptera: Labiduridae). *Ethology* **89,** 287–296.

Ralston, J. S. (1977). Egg guarding by male assassin bugs of the genus *Zelus* (Hemiptera: Reduviidae). *Psyche* 103–107.

Rankin, S. M., and Stay, B. (1985). Ovarian inhibition of juvenile hormone synthesis in the viviparous cockroach, *Diploptera punctata. Gen. Comp. Endocrinol.* **59,** 230–237.

Rankin, S. M., and Stay, B. (1987). Distribution of allatostatin in the adult cockroach, *Diploptera punctata* and effects on corpora allata *in vitro. J. Insect Physiol.* **33,** 551–558.

Rankin, S. M., Fox, K. M., and Stotsky, C. E. (1995a). Physiological correlates to courtship, mating, ovarian development and maternal behavior in the ring-legged earwig. *Physiol. Entomol.* **20,** 257–263.

Rankin, S. M., Palmer, J. O., Yagi, K. J., Scott, G. L., and Tobe, S. S. (1995b). Biosynthesis and release of juvenile hormone during the reproductive cycle of the ring-legged earwig. *Comp. Biochem. Physiol.* **110,** 241–251.

Reid, M. L., and Roitberg, B. D. (1994). Benefits of prolonged male residence with mates and brood in pine engravers (Coleoptera: Scolytidae). *Oikos* **70,** 140–148.

Robinson, G. E. (1987). Regulation of honey bee age polyethism by juvenile hormone. *Behav. Ecol. Sociobiol.* **20,** 329–338.

Robinson, G. E., Page, R. E., Strambi, C., and Strambi, A. (1989). Hormonal and genetic control of behavioral integration in honey bee colonies. *Science (Washington, D. C.)* **246,** 109–112.

Rollo, C. D. (1984). Resouce allocation and time budgeting in adults of the cockroach *Periplaneta americana:* The interaction of behavior and metabolic reserves. *Res. Pop. Ecol.* **26,** 150–187.

Röseler, P.-F., Röseler, I., Strambi, A., and Augier, R. (1984). Influence of insect hormones on the establishment of dominance hierarchies among foundresses of the paper wasp *Polistes gallicus. Behav. Ecol. Sociobiol.* **15,** 133–142.

Röseler, P.-F., Röseler, I., and Strambi, A. (1985). Role of ovaries and ecdysteroids in dominance hierarchy establishment among foundresses of the primitively social wasp *Polistes gallicus*. *Behav. Ecol. Sociobiol.* **18**, 9–13.

Rosenheim, J. A. (1987). Nesting behavior and bionomics of a solitary ground-nesting wasp, *Ammophila dysmica* (Hymenoptera: Sphecidae): Influence of parasite pressure. *Ann. Entomol. Soc. Am.* **80**, 739–749.

Roth, L. M., and E. R. Willis. (1960). The biotic associations of cockroaches. *Smithson. Misc. Collect.* **141**, 1–470.

Ruegg, D., and Rose, H. A. (1991). Biology of *Macropanesthia rhinoceros* Saussure (Dictyoptera: Balberidae). *Ann. Entomol. Soc. Am.* **84**, 575–582.

Ruttan, L. M. (1990). Experimental manipulations of dispersal in the subsocial spider, *Theridion pictum*. *Behav. Ecol. Sociobiol.* **27**, 169–173.

Ryan, M. J. (1990). Sexual selection, sensory systems, and sensory exploitation. *Oxford Surv. Evol. Biol.* **7**, 156–195.

Sato, H., and Hiramatsu, K. (1993). Mating behavior and sexual selection in the African ball-rolling scarab, *Kheper platynotus* (Bates) (Coleoptera: Scarabaeidae). *J. Nat. Hist.* **27**, 657–668.

Sato, H., and Immamori, M. (1987). Nesting behavior of a subsocial African ball-roller *Kheper platynotus* (Coleoptera: Scarabaeidae). *Ecol. Entomol.* **12**, 415–425.

Sato, H., and Immamori, M. (1988). Further observations on the nesting behavior of a subsocial ball-rolling scarab, *Kheper aegyptiorum*. *Kontyu* **56**, 873–878.

Scharrer, B. (1987). Insects as models in neuroendocrine research. *Annu. Rev. Entomol.* **32**, 1–16.

Schmidt, J. M., and Smith, J. J. B. (1987). Short interval time measurement by a parasitoid wasp. *Science (Washington, D. C.)* **237**, 903–905.

Schmitz, R. F. (1972). Behavior of *Ips pini* during mating, oviposition, and larval development (Coleoptera: Scolytidae). *Can. Entomol.* **104**, 1723–1728.

Schreiner, I. H., and Nafus, D. M. (1991). Evolution of sub-social behavior in the nymphalid butterfly *Hypolimnas anomala*. *Ecol. Entomol.* **16**, 261–264.

Schuster, J. C., and Schuster, L. B. (1985). Social behavior in passalid beetles (Coleoptera: Passalidae): Cooperative brood care. *Fla. Entomol.* **68**, 266–272.

Scott, M. P. (1990). Brood guarding and the evolution of male parental care in burying beetles. *Behav. Ecol. Sociobiol.* **26**, 31–39.

Scott, M. P. (1994). The benefit of paternal assistance in intra- and interspecific competition for the burying beetle, *Nicrophorus defodiens*. *Ethol. Ecol. Evol.* **6**, 537–543.

Scott, M. P., and Traniello, J. F. A. (1987). Behavioural cues trigger ovarian development in the burying beetle *Nicrophorus tomentosus*. *J. Insect Physiol.* **33**, 693–696.

Scott, M. P., and Traniello, J. F. A. (1990). Behavioral and ecological correlates of male and female parental care and reproductive success in burying beetles (*Nicrophorus* spp.). *Anim. Behav.* **39**, 274–283.

Seelinger, G., and Seelinger, U. (1983). On the social organisation, alarm and fighting in the primitive cockroach *Cryptocercus punctulatus* Scudder. *Z. Tierpsychol.* **61**, 315–333.

Shellman-Reeve, J. S. (1990). Dynamics of biparental care in the dampwood termite, *Zootermopsis nevadensis* (Hagen): Response to nitrogen availability. *Behav. Ecol. Sociobiol.* **26**, 389–397.

Shepard, M., Waddill, V., and Kloft, W. (1973). Biology of the predaceous earwig *Labidura riparia* (Dermaptera: Labiduridae). *Ann. Entomol. Soc. Am.* **66**, 837–841.

Sites, R. W., and McPherson, J. E. (1982). Life history and laboratory rearing of *Sehirus cinctus cinctus* (Hemiptera: Cydnidae), with descriptions of immature stages. *Ann. Entomol. Soc. Am.* **75**, 210–215.

Smith, C. C., and Fretwell, S. D. (1974). The optimal balance between size and number of offspring. *Am. Nat.* **108,** 499–506.
Smith, R. L. (1976). Male brooding behavior of the water bug *Abedus herberti* (Hemiptera: Belostomatidae). *Ann. Entomol. Soc. Am.* **69,** 740–747.
Smith, R. L. (1979). Repeated copulation and sperm precedence: paternity assurance for a male brooding water bug. *Science (Washington, D. C.)* **205,** 1029–1031.
Smith, R. L. (1980). Evolution of exclusive postcopulatory paternal care in the insects. *Fla. Entomol.* **63,** 65–78.
Smith, R. L. (in press). Evolution of paternal care in the giant water bugs (Heteroptera: Belostomatidae). *In* "Social Competition and Cooperation in Insects and Arachnids, Volume II, Evolution of Sociality" (B. J. Crespi and J. C. Choe, eds.), Princeton Univ. Press, Princeton, NJ.
Smith, R. L., and Larsen, E. (1993). Egg attendance and brooding by males of the giant water bug *Lethocerus medius* (Guerin) in the field (Heteroptera: Belostomatidae). *J. Insect Behav.* **6,** 93–106.
Stamp, N. E. (1980). Egg deposition patterns in butterflies: Why do some species cluster their eggs rather than deposit them singly? *Am. Nat.* **115,** 367–380.
Stamps, J. A. (1991). Why evolutionary issues are reviving interest in proixmate behavioral mechanisms. *Am. Nat.* **31,** 338–348.
Stay, B., and Coop, A. (1973). Developmental stages and chemical composition in embryos of the cockroach, *Diploptera punctata*, with observations on the effect of diet. *J. Insect Physiol.* **19,** 147–171.
Stay, B., and Lin, H. L. (1981). The inhibition of milk synthesis by juvenile hormone in the viviparous cockroach, *Diploptera punctata. J. Insect Physiol.* **27,** 551–557.
Stay, B., Ostedgaard, L. S., Tobe, S. S., Stambi, A., and Spaziani, E. (1984). Ovarian and haemolymph titres of ecdysteroid during the gonadotrophic cycle in *Diploptera punctata. J. Insect Physiol.* **30,** 643–651.
Stay, B., Tobe, S. S., and Bendena, W. G. (1994). Allatostatins: Identification, primary structures, functions and distribution. *Adv. Insect Physiol.* **25,** 267–337.
Stephens, D. W. (1989). Variance and the value of information. *Am. Nat.* **134,** 128–140.
Strambi, C., Strambi, A., de Reggi, M. L., and Delaage, M. A. (1984). Radioimmunoassays of juvenile hormones: State of the methods and recent data on validation. *In* "Biosynthesis, Metabolism and Mode of Action of Invertebrate Hormones" (J. Hoffman and M. Porchet, eds.), pp. 356–372. Springer-Verlag, Berlin.
Tachikawa, S., and Schaefer, C. W. (1985). Biology of *Parastrachia japonensis* (Hemiptera: Pentatomidae). *Ann. Entomol. Soc. Am.* **78,** 387–397.
Tallamy, D. W. (1982). Age specific maternal defense in *Gargaphia solani* (Hemiptera: Tingidae). *Behav. Ecol. Sociobiol.* **11,** 7–11.
Tallamy, D. W. (1984). Insect parental care. *BioScience* **34,** 2024.
Tallamy, D. W. (1986). Age specificity of "egg-dumping" in *Gargaphia solani* (Hemiptera: Tingidae). *Anim. Behav.* **34,** 599–603.
Tallamy, D. W. (1994). Nourishment and the evolution of paternal investment in subsocial arthropods. *In* "Nourishment and Evolution in Insect Societies" (J. H. Hunt and C. A. Nalepa, eds.), pp. 21–56. Westview Press, Boulder, CO.
Tallamy, D. W. (in press). Maternal care in the Hemiptera: ancestry, alternatives and current adapative value. *In* "Social Competition and Cooperation in Insects and Arachnids, Volume II, Evolution of Sociality" (B. J. Crespi and J. C. Choe, eds.), Princeton Univ. Press, Princeton, NJ.
Tallamy, D. W., and Denno, R. F. (1981). Maternal care in *Garagaphia solani. Anim. Behav.* **29,** 771–778.

Tallamy, D. W., and Denno, R. F. (1982). Life history tradeoffs in *Gargaphia solani* (Hemiptera: Tingidae): The cost of reproduction. *Ecology* **63,** 616–620.
Tallamy, D. W., and Dingle, H. (1986). Genetic variation in the maternal defensive behavior of the lace bug *Gargaphia solani*. In "Evolutionary Genetics of Invertebrate Behavior" (M. D. Huettel, ed.), pp. 135–143. Plenum Press, New York.
Tallamy, D. W., and Horton, L. A. (1990). Costs and benefits of the egg-dumping alternatives in *Gargaphia* lace bugs (*Hemiptera: Tingidae*). *Anim. Behav.* **39,** 352–359.
Tallamy, D. W., and Wood, T. K. (1986). Convergence patterns in subsocial insects. *Annu. Rev. Entomol.* **31,** 369–390.
Thornhill, R., and Alcock, J. (1983). "The Evolution of Insect Mating Systems." Harvard Univ. Press, Cambridge, MA.
Thorson, G. (1950). Reproductive and larval ecology of marine bottom invertebrates. *Biol. Rev.* **25,** 1–45.
Tobe, S. S., and Stay, B. (1980). Control of juvenile hormone biosynthesis during the reproductive cycle of a viviparous cockroach. *Gen. Comp. Endocrinol.* **40,** 89–98.
Tobe, S. S., and Stay, B. (1985). Structure and regulation of the corpus allatum. *Adv. Insect Physiol.* **18,** 305.
Trivers, R. L. (1972). Parental investment and sexual selection. In "Sexual Selection and the Descent of Man" (B. Campbell, ed.), pp. 136–179. Aldine, Chicago.
Trivers, R. L. (1974). Parent-offspring conflict. *Am. Nat.* **14,** 249–264.
Trumbo, S. T. (1990a). Reproductive benefits of infanticide in a biparental burying beetle, *Nicrophorus orbicollis*. *Behav. Ecol. Sociobiol.* **27,** 269–273.
Trumbo, S. T. (1990b). Brood size regulation in a burying beetle, *Nicrophorus tomentosus* (Silphidae). *J. Insect Behav.* **3,** 491–500.
Trumbo, S. T. (1991). Reproductive benefits and the duration of paternal care in a biparental burying beetle, *Nicrophorus orbicollis*. *Behaviour* **117,** 82–105.
Trumbo, S. T. (1992). Monogamy to communal breeding: Exploitation of a variable resource base in burying beetles. *Ecol. Entomol.* **17,** 289–298.
Trumbo, S. T. (1994). Interspecific competition, brood parasitism, and the evolution of biparental cooperation in burying beetles. *Oikos* **69,** 241–249.
Trumbo, S. T., and Eggert, A.-K. (1995). Beyond monogamy: Territory quality influences sexual advertisement in male burying beetles. *Anim. Behav.* **48,** 1043–1047.
Trumbo, S. T., and Fernandez, A. G. (1995). Cues employed to assess resource size by burying beetles and regulation of brood size by male parents. *Ethol. Ecol. Evol.* **7,** 313–322.
Trumbo, S. T., and Wilson, D. S. (1993). Brood discrimination, nestmate discrimination and determinants of social behavior in facultatively quasisocial beetles (*Nicrophorus* spp.). *Behav. Ecol.* **4,** 332–339.
Trumbo, S. T., Borst, D. W., and Robinson, G. E. (1995). Rapid elevation of JH titre during behavioural assessment of the breeding resource by the burying beetle, *Nicrophorus orbicollis*. *J. Insect Physiol.* **41,** 535–543.
Tyndale-Biscoe, M. (1984). Adaptive significance of brood care of *Copris diversus* Waterhouse (Coleoptera: Scarabaeidae). *Bull. Entomol. Res.* **74,** 453–461.
Valenzuela-González, J. (1993). Pupal cell-building behavior in passalid beetle (Coleoptera: Passalidae). *J. Insect Behav.* **6,** 33–41.
Vancassel, M. (1973). Rapports entre les comportements sexuel et parental chez *Labidura riparia* P. (Derm. Labiduridae). *Ann. Soc. Entomol. Fr.* **9,** 441–455.
Vancassel, M. (1977). Le développement du cycle parental de *Labidura riparis*. *Biol. Behav.* **2,** 51–64.
Vancassel, M., Foraste, M., Strambi, A., and Strambi, C. 1984. Normal and experimentally induced changes in hormonal hemolymph titers during parental behavior of the earwig *Labidura riparia*. *Gen. Comp. Endocrinol* **56,** 444–456.

Vancassel, M., Foraste, M., Quris, R., Strambi, A., and Strambi, C. (1987). The parental response of females *Labidura riparia* to young and its control. *Comp. Endocrinol.* **6,** 169–173.

Vancassel, M., Foraste, M., Strambi, C., Strambi, A., and Delbecque, J.-P. (1991). Analysis of hemolymph ecdysteroids in the female earwig: *Labidura riparia. Invertebr. Reprod. Dev.* **20,** 37–43.

van Doorn, A. (1987). Investigations into the regulation of dominance behavior and of the division of labour in bumblee colonies (*Bombus terrestries*). *Neth. J. Zool.* **37,** 255–276.

van Noordwijk, A. J., and de Jong, G. (1986). Acquisition and allocation of resources: Their influence on variation in life history tactics. *Am. Nat* **128,** 137–142.

Weaver, R. J., Pratt, G. E., and Finney, J. R. (1975). Cyclic activity of the corpus allatum related to gonadotrophic cycles in adult female *Periplaneta americana. Experientia* **31,** 597–598.

West, M. J., and Alexander, R. D. (1963). Sub-social behavior in a burrowing cricket *Anurogryllus muticus* (De Geer). *Ohio J. Sci.* **63,** 19–24.

Wilson, D. S., and Fudge, J. (1984). Burying beetles: Intraspecific interactions and reproductive success in the field. *Ecol. Entomol.* **9,** 195–204.

Wilson, D. S., and Knollenberg, W. G. (1984). Species packing and temperature dependent competition among burying beetles (Silphidae: *Nicrophorus*). *Ecol. Entomol.* **9,** 205–216.

Wilson, E. O. (1971). "The Insect Societies." Harvard Univ. Press, Cambridge, MA.

Wilson, E. O. (1975). "Sociobiology." Harvard Univ. Press, Cambridge, MA.

Wilson, K., and Lessels, C. M. (1994). Evolution of clutch size in insects. I. A review of static optimality models. *J. Evol. Biol.* **7,** 339–363.

Windsor, D. M. (1987). Natural history of a subsocial tortoise beetle, *Acromis sparsa* Boheman (Chysomelidae, Cassidinae) in Panama. *Psyche* **94,** 127–149.

Windsor, D. M., and Choe, J. C. (1994). Origins of parental care in chrysomelid beetles. *In* "Novel Aspects of the Biology of Chrysomelidae" (P. H. Jolivet, M. L. Cox, and E. Petitpierre, eds.), pp. 111–117. Kluver Academic Publishers, The Netherlands.

Wingfield, J. C., Ball, G. F., Duffy, A. M., Jr., Hegner, R. E., and Ramenofsky, M. (1987). Testosterone and aggression in birds. *Am. Sci.* **75,** 602–608.

Wingfield, J. C., Hegner, R. E., Duffy, A. M., and Ball, G. F. (1990). The "challenge hypothesis": Theoretical implications for patterns of testosterone secretion, mating systems and breeding strategies. *Am. Nat.* **136,** 829–846.

Wittenberger, J. F., and Tilson, R. L. (1980). The evolution of monogamy. *Annu. Rev. Ecol. Syst.* **11,** 197–232.

Wood, T. K. (1974). Aggregating behavior of *Umbonia crassicornis* (Homoptera: Membracidae). *Can. Entomol.* **106,** 169–173.

Wood, T. K. (1976). Alarm behavior of brooding female *Umbonia crassicornis* (Homoptera: Membracidae). *Ann. Entomol. Soc. Am.* **69,** 340–344.

Woodhead, A. P., Asano, W. Y., and Stay, B. (1993). Allatostatins in the haemolymph of *Diploptera punctata* and their effect *in vivo. J. Insect Physiol.* **39,** 1001–1005.

Wyatt, T. D. (1986). How a subsocial intertidal beetle, *Bledius spectabilis,* prevents flooding and anoxia in its burrow. *Behav. Ecol. Sociobiol.* **19,** 323–331.

Wyatt, T. D., and Foster, W. A. (1989a). Leaving home: Predation and the dispersal of the larvae from the maternal burrow of *Bledius spectabilis,* a subsocial intertidal beetle. *Anim. Behav.* **38,** 778–785.

Wyatt, T. D., and Foster, W. A. (1989b). Parental care in the subsocial intertidal beetle, *Bledius spectabilis,* in relation to parasitism by the ichneumonid wasp, *Barycnemis blediator. Behavior* **110,** 76–92.

Zeh, D. W., and Smith, R. L. (1985). Paternal investment by terrestrial arthropods. *Am. Zool.* **25,** 785–805.

# Cause and Effect of Parental Care in Fishes
## An Epigenetic Perspective

STEPHEN S. CRAWFORD AND EUGENE K. BALON

INSTITUTE OF ICHTHYOLOGY AND DEPARTMENT OF ZOOLOGY
UNIVERSITY OF GUELPH
GUELPH, ONTARIO
N1G 2W1, CANADA

## I. INTRODUCTION

> The time is ripe for a new, evolutionary history of reproduction. —*Herbert Wendt, 1965, in* The Sex Life of the Animals

In one of the earliest texts on parental care in fishes, Theodore Gill (1905) provided us with a direct connection back to the roots of our modern science as follows:

> the greatest naturalist of antiquity, Aristotle, told of a kind of fish, inhabiting the largest river of Greece, the Macedonian Achelous, which, in the person of the male parent, exerted the greatest care of both eggs and young (p. 403).

A few pages later he continued (p. 409):

> During the consideration of the social economy of these fishes the question must often recur, How did the parental instinct manifested originate? It was easy enough in olden time to give an answer which would be regarded as all sufficient in those days; it was a specific instinct implanted by an omnipotent creator in every case. In these days of evolutionary belief, however, such an answer is equivalent to no answer. The instinct itself must be regarded as a development of an aptitude inherent in the fish itself.

Even at the beginning of the twentieth century, researchers already recognized the importance of identifying cause-and-effect relationships in order to explain the existence of parental care in fishes.

When the *Advances in the Study of Behavior* series editors first approached us with the possibility of writing a review on parental care in fishes, Eugene indicated that our work deals primarily with the processes of ontogeny and phylogeny—rather than parental care in the strict sense. From our perspective, parental care is simply one component in an integrated system of possible evolutionary solutions to the demands of life. We suggested that perhaps it would be more appropriate to deal with somebody who had worked on the specific topic of parental care, but the editors insisted that they wanted the concept of parental care in fishes integrated within a larger evolutionary context. Given their intentions, we accepted the offer.

It would be appropriate at this point to provide some indication of what this article is and is not. Its general goal is to provide an analysis of concepts associated with parental care in fishes, using selected examples. We will not attempt to summarize or synthesize the tremendous primary literature. Rather, we will present the reader with an overview of historical trends in recent research, along with possible explanations of how these trends formed. Throughout the review we will attempt to focus on the mechanisms that may be responsible for perceived patterns associated with parental care. Throughout this historical perspective, we will present our ideas on the future of research on parental care in the broad sense.

Our specific objectives can be stated as follows: (1) Examine the diversity of parental care in fishes. It is important to have a general picture of the phenomenon in order to make sense of research that has been conducted to explain it. (2) Describe an evolutionary model of the life history of fishes, with emphasis on reproduction. As mentioned previously, we believe that parental care is so intimately tied with the processes of maintenance, development, and reproduction that it cannot be understood in isolation. (3) Present a simple concept of individual energy expenditures associated with reproduction, within the context of an integrated life-history model for fishes. This energy expenditure model describes the various options across which an individual parent could distribute energy resources to increase the probability of successful reproduction. (4) Analyze the definitions and assumptions that have been associated with parental care, within the context of the energy expenditure model mentioned previously. Previous work in this field has shaped, and has been shaped by, the different meanings that researchers have attributed to "parental care". A clear understanding of these meanings is required in order to evaluate their contributions. (5) Describe the focus of research on parental care in fishes over the past 15 years. It is possible to sample the primary literature in a manner that makes it possible to evaluate qualitatively the patterns of research within the context of our life-history model. (6) Develop hypotheses to describe causal mechanisms that could help to explain the evolution of reproductive styles, including parental care. Hypotheses like these are required to generate predictions that, in turn, can be tested to determine the confidence with which we may view piscine reproductive styles[1]. By addressing each of these objectives, we hope to provide a useful synthesis of information regarding the study of parental care in fishes.

## II. THE DIVERSITY OF PARENTAL CARE IN FISHES

According to estimates, there are more than 24,000 living fish species in the world, constituting more than one-half of all recognized vertebrate species (Nelson, 1994). Parental care of early offspring in the strict (i.e., behavioral) sense has been reported in at least 89 of 422 (21%) families of the bony fishes alone, representing an estimated 3000–5000 species (Blumer, 1982; but see Ridley, 1993 for a discussion of problems with such statistics). However, some form of parental care in the broad sense (as explained later in this chapter) is exhibited by all species of fishes. As might be expected, such a diversity of species is associated with a dazzling array of solutions for successful reproduction. The evolution of parental care, however, is not directly correlated with the evolutionary history of Linnaean taxa (Balon, 1975a,b; Blumer, 1982). Rather, reproductive styles seem to have evolved independently and repeatedly in the various forms of fishes (Balon, 1981a; see Section II,C).

The wide variety of parental care in fishes can be viewed as an operational blessing or curse, depending on one's perspective. On one hand, the observed diversity of reproductive patterns in fishes represents a rich "natural experiment" of evolutionary processes, but on the other, someone has to undertake the daunting task of developing an evolutionary classification system to encompass these patterns (Keenleyside, 1981). At the very least, a classification system should attempt to identify significant differences in the mechanisms employed to solve the problem of reproduction. The following text presents a brief, chronological description of previous attempts to classify reproductive patterns in fishes, including parental care.

### A. KRYZHANOVSKY'S ECOLOGICAL GROUPS

Sergei Grigoryevitch Kryzhanovsky (1949, p. 237, see Smirnov *et al.*, 1995) reasoned that

> adaptations of fishes for spawning and development reflect not only the essential ecological factors of the embryonic period, but also the essential factors of all the other intervals

[1] We use the term *style* as a neutral term instead of the often confusing *strategy* which we restrict to the context of game theory.

of life. These adaptations mark the biology of adults, and define the type of migrations, invasion abilities, and limits of distribution.

To construct natural groupings of fishes according to their broad ecological requirements for reproduction, Kryzhanovsky declared further that

> Two factors play leading roles during embryonic development: predators and the availability of oxygen. All other factors are associated with these two and together they create an extraordinary variety of adaptations associated with early development. However, the different reproductive styles and spawning grounds predetermine the respiratory conditions and the potential for defence against predators. Hence, to a considerable degree, they predetermine the nature of adaptations associated with early development. Therefore, the astounding multitude of adaptations associated with development reveal the ecological patterns that reflect the essential relations of fish in nature.

Kryzhanovsky's ecological groups were named after the different kinds of spawning grounds, e.g., lithophilous, phytophilous, psammophilous, pelagophilous, and ostracophilous, but were characterized by a complex suite of developmental, ecological, and behavioral attributes. We believe that a brief review of these essentially ecological groups will form a necessary, but often neglected, background for the classification of piscine parental care (see Balon, 1975a).

All fish species belonging to the lithophilous group deposit their eggs on a rock or gravel substrate, and their eggs and embryos develop on or within that substrate. This occurs in streams and rivers as well as oligotrophic lakes. Some embryos hatch early and have a photophobic escape reaction, enabling them to scatter and hide under stones. Embryonic respiratory organs in this group are moderately developed (Fig. 1). Segmental capillaries in the fin folds appear in all representatives of this group, while branches of the subintestinal vein are exhibited only in some species.

Fishes of the phytophilous group are adapted to develop on or within vegetation (alive or dead), above a muddy or silted bottom, and under conditions of very low available oxygen, especially at night. Free embryos and larvae of fishes belonging to this group are not photophobic and are much more pigmented than lithophils. Their embryonic respiratory organs are well developed, including enlarged ducts of Cuvier, a caudal vein network in the ventral fin fold, segmental vessels in the dorsal fin fold, and, in some taxa, capillary networks in gill covers and pectoral fins. Most species have egg envelopes with an adhesive chorion and/or offspring with cement glands that enable them to attach to plants above the anoxic bottom. Instead of hiding, the free embryos typically hang motionless on plants until sufficiently developed to swim and start their exogenous feeding as larvae.

Psammophilous fishes spawn on roots or grass above a sandy bottom or even on the sand itself. They are not photophobic and, in contrast to

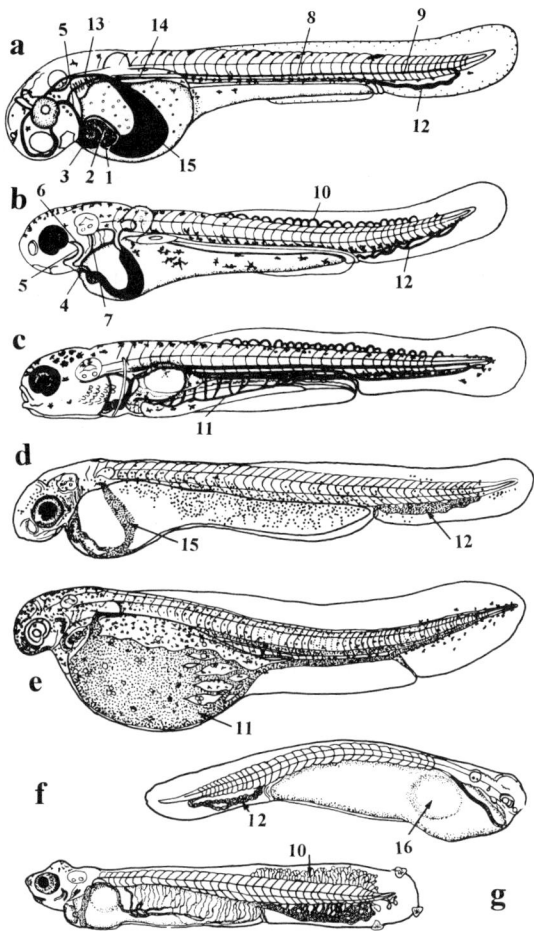

Fig. 1. The blood circulation and the **temporary respiratory networks** in late embryos and larvae of (a–c) phytophilous common carp (*Cyprinus carpio*), (d) lithophilous roach (*Rutilus frisi*), (e) phytophilous pike (*Esox reicherti*), (f) younger and (g) older embryo of the ostracophilous bitterling (*Rhodeus sericeus*). 1 = sinus venosus, 2 = atrium, 3 = ventricle, 4 = ventral aorta, 5 = mandibular and 6 = hyoid arches, 7 = gill cover, 8 = dorsal aorta, 9 = caudal artery, 10 = **segmental vessels,** 11 = **subintestinal vitelline vein,** 12 = **inferior caudal vein,** 13 = anterior and 14 = posterior caudal veins, 15 = **enlarged duct of Cuvier,** and 16 = yolk disk preventing premature expulsion by the mussel. (a–c after Makeyeva, 1992; d after Smirnova, 1957; and e–g after Kryzhanovsky *et al.*, 1951).

phytophilous fishes, lack cement glands. After hatching, the free embryos drop onto the sand. In response to moderately high oxygen availability, the embryonic respiratory organs are usually limited, and the gills become functional late, at a relatively advanced developmental state.

Fishes in the pelagophilous group produce free-floating eggs that are scattered in the water column. Neutral or positive buoyancy is achieved by a very large perivitelline space; single large or numerous small oil globules in the yolk; or by having watery, low-density yolk. Embryos and larvae are usually lightly pigmented, probably for cryptic protection against predation. Embryonic respiratory organs are absent in well-oxygenated waters. Many of the freshwater pelagophils are of marine origin and some, such as the burbot (*Lota lota*), display characteristics of a condition called semipelagophilous (e.g., a secondary loss of buoyancy in fresh water, which has a lower density than sea water).

The last of Kryzhanovsky's ecological groups referred to the ostracophilous fishes. This group encompasses specialized fishes that deposit their eggs in the gill cavity of mussels (Fig. 2). Not only does the adult female possess a special ovipositor for placing the eggs into the proper position within a live mollusc, but her eggs are elongated ovals with almost no perivitelline space. Consequently, more eggs can be deposited within the limited space between the gills (Balon, 1962; Nagata, 1985; Nagata and Nakata, 1988). Embryos are especially well prepared for the possibility of low oxygen concentrations in these confined environments by having dense respiratory networks that are formed mainly by the subintestinal vein on the yolk and by vessels in the dorsal and ventral fin folds (Fig. 1). Their development is very slow, and they retain the embryonic respiratory organs until ready to leave the mussel. Prior to this, and soon after the first appearance of the eye pigment, they become extremely photophobic. This is probably a protective mechanism that prevents premature expulsion from the protection of the mollusc, in addition to the earlier development of spines (unculi) or disks on the yolk sac.

The classification of ecological groups that Kryzhanovsky (see also Kryzhanovsky *et al.*, 1951) proposed is summarized in Table I. Note that Kryzhanovsky initially applied this classification system to fishes of a single drainage. Later it was expanded by some of his followers in Russia and elsewhere to include a wider range of species (Smirnov *et al.*, 1995).

B. Breder and Rosen's Compendium of Data

The compendium published by Charles M. Breder and Donn E. Rosen (1966) represents the first serious effort to compile and organize the data

FIG. 2. (a) Bitterling (*Tanakia lanceolata*) female with ovipositor and three eggs in it. (b) Bivalve mussel (*Pseudodon omiensis*) opened to show embryos (see arrows) of *Acheilognatus tabira* in its gill cavities. (Photographs by E. K. Balon, Futatsu-kawa at Yamagawa, Japan, 1994).

TABLE I
SUMMARY OF AN ECOLOGICAL CLASSIFICATION FOR FISHES PROPOSED BY KRYZHANOVSKY
(1948, 1949)[a]

| Major category | Ecological group | Description |
| --- | --- | --- |
| Fishes that scatter their eggs | Lithophilous | Having a predilection for stones |
| | Phytophilous | Having a predilection for plants |
| | Pelagophilous | Having a predilection for open water |
| | Semipelagophilous | Parents having a predilection for rock substrate, offspring having a predilection for open water |
| | Psammophilous | Having a predilection for sand |
| Fishes that hide their eggs | Ostracophilous | Having a predilection for ostraceans or shells |
| | Lithophilous | Having a predilection for stones |
| Fishes that guard their eggs | Lithophilous | Having a predilection for stones |
| | Phytophilous | Having a predilection for plants |
| | Psammophilous | Having a predilection for sand |
| | Nest builders | Construction in a hollow, or on the surface of the bottom |
| | Brooders | Holding or carrying offspring on the body surface or in an external pouch |
| | Live bearers | Carrying offspring within the body cavity |

[a] The classification is basically hierarchical in structure, with major categories (not his terms) that differentiate how parents deal with gametes, and ecological groups (his terms) that distinguish between predilections for spawning substrate.

on reproductive patterns in fishes. The authors recognized several categories of fishes, at the taxonomic level of families, in which each category was operationally defined as "... an assemblage of fishes that may conveniently be treated together from the standpoint of reproduction" (Breder and Rosen, 1966, p. 4). For each family of fishes, they provided information on nine aspects of reproduction: breeding season, breeding sites, migration, secondary sex characters, sex discrimination, competition for mates, courtship, mating, and parental care. The authors simply collected all available material from the published literature (unfortunately short on Russian and other non-English sources), and then presented a summary consisting of a brief discussion of their reproductive features. While their discussion focused on several general aspects of reproductive styles, including parental care, they did not provide a set of characteristics that clearly could be associated with each division. Thus, in spite of an extensive review, their work provides few, if any, insights into evolutionary mechanisms that might account for the diversity of reproductive styles among fishes. By ignoring the earlier work of Kryzhanovsky, and being unable to benefit from a cladistic approach to piscine systematics, Breder and Rosen (1966) missed an opportunity to synthesize their data into a classification that was meaningful from the perspective of evolutionary ecology.

### C. Balon's Reproductive Guilds

In 1975, Eugene K. Balon introduced the concept of reproductive guilds to serve as a framework for classifying the evolution of reproductive patterns in fishes, including parental care. The concept of reproductive guilds was a further development of Kryzhanovsky's (1948, 1949) idea of distinguishing groups of fishes on the basis of ecologically relevant variables associated with early ontogeny (Balon, 1964, 1965, 1975b, 1981c, 1984a, 1990). The first objective of the classification of reproductive guilds was to discriminate between observed patterns in reproductive styles on the basis of ethological, environmental, ecological, and ontogenetic variables (Table II). It was intimately tied to the development of hypotheses regarding the evolution of life-history styles in fishes, including parental care.

The highest level of this hierarchical system of classification (Ethological Section) served to distinguish reproductive styles on the basis of the relationship between parents and their offspring. The three sections at this level separated those species that do not guard their offspring (nonguarders) from those that guard their offspring; either separated by some actual distance from them (guarders) or by carrying them on or in their bodies (bearers).

The second level of this hierarchical system of classification (Ecological Group) served to distinguish reproductive styles on the basis of the specific environmental conditions under which the offspring exist. Within the nonguarders, a distinction was made between those species that release their gametes to exposed environments (open substratum egg scatterers) and those that deposit their eggs in less exposed environments (brood hiders). Within the guarders, a distinction was made between those species that protect their offspring without constructing a nest (clutch tenders) and those that protect their offspring by constructing a nest to hold the young (nesters). Finally, within the bearers, a distinction was made between those species that transfer and carry their offspring outside of the body cavity (external brooders) and those species that retain their offspring within their body cavities (live bearers).

The third level of this classification of reproductive patterns (Reproductive Guilds proper) served to distinguish specific styles within the Ecological

TABLE II
A Summary of Balon's (1975a, 1990) Classification of Reproductive Styles in Fishes, Using the Concept of the Reproductive Guild[a]

| Ethological section | Ecological group | Reproductive guild |
|---|---|---|
| A. Nonguarders | A.1 Open substratum egg scatterers | A.1.1 Pelagic spawners<br>A.1.2 Rock and gravel spawners with pelagic larvae<br>A.1.3 Rock and gravel spawners with benthic larvae<br>A.1.4 Nonobligatory plant spawners<br>A.1.5 Obligatory plant spawners<br>A.1.6 Sand spawners<br>A.1.7 Terrestrial spawners |
| | A.2 Brood hiders | A.2.1 Beach hiders<br>A.2.2 Cave hiders<br>A.2.3 Rock and gravel hiders<br>A.2.4 Hiders in live invertebrates<br>A.2.5 Annual fishes |
| B. Guarders | B.1 Clutch tenders | B.1.1 Pelagic tenders<br>B.1.2 Above water tenders<br>B.1.3 Rock tenders<br>B.1.4 Plant tenders |
| | B.2 Nesters | B.2.1 Froth nesters<br>B.2.2 Miscellaneous substrate nesters<br>B.2.3 Rock and gravel nesters<br>B.2.4 Gluemaking nesters<br>B.2.5 Plant material nesters<br>B.2.6 Sand nesters<br>B.2.7 Hole nesters<br>B.2.8 Anemone nesters |
| C. Bearers | C.1 External brooders | C.1.1 Transfer brooders<br>C.1.2 Auxiliary brooders<br>C.1.3 Mouth brooders without buccal feeding<br>C.1.4 Mouth brooders with buccal feeding of embryos<br>C.1.5 Gill-chamber brooders<br>C.1.6 Pouch brooders |
| | C.2 Internal live bearers | C.2.1 Facultative lecithotrophic live bearers<br>C.2.2 Obligate lecithotrophic live bearers<br>C.2.3 Embryonic cannibal live bearers<br>C.2.4 Histotrophic live bearers<br>C.2.5 Placentotrophic live bearers<br>C.2.6 Combined live bearers |

[a] The classification is hierarchical, with three levels of organization based on ethological, environmental, ecological, and ontogenetic variables.

Groups. Individual Reproductive Guilds were distinguished mainly according to ontogenetic attributes of the offspring, as well as the basis of environmental and ethological characteristics of parental behavior.

For example, while the names of most Reproductive Guilds are derived from the egg deposition substrate, the allocation of a species to any particular guild is based mainly on the existence of specific structures in eggs, embryos, and larvae (if present). Pelagic spawners will exhibit special buoyancy structures in their eggs and embryos (e.g., oil globules, large perivitelline space, watery yolk) and feeble embryonic respiratory networks. In contrast, plant spawners will exhibit well-developed respiratory networks (e.g., ducts of Cuvier, yolk sac and fin fold capillary plexuses, segmental vessel loops—see Fig. 1), in addition to cement gland cells that enable motionless attachment to plants above an anoxic bottom. Various absorptive structures and placental analogs characterize embryos of live bearers (see Balon, 1975a, 1990 for details). Thus, allocating a fish species to a Reproductive Guild solely on the basis of ethological characteristics or spawning-deposition substrate is incorrect or, at best, incomplete (e.g., Kottelat, 1990).

This classification recognized 36 different Reproductive Guilds, distributed within six Ecological Groups, and belonging to three general Ethological Sections (Table II). Given certain basic information regarding the reproductive style and early ontogeny of any particular species, it is possible to identify the appropriate Reproductive Guild to which it belongs. More importantly, however, the sequence of guilds within each section was arranged according to an ancestor-descendant progression using criteria of epigenetic relationships between ontogeny and phylogeny (e.g., Balon, 1981a, 1983, 1985b), which will be discussed in Section VII,A. After a review of the scarce data from the fossil record, it even proved possible to construct a putative model of the evolution of reproductive styles, including parental care (Fig. 3) (see also Balon *et al.*, 1977).

D. BLUMER'S ETHOLOGICAL FORMS

Lawrence S. Blumer published a pair of papers (1979, 1982) in which he distinguished 15 ethological forms of parental care that teleost fishes exhibited. These forms are not mutually exclusive, since parents belonging to a single species can exhibit a combination of different forms. On the basis of his definitions and survey of the published literature, Blumer (1979) assigned each form of parental care behavior to the 89 families of teleost fishes exhibiting parental care. These forms of parental care are presented in Table III (page 66), including the frequency of their occurrence in different families as calculated by Blumer (1979).

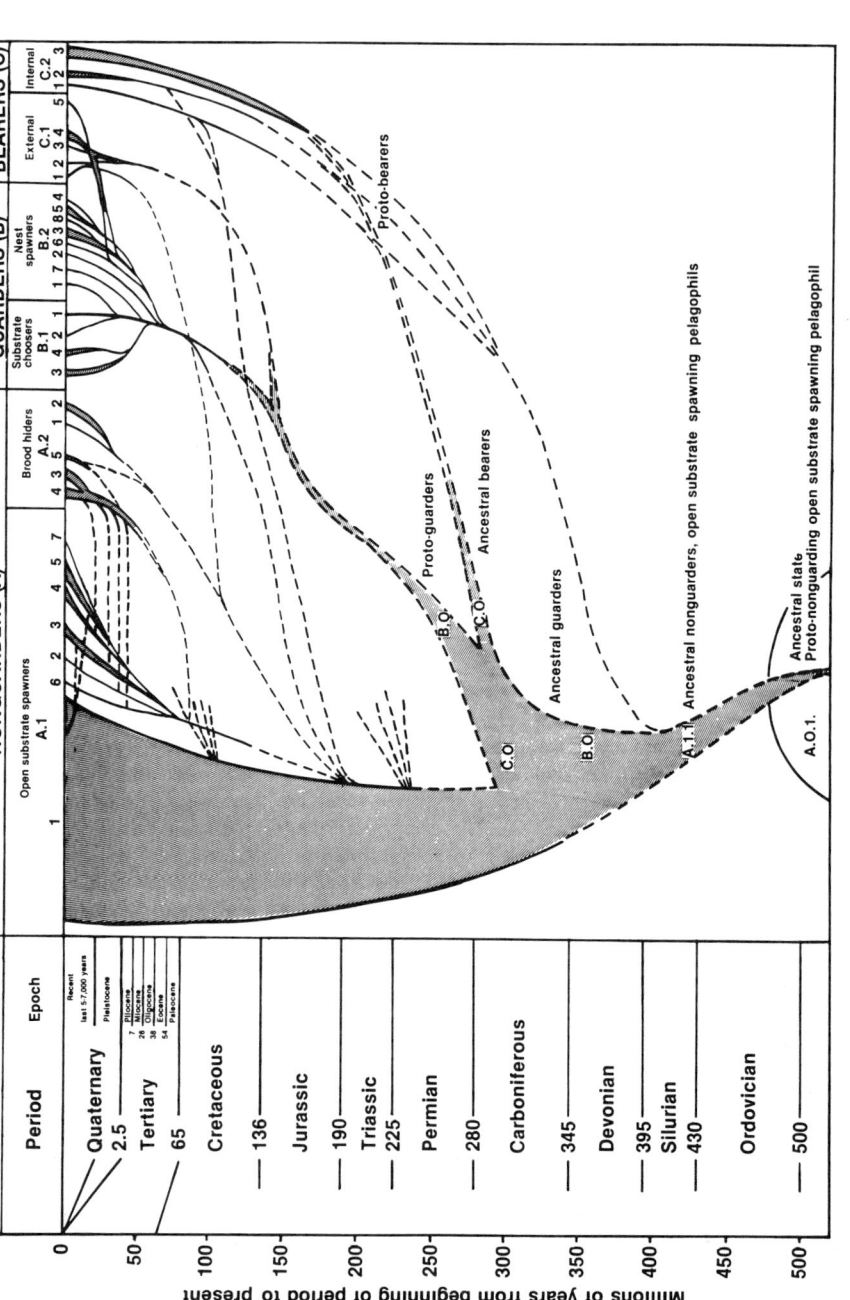

FIG. 3. Putative, schematic representation of the major trends in the evolution of reproductive styles (symbols and numerals of sections, groups, and guilds follow Table II except for ancestral groups A.0, B.0, and C.0. The derived styles may have risen more than once as marked by the dashed branches to the right of A.1.1. (From Balon, 1981a).

Blumer also provided information on several other characters associated with each of the families. These characters included the distinction between internal and external insemination, and between male and female care of offspring. Due to the length of these tables, they have not been included in this article, but the interested reader can refer to Blumer (1979, 1982) for additional details.

It should be noted that the primary focus in both of Blumer's papers was the distribution of parental care between the sexes. Notwithstanding this focus, Blumer's work takes its place with that of Breder and Rosen (1966) as one of the most comprehensive surveys of data on parental care behavior in fishes.

E. EVALUATION

The work of both Breder and Rosen (1966) and Blumer (1979, 1982) reflects the wealth of information already available in the published literature. However, the reader should be aware that the available information represents only a small fraction of fish species, both extant and extinct. The scarcity of such information is reflected in the fact that in both publications the information was summarized at the level of Linnaean families. This is troublesome, since much variation in parental care occurs below the level of families. It would seem that the development of a phylogenetic model of parental care in fishes is either not feasible at lower taxonomic levels (e.g., species, populations, forms), or is feasible, but only in the distant future. In either case, it seems unlikely that systems based on Linnaean classification will lead to the development of hypotheses regarding cause-and-effect relationships between parental care and other aspects of an organism's life history.

It should be stressed that the reviews by Breder and Rosen (1966) and Blumer (1979, 1982) were simply first steps toward describing the diversity of parental care behavior in fishes. As Winkler (1987, p. 526) noted: "Biologists have accumulated an enormous amount of information on the various patterns of parental care in animals (. . .) but current understanding of the determinants of the kind and amount of parental care extends little beyond a characterization of the typical pattern of care for a given species."

The Ecological Groups proposed by Kryzhanovsky (1948, 1949) and the Reproductive Guilds proposed by Balon (1975a, 1981c, 1984a, 1990) offered a different approach to the problem by providing a classification of parental care in the context of reproductive styles. At first, Kryzhanovsky was concerned with developing a classification system that would distinguish between functional components of fish communities, rather than constructing a model of evolutionary sequences that would explain the emergence of

TABLE III
Form, Codes, and Definitions of Parental Care Exhibited by Teleost Fishes, as Defined by Blumer (1979, 1982)[a]

| Form of parental care | Code | Definition | No. of families |
|---|---|---|---|
| Guarding | G | Displaying toward and/or actively chasing conspecific and heterospecific that approach the eggs or offspring, or the site where they are located. | 63 |
| Nest building and/or substrate cleaning | N | Nest: Digging a depression in the substrate, a burrow, or making an elevated mound with substrate materials. Assembling a cup or tube structure with pieces of vegetation. Blowing mucus-covered bubbles that form a floating mass. All these structures are used to hold eggs and/or offspring. Substrate: The removal of detritus, algae, and animals from the site where eggs are to be deposited. | 39 |
| Fanning | F | Moving the pectoral, pelvic, anal, or caudal fins over the egg mass or offspring, thereby aerating them and removing sediment. Aeration and removal of sediment may also be accomplished by forcing water over the eggs or offspring through the mouth or gill cavities. | 30 |
| Internal gestation | IG | Female retains eggs in her body and their development takes place inside the ovaries or oviducts. | 14 |
| Removal | RD | Dead or diseased eggs removed by mouth from the egg mass. | 12 |
| Oral brooding | ORB | Holding eggs and offspring in the mouth or gill cavities during their development, typically ending with the release of well-developed offspring. | 10 |
| Retrieval | R | Taking eggs or offspring that fall or stray from the nest or school into the mouth, and returning them to the nest or school. | 8 |
| Cleaning eggs | CE | Taking eggs into mouth, manipulating inside the buccal cavity, and returning them to the site from which they were taken. | 6 |

*(continues)*

TABLE III  *Continued*

| Form of parental care | Code | Definition | No. of families |
|---|---|---|---|
| External egg carrying | EEC | Eggs attached to the parent externally and carried until hatching. | 6 |
| Egg burying | EB | Depositing eggs beneath the substrate surface or covering eggs with substrate material. | 5 |
| Moving | M | Taking eggs or offspring by mouth from one location to another, often from one nest to another. | 5 |
| Coiling | GEC | The parent coils its body around the egg mass while guarding them. This guarding posture reduces the eggs' exposure to air at low tide, when the oviposition site is intertidal. | 4 |
| Ectodermal feeding | EF | A specialized mucus produced on the body surface of the parent is used as food by the young offspring. | 3 |
| Brood pouch | BP | Eggs held in a special saclike structure during their development. | 2 |
| Splashing | S | Splashing water on eggs deposited out of water or eggs exposed at times of low tide. | 2 |

[a] The total number of fish families with at least one species exhibiting a particular form of parental care is indicated in the final column (from Blumer, 1979, but see discussion in Section II,E). Blumer's use of the term *fry* has been replaced with the term *offspring*.

Ecological Groups. For this reason, he made reference to features associated with early ontogeny, but he did not concern himself with the evolutionary mechanisms that linked ontogeny and phylogeny. Balon extended the utility of Kryzhanovsky's work by developing a classification that integrated the concept of parental care within a framework that included the organism's structure, physiology, and ecology. This made it possible to focus on the evolutionary trajectories of reproductive styles, including parental care.

We recommend looking at parental care in fishes from a perspective based on the concept of Reproductive Guilds. This approach would incorporate the substantial efforts of many biologists to view reproduction as one component of ontogeny. It is for this reason that we now consider a general model of life history in fishes.

## III. A General Model of Life History in Fishes

In its simplest form, the term *life history* refers to ontogeny, whereby an individual progresses from activation to death. For the fishes, this process of ontogeny has been modeled by Balon (1975c, 1984b, 1986a,b, 1990; Flegler-Balon, 1989) as a hierarchical representation of morphological, physiological, and behavioral development over the course of time. At the heart of this model is the theory of saltatory ontogeny (Balon, 1979, 1986b) that describes a progression of stabilized and self-organizing energy states that are linked, successively one to the other, by thresholds of developmental reorganization. For present purposes, we will focus only on a few aspects of this life-history model. Readers who are interested in the applied aspects of the model should refer to the case studies published, for example, by Balon (1980), McElman and Balon (1979, 1980), Paine and Balon (1984a,b, 1986), Cunningham and Balon (1985, 1986a,b), Holden and Bruton (1992, 1994), Crawford and Balon (1994a,b,c), and Kováč (1995). Note that all of these studies arise from the study of ontogeny, rather than from the commonly cited "life-history theory" (e.g., Stearns, 1989; Harvey *et al.*, 1991; Smith and Wootton, 1995), which originates from population models in theoretical ecology. We believe that such life-history theory relies too heavily on simplistic quantitative parameters (e.g., egg size, mortality rate, growth, and maturation—Beverton, 1963; Pauly, 1980) and ignores important structural and functional characteristics that link organisms and their environments (e.g., yolk platelets, temporary respiratory organs, metamorphosis, and nest construction—Balon, 1983, 1989a,b; Beverton in Charnov and Berrigan, 1991; Hall, 1992).

One of the first requirements for any life-history model is to provide a classification of ontogenetic intervals that reflect a sequence of self-organizing units bordered by important thresholds in the developing organism. In the case of fishes, we recognize five possible periods of ontogeny: embryo, larva, juvenile, adult, and senescent. Each of these periods can be divided into several phases and numerous saltatory units, the steps (see Balon, 1975c, 1985a, 1986a,b, 1990).

The model is a natural consequence of the theory of saltatory ontogeny and it enables us to separate, by decisive thresholds and in a hierarchical fashion, the main periods of life. At the same time, it enables us to interpret the ecological and evolutionary significance of heterochrony. For example, a model of life-history intervals allows us to inquire about the importance of not having a larva, as well as the importance of having a larva despite the cost of metamorphosis (Flegler-Balon, 1989).

The embryo period of ontogeny is characterized by primarily endogenous feeding. The transition to exogenous feeding (i.e., the oral ingestion and intestinal digestion of nutrients) marks the beginning of the next period of life history: larva in case of indirect ontogeny, and juvenile in the case of direct ontogeny (see Table IV).

In general, larvae are restricted to small food particles by virtue of their small sizes. However, eggs with a smaller amount of low-density yolk can be produced in larger quantities, thus providing compensation for a lower survival rate of larvae. Being chiefly nutrient-gathering devices, larvae must obtain the required energy before a definitive phenotype can be formed. Also, most larvae exploit niches (e.g., planktonic environments) that are different from those of their definitive phenotypes, so the two do not usually

TABLE IV

Examples of Indirect (Metamorphic), Intermediate, and Direct (Ametamorphic) Ontogenies within the Proposed Life-History Model, and the Hierarchy of Intervals[a]

| Periods and phases of the life-history model with ontogenies | | |
|---|---|---|
| Indirect (altricial) | Transitory | Direct (precocial) |
| Embryo  Cleavage egg  Embryo  Free embryo | Embryo  Cleavage egg  Embryo  Free embryo | Embryo  Cleavage egg  Embryo  Free embryo |
| Larva  Fin fold larva  Fin-formed larva | Alevin (larval vestige)  Alevin | |
| Juvenile | Juvenile  Parr  Smolt  Juvenile | Juvenile |
| Adult | Adult | Adult |
| (Senescent) | Senescent | Senescent |

[a] Based on homeorhetic saltatory boundaries described in the footnote b (after Balon, 1990).

[b] Period = the longest interval of ontogeny separated by the strongest thresholds. Phase = the next longest interval into which periods are divided as morphological units for identification purposes, of lesser saltatory significance. Step = the shortest natural (homeorhetic) interval of ontogeny separated by thresholds; most important as an epigenetic stabilized state. Stage = an instantaneous state of ontogeny; should not be used to denote an interval.

compete for food. The benefits of early dispersal (e.g., Metz *et al.,* 1983) may be an additional incentive for indirect ontogeny with larvae (Garstang, 1929; Barlow, 1981; Doherty *et al.,* 1985).

Aside from high mortality, there is another price to be paid for having a larva. The numerous temporary structures of larvae, which allow survival in different habitats and niches, need to be remodeled into the permanent organs and shapes at some cost of energy. This process of remodeling—that is, metamorphosis—terminates the larva period (e.g., Fostner *et al.,* 1983). In some cases (e.g., elopomorphs, stomiidids), much of the body size gained during the larva period must be sacrificed for the remodeling (Fig. 4a), thus reducing the survival value of larger size. This provides clear circumstantial evidence that the main purpose of the larva is to acquire external nutrients when the endogenous supply is insufficient.

In contrast, when a sufficient endogenous food supply is provided at the cost of a lower number of eggs (Balon, 1984b), elimination of the vulnerable larva and its costly metamorphosis facilitates the direct development into a juvenile that is comparatively advanced at the time of first feeding—a clear survival advantage (Fig. 4b). In fishes, the combination of low egg number, larger volume and greater density of yolk (associated with negative buoyancy), prolonged development within egg envelopes, and sessile stages of free embryos all predispose the parent to contribute further protection through parental behavior (Balon, 1975a, 1981a,c, 1984a; Baylis, 1981; Blumer, 1982; Townsend and Stewart, 1985).

Direct ontogeny, therefore, appears to be the more specialized type of development, in the sense that it would result in a phenotype that is less variable than one resulting from indirect development. In theory, the evolutionary trend should proceed towards direct development in order to shorten the most vulnerable period, and to improve competitiveness when community structure becomes more diverse and complex (Balon, 1983, 1985b; Bruton, 1989). Thus, the degree of parental care should increase with the energy invested into individual offspring (see Section IV).

Finally, it is worth noting that the progression of an individual from activation to death is typically represented as a simple cycle of ontogeny (Fig. 5a); however, in reality this progression is better represented as a spiral trajectory in which the cycle of ontogeny is wound along the dimension of time (Fig. 5b). Moving up to the level of cohort, the individuals would be represented by a cable of trajectories moving along a similar spiral path. Similarly, individuals from different cohorts in a population would be represented by interwoven and possibly overlapping cables of spiral trajectories through the dimension of time.

Two aspects of this spiral metaphor are important for our purposes. First, the trajectory of ontogeny is not fixed, either within or between cohorts. Rather, it is a dynamic vector that responds to changes in abiotic and

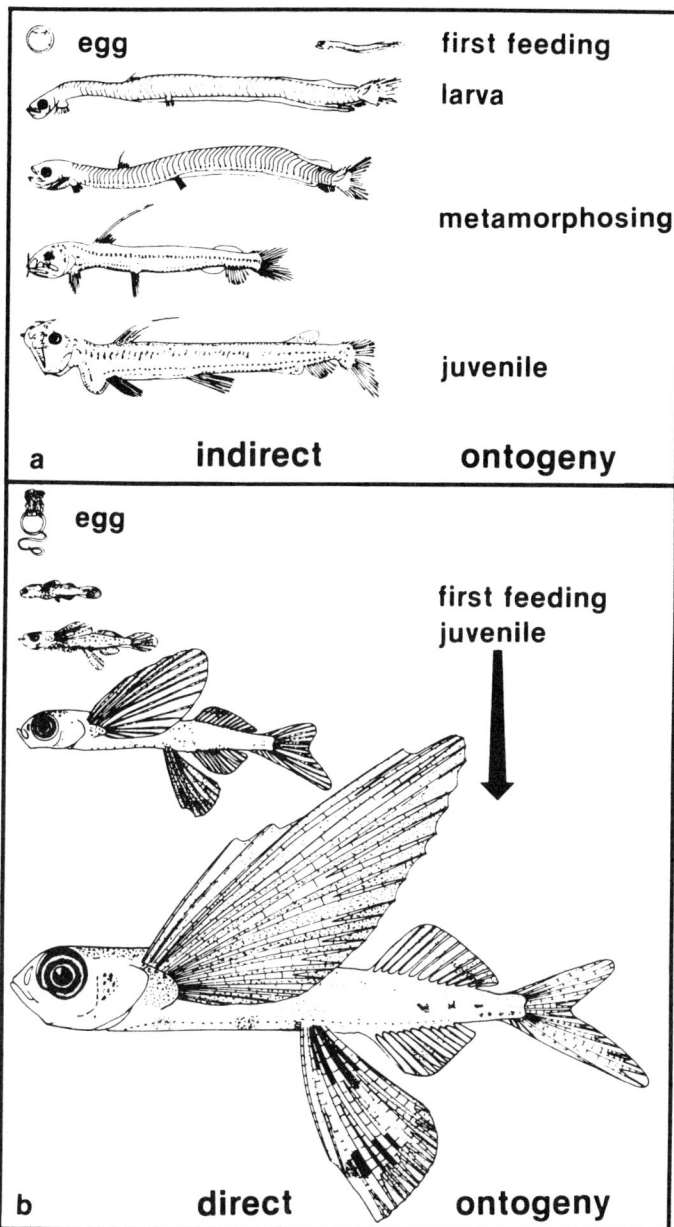

FIG. 4. Even among marine pelagic fishes, some species do not have larvae. Yolk densities, not egg sizes, determine the type of life history: (a) the mesopelagic viperfish (*Chauliodus sloani*) has indirect, metamorphic ontogeny with larvae that decrease in size during metamorphosis; (b) the epipelagic flying fish (*Hirundichthys rondeleti*) undergoes a direct (ametamorphic) ontogeny without larvae, beginning with eggs of the same size as the viperfish but of much higher yolk density. (After Balon, 1985a.)

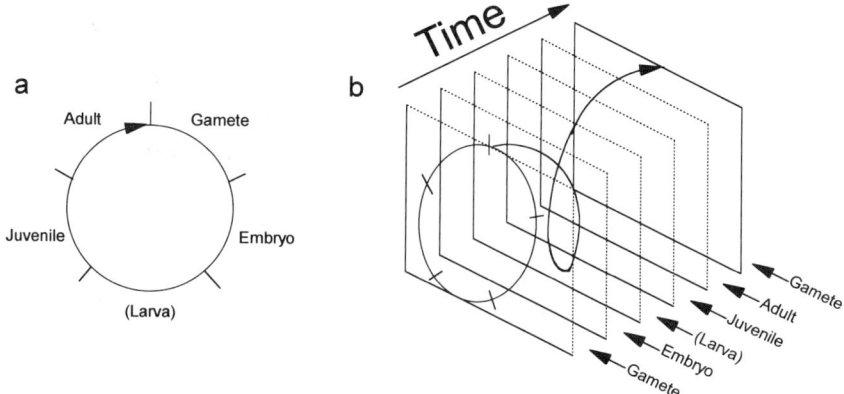

FIG. 5. Two different perspectives on the life history (ontogeny) in fishes. (a) The life history is typically represented as a circle in which an individual progresses through different periods of development: gamete(s), embryo (inside or outside of the egg envelopes), larva (in the case of indirect development), juvenile, adult (during which gametes for the next generation are formed). (b) A single ontogeny represented as a spiral through the dimension of time. The continuous, and possibly overlapping, nature of spiral trajectories from one generation to the next allows for changes in energy expenditures during one revolution to have effects on energy expenditures during the next revolution.

biotic components of the environment. For example, the rate at which an individual (a point on the trajectory) moves through ontogeny (the trajectory) could be affected by factors such as temperature, concentration of metabolic enzymes, and food supply. If one thinks of phylogeny as the extension of these spiral trajectories, it should also be clear that the trajectory characteristics of a lineage can change over evolutionary time. The second aspect of the model that is important for this article is the continuous, and possibly overlapping, nature of the spiral trajectories from one generation (cycle) to the next—the possibility that causal mechanisms initiated by energy expenditures during a parent's ontogeny (one revolution) can have dramatic effects (e.g., developmental rates) on the trajectory of the offspring's ontogeny (the next cycle).

## IV. Energy Requirements for Reproduction

Of the total energy produced in an ecosystem, a certain amount will be available to support the existence of an individual fish and its offspring. The finite amount of this available energy will be a function of environmental conditions (abiotic and biotic), as well as the performance of an individual in that environment. If we consider that any cause-and-effect relationship between parent and offspring requires some change in the expenditure of energy, we can classify these expenditures on the basis of where and when the energy is expended by the parent. One of the central tenets in biology is the need for an organism to distribute this finite amount of energy in a manner that will satisfy three qualitatively different needs: (1) maintenance—survival of the individual at its current state of ontogeny, (2) development—progression of the individual from its current state of ontogeny to the next state of ontogeny, and (3) reproduction—production of offspring (Fisher, 1930; Williams, 1966; Wootton, 1985, 1990; Sibly and Calow, 1986; Calow and Sibly, 1990; Hewett and Johnson, 1992). In this article, we are interested in the energy requirements for reproduction. However, we also recognize that this category of energy demand is intimately associated with the other two categories (i.e., energy expended in maintenance or development detracts from the total energy available for reproduction).

In order to understand the various energy expenditures that may be demanded by reproduction, it would be helpful to map out these potential demands in a hierarchical model. We offer a classification of energy distribution by a potential parent (male or female) as depicted in Fig. 6. This classification is based on the distinction of qualitative and quantitative differences in the expenditure of energy towards reproduction by an individual fish (see also Wootton, 1990). Qualitative differences reflect the possibility that some processes of expenditure may be exhibited by an individual fish, while others are not (e.g., nest-building behavior, no nest-building behavior). Quantitative differences reflect the possibility of variable energy expenditure within a particular process of an expenditure category (e.g., where nest-building behavior is exhibited, it could range from simple substrate manipulation to the construction and maintenance of elaborate structures). In some cases, there is a necessary sequence of energy expenditures that is required to achieve a final goal (e.g., for gamete production: develop gonads, maintain gonads, form gametes, store gametes). In other cases, the different energy expenditures can be considered to be options that do not necessarily require other options (e.g., defense of offspring: nest cleaning, offspring feeding). It should be noted that this conceptual model differs from mathematical models that have related energy expenditures and parental care (e.g., the dynamic programming model of Sargent and Gross, 1993), by attempting to more fully account for specific expenditures of parental energy into different forms of parental care.

Energy committed by an individual to reproduction can be divided into different classes, based on the fate of that energy (Wootton, 1990). Specifi-

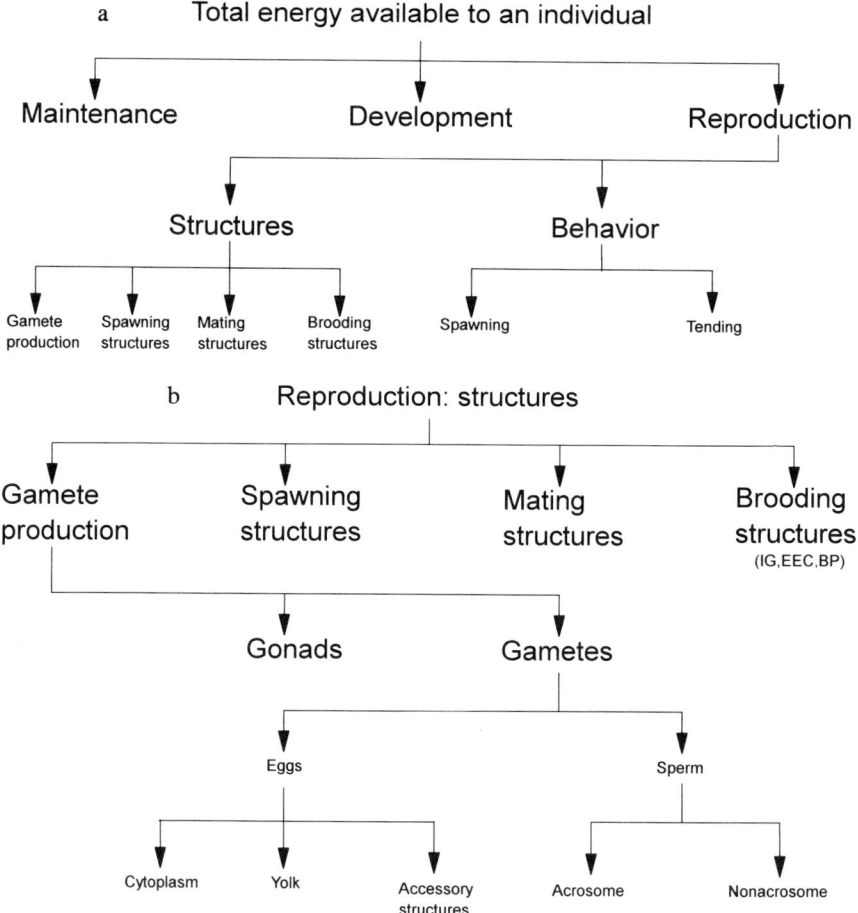

FIG. 6. A general hierarchical model for the expenditure of total available energy by an individual fish. Arrows indicate possible channels for the distribution of energy, with a breakdown of specific components of (a) total energy available to an individual, (b) reproduction: structures, and (c) reproduction: behavior. Alphabetic codes refer to energy expenditures associated with forms of parental care defined by Blumer (1979, 1982): BP, brood pouch; CE, cleaning eggs; EB, egg burying; EEC, external egg carrying; EF, ectodermal feeding; F, fanning; G, guarding; GEC, coiling, IG, internal gestation; M, moving; N, nest building and/or substrate cleaning; ORB, oral brooding; R, retrieval; RD, removal; S, splashing.

cally, we can distinguish between two classes of energy expenditure in the reproductive patterns of fishes: (1) structures associated with reproduction, and (2) behaviors associated with reproduction (Fig. 6a). A brief description of this hierarchical classification is given further. We have purposefully

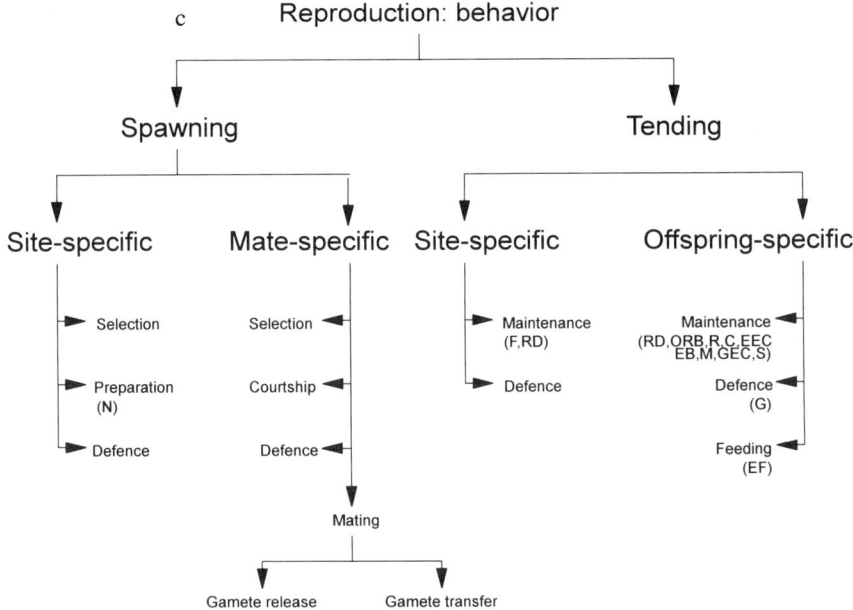

FIG. 6 *Continued*

painted this picture with broad strokes, our intention being that it will serve as a useful tool for understanding previous research on parental care in fishes. As an example, the various forms of parental care that Blumer (1979, see previous Section II,D) defined have been indicated on Fig. 6b,c with the appropriate codes from Table III.

Reproductive energy can be allocated to the development and maintenance of structures. For the purpose of this article, we distinguish among four different categories of structures directly associated with reproduction: structures associated with gamete production, structures associated with spawning in general, structures associated with mating in particular, and structures associated with the brooding of offspring (Fig. 6b). In the process of gamete production, energy can be allocated to the gonadal tissue that produces the gametes, or to the gametes themselves (Beamish, 1979; Beamish *et al.*, 1979; Wootton, 1990). In particular, the energy allocation to female gametes must be further partitioned to the cytoplasm, yolk energy reserves, or other accessory structures that may play a role in determining the survival of offspring (e.g., adhesive chorionic fibrils). Structures associated with spawning include special organs required for migration, establishment of spawning territory, competition for mates, and construction of a suitable nest. In contrast, energy allocated to mating structures serves to support the development and operation of any organs that facilitate the transfer of gametes from the parent (e.g., the ovipositor depicted in Fig. 2a). Finally, any special structures that are required for brooding offspring (e.g., brood pouches) can be considered to draw on energy that is available for reproductive structures (Wetzel and Wourms, 1995).

Reproductive energy can also be allocated to the behaviors associated with reproduction, either before or after gamete union. In this case, we distinguish between spawning-specific and tending-specific behaviors (Fig. 6c). Within spawning-specific behaviors, an individual parent may expend energy in a site-specific or mate-specific manner. Site-specific behaviors could include site selection (as well as migration), site preparation (e.g., nest construction), and the defense of the spawning site from conspecific competitors. Mate-specific behaviors could include the processes of mate selection, courtship rituals (see Balon, 1982), and mate defense before and after mating. In this category, we have also included the energy required for mating itself—a cost that may not be trivial in some species.

Similarly, tending-specific behaviors can be divided into site-specific and offspring-specific behaviors. Energy can be expended on direct maintenance of the site (e.g., nest repair) or on defense of the spawning site after mating. Energy can also be allocated to behaviors that are directed at the offspring, regardless of spawning site—this is especially true when offspring leave the spawning site *en masse*. Within offspring-specific tending behaviors, we recognize the possibility of parental energy expenditures on offspring maintenance (e.g., moving or fanning the young), offspring defense, and offspring feeding. Although tending behaviors such as maintenance and defense often apply to both sites and offspring, we have made a distinction between the two for heuristic reasons.

Two important theoretical implications are associated with this model of parental energy expenditures: one dealing with the characteristics of parental expenditures, and another dealing with the manner in which those expenditures have effects on the offspring. First, with a finite amount of energy devoted to reproduction, expenditure on one component (e.g., gamete production) necessarily precludes the expenditure of that energy on another component (e.g., tending). In order to achieve increased energy expenditures in both components, a decision must be made at a higher level of the hierarchy to increase the total energy expenditure on reproduction. This, in turn, would require a decrease in energy expenditure for maintenance and/or growth, or an increase in total energy available to the individual. In a dynamic system, it might be possible for an organism to achieve an increase of total energy input by increasing energy expenditures to obtain more energy (e.g., movement to better foraging areas, increased feeding activity).

The second implication of this model stems from the fact that we have been dealing exclusively with the expenditures of energy by the parent. At this point, we have not begun to consider the cause-and-effect relationships between these parental energy expenditures and the results they have on the offspring (see Kamler, 1992). Recognition of such interactions is vital to the integration of reproduction (and parental care) within the context of a complete life-history model. We will return to this topic in Section VII.

## V. Definitions of Parental Care

The parental energy expenditure and parental care have been discussed widely in the biological literature—in which they often are treated as well-defined phenomena. Our survey of this literature, however, has shown that many authors do not explicitly discriminate between parental care and other life-history attributes. As a result, the term *parental care* is often used without necessarily meaning one thing or another (Clutton-Brock and Godfray, 1991). In some studies, authors have attempted to define the phenomenon of parental care by explicitly recognizing certain criteria that must be met. The purpose of this section is to review such definitions, and to evaluate them within the context of the general life-history model presented previously.

### A. Parental Investment (Trivers, 1972)

Theories associated with parental care have been closely linked to theories regarding parental investment in general. Perhaps the most frequently cited definition of *parental investment* is the one by Trivers (1972, p. 139): "I first define parental investment as any investment by the parent in an individual offspring that increases the offspring's chance of surviving (and hence reproductive success) at the cost of the parent's ability to invest in other offspring." According to this definition, parental investment would be the sum of the various components of reproduction in Fig. 6a, including expenditures associated with both form and function. However, most authors who cite this definition ignore Trivers' further comments that directly follow the previous sentence (Trivers, 1972, p. 139):

> So defined, parental investment includes the metabolic investment in the primary sex cells but refers to any investment (such as feeding or guarding the young) that benefits the young. It does not include effort expended in finding a member of the opposite sex or in subduing members of one's own sex in order to mate with a member of the opposite sex, since such effort (except in special cases) does not affect the survival chances of the resulting offspring and is therefore not parental investment.

Thus, it appears that what Trivers really meant by parental investments were the gamete production components of reproductive structures and the tending components of reproductive behavior (Fig. 6a).

But what about the other components of reproductive structures and behavior related to reproduction? At this point, it may be helpful to view the allocation of reproductive energy from another perspective. Table V is a contingency table that separates the target of parental energy expenditures (offspring or environment) from the timing of parental energy expenditures (before spawning or after spawning). Thus, four qualitatively different forms of parental energy expenditures can be distinguished: Type I expenditures are directed toward the offspring before spawning, Type II expenditures are directed toward the environment before spawning, Type III expenditures are directed toward the offspring after spawning, and Type IV expenditures are directed toward the environment after spawning. The various parental energy expenditures now can be seen as having different kinds of effects on the survival of the offspring.

It is clear that the definition of parental investment that Trivers (1972) offered corresponds only to Types I, III, and IV of the classification in Table V. Trivers claimed that Type II expenditures do not affect the survival of the offspring, but we believe that here Trivers has painted himself into a corner. In the first place, we consider parental energy expenditures such

TABLE V
A Classification of Effects Associated with Energy Expenditures from a Parent to an Offspring[a]

| Timing of parental energy expenditures | Target of parental energy expenditures | |
| --- | --- | --- |
| | Offspring | Environment |
| Before spawning | Type I | Type II |
| | Sperm | Spawning site selection |
| | Egg cytoplasm | Spawning site defense |
| | Egg yolk | Nest construction |
| After spawning | Type III | Type IV |
| | Skin feeding | Nest maintenance |
| | Filial cannibalism | Nest defense |
| | Placental transfers | Brooding |
| | Histotrophic production | Bearing |

[a] Parental energy expenditures are either targeted directly at the offspring, or indirectly at the offspring by changing the offspring's environment. The timing of parental energy expenditures is classified as either before spawning or after spawning. Examples of Types I–IV of parental energy expenditures are given in the appropriate parts of the table.

as mate selection to be preconditions of successful spawning that could indeed "affect the survival chances of the resulting offspring," for example, by affecting the quality of behavioral care contributed by the mate. Secondly, Trivers' definition of parental investment fails to consider the balanced relationships among the various components of parental energy expenditures; that is, given a finite amount of energy available to reproduction, energy that is distributed in one direction (e.g., reproductive structures) cannot be distributed in another (e.g., reproductive behavior). Thirdly, Trivers' definition does not recognize the possibility of cause-and-effect relationships between components of parental energy expenditure that may cross over his arbitrary distinctions.

Given this analysis, it is interesting that Trivers (1972, p. 136) recognized a conceptual error in one of his more famous predecessors: "Charles Darwin's (1871) treatment of the topic of sexual selection was sometimes confused because he lacked a general framework within which to relate the variables he perceived to be important (. . .). This confusion permitted others to attempt to show that Darwin's terminology was imprecise, that he misinterpreted the function of some structures, and that the influence of sexual selection was greatly overrated". Ironically, it is exactly the lack of a "general framework within which to relate the variables" and "imprecise terminology" that constitute the major weaknesses of Trivers' definition of parental investment.

It is likely that Trivers' simplistic definition of parental investment was a reflection of his desire to satisfy another theoretical goal—the calculation of "typical parental investment by an individual per offspring" for a model that "assumes that natural selection has favoured the total parental investment that leads to maximum net reproductive success" (Trivers, 1972, p. 139). Perhaps if Trivers had placed more importance on the actual nature of reproduction, he would have been able to provide a more realistic model of the underlying evolutionary processes.

B. Parental Care (Blumer, 1979, 1982)

In Blumer's (1979) paper on the origins of paternal care in teleost fishes, he also criticized Trivers' definition of parental investment (Blumer, 1979, p. 149):

> Trivers' definition excludes behaviors that serve both purposes such as nest building and territoriality. A male may build a nest or defend an oviposition site in order to attract or to control mates. Yet, the deposition of eggs in an area where egg-eating conspecifics and heterospecifics have been excluded may increase the eggs' chances of surviving. My own operational definition of parental care goes beyond Trivers' definition for this reason.

As discussed previously, we agree with this criticism (see also Clutton-Brock and Godfray, 1991), yet Blumer's own definition of parental care also was limited: "I define parental care as non-gametic contributions that directly or indirectly contribute to the survival and reproductive success of the offspring. Although the gametes constitute a real contribution by the parents to the offspring, I wish to focus attention on those parental contributions that do not characterize all gamete producing organisms" (Blumer, 1979, p. 149), or, put more succinctly: "non-gametic contributions that enhance the survival and reproductive success of the care-giver's offspring" (Blumer, 1982, p. 1).

Thus, according to this definition of parental care, Blumer (1979, 1982) was referring to all behavioral components of reproduction, including both spawning and tending behaviors (Fig. 6a). The status of reproductive structures, however, is not as clear. We do not know whether Blumer would consider energy expenditures on spawning, mating, or brooding structures in his definition of parental care. In contrast, Blumer's opinion on gametes is perfectly clear, that is, energy expended on gametes is definitely not included in his definition of parental care. But why would Blumer have recognized Types II, III, and IV of parental energy expenditures (Table V), and not Type I expenditures? From an evolutionary perspective, this seems to be an unjustified separation of energy transfers that are intimately linked to one another.

### C. PARENTAL CARE (SARGENT ET AL., 1987)

The next major attempt to define parental care was by Sargent *et al.* (1987, p. 33), in a paper on parental care and gamete size:

> For the purposes of this paper, we define parental care as any investment by a parent, other than egg yolk, that increases the survival of the offspring until they are independent of all parental resources. Thus, parental care includes the selection or building of an oviposition site and any guarding or feeding that may occur after fertilization. Note that although parental care is generally defined as investment in offspring after fertilization (e.g., Gross and Sargent, 1985; Sargent and Gross, 1986), we have broadened our definition of parental care here to include investment before fertilization that affects offspring survival after fertilization (see also Blumer, 1982).

This definition of parental care was similar to the definition that Blumer offered, with a few twists of its own. First, Sargent *et al.* (1987) arbitrarily excluded the energy expended on gametes (focusing exclusively on egg yolk for some undefined reason) from the realm of parental care. Second, they attempted to anchor the definition of parental care in some ontogenetic context by limiting it to those expenditures that would have an effect on offspring survival, up until the state when the offspring became independent of all parental resources. We assume in this case that they were referring to the beginning of the larva or juvenile period (see Section III, Table IV); however, their acceptance of Shine's (1978) ontogenetic model leaves room for some doubt. Third, Sargent *et al.* (1987) also extended the coverage of parental care expenditures to the period before spawning, as suggested by Blumer (1979, 1982; but see Sargent and Gross, 1993). Therefore, the definition of parental care by Sargent *et al.* (1987) is generally similar to that of Blumer in that it concerns parental energy expenditures of Types II, III, and IV. Obviously, the same questions that we have raised about Blumer's definition of parental care (see previously mentioned) apply to that of Sargent *et al.* (1987).

D. EVALUATION

There has been little consistency within or between the various definitions of parental investment or parental care that have been offered in the past. In our opinion, these definitions have ignored the integrated nature of the energy expenditures associated with reproduction and have placed artificial limits on what constitutes parental care. There has been a tendency to isolate behavior from other components of reproductive energy expenditures, even though individual cause-and-effect relationships have been suggested many times. This isolation is probably a reflection of the fact that many of the researchers have had specialized training in ethology or behavioral ecology, rather than whole animal life history per se.

In Section VI, we review some of the major historical trends in the study of piscine parental care. In particular, we will focus on the cause-and-effect relationships that have been hypothesized to exist between various aspects of life history. We will attempt to pull together many of the disparate concepts associated with parental care in fishes, within the structure of an integrated life-history model. It should be noted that when we use the term *parental care* in Section VI, we refer to the entire class of reproductive energy expenditures (see Fig. 6).

VI. A REVIEW OF RECENT WORK ON PARENTAL CARE IN FISHES

Having presented a conceptual framework for understanding parental care in fishes, it would be appropriate for us to describe the distribution of recent work in this field. For this reason, we sampled the primary literature for references during the previous 15 years. Our intention was not to present a comprehensive bibliography and review of this literature, but rather to detect general trends in the field. To obtain a relatively unbiased sample of the recent literature, we searched the Science Citation Index (Compact Disk Edition, Institute for Scientific Information Inc.) for the period 1980 to March 1995, inclusive. The search condition used in this sampling program was the occurrence of the words *parent* (which indicates a wildcard to include associated words such as *parents* and *parental*) and *care* in the title of the publication. When the search was completed, we selected those references (notes or full papers) that focused primarily on parental care in fishes. We recognize that our limited search parameters may have excluded some attempts to place parental care in a greater life history context; however, we were most interested in studies with a primary focus on parental care in fishes. We believe that our methods resulted in a representative sampling of such studies that have been published in the primary literature over the past 15 years.

We reviewed each of the selected papers, and summarized the work of the authors in two different ways: (1) we classified the reproductive energy expenditures that were considered by the authors (Table VI), using the hierarchical model presented in Fig. 6; and (2) we classified the major factors associated with parental care that were investigated by the authors (Table VII). The following is a description of the trends that we observed.

### A. Reproductive Energy Expenditures

The most obvious trend in recent research on energy expenditures associated with parental care was the distinction between energy expended on structure versus behavior. The vast majority of papers focused on spawning and/or tending behavior, rather than on the tissues or organs related to gamete production, mating, spawning, or brooding (Table VI).

Of the papers that focused on energy expenditures in structure (12/40 = 30%), most referred to the properties of female gametes. Twenty percent of the papers in our sample discussed the relationship between parental care and some measure of yolk investment, usually referred to as *egg size*. We present a more detailed interpretation of the work on parental care and female gamete expenditures in Section VI,C. The only other component of reproductive structure (morphology) that received much attention from the parental care researchers was the energy associated with parental body structures used to brood offspring. The remaining components of morphological energy expenditures each received attention from less than 10% of the papers in our sample.

All of the papers in our sample focused on parental energy expenditures in behavior of various kinds. The most frequent topics were associated with tending behaviors, site-specific defense, offspring-specific maintenance, and

TABLE VI

A Summary of Major Concepts Associated with Published Research on Parental Care in Fishes (1980–1995), According to the Reproductive Energy Expenditures by the Parents[a]

| | Hierarchical order of reproductive energy expenditures | | | | References | |
|---|---|---|---|---|---|---|
| Level I | Level II | Level III | Level IV | Level V | Code | Frequency |
| Structure | Gamete production | Gonads | | | 2,7,39 | 3 |
| | | Gametes | Eggs | Cytoplasm | | 0 |
| | | | | Yolk | 2,14,15,29,31,33,35,36 | 8 |
| | | | | Accessory structures | 30,36 | 2 |
| | | | Sperm | | 2 | 1 |
| | Spawning structures | | | | 31 | 1 |
| | Mating structures | | | | | 0 |
| | Brooding structures | | | | 15,26,30,35,39 | 5 |
| Behavior | Spawning | Site-specific | Selection | | 2,4,12,14,19,33,39 | 7 |
| | | | Preparation | | 1,2,12,14,16,18,19,33,37 | 9 |
| | | | Defense | | 2,4,12,14,16,20,33,37,38 | 9 |
| | | Mate-specific | Selection | | 2,4,7,9,12,14,16,20 | 8 |
| | | | Courtship | | 4,9,12,14,16,19,20 | 7 |
| | | | Defense | | 2,4,9,12 | 4 |
| | | | Mating | Gamete release | 12,14,31 | 3 |
| | | | | Gamete transfer | 17,31 | 2 |
| | Tending | Site-specific | Maintenance | | 5,9,13,14,16,25,33,37,38 | 9 |
| | | | Defense | | 1,2,4,5,6,9,10,11,13,14,15,16,17,19,20,22,23,24, 25,32,33,37,38 | 23 |
| | | Offspring-specific | Maintenance | | 3,5,6,7,12,13,14,22,24,25,26,27,28,29,30,31,33, 34,35,36,39,40 | 22 |
| | | | Defense | | 1,2,3,4,5,6,7,8,9,10,11,12,13,14,15,17,19,20,21, 22,23,25,26,27,28,29,31,32,33,34,37,39 | 32 |
| | | | Feeding | | 5,17 | 2 |

[a] The first five columns represent the hierarchical organization of energy expenditures on reproduction, as presented in Figure 6. The last two columns list the identifying reference code of each research paper (see bracketed numbers at the end of entries in References) and the frequency of research papers from a sample of $N = 40$, during the period 1980–1995 (see text for details).

TABLE VII
A Summary of Major Factors Associated with Research on Parental Care in Fishes (1980–1995), According to Hypotheses Discussed by the Authors[a]

| Category | Description | References Codes | Frequency |
|---|---|---|---|
| Environment | Spawning substrate | 1,10,25 | 3 |
| | Dissolved oxygen | 6,40 | 2 |
| | Water temperature | 32 | 1 |
| | Food supply | 21 | 1 |
| | Wild versus aquaria | 34 | 1 |
| | Marine versus fresh water | 10 | 1 |
| | Pelagic versus benthic | 10 | 1 |
| Parent | Growth | 4,7,9,14,15,21,24,32,33,34,35,38,39 | 13 |
| | Feeding | 7,21,24,34,35,38 | 6 |
| | Survival | 15,24,33,35,38 | 5 |
| | Fecundity–egg size | 2,15,33,35 | 4 |
| | Future reproduction | 15,29,33,38 | 4 |
| | Predation risk | 23 | 1 |
| Mating | Season, timing | 7,9,12,14,16,29,32,35,38,39 | 10 |
| | Nest construction | 2,6,14,16,18 | 5 |
| | Brood size | 9,20,27,29,32 | 5 |
| | Mate choice | 4,9,16,20 | 4 |
| | Fertilization mechanisms | 15,17,31 | 3 |
| | Migration | 12 | 1 |
| Tending | Gender bias | 2,5,7,11,12,15,21,22,26,29,31,34,35 | 13 |
| | Division of labor | 5,6,7,11,12,22,25,26,27,34,36,39 | 12 |
| | Ethological components | 5,6,9,14,19,22,31,40 | 8 |
| | Intensity of tending | 4,16,18,21,22,34 | 6 |
| | Duration | 8,24,26,29,32,39 | 6 |
| | Motivational drive | 16,19,34,40 | 4 |
| | Abandonment | 7,15,29,34 | 4 |
| | Mouthbrooding | 26,35,39 | 3 |
| | Physiological mechanisms | 18,19,40 | 3 |
| | Offspring cannibalism | 24,26 | 2 |
| | External brooding | 30 | 1 |
| | Territorial characteristics | 20 | 1 |
| | Fungal infection | 6 | 1 |
| | Offspring discrimination | 3 | 1 |
| Offspring | Development | 6,9,13,24,26,27,28,32,33,39,40 | 11 |
| | Survival | 1,6,8,13,15,20,21,27,33,40 | 10 |
| | Predation risk | 1,6,8,13,14,23,27,34 | 8 |
| | Growth | 8,21,27,32,33 | 5 |
| | Behavior | 3,8,9,32 | 4 |
| | Feeding | 17,21 | 2 |
| | Morphology | 3 | 1 |

[a] The last three columns indicate the reference code of each research paper (see References) and the frequency of research papers from a sample of $N = 40$, during the period 1980–1995 (see text for details).

offspring-specific defense, each receiving strong interest from more than half of the sampled papers. The relationships between parental care and spawning behaviors (site-specific and mate-specific) were not discussed frequently. Although the topic of offspring feeding was rarely discussed in our sample of papers (2/40 = 5%), this statistic serves to remind us that the frequency of papers discussing a particular aspect of energy expenditure is confounded by the absolute and relative frequency of occurrence in the reproductive styles of the fishes (e.g., compare the scientific interest in mouthbrooding with the actual frequency of mouthbrooding among fishes). This consideration should be kept in mind when interpreting all of the analyses and interpretations of the literature sampled for this article.

In summary, we found that recently published research on parental care exhibited a bias toward studying behavioral rather than structural aspects of reproductive energy expenditures. Where structural components were investigated, there was a research bias toward female, rather than male, gamete production. Both structural and mate-specific behavioral aspects of reproductive energy expenditures require additional attention.

B. Major Factors Investigated

For the purpose of convenience, we categorized the major factors investigated by researchers in our sample of papers on parental care in fishes (Table VII). Taking our cues from the work of the authors themselves, we recognized five general categories of factors relating to parental care: environmental conditions, the characteristics of parent(s), mating, tending, and offspring. While the reader may notice some similarities between this categorization and the classification of reproductive energy expenditures (see previously mentioned), our primary purpose was to summarize trends that the authors themselves considered to be important.

Environmental conditions were considered to be important factors relating to parental care in 8 of the 40 (20%) papers in our sample. Among these, basic environmental characteristics, such as spawning substrate or the physical properties of water, were discussed. In a few instances, authors attempted to compare the properties of parental care in contrasting environments (e.g., in the wild versus aquaria, marine versus fresh water, pelagic versus benthic habitats). It is interesting to remember that almost half a century ago Kryzhanovsky (1948, 1949) recognized that environmental conditions were a major determinant of the reproductive style (including all aspects of parental care) in fishes. From our sample of the literature, it seems that few investigators have heeded Kryzhanovsky's advice to study the life history of their subjects within the context of the environment in which they evolved.

In 16 (40%) of the papers in our sample, the characteristics of the parent were considered as major factors relating to the quality and quantity of parental care. Among these papers, most authors focused on the relationship between parental body growth (size) and parental care. Other parental characteristics that were recognized as important characters included feeding, fecundity, the probability of future survival, and reproduction. It should be noted that the relationship of fecundity–egg size to parental care in this category involved explicit statements of causal relationships. In contrast, the yolk energy expenditure in Table VI (Level V) did not require any discussion of a causal relationship with other aspects of parental care.

The relationship between mating characteristics and parental care was discussed in 17 different papers (42.5%) with more than half of them focusing on the seasonality or timing of mating events. Others papers considered the interrelationships between nest construction, mate choice or brood size, and other components of parental care. It should be noted that in a few papers, the subect of "fertilization" mechanisms (better called insemination, see Balon, 1985a, 1990)—especially the phenomenon of internal insemination—was considered as an important factor in parental care.

The characteristics of tending behavior were clearly the most popular topics of research in our sample of the literature. A total of 31 different papers (77.5%) discussed tending behavior as a major factor associated with parental care. Within this category, the issues of gender bias and division of labor between parents were by far the most common topics of discussion. Other aspects of tending behavior that received attention included ethological analysis, the quality and quantity of tending behavior, and the physiological basis for behavior.

Finally, the relationship between parental care and characteristics of offspring were discussed as major factors in 21 (52.5%) of the papers in our sample. More than half of these papers referred to the development of the offspring (usually to specific events that take place during ontogeny, like hatching) and/or the relationship between parental care and offspring survival. Only a few papers attempted a detailed analysis of the relationship between aspects of parental care and the physiological, morphological, and ethological life-history characteristics of the offspring. It is precisely for this reason that we advocate a new approach to the study of parental care—within an integrated life-history model that focuses on the interrelationships between parents, offspring, and their environment (nonliving and living). The following text provides an example of how we would interpret work on parental care from the perspective of such a life-history model.

### C. The Relationship between Egg Size and Parental Care

From a structural perspective, much of the work that has taken place in the study of parental care in fishes has focused on the effect of gamete production on other components of reproductive energy expenditures (Table VI). Past research has devoted virtually all of its attention on female gamete (egg) production, implicitly assuming that the total energy expenditure in male gametes (sperm) is insignificant when compared with that of the female (but see Blumer, 1979; Jamieson, 1991). As a result, the concept of egg size has become widely distributed throughout the literature in this field. It should be noted that the phrase *egg size* is actually an overly simplistic and often incorrect representation of the quantity and quality of energy expended by a female into the cytoplasm, yolk, and accessory structures of a gamete (Fig. 6, and see Crawford and Balon, 1994c). By presenting a re-evaluation of issues associated with research in this field, we hope to show how future work in the study of parental care can be conducted within the context of an integrated life-history model.

Unfortunately, very few authors have recognized the fact that egg size is often not a direct correlate of energy expenditure—thus exposing themselves to first-order failures in logic (see Fig. 4). For example, consider the premises of the paper published by Sargent *et al.* (1987, p. 34):

> We define egg size as the amount of maternal yolk reserves allotted per offspring (in units of mass). Although other factors may contribute to egg size variation (e.g., chemical composition of yolk, oil droplets, hydration), we assume that natural selection maximizes maternal fitness and that egg size is "optimized" with respect to maximizing the number of offspring that survive to maturity. This measure of fitness assumes that a mature individual's reproductive success is independent of the size of the egg from which it hatched. Our approach is to assume an optimal egg size and to analyze how this optimum depends on one variable, the quality of parental care. In our models, parental care subsumes all parental behaviors that reduce instantaneous mortality during the egg stage.

Other technical considerations aside, the assumption that natural selection maximizes maternal fitness and optimizes egg size (see also Gross and Sargent, 1985; Sargent and Gross, 1986, 1993) can lead to mistaken interpretations of patterns in gametic investment. Given that energy supplies can be allocated in different densities, other factors (e.g., the surface area requirements for embryonic respiratory structures, see Crawford and Balon, 1994c) may account for more variation in egg size than the energy bundled in the packet. Most authors fail to consider that egg size and energy expenditures are dynamic conditions—both between and within individuals of a particular species (e.g., Geist, 1971; Balon, 1986a, 1989b). Notwithstanding these criticisms, for the purpose of argument, we will use the term *egg size* to represent some measure of the energy expended by a female into the gamete (i.e., measure of energy associated with endogenous nutrients).

Shine (1978) commented on the relationship between the degree of parental care (undefined) exhibited by different species of organisms including fishes, and their respective egg size and number: "Why do females of some species produce many small eggs while related species produce a few large eggs? The evolution of propagule size is a major unsolved problem in life-history theory, despite much theoretical work (. . .). The present paper approaches this problem from a new viewpoint: analysis of an empirical correlation between propagule size and parental care" (Shine, 1978, p. 417). To his credit, Shine (1978) distinguished between correlation and causation in the following manner: "The pattern is this: species that protect their offspring usually produce larger (and hence fewer) propagules than do related species that do not protect their offspring. This correlation is shown by almost all poikilothermic groups with parental care, ranging from sponges to reptiles (Table 1)" (Shine, 1978, p. 419). If we assume that there are empirical data to support the claim of this general correlation (the data for fishes have never been presented in a comprehensive manner)—so far, so good! We have an inverse relationship between egg size and egg number (e.g., Smith and Fretwell, 1974; Peters and Berns, 1982; Elgar, 1990), and a direct relationship between egg size and parental care: "Species that protect their offspring (by egg-brooding or live-bearing) produce larger (and hence fewer) propagules than do related species that do not protect their offspring" (Shine, 1978, p. 417).

It now remains for researchers to develop hypotheses that would link these variables in cause-and-effect relationships. Shine (1978) offered the following transition from correlation to causation: "What the correlation means, in effect, is that the evolution of parental care is usually accompanied by the evolution of a mechanism to increase propagule size. The actual mechanism varies: some groups merely produce larger eggs (. . .) while others retain and feed the young after the egg hatches (. . .). A several-fold size difference is often seen between propagules of related species with and without parental care . . . ." (Shine, 1978, p. 418). Indeed, a hypothesis that suggested a cause-and-effect relationship between parental care and egg size would have been perfectly reasonable at this point. We searched through Shine's (1978) paper for a passage of text that would describe such a hypothesis, but were unsuccessful. In its place, we present a description of Shine's (1978) safe harbor hypothesis that was written by Gross and Sargent (1985, p. 816):

> When offspring survivorship is increased by parental care, selection may favor the female making larger (and fewer) eggs. Since development time lengthens with increasing egg size (Steele 1977, Noakes and Balon, 1982) a female producing larger eggs will ensure that her offspring spend more of their life history in a relatively "safe" place. Moreover, because egg size determines initial juvenile size, offspring that hatch from large eggs will also have higher survivorships over the juvenile stage of development. Thus egg size should increase until the proportional gain in surviving offspring equals the female's

costs in decreased fecundity due to large eggs. Shine (1978) termed this the "safe harbor" hypothesis.

There are several points that we would like to make about this hypothesis. First, it implicitly or explicitly makes several assumptions: (1) natural selection is a major factor in determining egg size, (2) development time increases with egg size, (3) mortality of offspring within the egg envelopes is lower than mortality after hatching, and (4) mortality of offspring (especially juveniles) decreases with egg size. These assumptions have not been demonstrated as general truths and thus should not be considered as general representations of life history in fishes. Thus, it should be clear that the safe harbor hypothesis is more a series of possible (not necessarily observed) correlations, brought together without explicitly specifying cause-and-effect relationships between gametic energy expenditure and the intensity of parental care (in any form). In fact, Shine reveals the fundamental weakness of his hypothesis later in his paper: "changes in propagule survivorship are the most consistent correlates of, and hence the most likely cause of, the parental care-propagule size correlation" (Shine, 1978, p. 421). We suspect that the actual causal mechanisms in this field have been swept into a black box with the words *natural selection* embossed on the lid. These same difficulties were inherited by subsequent theoretical works on the relationship between parental care and egg size (e.g., Gross and Sargent, 1985; Sargent *et al.,* 1987; Sargent and Gross, 1993).

One final observation is warranted about past research on relationships between parental care and gamete energy expenditures. We commented previously that a hypothesized cause-and-effect relationship was appropriate to serve as a potential explanation of correlation. Shine (1978) and Gross and Sargent (1985) implicitly assumed that changes in parental care led to subsequent changes in physiology, which in turn altered egg size. We have been unable to find any clear indication that they considered the possibility that changes in physiology, which in turn altered egg size, might have led to subsequent changes in parental care that have been documented previously (e.g., Balon, 1975a, 1978, 1980, 1985b; Matsuda, 1987). Although we recognize both possibilities, we tend to think that the latter epigenetic relationship is more probable from an evolutionary point of view (see Balon, 1988, 1989b). Even within the same species, better feeding conditions of the female at the time of vitellogenesis can result in more yolk per egg relative to less yolk under poorer feeding conditions (see also Geist, 1971; Balon, 1989b).

In summary, research on the relationship between egg size and parental care has suffered from several fundamental weaknesses. First, investigators have commonly been uncritical with their use of egg size as a measure of reproductive energy expenditure. Second, there has been a lack of distinction between correlation and causation in the relationship between reproductive energy expenditure and the degree of parental care exhibited, both within and between species. Third, there has been a tendency for *a priori* application of other theories (e.g., natural selection) without questioning or empirical testing of predictions generated by these theories. All of these weaknesses suggest a lack of desire to consider alternative life-history hypotheses that may account for parental care phenomena in a cause-and-effect manner. Our analysis of arguments forwarded on this issue suggests that researchers would benefit greatly from developing and using life-history models that place more emphasis on the early ontogeny and environment of the offspring.

## VII. The Epigenesis of Piscine Parental Care

[I]t is unproductive to defend the status quo of the evolutionary synthesis by denying that objections to it are, in fact, really objections.—*P. H. Greenwood, 1995, p. x, in a Foreword to* Speciation and the Recognition Concept *(Lambert and Spencer, 1995)*

### A. Neo-Darwinism versus Epigenetics

In the preceding text we have suggested some cause-and-effect mechanisms that are not consistent with the traditional neo-Darwinian views of evolution. Therefore, it would be appropriate for us to discuss these mechanisms more fully, especially as they relate to the evolution of parental care (structural and behavioral) in fishes. It should be noted that this discussion is intended to be an example of how epigenetic theory can be applied to the issue of parental care, rather than a treatise on epigenetic theory itself. For a more complete treatment of these ideas, we direct interested readers to works by Løvtrup (1974, 1982), Balon (1983, 1988, 1989b), Bruton (1990), Hall (1992), and references therein.

Currently, there are two general theories in evolutionary biology that compete with each other by proposing alternative hypotheses of cause-and-effect relationships: (1) neo-Darwinism, and (2) epigenetics. The following is a very brief summary of the differences between these competing theories, as they relate to the ontogeny and phylogeny of concepts such as parental care.

Neo-Darwinian theory has been a champion in evolutionary biology for the past 50 years. This complex theory is a hybrid of the Darwinian theory of natural selection and the genetic theory of inheritance. Hypotheses that stem from this theory share a common dependence on creative factors that are intrinsic to an organism, and that are physically inherited by its offspring. These hypotheses explain the development of features by suggesting the existence and action of predetermined (i.e., genetic) instructions that are inherited with the possibility of random variation, from parent to offspring. Organisms in a population become adapted to local environmental conditions over time, through differential survival–reproduction, and hence exhibit differential transmission of the instructions to their offspring. According to this theory, all functional characteristics of organisms can be linked directly to specific units of inheritance (i.e., genes in the genotype) that account for particular phenomena (structure or behavior in the phenotype) in a 1:1 manner. Currently, neo-Darwinian theory is common, well defined, and firmly entrenched in the field of evolutionary biology.

Epigenetic theory has emerged as a challenger in modern evolutionary biology (Løvtrup, 1974, 1982, 1987; Balon, 1983, 1985b; Hall, 1992). The major difficulty in defining current epigenetic theory results from its history. The theory actually began as an assortment of specific challenges to neo-Darwinism; an assortment that lacked a general, common basis. In this sense, epigenetic theories were simply bound by the fact that they recognized the limitations of genetic mechanisms in accounting for phenomena in the "real" world (i.e., they were considered antigenetic or nongenetic in their interpretation). Over time, these challenges to neo-Darwinism began to gravitate towards each other in a manner that was similar to the effect of "strange attractors" in chaos theory (e.g., Gleick, 1987). In this case, the various challenging theories came together on the principle of extrinsic, nonpredetermined (i.e., epigenetic) forces that are associated with environmental conditions under which an organism develops and survives.

The greatest irony about the perceived champion-challenger relationship between neo-Darwinism and epigenesis is that they are simply the most recent installments in an ongoing, historic tale of struggle between two basic kinds of biological theories: (1) those theories that explain ontogeny and phylogeny by reference to the predetermined action of intrinsic factors (i.e., preformation theories), and (2) those theories that explain ontogeny and phylogeny by reference to the novel action of extrinsic factors (i.e., epigenetic theories). A review of the historical literature shows clearly that the struggle for popular acceptance in the scientific community has shifted between the preformation and epigenetic camps at least five times since the reported origin of the contest between Plato and Aristotle (see Huxley and de Beer, 1934; Adelmann, 1966; Gasking, 1967; Gould, 1977; Magner, 1979). The ongoing dynamic struggle between preformation and epigenesis has always been a balance of forces that results in a continuum of conceptual change, not unlike the movement of a pendulum. However, at a more proximate level, the struggle has more to do with the discrete question of how individual minds perceive the task at hand—does one tend to explain their everyday world with handy, prepackaged hypotheses, or does one tend to search for exceptions that require new expressions of creative forces? In this case, science is clearly a reflection of the human condition.

As in any field in which evolutionary forces are at work, parental care cannot be understood, its origin and mechanisms revealed, without an understanding of developmental biology. A change in the generation of parental care within an individual life history (ontogeny) leads to a change in the generation of parental care within the life history of a species lineage (phylogeny). The trick for evolutionary biologists is to distinguish between genetic and epigenetic mechanisms that combine to result in changes of parental care. Such distinctions are at the heart of understanding cause-and-effect relationships in the evolution of piscine parental care.

## B. The Evolution of Reproductive Styles

The following is our view of the major ontogenetic factors that have caused the evolution of diverse quality and quantity of parental care among fishes, within the general context of reproductive guilds (see Sections II,C and II,E). We have highlighted features of this evolutionary history that are likely to be associated with epigenetic factors—that is, factors that are not encoded within an organism and inherited directly from parent to offspring. Our purpose for introducing this epigenetic bias is not to repudiate genetic mechanisms, but rather to bring to the reader's attention possible epigenetic mechanisms that have been largely ignored by research in recent times (see Section VI).

Parental care in its most basic form started when nonguarding fish sought spawning grounds on which to deposite gametes. The parents must have cared in advance where to deposit their gametes, for otherwise the progeny would have likely not survived. Parents even of the nonguarding egg-scattering kind, displayed some form of parental care. Offspring of each species developed structures and functions suited best for the particular environment in which the parents had deposited them. Hiding eggs within the selected spawning grounds was obviously a natural step in enhancing the survival of progeny, be it in gravel pits and rock crevices as in some salmonids, or in living mussels as in bitterlings. In some cases, such as species with extensive spawning migrations, this preparatory care could have cost parents much more than their direct care for offspring.

The evolutionary sequence of transition from nonguarding to guarding can be understood by making reference to particular characteristics of the eggs and embryos. When increased endogenous nutrient deposition in eggs was feasible, despite decreased fecundity, heavy eggs had a tendency to form a compact structure that could have elicited extended care from the parent(s) beyond simple substrate selection. If high fecundity was required for successful dispersal, limited and watery yolk production would have been retained. The presence of a large amount of yolk, in itself, formed a perfect surface for the development of temporary respiratory networks, enabling not only more effective hiding of the egg but also expanding the utilization of low-oxygen areas where predators were less common. Fanning during guarding helped in this respect. Cement gland cell development, to facilitate the embryos' attachment above an anoxic muddy bottom after hatching, was a feature most obligatory plant spawners exhibited, irrespective of their offspring being abandoned or guarded.

The evolutionary sequence of transition from substrate guarding and nesting to mouthbrooding can be illustrated by reference to the cichlids (Fig. 7). Substrate or nest-tending cichlids produced much smaller eggs than mouthbrooding ones, and the lesser amount of yolk resulted in indirect development. After hatching, the embryos of the former were guarded and fanned, each one attached to the substrate by fibers from cement glands on the head in a more or less compact clutch. As feeding larvae, they arose from the substrate in a compact school and were guarded by the parent(s) until metamorphosis into juveniles or even longer. In most species, the eggs were not only fanned but also were mouthed-cleaned and often were picked up into the parent's mouth and transferred onto a clean substrate or pit. These transfers became more frequent, and the eggs and embryos were retained in the buccal cavity for a longer and longer time. This tendency was associated with fewer, more yolky eggs and larger larvae at first feeding, and ultimately, the elimination of the larva period entirely (i.e., direct development).

From this state, it is easy to see how the evolution of mouthbrooding could have occurred. In fact, we know of examples of cichlids exhibiting almost every possible intermediate from ancestral substrate tending to advanced oral brooding (Fig. 7). There is no limit to the creation of life histories under the influence of heterochrony. For example, the mouthbrooding cichlid *Cyphotilapia frontosa* releases unusually large juveniles directly into the adult habitat of the Lake Tanganyika hypolimnion by producing very large eggs, as well as by having embryos that hatch after a very brief incubation. Offspring in the female's buccal cavity, while still carrying a large yolk supply, begin feeding on external food inhaled by the female. This double delivery of nutrients from both yolk and exogenous feeding accelerates the direct development into a large juvenile. Upon release, the offspring do not require a separate nursery feeding ground of shallow water (see Balon, 1990). After the publication of these laboratory

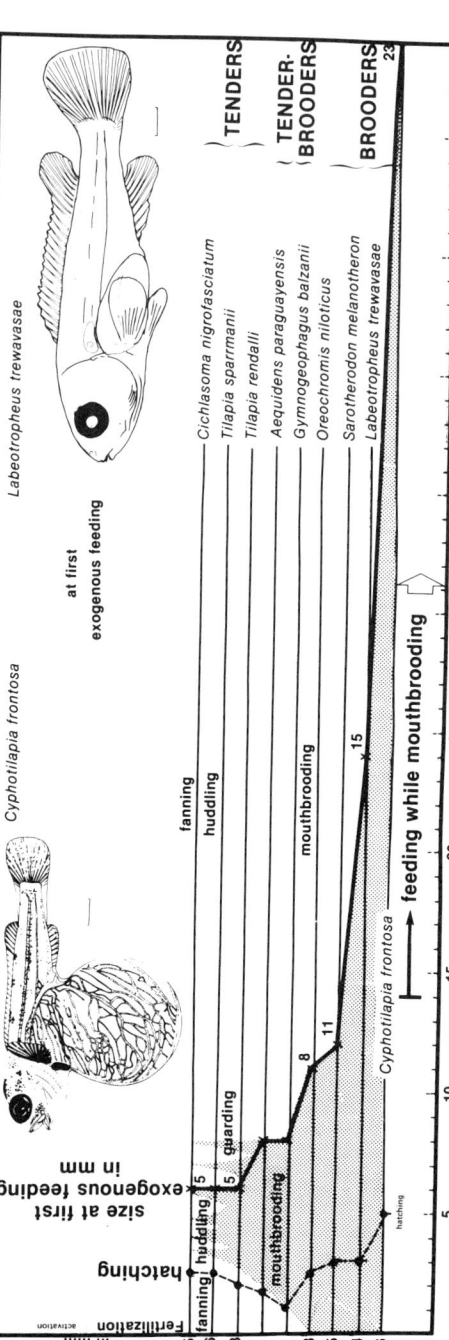

FIG. 7. The evolution of parental care in cichlid fishes from clutch tenders on a substratum, via intermediate initial substrate tenders and subsequent mouthbrooders, to mouthbrooders with immediate uptake of ova after or prior to insemination. Fish sketches represent the size and form of *Cyphotilapia frontosa* (left) and *Labeotropheus trewavasae* (right) at the time of first exogenous feeding. In the diagram, tenders are represented by *Cichlasoma nigrofasciatum*, a guarding species that tends and fans the embryos and transfers them several times into new pits excavated in the bottom using its mouth (huddling). The two nest guarding tilapias (*Tilapia sparrmanii* and *Tilapia rendalli*) do the same but retain the free embryos longer in their buccal pouches during such transfers; *Aequidens paraguayensis* and *Gymnogeophagus balzanii* guard and fan the clutch in a nest until hatching, at which time the embryos are collected into the mouth and brooded continuously in the buccal cavity until their final release, which coincides with the offspring's first exogenous feeding. *Oreochromis niloticus*, *Sarotherodon melanotheron*, and *Labeotropheus trewavasae* represent species with continuous mouthbrooding, but of different duration, depending on the yolk supply reflected in the egg size (left scale). *Cyphotilapia frontosa* represents the most advanced care with exogenous feeding already started in early embryos during mouthbrooding. For most of the brooding interval the young survive on mixed feeding, that is yolk, and particles inhaled by the female into her buccal cavity. The major goal of all these energy expenditures is the creation of more viable young, reflected in an increase of size and completion of the definitive form at the time of release from the protection of the parental body (bold line). Hatching is at first accelerated with the prolongation of brooding time, but an increase in yolk ultimately enables the incorporation of more carotenoid pigments with the ability to provide an endogenous oxygen supply, and so hatching can be delayed (dashed line). Buccal uptake and brooding until first release of offspring are represented by the finely dotted screen. (From Balon, 1990.)

observations (Balon, 1985a), the same life-history pattern was not only confirmed in the field but was also discovered in other cichlids (Yanagisawa and Sato, 1990; Yanagisawa and Ochi, 1991; Yanagisawa, 1993). It proves the earlier contention that the modification of behavioral care must have occurred after changes in ontogeny!

One advantage of internal insemination is the elimination of gamete exposure to environmental influences like predation or physical and chemical shocks. Although relatively few oviparous species of fishes practice internal insemination, the oral-brooding cichlids have minimized the exposure of their gametes by: (1) shortening the intervals between egg laying, sperm release, and engulfing of the eggs by a parent; (2) releasing and retrieving the eggs in midwater instead of on a substrate; and (3) having the eggs inseminated in the female's mouth (Balon, 1977, 1978; Eccles and Lewis, 1981). None of these behaviors appears to have served as an intermediate state for any other.

Viviparous fishes exhibit styles of live bearing and parental nutrition of embryonic young ranging from lecithotrophy (dependence on yolk) to a multitude of placental analogs (some of which are more complex than in mammals), as well as the packaging of sperm into spermatophores or spermatozeugmata to facilitate transfer into the female genital tract (Wourms *et al.,* 1988; Balon, 1991). It is noteworthy that none of these shows any structural or behavioral connection with each other. That is, each one of 6 different guilds of viviparity appears to have independently evolved (Wourms and Lombardi, 1992).

C. EVOLUTIONARY MECHANISMS REVISITED

At this point, we would like to discuss the factors affecting quantity and quality of parental care for a particular species (or population), within an evolutionary context that focuses on speciation. We use our general life-history model, and associated theories related to the bifurcation of altricial and precocial forms, to demonstrate how parental care can be both a cause and an effect in the evolution of new life-history patterns (see Balon, 1983, 1985, 1989a,b).

As an alternative to the suspect concepts of Darwinian or neo-Darwinian natural selection, fitness, and adaptation (Krimbas, 1984; Reid, 1985; Løvtrup, 1987) that provide the *modus operandi* behind most previous explanations of parental care (e.g., Harvey *et al.,* 1991), two closely related theories have been formulated to explain the epigenetic mechanisms responsible for the evolution of organisms, including their styles of reproduction. The theory of saltatory ontogeny (Balon, 1981b, 1986b) postulates that development from activation of gametes to death of an individual proceeds not in a gradual fashion but in saltatory steps. Each step is a stabilized interval of ontogeny with self-organizing abilities in terms of nonequilibrium thermodynamics, separated by relatively brief, unstable thresholds. During these steps, the organism resists destabilizing variation, but during thresholds, novelties can be introduced and reorganization can be initiated. Inasmuch as we can hardly expect the newly-formed embryo to be the same stabilized entity as the adult, the interpretation of ontogeny as saltatory is not only logical, it is also supported by empirical evidence that is being increasingly recognized (e.g., Lampl et al., 1992; Wray, 1995).

The life-history model to which we have referred is a direct consequence of saltatory ontogeny. The hierarchical intervals (periods, phases, and steps) are each clearly defined by threshold boundaries. Changes in life history are caused, for example, by changes in nutrient deliveries (more or less yolk, earlier or later transition to exogenous feeding, switching to new modes of absorptive feeding). These changes are reflected in the timing of thresholds (heterochronies), including the addition and subtraction of developmental steps. Ultimately, life histories can be readily modified in both directions along the continuum from indirect to direct development (Fig. 8).

The second theory, nicknamed *alprehost* (altricial ⇄ precocial homeorhetic states), attempts to explain the mechanism responsible for generating alternative altricial and precocial states in ontogeny, and by extension, the process of speciation (see Balon, 1981b, 1985b, 1990; Bruton, 1989, 1990). Saltatory ontogeny is an indispensable mechanism for the introduction of changes during a threshold between two stabilized states. The earlier in ontogeny the change, the more effective and extensive it may be. These changes occur in an individual, but may be synchronized to occur similarly within an entire clutch. The oxygen-dependent timing of hatching is one example of this synchronization. Not only will the same cue (i.e., change in oxygen concentration) initiate the event in a group of individual embryos, but if eggs are deposited in clusters, the hatching enzymes of the first embryo that has broken free will induce hatching of the adjacent embryos. Hence, both the environmental cue and the "message" from the first individual will stimulate the group to develop in a synchronized manner, with ultimate consequences for their entire ontogenies.

Other environmental cues, such as cellular interactions and positional activations, will have similar effects on various developmental events, as experiments on temperature and synchrony of skeletal calcification have shown (Balon, 1980). In no instance, however, can such synchrony encompass an entire population, even if it is restricted to a single nesting colony. The synchrony of developmental changes requires close proximity in the case of both exogenous and endogenous cues. An introduction of alternative

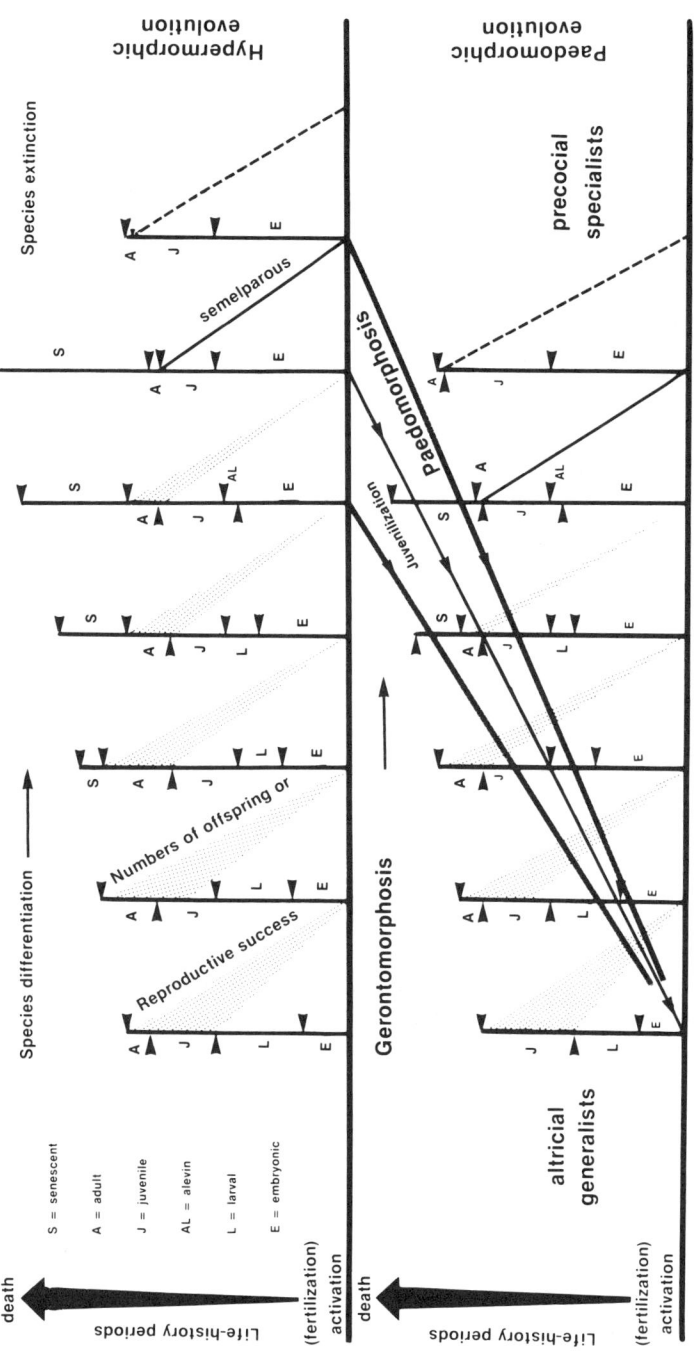

FIG. 8. A scheme illustrating a succession of possible ontogenies, that is, life histories (vertical lines), in a sequence either of increased phenotypic specialization (from left to right) or, under certain conditions, of phenotypic despecialization by paedomorphosis (from upper right to lower left). The relative duration of ontogenetic periods (arrowheads represent thresholds) changes during specialization from altricial (left) towards precocial (right). Concomitantly, reproductive success, that is, number of offspring (dotted areas), is reduced and paralleled by truncation of the adult period, elimination of the larva period, and prolongation of senescence period. Any two of the ontogenies (vertical lines) can represent the dichotomous, steady conditions named "altricial ⇄ precocial homeorhetic states" (alprehost). In both the upper, hypermorphic, and the lower, paedomorphic trajectories, specialization leads to gerontomorphosis. (From Balon, 1985b.)

changes still remains more likely, and ultimately more effective, as an evolutionary driving force.

Even when zygotes are in close proximity to each other (e.g., within a single clutch from one parental pair where peripheral ova are often last to be deposited), differences in location within the clutch will suffice for bifurcations to occur. Ultimately, such bifurcations will result in the formation of two distinct trajectories of stabilized states: one group of relatively altricial offspring, and another group of relatively precocial offspring. Depending on the "strength" of the cue, or the "size" of the activated field, the two forms could be very similar, or quite different, in their life-history attributes. Often only one form will survive to maturity, but it will again produce offspring of both forms. We argue that many of the alternative behavioral "tactics" may have their foundations in such epigenetic bifurcations, rather than in natural selection (cf., Gross, 1984; Caro and Bateson, 1986).

For the most part, the initial differences between altricial and precocial forms in ontogeny are small (Nice, 1962; Ricklefs, 1979). The main attributes of the two forms can be distinguished as an altricial form with smaller, incompletely developed young receiving little parental care, and a precocial form with larger, more completely developed young receiving more parental care. In extreme cases, the definitive phenotype of the altricial form is reached through a slow differentiation and remodeling (metamorphosis) of the organism during a temporary nutrient-gathering interval (e.g., caterpillar, larva, tadpole). In contrast, the definitive phenotype of the precocial form differentiates directly as a result of sufficient endogenous food supplies (e.g., yolk, trophodermy, and placentotrophy), and develops into a nearly definitive phenotype.

Large differences between altricial and precocial forms will appear only when the same theoretical concept is applied between various taxa: for example, if altricial substrate-nesting cichlids were compared with precocial mouthbrooders, or if altricial marsupial mammals were compared with precocial placentals. Because both altricial and precocial states may be created by the same epigenetic mechanisms, a broad usage of the terms is justified. As the ontogeny of each taxon is created in every generation lineage—through a sequence of alternative altricial ⇌ precocial homeorhetic states—so too may different taxa be formed by a similar mechanism.

## VIII. Conclusions

There are several general conclusions that can be drawn from this review of piscine parental care, based on the objectives stated at the outset. It should be clear that the fishes exhibit tremendous variation in their styles of parental care—variation that can not be effectively classified by reference to taxonomy or behavior of the parent(s) in isolation from other aspects of the organism's life history. For this reason, we recommend the examination of parental care in fishes using the concept of Reproductive Guilds, where reproduction is considered to be one component of ontogeny.

Given the need for a life-history perspective on parental care, we describe a general model of life history in fishes that serves to identify the different states of development, and the processes through which an individual progresses from activation to death. This model is useful because it emphasizes the continuous nature of ontogeny through phylogeny.

There is an urgent need for researchers to consider parental care in terms of energy allocated to the general phenomenon of reproduction, including both structural and behavioral components of parental energy allocation. We offer a general model of these energy allocations, with the intention of describing the various options across which an individual parent could distribute energy resources to increase the probability of successful reproduction. In this model, parental care can be interpreted as being much more than parental behavior.

We analyze previous definitions and assumptions that have been associated with parental care, and conclude that these definitions have shown little consistency. In our opinion, these definitions have ignored the integrated nature of energy expenditures associated with reproduction, and have placed artificial limits on what constitutes parental care. In particular, there has been a tendency to isolate parental behavior from other components of reproductive energy expenditure.

Our review of recent trends in the study of piscine parental care indicates that a bias toward parental behavior was indeed apparent in the primary literature. In order to demonstrate how parental care can be viewed in the context of a general life-history model, we select the often-cited relationship between egg size and parental care, and reinterpret hypotheses that have been proposed to account for this relationship. We find that investigators have commonly been uncritical with their use of egg size as a measure of reproductive energy expenditure, and that there has been a lack of distinction between correlation and causation between the two variables. We conclude that researchers would greatly benefit from developing and using life-history models that place more emphasis on the early ontogeny and environment of the offspring.

Finally, we outline the basic conflict between neo-Darwinistic and epigenetic explanations of evolutionary phenomena, and we recommend that researchers consider the development of nontraditional explanations for the evolution of parental care. Focusing on epigenetic processes, we offer an account for the evolution of reproductive styles in fishes, and discuss the bifurcation of altricial and precocial forms as possible mechanism that could account for the evolution of fishes, including their parental care.

### Acknowledgments

We would like to extend our thanks to several people who read early draft manuscripts and provided suggestions that greatly improved this article. Patrice Baker carefully read the manuscript for errors in logic and format. Kevin McCann gave useful comments on the need for changes in conventional life-history investigations, and helped us develop the spiral metaphor to describe ontogeny. David Noakes and Gene Helfman encouraged temperance, and directed us to useful references in the behavioral literature. The editors, Charles Snowdon and Jay Rosenblatt, made constructive comments on the general structure of this paper and highlighted key issues for discussion. Finally, we express special thanks to Jim Atz, who gave exhaustive comments on logic and clarity that greatly improved the manuscript.

### References

(Numbers in brackets at the end of a reference refer to the codes used in Tables VI and VII.)

Adelmann, H. B. (1966). "Marcello Malpighi and the Evolution of Embryology." Cornell Univ. Press, Ithaca, NY.

Bain, M. B., and Helfrich, L. A. (1983). Role of male parental care in survival of larval bluegills. *Trans. Am. Fish. Soc.* **112,** 47–52. [1]

Balon, E. K. (1962). Note on the number of Danubian bitterling developmental stages in mussels. *Věstn. Česk. Spol. Zool.* **26,** 250–256.

Balon, E. K. (1964). Spis i ekologiczna charakterystyka stodkowodnych krągtoustych i ryb Polski (A list and ecological characteristics of the Polish freshwater cyclostomes and fishes). *Pol. Arch. Hydrobiol.* **12,** 233–251. (In Polish.)

Balon, E. K. (1965). Ekologické skupiny rýb a poznámky k ochrane našej ichtyofauny (Ecological groups of fishes and notes on the conservation of our ichthyofauna). *Česk. Ochrana Prírody* **2,** 135–160. (In Slovak.)

Balon, E. K. (1975a). Reproductive guilds of fishes: a proposal and definition. *J. Fish. Res. Board Can.* **32,** 821–864.

Balon, E. K. (1975b). Ecological guilds of fishes: a short summary of the concept and its application. *Verh. Int. Ver. Limnol.* **19,** 2430–2439.

Balon, E. K. (1975c). Terminology of intervals in fish development. *J. Fish. Res. Board Can.* **32,** 1663–1670.

Balon, E. K. (1977). Early ontogeny of *Labeotropheus* Ahl, 1927 (Mbuna, Cichlidae, Lake Malawi), with a discussion on advanced protective styles in fish reproduction and development. *Environ. Biol. Fishes* **2,** 147–176.

Balon, E. K. (1978). Reproductive guilds and the ultimate structure of fish taxocenes: Amended contribution to the discussion presented at the mini-symposium. *Environ. Biol. Fishes* **3,** 149–152.

Balon, E. K. (1979). The theory of saltation and its application in the ontogeny of fishes: Steps and thresholds. *Environ. Biol. Fishes* **4,** 97–101.

Balon, E. K. (ed.) (1980). "Charrs: Salmonid Fishes of the Genus *Salvelinus.*" Perspectives in Vertebrate Science 1, Dr W. Junk Publishers, The Hague, The Netherlands.
Balon, E. K. (1981a). About processes which cause the evolution of guilds and species. *Environ. Biol. Fishes* **6,** 129–138.
Balon, E. K. (1981b). Saltatory processes and altricial to precocial forms in the ontogeny of fishes. *Am. Zool.* **21,** 573–596.
Balon, E. K. (1981c). Additions and amendments to the classification of reproductive styles in fishes. *Environ. Biol. Fishes* **6,** 377–389.
Balon, E. K. (1982). About the courtship ritual in fishes, but also about a false sense of security given by classification schemes, 'comprehensive' reviews and committee decisions. *Environ. Biol. Fishes* **7,** 193–197.
Balon, E. K. (1983). Epigenetic mechanisms: Reflections on evolutionary processes. *Can. J. Fish. Aquat. Sci.* **40,** 2045–2058.
Balon, E. K. (1984a). Patterns in the evolution of reproductive styles in fishes. *In* "Fish Reproduction: Strategies and Tactics" (G. W. Potts and R. J. Wootton, eds.), pp. 35–53. Academic Press, London.
Balon, E. K. (1984b). Reflections on some decisive events in the early life of fishes. *Trans. Am. Fish. Soc.* **113,** 178–185.
Balon, E. K. (1985a). The theory of saltatory ontogeny and life history models revisited. *In* "Early Life Histories of Fishes: New Developmental, Ecological and Evolutionary Perspectives, Dev. in Environ. Biol. Fishes 5" (E. K. Balon, ed.), pp. 13–30. Dr W. Junk Publishers, Dordrecht, The Netherlands.
Balon, E. K. (1985b). Reflections on epigenetic mechanisms: Hypotheses and case histories. *In* "Early Life Histories of Fishes: New Developmental, Ecological and Evolutionary Perspectives, Dev. in Environ. Biol. Fishes 5" (E. K. Balon, ed.), pp. 239–270. Dr W. Junk Publishers, Dordrecht, The Netherlands.
Balon, E. K. (1986a). Types of feeding in the ontogeny of fishes and the life-history model. *Environ. Biol. Fishes* **16,** 11–24.
Balon, E. K. (1986b). Saltatory ontogeny and evolution. *Riv. Biol./Biol. Forum* **79,** 151–190. (In English and Italian.)
Balon, E. K. (1988). Tao of life: universality of dichotomy in biology. 2. The epigenetic mechanisms. *Riv. Biol./Biol. Forum* **81,** 339–380. (In English and Italian.)
Balon, E. K. (1989a). The Tao of life: From the dynamic unity of polar opposites to self-organization. *In* "Alternative Life-History Styles of Animals, Perspectives in Vertebrate Science 6" (M. N. Bruton, ed.), pp. 7–40. Kluwer Academic Publishers, Dordrecht, The Netherlands.
Balon, E. K. (1989b). The epigenetic mechanisms of bifurcation and alternative life-history styles. *In* "Alternative Life-History Styles of Animals, Perspectives in Vertebrate Science 6" (M. N. Bruton, ed.), pp. 467–501. Kluwer Academic Publishers, Dordrecht, The Netherlands.
Balon, E. K. (1990). Epigenesis of an epigeneticist: The development of some alternative concepts on the early ontogeny and evolution of fishes. *Guelph Ichthyol. Rev.* **1,** 1–42.
Balon, E. K. (1991). Probable evolution of the coelacanth's reproductive style: Lecithotrophy and orally feeding embryos in cichlid fishes and in *Latimeria chalumnae. Environ. Biol. Fishes* **32,** 249–265.
Balon, E. K., Momot, W. T., and Regier, H. A. (1977). Reproductive guilds of percids: results of the paleogeographical history and ecological succession. *J. Fish. Res. Board Can.* **34,** 1910–1921.
Barlow, G. W. (1981). Patterns of parental investment, dispersal and size among coral-reef fishes. *Environ. Biol. Fishes* **6,** 65–85.

Baylis, J. R. (1981). The evolution of parental care in fishes, with reference to Darwin's rule of male sexual selection. *Environ. Biol. Fishes* **6,** 223–251. [2]

Beamish, F. W. H. (1979). Migration and spawning energetics of the anadromous sea lamprey, *Petromyzon marinus*. *Environ. Biol. Fishes* **4,** 3–7.

Beamish, F. W. H., Potter, I. C., and Thomas, E. (1979). Proximate composition of the adult anadromous sea lamprey, *Petromyzon marinus*, in relation to feeding, migration and reproduction. *J. Anim. Ecol.* **48,** 1–19.

Bernstein, J. W. (1980). Parental care behavior in the cichlid fish *Cichlasoma nigrofasciatum*. *Copeia* **1980,** 682–686. [3]

Beverton, R. J. H. (1963). Maturation, growth and mortality of cluepid and engraulid stocks in relation to fishing. *Rapp. P.V. Réun. Cons. Int. Explor. Mer* **154,** 44–67.

Bisazza, A., and Marconato, A. (1988). Female mate choice, male-male competition and parental care in the river bullhead, *Cottus gobio* L. (Pisces, Cottidae). *Anim. Behav.* **36,** 1352–1360. [4]

Blumer, L. S. (1979). Male parental care in the bony fishes. *Q. Rev. Biol.* **54,** 149–161.

Blumer, L. S. (1982). A bibliography and categorization of bony fishes exhibiting parental care. *Zool. J. Linn. Soc.* **76,** 1–22. [5]

Blumer, L. S. (1986a). The function of parental care in the brown bullhead, *Ictalurus nebulosus*. *Am. Midl. Nat.* **115,** 234–238. [6]

Blumer, L. S. (1986b). Parental care sex differences in the brown bullhead, *Ictalurus nebulosus* (Pisces, Ictaluridae). *Behav. Ecol. Sociobiol.* **19,** 97–104. [7]

Breder, C. M., and Rosen, D. E. (1966). "Modes of Reproduction in Fishes." Natural History Press, Garden City, NY. (Reprinted by T.F.H. Publications, Jersey City, NJ, with the title "Modes of Reproduction in Fishes: How Fishes Breed.")

Brown, J. A. (1984). Parental care and the ontogeny of predator-avoidance in two species of centrarchid fish. *Anim. Behav.* **32,** 113–119. [8]

Bruton, M. N. (1989). The ecological significance of alternative life-history styles. *In* "Alternative Life-History Styles of Animals" (M. N. Bruton, ed.), pp. 503–553. Kluwer Academic Publishers, Dordrecht, The Netherlands.

Bruton, M. N. (ed.) (1990). "Alternative Life-History Styles of Fishes. Developments in Environmental Biology of Fishes 10" Kluwer Academic Publishers, Dordrecht, The Netherlands.

Calow, P., and Sibly, R. M. (1990). A physiological basis of population processes: Ecotoxicological implications. *Funct. Ecol.* **4,** 283–288.

Caro, T. M., and Bateson, P. (1986). Organization and ontogeny of alternative tactics. *Anim. Behav.* **34,** 1483–1499.

Charnov, E. L., and Berrigan, D. (1991). Dimensionless members and the assembly rules for life histories. *In* "The Evolution of Reproductive Strategies" (P. H. Harvey, L. Partridge, and T. R. E. Southwood, eds.), pp. 41–48. The Royal Society, London. (Discussion by R. J. H. Beverton, p. 48.)

Clutton-Brock, T., and Godfray, C. (1991). Parental investment. *In* "Behavioural Ecology: An Evolutionary Approach, Third Edition" (J. R. Krebs and N. B. Davies, eds.), pp. 234–262. Blackwell Scientific Publications, London.

Coleman, R. M. (1992). Reproductive biology and female parental care in the cockscomb prickleback, *Anoplarchus purpurescens* (Pisces, Stichaeidae). *Environ. Biol. Fishes* **35,** 177–186. [9]

Crawford, S. S., and Balon, E. K. (1994a). Alternative life histories of the genus *Lucania*: 1. Early ontogeny of *L. parva*, the rainwater killifish. *Environ. Biol. Fishes* **40,** 349–389.

Crawford, S. S., and Balon, E. K. (1994b). Alternative life histories of the genus *Lucania*: 2. Early ontogeny of *L. goodei*, the bluefin killifish. *Environ. Biol. Fishes* **41,** 331–368.

Crawford, S. S., and Balon, E. K. (1994c). Alternative life histories of the genus *Lucania:* 3. An ecomorphological explanation of altricial (*L. parva*) and precocial (*L. goodei*) species. *Environ. Biol. Fishes* **41**, 369–402.

Cunningham, J. E. R., and Balon, E. K. (1985). Early ontogeny of *Adinia xenica* (Pisces, Cyprinodontiformes): 1. The development of embryos in hiding. *Environ. Biol. Fishes* **14**, 115–166.

Cunningham, J. E. R., and Balon, E. K. (1986a). Early ontogeny of *Adinia xenica* (Pisces, Cyprinodontiformes): 2. Implications of embryonic resting interval for larval development. *Environ. Biol. Fishes* **15**, 15–45.

Cunningham, J. E. R., and Balon, E. K. (1986b). Early ontogeny of *Adinia xenica* (Pisces, Cyprinodontiformes): 3. Comparison and evolutionary significance of some patterns in epigenesis of egg-scattering, hiding and bearing cyprinodontiforms. *Environ. Biol. Fishes* **15**, 91–105.

Doherty, P. J., Williams, D. McB., and Sale, P. F. (1985). The adaptive significance of larval dispersal in coral reef fishes. *Environ. Biol. Fishes* **12**, 81–90.

Eccles, D. H., and Lewis, D. S. C. (1981). Midwater spawning in *Haplochromis chrysonotus* Boulenger (Telostei: Cichlidae) in L. Malawi. *Environ. Biol. Fishes* **6**, 201–202.

Elgar, M. A. (1990). Evolutionary compromise between a few large and many small eggs: Comparative evidence in teleost fish. *Oikos* **59**, 283–287.

Fisher, R. A. (1930). "The Genetical Theory of Natural Selection." Dover Books, New York.

Flegler-Balon, C. (1989). Direct and indirect development in fishes—examples of alternative life-history styles. *In* "Alternative Life-History Styles of Animals" (M. N. Bruton, ed.), pp. 71–100. Kluwer Academic Publishers, Dordrecht, The Netherlands.

Fostner, H., Hinterleitner, S., Mahr, K., and Wieser, W. (1983). Towards a better definition of 'metamorphosis' in *Coregonus* sp.: Biochemical, histological, and physiological data. *Can. J. Fish. Aquat. Sci.* **40**, 1224–1232.

Garstang, W. (1929). The origin and evolution of larval forms. *Br. Assoc. Adv. Sci. Rep.* **96**, 77–98.

Gasking, E. B. (1967). "Investigations into Generation 1651–1828." The Johns Hopkins Press, Baltimore.

Gebhardt, M. D. (1987). Parental care—a freshwater phenomenon? *Environ. Biol. Fishes* **19**, 69–72. [10]

Geist, V. (1971). "Mountain Sheep. A Study in Behavior and Evolution." Univ. of Chicago Press, Chicago.

Gill, T. N. (1905). Parental care among fresh-water fishes. *Annu. Rep. Smithson. Inst.* **1905**, 403–531.

Gittleman, J. L. (1981). The phylogeny of parental care in fishes. *Anim. Behav.* **29**, 936–941. [11]

Gladstone, W. (1994). Lek-like spawning, prenatal care and mating periodicity of the trigger-fish, *Pseudobalistes flavimarginatus* (Balistidae). *Environ. Biol. Fishes* **39**, 249–257. [12]

Gleick, J. (1987). "Chaos, Making a New Science." Penguin Books, New York.

Goto, A. (1989). Effects of parental care on egg survival estimated from an experimental removal of nest-guarding males in the river sculpin *Cottus amblystomopsis. Jpn. J. Ichthyol.* **36**, 281–284. [13]

Gould, S. J. (1977). "Ontogeny and Phylogeny." Harvard Univ. Press, Cambridge, MA.

Goulet, D., Green, J. M., and Shears, T. H. (1986). Courtship, spawning, and parental care behavior of the lumpfish, *Cyclopterus lumpus* L., in Newfoundland. *Can. J. Zool.* **64**, 1320–1325. [14]

Gross, M. R. (1984). Sunfish, salmon, and the evolution of alternative reproductive strategies and tactics in fishes. *In* "Fish Reproduction: Strategies and Tactics" (G. W. Potts and R. J. Wootton, eds.), pp. 55–75. Academic Press, London.

Gross, M. R., and Sargent, R. C. (1985). The evolution of male and female parental care in fishes. *Am. Zool.* **25,** 807–822. [15]

Hall, B. K. (1992). "Evolutionary Developmental Biology." Chapman and Hall, London.

Harvey, P. H., Partridge, L., and Southwood, T. R. E. (1991). "The Evolution of Reproductive Strategies." The Royal Society, London.

Hewett, S. W., and Johnson, B. L. (1992). Fish bioenergetics model 2: an upgrade of a generalized bioenergetics model of fish growth for microcomputers. UW Sea Grant Technical Report No. WIS-SG-92-250. University of Wisconsin Sea Grant Institute, Madison, WI.

Holden, K. K., and Bruton, M. N. (1992). A life-history approach to the early ontogeny of the Mozambique tilapia *Oreochromis mossambicus* (Pisces, Cichlidae). *S. Afr. J. Zool.* **27,** 173–191.

Holden, K. K., and Bruton, M. N. (1994). The early ontogeny of the southern mouthbrooder, *Pseudocrenilabrus philander* (Pisces, Cichlidae). *Environ. Biol. Fishes* **41,** 311–329.

Huxley, J. S., and G. R. de Beer. (1934). "The Elements of Experimental Embryology." Cambridge Univ. Press, New York.

Jamieson, B. G. M. (1991). "Fish Evolution and Systematics: Evidence From Spermatozoa." Cambridge Univ. Press, Cambridge, MA.

Jamieson, I. G., Blouw, D. M., and Colgan, P. W. (1992). Parental care as a constraint on male mating success in fishes: A comparative study of threespine and white sticklebacks. *Can. J. Zool.* **70,** 956–962. [16]

Kamler, E. (1992). "Early Life History of Fish: An Energetics Approach." Chapman and Hall, London.

Keenleyside, M. H. A. (1981). Parental care patterns of fishes. *Am. Nat.* **117,** 1019–1022. [17]

Kindler, P. M., Bahr, J. M., Gross, M. R., and Philipp, D. P. (1991a). Hormonal regulation of parental care behavior in nesting male bluegills: Do the effects of bromocriptine suggest a role for prolactin? *Physiol. Zool.* **64,** 310–322. [18]

Kindler, P. M., Bahr, J. M., and Philipp, D. P. (1991b). The effects of exogenous 11-ketotestosterone, testosterone, and cyproterone acetate on prespawning and parental care behaviors of male bluegill. *Horm. Behav.* **25,** 410–423. [19]

Knapp, R. A., and Warner, R. R. (1991). Male parental care and female choice in the bicolar damselfish, *Stegastes partitus:* bigger is not always better. *Anim. Behav.* **41,** 747–756. [20]

Kottelat, M. (1990). Synopsis of the endangered buntingi (Osteichthyes: Adrianichthyidae and Oryziidae) of Lake Poso, Central Sulawesi, Indonesia, with a new reproductive guild and description of three new species. *Ichthyol. Explor. Freshwater* **1,** 49–67.

Kováč, V. (1995). Reproductive behavior and early development of the European mudminnow, *Umbra krameri. Folia Zoologica* **44,** 57–80.

Krimbas, C. B. (1984). On adaptation, neo-Darwinian tautology, and population fitness. *In* "Evolutionary Biology 17" (M. K. Hecht, B. Wallace, and G. T. Prance, eds.), pp. 1–57. Plenum Press, New York.

Kryzhanovsky, S. G. (1948). Ecological groups of fishes and the principles of their development. *Izv. Tikhookean. Nauchno-Issled. Inst. Rybn. Khor. Okeanogr.* **2,** 3–114. (In Russian.)

Kryzhanovsky, S. G. (1949). Eco-morphological principles of development among carps, loaches, and catfishes. *Tr. Inst. Morph. Zhivotn. Severtsova* **1,** 5–332. (In Russian.) (Part II, Ecological groups of fishes and patterns of their distribution, pp. 237–331. Translated from Russian by *Fish. Res. Board Can. Transl. Ser. No. 2945*, 1974).

Kryzhanovsky, S. G., Smirnov, A. J., and Soin, S. G. (1951). Data on the development of Amur River fishes. *In* "Trudy Amurskoy ichtiol. exped. 1945–1949" (S. G. Kryzhanovsky, ed.), pp. 5–222. Vol. 2, Moscow Society of Naturalists Press, Moscow. (In Russian.)

Lambert, D. M., and Spencer, H. G. (eds.) (1995). "Speciation and the Recognition Concept. Theory and Application." The Johns Hopkins Univ. Press, Baltimore, MD.
Lampl, M., Veldhuis, J. D., and Johnson, M. L. (1992). Saltation and stasis: A model of human growth. *Science (Washington, D.C.)* **258,** 801–803.
Lavery, R. J., and Kieffer, J. D. (1994). Effects of parent and offspring food rations on parental care in the convict cichlid fish (Pisces, Cichlidae). *Behavior* **129,** 63–77. [21]
Lavery, R. J., and Reebs, S. G. (1994). Effect of mate removal on current and subsequent parental care in the convict cichlid (Pisces: Cichlidae). *Ethology* **97,** 265–277. [22]
Løvtrup, S. (1974). "Epigenetics—A Treatise on Theoretical Biology." John Wiley and Sons, London.
Løvtrup, S. (1982). The four theories of evolution. *Rivista di Biologia* **75,** 53–66, 231–272, 385–409.
Løvtrup, S. (1987). "Darwinism: The Refutation of a Myth." Croom Helm, London.
Magner, L. N. (1979). "A History of the Life Sciences." Dekker, New York.
Magnhagen, C. (1992). Parental care and predation risk in fish. *Ann. Zool. Fenn.* **29,** 227–232. [23]
Makeyeva, A. P. (1992). "Embryology of Fishes." Moscow Univ. Press, Moscow. (In Russian.)
Marconato, A., Bisazza, A., and Fabris, M. (1993). The cost of parental care and egg cannibalism in the river bullhead, *Cottus gobio* L. (Pisces, Cottidae). *Behav. Ecol. Sociobiol.* **32,** 229–237. [24]
Matsuda, R. (1987). "Animal evolution in changing environment with special reference to abnormal metamorphosis." John Wiley & Sons, New York.
McElman, J. F., and Balon, E. K. (1979). Early ontogeny of walleye, *Stizostedion vitreum*, with steps of saltatory development. *Environ. Biol. Fishes* **4,** 309–348.
McElman, J. F., and Balon, E. K. (1980). Early ontogeny of white sucker, *Catostomus commersoni*, with steps of saltatory development. *Environ. Biol. Fishes* **5,** 191–224.
Metz, J. A. J., de Jong, T. J., and Klinkhamer, P. G. L. (1983). What are the advantages of dispersing: A paper by Kuno explained and extended. *Oecologia* **57,** 166–169.
Mrowka, W. (1982). Effect of removal of the mate on the parental care behavior of the biparental cichlid. *Aequidens paraguayensis. Anim. Behav.* **30,** 295–297. [25]
Mrowka, W. (1984). Is the parental-care behavior of *Aequidens paraguayensis* (Cichlidae) optimal? *Behavior* **89,** 128–146. [26]
Nagata, Y. (1985). Estimation of population fecundity of the bitterling, *Rhodeus ocellatus*, and ecological significance of its spawning habit into bivalves. *Jpn. J. Ichthyol.* **32,** 324–334. (In Japanese.)
Nagata, Y., and Nakata, Y. (1988). Distribution of six species of bitterlings in a Creek in Fukuoka Prefercture, Japan. *Jpn. J. Ichthyol.* **35,** 320–331.
Nagoshi, M. (1987). Survival of broods under parental care and parental roles of the cichlid fish, *Lamprologus toae*, in Lake Tanganyika. *Jpn. J. Ichthyol.* **34,** 71–75. [27]
Nagoshi, M., and Gashagaza, M. M. (1988). Growth of the larvae of a Tanganyikan cichlid. *Lamprologus attenuatus*, under parental care. *Jpn. J. Ichthyol.* **35,** 392–395. [28]
Nelson, J. S. (1994). "Fishes of the World, 3rd Edition." John Wiley and Sons, Inc., Toronto.
Nice, M. M. (1962). Development of behavior in precocial birds. *Trans. Linn. Soc. New York* **8,** 1–211.
Ochi, H. (1985). Termination of parental care due to small clutch size in the temperate damselfish, *Chromis notata. Environ. Biol. Fishes* **12,** 155–160. [29]
Paine, M. D., and Balon, E. K. (1984a). Early development of the northern logperch, *Percina caprodes semifasciata*, according to the theory of saltatory ontogeny. *Environ. Biol. Fishes* **11,** 173–190.

Paine, M. D., and Balon, E. K. (1984b). Early development of the rainbow darter, *Etheostoma caeruleum*, according to the theory of saltatory ontogeny. *Environ. Biol. Fishes* **11,** 277–299.

Paine, M. D., and Balon, E. K. (1986). Early development of johnny darter, *Etheostoma nigrum*, and fantain darter, *E. flabellare*, with a discussion of its ecological and evolutionary aspects. *Environ. Biol. Fishes* **15,** 191–220.

Pauly, D. (1980). On the interrelationships between natural mortality, growth parameters, and mean environmental temperature in 175 fish stocks. *J. Cons. Cons. Int. Explor. Mer* **39,** 175–192.

Peters, H. M., and Berns, S. (1982). Die Maulbrutpflege der Cichliden. Untersuchungen zur Evolution eines verhaltensmusters. *Z. Zool. Syst. Evolutions forsch.* **20,** 18–52.

Pietsch, T. W., and Grobecker, D. B. (1980). Parental care as an alternative reproductive mode in an antennariid anglerfish. *Copeia* **1980,** 551–553. [30]

Ragland, H. C., and Fischer, E. A. (1987). Internal fertilization and male parental care in the scalyhead sculpin, *Artedius harringtoni*. *Copeia* **1987,** 1059–1062. [31]

Reid, R. G. B. (1985). "Evolutionary Theory: The Unfinished Synthesis." Croom Helm, London.

Ricklefs, R. E. (1979). "Ecology, 2nd Edition." Thomas Nelson, Sunbury-on-Thames, England.

Ridgway, M. S., and Friesen, T. G. (1992). Annual variation in parental care in smallmouth bass, *Micropterus dolomieu*. *Environ. Biol. Fishes* **35,** 243–255. [32]

Ridley, M. (1993). "Evolution." Blackwell Scientific Publications, Boston.

Sargent, R. C., and Gross, M. R. (1986). William's principle: An explanation of parental care in teleost fishes. *In* "The Behavior of Teleost Fishes" (T. J. Pitcher, ed.), pp. 275–293. Chapman and Hall, London.

Sargent, R. C., and Gross, M. R. (1993). William's principle: An explanation of parental care in teleost fishes. *In* "Behaviour of Teleost Fishes, 2nd Edition" (T. J. Pitcher, ed.), pp. 333–361. Chapman and Hall, London.

Sargent, R. C., Taylor, P. D., and Gross, M. R. (1987). Parental care and the evolution of egg size in fishes. *Am. Nat.* **129,** 32–46. [33]

Schwanck, E. J. (1989). Parental care of *Tilapia mariae* in the field and in aquaria. *Environ. Biol. Fishes* **24,** 251–265. [34]

Shine, R. (1978). Propagule size and parental care: The "safe harbor" hypothesis. *J. Theor. Biol.* **75,** 417–424.

Sibly, R., and Calow, P. (1986). Why breeding earlier is always worthwhile. *J. Theor. Biol.* **123,** 311–319.

Smirnov, S. A., Makeyeva, A. P., and Smirnov, A. I. (1995). Development of ecomorphology of fishes in Russia. *Environ. Biol. Fishes* **44,** 23–33.

Smirnova, E. N. (1957). Morpho-ecological features of development of *Rutilus frisii* (Nordmann). *Trudy Inst. Morph. Zhiv. Severtsova* **20,** 95–120. (In Russian.)

Smith, C. C., and Fretwell, S. D. (1974). The balance between size and number of offspring. *Am. Nat.* **108,** 677–699.

Smith, C., and Wootton, R. J. (1994). The cost of parental care in *Haplochromis 'argens'* (Cichlidae). *Environ. Biol. Fishes* **40,** 99–104. [35]

Smith, C., and Wootton, R. J. (1995). The cost of parental care in teleost fishes. *Rev. Fish Biol. Fish.* **5,** 7–22.

Stearns, S. C. (1989). Trade-offs in the life history evolution. *Funct. Ecol.* **3,** 259–268.

Stiassny, M. L. J., and Gertner, C. L. (1992). The parental care behavior of *Paratilapia polleni* (Perciformes, Labroidei), a phylogenetically primitive cichlid from Madagascar, with a discussion of the evolution of maternal care in the family Cichlidae. *Environ. Biol. Fishes* **34,** 219–233. [36]

Townsend, D. S., and Stewart, M. M. (1985). Direct development in *Eleutherodactylus coqui* (Anura: Leptodactylidae): A staging table. *Copeia* **1985,** 423–436.

Trivers, R. L. (1972). Parental investment and sexual selection. *In* "Sexual Selection and the Descent of Man 1871-1971" (B. Campbell, ed.), pp. 136–179. Aldine, Chicago.
van den Berghe, E. P. (1990). Variable parental care in a labrid fish—how care might evolve. *Ethology* **84,** 319–333. [37]
van den Berghe, E. P. (1992). Parental care and the cost of reproduction in a Mediterranean fish. *Behav. Ecol. Sociobiol.* **30,** 373–378. [38]
Wendt, H. (1965). "The Sex Life of the Animals." Simon and Schuster, New York.
Wetzel, J., and Wourms, J. P. (1995). Adaptations for reproduction and development in the skin-brooding ghost pipefishes, *Solenostomus*. *Environ. Biol. Fishes* **44,** 363–384.
Williams, G. C. (1966). "Adaptation and Natural Selection." Princeton Univ. Press, Princeton, NJ.
Winkler, D. W. (1987). A general model for parental care. *Am. Nat.* **130,** 526–543.
Wootton, R. J. (1985). Introduction: Strategies and tactics in fish reproduction. *In* "Fish Reproduction: Strategies and Tactics" (G. W. Potts and R. J. Wootton, eds.), pp. 1–12. Academic Press, New York.
Wootton, R. J. (1990). "Ecology of Teleost Fishes." Chapman and Hall, London.
Wourms, J. P., and Lombardi, J. (1992). Reflections on the evolution of piscine viviparity. *Am. Zool.* **32,** 276–293.
Wourms, J. P., Grove, B. D., and Lombardi, J. (1988). The maternal-embryonic relationship in viviparous fishes. *In* "Fish Physiology, Vol. 11b" (W. S. Hoar and D. J. Randall, eds.), pp. 1–134. Academic Press, San Diego.
Wray, G. A. (1995). Punctuated evolution of embryos. *Science (Washington, D.C.)* **267,** 1115–1116.
Yanagisawa, Y. (1986). Parental care in a monogamous mouthbrooding cichlid *Xenotilapia flavipinnis* in Lake Tanganyika. *Jpn. J. Ichthyol.* **33,** 249–261. [39]
Yanagisawa, Y. (1993). Long-term territory maintenance by female *Tropheus duboisi* (Cichlidae) involving foraging during the mouth-brooding period. *Ecol. Freshwater Fish* **1993,** 1–7.
Yanagisawa, Y., and Ochi, H. (1991). Food intake by mouthbrooding females of *Cyphotilapia frontosa* (Cichlidae) to feed both themselves and their young. *Environ. Biol. Fishes* **30,** 353–358.
Yanagisawa, Y., and Sato, T. (1990). Active browsing by mouthbrooding females of *Tropheus duboisi* and *Tropheus moorii* (Cichlidae) to feed the young and/or themselves. *Environ. Biol. Fishes* **27,** 43–50.
Zoran, M. J., and Ward, J. A. (1983). Parental egg care behavior and fanning activity for the orange chromide, *Etroplus maculatus*. *Environ. Biol. Fishes* **8,** 301–310. [40]

# Parental Care among the Amphibia

Martha L. Crump

DEPARTMENT OF BIOLOGICAL SCIENCES
NORTHERN ARIZONA UNIVERSITY
FLAGSTAFF, ARIZONA 86011

## I. Introduction

Amphibians are the most diverse of all the terrestrial vertebrates in their modes of reproduction, no doubt a reflection of various stages in their evolution towards greater terrestriality and independence from standing water. Land, however, represents an inhospitable environment for anamniotic eggs. Perhaps it is in response to this harsh environment that some amphibians have evolved unique forms of parental care that enhance offspring survivorship. For example, parents transport tadpoles from terrestrial oviposition sites to water; parents attend direct-developing eggs under logs and moss and inside tunnels; adults carry eggs and tadpoles, throughout development to metamorphosis, in dorsal pouches, vocal sacs, and even in the stomach; females of some species provide eggs to their tadpoles for food; some species have internal gestation within the oviduct. Despite the fascinating parental care behaviors exhibited by amphibians, relatively few quantitative studies have been published on this class in comparison to birds and mammals (see chapters and references cited therein, this volume). In this article I hope to convey that parental care in amphibians represents an exciting field for future research.

What should be and what should not be considered a parental care behavior in amphibians? Does the presence of a female salamander next to a clutch of eggs constitute parental care? In order to answer this question we need to define some terms. The term *parental investment* was introduced by Trivers (1972), who defined it as "any investment by the parent in an individual offspring that increases the offspring's chance of surviving (and hence reproductive success) at the cost of the parent's ability to invest in other offspring." The term *parental care, sensu stricto,* has been used in at least two ways. Some authors refer to parental care as a subset of parental investment. In this sense, parental care includes all nongametic investments in offspring following fertilization, and thus entails a cost to the parent (e.g., Wittenberger, 1981). Other authors refer to parental care as a behavior that increases survivorship of young, but without implications of cost to the parent (e.g., Clutton-Brock, 1991).

In this article I use the terminology of Wittenberger (1981). The female salamander found with a clutch of eggs is generally assumed to be exhibiting parental care, based on data that in many species the presence of a female definitely increases offspring survivorship, and the assumption that the female incurs a cost relative to the alternative behavior of egg abandonment. For many species, however, we lack hard data on benefits and especially costs for adults that have been recorded in association with offspring. Thus, some presumed examples of parental care may, in fact, not be parental care. For example, the female could merely be "resting" after oviposition, to depart as soon as the observer leaves; if no benefits or costs were involved, the female's behavior would not constitute parental care. This example points out a critical weakness in the study of parental care in amphibians. Analysis of benefits and costs of parental behaviors towards offspring, more than simple documentation of the behavior, is a crucial field for future investigation.

Some discussions of parental care omit live-bearing modes of reproduction (viviparity and ovoviviparity) from consideration, and instead focus on behaviors and associations between parents and their offspring that are "voluntary" on the part of the parent (e.g., McDiarmid, 1978; Shine, 1988; Crump, 1995; Townsend, 1996). In this article I include internal gestation within the oviduct for "completeness sake" because it certainly satisfies the definition of parental care. Those behaviors that can be considered as voluntary are referred to as *postoviposition* parental care, as distinct from internal gestation in the oviduct.

Three additional terms need to be defined because of inconsistent use in the literature: brooding, attendance, and guarding. Here I restrict use of the term *brooding* to anurans that carry eggs and/or larvae on or within their bodies (other than inside the oviduct). *Attendance* refers to the behavior of salamanders, caecilians, and anurans that remain with their eggs at a fixed site. *Guarding* refers to the behavior of a parent actively defending its offspring.

This article represents an update to Crump (1995), and includes 34 additional species known to exhibit parental care (1 *Centrolene,* Burrowes, 1987; 1 *Colostethus,* Juncá *et al.,* 1994; 28 *Eleutherodactylus,* Townsend, 1996; and 4 *Epipedobates,* Rodríguez and Myers, 1993; Haddad and Martins, 1994). Furthermore, amphibian systematics has undergone major changes. Instead of 21 anuran families, 27 are currently recognized (Duellman, 1993; Ford and Cannatella, 1993); 10 families of salamanders are now recognized instead of 9 (Good and Wake, 1992). Tables here reflect the revised systematics. Percentages of species with parental care in this article differ slightly from those in Crump (1995), reflecting the addition of named species (Duellman, 1993) and the 34 additional species that exhibit parental care. In this contribution I have included internal gestation (viviparity and ovoviviparity) as a sixth mode of parental care. Finally, I present a geographical analysis of the distribution of parental care. The reader is referred to Crump (1995) for a more extensive list of references on parental care in amphibians and for an appendix of species that exhibit parental care.

## II. Phylogenetic Distribution of Parental Care

Parental care is found in all three orders of the class Amphibia: Gymnophiona (caecilians), Caudata (salamanders), and Anura (frogs and toads). Parental care has been documented for relatively few species, but these are distributed in many different families. Documented reports of postoviposition parental care exist for only about 5% of caecilians (8 of 163 species), 18% of salamanders (72 of 409 species), and 6% of anurans (240 of 3967 species). These are conservative estimates, however, as many additional species may also exhibit parental care (see as follows).

In caecilians, postoviposition parental care consists of the female lying coiled around the eggs (Table I). Wake (1986) speculated that all oviparous species attend their eggs, and, in fact, all caecilian eggs that have been found have had attendant females (R. A. Nussbaum, personal communication). Although oviparity is known or presumed to occur in four families, egg attendance has been reported only in two (Table I). About 25% of caecilian species are assumed to lay eggs, so perhaps some 40 species attend their eggs. The remaining species are known or presumed to have internal gestation in the oviduct (viviparity). Thus, it is possible that all caecilians provide parental care to their offspring. Other than recognizing that the behavior exists, we know very little about parental care in caecilians. Therefore, this article deals primarily with ecological and evolutionary aspects of parental care in salamanders and anurans.

In salamanders, postoviposition parental care has been reported from 8 of 10 families (all families except Rhyacotritonidae) and Salamandridae; Table I). Plethodontids, which comprise about 63% of recognized species of salamanders (Duellman and Trueb, 1986), provide 79% of the cases. Nussbaum (1985) predicted that further study will reveal many additional examples of parental care in salamanders, especially in stream-adapted species and in many of the tropical, terrestrially breeding plethodontids whose reproductive biology is currently unknown. In addition, internal gestation in the oviduct occurs in four species of Salamandridae.

TABLE I
OCCURRENCE OF EGG ATTENDANCE IN CAECILIANS AND SALAMANDERS

| Taxon[a] | No. species | Sex of caregiver[b] | Oviposition site | No. eggs[c] | Fert.[d] |
|---|---|---|---|---|---|
| Gymnophiona | | | | | |
| Caeciliidae | | | | | |
| Afrocaecilia | 1 | F | Terrestrial | I | N |
| Grandisonia | 3 | F | Terrestrial | I | N |
| Hypogeophis | 1 | F | Terrestrial | I | N |
| Idiocranium | 1 | F | Terrestrial | I | N |
| Praslinia | 1 | F | Terrestrial | I | N |
| Ichthyophiidae | | | | | |
| Ichthyophis | 1 | F | Terrestrial | I | N |
| Caudata | | | | | |
| Ambystomatidae | | | | | |
| Ambystoma | 1 | F | Terrestrial | II | N |
| Amphiumidae | | | | | |
| Amphiuma | 2 | F | Terrestrial | I–II | N |
| Cryptobranchidae | | | | | |
| Andrias | 1 | M | Aquatic | III | X |
| Cryptobranchus | 1 | M | Aquatic | III | X |
| Dicamptodontidae | | | | | |
| Dicamptodon | 3 | F | Aquatic | II | N |
| Hynobiidae | | | | | |
| Hynobius | 3 | M | Aquatic | II | X |
| Plethodontidae | | | | | |
| Aneides | 5 | 3F; 2M + F? | Terrestrial | I | N |
| Batrachoceps | 3 | F | Terrestrial | I | N |
| Bolitoglossa | 8 | 6F; 1F/M; 1NR | Terrestrial | I | N |
| Chiropterotriton | 1 | F | Terrestrial | I | N |
| Desmognathus | 11 | F | 3 Aquatic; 8 Terrestrial | I | N |
| Ensatina | 1 | F | Terrestrial | I | N |
| Eurycea | 3 | F | Aquatic | I | N |
| Gyrinophilus | 1 | F | Aquatic | I | N |
| Hemidactylium | 1 | F | Terrestrial | I | N |
| Hydromantes | 3 | F | Terrestrial | I | N |
| Leurognathus | 1 | F | Aquatic | I | N |
| Lineatriton | 1 | NR | Terrestrial | I | N |
| Parvimolge | 1 | F | Terrestrial | I | N |
| Plethodon | 10 | F | Terrestrial | I | N |
| Pseudoeurycea | 5 | F | Terrestrial | I | N |
| Pseudotriton | 1 | F | Aquatic | I | N |
| Stereochilus | 1 | F | Terrestrial | I | N |
| Proteidae | | | | | |
| Necturus | 2 | 1F; 1M | Aquatic | I–II | N |
| Proteus | 1 | F and/or M? | Aquatic | I | N |
| Sirenidae | | | | | |
| Siren | 1 | F | Aquatic | II–III | X? |

[a] Systematics follow Duellman and Trueb, 1986.
[b] Sex of caregiver: M = male, F = female, M + F = both sexes, M/F = either sex, NR = not reported.
[c] Clutch size (no. eggs) rankings represent averages from the literature: I = 1–50 eggs, II = 51–200 eggs, III = 201–1000 eggs.
[d] Fertilization (Fert.): X = external, N = internal.

Among frogs and toads, at least 17 of the 27 families have postoviposition parental care (Table II on page 110). In addition, five species of frogs (one Leptodactylidae and four Bufonidae) have internal gestation in the oviduct. McDiarmid (1978) speculated that about 10% of anurans will be found to have parental care. This figure seems reasonable, considering that we have no information on the reproductive biology of more than half the species of dart-poison frogs (Dendrobatidae) or on many other species of anurans suspected to have parental care (e.g., some egg-brooding tree frogs, *Pipa, Eleutherodactylus,* and many Australo-Papuan microhylids).

Among amphibian families there is a considerable range in the percentage of species that exhibit parental care, in large part due to the wide range of sizes of families. For example, within some of the families composed of only a few species, all or most have parental care (e.g., salamanders: Amphiumidae, 2 of 3 species; Cryptobranchidae, 2 of 3 species; Dicamptodontidae, 3 of 4 species; anurans: Leiopelmatidae, 3 of 3 species; Rhinodermatidae, 2 of 2 species; Sooglossidae, 2 of 3 species). In the only large family of salamanders, Plethodontidae, at least 23% of species (57 of 251) exhibit parental care. In contrast, parental care is rare, recorded for 1% or fewer of the species, in some of the large families of anurans (e.g., Hyperoliidae, 1 of 226 species; Ranidae, 8 of 625 species; Rhacophoridae, 1 of 204 species). Even in the large anuran families with the highest percentages of species with parental care recorded, the behavior is still not common (Leptodactylidae, 56 of 850 species, or about 7%; Hylidae, 55 of 719 species, or about 8%; Microhylidae, 27 of 313 species, or about 9%). These estimates are conservative, however, as additional species are presumed to provide parental care.

As far as we know, most anuran genera have never evolved parental care. Within some speciose genera, some species exhibit parental care, some do not (e.g., the anurans *Centrolene, Eleutherodactylus,* and *Hyla*), suggesting that presence of the behavior is a derived character state. In other speciose genera, all species may be found to provide parental care (e.g., *Dendrobates* and *Gastrotheca*), suggesting that this behavior is a primitive character state. Some specialized behaviors have evolved only within a single lineage (e.g., brooding of offspring in the stomach, vocal sacs, or inguinal pouches). In contrast, some parental care behaviors have evolved independently in different families of amphibians (e.g., attendance of terrestrial, direct-developing eggs; transport of tadpoles to water; attendance of nonfeeding, terrestrial tadpoles; and internal gestation in the oviduct).

### III. Modes of Parental Care: Occurrence and Function

I recognize six modes of parental care within the Amphibia: egg attendance, egg transport, tadpole attendance, tadpole transport, tadpole feed-

TABLE II
Occurrence of Postoviposition Parental Care within Anurans

| Taxon[a] | No. species | Sex of caregiver[b] | Oviposition site | No. eggs[c] |
|---|---|---|---|---|
| Arthroleptidae | | | | |
|   *Trichobatrachus* | 1 | M | Aquatic | ? |
| Bufonidae | | | | |
|   *Nectophryne* | 1 | M | Aquatic | I |
|   *Nectophrynoides* | 1 | M | Terrestrial | I |
|   *Oreophrynella* | 1 | F? | Terrestrial | I |
| Centrolenidae | | | | |
|   *Centrolene* | 4 | M | Terrestrial | I–II |
|   *Cochranella* | 1 | M | Terrestrial | I |
|   *Hyalinobatrachium* | 4 | M | Terrestrial | I–II |
| Dendrobatidae | | | | |
|   *Colostethus* | 16 | 13M; 2F; 1NR | Terrestrial | I |
|   *Dendrobates* | 12 | 3M; 4F; 3M + F; 1M/F; 1NR | Terrestrial | I |
|   *Epipedobates* | 18 | 16M; 2M/F | Terrestrial | I |
|   *Minyobates* | 3 | M | Terrestrial | I |
|   *Phyllobates* | 3 | M | Terrestrial | I |
| Discoglossidae | | | | |
|   *Alytes* | 2 | M | Terrestrial | I–II |
| Hemisotidae | | | | |
|   *Hemisus* | 2 | F | Terrestrial | II–IV |
| Hylidae | | | | |
|   *Cryptobatrachus* | 2 | F | Terrestrial | I |
|   *Flectonotus* | 5 | F | Terrestrial | I |
|   *Gastrotheca* | 38 | F | Terrestrial | II–III |
|   *Hemiphractus* | 5 | F | Terrestrial | I |
|   *Hyla* | 1 | M | Aquatic | IV |
|   *Osteopilus* | 1 | F | Aquatic | II–III |
|   *Stefania* | 3 | F | Terrestrial | I |
| Hyperoliidae | | | | |
|   *Alexteroon* | 1 | F | Terrestrial | I |
| Leiopelmatidae | | | | |
|   *Leiopelma* | 3 | M | Terrestrial | I |
| Leptodactylidae | | | | |
|   *Cyclorhamphus* | 1 | F | Terrestrial | I |
|   *Eleutherodactylus* | 48 | 19M; 27F; 2NR | Terrestrial | I–II |
|   *Geobatrachus* | 1 | F | Terrestrial | I |
|   *Leptodactylus* | 4 | 3F; 1M + F | 2 Aquatic; 2 terrestrial | I–IV |
|   *Thoropa* | 1 | M | Terrestrial | I |
|   *Zachaenus* | 1 | F | Terrestrial | I |
| Microhylidae | | | | |
|   *Anodonthyla* | 1 | M | Aquatic | I |

(*continues*)

TABLE II Continued

| Taxon[a] | No. species | Sex of caregiver[b] | Oviposition site | No. eggs[c] |
|---|---|---|---|---|
| Barygenys | 1 | M | Terrestrial | I |
| Breviceps | 3 | 2F; 1M + F | Terrestrial | I–II |
| Cophixalus | 8 | 5M; 1F; 1M/F; 1NR | Terrestrial | I |
| Myersiella | 1 | F | Terrestrial | I |
| Oreophryne | 3 | 1M; 1M + F; 1NR | Terrestrial | I |
| Phrynomantis | 2 | M | Terrestrial | I |
| Platypelis | 1 | M | Aquatic | II |
| Plethodontohyla | 2 | 1M; 1F | 1 Aquatic; 1 terrestrial | II |
| Probreviceps | 1 | F | Terrestrial | I |
| Sphenophryne | 3 | 1M; 1F; 1NR | Terrestrial | I |
| Synapturanus | 1 | M | Terrestrial | I |
| Myobatrachidae | | | | |
| Adelotus | 1 | M | Aquatic | III |
| Assa | 1 | M + F | Terrestrial | I |
| Kyarranus | 1 | F | Terrestrial | I |
| Pseudophryne | 6 | 5M; 1F | Terrestrial | I–II |
| Rheobatrachus | 2 | F | Aquatic | I |
| Pipidae | | | | |
| Pipa | 5 | F | Aquatic | II–III |
| Ranidae | | | | |
| Petropedates | 3 | M | Terrestrial | I |
| Pyxicephalus | 1 | M | Aquatic | IV |
| Phrynodon | 1 | F | Terrestrial | ? |
| Rana | 3 | M | 1 Aquatic; 2 terrestrial | I, III |
| Rhacophoridae | | | | |
| Chiromantis | 1 | F | Terrestrial | II |
| Rhinodermatidae | | | | |
| Rhinoderma | 2 | M | Terrestrial | I |
| Sooglossidae | | | | |
| Sooglossus | 2 | F | Terrestrial | I |

[a] Systematics follow Duellman (1993), and Ford and Cannatella (1993).

[b] Sex of caregiver: M = male, F = female, M + F = both sexes, M/F = either sex, NR = not reported.

[c] Clutch size (no. eggs) rankings represent averages from the literature: I = 1–50 eggs, II = 51–200 eggs, III = 201–1000 eggs, IV = >1000 eggs. All of the species included in this table have external fertilization with the exception of one *Eleutherodactylus* (*E. coqui*), which has internal fertilization.

ing, and internal gestation in the oviduct (viviparity and ovoviviparity). Of these, caecilians and salamanders exhibit only egg attendance and internal gestation in the oviduct. In contrast, anurans exhibit considerable diversity in modes of parental care even within some families (Table III); many species exhibit more than one mode.

## A. Egg Attendance

Egg attendance refers to a parent remaining (full or part time) with the eggs at a fixed location. Egg attendance occurs in at least 8 species of caecilians, at least 72 species of salamanders, and at least 176 species of anurans from 15 families (Tables I and III; Crump, 1995). Some anurans attend their eggs and subsequently attend or transport their tadpoles. These examples will be discussed as follows.

What are the apparent functions of egg attendance? (1) *Protection against predators.* Observations and staged encounters, in the field and in the laboratory, show that some species actively defend their eggs against po-

TABLE III
Diversity of Modes of Parental Care within Anuran Families

| Family[a] | Egg attendance | Egg transport | Tadpole attendance | Tadpole transport | Tadpole feeding | Int. gest.[b] |
|---|---|---|---|---|---|---|
| Arthroleptidae | X | | | | | |
| Bufonidae | X | | X | | | X |
| Centrolenidae | X | | | | | |
| Discoglossidae | | X | | | | |
| Dendrobatidae | X | | | X | X | |
| Hemisotidae | X | | X | | | |
| Hylidae | X | X | X | X | X | |
| Hyperoliidae | X | | | | | |
| Leiopelmatidae | X | | X | | | |
| Leptodactylidae | X | | X | X | | X |
| Microhylidae | X | | X | | | |
| Myobatrachidae | X | X | | X | | |
| Pipidae | | X | | | | |
| Ranidae | X | | X | X | | |
| Rhacophoridae | X | | | | | |
| Rhinodermatidae | X | | | X | | |
| Sooglossidae | X | | | X | | |

[a] See Crump (1995) for a list of species within each family that exhibit each mode of parental care, and for references.
[b] Internal gestation within the oviduct (ovoviviparity or viviparity).

tential predators, including conspecifics (Highton and Savage, 1961; Vaz-Ferriera and Gehrau, 1975; Kluge, 1981; Townsend *et al.,* 1984). Several field studies on egg-attending species have documented that eggs deprived of attendant parents quickly disappear (Tilley, 1972; Forester, 1978, 1979b; Simon, 1983). (2) *Protection against pathogens* (especially fungi). In some species the presence of the female decreases the incidence of fungal infestation (e.g., Simon, 1983), in either or both of two ways. First, in certain terrestrial breeders, the attendant parent rolls the eggs around; these movements presumably destroy fungal hyphae (Salthe and Mecham, 1974). Second, oophagy is common in terrestrial breeders; the parent eats dead eggs before fungal growth becomes established within the clutch (Forester, 1979b). (3) *Aeration of aquatic eggs.* Some aquatic species rock back and forth near their eggs, beat their gills near the eggs, or swim in place (Durand and Vandel, 1968; Scheel, 1970; Salthe and Mecham, 1974). These behaviors jostle the eggs and presumably increase the flow of oxygen over them. (4) *Hydration of terrestrial eggs.* In some species, the parent actively expels its bladder contents over the eggs (Weygoldt, 1980, 1987). In other species, the parent reduces the rate of water loss by decreasing the exposed surface area of the egg mass (Forester, 1984). (5) *Prevention of developmental abnormalities.* Manipulation of eggs by the parent (often resulting simply from the adult moving around) may help to prevent developmental adhesions and yolk layering (Organ, 1960; Forester, 1979b; Simon, 1983).

B. Egg Transport

Egg transport (*egg brooding*) involves a parent transporting eggs either in or on its body. In some species the eggs develop directly, bypassing a free-swimming larval stage. In others, adults carry the eggs until they hatch, and then deposit the tadpoles in water. Egg transport occurs in four families of anurans (Table III).

Males of two species of midwife toads, *Alytes* (Discoglossidae), carry their eggs entwined on their hind legs until the eggs are ready to hatch, at which point the male enters water and the tadpoles swim off and undergo aquatic development. Because a male may deposit his offspring in several different ponds, this form of parental care may function in dispersing the offspring (Salthe and Mecham, 1974), and many increase tadpole survivorship if the ponds vary in quality and duration.

Some tropical tree frogs (Hylidae: Hemiphractinae) carry their eggs on their backs. Three groups of egg-brooding tree frogs are distinguished based on where the eggs develop (within a pouch or not) and on the morphology of the pouch (Duellman and Maness, 1980; del Pino, 1980; Duellman and Gray, 1983; Duellman and Hoogmoed, 1984). (1) In

*Cryptobatrachus* (2 species), *Stefania* (3 species), and *Hemiphractus* (5 species), the eggs are carried exposed on the dorsum of the female. Mucous glands of the dorsal integument secrete a matrix of material that glues the eggs to the female's back (del Pino, 1980). In all of these species the young hatch as froglets. (2) In *Flectonotus* (5 species), the eggs develop in a maternal dorsal pouch that opens through a longitudinal midline slit (del Pino, 1980). Eggs hatch as advanced tadpoles, which are transported by the female to water-filled bromeliads or bamboo where they continue development. The tadpoles contain large amounts of yolk and do not feed on external sources of food (Duellman and Gray, 1983). (3) Female *Gastrotheca* (38 species) brood their eggs in a closed, dorsal pouch that opens only at the posterior of the body. Some species release froglets, others transport larvae to aquatic developmental sites (Duellman and Maness, 1980). Presumably other species in these genera of egg-brooding hylids will be found to have similar parental behaviors.

Some aquatic species brood their eggs. For example, two species of *Rheobatrachus* (Myobatrachidae) exhibit gastric brooding (Corben *et al.*, 1974; McDonald and Tyler, 1984). The female swallows her late-stage eggs (or early-stage larvae), and broods them in her stomach. Following metamorphosis, the young emerge from the female's mouth as froglets (see also Section III,D). Female *Pipa* (Pipidae) carry their eggs embedded in their dorsum (Rabb and Snedigar, 1960). Although this mode has been documented for only five species, presumably the other two brood in this manner. In two species the eggs hatch as tadpoles, which leave the female and undergo a larval stage; the other three species have direct development.

C. TADPOLE ATTENDANCE

Attendance of aquatic larvae has been recorded in the anuran families Bufonidae, Leptodactylidae, and Ranidae, but only for a few species. Eggs of the bufonid *Nectophryne afra* are deposited in small, terrestrial cavities filled with water. The male attends the eggs and remains with the tadpoles for 1 to 2 weeks after they hatch; throughout this time, he periodically swims in place and rapidly kicks his hind legs (Scheel, 1970), presumably helping to aerate the water. In *Leptodactylus ocellatus,* the eggs are contained within a large, floating foam nest. The female (and sometimes the male) remains near the nest and aggressively defends the eggs against intruders (Vaz-Ferreira and Gehrau, 1974, 1975). After hatching, the tadpoles move about in a dense mass. The female (but not the male) stays with them throughout development to metamorphosis 40–50 days later, and will attack potential predators (e.g., wading birds). In *Leptodactylus bolivianus*, females not only attend their densely packed masses of tadpoles, but also they appear to lead them around in the water (Wells and Bard, 1988). The function of this attendance is unclear, but it could be to guide the tadpoles to sites ideal for optimal development (i.e., the female could be making choices about water depth or temperature) or to protect the tadpoles from predators. In the African ranid, *Pyxicephalus adspersus,* males accompany schools of tadpoles and actively defend them against predators until the tadpoles metamorphose (Balinksy and Balinksy, 1954). Males construct channels between small pools (where tadpoles are isolated) and larger bodies of water, thus allowing the tadpoles to escape entrapment (Kok *et al.,* 1989).

Attendance of terrestrial tadpoles has been documented in four families. In *Hemisus* (Hemisotidae) the female remains near her subterranean eggs; after they hatch, she digs a tunnel into a nearby water body, providing a path for the tadpoles to enter the water, where they undergo further development. The eggs and larvae of *Synapturanus salseri* (Microhylidae) undergo terrestrial development; the male remains with the eggs and nonfeeding larvae in an underground burrow until the offspring metamorphose (Pyburn, 1975). Likewise, in the three species of *Leiopelma* (Leiopelmatidae) after attending terrestrial eggs, the males remain near their terrestrial, nonfeeding tadpoles until they metamorphose (Bell, 1985). In *Leptodactylus fallax* (Leptodactylidae the female stays with her terrestrial foam nest, in which the tadpoles develop (Lescure, 1979).

D. TADPOLE TRANSPORT

Frogs from seven families of anurans transport their tadpoles (Table III). Some species carry them on their dorsum. In *Sooglossus seychellensis* (Sooglossidae), the female remains with her terrestrial eggs in leaf litter. After the eggs hatch, the tadpoles wriggle up onto her dorsum, where they remain without feeding until they metamorphose (R. A. Nussbaum, personal communication). Likewise, female *Cyclorhamphus stejnegeri* (Leptodactylidae) carry tadpoles on their dorsum; the larvae have large yolk stores, and thus presumably are nonfeeding (Heyer and Crombie, 1979). Tadpoles of *Colostethus degranvillei* (Dendrobatidae) also have large yolk stores and complete their larval development while carried on the dorsum of the male parent (Lescure, 1984), a pattern that contrasts with that of other dendrobatids, discussed as follows.

Many anurans that attend terrestrial eggs transport their newly hatched tadpoles to water. Dendrobatids deposit their eggs out of water. In one species, *Colostethus stepheni,* the tadpoles complete their development in a terrestrial nest (Juncá et al., 1994). In all the remaining species for which the reproductive biology is known (51 species), either the male or the female attends the eggs. After hatching, the tadpoles wriggle onto the back of the parent, and are subsequently carried to water. Some species transport their entire complement of tadpoles at once, usually to streams; others transport tadpoles individually to small bodies of still water such as tree holes or water-filled leaf axils of bromeliads. In some of the latter species, the female returns periodically and feeds unfertilized eggs to her young (Section III,E). The time the tadpoles spend on the parent's dorsum is variable. In some species, the tadpoles are released within a few hours (e.g., *Dendrobates tricolor,* Myers and Daly, 1983). In others, the tadpoles remain on the parent's dorsum for several days or even more than 1 week before being released (e.g., *Colostethus inguinalis,* Wells, 1980). Males of *Rana finchi* and *Rana palavanensis* (Ranidae) from Borneo also carry their tadpoles on their backs (Inger and Voris, 1988). Presumably, as with dendrobatids, the eggs are deposited on land and the male subsequently transports the tadpoles to a suitable aquatic site.

Other anurans also transport tadpoles but not exposed on the dorsum. Some egg-brooding hylids transport their newly hatched tadpoles in dorsal pouches to water, where they release them for subsequent development (e.g., *Flectonotus* and some *Gastrotheca*). In *Rhinoderma* (Rhinodermatidae), the eggs are deposited on moist ground. The male remains close to the eggs, and once the embryos show signs of movement, the male takes the eggs into his mouth and places them in his vocal sac. In *Rhinoderma rufum,* the male carries tadpoles from the terrestrial nest site to a water body, where he releases them (Formas et al., 1975).

Some species carry tadpoles within their bodies until the young metamorphose without ever passing through an aquatic larval stage. Male *Rhinoderma darwinii* house their tadpoles in the vocal sac, but unlike *R. rufum,* young are retained there until they metamorphose and leave the male's body about 52 days later (Formas et al., 1975). In *Assa darlingtoni* (Myobatrachidae), the female attends her terrestrial eggs while the male remains close by (Ehmann and Swan, 1985). About 11–12 days after oviposition, the male climbs into the egg mass, rupturing the capsules around the embryos. The newly hatched larvae then wriggle, aided by the male's movements, into bilateral brood pouch openings in the male's inguinal region. The male carries his nonfeeding tadpoles in these pouches until metamorphosis (Ingram et al., 1975; Ehmann and Swan, 1985).

Perhaps the most unique form of tadpole transport is found in two species of *Rheobatrachus* (Myobatrachidae) that exhibit gastric brooding. As mentioned previously, the female swallows her eggs or early-stage larvae. Feeding and digestion stop, and the tadpoles develop in the stomach (Corben et al., 1974; McDonald and Tyler, 1984). The tadpoles secrete prostaglandin

$E_2$, which inhibits gastric acid secretion by the mother (Tyler *et al.,* 1983). The tadpoles rely exclusively on their yolk during the approximately 8-week developmental period. Following metamorphosis, the young emerge from the female's mouth over a period of several days (Tyler and Carter, 1981). Within a few days after expelling her young, a female's stomach resumes normal digestive functions.

E. TADPOLE FEEDING

Female *Osteopilus brunneus* (Hylidae) oviposit in water-filled leaf axils of bromeliads, where food for tadpoles may be limited. These tree frogs display an elaborate maternal care wherein the female returns and supplies about 250 small eggs to her offspring (as many as 170 tadpoles in a bromeliad) every few days throughout the period of larval development (Thompson, 1992). Initially, the female provides fertilized eggs. About the time the first tadpoles metamorphose, however, the female begins to deposit unfertilized eggs into the leaf axil. The remaining larvae feed on unfertilized eggs until they metamorphose. Provision of eggs appears to be obligatory for reproductive success; clutches that never receive eggs always fail, and in those clutches where a female stops providing eggs, the larvae stop growing and eventually die unless they are already close to metamorphosis (Thompson, 1992). Lannoo *et al.,* (1987) found that large tadpoles can hold more than 180 eggs in their stomachs.

Females of six species of dart-poison frogs, *Dendrobates,* have been observed in captivity to provide unfertilized eggs to their developing larvae (Weygoldt, 1980, 1987). One of these species (*Dendrobates pumilio*), at least, does the same under natural field conditions (Brust, 1993). Field data also suggest that *Dendrobates granuliferus* provide eggs to their tadpoles (van Wijngaarden and Bolaños, 1992). As in *Osteopilus brunneus,* egg provisioning by these seven dendrobatids is associated with larval development in presumed food-limited habitats (water-filled bromeliads).

Indirect feeding of the young may occur in Darwin's frog, *Rhinoderma darwinii.* Goicoechea *et al.* (1986) used peroxidase as a tracer to show that substances can move from paternal tissue in the vocal sac to larval tissues. These data suggest that nutrients provided by the male enter through the tadpole's skin during early stages of development. Once the tadpole's yolk has been absorbed and the intestinal epithelium has become differentiated, the nutrients could instead enter directly through the mouth into the digestive system.

F. INTERNAL GESTATION IN THE OVIDUCT

Retention of developing embryos in the oviduct (*live-bearing*) is an advanced mode of reproduction found in all three orders of amphibians.

Ovoviviparity (the young rely solely on their own egg yolk for development) occurs in three species of anurans (*Eleutherodactylus jasperi, Nectophrynoides tornieri,* and *Nectophrynoides viviparus*) and in two species of salamanders (*Salamandra salamandra* and *Mertensiella caucasica*) (Duellman and Trueb, 1986); this mode of reproduction is unknown for caecilians. Viviparity (the young receive nourishment from the female after exhausting their yolk supplies) occurs in two species of anurans (*Nectophrynoides liberiensis* and *Nectophrynoides occidentalis*), two species of salamanders (*Salamandra atra* and *Mertensiella luschani antalyana*), and probably about 75% of all caecilians (Duellman and Trueb, 1986). For reviews of live-bearing modes in amphibians, see Wake (1977, 1982), Salthe and Mecham (1974), Lamotte and Lescure (1977), Duellman and Trueb (1986), and Guillette (1987).

Wake (1982) emphasized that within the three orders of amphibians the functional morphology of parental and fetal adaptations for retention of young in the oviduct shows strong patterns of convergence. Reduced clutch size is a correlate of live-bearing modes of reproduction. Furthermore, viviparous–ovoviviparous members of all three orders possess an endocrine system of preparation and maintenance of the "pregnancy" involving both estrogen and progesterone. Fetal modifications also show convergent patterns. For example, functional gills are present very early in development; later on during development the fetuses make use of their entire body skin for gaseous exchange. In the viviparous species of all three orders, the fetuses feed on the secretory oviducal epithelium.

### IV. Geographic Distribution of Parental Care

Can we identify geographical "hot spots" for occurrence of parental care in amphibians? If so, then perhaps future investigators should focus on these areas in an effort to identify possible environmental factors that may have favored the evolution of parental care. In order to examine this question, we must consider the geographic distribution of caecilians, salamanders, and anurans.

Caecilians are pantropical in distribution, occurring in Southeast Asia, tropical Africa, the Seychelles, and Central and South America (Duellman and Trueb, 1986). Of the eight species known to attend their eggs, five occur only in the Seychelles, two in central Africa, and one in Sri Lanka. We know so little regarding egg attendance in caecilians that it is useless to speculate why the behavior has not been observed in species from the New World tropics, except to point out that most of the New World species may be viviparous.

Nine of the 10 familes of salamanders, all except the largest (Plethodontidae), occur almost exclusively in North Temperate regions. Although plethodontids are most diverse in North America, they have also invaded the New World tropics, to southern Brazil and central Bolivia (Duellman and Trueb, 1986). Because most of the salamanders that attend their eggs are plethodontids, it is no surprise that most parental care in this order is found in regions where plethodontids are most common: 49 species in the United States (1 ambystomatid, 2 amphiumids, 1 cryptobranchid, 3 dicamptodontids, 39 plethodontids, 2 proteids, and 1 sirenid) and 15 species in Mexico and Central America (all plethodontids). One plethodontid that attends eggs occurs in northern South America, The other 7 species that attend eggs are found in the Old World: 4 in Asia (3 hynobiids, 1 cryptobranchid), and 3 in Europe (2 plethodontids, 1 proteid). The 4 ovoviviparous–viviparous species occur in southern Europe and western Asia.

Species richness of anurans is higher in the Southern Hemisphere than in the Northern Hemisphere, and is highest in tropical wet forests. The Neotropics (Central and South America) boast the highest species richness of anurans in the world. Consequently, by far the greatest anuran diversity of modes of postoviposition parental care (all five modes) exists in Central and South America (Table IV; Fig. 1). One can even find all five modes within an area of a few square kilometers. The lowest diversites (one mode) occur in the North Temperate areas of North America, Europe, and Asia, in southern Asia, Papua–New Guinea, and New Zealand. No species with parental care has been recorded from Africa north of the Tropic of Cancer (Region G, Fig. 1), nor from southern Indonesia (Region O, Fig. 1). The ovoviviparous leptodactylid is found in Puerto Rico, and the four bufonids that are ovoviviparous or viviparous occur in central Africa.

Is there an obvious difference in occurrence of parental care within tropical anurans among the major regions of the world at the family level? Between the Tropic of Cancer and the Tropic of Capricorn we find: (1) Central and South America (Regions B, C, and D, Fig. 1): species from seven families exhibit parental care; (2) Africa, the Seychelles, and Madagascar (Regions H, I, K, and L, Fig. 1): eight families exhibit parental care; and (3) southern Asia, Indonesia, Papua–New Guinea, and the northern half of Australia (Regions N, O, P, and Q, Fig. 1): three families exhibit parental care (Table IV). The only family that exhibits parental care in all three geographic areas is Microhylidae.

How does the picture look at the species level? As discussed by Brooks and McLennan (1991) and Harvey and Pagel (1991), one must realize that when an association is analyzed at the species level, one introduces potential phylogenetic bias. For example, there are 48 species of *Eleutherodactylus* (New World tropics) that exhibit egg attendance as compared to one *Chiro-*

TABLE IV
TAXONOMIC DISTRIBUTION OF ANURANS WITH POSTOVIPOSITION PARENTAL CARE BY GEOGRAPHIC AREA

| Geographical area[a] | Modes of parental care | | | | |
|---|---|---|---|---|---|
| | Egg attendance[b] | Egg transport[b] | Tadpole attendance[b] | Tadpole transport[b] | Tadpole feeding[b] |
| A | Leptodactylidae 1 | | | | |
| B | Centrolenidae 3<br>Hylidae 1<br>Leptodactylidae 38 | Hylidae 3 | Leptodactylidae 2 | Dendrobatidae 6 | Dendrobatidae 3<br>Hylidae 1 |
| C | Bufonidae 1<br>Centrolenidae 8<br>Hylidae 1<br>Leptodactylidae 14 | Hylidae 27<br>Pipidae 2 | Leptodactylidae 1<br>Microhylidae 1 | Dendrobatidae 25 | Dendrobatidae 3 |
| D | Centrolenidae 1<br>Dendrobatidae 1<br>Leptodactylidae 4<br>Microhylidae 1 | Hylidae 34<br>Pipidae 4 | Leptodactylidae 2 | Dendrobatidae 19<br>Leptodactylidae 1 | Dendrobatidae 2 |
| E | Leptodactylidae 1<br>Microhylidae 1 | Hylidae 4 | Leptodactylidae 1 | Rhinodermatidae 2 | |
| F | | | | | |
| G | | Discoglossidae 2 | | | |

| | | |
|---|---|---|
| H | Arthroleptidae 1 | Bufonidae 1 |
| | Bufonidae 1 | Hemisotidae 1 |
| | Hyperoliidae 1 | Ranidae 1 |
| | Ranidae 4 | |
| I | Arthroleptidae 1 | Hemisotidae 1 |
| | Microhylidae 1 | Ranidae 1 |
| | Rhacophoridae 1 | |
| J | Microhylidae 3 | Hemisotidae 1 |
| | | Ranidae 1 |
| K | Sooglossidae 1 | Sooglossidae 1 |
| L | Microhylidae 3 | Microhylidae 1 |
| M | Ranidae 1 | Ranidae 2 |
| N | | |
| O | | |
| P | Microhylidae 13 | |
| Q | Microhylidae 4 | Myobatrachidae 1 |
| | Myobatrachidae 3 | |
| R | Myobatrachidae 7 | Myobatrachidae 2 |
| S | | Leiopelmatidae 3 |

[a] Letters for geographical areas correspond to the letters in Fig. 1.
[b] Numbers following family names indicate number of species. For those species that exhibit more than one mode of parental care (e.g., egg attendance and tadpole transport), only the behavior exhibited latest in the offspring's development is included here (in this case, tadpole transport); thus, a species is represented only once within a geographical area. Because some species are found in more than one geographical area, the column totals are greater than the actual number of anurans known to exhibit each mode of parental care.

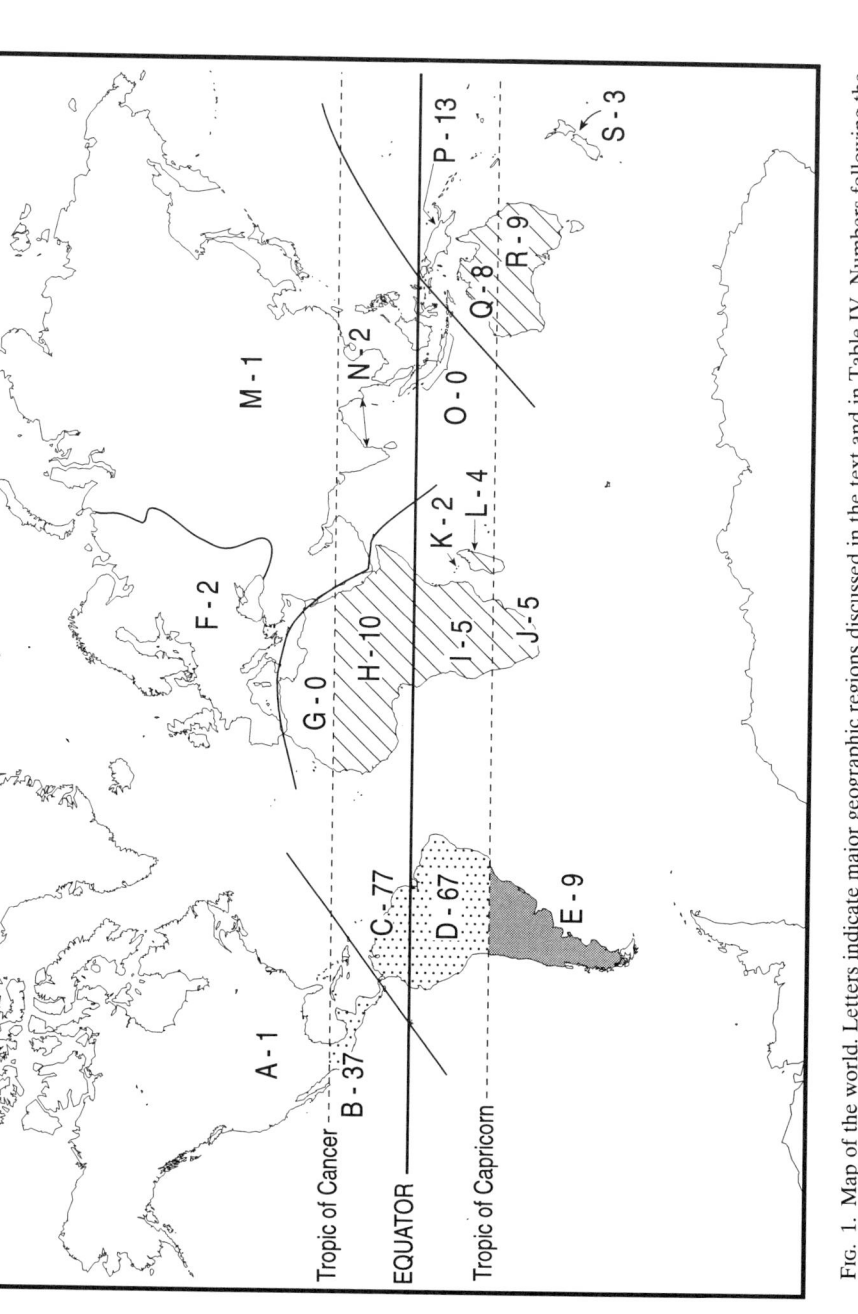

FIG. 1. Map of the world. Letters indicate major geographic regions discussed in the text and in Table IV. Numbers following the letters indicate the number of species of anurans known to provide parental care that occur within that particular geographic region. The regions indicated by pattern (stippled, shaded, and crosshatched) represent those with the highest diversity of modes of parental care—stippled, five modes; shaded, four modes; crosshatched, two modes. The Seychelles (Region K) also is represented by two

*mantis* (tropical Africa) that does the same. The superabundance of parental care in *Eleutherodactylus* will greatly inflate our species total figure for the New World tropics. The bias results because not all genera are equally speciose, nor do they all have equal geographic distributions. Nonetheless, accepting this bias for the moment, we find that the number of anuran species that exhibit postoviposition parental care is much greater in Central America and the Caribbean, and in South America than anywhere else (Table IV; Fig. 1). For example, a comparison of the world's south tropical regions (between the equator and the Tropic of Capricorn) reveals the following: (1) South America (Region D, Fig. 1): egg attendance 7 species, egg transport 38, tadpole attendance 2, tadpole transport 20, tadpole feeding 2, for a total of 69; (2) Africa, the Seychelles, and Madagascar (Regions I, K, and L, Fig. 1): egg attendance 7 species, tadpole attendance 3, tadpole transport 1, for a total of 11 species; and (3) southern Indonesia, Papua–New Guinea, and the northern half of Australia (Regions O, P, and Q, Fig. 1): egg attendance 20 species, tadpole transport 1, for a total of 21 species that exhibit parental care.

The only obvious conclusion regarding distribution of parental care in amphibians is that, as expected, the highest diversity of modes of parental care and the highest number of species exhibiting parental care are found in regions of highest species richness. No one has examined the selective factors that influence the evolution of parental care, so we do not know how important a role environmental conditions play. Maybe they are not very important. It is true that within one habitat one finds closely related species that provide care and others that do not. It would be interesting to obtain quantitative measures for how "successful" each of these species is. Furthermore, viviparity is found in cold, montane areas as well as lowland, tropical sites. Frogs transport their tadpoles at wet sites, dry sites, and intermediate sites. I offer this descriptive picture of geographic distribution of parental care in amphibians in the hope that it may serve as a springboard for analysis of correlations between occurrence of the behavior and environmental factors on a finer scale.

## V. Distribution of Parental Care between the Sexes

Which sex does the caring, and why? Mode of fertilization (external versus internal) may be implicated (Ridley, 1978). Within ectotherms, female care is most often associated with internal fertilization and male care with external fertilization. Three hypotheses have been proposed to explain this association: gamete order (Dawkins and Carlisle, 1976), reliability of paternity (Alexander, 1974), and association of a parent with embryonic stages (Williams, 1975).

Does a correlation exist between mode of fertilization and sex of the caregiver for amphibians? For this discussion I will not consider the species that have internal gestation in the oviduct because of the bias inclusion would introduce since in all of these species the caring parent is necessarily the female. As expected, paternal care occurs in only 3 families with internal fertilization and in 14 families with external fertilization. Maternal care, however, occurs in 7 families with internal fertilization and 12 with external fertilization (Tables I and II; Crump, 1995). Within the class the correlation exists for caecilians and salamanders but not for anurans. All caecilians are presumed to have internal fertilization and the only care is maternal. Almost all instances of paternal care in salamanders are those that have external fertilization, whereas nearly all maternal care is found in salamanders with internal fertilization (Table I). Very few anurans have internal fertilization, and only one of these exhibits parental care: *Eleutherodactylus coqui*, which is oviparous and has paternal care (Townsend *et al.*, 1981). Of the families with external fertilization and parental care, paternal care is found in 12 families and maternal care in 12 families.

Despite a phylogenetic bias when the association is analyzed at the species level (Brooks and McLennan, 1991; Harvey and Pagel, 1991), nonetheless it is revealing that of the 221 species of anurans with external fertilization that also provide uniparental care, 47% (104 species) have paternal care and 53% (117 species) have maternal care (Crump, 1995). Why do so many of these species provide maternal care, when theory suggests that male care should dominate? Whether the male or female provides care may depend on the cost of care to the parent (to the female, in terms of the effect on future egg production; to the male, in terms of ability to mate with additional females; Section VII, B).

## VI. Correlates of Parental Care

### A. Large Egg Size

Generally, egg size is larger in amphibians that exhibit parental care than in those lacking it (Salthe, 1969; Salthe and Duellman, 1973; Salthe and Mecham, 1974), although exceptions exist (e.g., the salamander *Ambystoma opacum* and the tree frogs *Hyla rosenbergi* and *Osteopilus brunneus*). Within a family, species that provide parental care often have eggs that are four to five times larger than those of species that abandon their eggs. In most cases, these large eggs are deposited on land (see as follows). Large eggs have a smaller surface-area-to-volume relationship than small eggs, and thus are less prone to problems of water loss. This attribute is an advantage under potentially desiccating circumstances, such as in terrestrial environments where parental care often occurs. Furthermore, because large eggs result in large offspring at hatching, large egg size may be a means of placing more robust juveniles into the environment (Salthe and Duellman, 1973; Nussbaum, 1987). The question of which came first, large egg size or parental care, has stimulated considerable discussion and speculation (Shine, 1978, 1989; Nussbaum, 1985, 1987; Nussbaum and Schultz, 1989).

### B. Small Clutch Size

Most species that exhibit parental care have small clutch sizes. Of the 72 species of salamanders that provide care, only 14 (19%) produce a clutch of more than 50 eggs (Table I), whereas many aquatic-breeding salamandrids and ambystomatids that lack parental care produce clutches of over 200 eggs. The same pattern exists for anurans (Table II). Few species with parental care produce more than 200 eggs (those that do most often care for small, aquatic eggs), whereas species that abandon their eggs generally produce hundreds or even thousands of eggs.

At least three nonmutually exclusive hypotheses may explain this observation. First, large eggs can be produced only at the expense of lowered fecundity, since a female has a fixed amount of energy. Since parental care usually is associated with large eggs, few can be produced. Second, constraints on fecundity are imposed by female size. Generally, within a mode of reproduction, large females produce large clutch sizes (Salthe and Duellman, 1973). Most amphibians that exhibit parental care are small relative to noncaring relatives, and they have smaller clutch sizes as well. Third, the cost–benefit ratio of parental care may vary with clutch size. A parent may be able to increase survivorship of its offspring economically if there are just a few young for which to care, but the costs may outweigh the benefits if the parent must care for many offspring.

### C. Terrestrial Modes of Reproduction

Most amphibians that provide parental care lay their eggs out of water, and most of these species occur in the humid tropics. All eight caecilians known to attend their eggs are terrestrial breeders (Table I). Of the salamanders that attend their eggs, 71% (51 species) have terrestrial eggs (Table I). Of the anurans that provide parental care, 92% (221 species) deposit their eggs out of water (a few of these subsequently care for aquatic larvae) (Table II).

## VII. EVOLUTION OF PARENTAL CARE

### A. EVOLUTIONARY TRENDS IN TERRESTRIAL BREEDING AND PARENTAL CARE

A broad perspective of overall reproductive biology is important when contemplating the evolution of parental care in amphibians. All three orders of amphibians have followed an evolutionary trend towards greater terrestriality of reproduction (Crump, 1974; Lamotte and Lescure, 1977; McDiarmid, 1978; Duellman, 1985; Duellman and Trueb, 1986). The selective pressures for this move from water to land may have been high levels of predation in aquatic sites and unpredictable duration of small bodies of water. Amphibian eggs are anamniotic and they lack a protective shell. Because the layers of jelly surrounding the embryo are extremely permeable and prone to desiccation, terrestrial environments are potentially inhospitable. Parental care may have evolved in response to the inhospitable terrestrial environment, and the behavior presumably has been maintained because it increases offspring survivorship.

Major trends in the evolution of reproductive biology in caecilians include elimination of the larval stage and protection of the embryos, either through maternal egg attendance of direct-developing eggs or by viviparity (Wake, 1977). Oviparity with aquatic larvae is considered the primitive mode of reproduction for caecilians, represented in the families Ichthyophiidae, Rhinatrematidae, and some Caeciliidae (Wake, 1977). Many of these species lay their eggs on land near water either in burrows or under vegetation (Wake, 1982). They characteristically produce few large eggs, around which the female coils. After hatching, the larvae wriggle into the water, where they continue development. The intermediate mode in the presumed evolutionary sequence is oviparity with direct development. Again, the female attends the eggs. The most advanced mode is viviparity, thought to be exhibited by 75% of all caecilian species.

External fertilization of aquatic eggs is the primitive condition in salamanders, characteristic of the cryptobranchids, hynobiids, and presumably the sirenids (Duellman and Trueb, 1986). The 90% of salamanders that have internal fertilization (the remaining seven families) progress from (1) aquatic eggs and larvae to (2) terrestrial eggs and aquatic larvae to (3) terrestrial eggs with direct development and eventually to (4) retention of eggs or young in the oviduct (ovoviviparity and viviparity).

Among salamanders that lay eggs, three major modes of reproduction are recognized (Salthe, 1969; Nussbaum, 1985). Mode I salamanders deposit many small eggs in exposed sites in still water and provide no parental care. The larvae hatch at small sizes and at relatively early stages of development. Examples include most *Ambystoma* and most newts (Salamandridae). Mode II species deposit relatively few, large eggs in hidden nest sites in running water. The eggs hatch at a more developed stage than those of Mode I species, and the larvae are larger at hatching. Many Mode II species provide parental care; examples include two cryptobranchids, some hynobiids, two species of *Necturus, Proteus,* and many plethodontids. Mode III species deposit direct-developing terrestrial eggs beneath cover such as logs or rocks. Fewer, larger eggs are produced than in either Mode I or Mode II species. So far as is known, almost all Mode III species provide parental care, and all are plethodontids. Parental care presumably enabled these animals to exploit the terrestrial habitat more fully, and in part may be responsible for their widespread distribution in the tropics.

Nussbaum (1985, 1987) hypothesized that parental care evolved among Mode II salamanders and not Mode I species as a consequence of basic differences in the trophic structures of ponds and streams. Salamander larvae in ponds feed primarily on zooplankton, whereas in streams they feed mainly on immature insects. Nussbaum reasoned that selection should adjust egg size in such a way that food supply for the hatchlings is maximized. Hatchlings of stream-breeding species must be large to take advantage of their food source. The only way to have a large hatchling is to have a large egg, which entails a long developmental time (Salthe and Duellman, 1973; Salthe and Mecham, 1974). The long developmental time translates into a longer time of exposure to predators than for a small egg, laying the groundwork for parental care. In contrast, in pond environments selection should favor small eggs due to the small size of the primary food source (plankton).

What might be the hypothetical steps in the evolutionary progression of anurans from the primitive mode of reproduction (aquatic eggs and larvae, no parental care) to the most derived mode (complete independence of standing water, parental care) (McDiarmid, 1978; Duellman and Trueb, 1986; Crump, 1995)? Initial steps may have involved laying eggs on land, but close enough to the water so that the tadpoles could wriggle, fall into, or be washed into the water to complete larval development; the eggs presumably were abandoned by the parents. Many extant species exhibit these modes of reproduction. Subsequent steps in the evolutionary progression toward greater terrestriality may have involved parental care ranging from terrestrial egg attendance but subsequent aquatic larval development, to care of offspring until metamorphosis (by brooding, internal gestation in the oviduct, or attendance of direct-developing eggs) and thus complete independence of standing water, as in the extant examples of species with these parental behaviors discussed in Section III.

Different explanations apply to aquatic-breeding anurans that have evolved parental care. Some of these breed in predator-rich environments of open bodies of water where parental care may enhance survivorship (e.g., *Pyxicephalus, Rheobatrachus, Pipa, Adelotus,* and *Leptodactylus*). Other species that have retained aquatic larvae escape these predator-rich sites by ovipositing in aquatic, arboreal sites or in water-filled terrestrial crevices that would be inhospitable sites without parental intervention. Parental care in the form of aeration or provision of eggs as food allows utilization of such aquatic sites (e.g., *Nectophryne afra, Osteopilus brunneus,* and some dendrobatids).

As suggested for reptiles (Shine, 1988), parental care in amphibians may have evolved from other behaviors. Parental care in some species may have begun with the female being "exhausted" from oviposition and simply remaining in a hidden nest site for a period of time. An early stage in this association between parent and young may have been simply the physical presence of the female with the eggs. The addition of behaviors such as clutch defense by the female or hydration of the eggs would have led eventually to a direct benefit of prolonged nest attendance. Such stages, from brief (a few days) to longer (several months) association, and from a facultative to obligatory relationship, are seen in living amphibians (Crump, 1995). Another possible origin of parental care is the parent returning intermittently to a good refuge spot or to a site it defended for mating purposes, and at which its eggs happen to be located. Although the original fidelity might have involved the site, not the eggs, through time, selection may have favored more persistent attendance and active protection of the eggs. The intermediate stages may be expressed today in the variability of the strength of the bond between the parent and the offspring (Crump, 1995). Either or both of these two alternatives may, in fact, explain some cases of amphibians observed with egg clutches if we find no evidence of costs or benefits.

B. Costs and Benefits of Parental Care

*1. What Are the Costs and Benefits of Parental Care for Amphibians?*

The benefit of parental care is increased survivorship of the offspring (Trivers, 1972). Presence of an attendant parent has been shown to increase egg survivorship in a number of species of anurans and salamanders (e.g., Tilley, 1972; McDiarmid, 1978; Forester, 1979b; Harris and Gill, 1980; Simon, 1983; Townsend *et al.*, 1984; Juterbock, 1987; Jackson *et al.*, 1989). Benefits of egg and tadpole transport, tadpole attendance, and tadpole feeding have received less attention, but in many cases these behaviors are obligatory in the sense that without them survivorship would be zero.

The major cost of parental care is reduced future survival or reduced reproductive success of the caring parent (Trivers, 1972). Unfortunately,

costs are difficult to measure and few data are available. The ultimate cost of reduced future survival or reproductive success may come about in any one of the following ways, and may differ between males and females.
(1) *Increased vulnerability to predation.* Because parental care restricts an animal's mobility, the caring parent may be more vulnerable to predation. Either sex might experience decreased or zero future reproductive success if attacked by a predator. This assumption, however, needs to be validated through field observations and experiments. For certain species (e.g., species that attend eggs in an underground or arboreal nest site and do not leave to forage), parental care may actually reduce vulnerability to predation.
(2) *Decreased food intake.* Many species eat little while providing parental care (e.g., Simon, 1983; Townsend, 1986; Organ, 1961; Krzysik, 1980; Forester, 1981; Hom, 1987; Juterbock, 1987; Ng and Wilbur, 1995). The potential cost of less food presumably is greater for females than for males if it delays production of subsequent clutches. On the other hand, certain species may not experience such a cost (Tilley, 1972; Kaplan and Crump, 1978).
(3) *Reduced mating opportunities.* Time and energy spent on parental care may reduce opportunities for additional matings. This cost is probably more severe for males than for females (e.g., Kluge, 1981; Townsend, 1986). Townsend's studies emphasize the need to consider both the benefits and the costs of parental care; although male *Eleutherodactylus coqui* miss additional mating opportunities, they still enhance their fitness by attending their eggs (Townsend, 1986).

2. *Can Costs and Benefits Help Explain Presence and Type of Parental Care?*

Why do some amphibians lack parental care, others exhibit uniparental care, and a few provide biparental care? This complex question is difficult to answer because so few quantitative data exist on the benefits and costs of parental care.

Parental care is unlikely to evolve if neither parent can increase offspring survival sufficiently to offset the costs involved (Perrone and Zaret, 1979). This may be the case when either the benefits are low or the costs are high. In many amphibians the benefits of increased offspring survivorship through parental care may not offset the costs to males of being unable to fertilize additional clutches, nor the costs to females of being unable to replenish energy stores for future reproduction. Other constraints may hinder the evolution of parental care. For example, adult body size may morphologically constrain egg guarding; adults may not be large enough relative to their potential predators to defend their young effectively. Oviposition pattern may also constrain egg attendance. Attendance is associated with species that lay their eggs in one place, often in a compact mass with the eggs adhering to each other. Conversely, dispersion of the clutch among several sites, exhibited by many species, presumably precludes effective egg attendance. Another constraint may be foraging behavior. Many salamanders and anurans are active foragers, seeking dispersed arthropod prey. Thus, an individual might not be able to secure sufficient food by remaining in one place.

Parental care is expected to evolve in species and in environments where the benefit to the offspring is the greatest relative to the cost (Shine, 1988; Clutton-Brock, 1991), for instance in physically harsh environments. As discussed, parental care predominates in amphibians with terrestrial modes of reproduction (Tables I and II). Likewise, parental care is especially beneficial where predation pressure (including from conspecifics) on the egg or larval stage is heavy.

Alternatively, parental care is expected to evolve in species and in environments where the costs to the parent are minimal (Shine, 1988; Clutton-Brock, 1991), assuming that benefits exist. For example, parental care is more likely to evolve in species where parents attend their young in hidden nest sites than in species that oviposit in open, exposed environments, all else equal. Observations of parental care in amphibians support this prediction; most amphibians that attend their eggs do so in well-protected sites such as subterranean burrows, tree holes, and hidden cavities.

Can costs and benefits help explain which sex provides parental care? For which sex are the benefits of parental care greater? For which sex is parental care least likely to decrease adult survivorship or interfere with future reproductive success? Presumably, parental care evolves and is maintained in the sex for which the fitness benefits are higher relative to the costs.

Probably few cases exist where parental care by one sex would result in higher survivorship of the offspring than would come by the other. One case might be the tree frog, *Hyla rosenbergi,* in which males are equipped with prepollical spines used to defend eggs against intruding conspecifics. Females lack these spines. In some dart-poison frogs (Dendrobatidae) with one territorial sex, transport of the tadpoles may have evolved in the opposite sex because of the danger of the tadpoles being dislodged from the aggressive sex during territorial encounters (Wells, 1981). In most species of amphibians, however, either sex may be able to do the job of caring equally well. If so, then on average, males and females probably derive similar benefits to their fitness from providing parental care.

Do males and females differ in the cost of parental care to their future survivorship? No evidence exists that one sex would experience greater vulnerability to predation than would the other as a result of providing parental care. Thus, the major difference in potential cost between the sexes may lie in future reproductive success.

Generally, the factor that limits reproductive success of males is the number of mates they can acquire. Any behavior that restricts that potential may decrease a male's reproductive success. If additional females lay eggs in the territory of a male already attending offspring, the cost of care to that male may be reduced and thus paternal care might be favored. On the other hand, if males increase their reproductive success by actively seeking additional mates over a large area, the cost of remaining sedentary and caring for young would be high. Thus, we would not expect to find paternal care in these species.

For females, the limitation to reproductive success is generally egg production. Thus, maternal care is expected to evolve when the costs of care are minimal in terms of future reproduction. Fecundity increases with body size. Thus, any behavior that interferes with the ability to increase body size will limit reproductive success.

As pointed out earlier, in 53% (117 species) of the anurans with external fertilization that provide uniparental care, the caregiver is female, yet theory would predict male care (Ridley, 1978). Why then is maternal care so prevalent? One scenario with limited cost to the female would be where the female can continue to feed while providing care (Wells, 1981). Examples might include species that do not attend their offspring at a fixed location, but instead carry their young and are free to forage widely (e.g., egg-brooding hylids, *Pipa,* and dendrobatids). We need data on feeding rates of parental versus nonparental females to test this prediction. This nondieting scenario may apply to about 75% of the 117 species with maternal care. Two other hypotheses may explain the other 25%, species in which the female remains in one place with the eggs or young and thus may experience decreased food intake. First, even in these species, if the duration of parental care is short relative to the time required to produce an additional clutch, then the cost to the female would be minimal (Wells, 1981). For most species, though, we do not know how often individual females lay eggs, or the cost of parental care behavior on subsequent vitellogenesis. Second, cost might also be minimal where the cost of vitellogenesis is so great or the breeding season so short that even without parental care a female could not produce a second clutch in the same season. We need manipulative field experiments, and studies of food resource availability and length of the breeding season, in order to test this idea.

Biparental care is expected to evolve only if offspring survival is greatly enhanced when a second parent assists (Perrone and Zaret, 1979). According to theory, one parent should not help the other unless the incremental increase in survival of the offspring surpasses possible reproductive benefits gained from other behaviors, such as fertilizing additional clutches or replenishing energy for future reproduction. Very few cases of biparental care have been documented for amphibians, suggesting that one parent may generally be as effective as two. No well-documented cases of biparental care exist for salamanders (Nussbaum, 1985). The few cases of biparental care in anurans generally involve care for both egg and larval stages. Perhaps in these species the care process is either so prolonged or so demanding that two parents are required for offspring survivorship to be enhanced substantially. For example, in *Assa darlingtoni,* the female attends the eggs, after which the male broods the tadpoles in his inguinal pouches (Ehmann and Swan, 1985). Biparental care occurs in some of the dendrobatids that provide tadpole attendance and feeding (Zimmermann and Zimmermann, 1984; Weygoldt, 1987). In *Leptodactylus ocellatus,* both parents sometimes defend their eggs against intruders (Vaz-Ferreira and Gehrau, 1975). In *Breviceps gibbosus,* both the female and male attend their eggs, which hatch into nonfeeding tadpoles that complete development within the terrestrial nest (Rose, 1962; Wager, 1965).

## VIII. Flexibility in Parental Care

Models of parental care generally assume that a parent should be flexible in the extent of care given to particular offspring in order to adjust expenditure relative to proximate variation in levels of benefits and costs (Winkler, 1987). The level of parental care should increase as benefits to the particular offspring increase, and the level should decline as the costs to the parents' lifetime fitness increase. Although we are beginning to understand patterns of flexibility in parental care in fish, birds, and mammals (Clutton-Brock, 1991), relatively little is known concerning flexibility in amphibians.

### A. Variation in Care Relative to Benefits

Offspring should derive the greatest benefit from parental care when environmental conditions are worst. Therefore, in species where parental care is facultative, the behavior might vary as ecological conditions change (Wells, 1981). For example, a species might attend eggs only when environmental conditions are drier than usual or when predators are more abundant. In support of this idea, more male *Hyla rosenbergi* guard their nests against conspecifics when population density is high (thus increasing the likelihood of intrusion by conspecifics) than when population density is lower (Kluge, 1981). Two species of Salamandridae are oviparous and have aquatic larvae when conditions are less harsh, but retain the eggs in the oviduct for varying periods of time under harsher conditions (Duellman and Trueb, 1986). Wake (1982), in reviewing viviparity in amphibians, noted that in *Nectophrynoides* and *Salamandra* the length of time the young are retained in the oviduct varies with the harshness of the environment; young are retained longer during years with long, cold winters and short summer activity periods. In contrast, Juterbock's data on nest site tenacity in the terrestrially breeding salamander, *Desmognathus fuscus,* do not support the prediction that females increase care under stressful conditions (Juterbock, 1987). Females were equally likely to abandon their eggs following disturbance regardless of ground moisture conditions, cover type under which the eggs were deposited, or nest placement relative to the median height above water level.

In theory, the level of parental care should increase with clutch size both in situations where the care is depreciable (i.e., where the benefits to individual offspring decline as clutch size increases, e.g., provision of food) and nondepreciable (i.e., where the benefits to individual offspring do not decline with increasing clutch size, e.g., parental vigilance) (Lazarus and Inglis, 1986; Winkler, 1987; Clutton-Brock, 1991). Some data for amphibians support this hypothesis. For example, female *Osteopilus brunneus* seem to increase parental care with larger clutches; the number of egg provisionings increases with increasing clutch size, presumably because larger clutches take longer to develop to metamorphosis (Thompson, 1992). In contrast, Juterbock (1987) found no correlation in *Desmognathus fuscus* between clutch size and tendency of the female to remain with her eggs after disturbance.

Clutton-Brock (1991) noted that it is difficult to predict how parental care should change with age of the offspring. As offspring mature, their reproductive value presumably increases, and thus we might expect parents to increase the level of care for older offspring. On the other hand, because younger stages are more vulnerable, they stand to benefit the most from parental care; by this reasoning, we might expect a decrease in the level of care as the offspring mature. Indeed, both patterns are found in amphibians. Within anurans, for example, male *Hyla rosenbergi* decrease their responsiveness with increasing age of the embryos (Kluge, 1981). Likewise, Thompson (1992) found that female *Osteopilus brunneus* fed their tadpoles significantly more frequently during the first half of development than during the second half. The opposite pattern exists in other species, for example, in the salamander *Desmognathus ochrophaeus.* The strength of the bond between a female and her eggs is correlated positively with developmental stage of the eggs and with the time she has spent attending them (Forester, 1979a). Furthermore, attending females are significantly more aggressive toward potential predators (conspecific males) later in the incubation period than earlier (Forester, 1983).

In other species no obvious pattern has been detected. For example, Townsend *et al.* (1984) found that male *Eleutherodactylus coqui* were as likely to attend late-stage embryos as early-stage ones. After the eggs hatched, however, the level of attendance decreased significantly. Female *Dendrobates pumilio* do not increase the number of eggs provided per meal as their tadpoles grow (Brust, 1993). In *Desmognathus fuscus,* no consistent pattern was found between tendency of the female to remain with the eggs following disturbance and time already spent attending eggs.

B. Variation in Care Relative to Costs

Theoretically, a parent should increase the level of care when the costs to itself are low and decrease the level of care when the costs are high. For example, in some species parental care is reduced when resource availability is low (reviewed in Clutton-Brock, 1991). There are no data available for amphibians. Another general prediction is that parental care by males should decline as access to additional mates increases (Clutton-Brock, 1991), but again data for amphibians are not available. Dendrobatids would be an ideal group for which to test this prediction, especially in those species that exhibit polygyny and mate guarding (Summers, 1989). Members of this family are diurnal, and for many species it is relatively easy to observe social interactions and mating behavior.

Some evidence for other vertebrates suggests that older parents invest more heavily in parental care than do younger individuals (Clutton-Brock, 1991), but this pattern is not supported by the limited data available for salamanders. For example, larger (older?) female *Desmognathus fuscus* did not exhibit a greater tendency to remain with the clutch following disturbance than did smaller (younger?) females (Juterbock, 1987). Interpretation of the lack of a positive correlation may be confounded, however, if age and body size are not perfectly correlated (i.e., large size may not always reflect older age).

IX. Summary

Six modes of parental care are recognized: egg attendance, egg transport, tadpole attendance, tadpole transport, tadpole feeding, and internal gestation in the oviduct. Whereas egg attendance and internal gestation in the oviduct are the only modes of parental care known for caecilians and salamanders, anurans display all six modes. Although parental care is widely distributed phylogenetically within the Amphibia, the percentage of species providing care is low. Postoviposition parental care has been documented in only about 5% of caecilians (8 species, 2 families), 18% of salamanders (72 species, 8 families), and 6% of anurans (240 species, 17 families). Internal gestation in the oviduct probably occurs in about 75% of all caecilians, in 4 species of salamanders, and in 5 species of anurans. Parental care in amphibians is most commonly found in geographical areas of correspondingly high species richness.

All caecilians that provide care have maternal care and internal fertilization. In salamanders, maternal care is associated with internal fertilization, paternal care with external fertilization. This association does not hold up for anurans. The only anuran with internal fertilization and parental care has paternal rather than maternal care. In the 221 species with external fertilization that provide uniparental care, 47% have paternal care and 53% have maternal care. Large egg size, small clutch size, and terrestrial breeding habits are all correlated with parental care. In each of the three orders of amphibians, there has been an evolutionary trend towards increasing terrestriality of breeding activity. In many groups, parental care probably evolved as a means of enhancing offspring survivorship in the potentially harsh terrestrial environment.

Increased survivorship of the offspring is the main benefit of parental care, as documented quantitatively by numerous studies. Reduced fitness to the parent, measured by reduced future survival or reproductive success, is the major cost of parental care. This cost may come about through increased vulnerability to predation, decreased food intake, or reduced mating opportunities, and may differ between males and females. Unfortunately, costs are difficult to quantify and thus few data are available.

More quantitative natural history data are needed. For most species, we do not know the function of the parental behavior or how critical the care is. Long-term mark and recapture studies are needed to determine how often a given female breeds in a season, whether or not parental care precludes additional matings, and what determines reproductive success—number of matings or parental care. Also needed are experimental tests of the costs and benefits of parental care. Benefits can be quantified by removing the parent and observing the fate of the clutch. Experimental field manipulations involving removal of clutches need to be done in order to quantify the differences in vulnerability of caring versus noncaring adults to predation, food intake by adults, rate of egg production, growth rate of females, and mating success of males.

**Acknowledgments**

I thank W. E. Duellman and especially P. Feinsinger for helpful comments and suggestions on the manuscript. Robyn O'Reilly drew the map.

## References

Alexander, R. D. (1974). The evolution of social behavior. *Annu. Rev. Ecol. Syst.* **5**, 325–383.
Balinsky, B. I., and Balinsky, J. B. (1954). On the breeding habits of the South African bullfrog, *Pyxicephalus adspersus. S. Afr. J. Sci.* **51**, 55–58.
Bell, B. D. (1985). Development and parental-care in the endemic New Zealand frogs. *In* "The Biology of Australasian Frogs and Reptiles" (G. Grigg, R. Shine, and H. Ehmann, eds.), pp. 269–278. Royal Zool. Society of New South Wales, Australia.
Brooks, D. R., and McLennan, D. A. (1991). "Phylogeny, Ecology, and Behavior." The Univ. of Chicago Press, Chicago.
Brust, D. G. (1993). Maternal brood care by *Dendrobates pumilio:* A frog that feeds its young. *J. Herpetol.* **27**, 96–98.
Burrowes, P. A. (1987). An ecological study of a cloud forest herpetofauna in southern Colombia. Unpublished Masters Thesis, Univ. of Kansas.
Clutton-Brock, T. H. (1991). "The Evolution of Parental Care." Princeton Univ. Press, Princeton, NJ.
Corben, C. J., Ingram, G. J., and Tyler, M. J. (1974). Gastric brooding: Unique form of parental care in an Australian frog. *Science (Washington, D. C.)* **186**, 946–947.
Crump, M. L. (1974). Reproductive strategies in a tropical anuran community. *Misc. Publ. Univ. Kansas Mus. Nat. Hist.* **61**, 1–68.
Crump, M. L. (1995). Parental care. *In* "Amphibian Biology" Vol. II (H. Heatwole and B. K. Sullivan, eds.), pp. 518–567. Surrey Beatty & Sons Pty Ltd, Chipping Norton NSW, Australia.
Dawkins, R., and Carlisle, T. R. (1976). Parental investment, mate desertion and a fallacy. *Nature (London)* **262**, 131–133.
del Pino, E. M. (1980). Morphology of the pouch and incubatory integument in marsupial frogs (Hylidae). *Copeia* **1980**, 10–17.
Duellman, W. E. (1985). Reproductive modes in anuran amphibians: Phylogenetic significance of adaptive strategies. *S. Afr. J. Sci.* **81**, 174–178.
Duellman, W. E. (1993). Amphibian species of the world: Additions and corrections. *Spec. Publ. Mus. Nat. Hist. Univ. Kansas* **21**, 1–372.
Duellman, W. E., and Gray, P. 1983. Developmental biology and systematics of the egg-brooding hylid frogs, genera *Flectonotus* and *Fritziana. Herpetologica* **39**, 333–359.
Duellman, W. E., and Hoogmoed, M. S. (1984). The taxonomy and phylogenetic relationships of the hylid frog genus *Stefania. Misc. Publ. Univ. Kansas Mus. Nat. Hist.* **75**, 1–39.
Duellman, W. E., and Maness, S. J. (1980). The reproductive behavior of some hylid marsupial frogs. *J. Herpetol.* **14**, 213–222.
Duellman, W. E., and Trueb, L. (1986). "Biology of Amphibians." McGraw-Hill Book Co., New York.
Durand, J., and Vandel, A. (1968). *Proteus:* an evolutionary relict. *Sci. J.* **1968**, 44–49.
Ehmann, H., and Swan, G. (1985). Reproduction and development in the marsupial frog, *Assa darlingtoni* (Leptodactylidae, Anura). *In* "Biology of Australasian Frogs and Reptiles" (G. Grigg, R. Shine, and H. Ehmann, eds.), pp. 279–285. Royal Zool. Society of New South Wales, Australia.
Ford, L. S., and Cannatella, D. C. (1993). The major clades of frogs. *Herpetol. Monogr.* **7**, 94–117.
Forester, D. C. (1978). Laboratory encounters between attending *Desmognathus ochrophaeus* (Amphibia, Urodela, Plethodontidae) females and potential predators. *J. Herpetol.* **12**, 537–541.
Forester, D. C. (1979a). Homing to the nest by female mountain dusky salamanders (*Desmognathus ochrophaeus*) with comments on the sensory modalities essential to clutch recognition. *Herpetologica* **35**, 330–335.

Forester, D. C. (1979b). The adaptiveness of parental care in *Desmognathus ochrophaeus* (Urodela: Plethodontidae). *Copeia* **1979,** 332–341.
Forester, D. C. (1981). Parental care in the salamander *Desmognathus ochrophaeus:* Female activity pattern and trophic behavior. *J. Herpetol.* **15,** 29–34.
Forester, D. C. (1983). Duration of the brooding period in the mountain dusky salamander (*Desmognathus ochrophaeus*) and its influence on aggression toward conspecifics. *Copeia* **1983,** 1098–1101.
Forester, D. C. (1984). Brooding behavior by the mountain dusky salamander: Can the female's presence reduce clutch desiccation? *Herpetologica* **40,** 105–109.
Formas, R., Pugin, E., and Jorquera, B. (1975). La identidad del batracio chileno *Heminectes rufus* Philippi, 1902. *Physis* (Buenos Aires) (*Sección C*) **34,** 147–157.
Goicoechea, O., Garrido, O., and Jorquera, B. (1986). Evidence for a trophic paternal-larval relationship in the frog *Rhinoderma darwinii. J. Herpetol.* **20,** 168–178.
Good, D. A., and Wake, D. B. (1992). Geographic variation and speciation in the torrent salamanders of the genus *Rhyacotriton* (Caudata: Rhyacotritonidae). *Univ. Calif. Publ. Zool.* **126,** 1–91.
Guillette, L. J., Jr. (1987). The evolution of viviparity in fishes, amphibians and reptiles: an endocrine approach. *In* "Hormones and Reproduction in Fishes, Amphibians, and Reptiles" (D. O. Norris and R. E. Jones, eds.), pp. 523–562. Plenum Publishing Corp., New York.
Haddad, C. F. B., and Martins, M. (1994). Four species of Brazilian poison frogs related to *Epipedobates pictus* (Dendrobatidae): Taxonomy and natural history observations. *Herpetologica* **50,** 282–295.
Harris, R. N., and Gill, D. E. (1980). Communal nesting, brooding behavior, and embryonic survival of the four-toed salamander *Hemidactylium scutatum. Herpetologica* **36,** 141–144.
Harvey, P. H., and Pagel, M. D. (1991). "The Comparative Method in Evolutionary Biology." Oxford Univ. Press, New York.
Heyer, W. R., and Crombie, R. I. (1979). Natural history notes on *Craspedoglossa stejnegeri* and *Thoropa petropolitana* (Amphibia: Salientia, Leptodactylidae). *J. Wash. Acad. Sci.* **69,** 17–20.
Highton, R., and Savage, T. (1961). Functions of the brooding behavior in the female redbacked salamander, *Plethodon cinereus. Copeia* **1961,** 95–98.
Hom, C. L. (1987). Reproductive ecology of female dusky salamanders, *Desmognathus fuscus* (Plethodontidae), in the southern Appalachians. *Copeia* **1987,** 768–777.
Inger, R. F., and Voris, H. K. (1988). Taxonomic status and reproductive biology of Bornean tadpole-carrying frogs. *Copeia* **1988,** 1060–1062.
Ingram, G. J., Anstis, M., and Corben, C. J. (1975). Observations on the Australian leptodactylid frog, *Assa darlingtoni. Herpetologica* **31,** 425–429.
Jackson, M. E., Scott, D. E., and Estes, R. A. (1989). Determinants of nest success in the marbled salamander (*Ambystoma opacum*). *Can. J. Zool.* **67,** 2277–2281.
Juncá, F. A., Altig, R., and Gascon, C. (1994). Breeding biology of *Colostethus stepheni,* a dendrobatid frog with a nontransported nidicolous tadpole. *Copeia* **1994,** 747–750.
Juterbock, J. E. (1987). The nesting behavior of the dusky salamander, *Desmognathus fuscus.* II. Nest site tenacity and disturbance. *Herpetologica* **43,** 361–368.
Kaplan, R. H., and Crump, M. L. (1978). The non-cost of brooding in *Ambystoma opacum. Copeia* **1978,** 99–103.
Kluge, A. G. (1981). The life history, social organization, and parental behavior of *Hyla rosenbergi* Boulenger, a nest-building gladiator frog. *Misc. Publ. Mus. Zool. Univ. Mich.* **160,** 1–170.

Kok, D., du Preez, L. H., and Channing, A. (1989). Channel construction by the African bullfrog: another anuran parental care strategy. *J. Herpetol.* **23**, 435–437.

Krzysik, A. J. (1980). Trophic aspects of brooding behavior in *Desmognathus fuscus fuscus*. *J. Herpetol.* **14**, 426–428.

Lamotte, M., and Lescure, J. (1977). Tendances adaptives a l'affranchissement de milieu aquatique chez les amphibiens anoures. *Terre Vie* **31**, 225–312.

Lannoo, M. J., Townsend, D. S., and Wassersug, R. J. (1987). Larval life in the leaves: Arboreal tadpole types, with special attention to the morphology, ecology, and behavior of the oophagous *Osteopilus brunneus* (Hylidae) larva. *Fieldiana Zool.* **38**, 1–31.

Lazarus, J., and Inglis, I. R. (1986). Shared and unshared parental investment, parent-offspring conflict and brood size. *Anim. Behav.* **34**, 1791–1804.

Lescure, J. (1979). Étude taxonomique et éco-éthologique d'un amphibien des petites Antilles *Leptodactylus fallax* Muller, 1926 (Leptodactylidae). *Bull. Mus. Nat. Hist. Nat. Paris* **1**, 757–774.

Lescure, J. (1984). Las larvas de Dendrobatidae II. *Reunión Iberamer. Cons. Zool. Vert.* 37–45.

McDiarmid, R. W. (1978). Evolution of parental care in frogs. *In* "The Development of Behavior: Comparative and Evolutionary Aspects" (G. M. Burghardt and M. Bekoff, eds.), pp. 127–147. Garland STPM Press, New York.

McDonald, K. R., and Tyler M. J. (1984). Evidence of gastric brooding in the Australian leptodactylid frog *Rheobatrachus vitellinus*. *Trans. R. Soc. South Aust.* **108**, 226.

Myers, C. W., and Daly, J. W. (1983). Dart-poison frogs. *Sci. Am.* **248**, 120–133.

Ng, M. Y., and Wilbur, H. M. (1995). The cost of brooding in *Plethodon cinereus*. *Herpetologica* **51**, 1–8.

Nussbaum, R. A. (1985). The evolution of parental care in salamanders. *Misc. Publ. Mus. Zool. Univ. Mich.* **169**, 1–50.

Nussbaum, R. A. (1987). Parental care and egg size in salamanders: An examination of the safe harbor hypothesis. *Res. Popul. Ecol.* **29**, 27–44.

Nussbaum, R. A., and Schultz, D. L. (1989). Coevolution of parental care and egg size. *Am. Nat.* **133**, 591–603.

Organ, J. A. (1960). Studies on the life history of the salamander, *Plethodon welleri*. *Copeia* **1960**, 287–297.

Organ, J. A. (1961). Studies on the local distribution, life history, and population dynamics of the salamander genus *Desmognathus* in Virginia. *Ecol. Monogr.* **31**, 189–220.

Perrone, M., Jr., and Zaret, T. M. (1979). Parental care patterns of fishes. *Am. Nat.* **113**, 351–361.

Pyburn, W. F. (1975). A new species of microhylid frog of the genus *Synapturanus* from southeastern Colombia. *Herpetologica* **31**, 439–443.

Rabb, G. B., and Snedigar, R. (1960). Observations on breeding and development of the Surinam toad, *Pipa pipa*. *Copeia* **1960**, 40–44.

Ridley, M. (1978). Paternal care. *Anim. Behav.* **26**, 904–932.

Rodríguez, L., and Myers, C. W. (1993). A new poison frog from Manu National Park, southeastern Peru (Dendrobatidae, *Epipedobates*). *Am. Mus. Novit.* **3068**, 1–15.

Rose, W. (1962). "The Reptiles and Amphibians of Southern Africa." Maskew Miller, Capetown, Africa.

Salthe, S. N. (1969). Reproductive modes and the numbers and sizes of ova in the urodeles. *Am. Midl. Nat.* **81**, 467–490.

Salthe, S. N., and Duellman, W. E. (1973). Quantitative constraints associated with reproductive mode in anurans. *In* "Evolutionary Biology of the Anurans" (J. L. Vial, ed.), pp. 229–249. University of Missouri Press, Columbia, MO.

Salthe, S. N., and Mecham, J. S. (1974). Reproductive and courtship patterns. *In* "Physiology of the Amphibia" Vol. II (B. Lofts, ed.), pp. 309–521. Academic Press, New York.

Scheel, J. J. (1970). Notes on the biology of the African tree-toad, *Nectophryne afra* Buchholz & Peters, 1875, (Bufonidae, Anura) from Fernando Poo. *Rev. Zool. Bot. Afr.* **81,** 225-236.
Shine, R. (1978). Propagule size and parental care: The "safe harbor" hypothesis. *J. Theor. Biol.* **75,** 417-424.
Shine, R. (1988). Parental care in reptiles. *In* "Biology of the Reptilia" Vol. 16 (C. Gans and R. B. Huey, eds.), pp. 275-329. Alan R. Liss, Inc., New York.
Shine, R. (1989). Alternative models for the evolution of offspring size. *Am. Nat.* **134,** 311-317.
Simon, M. P. (1983). The ecology of parental care in a terrestrial breeding frog from New Guinea. *Behav. Ecol. Sociobiol.* **14,** 61-67.
Summers, K. (1989). Sexual selection and intra-female competition in the green dart-poison frog, *Dendrobates auratus. Anim. Behav.* **37,** 797-805.
Thompson, R. L. (1992). Reproductive behaviors and larval natural history of the Jamaican brown frog, *Osteopilus brunneus* (Hylidae). Unpublished Masters Thesis, Univ. of Florida.
Tilley, S. G. (1972). Aspects of parental care and embryonic development in *Desmognathus ochrophaeus. Copeia* **1972,** 532-540.
Townsend, D. S. (1986). The costs of male parental care and its evolution in a neotropical frog. *Behav. Ecol. Sociobiol.* **19,** 187-195.
Townsend, D. S. (1996). Patterns of parental care in frogs of the genus *Eleutherodactylus. In* "Contributions to West Indian Herpetology: A Tribute to Albert Schwartz" (R. Powell and R. W. Henderson, eds.), pp. 229-239. Society for the study of Amphibians and Reptiles, Ithaca, NY. Contributions to *Herpetology* **12.**
Townsend, D. S., Stewart, M. M., Pough, F. H., and Brusard, P. F. (1981). Internal fertilization in an oviparous frog. *Science (Washington, D. C.)* **212,** 469-471.
Townsend, D. S., Stewart, M. M., and Pough, F. H. (1984). Male parental care and its adaptive significance in a neotropical frog. *Anim. Behav.* **32,** 421-431.
Trivers, R. L. (1972). Parental investment and sexual selection. *In* "Sexual Selection and the Descent of Man" (B. Campbell, ed.), pp. 136-179. Aldine Press, Chicago.
Tyler, M. J., and Carter, D. B. (1981). Oral birth of the young of the gastric brooding frog *Rheobatrachus silus. Anim. Behav.* **29,** 280-282.
Tyler, M. J., Shearman, D. J. C., Franco, R., O'Brien, P., Seamark, R. F., and Kelly, R. (1983). Inhibition of gastric acid secretion in the gastric brooding frog, *Rheobatrachus silus. Science (Washington, D. C.)* **220,** 609-610.
van Wijngaarden, R., and Bolaños, F. (1992). Parental care in *Dendrobates granuliferus* (Anura: Dendrobatidae), with a description of the tadpole. *J. Herpetol.* **26,** 102-105.
Vaz-Ferreira, R., and Gehrau, A. (1974). Proteccion de la prole en leptodactylidos. *Rev. Biol. Uruguay* **2,** 59-62.
Vaz-Ferreira, R., and Gehrau, A. (1975). Comportamiento epimeletico de la rana comun, *Leptodactylus ocellatus* (L.) (Amphibia, Leptodactylidae) 1. Atencion de la cria y actividades alimentaris y agresivas relacionadas. *Physis* (Buenos Aires) *(Sección B)* **34,** 1-14.
Wager, V. A. (1965). "The Frogs of South Africa." Purnell & Sons, Capetown, Africa.
Wake, M. H. (1977). The reproductive biology of caecilians: an evolutionary perspective. *In* "The Reproductive Biology of Amphibians" (D. H. Taylor and S. I. Guttman, eds.), pp. 73-101. Plenum Press, New York.
Wake, M. H. (1982). Diversity within a framework of constraints. Amphibian reproductive modes. *In* "Environmental Adaptation and Evolution" (D. Mossakowski and G. Roth, eds.), pp. 87-106. Gustav Fischer, New York.
Wake, M. H. (1986). Caecilians. *In* "The Encyclopedia of Reptiles and Amphibians" (T. R. Halliday and K. K. Adler, eds.), pp. 16-17. Equinox-Facts on File, Inc., New York.

Wells, K. D. (1980). Evidence for growth of tadpoles during parental transport in *Colostethus inguinalis. J. Herpetol.* **14,** 428–430.

Wells, K. D. (1981). Parental behavior of male and female frogs. *In* "Natural Selection and Social Behavior" (R. D. Alexander and D. W. Tinkle, eds.), pp. 184–197. Chiron Press, New York.

Wells, K. D., and Bard, K. M. (1988). Parental behavior of an aquatic-breeding tropical frog, *Leptodactylus bolivianus. J. Herpetol.* **22,** 361–364.

Weygoldt, P. (1980). Complex brood care and reproductive behavior in captive poison-arrow frogs, *Dendrobates pumilio* O. Schmidt. *Behav. Ecol. Sociobiol.* **7,** 329–332.

Weygoldt, P. (1987). Evolution of parental care in dart poison frogs (Amphibia: Anura: Dendrobatidae). *Z. Zool. Syst. Evolutionsforsch.* **25,** 51–67.

Williams, G. C. (1975). "Sex and Evolution." Princeton Univ. Press, Princeton, NJ.

Winkler, D. W. (1987). A general model for parental care. *Am. Nat.* **130,** 526–543.

Wittenberger, J. F. (1981). "Animal Social Behavior." Duxbury Press, Boston.

Zimmermann, H., and Zimmermann, E. (1984). Durch Nachzucht erhalten: Baumsteigerfrosche, *Dendrobates quinquevittatus* und *D. reticulatus. Aquarium Mag.* **1984,** 35–41.

# An Overview of Parental Care among the Reptilia

CARL GANS

DEPARTMENT OF BIOLOGY
UNIVERSITY OF MICHIGAN
ANN ARBOR, MICHIGAN 48109-1048

## I. INTRODUCTION

*Parental care* has been defined as "any kind of parental behavior that appears likely to increase the fitness of the parent's offspring" (Clutton-Brock, 1991). The term parental care is usually restricted to care expended after the deposition of eggs or live birth; otherwise the term might include such investments as devotion of energy to egg size or quality, or their retention until hatching. Other terms of significance to this topic are parental expenditure, absolute and relative, and parental investment, which takes into account the expenditure-associated reduction of capacity to invest in future offspring.

Obviously, successful parental care, no matter how it is defined, represents an adaptation of the organism, which is assumed to have a positive effect on the *fitness of the organism,* which may in turn be defined as the production of offspring that will also be fit. Such hypotheses of adaptation require test by comparative study. Adaptation can be defined in historical terms; then morphological, physiological, or behavioral traits can only be claimed to represent an adaptation if they arose at a time at which the then-current circumstances had favored their development. Consequently, proof of adaptation must document that the process leading to the new phenotype better matched the past environment (and one could argue that such a past adaptation would retain a status of adaptation, even if it no longer remained advantageous in the present). Obviously, this imposes major problems, even if possible at all (Leroi *et al.,* 1994). However, adaptation may also be defined in a current sense, making it more easily testable. The more operational definition, which is used more commonly in physiological studies, considers adaptive any phenotypic aspect of the parents that demonstrably increases the fitness of the offspring under current circumstances relative to other members that lack it.

Obviously, the decision whether a phenotypic aspect represents parental care would differ depending on the definition of adaptation. Here, I choose a contemporary and more operational definition of adaptation (Gans, 1988, 1993). Also, I choose to stress that parallel origins, although they may represent several kinds of investment in the offspring, represent stages of a continuum rather than discrete end states. Differences in investment patterns probably are due more to our limited data and understanding than to the presumptive genetic or morphological bases or costs to the animal.

Many aspects of reptilian reproductive patterns prove to be vagile among the vertebrates. Thus, the most heavily yolked or "telolecithal" eggs (such as seen in birds), viviparity, and internal fertilization have evolved and reversed repeatedly. Reversals complicate, and may even invalidate, the characterization of broad trends. Furthermore, the 7000 species of reptiles show dozens of modes that seem to enhance the fitness of their offspring, thereby providing a vast opportunity of testing the reality of these adaptations. Certain aspects of parental care, including reptilian patterns, have been reviewed extensively during the last decade (Packard and Packard, 1987; Shine, 1985, 1987; Somma, 1990). The preceding studies provide a most useful resource but also complicates the preparation of this brief summary. I have considered it more critical to transcend the tabulation of the many kinds of reptilian parental investments, and instead to document their interaction and to consider the ways in which the observed mechanisms might uniquely reflect the reptilian condition.

## II. The Reptilian Diversity

The Reptilia are paraphyletic; they do not include the descendant groups Mammalia and Aves. Hence the group of reptiles cannot be defined by derived characteristics that reptiles share without including their descendants, birds and mammals, in the definition. However, the extant species of reptiles represent three major clearly distinct and monophyletic lineages: (1) Testudines (turtles, tortoises, and terrapins), (2) lepidosaurians (sphenodontids and squamates, including snakes, lizards, and amphisbaenians), and (3) crocodylians. All of these share the derived key characteristic (synapomorphy) of the amniotic egg (or mode of reproduction), but have lost metamorphosis (a drastic change between the juvenile and adult phenotypes, which represents a set of plesiomorphies present in most fishes and amphibians), yet retained ectothermy (a symplesiomorphy). As shown in Fig. 1, extant reptiles represent a grade of similar evolutionary specialization that appears in several parallel evolutionary lineages (i.e., clades or evolutionary lines).

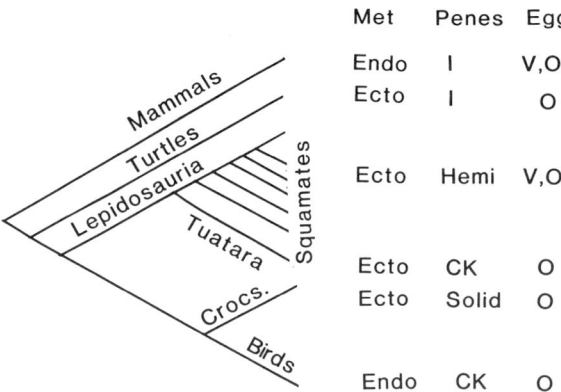

FIG. 1. Phylogenetic relationships among the amniotes showing the distribution of various reproductive structures and mechanisms. The first two columns indicate why it is impossible to define reptiles as a monophyletic category. The term Squamates bridges the lines for the various squamates (including lizards, snakes, and amphisbaenians). The codes for the columns (and items below these) stand for: Met, metabolism (Endo, endothermy; Ecto, ectothermy); Penes, sperm transfer system (CK, cloacal kiss; Hemi, double inflatable and reversing hemipenes; I, single inflatable; Solid, single rigid structure); Egg, offspring deposition pattern (O, oviparous; V, viviparous).

The extant Reptilia evolved from the Amphibia, and various fossil amphibians comprise a sister group for the entire assemblage. We assume that the amniotic egg (not observable in most fossils) represents a key synapomorphy for reptiles. However, because Recent amphibians, including salientians (frogs and toads), urodeles, and caecilians, are highly derived relative to the stem amphibians, early lineages that gave rise to the reptiles, we cannot equate their structural or functional features with the primitive states for the early reptiles. In the present context, Recent amphibians show enormous diversity in their patterns of egg deposition, larval state, and metamorphosis (Duellman and Trueb, 1986; Crump, this volume). Amphibian diversity may be seen as a by-product of the initial invasion of terrestrial environments, which led to an experimental radiation in modes of reproduction (Gans, 1990). The fundamental difference between the early stem amphibians and those surviving today discounts the latter from consideration as ancestral to reptiles.

This set of considerations forces us to recognize that the amniotic egg and amniotic developmental patterns are the first of the explicitly reptilian parental specializations. The amniotic condition allowed reproduction to proceed in the absence of standing water and may have proceeded in parallel with absolute reliance on aerial respiration, the development of a water-resistant integument, and a modified kidney system. This suite of characteristics allowed occupation of the more xeric parts of the terrestrial environment. Adults and offspring were freed from reliance on free water (rather than zones of temporally high humidity).

Parental investment is a key component to these changes, and explicitly involves the reconstruction of the reproductive tract, including the development of a new set of reproductive tubes. Even more important is the recognition that the amniotic egg was combined with the development of obligate internal fertilization, unique sperm transfer systems, and sperm storage in females. The three major lines of extant reptiles each show at least one distinct mode of sperm transfer (inflatable penis in Testudines, generally reversing hemipenes in squamates, and solid penis in crocodylians, Raynaud and Pieau, 1985), suggesting that we are observing the results of another experimental radiation.

### III. Structural Types of Parental Investment

Whereas any aspect of reproduction represents an investment on the part of the parents, we can begin the present analysis by noting the pattern of egg formation. Rather than starting with a cytoplasmic fertilized ovum, the precursor egg likely contained much yolk, which was provided in the ovary, whereas other materials are provided in the oviduct. Consequently, the telolecithal condition is the plesiomorphic state in reptiles (witness the development of isolecithal eggs (with minimal and evenly distributed yolk) as an apomorphy among individual lizards). Isolecithal eggs are only possible in species that already possess protoadaptive modifications (Gans, 1979) of the oviduct and some aspects of placentation. Whereas the initial demands on protoplacentation reflect only the provision of oxygen and water, as well as the discharge of carbon dioxide, advanced placentation involves the provision of many other substances, including amino acids, which must pass the uterine lining to supply the embryo (Packard and Packard, 1987; Blackburn, 1995). All of these were present in some oviparous species and could be co-opted as part of the process in which the ovum is no longer provided with a substantial mass of yolk.

The oviductal modifications allowing placentation and hence vivipary are spectacular; as spectacular is the vagility of the reptilian (particularly the lepidosaurian) system (Guillette, 1993). There are several genera that have both viviparous and oviparous species, and even species that have both viviparous and oviparous populations, suggesting that there may have been recent shifts; indeed, analysis of the system indicates that the pattern often reflects environmental conditions and with this that there may have been evolutionary reversals.

A second, but less far-reaching, suite of egg-forming specializations is required because the opportunities each for mating and insemination, for embryonic growth, and for egg laying or viviparous birth show temporal differences. Thus, the encounters (mating opportunities) among members of a species not only reflect the mean density of the species within the occupied region, but also the distribution of its members; the frequency of encounters is affected by clustering or patchiness of distribution. Seasonal restriction, such as that associated with particular hibernacula, or even of feeding or egg-deposition sites, hence allows an increase in encounter frequency.

The density affects the opportunity for initial mating of reproductively active individuals. The potential temporal differences between the male and female readiness (Crews, 1984) may be bridged by storage of sperm in the oviduct. Sperm storage may proceed as a function of preservative compounds in oviductal crypts, or by the occurrence of sperm storage cells analogous to the Sertoli cells of males (Saint Girons, 1985).

Both the size of the clutch and the number of offspring have a potential relation to fitness. Also, they tend to increase with the size of the adult. Adult size and growth in reptiles are complex topics (Andrews, 1982). Many reptiles do not show terminal growth, but rather grow incrementally even after becoming sexually mature, hence they produce larger clutches with age (Wilbur and Morin, 1987; Dunham *et al.*, 1987). Consequently, members of such species face a selective decision whenever the cost of the current parental effort may affect their survival and potential capacity to reproduce in subsequent seasons. The need to consider the future certainly affects the likelihood of parental defense of the brood by smaller species and by those that lack obvious defensive devices.

On the other hand, there is a limit on the mass (volume) of offspring. Very small arboreal species, particularly those that propel themselves among the leaves of the canopy by a saltatorial locomotor pattern, may encounter adhesive limitations here; thus the anoles lizards and some geckonids constrain their reproduction to a single young at a time (Andrews and Rand, 1974). Whenever the parents are capable of transporting a larger mass, the species encounters the potential decision between the production of many small and fewer larger eggs. The reverse effect is seen in burrowing squamates in which the offspring often lie in a single series on a single, rather than a bilateral, oviduct (Rajendran, 1985).

Sinervo and colleagues (Sinervo, 1993) have used experimental techniques on fence lizards to change the number of offspring per clutch and to manipulate the amount of yolk per egg. Both approaches show that the resulting offspring differ in their size and also in their performance. They also show that the effects show species and population specificity. For instance, miniaturized southern hatchlings have the same locomotor speed as northern hatchlings of the same size; however, they will differ in stamina. The analysis points to the possible influence of maternal effects on egg quality and the complications in deriving conclusions about the causes of observed differences.

## IV. Physiological Types of Parental Investment

Reptiles are able to conform to environmental circumstances. Indeed, since Benedict (1932), we have recognized that ectothermy facilitates conforming (Gans and Pough, 1982). However, we know that change of temperature affects metabolic rate and hence the cost of maintenance. Furthermore, the ability to conform does not imply the absence of hidden costs. Whereas the incubation temperature differs among species, changes in incubation time, besides affecting the sex of the offspring, may lead to other modifications. Examples are change of vertebral number (Fox *et al.*, 1961), growth pattern (Tousignant and Crews, 1995; Tousignant *et al.*, 1995), behavior (Flores and Crews, 1995; Flores *et al.*, 1994), endocrine physiology (Crews and Gans, 1992; Tousignant *et al.*, 1995), and brain organization (Crews *et al.*, 1996); also, it may have teratological effects (Vinegar, 1974; Branch and Patterson, 1975). Similar effects are seen in response to the modification of hydration (Packard and Packard, 1987).

The first aspect of the physiology underlying care then appears in the selection of the egg-deposition site, its patency (protection from collapse), depth, humidity conditions, and likely temperature regime. Here, as in some of the previously noted aspects, the parent has to predict future climatic (and microclimatic) conditions because these will affect the merit of different strategies. The ability to engage in successful predictions clearly is part of the parental investment. The cues and mechanisms by which the predictions are made is an obvious topic for future analysis.

The preparation of the egg chamber represents a further step. Certain crocodylians (i.e., *Crocodylus palustris*) not only build a simple egg-deposition pit, but equip the pit with a laterally branching side chamber, into which the relatively large eggs are lifted by their hind feet (C. Gans, personal observation). This lateral displacement of the egg mass reduces the risk of crushing once the excavated soil is replaced into the shaft during closure of the nest. The capacity of other crocodylians to assemble masses of decomposing vegetation and then to deposit their eggs in these has been supposed to maintain both a zone of elevated temperature and an enhanced humidity regime (Greer, 1971). Marine turtles discharge the contents of their bladder over the nest during closure (Obst, 1988); urination may affect the soil moisture in the next chamber, but has also been claimed to serve as a source of olfactory confusion, keeping potential predators from locating the egg chamber, and as an antifungal material. The parent may also use a modification of the egg shell to maintain water balance. Many species of geckos glue their eggs to vertical rock surfaces or the surfaces of tree trunks. Site selection affects the temperature to which the embryos will be exposed; however, the leathery egg shell must be able to conform to the general humidity regime, including wind effects on evaporation.

In the animals depositing clutches of multiple eggs, there is often an advantage to synchronization of the developmental process and the arrest of development has been referred to as egg diapause (Ewert, 1985). Diapause occurs in a highly complex mode in the testudines, in which it permits deposit of the clutch to be delayed by increasing the time from fertilization to hatching (Ewert, 1985); developmental arrest has been shown to allow the eggs of a clutch to survive weeks of flooding (Kennett *et al.,* 1993). The timing of hatching may then be adjusted to occur under circumstances that match the needs of the hatchlings, making it possible to change the period of development and allowing the animals to obtain feedback from climatic conditions, such as from shifting weather patterns. Whereas we are still dealing with morphological–developmental adaptations, these directly involve behavior affecting the time and place of egg deposition. This confirms that the parental investment is not easily divisible into morphological, physiological, and behavioral compartments.

It has been demonstrated that females of certain (but not all) boid snakes incubate their eggs, coiling around the mass and regularly twitching their muscles, thus raising the egg temperature (Vinegar *et al.,* 1970). Females of other snakes (such as king cobras) and some lizards coil around the egg mass providing a passive thermal barrier, and transporting heat and moisture whenever the adults enter after basking on the surface.

In a variety of species, the brooding or egg-associated adults eat or otherwise remove nonviable eggs or aborted ova (see Somma, 1990 for citations). Removal of nonviable materials has been claimed to protect the viable eggs from infection or from spread of fungal growth, which is commonly seen on eggs that died during attempts to incubate them in captivity. Besides controlling the spread of infection, the removal of nonviable materials also removes the source of odors that could otherwise attract predators.

The final kind of investment that deserves attention here is temperature-dependent (rather than chromosomal) sex determination (Raynaud and Pieau, 1985; Crews, 1984; Crews *et al.,* 1994). Sensitivity to the thermal condition is restricted to specific temporal moments during development. As different sex ratios are likely to be advantageous under distinct circumstances, environmental sex determination potentially represents another aspect for matching environmental changes; however, we know far more cases of the phenomenon than we have documentations of its actual advantage. Temperature-dependent sex determination clearly requires certain behaviors, as the adults must select places likely to be exposed to particular thermal regimes. It remains unclear whether the incubation temperature of the parents will affect their temperature preference for egg-laying sites.

## V. Behavioral Types of Parental Investment

In a number of species of lizards and snakes, the egg laying is communal and more than a single parent may be associated with the clutch (Shine, 1987). It is possible that some of the physiological merits of brooding are enhanced here; on the other hand, there may be a benefit in terms of deterring predators.

Whereas some lepidosaurians remain in the vicinity of the eggs, many others do not. Testudines tend to abandon their nest sites, although some marine species tear up the soil in the vicinity of the nest, complicating the task of predators in discovering the specific nesting site. Several marine turtles and some South American freshwater species, as well as the beach nesting lizard *Iguana,* show temporally and spatially restricted nesting patterns. Referred to as *arribadas,* waves of pregnant females reach a specific portion of the beach and each begins its excavation pattern, seemingly without consideration of the existence of earlier nests. Commonly, the later arrivals will displace and destroy the nests of earlier arrivals. The relative advantages of the several tactics during mass nesting remain to be worked out, although there are clear similarities to mast fruiting patterns or synchronous birthing of herd animals.

Some large crocodylians stay in the vicinity of the nesting site; the females may even lie on top of the mound. Whenever members of such a species engage in communal nesting along the water's edge as, for instance, do the Nile crocodiles, the efforts of adjacent individuals are likely synergistic as a predator deterrent.

The patterns of egg laying and protection grade into the issue of viviparity, which does not occur among Testudines and Crocodylia. Among the various lepidosaurs one sees a range of egg-retention regimens from the deposit of eggs at very early developmental stages to their retention to term (Shine, 1985; de Fraipont *et al.,* 1996). Animals with intermediate development likely retain the shell and embryonic membranes, whereas true live bearers never form these. The observed diversity has thus far precluded generalizations.

Even the live bearers rarely show particular attention to their offspring. Feeding of young has not been shown and Somma (1990) only cites a unique report in which a brooding female lizard avoided feeding until its young had eaten. However, there may be more subtle effects. Saint Girons (1992) notes that in northern zones in which the lizard population size is relatively small in comparison with that of the snakes, adult European vipers avoid predation on the local lizards. The adult vipers seemingly restrict themselves to feeding on mammals and reserve the lizards for the juveniles of their species that cannot ingest rodents. This system deserves further test on the reality of the switch in feeding preference.

Perhaps the most spectacular instance of reptilian parental care occurs in crocodylians, in which both members of the mating pair will respond to the call of the hatchlings, although the specificity of the response needs further study. The call is recognized by both sexes and even by members of other crocodylian species. However, it is unclear whether the response sequence pertains only to their own young. The hatchlings vocalize underground while emerging from the eggs, causing the adults to dig them up, and carry them to the water within the relatively enormous buccal pouch (Pooley, 1977). Not only are the young then washed, but they stay in association with the adults for a relatively brief period after which their response to large crocodylians reverses. Thereafter the juveniles then spread suddenly into small adjacent channels and may even dig themselves tunnels here. In view of the fact that crocodylians may be cannibalistic, there seems to be both an inhibition of cannibalism in the parents and an inhibition of a possible adult avoidance reaction in the neonates. There are obvious parallels to mouth-brooding fishes. Kin and offspring recognition reactions are topics deserving further study.

## VI. Opportunities for Study

The adaptations seen in a major group such as the lepidosaurians may represent a single suite reflecting the response to a particular situation. Indeed, we would expect to find either a unique solution throughout or a series of parallel modifications that solve a single problem. A single solution is clearly not what is observed here, although Testudines and Crocodylia each show a single pattern. In contrast, Lepidosauria exhibit diversity not just in solutions, but seemingly also in the situations addressed. This produces a spectacular opportunity for studying diversity in a complex of radiations.

Understanding such a situation clearly requires a taxonomic basis. Furthermore, one needs a catalog of the observable modifications, in this case of the many putative kinds of parental care. (Phenomena still need to be discovered and described; witness that an analysis of viviparity (de Fraipont *et al.,* 1996) notes that the database used differed significantly from all previous ones, including those assembled less than a decade ago.) The several sets of reproductive characters may then be studied relative to these cladograms, letting one compare the observed sequences and search for possible trends. Beyond discovering these trends, one should be able to determine the tasks that each of them addresses.

The multiplicity of possible trends must take into account the potential of reversals in the evolution of characteristics of the reproductive system, as well as of their frequency (de Fraipont *et al.,* 1996). After all, the existence of diversity is an indication of the absence of a unique solution, in the present case to a demand on a unique type of aid from the parents. Although some characteristics occur in clear association with phylogenetic patterns, others are more environmentally vagile. Fine-grained analysis of the phenomena is most likely to disclose such distinctions.

What is always of interest would be a way of evaluating the costs and benefits of the multiple reproductive solutions observed. Neither item is obvious. It is simple to evaluate the costs of parental investment in calories or in time, or in terms of an increased exposure to disease or predation. One may characterize benefits in terms of the number of offspring produced; as important is the relative success these offspring incur in their own reproduction. Most important is that the benefit probably should be expressed in terms of the capacity of the species not just for occupying and surviving in, but for reproducing in particular environments. Temperate and tropical zones, deserts, oceans, forests, and alpine environments, all provide distinct problems. It is critical that we develop mechanisms for evaluating and perhaps for quantifying these kinds of benefits.

## VII. Summary

Reptiles are a paraphyletic group and the three extant lines differ in structure and function. The care extended by the parent generation appears as structural aspects, as modification of physiological processes, and aspects of behavior. Whereas reptilian parental care represents evolution's first experiment with the amniotic egg, the condition in snakes, lizards, and their relatives shows much variation between egg laying and viviparity. There are many examples of egg guarding, moisture and heat provision, and antipredator devices. However, for most species we lack clear documentation of the magnitude of the cost to the parent and of the reality and magnitude of the putative selective advantage. In view of the fact that the several advantages are established by distinct environmental conditions, the provision of parental effort requires decisions that must be based on a predictive capacity by the parent, which in turn represents its investment.

### Acknowledgments

This account was prepared with assistance of a grant from the Leo Leeser Foundation. I thank David Crews, John Denver, Ernie Liner, Bradley Moon, and Charles T. Snowdon for comments on a draft manuscript.

### References

Andrews, R. (1982). Patterns of growth in reptiles. *In* Vol. 13 (C. Gans and F. H. Pough, eds.), "Biology of the Reptilia," pp. 273–320. Academic Press, London.
Andrews, R., and Rand, A. S. (1974). Reproductive effort in anoline lizards. *Ecology* **55**(6), 1317–1327.
Benedict, F. G. (1932). The physiology of large reptiles. Carnegie Institution of Washington, Publication 425.
Blackburn, D. G. (1995). Saltationist and punctuated equilibrium models for the evolution of viviparity and placentation. *J. Theoret. Biol.* **174**(2), 199–216.
Branch, W. R., and Patterson, R. W. (1975). Notes on the development of embryos of the African rock python, *Python sebae. J. Herpetol.* **9**, 243–248.
Clutton-Brock, T. H. (1991). "The Evolution of Parental Care." Princeton Univ. Press, Princeton, NJ.
Crews, D. (1984). Gamete production, sex hormone secretion and mating behavior uncoupled. *Horm. Behav.*, **18**, 22–28.
Crews, D., and Gans, C. (1992). The interaction of hormones, brain, and behavior; An emerging discipline in herpetology. *In* "Biology of the Reptilia," Vol. 18 (C. Gans and D. Crews, eds.), pp. 1–23. Univ. of Chicago Press, Chicago.
Crews, D., Bergeron, J. M., Flores, D., Bull, J. J., Skipper, J. K., Tousignant, A., and Wibbels, T. (1994). Temperature-dependent sex determination in reptiles: Proximate mechanisms, ultimate outcomes, and practical applications. *Dev. Genet.* **15**, 297–312.
Crews, D., Coomber, P., and Gonzales-Lima, F. (1996). Independent effects of gonadal sex, and incubation temperature on the volume and metabolic capacity of brain nuclei in the leopard gecko (*Eublepharis macularius*), a lizard with temperature-dependent sex determination. *J. Comp. Neurol.* accepted.
de Fraipont, M., Clobert, J., and Barbault, R. (1996). The evolution of oviparity with egg guarding and viviparity in lizards and snakes. *Evolution* **50**(1), 391–400.
Duellman, W. E., and Trueb, L. (1986). "Biology of Amphibians." McGraw-Hill Book Co., New York.
Dunham, A. E., Miles, D. B., and Reznick, R. N. (1987). Life history patterns in squamate reptiles. *In* "Biology of the Reptilia," Vol. 16 (C. Gans and R. B. Huey, eds.), pp. 441–522. A. Liss, New York; 1994 reprint Branta Books, Ann Arbor, MI.
Ewert, M. (1985). Embryology of turtles. *In* "Biology of the Reptilia," Vol. 14 (C. Gans, F. Billett, and P. F. A. Maderson, eds.), pp. 269–328. John Wiley & Sons, New York.

Flores, D. L., and Crews, D. (1995). Effects of hormonal manipulation on sociosexual behavior in adult female leopard geckos (*Eublepharis macularius*), a species with temperature-dependent sex determination. *Horm. Behav.* **29,** 458–473.

Flores, D. L., Tousignant, A., and Crews, D. (1994). Incubation temperature affects the behavior of adult leopard geckos (*Eublepharis macularius*). *Physiol. Behav.* **55,** 1067–1072.

Fox, W., Gordon, C., and Fox, M. H. (1961). Morphological effects of low temperatures during the embryonic development of the garter snake, *Thamnophis elegans. Zoologica (New York)* **46**(2), 57–71.

Gans, C. (1979). Momentarily excessive construction as the basis for protoadaptation. *Evolution* **33**(1), 227–233.

Gans, C. (1988). Adaptation and the form-function relation. *Am. Zool.* **28**(2), 681–697.

Gans, C. (1990). Adaptations and conflicts. *In* "Evolutionary Biology: Theory and Principles" Proc. of an Int. Symp., Plzen, 1988, pp. 23–31. Czechoslovak Acad. Sci., Praha, Czeck Republic.

Gans, C. (1993). On the merits of adequacy. *Am. J. Sci. Ser. A* **293,** 391–406.

Gans, C., and Pough, F. H. (1982). Physiological ecology: Its debt to reptilian studies, its value to students of reptiles. *In* "Biology of the Reptilia," Vol. 12 (C. Gans and F. H. Pough, eds.), pp. 1–13. Academic Press, London.

Greer, A. E., Jr. (1971). Crocodilian nesting habits and evolution. *Fauna* **2,** 20–28.

Guillette, L. J., Jr. (1993). The evolution of viviparity in lizards. *BioScience* **43**(11), 742–751.

Kennett, R., Georges, A., and Palmer-Allen, M. (1993). Early developmental arrest during immersion of eggs of a tropical freshwater turtle, *Chelodina rugosa* (Testudinata: Chelidae), from northern Australia. *Austr. J. Zool.,* **41,** 37–45.

Leroi, A. M., Rose, M. R., and Lauder, G. V. (1994). What does the comparative method reveal about adaptation? *Am. Nat.* **143**(3), 381–402.

Obst, F. J. (1988). "Turtles, Tortoises and Terrapins." St. Martin's Press, New York.

Packard, G. C., and Packard M. J. (1987). The physiological ecology of reptilian eggs and embryos. *In* "Biology of the Reptilia," Vol. 16 (C. Gans and R. B. Huey, eds.), pp. 523–605. A. Liss, New York; 1994 reprint Branta Books, Ann Arbor, MI.

Pooley, A. C. (1977). Nest opening response of the Nile crocodile *Crocodylus niloticus. J. Zool. (London)* **182**(1), 17–26.

Rajendran, M. V. (1985). "Studies in Uropeltid Snakes." Madurai Karamaraj Univ., Tamil Nadu, India.

Raynaud, A., and Pieau, C. (1985). Embryonic development of the genital system. *In* "Biology of the Reptilia," Vol. 15 (C. Gans and F. Billett, eds.), pp. 149–300. John Wiley & Sons, New York.

Saint Girons, H. (1985). Comparative data on lepidosaurian reproduction and some time tables. *In* "Biology of the Reptilia," Vol. 15 (C. Gans and F. Billett, eds.) pp. 35–58. John Wiley & Sons, New York.

Saint Girons, H. (1992). Strategies reproductrices des Vipéridae dans les zones temperées fraiches et froides. *Bull. Soc. Zool. France* **117**(3), 267–278.

Shine, R. (1985). The evolution of viviparity in reptiles: An ecological analysis. *In* "Biology of the Reptilia," Vol. 15 (C. Gans and F. Billett, eds.), pp. 605–694. John Wiley & Sons, New York.

Shine, R. (1987). Parental care in reptiles. *In* "Biology of the Reptilia," Vol. 16 (C. Gans and R. B. Huey, eds.), pp. 275–329. A. Liss, New York; 1994 reprint Branta Books, Ann Arbor, MI.

Sinervo, B. (1993). The effect of offspring size on physiology and life history. *BioScience* **43**(3), 210–218.

Somma, Louis A. (1990). A categorization and bibliographic survey of parental behavior in lepidosaurian reptiles. *Smithson. Herpetol. Infor. Serv.* **81,** 1–53.

Tousignant, A., and Crews, D. (1995). Incubation temperature and gonadal sex affect growth and physiology in the leopard gecko (*Eublepharis macularius*), a lizard with temperature/dependent sex determination. *J. Morphol.* **224,** 159–170.

Tousignant, A., Viets, B., Flores, D., and Crews, D. (1995). Ontogenetic and social factors affecting the endocrinology and timing of reproduction in the female leopard gecko (*Eublepharis macularius*). *Horm. Behav.* **29,** 458–473.

Vinegar, A. (1974). Evolutionary implications of temperature induced anomalies of development in snake embryos. *Herpetologica* **30**(1), 72–74.

Vinegar, A., Hutchison, V. H., and Dowling, H. G. (1970). Metabolism, energetics, and thermoregulation during brooding of snakes of the genus *Python* (Reptilia, Boidae). *Zoologica* **55**(2), 19–48.

Wilbur, H. M., and Morin, P. J. (1987). Life history evolution in turtles. *In* "Biology of the Reptilia," Vol. 16 (C. Gans and R. B. Huey, eds.), pp. 387–439. A. Liss, New York; 1994 reprint Branta Books, Ann Arbor, MI).

# PART II

# ENDOCRINE, NEURAL, AND BEHAVIORAL FACTORS GOVERNING PARENTAL CARE AMONG MAMMALS AND BIRDS

# Neural and Hormonal Control of Parental Behavior in Birds

JOHN D. BUNTIN

DEPARTMENT OF BIOLOGICAL SCIENCES
UNIVERSITY OF WISCONSIN-MILWAUKEE
MILWAUKEE, WISCONSIN 53201

## I. Introduction

Birds exhibit a wide spectrum of parental care patterns ranging from brood parasitism, in which no parental behavior is displayed, to full-time attentiveness to the eggs and young by one or both parents. More than 99% of avian species exhibit some form of parental care, but the type and amount of care that is delivered and the nature and extent of male and female contribution to the parental effort varies widely across avian taxa. These differences correlate with mating system and with the mode of development of the young, although several other factors, such as habitat, diet, and clutch characteristics, may also be linked to parental behavior variation (Silver et al., 1985). As this article documents, the physiological underpinnings of this diversity in parental behavior are just beginning to be explored, and relatively few species have been systematically investigated. Accordingly, there is as yet no clear understanding of how diversity in avian parental care strategies arises from differences in underlying physiological mechanisms.

Galliform and columbiform species dominate the experimental literature on physiological regulation of parental behavior in birds, but both of these groups are rather atypical representatives of the Class Aves. Galliform birds are precocial at birth, and parental care in most members of this group is delivered exclusively by the female (Kendeigh, 1952). In contrast, 85–90% of extant avian species exhibit some form of biparental care (Kendeigh, 1952) and more than two-thirds of avian subfamilies rear altricial young (Silver et al., 1985). While columbiform birds are more typical in these respects, the parental care system that has evolved in this avian group is highly specialized, as pigeons and doves are unusual in their ability to manufacture food for the young in the form of crop "milk." Although investigations of parental behavior regulation in more typical avian species are urgently needed, studies of avian representatives with atypical or specialized parental care patterns are also advantageous, since similarities in the hormonal and neural determinants of parental behavior across such groups could suggest some physiological characteristics of avian parental care that are shared among a wide range of avian species.

In an effort to identify both differences and similarities in the physiology of parental care across various avian species, the neural and hormonal mechanisms that regulate parental care expression in galliform and columbiform species will be discussed in detail in this review, and attempts will be made where possible to relate the evidence to that obtained in other avian species that have received less experimental attention. For purposes of this discussion, *parental behavior* will be defined as those activities directed toward care of eggs or young. However, it is important to emphasize that these behaviors arise from antecedent events. At the risk of minimizing the importance of physiological and stimulus continuity in promoting transitions from one form of parental activity to another during the breeding cycle, the neural and hormonal regulation of parental activities exhibited prior to hatching (i.e., incubation behavior) will be discussed separately from those occurring after hatching (brooding and feeding the young). Interposed between these two sections will be a discussion of the physiological determinants of nest defense behavior, which is displayed both before and after hatching in many species.

## II. Incubation Behavior

### A. Hormonal Mechanisms

More than 60 years ago, Leinhart (1927) observed that nonincubating fowl could be induced to sit on eggs by injecting them with serum from incubating hens. Subsequent studies by Riddle *et al.* (1935) implicated the pituitary hormone prolactin as a key humoral agent in promoting this sitting response, and later measurements of prolactin in blood supported this conclusion. Nevertheless, studies on other birds soon revealed that the relationship between prolactin secretion and incubation behavior is not uniform across species and often involves complex interactions with other hormones.

#### 1. Fluctuations in Plasma Prolactin during Incubation: Correlations with Sitting Activity

Measurements of plasma prolactin during the breeding cycle suggest that prolactin is somehow linked to incubation activity in a variety of avian taxa. In avian species in which incubation duties are shared extensively between the sexes, it is generally the case that both members of the breeding pair show a substantial increase in plasma prolactin concentration during the incubation period. Species that conform to this pattern include the ring dove (*Streptopelia risoria;* Goldsmith *et al.*, 1981; Cheng and Burke, 1983; Ramsey *et al.*, 1985; Lea *et al.*, 1986); cockatiel (*Nymphicus hollandicus;* Myers *et al.*, 1989); semipalmated sandpiper (*Calidris pusilla;* Gratto-Trevor *et al.*, 1990); Bengalese finch (*Lonchura striata;* Seiler *et al.*, 1992); Australian black swan (*Cygnus atratus;* Goldsmith, 1982b); macaroni and gentoo penguin (*Eudyptes chrysolophus* and *Pygoscelis papua,* respectively; Williams and Sharp, 1993); king penguin (*Aptenodytes patagonicus;* Cherel *et al.*, 1994); wandering, grey-headed, and black-browed albatross (*Diomedea exultans, D. chrysostoma,* and *D. melanophris,* respectively; Hector and Goldsmith, 1985); and Cape gannet (*Sula capensis;* Hall, 1986). Conversely, in species in which incubation is predominantly or exclusively performed by one sex, plasma prolactin levels tend to be substantially higher in the incubating sex. This pattern is exhibited by several species in which the female normally sits, such as the mallard duck (*Anas platyrhynchos;* Goldsmith and Williams, 1980), bar-headed goose (*Anser indicus;* Dittami, 1981), canary (*Serinus canaria;* Goldsmith, 1982a), and white-crowned sparrow (*Zonotrichia leucophrys;* Hiatt *et al.*, 1987), as well as species such as Wilson's phalarope (*Phalaropus tricolor;* Oring *et al.*, 1988), red-necked phalarope (*Phalaropus lobatus;* Gratto-Trevor *et al.*, 1990), and spotted sandpiper (*Actitis macularia;* Oring *et al.*, 1986a), in which males assume most or all of the incubation duties. Documented reductions in plasma prolactin following interruptions in incubation (El Halawani *et al.*, 1980a; Lea and Sharp, 1989; Ramsey *et al.*, 1985; Hall and Goldsmith, 1983) and the subsequent increase in plasma prolactin following its reinstatement (El Halawani *et al.*, 1980a; Goldsmith *et al.*, 1984; Sharp *et al.*, 1988) also suggest an association between sitting activity and elevated prolactin secretion.

Although incubation activity is typically accompanied by high titers of plasma prolactin, the relationship between these two events is complex. For example, elevated plasma prolactin levels persist during periods off the nest in several species in which incubation is shared between the breeding partners (ring dove: Goldsmith *et al.*, 1981; Lea *et al.*, 1986; Cape gannet: Hall, 1986), even when sitting recesses last several days (Hector and Goldsmith, 1985). Furthermore, sex differences in prolactin secretion patterns do not always correlate well with sex differences in sitting activity. For example, plasma prolactin concentrations in female wandering albatross are significantly higher than those of their male partners during incubation despite extensive sharing of incubation, as characterized by protracted incubation shifts by both sexes (Hector and Goldsmith, 1985). In the European starling (*Sturnus vulgaris*), males paired with one female (monogamous males) typically share in incubation, while those paired with several females (polygynous males) usually do not (Pinxten *et al.*, 1993). Nevertheless, plasma prolactin levels are elevated in males while their breeding partners are sitting on eggs, even if the males take no part in incubation behavior (Dawson and Goldsmith, 1982). This pattern has also been reported in nonincubating males of a variety of other species (lesser snow goose, *Anser caerulescens caerulescens;* Campbell *et al.*, 1981; song sparrow, *Melospiza melodia;* Wingfield and Goldsmith, 1990; white-crowned sparrow, Hiatt *et al.*, 1987; pied flycatcher, *Ficedula hypoleuca,* Silverin and Goldsmith, 1983). In some cases, the increase in plasma prolactin in the nonincubating male has been attributed to his participation in other parental activities, such as provisioning the incubating female, or as a prelude to later parental contributions, such as feeding the nestlings (Silverin and Goldsmith, 1983; Dawson and Goldsmith, 1982). However, factors unrelated to breeding activity may also contribute to these changes. For example, increasing daylength, which accompanies breeding activity in temperate zones, stimulates increased plasma prolactin concentrations in males of a variety of passerine species (starling: Ebling *et al.*, 1982; Dawson and Goldsmith, 1983; white-crowned sparrow: Hiatt *et al.*, 1987; song sparrow: Wingfield and Goldsmith, 1990). Photoperiodic cues may also be responsible for the increase in plasma prolactin that reportedly occurs in brown-headed cowbirds (*Molothrus ater*) of both sexes during the breeding season (Dufty *et al.*, 1987). The elevation in prolactin concentration in breeding cowbirds is particularly interesting, as these birds are brood parasites that exhibit no incubation behavior or care of young. Nest removal and replacement studies also raise questions about the nature of the linkage between prolactin and sitting activity, since an interest in sitting can persist for 1–3 days in birds receiving anti-prolactin antiserum (bantam hen: Lea *et al.*, 1981) or in birds in which circulating prolactin levels have declined significantly as a result of nest deprivation (canary: Goldsmith *et al.*, 1984; ring dove: Ramsey *et al.*, 1985; Lea and Sharp, 1989). Furthermore, the resumption of sitting activity following a period of nest deprivation is not always preceded by or associated with an increase in plasma prolactin concentration (ring dove: Lea and Sharp, 1989; turkey: El Halawani *et al.*, 1980a,b).

Collectively, these findings suggest that a variety of social, environmental, and physiological factors act to modulate the relationship between prolactin and sitting behavior during the normal breeding cycle. Such complexities underscore the need for experimental investigations to help parcel out how much of the apparent correlation between prolactin levels and incubation behavior is due to sitting-induced prolactin release, to prolactin-induced sitting activity, and to the presence of other temporally coincident factors that influence one or both of these events.

## 2. Direct Evidence for Prolactin and Steroid Hormone Involvement in Incubation Behavior

*a. Galliformes.* The fact that plasma prolactin concentrations increase in several galliform species during egg laying and the onset of incubation (Etches *et al.*, 1979; Lea *et al.*, 1982; Burke and Dennison, 1980) suggests that prolactin may be responsible for the initiation of sitting in this avian order (Fig. 1). In early studies on domestic fowl (*Gallus domesticus*), sitting behavior was reportedly induced by systemic (Riddle *et al.*, 1935; Saeki and Tanabe, 1955) or intracranial (Opel, 1971) administration of mammalian prolactin preparations. However, the fact that hens must be in laying condition in order for prolactin to be successful in this regard suggested that the efficacy of prolactin may depend on its synergistic interactions with gonadal steroids. This was later confirmed in a definitive study by El Halawani *et al.* (1986), who reported that the hormonal requirements for onset of nesting activity and for full induction of incubation behavior in ovariectomized turkeys (*Meleagris gallopavo*) were quite specific and paralleled normal

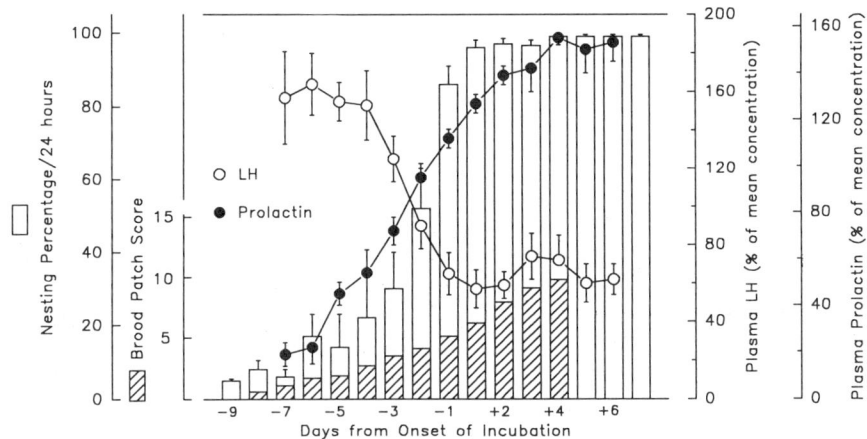

FIG. 1. Changes in plasma prolactin, plasma luteinizing hormone (LH), percent time spent on the nest, and brood patch development in bantam hens ($n = 5$) in relation to incubation onset. Brood patch development was recorded on a 15-point scale as described by Etches *et al.* (1979). Plasma prolactin and LH values were converted to a percentage of the mean hormone concentration for each bird before group means were computed. Values shown are group mean ± SEM (The figure is modified from Lea *et al.*, 1981 with the permission of the *Journal of Endocrinology Ltd.*)

changes in hormone secretion during the laying period. While neither estradiol, progesterone, or ovine prolactin alone effectively induced the turkey hens to occupy the nest box or to display incubation behavior, initial nesting activity was induced by a period of estrogen priming followed by progesterone. These results reinforced earlier findings of estrogen and progesterone involvement in nesting onset in ovariectomized chickens (Wood-Gush and Gilbert, 1973). Sustained nesting (i.e., full incubation), by contrast, required the addition of prolactin to the steroid regimen. Despite the effectiveness of this treatment protocol, there are apparently crucial aspects of the hormonal milieu necessary for incubation induction that remain to be identified, since Opel and Proudman (1980) reported that prolactin injections failed to induce incubation in intact chickens or turkeys that were actively laying eggs. Maintenance of elevated plasma prolactin levels over a prolonged period may be one important requirement for prolactin-induced incubation, as the El Halawani group employed polyvinylpyrrolidone, which extends the half-life of injected peptides, whereas Opel and Proudman (1980) used a conventional saline vehicle.

As Lea (1987) has pointed out, this two-stage process of incubation induction may be particularly well suited to galliform species, which lay a large clutch of eggs over an extended period before functional incubation begins. As gonadal steroids must be elevated throughout the extended egg-laying period to support ovulation, they cannot directly signal incubation onset, although it would be advantageous for these hormones to promote the initial stage of nesting activity that is associated with oviposition. In contrast, the subsequent increase in prolactin secretion, which is stimulated by rising gonadal steroids and steroid-facilitated nesting activity (El Halawani *et al.*, 1983; Book *et al.*, 1991), would be an effective stimulus for persistent nest occupation and full incubation onset. The effectiveness of estrogen and progesterone in promoting nest occupation in ovariectomized chickens, and the close association between prolactin secretion with nesting activity in bantam hens during the egg-laying period (Lea *et al.*, 1981), imply that the physiological mechanisms for the induction of incubation in chickens are similar to those in turkeys. However, whether incubation induction in other galliform species conforms to this two-stage process remains unclear due to a lack of systematic comparative studies. In Japanese quail (*Coturnix coturnix japonica*) housed under short days, combined estrogen and prolactin treatment reportedly failed to induce incubation behavior, although the effectiveness of a treatment regimen consisting of estradiol, progesterone, and prolactin administration was not investigated (Hohn, 1981).

Studies generally reinforce the view that prolactin plays an important role in incubation onset in galliform birds. In bantam hens, a reduction in the incidence of incubation behavior or a delay in its onset was observed following active immunization against chicken prolactin (March *et al.*, 1994). In laying turkeys, intracerebroventricular (ICV) infusions of ovine prolactin effectively induce incubation behavior (Youngren *et al.*, 1991; Pitts *et al.*, 1994) and active immunization against vasoactive intestinal polypeptide (VIP), a potent prolactin-releasing neuropeptide in birds (El Halawani *et al.*, 1990; Macnamee *et al.*, 1986; Lea and Vowles, 1986), strongly inhibits incubation onset in this species (El Halawani *et al.*, 1995). Contrary to expectations, however, Pitts *et al.* (1994) reported that ICV-infused porcine VIP failed to mimic the facilitatory action of ICV-infused prolactin on incubation onset in laying turkeys. At present, this inconsistency remains unexplained.

Plasma levels of gonadotropins (Fig. 1) and gonadal steroids decline at the end of egg laying and remain at low concentrations for the remainder of the incubation and posthatching periods (Sharp *et al.*, 1979, 1988; Lea *et al.*, 1981; Zadworny *et al.*, 1988). These changes are likely to be due in part to rising prolactin concentrations in blood (Lea *et al.*, 1981; Camper and Burke, 1977). The low plasma levels of gonadal steroids that are present during incubation are apparently unnecessary for incubation maintenance, since sitting behavior and elevated titers of prolactin persist in incubating bantam hens after ovariectomy (Sharp *et al.*, 1986) and in incubating, ovariectomized turkey hens after exogenous hormone treatments are discontinued (El Halawani *et al.*, 1986). At the same time, sitting behavior and prolactin secretion appear to be mutually reinforcing during the incubation period, with prolactin sustaining sitting, and sitting activity in turn promoting further prolactin release. This conclusion is supported by the decline in plasma prolactin levels that follows nest removal and the increase that typically accompanies reinstatement of sitting when the nest and eggs are returned (El Halawani *et al.*, 1980a; Sharp *et al.*, 1988). Furthermore, exogenous prolactin maintains readiness to incubate during an extended (72 h) period of nest deprivation in bantam hens (Sharp *et al.*, 1988). In addition, injections of antibodies raised against the prolactin-releasing peptide VIP resulted in nest abandonment in all incubating bantam hens that were given this treatment, while, in contrast, no significant inhibitory effects were observed in immunized females given coinjections of ovine prolactin (Sharp *et al.*, 1989). Collectively, these results provide strong evidence for a role of prolactin in incubation maintenance in galliform species. However, the mechanisms by which prolactin influences sitting behavior have not been well characterized. Richard-Yris *et al.* (1995) noted that incubating hens that were nest deprived for 72 h were more likely to resume incubation when the nest was returned if they had been exposed to chicks during the period of nest deprivation. While exposure to chicks did not prevent the characteristic decline in prolactin secretion that resulted from nest deprivation, it did prevent the rise in plasma luteinizing hormone (LH) and estradiol that normally accompanies this manipulation. These results, together with evidence that normal incubation or prolactin-induced incubation is disrupted in domestic hens or turkey hens given high doses of gonadal steroids (Collias, 1950; van Tienhoven, 1958), suggest that part of the prolactin-induced facilitation could be an indirect result of the hormone's antigonadal (Camper and Burke, 1977) or antigonadotropic action (Opel and Proudman, 1980; Lea *et al.*, 1981).

*b. Columbiformes.* In contrast to the pattern observed in Galliformes, the onset of incubation behavior in the best-studied columbiform species, the ring dove, is not accompanied by a marked and sustained increase in plasma prolactin levels. In fact, a detectable increase in prolactin levels in the blood is not observed until the birds have been sitting on eggs for 5–7 days (Goldsmith *et al.*, 1981; Lea *et al.*, 1986). Doves also differ from galliform species in that prolactin concentrations during the second half of incubation increase to reach peak levels during the early posthatching period. This increase is observed in both sexes, which share in the responsibilities of incubating eggs and feeding the young. The incubation phase increase in plasma prolactin is responsible for and closely parallels the marked proliferation of the crop sac and the production of crop milk, which is regurgitated by the parents to the newly hatched young. Because the extensive crop sac development and crop milk formation that is seen in pigeons and doves is unique to this avian group, and because maximal prolactin-induced crop sac growth and crop milk production must be closely synchronized with hatching in order to be effective, it is tempting to hypothesize that the prolactin secretion profile that underlies these changes during incubation is also unique. While this may be true in the strict sense, there are features of incubation-phase prolactin changes in other species that resemble the dove pattern (Section II,A,2,c, which follows).

In female ring doves, circulating estradiol and progesterone increase during the prelaying period (Korenbrot *et al.*, 1974; Silver *et al.*, 1974) in response to male courtship activity and the stimulatory effects that such activity has on the female's own behavior (Cheng, 1992). Studies on ovariectomized female doves indicate that synergistic interactions between these two hormones are sufficient to promote courtship behavior, nest-oriented activities, and onset of incubation (Cheng and Silver, 1975). Nevertheless, a modest and transient increase in plasma prolactin levels has been reported in intact females during ovulation and egg laying when functional incubation begins (Cheng and Burke, 1983; Lea *et al.*, 1986). At present it is difficult to evaluate the possibility that this periovulatory elevation in prolactin facilitates incubation onset, since the incubation-promoting effects of combined estrogen, progesterone, and prolactin treatment have yet to be characterized. Nevertheless, this possibility is unlikely for several reasons. First, most studies indicate that, apart from the brief periovulatory surge of prolactin, sustained elevations in circulating prolactin above baseline levels are not seen in ring doves of either sex until several days into the incubation period (Goldsmith *et al.,* 1981; Lea *et al.,* 1986). Second, Lehrman *et al.* (1961) demonstrated that doves will sit on eggs if they are given to them during the period of nest-building activity prior to egg laying (and presumably prior to the periovulatory prolactin surge). Third, the work of Lehrman (1958) and Lehrman and Brody (1961) clearly indicate that progesterone is a more effective agent than prolactin in inducing incubation behavior in intact, nonbreeding doves. Collectively, these data suggest important differences between Columbiformes and Galliformes in the hormonal determinants of incubation onset. Lea (1987) has suggested that the limited clutch size (two) and short laying interval in columbiform birds, and the abbreviated period of elevated gonadal steroid levels that is associated with these features, permit a one-stage process of incubation onset in the female. In this view, incubation is induced in the female by the same changes in gonadal steroids that are necessary for nest building, ovulation, and nest attachment associated with successful oviposition. Whether this mode of incubation induction applies to other birds laying small clutches remains to be evaluated, but the available evidence suggests that it may not. Studies in the black-browed albatross and the grey-headed albatross, which lay only one egg, suggest some role for prolactin in incubation onset, as plasma prolactin levels begin to rise prior to egg laying in these species (Hector and Goldsmith, 1985).

The physiological mechanisms responsible for the induction of incubation behavior in male doves may be more complicated, as there is no indication that progesterone titers in blood rise at the onset of the incubation period (Silver *et al.,* 1974; Lea *et al.,* 1986). In addition, the onset of incubation in breeding male doves is not disrupted by the apparent suppression of adrenal progesterone secretion (Silver and Buntin, 1973) coupled with removal of gonadal progesterone (Silver and Feder, 1973), although the protocols used in these studies did not conclusively rule out the possibility that facilitatory effects of progesterone on sitting readiness had already occurred before the experimental manipulations were imposed (Cheng, 1975). Stimuli associated with the nest and the interactions with the mate that occur prior to egg laying are clearly important determinants of incubation onset in male doves (Silver *et al.,* 1973). Nevertheless, long-term castrates show little or no incubation when paired with suitable breeding partners (Patel, 1936; Cheng, 1975; Ramos and Silver, 1992). This suggests that hormonal state is also an important determinant, although it is not clear if hormones do more than simply promote behavioral interactions between mates that in turn generate incubation-promoting stimuli. Cheng (1975) has suggested that hormonal and environmental cues act in complementary fashion to promote the normal onset of incubation in this species, such that effective hormonal stimulation can substitute for inadequacies in situational cues. Supporting this view are several reports that progesterone stimulates a rapid onset of incubation behavior in male doves that were housed in visual isolation from conspecifics during the progesterone treatment period and subsequently were provided with a nest and eggs in the presence of a female partner (Lehrman, 1958; Stern and Lehrman, 1969; Stern, 1974; Cheng, 1975; Komisaruk, 1967). Progesterone's efficacy in this regard is markedly enhanced by testosterone or estradiol (Stern and Lehrman, 1969; Stern, 1974) and by prior experience with nesting (Michel and Moore, 1985). Interestingly, a later study revealed that the concentration of progestin receptors in the anterior and posterior hypothalamus of male doves are significantly increased by testosterone (Balthazart et al., 1980). Since plasma testosterone levels increase in the male during the courtship phase of the breeding cycle (Feder et al., 1977), these results leave open the possibility that the male brain becomes more sensitive to progesterone as the prelaying period progresses and it is this change, rather than an elevation in plasma progesterone concentration, that normally promotes the induction of sitting behavior.

While there is little evidence that prolactin promotes the onset of incubation in ring doves, the fact that prolactin levels increase in blood during the last half of the incubation period suggests a possible role of the hormone in incubation maintenance. In support of this idea, Lehrman and Brody (1964) observed that a majority of male and female doves with previous breeding experience maintained their interest in sitting on eggs during a prolonged (12 day) period of nest deprivation. Janik and Buntin (1985) later confirmed these findings, although the effect of prolactin was not as robust as that reported in the earlier study and the number of birds showing incubation behavior following nest return did not increase uniformly with the dose of prolactin administered. This raised questions about the degree to which prolactin acts directly on the neural circuitry underlying nest attentiveness, and these concerns were reinforced by the subsequent finding the ICV injections of prolactin failed to mimic the effects of subcutaneous injections on readiness to incubate (Buntin and Tesch, 1985). Together, these results suggest that prolactin acts indirectly to maintain incubation readiness during periods of nest deprivation, although the peripheral target tissues and specific modes of prolactin action that are presumed to mediate this response remain undefined.

The question of whether prolactin maintains ongoing incubation behavior in ring doves has received less experimental attention than has the related but separate question of whether prolactin maintains an interest in sitting during a period of nest deprivation. Administration of ovine prolactin significantly prolongs the period during which ring dove pairs will sit on infertile eggs. However, the extension of incubation is limited and corresponds to the period that normal breeding pairs continue to brood their nestlings after hatching (Lea *et al.*, 1986). In normal pairs, this posthatching period of sitting activity is presumably dependent upon the prolactin released in response to stimulation from the newly hatched young (E. W. Hansen, 1966; Buntin, 1979; Lea and Sharp, 1991). The inability of exogenous prolactin to sustain incubation indefinitely in birds sitting on infertile eggs could simply reflect an immunological response to the prolonged administration of a mammalian prolactin preparation or, alternatively, a spontaneous change in sensitivity to prolactin. The latter possibility resembles the endogenous timing mechanism that has been invoked to explain the observation that the total length of the sitting period, and the period over which plasma prolactin remains elevated, is surprisingly constant in doves sitting on infertile eggs (Allen and Erickson, 1982; Vowles and Lea, 1986). Interestingly, photoperiodic manipulations further implicate a circadian component to the putative clock that underlies this response (Vowles and Lea, 1986). Studies on the pied flycatcher (Silverin and Goldsmith, 1984) and the grey-headed and black-browed albatross (Hector and Goldsmith, 1985) also suggest the existence of an endogenous timing mechanism controlling the duration of prolactin secretion and incubation behavior. In other species, however, the incubation period can be extended considerably if hatching is delayed, which suggests that this mechanism is absent (Wilson's phalarope: Oring *et al.*, 1988; bantam hen: Lea and Sharp, 1982; canary: Goldsmith, 1982a).

Unlike the situation in bantam hens (Sharp *et al.*, 1988), turkey hens (Book *et al.*, 1991), and female domestic mallard ducks (Hall and Goldsmith, 1983; Hall, 1987), in which elevated prolactin levels depend upon sustained contact with the nest and eggs, plasma prolactin levels normally remain elevated in incubating ring doves of both sexes during incubation recesses of several hours (Goldsmith *et al.*, 1981; Lea *et al.*, 1986). This implies differences in the stimulus requirements for prolactin release during incubation in Galliformes and Anseriformes on the one hand and Columbiformes on the other. There is convincing evidence that visual stimuli from the incubating partner sustain prolactin release during incubation in pigeons and doves of both sexes, although prior experience with incubation is an important requirement for these cues to be effective (Patel, 1936; Friedman and Lehrman, 1968; Michel and Moore, 1986).

Sex steroid concentrations in blood are relatively low in both ring dove sexes after egg laying (Feder *et al.*, 1977; Korenbrot *et al.*, 1974; Silver *et al.*, 1974; Lea *et al.*, 1986). In the case of progesterone, the decline may be functionally important, as incubation behavior and crop sac development are reportedly reduced in male doves given daily injections of progesterone during the incubation period (Silver and Feder, 1973). In contrast, sitting behavior is not disrupted in testosterone-treated intact male pigeons, *Columba livia* (Collias, 1950), or castrated male doves given daily testosterone injections during incubation (Silver and Feder, 1973). Yet to be resolved, however, is the question of whether the low plasma concentrations of sex steroids that are present during the incubation period are important for the maintenance of sitting activity. Incubation is apparently maintained in male pigeons that are castrated at the onset of incubation (Patel, 1936) and in testosterone-treated, castrated male ring doves in which hormone replacement is discontinued at the time of pairing with a female breeding partner (Silver and Feder, 1973). Similarly, the incubation behavior that is induced in ovariectomized female doves by estradiol and progesterone administration is maintained when injections are discontinued at the onset of incubation (Cheng and Silver, 1975). These data suggest that sex steroids are not required for incubation maintenance in either sex. However, it is difficult, in the absence of clearance data, to estimate the period over which sitting behavior persisted in these studies after the hormones had disappeared from the circulation.

Although gonadal steroids may not contribute significantly to the maintenance of sitting activity in doves, they may play other roles during this breeding stage. Data collected by Ramos and Silver (1992) raise the interesting possibility that sex steroids are important determinants of the daily rhythm of sitting activity in the two sexes, which is characterized by male nest occupation during a 6- to 8-h period during the middle of the day and a lengthy incubation shift by the female during the remaining evening, night, and early morning hours. Gonadectomized males and females given heterotypical gonadal hormone replacement in this study were reported to exhibit incubation behavior when paired with untreated, same-sex breeding partners and, most interestingly, to assume daily incubation shifts that were characteristic of the opposite sex. In addition to this possible function, gonadal steroids may also prime the incubating pair so that they later respond to newly hatched squabs by showing prolactin release (Lea and Vowles, 1985).

*c. Other Avian Species.* Cross-species comparisons reveal at least three distinct patterns in the temporal relationship between plasma prolactin fluctuations and incubation onset, as shown in Fig. 2 (see also Goldsmith, 1983). In female bantam chickens (Lea *et al.*, 1981), ruffed grouse (*Bonasa umbellus;* Etches *et al.,* 1979), turkeys (Burke and Dennison, 1980), domestic and wild mallard ducks (Goldsmith and Williams, 1980; Hall and Goldsmith, 1983), and song sparrows (Wingfield and Goldsmith, 1990), circulating prolactin levels increase abruptly around the time of egg laying and incubation onset, and remain elevated throughout the incubation period. In other birds, prolactin levels rise more slowly, however, and do not plateau or reach peak levels until later in incubation. Although plasma prolactin levels may not be at peak concentrations during early incubation in these species, they are typically elevated above prelaying values at this stage. Female Australian black swans (Goldsmith, 1982b), pied flycatchers (Silverin and Goldsmith, 1983), canaries (Goldsmith, 1982a), white-crowned sparrows (Hiatt *et al.,* 1987), and great tits (*Parus major;* Silverin, 1991a) are examples of this pattern. Male and female wandering albatrosses (Hector and Goldsmith, 1985), ring doves (Goldsmith *et al.,* 1981), and possibly macaroni penguins (Williams and Sharp, 1993) are unusual among the birds that have been studied to date in exhibiting a third pattern characterized by a delay in the sustained increase in plasma prolactin until incubation is already well established (Hector and Goldsmith, 1985; Goldsmith *et al.,* 1981). Interestingly, a transient spike in plasma prolactin levels has been reported in females of all three of these species at the time of egg laying (Cheng and Burke, 1983; Lea *et al.,* 1986; Hector and Goldsmith, 1985; Williams and Sharp, 1993) although, as previously discussed, the functional significance of this event remains to be demonstrated. Experimental studies on a variety of other species are clearly needed in order to determine if the specific contribution of prolactin to the initiation and maintenance of incubation is correlated with its pattern of secretion during the incubation period.

The hormonal requirements for incubation induction and maintenance have been largely unexplored in most avian orders. Accordingly, it is not yet possible to determine which aspects of the patterns observed in galliform and columbiform birds are generalizable features. Prolactin reportedly failed to promote incubation behavior in female mallard ducks when administered after a period of gonadal steroid priming (Hohn, 1971). However, the small sample sizes used in this pilot study, and the fact that hormone dosage and temporal order of hormone administration was not varied renders these results inconclusive. In contrast to these negative results, prolactin has been shown to be an effective agent in promoting extended nest box occupation activity leading to incubation behavior in ovariectomized female budgerigars (*Melopsittacus undulatus*), but only when it is administered in conjunction with estradiol (Hutchison, 1975). This resembles the steroid priming that is necessary for prolactin-induced onset of incubation in female turkeys (El Halawani *et al.,* 1986) and for prolactin-induced development of the brood patch, which facilitates heat transfer

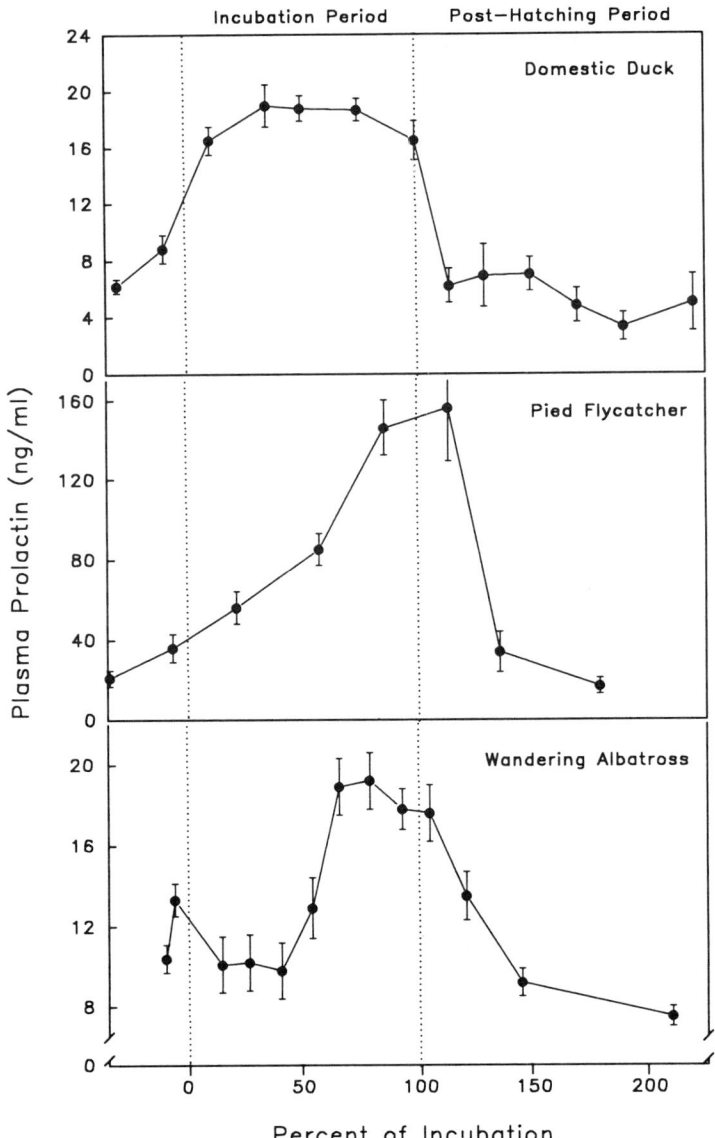

from the parent to the eggs or young in a variety of avian species (Jones, 1971). Estradiol-plus-prolactin treatment was also more effective than the administration of estradiol and progesterone in inducing advanced nesting and the onset of incubation. However, the effects of a combined regime consisting of estradiol, progesterone, and prolactin, which proved highly effective in inducing incubation in ovariectomized turkeys (El Halawani *et al.*, 1986), was not examined in this study.

Studies in nonparental brood parasites are potentially useful in establishing the physiological basis of parental responsiveness. In one such species, the brown-headed cowbird, nest building and incubation could not be induced by estradiol, progesterone, or prolactin when administered alone or in various combinations (Selander and Kuich, 1963; Selander and Yang, 1966). It is of particular interest that plasma prolactin levels are elevated during the breeding season in cowbirds even though no parental behavior is displayed (Dufty *et al.*, 1987). This suggests that these birds have evolved some degree of target tissue insensitivity to prolactin. The failure of gonadal hormones and prolactin to promote brood patch formation in cowbirds is consistent with the tissue insensitivity hypothesis (Selander and Kuich, 1963), and further support is provided by an autoradiographic study which indicated that prolactin binding activity in the preoptic area (POA), a region implicated in prolactin-induced regulation of parental behavior, is lower in cowbirds than in two parental songbird species (Ball *et al.*, 1988, 1990; Section IV,B,1).

The degree to which gonadal steroids modulate ongoing incubation behavior has been experimentally investigated in only a few avian representatives outside the orders Galliformes and Columbiformes, but the available evidence points to important species differences in this regard. Incubation

---

FIG. 2. Examples of three distinct patterns of circulating prolactin in relation to the incubation and posthatching periods of the avian breeding cycle. The female domestic duck (top panel) exhibits a pattern characterized by an abrupt increase in plasma prolactin in association with incubation onset. (Data reproduced with permission from Hall and Goldsmith, 1983.) The female pied flycatcher (middle panel) illustrates a second pattern characterized by a gradual increase in prolactin levels during the egg-laying and early incubation periods, with peak prolactin titers seen later in incubation. (Data from Silverin and Goldsmith, 1983 by permission of Oxford Univ. Press, and by permission from Silverin and Goldsmith, 1984.) The female wandering albatross (bottom panel) typifies a third pattern in which incubation behavior is well established before sustained elevations in plasma prolactin are observed. The transient increase in plasma prolactin during egg laying seen here is also typical of females of other species that exhibit this delayed pattern of prolactin secretion. (Data reproduced with permission from Hector and Goldsmith, 1985.)

behavior and care of young are not impaired in female song sparrows, white-crowned sparrows, or pied flycatchers given estrogen implants even though these females, unlike untreated subjects, reportedly remain sexually receptive after egg laying (Moore, 1982, 1983; Wingfield et al., 1989; Silverin, 1991b). In contrast, female starlings must experience the normal decline in gonadal steroid secretion in order for incubation behavior to be fully maintained (Ball and Wingfield, 1988). To add further complexity, there is evidence that supplemental estrogen prolongs incubation in female Bengalese finches, while exogenous progesterone leads to premature termination of incubation (Eisner, 1969). Similar species differences have emerged in analogous studies on incubating male birds. The decline in androgen secretion that normally occurs after egg laying in the polyandrous male spotted sandpiper (Fivizzani and Oring, 1986) is typical of incubating and nonincubating males of many species (Wingfield et al., 1990). This decline appears to be functionally significant in male sandpipers, as Oring and Fivizzani (1991) found that administration of the antiandrogen flutamide during the egg-laying period led to premature onset of incubation. In addition, ongoing incubation was disrupted in male sandpipers given testosterone implants even though the treatment did not interfere with the high circulating levels of plasma prolactin that accompany sitting activity (Oring et al., 1989). Interestingly, androgen-treated male sandpipers that were intensively courted by females were most likely to desert their nests, while those that nested later in the breeding season when courtship activity is reduced generally continued to incubate (Oring et al., 1989). This implies that the disruptive effects of testosterone are due in part to an increase in sexual activity exhibited by treated males in the presence of females, which is incompatible with sitting activity. These results contrast with maintenance of normal incubation in testosterone-treated male ring doves housed with their breeding partners in visual isolation from other birds (Section II,A,2,b). Given the fact that the disruption of incubation in androgen-treated spotted sandpipers is situation dependent, however, and in view of fact that ring doves are monogamous rather than polygamous, it would be of interest to determine if testosterone also fails to disrupt incubation in male ring doves breeding in social groups.

B. NEURAL MECHANISMS

In turkey hens and female domestic mallard ducks, sensory input from the brood patch, which develops in response to the same hormonal milieu that promotes incubation onset (Jones, 1971; see Fig. 1), is an important stimulus for prolactin release during the nesting and incubation periods (Book et al., 1991; Hall and Goldsmith, 1983; Hall, 1987). In turkey hens, nesting-induced prolactin secretion is disrupted by denervation of the brood patch and, as a result, denervated hens do not initiate incubation behavior (Book et al., 1991). In other species, however, there is a less obvious connection between incubation behavior and the tactile cues from the brood patch or ventral apterial surface that contacts the eggs (Medway, 1961; Hall, 1987; Hall and Goldsmith, 1983).

The neural substrate underlying incubation behavior itself has not been characterized in detail in any avian species, but the avian diencephalon, and the preoptic–hypothalamic continuum in particular, appears to be a critically important component. In studies on turkey hens, Youngren et al. (1989) observed that lesions in the POA prevented the onset of incubation and the rise in plasma prolactin that accompanies it. Similar results were obtained following lesions of the ventromedial hypothalamic nucleus (VMN) and lateral hypothalamus, which disrupted components of the putative efferent pathway to the median eminence that mediates the prolactin release induced by electrical stimulation of the POA. Lesions of the VMN also proved to be effective in disrupting ongoing incubation behavior and prolactin secretion in this experiment. However, this study did not test the ability of lesioned hens to exhibit incubation behavior in response to exogenous hormone treatment. As a result, it is difficult to determine if the lesions were effective in disrupting incubation solely because they interfered with nesting-related prolactin release or because they also destroyed a component of the neural circuitry underlying the expression of the behavior. The possibility that the POA represents a component of the neural substrate for sitting behavior is supported by two hormone administration studies. Opel (1971) successfully induced incubation behavior in laying chickens following implants of prolactin into the POA, anterior hypothalamic area, and lateral forebrain bundle. In a large-scale implantation study on ring doves, in which a variety of diencephalic sites were investigated, Komisaruk (1967) observed that the POA was the most effective site of progesterone action in promoting incubation, although implants in a more caudal and lateral region encompassing the supraoptic decussation were also quite effective in this regard.

In view of the prominent role that prolactin is assumed to play in the expression of incubation behavior in birds, and in view of the fact that the blood-brain barrier prevents proteins, such as prolactin, from gaining direct access to neurons in the brain, it is important to determine how and to what extent prolactin acts directly on the central nervous system (CNS) to promote behavioral changes. Intracranial administration of ovine prolactin in small doses is effective in inducing incubation behavior in turkey hens when infused directly into the cerebrospinal fluid (CSF) even though these infusions suppress endogenous prolactin secretion (Youngren et al., 1991).

While these results and those of Opel (1971) suggest that prolactin can act directly on the brain to promote sitting behavior in galliform birds, there is currently no information on the distribution of prolactin-sensitive neurons in the CNS and the mechanisms by which blood-borne prolactin gains access to the brain in this avian group. However, specific binding sites for $^{125}$I-ovine prolactin have been detected in the brains of several other avian species using conventional *in vitro* ligand binding assays and autoradiographic–densitometric mapping techniques (Section IV,B,1). Interestingly, the preoptic region contains specific binding sites for prolactin in all avian species examined to date, which strengthens the conclusion that this area is a potential target of prolactin action in the regulation of incubation behavior. Indirect support for this view is also provided by work in Wilson's phalarope, a species in which incubation and care of young is delivered exclusively by the male and, correspondingly, plasma prolactin levels are higher in males than in females during incubation (Oring et al., 1988). Autoradiographic comparisons of $^{125}$I-ovine prolactin binding in 11 brain regions of female phalaropes and incubating and nonincubating males revealed few significant differences in specific binding activity, although there was a clear overall trend for incubating males to exhibit lower binding activity than nonincubating males or females in most brain regions (Fivizzani et al., 1993). Because plasma prolactin levels were elevated in these incubating males, it is likely that the associated reduction in binding activity was the result of an increase in prolactin receptor occupancy by the endogenous hormone, which decreased the number of unoccupied prolactin receptors available to bind to the ovine prolactin tracer. This pattern, however, was not observed in the POA and the lateral septum, where mean specific binding values were higher in incubating males than in nonincubating males or females. These results are noteworthy because they suggest a possible up-regulation of the prolactin receptor and a resulting increase in prolactin sensitivity in select areas of the phalarope brain, such as the POA, that are presumed, based on evidence in other avian species, to be involved in regulation of incubation behavior and other parental responses. Nevertheless, these results should be cautiously interpreted, since the observed differences were not statistically significant and there is as yet no experimental evidence that directly links elevated prolactin to the expression of parental behavior in this species.

If the rise in plasma prolactin that occurs during incubation influences behavior through direct interactions with receptors in the CNS, then the presence of the blood-brain barrier dictates that specific transport mechanisms must exist to ferry the hormone in the plasma to specific prolactin-sensitive target cells in the brain. Short-term intravenous injection studies in ring doves reveal that radiolabeled ovine prolactin in the blood gains access to prolactin-sensitive target cells in the brain (Buntin *et al.*, 1993). Furthermore, long-term pretreatment with ovine prolactin causes a marked reduction in specific binding of $^{125}$I-ovine prolactin in several prolactin-sensitive dove brain regions (Buntin and Ruzycki, 1987; Buntin *et al.*, 1993), presumably because of increased receptor occupancy by the exogenous hormone. These observations support the conclusion that blood-borne prolactin interacts directly with prolactin receptors in the CNS.

Although several mechanisms of prolactin uptake are possible, the CSF could be an important conduit by which circulating prolactin gains access to the brain. Three lines of evidence support this conclusion. First, the choroid plexus has been shown to possess specific binding sites for prolactin that are part of a blood-to-CSF transport mechanism in the rat (Walsh *et al.*, 1987). Second, the ring dove choroid plexus contains prolactin receptors that exhibit saturable binding to prolactin *in vivo* and *in vitro* (Buntin and Walsh, 1988). Third, prolactin receptors have been detected in the choroid plexus of a variety of other species (Posner *et al.*, 1983; Muccioli *et al.*, 1988, 1990). While these data do not provide direct evidence that the choroid plexus could function in promoting prolactin uptake in the avian brain, they are at least consistent with this possibility.

In addition to interacting with blood-borne prolactin, CNS prolactin receptors could conceivably bind to prolactin-like molecules of CNS origin. Immunocytochemical evidence for the existence of such molecules is available for several vertebrates (B. L. Hansen and Hansen, 1982; Wright, 1986; Harlan *et al.*, 1989), including one avian species, the Japanese quail (Berghman *et al.*, 1992). Nevertheless, the molecular identity of this material, and the degree to which it resembles pituitary prolactin, is a matter of some debate (Dutt *et al.*, 1994). The existence of prolactin-like substances in the avian brain, and the possibility that such substances are regulated by steroid hormones (DeVito *et al.*, 1992; Shivers *et al.*, 1989) would have several exciting implications for our understanding of prolactin action in regulating incubation behavior and related parental activity. Nevertheless, evidence that such molecules are synthesized in the brains of several avian species is needed before such speculation is warranted.

### III. Defense of the Nest and Young

#### A. Hormonal Mechanisms

Most birds that are incubating eggs or brooding young exhibit some form of defensive behavior when challenged by a potential intruder. While few studies have addressed the physiological basis of this behavior, it is reasonable to hypothesize that it is regulated by the same physiological milieu that promotes other parental activities. This hypothesis is supported by a series of studies of defensive behavior (wing raising and feather erection) and aggressive behavior (wing slapping and pecking) in male and female ring doves presented with a large toy spider that is attached to the end of a stick (Vowles and Harwood, 1966; Lea *et al.,* 1986). As one might expect, breeding doves are more likely to show defensive or aggressive behavior to the intruding stimulus when they are occupying the nest. Nevertheless, males and females exhibit different patterns of nest defense toward the spider at different breeding stages. Lea *et al.* (1986) reported that males exhibit a progressive increase in incidence of defensive and aggressive activity toward the intruding stimulus throughout incubation to reach peak levels after hatching, while females show an earlier rise that is maintained at the same level throughout the incubation behavior but declines thereafter (Fig. 3). Because progesterone reportedly enhances nest defense in the ring dove (Vowles and Harwood, 1966), Lea *et al.* (1986) suggested that early rise in aggressive responses toward the spider by the female is a reflection of the high titer of progesterone that is present in females, but not males, at the time of egg laying. Moreover, there is evidence to suggest that the heightened response seen during midincubation in the female represents a delayed effect of progesterone (Lea and Vowles, 1985). Although its efficacy in promoting nest defense behavior remains to be conclusively established (Buntin *et al.,* 1991), there is evidence that prolactin facilitates these responses in ring doves (Vowles and Harwood, 1966). Accordingly, the rise in prolactin secretion during late incubation and early posthatching is presumably responsible for the elevated responsiveness seen in both sexes at this stage. The differential response of the male and female at 7–9 days after hatching, however, remains unexplained (Fig. 3).

The physiological regulation of nest defense in galliform birds has received much less attention. In domestic fowl and in turkeys, defensive behavior usually accompanies incubation behavior, and because prolactin is strongly implicated in the expression of sitting activity, it is assumed that defensive behavior is also prolactin-dependent. However, this hypothesis has never been directly tested. During the posthatching period, domestic hens with chicks exhibit heightened aggression toward intruding conspecifics (Kent, 1992). Nevertheless, prolactin concentrations in blood decline during the posthatching period in domestic galliform species (Leboucher *et al.,* 1993; Sharp *et al.,* 1988; Opel and Proudman, 1988), thus raising questions about the involvement of prolactin in the expression of the behavior at this stage. In one galliform species, however, there is direct evidence for prolactin-induced facilitation of defensive behavior. The female willow ptarmigan (*Lagopus l. lagopus*) exhibits an elaborate injury-feigning distrac-

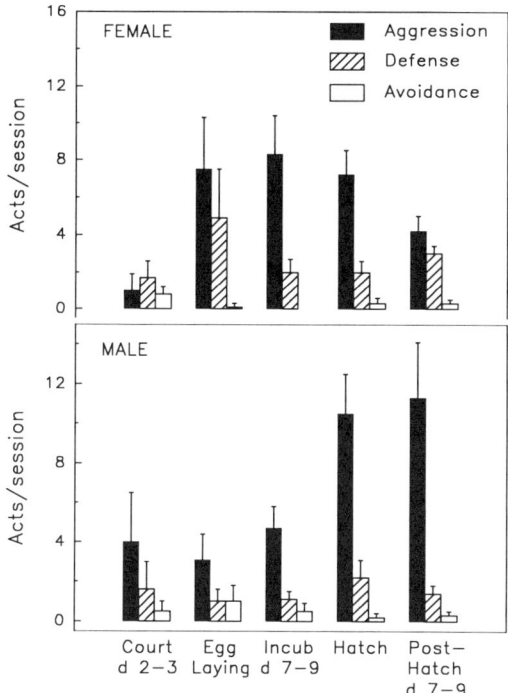

FIG. 3. Sex-specific patterns of defensive (feather fluffing or wing raising) and aggressive (pecking or wing slapping) reactions to a black model spider "intruder" in ring doves tested at different stages of the breeding cycle. (Redrawn from Lea *et al.*, 1986 with the permission of the *Journal of Endocrinology Ltd.*)

tion display in defense of her brood. In a remarkable study, Pedersen (1989) observed that the incidence of this distraction displays in free-living females was increased following continuous infusion of ovine prolactin. Prolactin-treated females also required more intrusive stimulation before they would leave the brood and traveled a shorter distance from the brood when flushed. Interestingly, this prolactin-induced increase in defensiveness and attentiveness also increased reproductive success, as chick survival in the prolactin-treated group was more than twice that seen in the vehicle-injected control group over the first 7 days of the posthatching period.

B. NEURAL MECHANISMS

While a variety of experimental techniques have been enlisted to identify the neural mechanisms involved in the expression of nest defense behavior in the ring dove, much less information is available in other avian species. Electrical stimulation, lesion, and hormone injection studies implicate the medial portion of the paleostriatum augmentatum (PA) and adjacent anterior hypothalamic area (AH) as important components of the neural substrate underlying the expression of aggressive and defensive responses during the ring dove breeding cycle (Harwood and Vowles, 1967; Lea, 1987; Vowles and Beazley, 1974). This neural substrate, which is extensive, is contiguous with that underlying escape behavior and resembles that mapped using electrical stimulation in other avian species (pigeon: Åkerman, 1965; mallard duck: Phillips, 1964; chicken: Phillips and Youngren, 1971). Although defensive–aggressive behavior can be elicited in ring doves by electrical stimulation of the medial PA and AH in the absence of environmental stimuli that normally evoke the response, the response is magnified in the presence of such stimuli (Vowles and Beazley, 1974). Interestingly, this area appears to be responsive to hormonal signals that are implicated in the expression of nest defense during the normal breeding cycle. Lesions of this region reportedly interfere with the ability of prolactin to facilitate defensive–aggressive responses in ring doves and intramuscular injections of prolactin reportedly increase defensive–aggressive responses to electrical stimulation in this area (work by D. M. Vowles, cited by Lea, 1987). Direct interaction of prolactin with neurons in this region is unlikely, however, as this area does not contain high concentrations of prolactin receptors in the ring dove brain (Fechner and Buntin, 1989; Buntin *et al.*, 1993). In addition, the prolactin-modulated neural mechanisms underlying the expression of parental behavior in the ring dove may be somewhat distinct from those involved in nest defense, as lesions of the preoptic region in nonbreeding, prolactin-injected birds disrupted parental regurgitation feeding activity towards nestlings without significantly attenuating the incidence of defensive activity toward an intruding stimulus (Slawski and Buntin, 1995).

## IV. Parental Responses toward Young

### A. Hormonal Mechanisms

#### 1. *Species with Precocial Young*

In the domestic hen, as in other Galliformes, the mother erects her feathers and crouches over her chicks to keep them warm. In addition to this brooding response, the hen attracts the chicks to her and leads them to food sources by emitting clucks and "tidbit" calls (Richard-Yris *et al.*, 1983). Brooding of the chicks declines during the first week after hatching as the young become more independent and as thermoregulatory ability improves (Sherry, 1981), and is paralleled by a rapid decline in plasma prolactin levels (e.g., domestic mallard duck: Hall and Goldsmith, 1983; chicken: Leboucher et al., 1990; Sharp et al., 1988; turkey: Opel and Proudman, 1989; Australian black swan: Goldsmith, 1982b). The precise cause of this decline in prolactin, however, is not clear. There is evidence that physical contact with chicks is required for the decline to occur (Opel and Proudman, 1988, 1989), but the close association between changes in brooding activity and changes in plasma prolactin suggests that the reduction in prolactin is not caused by a direct inhibitory effect of these stimuli on prolactin release. Instead, it appears to result from disruptive effects of these contact cues on the brooding activity of the hen (Lea et al., 1982; Sharp et al., 1988; Leboucher et al., 1993). The gradual posthatching decline in prolactin seen in several shorebid species represents an exception to the general pattern seen in other avian species with precocial young (spotted sandpiper: Oring et al., 1986b; Wilson's phalarope: Oring et al., 1988; red-necked phalarope: Gratto-Trevor et al., 1990). Nevertheless, data from this avian order support the concept that the rate of decline in prolactin levels parallels the rate of decline in brooding activity (Maxson and Oring, 1980; Gratto-Trevor et al., 1990). A notable exception, however, is the pattern seen in semipalmated sandpipers (Gratto-Trevor et al., 1990). Although brooding gradually declines during the first week after hatching in this species, plasma prolactin levels remain elevated throughout the posthatching period. Because this species breeds at high latitudes, it has been suggested that the sustained elevation of prolactin is related to other physiological events that coincide with the short breeding season, such as molt, which begins during incubation, or premigratory fattening, which begins soon after the parents leave the young (Gratta-Trevor et al., 1990). However, the pied flycatcher, an altricial species with similar breeding constraints (Silverin, 1981), shows a pronounced decline in prolactin release during the early posthatching period (Silverin and Goldsmith, 1983).

While prolactin levels decline when chicks are substituted for eggs in galliform species, the rate at which prolactin levels decline during the posthatching period is slower in hens raising chicks than in nest-deprived hens (Sharp et al., 1988; Leboucher et al., 1993). Field comparisons of bar-headed geese that successfully hatched chicks and those that did not also revealed an attenuating effect of chicks on the posthatching decline of prolactin in the female (Dittami, 1981). Nevertheless, questions could be raised about the functional significance of these effects of chicks on prolactin profiles, since they are superimposed on an overall decline in prolactin secretion. In addition, studies in another strain of domestic fowl failed to confirm these results, thereby suggesting that the effect could be specific to certain genotypes (Richard-Yris *et al.,* 1995).

It is difficult to reconcile the fact that prolactin levels decline sharply during the posthatching period with earlier reports that exogenous prolactin facilitates the display of parental behavior toward chicks in several galliform species (female turkeys: Crispens, 1957; Cain *et al.,* 1978; female pheasants, *Phasiamus colchicus torquatus:* Crispens, 1956; domestic cocks: Nalbandov and Card, 1945), and enhances the attentiveness of the willow ptarmigan hen toward her brood (Pedersen, 1989). It is clear that prolactin is not necessary to induce maternal behavior in domestic hens, as forced confinement of hens with chicks is effective in inducing maternal responsiveness in nonlaying hens in the absence of any detectable change in pituitary (Burrows and Byerly, 1938; Saeki and Tanabe, 1955) or plasma prolactin (Richard-Yris *et al.,* 1987a). However, onset of maternal responsiveness is not always immediate under these conditions (Burrows and Byerly, 1938; Richard-Yris and Leboucher, 1987a) and these findings do not rule out a stimulatory action of prolactin on parental responsiveness under normal conditions.

Interestingly, exposure to chicks not only retards the posthatching decline in plasma prolactin, but it also restrains gonadal steroid secretion in incubating domestic hens (Richard-Yris *et al.,* 1987a, 1995; Sharp *et al.,* 1988). Suppression of gonadal activity is also seen in hens brooding their own chicks (Sharp *et al.,* 1979) and in nonincubating hens confined with chicks (Richard-Yris *et al.,* 1983, 1987a,b). Notably, this gonadal response is not dependent upon elevated prolactin levels (Richard-Yris *et al.,* 1987a, 1995). High concentrations of gonadal steroids reportedly disrupt broody care of the chicks (Collias, 1950) and may delay the onset of maternal behavior, since the latency to the onset of parental responsiveness after forced confinement with chicks is shorter in nonlaying hens with low titers of gonadal steroids than in laying hens with high concentrations of these hormones (Richard-Yris *et al.,* 1983). However, a direct test of this hypothesis using nonlaying hens given testosterone or estradiol and subjected to forced confinement with chicks indicated that the clucking call was the only component of the maternal repertoire that was disrupted or delayed by exogenous steroid treatment (Richard-Yris and Leboucher, 1987b).

While the role of reduced gonadal steroid secretion in parental behavior expression remains to be resolved in the domestic hen, studies in other species with precocial young are consistent with the hypothesis that parental behavior is incompatible with high concentrations of circulating gonadal steroids. In the male spotted sandpiper, androgen treatment results in reduced parental attentiveness towards chicks even though prolactin levels remain elevated (Oring *et al.,* 1989). Similarly, Vleck and Dobrott (1993) observed a decrease in plasma testosterone in male bobwhite quail (*Colinus virginianus texanus*) that developed parental behavior toward foster chicks that were housed with them, while, in contrast, males that failed to develop parental responsiveness exhibited an increase in plasma androgen concentration. Nevertheless, the probability of successful induction of parental behavior was not increased by the administration of the antiandrogen flutamide, or flutamide in concert with an aromatase inhibitor to block testosterone conversion to estradiol. This suggests either that the fall in circulating androgen is a necessary but not a sufficient condition to promote parental responsiveness or that it is a consequence of parental behavior induction rather than a causal agent in its onset.

Clearly, more data are needed to determine the degree to which changes in circulating prolactin and steroid hormones support the normal posthatching expression of parental behavior toward chicks in galliform birds and in other species with precocial young. Nevertheless, it is potentially significant that maternal behavior can be induced and maintained in domestic hens by forced confinement with chicks. Although these manipulations are not without their hormonal consequences, they raise the possibility that nonhormonal cues may be significant to maintain parental activity during the normal posthatching period in the face of declining hormonal support. Conceptually, this is analogous to the transition between hormonal and nonhormonal regulation of parental behavior that is hypothesized to occur in lactating rodents during the postpartum period (Rosenblatt, 1967; Fleming and Rosenblatt, 1974).

### 2. Species with Altricial and Semialtricial Young

By necessity, the parental care of species with altricial young differs qualitatively from that seen in species with precocial young. The poor thermoregulatory and locomotor abilities of altricial nestlings dictate that they must be actively brooded for extended periods and must have food delivered to them by their parents. While semialtricial young are more mobile at hatching and possess better thermoregulatory ability, they remain dependent upon their parents for food over an extended period. The protracted period of parental care in altricial and semialtricial species is usually accompanied by a prolonged period of elevated prolactin secretion (Goldsmith, 1991) and reduced secretion of gonadal hormones (Wingfield *et al.*, 1990). The available evidence, while not conclusive, suggests that both of these hormonal features contribute to the normal expression of parental behavior during the posthatching period.

*a. Steroid Hormone Effects.* In many passerine species, males do not incubate eggs or brood the nestlings, but they do feed their young and sometimes provision their mates (Kendeigh, 1952). Plasma androgen levels remain low while these behaviors are displayed, and experimental studies suggest that this may be an important prerequisite for parental responsiveness. Parental feeding of young is markedly reduced in free-living male songbirds given exogenous testosterone (pied flycatcher: Silverin, 1980; house sparrow, *Passer domesticus;* Hegner and Wingfield, 1987; dark-eyed junco, *Junco hyemalis,* Ketterson *et al.,* 1992). Conversely, parental feeding rates are reported to rise in male house sparrows following treatment with flutamide, an antiandrogen (Hegner and Wingfield, 1987). Indirect evidence collected in free-living male song sparrows further suggests that an elevation in endogenous testosterone during the incubation and posthatching phases reduces parental feeding without influencing circulating prolactin levels (Wingfield *et al.,* 1989). The parental behavior deficits in androgen-treated males apparently result from an increase in activities associated with territorial defense and mate guarding, which are incompatible with provisioning the young (Wingfield *et al.,* 1990). It is noteworthy that in house sparrows and pied flycatchers, a reduction in fledgling survival accompanied the androgen-induced reduction in the male's parental feeding activity. However, this was not observed in juncos (Ketterson *et al.,* 1992). The possible linkage between high gonadal steroid titers and lowered reproductive success may help explain the observation that plasma testosterone levels increase in multiple brooding male song sparrows during the period prior to laying of the first clutch but remain at low concentrations prior to laying of the second clutch when the males are provisioning the fledglings (Wingfield and Goldsmith, 1990).

Apart from evidence that corticosterone may influence parental feeding efficiency in pied flycatchers, the possible effects of other steroid hormones on care of young have not received systematic attention. Plasma corticosterone is normally low in pied flycatchers during the posthatching period, but levels do rise when parents are feeding large broods and when females are provisioning the young without the assistance of their mates (Silverin, 1982). Because these birds increase their foraging activity to meet the nutritional demands of their growing young, it has been suggested that moderate, stress-induced elevations in corticosterone may enhance parental feeding activity (Silverin, 1990). However, higher corticosterone concentrations can lead to a decrease in the amount of food given to the nestlings due to an increase in food consumption by the parent, as Silverin (1986) has documented in free-living birds receiving corticosterone implants. At even higher doses, corticosterone caused the parents to abandon the nest altogether and to leave the breeding area. Based on the hyperphagia seen in captive flycatchers given the same high dose of corticosterone, it is likely that these free-living birds shifted their effort away from reproductive activity to engage in feeding (Silverin, 1990). Conflicting evidence precludes an answer to the question of whether corticosterone increases food intake in species with precocial young (Astheimer *et al.,* 1992), and negative results have been reported in two species with altricial young (junco: Gray *et al.,* 1990; pigeon: Bates *et al.,* 1962). It is of interest that the effects of corticosterone in white-crowned sparrows appear to vary with feeding history. Specifically, corticosterone implants had no effect on feeding rates in freely fed subjects but augmented the ingestive behavior of previously food-deprived birds (Astheimer *et al.,* 1992). In view of these conflicting data, it is difficult at present to assess the degree to which other altricial species resemble the pied flycatcher in their pattern of parental and ingestive responses to corticosterone.

*b. Patterns of Prolactin Secretion during the Posthatching Period.* The period of parental care in species with altricial young is usually longer than that seen in species with precocial young and is typically accompanied by a prolonged period of elevated prolactin (Goldsmith, 1991). An extreme example of this pattern is seen in male and female king penguins, which display elevated prolactin concentrations throughout an extended 11-month period of parental care during which they continue to feed the chicks at infrequent intervals (Cherel *et al.,* 1994). The possibility that parental behavior is dependent upon a period of elevated prolactin during the posthatching period is strongly suggested by studies in two albatross species of the genus *Diomedea* (Hector and Goldsmith, 1985). In these experiments, birds in which the incubation period had been artificially shortened by early introduction of the chick showed a longer than normal postincubatory period of elevated prolactin secretion and a corresponding prolongation of the period over which the chick was brooded. Conversely, birds that experienced an artificially prolonged incubation period displayed a truncated period of elevated prolactin after hatching and abandoned their chicks prematurely. In other altricial species, however, the relationship between prolactin concentrations and parental responses toward young may be less direct. Some aspects of parental care may persist in the face of low or declining prolactin levels during the posthatching period (e.g., ring dove: Horseman and Buntin, 1995; pied flycatcher: Silverin and Goldsmith, 1984; Silverin and Goldsmith, 1990). It is possible that nonhormonal stimuli become increasingly more effective in promoting these activities as the posthatching period proceeds, as discussed previously in the case of species with precocial young (Section IV,A,1). However, more experimental studies on the behavioral effects of prolactin will be needed in altricial species before this hypothesis can be meaningfully evaluated.

In some species, parents are assisted in feeding young by other members of the social group. Little is known about the hormonal correlates, if any, of this interesting phenomenon. It could be significant, however, that prolactin levels increase in the blood of nonbreeding adult male Harris' hawks (*Parabuteo unicinctus*) when they begin to assist the parents in feeding the young (Vleck et al., 1991). Indeed, prolactin levels during this period are higher in nonbreeding adult male helpers than in the male parents. Interestingly, this correlates with the observation that adult male helpers feed the young more frequently than the breeding males at this stage. A study by Schoech et al. (1996) in Florida scrub jays (*Aphelocoma c. coerulescens*) provides stronger evidence for a relationship between prolactin titers and parental behaviors by nonbreeding helpers at the nest. In this species, as well as in the related Mexican jay (*Aphelocoma ultramarina;* Vleck and Brown, 1992), plasma prolactin levels in nonbreeding helpers are lower than those of breeding birds, but are nonetheless elevated during incubation (scrub jay) and during the period of nestling care (both species). Significantly, Schoech et al. (1996) reported that nonbreeding scrub jays that fed nestlings had higher prolactin levels than those that did not, thereby suggesting a relationship between parental feeding and prolactin concentrations in blood.

*c. Role of the Young in Maintaining Elevated Prolactin during the Post-hatching Period.* In some altricial species, as in precocial birds (Section IV,A,1), stimuli from the young or cues associated with parent–young interactions may promote or sustain elevated prolactin titers. In free-living juncos, for example, males that are beginning to provision their young at the time of hatching were found to have higher levels of prolactin in blood than parental males that were removed from the breeding area at the end of incubation and housed in aviaries (Ketterson et al., 1990). Males that were attending to young also had higher plasma prolactin levels than the new occupants of the territories of the replaced males. In female pied flycatchers, prolactin reaches maximal concentrations in plasma at the end of incubation and remain elevated for the first 3 days of the posthatching period when newly hatched young are being intensely brooded (Silverin and Goldsmith, 1984, 1990). Thereafter, prolactin levels decline steeply. Exposure to newly hatched young clearly stimulates prolactin release in females in which prolactin levels had spontaneously declined during an experimentally extended incubation period (Silverin and Goldsmith, 1984) and constant exposure to 1- to 3-day-old nestlings extends the normal period of elevated prolactin by approximately 9 days (Silverin and Goldsmith, 1990). In contrast, substitution of newly hatched young with 3-day-old nestlings results in a premature decline in prolactin (Silverin and Goldsmith, 1990). The fact that male flycatchers do not respond to newly hatched young in this fashion suggests that the response is associated with brooding the young, since the male's contribution to parental care is restricted to feeding the nestlings. Nevertheless, prolactin levels eventually decline in the presence of newly hatched young even though females continue to brood (Silverin and Goldsmith, 1990). This implicates an endogenous timing mechanism that limits the maximal period over which prolactin is sustained in this species (Silverin and Goldsmith, 1984, 1990).

Stimulation of prolactin secretion by the young is also well documented in the ring dove (Buntin, 1979; Lea and Sharp, 1991). However, this relationship differs in at least two respects from that observed in the pied flycatcher. First, squab-induced release of prolactin occurs in both sexes, which reflects the fact that males and females share the brooding and feeding duties. Second, repeated exposure to newly hatched young, while effective, is no more effective in promoting prolactin-dependent crop sac growth than exposure to normally developing young (Hansen, 1971). Male and female doves differ in their stimulus requirements for squab-induced prolactin secretion, as assessed by crop sac growth responses (Buntin, 1977). Nevertheless, both sexes exhibit a greater growth response when exposed to food-deprived young, which evoke high rates of parental feeding activity, than when exposed to recently fed squab, which elicit little parental feeding (Buntin et al., 1977). This apparent elevation of prolactin secretion by stimuli associated with parental feeding interactions provides an effective mechanism for gearing crop milk production to the food demands of the young.

The hypothesis that prolactin release is facilitated by stimuli associated with parental feeding activity is most strongly supported by laboratory studies in ring doves. Correlative evidence obtained by Myers et al. (1989) suggests a similar relationship in male cockatiels, although it is unclear if the elevated prolactin is a cause or a consequence of elevated parental feeding rates. Field studies in the cooperatively breeding Florida scrub jay clearly indicate that parental feeding activity is associated with elevated prolactin secretion in nonbreeding helpers, although somewhat curiously, no such relationship could be demonstrated among breeding birds (Schoech et al., 1996). In other species, however, a close relationship between prolactin levels and parental feeding activity is not apparent. In free-living pied flycatchers and juncos, for example, plasma prolactin profiles of females that are feeding young at high rates without male assistance are essentially the same as those of assisted females despite marked differences in parental feeding rates (Ketterson et al., 1990; Silverin and Goldsmith, 1984).

 d. *Experimental Evidence that Prolactin Promotes Parental Behavior during the Posthatching Period.* The results presented in Section IV,A,2,b and c indicate that elevated prolactin and the display of parental activity are temporally contiguous during at least part of the posthatching period in a variety of altricial species, and that one mechanism by which these elevated levels of prolactin are maintained is through neuroendocrine reflexes triggered by stimuli from young or cues generated during parent–young interactions. While it is generally assumed that these elevated levels of prolactin are important in promoting parental responsiveness toward nestlings, direct evidence for such a relationship is very limited in altricial birds. In fact, the ring dove is the only altricial species in which this question has been experimentally addressed using direct hormone administration.

In ring doves, prolactin facilitates parental care during the posthatching period in several different ways. First, it may facilitate the transition from incubation behavior to posthatching parental care. Such a role is suggested by the observation that prolactin increases the relative attractiveness of newly hatched young, as assessed in a squab–egg choice test (Moore, 1976). Second, it generates an essential source of food for the newly hatched young by stimulating crop sac development during the incubation and early posthatching period, which results in the production of crop milk. Third, as discussed below, prolactin facilitates the regurgitation behavior required to transfer the crop milk to the offspring. Finally, prolactin stimulates feeding, which provides seed and other nutrients that are regurgitated to the rapidly growing young in increasing amounts as crop milk production declines (Fig. 4). In a sense, therefore, this hormone can be thought of as an integrating agent that acts in a variety of ways to ensure adequate provisioning of the young during the period that they are dependent upon their parents for nourishment.

*1. Regurgitation feeding of young in ring doves.* Lehrman (1955) reported that a 7-day period of intradermal prolactin administration was effective in facilitating the display of parental regurgitation feeding behavior in isolated, nonbreeding doves with previous breeding experience. Lehrman also demonstrated that the experiential background of the birds was important in this regard, as prolactin was ineffective in promoting parental regurgitation feeding activity in inexperienced subjects. The stimulatory effect of prolactin on parental responsiveness in experienced birds was confirmed in a later study (Buntin *et al.,* 1991) and, in inexperienced birds, Lott and Comerford (1968) subsequently showed that while neither prolactin nor progesterone alone stimulated parental regurgitation, the two hormones were effective when administered in combination. It is perhaps significant that nest occupation was not observed in prolactin-injected birds during the behavior test, but it was seen in birds receiving progesterone injections and those receiving the combined treatment. This implies that nest attachment or proximity to the foster squab is an important precondition for prolactin-induced regurgitation feeding to be displayed.

Lehrman originally hypothesized that prolactin acts to promote parental regurgitation feeding activity through its peripheral action on the crop sac. This hypothesis was suggested by the observation that crop anesthetization

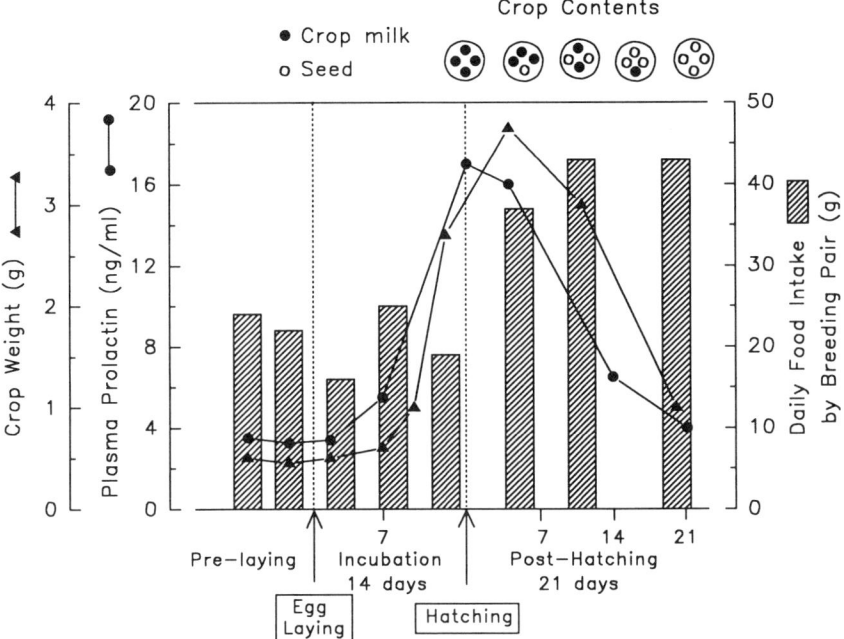

FIG. 4. Changes in plasma prolactin, crop development, crop contents, and parental food consumption during the ring dove breeding cycle. The pattern of plasma prolactin and crop weight changes that is depicted represents both sexes, although females typically exhibit higher peak values than males. (Taken from Horseman and Buntin, 1995 and represents data compiled from several different sources.) Reproduced, with permission, from the *Annual Review of Nutrition,* Volume 15, © 1995, by Annual Reviews Inc.)

dramatically reduces the incidence of prolactin-induced parental regurgitation behavior in experienced, nonbreeding doves (Lehrman, 1955). Specifically, Lehrman suggested that prolactin-induced crop sac growth and engorgement with milk sensitized the bird to the tactile cues provided by the begging squab as it moved its head against the parent's chest. However, the fact that 33% of the birds with anesthetized crops still exhibited regurgitation behavior during the test suggested that a developed and distended crop is not a prerequisite for the behavior to be displayed. This is consistent with the results of later studies, which demonstrated that doves given foster young during early stages of incubation before significant crop sac growth had occurred would nevertheless attempt to feed the squab by regurgitation (E. W. Hansen, 1971; Klinghammer and Hess, 1964).

In addition to a peripheral (i.e., crop sac) site of action, prolactin may also act directly on the CNS to promote parental regurgitation behavior.

In a study (Buntin *et al.,* 1991), nonbreeding birds with previous breeding experience were tested for parental behavior following a series of twice-daily ICV injections of prolactin at a dose that was too low to have any discernible effects on crop sac development (Fig. 5). Because prolactin also

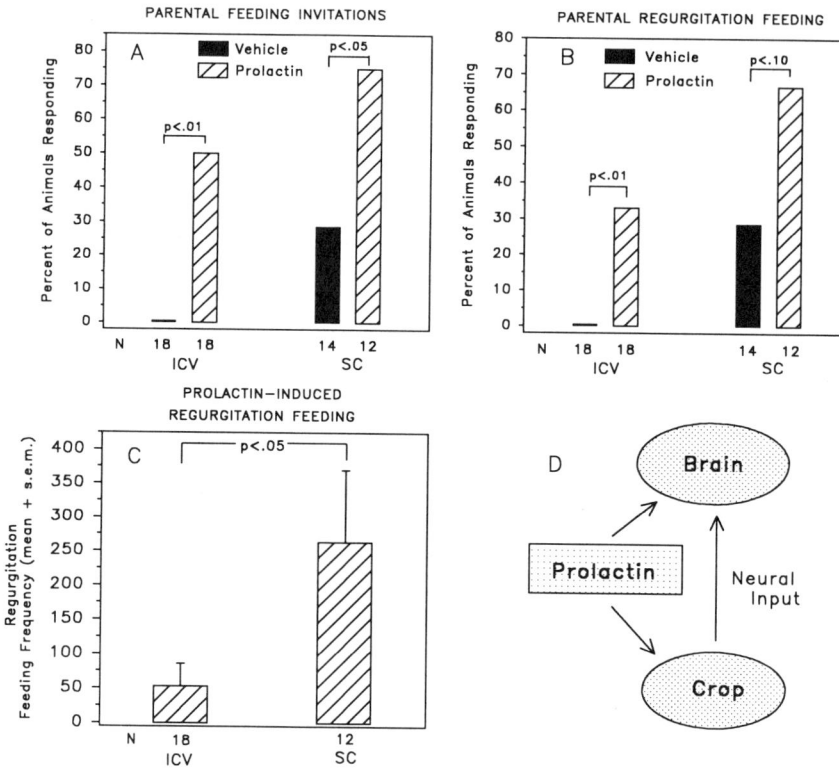

FIG. 5. Differences in the stimulatory effects of central (intracerebroventricular, ICV) and peripheral (subcutaneous, SC) administration of ovine prolactin on the incidence of parental regurgitation feeding invitations (A), incidence of parental regurgitation feeding (B), and frequency of parental regurgitation (C) of foster young in nonbreeding ring doves with previous breeding experience. Males and females received 7 days of twice-daily injections of prolactin or vehicle prior to a 2.5-h test with a hungry foster squab (6–8 days old). Birds receiving ICV injections were food deprived prior to the test to minimize the confounding effects of crop engorgement with seed on regurgitation behavior. Data (group mean ± SEM) from males and females were pooled, as no sex differences were apparent. D depicts a dual target site model of prolactin action that is consistent with the results shown in A–C (see text). (Modified from Horseman and Buntin, 1995 and represents data from Buntin *et al.,* 1991. Reproduced, with permission, from the *Annual Review of Nutrition,* Volume 15, © 1995, by Annual Reviews Inc.)

stimulates feeding behavior (Section IV,A,2), birds in the study were also food deprived prior to the test to eliminate the possibility that crop sac distension from the ingested food could generate cues that might facilitate the parental regurgitation feeding response. Although they were less responsive than prolactin-treated doves with developed crop sacs, birds receiving intracranial prolactin in the absence of any crop sac stimulation showed a significantly higher incidence of parental regurgitation behavior and parental feeding "invitations" during the test than did vehicle-injected controls (Fig. 5). These results suggest that central actions of prolactin may be sufficient to promote parental motivation. At the same time, however, normal parental regurgitation feeding rates may depend upon input from a developed or engorged crop.

*2. Parental hyperphagia in ring doves.* As the posthatching period proceeds, the production of crop milk declines and the young are fed increasingly on seed and other food items obtained by the parents (Fig. 4). The marked increase in parental food intake required to supply the increasing energy needs of the growing young begins very early after hatching (Lea *et al.*, 1992). Although the increase is progressive, it does not exceed prelaying levels of food intake until the young are several days old (Brisbin, 1969; Lea *et al.*, 1992). The sustained elevation in parental food intake that occurs throughout the posthatching period appears to be characteristic of the Columbidae, as it is also seen in pigeons (Mondloch and Timberlake, 1991). It is significant that the hyperphagia exhibited by the breeding pair at this stage is not accompanied by an increase in parental body weight. In fact, parental body mass declines significantly during this period, which suggests that the additional ingested food is being regurgitated to the young (Mondloch and Timberlake, 1991; ten Cate *et al.*, 1993).

Because prolactin levels are elevated during the early posthatching period when parental food intake is rapidly increasing, it is reasonable to hypothesize that prolactin may facilitate feeding activity. This is supported by a wealth of data, which collectively demonstrate a potent stimulatory effect of prolactin on food intake in columbid birds following either subcutaneous (Schooley *et al.*, 1941; Bates *et al.*, 1962; Buntin and Figge, 1988) or ICV administration of the hormone (Buntin and Tesch, 1985; Buntin and Figge, 1988; Buntin, 1989). Studies in male doves reveal that the food intake response to daily ICV injections of ovine prolactin is strongly dose dependent, and a virtual doubling of daily food intake is observed at doses that are too low to have any detectable effects on peripheral target tissues (Buntin and Figge, 1988). An interesting feature of the hyperphagia is that it is sexually dimorphic, with males exhibiting a food intake response that is approximately twice that seen in females (Buntin and Tesch, 1985; Buntin and Figge, 1988). The physiological basis of this sex difference is not yet clear, although it is conceivable, based on previous work (Pietras and Wenzel, 1974), that circulating androgens act synergistically with prolactin to enhance the male's hyperphagic response. The division of labor dictated by the opportunistic breeding strategy employed by Columbiformes suggests that this sex difference may be functionally significant. When conditions are favorable, female pigeons and doves typically lay a new clutch of eggs before their young are old enough to forage on their own, and with the female occupied with the new clutch of eggs, the male assumes most of the responsibility for provisioning the nestlings (Burley, 1980; Mondloch and Timberlake, 1991). Sex differences in prolactin-induced feeding are therefore consistent with differences in parental roles assumed by the breeding partners under these conditions.

Although there is evidence that prolactin can act directly on brain sites that have been implicated as components of the neural circuitry underlying feeding behavior (Section IV,B,2), the delayed onset and protracted duration of the hyperphagia induced by ICV injections of prolactin suggests a complex mode of action in the CNS (Buntin, 1989). Complexities are also suggested by the fact that the rise in prolactin secretion that occurs during late incubation in doves is not accompanied by an increase in feeding activity. In part, this could be due to the fact that incubation behavior is simply incompatible with foraging. Threshold differences in the level of prolactin necessary to promote incubation activity and feeding activity may also be a factor in this regard, as food intake in the parents does not begin to increase until the early posthatching period when peak prolactin levels are attained. However, other factors may also come into play during the posthatching period to predispose the parents to increase their food consumption in response to high circulating titers of prolactin. One interesting, although speculative, possibility is that prolactin activates the hypothalamo–pituitary–adrenal axis, and the resulting elevation in corticosterone synergizes with prolactin to promote feeding behavior. This notion is suggested by earlier work with hypophysectomized pigeons, which suggested that corticosterone amplifies the the hyperphagic action of prolactin but does not by itself influence food intake (Miller and Riddle, 1943; Bates *et al.*, 1962). The possibility that prolactin and corticosterone could work together to promote feeding activity during the posthatching period is strengthened by the observation that plasma corticosterone levels begin to increase in parent ring doves after their eggs have hatched (Lea *et al.*, 1992). As yet there is no direct evidence that prolactin stimulates corticosterone secretion in doves, although evidence in the rat suggests that prolactin can act directly on the brain to activate the hypothalamo–pituitary–adrenal axis (deGreef *et al.*, 1995).

*3. Is prolactin essential for the expression of parental behavior in ring doves?* While there is compelling evidence that prolactin stimulates parent doves to regurgitate food to their nestlings and to increase their own food intake for the presumed purpose of provisioning the young, these studies do not address the question of whether prolactin is required for the expression of parental behavior during the normal breeding cycle. The obligatory nature of prolactin's contribution to parental responsiveness has been called into question by Lea *et al.* (1991), who reported that passive immunization against the prolactin-releasing neuropeptide VIP did not disrupt incubation behavior or prevent the display of parental feeding of young in breeding ring doves, even though plasma prolactin levels and crop sac development were significantly depressed. While these results suggest that elevated prolactin levels are not required for parental behavior expression, a cautious interpretation is warranted for several reasons. First, the anti-VIP antiserum did not completely suppress the incubation-associated rise in plasma prolactin, and the suppression itself was short-lived due to an immune response mounted by the doves against the exogenously administered antibodies. Thus, it is possible that parental behavior could persist in the face of transient and/or partial decrements in plasma prolactin levels. Second, it is possible that prolactin may be more important in facilitating parental responsiveness under free-living conditions, where food availability may be reduced and forays away from the nest area are not restricted as they are in birds breeding in small cages in the laboratory. Third, it is possible, as Moore (1976) proposed, that complementarity exists between the internal state and breeding experience in the regulation of parental responsiveness, such that inexperienced subjects may be more dependent upon prolactin for the normal expression of parental behavior during the breeding cycle than are experienced birds. Because Lea *et al.* (1991) only used experienced subjects in their passive immunization study, this hypothesis has yet to be directly tested.

Despite the strong possibility that prolactin exerts some stimulatory effect on the regurgitation feeding of young and the parental hyperphagia exhibited by normally breeding birds, it is likely that its influence wanes during later stages of the posthatching period. This conclusion is based on the fact that parental feeding of young and the parental foraging activity that accompanies it persist at high levels as plasma prolactin concentrations decline to prelaying levels during later stages of the posthatching period (Fig. 4). As discussed previously, this dissociation between prolactin levels in blood and parental responses during the posthatching period is seen in other altricial species (e.g., pied flycatcher: Silverin and Goldsmith, 1984, 1990), as well as in the precocial Galliformes (Section IV,A,1). Although direct support for the concept is not yet available, these observations imply a transition from hormonal to nonhormonal regulation of parental behavior during the posthatching period, which resembles that reported in several mammalian species during the postpartum period (Rosenblatt, 1992).

B. NEURAL MECHANISMS

1. Sites of Prolactin Action in the Avian Brain

Specific binding sites for prolactin have been detected in the brain of several avian species, including pigeons (Muccioli *et al.*, 1988), ring doves (Buntin and Ruzycki, 1987; Fechner and Buntin, 1989; Buntin *et al.*, 1993), Wilson's phalarope (Fivizzani *et al.*, 1993), and several songbird species (redwing blackbird, *Agelaius phoeniceus*, European starling, brown-headed cowbird: Ball *et al.*, 1988, 1990; dark-eyed junco, Deviche and Buntin, 1992). However, characterization of these receptors is most complete in the ring dove. High-affinity (Kd = $10^{-10}$M) binding of $^{125}$I-ovine prolactin has been detected in membrane fractions prepared from dove brain homogenates (Buntin and Ruzycki, 1987) as well as in a variety of intact brain regions (Buntin *et al.*, 1993). Studies involving the use of unlabeled hormone competitors in the prolactin–growth-hormone–placental lactogen family indicate that these sites bind prolactin with high specificity, although they exhibit a much lower affinity for turkey prolactin than for ovine prolactin (Buntin and Ruzycki, 1987). Interestingly, this corresponds to the finding that turkey prolactin is less potent than ovine prolactin in suppressing gonadotropin secretion and promoting ingestive behavior in ring doves when administered via ICV injection. This implies that avian prolactins and prolactin receptors are structurally and functionally heterogeneous.

Detailed mapping studies of prolactin receptors in the ring dove CNS have been carried out through the incubation of $^{125}$I-ovine prolactin with slide-mounted dove brain sections and quantification of specific binding by film autoradiography and densitometry (Fechner and Buntin, 1989; Buntin *et al.*, 1993). These studies reveal a nonuniform distribution of prolactin binding activity, with highest concentrations of receptors in several diencephalic sites, including the POA; the lateral hypothalamic region; the tuberal hypothalamus; and the ventromedial, suprachiasmatic, and paraventricular hypothalamic nuclei. Binding was also detected in a variety of extrahypothalamic brain regions such as the medial habenula, lateral septum, and nucleus accumbens.

Although species differences are apparent in the regional distribution of prolactin receptors in the CNS, prolactin binding activity in the POA and the hypothalamus is generally high in those species that have been examined. Binding patterns in the POA are particularly interesting in view of evidence that this region may support the expression of incubation behavior and/or care of young in galliform and columbiform species (Section II,B and IV,B,2). Ball *et al.*, (1988, 1990) compared patterns of prolactin binding activity in the brains of two species of songbirds that exhibit parental care (starling, redwing blackbird) and one species that does not (brown-headed cowbird). All three species exhibited specific binding sites for prolactin in the CNS, and although differences in binding activity were observed, the general pattern of binding site distribution was similar across species. Interestingly, the POA emerged as the only area in which species differences in binding activity consistently correlated with differences in reproductive strategy, with both parental songbird species exhibiting significantly higher binding activity than the nonparental cowbird. Although a central role of prolactin in incubation and care of young has yet to be established in starlings and redwing blackbirds, these results are at least consistent with other evidence that the POA is an important site of prolactin action in regulating these activities.

## 2. *Neural Mechanisms Underlying Specific Parental Activities*

The ring dove is the only avian species in which the neural substrate(s) underlying the expression of parental responses toward young have been directly investigated. Work in this species suggests that the neural mechanisms underlying parental regurgitation feeding of young may differ somewhat from the underlying the prolactin-facilitated hyperphagia that is assumed to be involved in parental provisioning of the nestlings. Based on previous work implicating the POA in maternal behavior onset in mammals (Numan, 1994) and the display of incubation behavior in birds (Section II,B), Slawski and Buntin (1995) examined the effects of POA destruction on these two types of parental activities. Axon-sparing lesions of the POA profoundly disrupted parental regurgitation feeding behavior and related activities in isolated, nonbreeding doves with previous breeding experience when tested with foster young after a 7-day period of subcutaneous prolactin administration (Slawski and Buntin, 1995; Fig. 6). Furthermore, the magnitude of the behavioral deficits observed in these prolactin-treated birds was significantly correlated with the extent of POA damage. In marked contrast to this pattern, POA lesions did not attenuate the hyperphagia induced by prolactin administration. Thus, prolactin-induced hyperphagia and prolactin-induced regurgitation feeding of young differ in their dependence upon the POA for their expression. Nevertheless, food consumption is increased in birds given microinjections of prolactin into the POA (see as follows). This suggests that the POA may have some role to play in prolactin-induced feeding responses. An attractive but as yet untested hypothesis is that the POA synchronizes parental food procurement with the expression of the

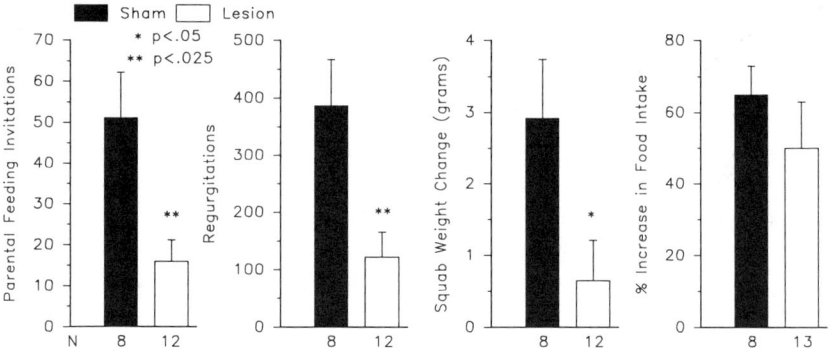

FIG. 6. Disruptive effects of the preoptic area (POA) destruction on parental feeding invitations and total parental regurgitations by prolactin-injected ring doves tested with a hungry foster squab (6–8 days old) during a 2.5-h parental behavior test. Data (group mean ± SEM) from males and females were pooled, as no sex differences were apparent. The average amount of weight gained by the foster squab during the test and the percentage increase in average daily food consumption over pretreatment values during the 7-day prolactin treatment period that preceded the test is also depicted. Note that the POA lesions did not suppress the prolactin-induced hyperphagic response. (Data reproduced with permission from Slawski and Buntin, 1995.)

parental regurgitation behavior that is required to actually transfer the food to the young.

The neural substrates responsible for prolactin-induced hyperphagia have also been examined in some detail in male ring doves. Doses of prolactin that are too low to induce detectable hyperphagia when administered ICV were nonetheless effective in promoting feeding activity if they were microinjected into the prolactin-sensitive POA, VMN, or tuberal hypothalamic region (Hnasko and Buntin, 1993; Fig. 7). In contrast, no significant elevation in food intake was observed when prolactin was administered to other prolactin-sensitive areas of the brain, including some that have been strongly linked to food intake regulation in birds and mammals, such as the paraventricular nucleus of the hypothalamus and the lateral hypothalamic area (Kuenzel, 1994). Based on the magnitude of the hyperphagia observed, the VMN was the most effective of the three sites that effectively supported prolactin-induced feeding responses (Fig. 7). Previous work has demonstrated an important role of the VMN in regulation of feeding behavior (Kuenzel, 1994; King, 1991), and accordingly, a direct action of prolactin on the neural circuitry underlying feeding activity is possible. Li *et al.* (1995) have provided evidence that prolactin receptors mediate, at least in part, the hyperphagia that is evoked by prolactin injections into the VMN. In

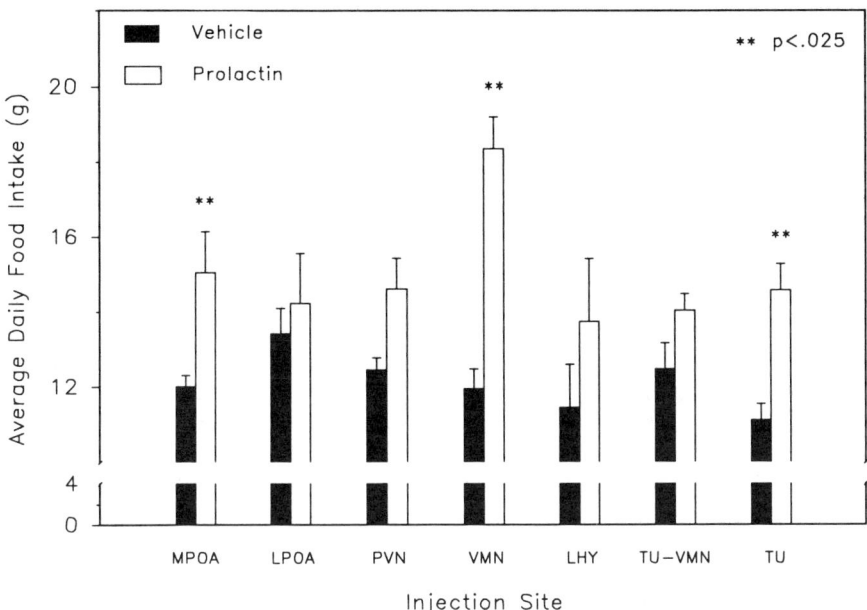

FIG. 7. Effects of injection site location on prolactin-induced food intake in male ring doves. Injections (50 ng/10 nl) were delivered unilaterally to single brain site twice daily for 4 days in a Latin square design, with a 6- to 12-day recovery period interposed between the prolactin and vehicle treatment periods. MPOA = medial preoptic area; LPOA = lateral preoptic area; PVN = paraventricular hypothalamic nucleus; VMN = ventromedial hypothalamic nucleus; TU = tuberal hypothalamus; TU-VMN = area between TU and VMN. (Data modified with permission from Hnasko and Buntin, 1993.)

this study, anti-prolactin receptor antibodies proved effective in binding to prolactin receptors in the dove brain and reduced the prolactin-induced elevation in food consumption by 50% when injected in tandem with prolactin into the VMN.

## V. Summary

Parental behavior patterns are remarkably diverse in birds, and the physiological factors that give rise to these differences are only beginning to be understood. However, several broad themes in the physiological regulation of parental behavior are emerging that may be applicable to a range of avian species. These themes are presented here with the hope that they will stimulate future research designed to test their durability as valid generalizations.

A. STEROID HORMONES INFLUENCE THE EXPRESSION OF PARENTAL ACTIVITY

Evidence that steroid hormones play an important role in the display of parental behavior is now indisputable. However, the specific contributions of steroid hormones in this regard may vary with sex, species, and breeding stage. Gonadal steroid hormones are important components of the hormonal milieu required for incubation induction in many species (e.g., female budgerigar, turkey, and ring dove). In particular, these hormones may serve to promote early nesting activity and, in some species, to serve a priming function to enhance prolactin secretion or responsiveness to prolactin for full incubation onset. Nevertheless, in some birds, a subsequent decline in gonadal steroid concentrations in blood may be required in order for incubation behavior itself to be expressed or maintained in response to elevated prolactin concentrations (e.g., male spotted sandpiper). In males of many species (e.g., spotted sandpiper, song sparrow, house sparrow, pied flycatcher), high concentrations of androgens are incompatible with the maintenance of incubation or parental feeding of young, while, in contrast, females of most species maintain parental responsiveness when given exogenous estrogen. While this suggests a possible sex difference in gonadal hormone contribution to parental activity, exceptions to these patterns have been documented (e.g., ring dove males, starling females), and the effects of other gonadal steroids, such as progesterone, have not been sufficiently well studied to formulate generalizations. The physiological basis for these differences are important issues that remain to be elucidated.

B. PROLACTIN FACILITATES PARENTAL BEHAVIOR EXPRESSION

An impressive body of correlational data indicates that prolactin is closely linked to the display of parental behavior in a variety of avian species. In addition, there is now direct evidence from several species for a causal relationship between elevated prolactin and various parental responses, including incubation behavior (turkey, chicken, ring dove, budgerigar), nest defense (ring dove, ptarmigan), and parental responses toward young (chicken, turkey, pheasant, ring dove). Nevertheless, the number of studies that provide correlational evidence for prolactin involvement in parental behavior far outweigh the number of studies that provide direct evidence for such a relationship. There is a particularly acute need to expand the number of avian species in which the behavioral effects of exogenous prolactin administration have been explored. In this context, the use of implantable pumps for the continuous administration of prolactin in the field (e.g., Pedersen, 1989) may provide a particularly valuable yet minimally invasive method for establishing the importance of prolactin in the expression of parental behavior in free-living populations. Experimental studies are also needed to characterize the neural substrate(s) for parental behavior expression. While changes in steroid hormones may modulate or perhaps even mediate prolactin-induced changes in parental responsiveness, the neural mechanisms underlying these interactions are completely unknown.

Despite the fact that elevations in plasma prolactin are usually associated with the display of parental behavior, there are many examples of situations in which these two events are uncoupled. There may be several reasons for these discrepancies. In some instances, they may reflect the confounding influence of other environmental factors the regulate prolactin secretion, such as photoperiod, or changes in other hormone synergists that may be important for prolactin-induced behavioral expression. The persistence of parental behavior in the face of low or declining prolactin levels could indicate that parental activity is under the control of several redundant physiological mechanisms, and is therefore not dependent upon prolactin for its expression (Lea *et al.*, 1991). It is also likely that experiential factors modulate the coupling between various physiological factors, including prolactin, and parental behavior expression. Indeed, with increasing parental experience with young acquired during the posthatching period, a transition from a hormonal to a nonhormonal mode of parental behavior control may occur, with stimuli from young or cues associated with parent–young interactions assuming pre-eminence as behavioral regulators.

Under some circumstances, short-term dissociations between plasma prolactin levels and parental responses could reflect the protracted time course of prolactin-induced changes in the brain or peripheral target organs. Differences in the clearance rate of blood-borne prolactin and prolactin that is transported from blood to CSF may also contribute to this uncoupling. As a result of such factors, prolactin-induced behavioral changes could conceivably persist for some period after plasma titers of the hormone have declined. The fact that some prolactin-induced behavioral alterations do persist for at least 48 h after a single injection (Buntin, 1989) lends some credibility to this hypothesis. An interesting but speculative possibility is that prolactin-like molecules of brain origin may act in tendem with circulating prolactin to support the display of parental behavior. If this were true, a dissociation between plasma prolactin and parental activity would not necessarily imply that behavioral expression is independent of prolactin.

C. Prolactin-Induced Changes in Parental Behavior Involve both Central and Peripheral Actions

Although we have much to learn about precisely how and where prolactin acts to promote parental activity, it seems clear that several sites and modes of prolactin action are involved. The CNS is a likely target of prolactin action based on evidence for prolactin receptors in the avian brain. Furthermore, evidence for ventromedial hypothalamic involvement in prolactin-induced hyperphagia in doves indicates that prolactin could conceivably interact directly with prolactin-sensitive neural substrates for specific parental activities. Circulating prolactin could presumably access these target sites through specific uptake mechanisms, such as the blood-to-CSF transport of prolactin that has been documented in the mammalian choroid plexus. Another possibility is that prolactin activates neural mechanisms that have indirect effects on parental behavior expression. For example, the hormone is capable of acting on the brain to suppress the hypothalamo–pituitary–gonad axis (Buntin *et al.,* 1988; Juss and Goldsmith, 1992), which in turn could promote the expression of parental activity through a reduction in gonadal steroid secretion. Finally, parental behavior may be facilitated by prolactin actions on peripheral target tissues. Possible examples of such actions include direct gonadal suppression by prolactin (Camper and Burke, 1977); stimulation of brood patch development and resulting changes in tactile sensitivity to the nest and eggs or young (Jones, 1971); and in columbiform species, enhancement of parental regurgitation toward young through stimuli generated by prolactin-induced changes in the crop sac (Lehrman, 1955). In addition to these mechanisms, which involve prolactin released by the pituitary, there is also the possibility that prolactin-like molecules of brain origin may promote the expression of parental behavior in birds. Because the existence of one or more prolactinergic neuronal systems could have far-reaching implications for the regulation of parental activity, the detection and characterization of such molecules in the avian brain should be a high priority for future investigations.

#### Acknowledgments

I wish to thank Linda Buntin, Bob Lea, Charles Snowdon, Jay Rosenblatt, and Charles Weise for their comments and suggestions on the manuscript. I am deeply indebted to the many technicians, undergraduates, graduate students, and collaborators who have made important research contributions to my work. I am also grateful for the research support provided by National Institute of Mental Health (MH 41447) and the National Science Foundation (DCB 8303026), which made these contributions possible.

This article is dedicated to the memory of Daniel S. Lehrman, who served as my Ph.D. advisor before his untimely death in 1972. In addition to being an extraordinarily gifted teacher, Danny was a true naturalist, an insightful experimentalist, and an influential theorist whose keen understanding of the interactive nature of biological systems left an indelible mark on the fields of ethology, developmental psychobiology, and behavioral endocrinology.

#### References

Åkerman, B. (1965). Behavioral effects of electrical stimulation in the forebrain of the pigeon I. reproductive behavior. *Behavior* **26,** 323–338.

Allen, T. O., and Erickson, C. J. (1982). Social aspects of the termination of incubation behavior in the ring dove (*Streptopelia risoria*). *Anim. Behav.* **30,** 345–351.

Astheimer, L. B., Buttemer, W. A., and Wingfield, J. C. (1992). Interactions of corticosterone with feeding, activity and metabolism in passerine birds. *Ornis Scand.* **23,** 355–365.

Ball, G. F., and Wingfield, J. C. (1988). Effects of long lasting estradiol and progesterone on female incubation behavior in free-living starlings. Presented at the XIXth International Ornithological Congress. Ottawa, Canada, June 22–29, 1986. University of Ottawa Press, Ottawa, Ontario.

Ball, G. F., Dufty, A. M., Goldsmith, A. R., and Buntin, J. D. (1988). Autoradiographic localization of brain prolactin receptors in a parental and non-parental songbird species. *Soc. Neurosci. Abstr.* **14,** 88.

Ball, G. F., Dufty, A. M., Johnson, A. E., and Buntin, J. D. (1990). Autoradiographic localization of brain prolactin receptors in three songbird species. Presented at the 22nd Annual Conference on Reproductive Behavior. Atlanta, GA, June 8–11, 1990.

Balthazart, J., Blaustein, J. D., Cheng, M.-F., and Feder, H. H. (1980). Hormones modulate the concentration of cytoplasmic progestin receptors in the brain of male ring doves (*Streptopelia risoria*). *J. Endocrinol.* **86,** 251–261.

Bates, R. W., Miller, R. A., and Garrison, M. M. (1962). Evidence in the hypophysectomized pigeon of a synergism among prolactin, growth hormone, thyroxine and prednisone upon weight of the body, digestive tract, kidney and fat stores. *Endocrinology* **71,** 345–360.

Berghman, L. R., Grauwels, L., Vanhamme, L., Proudman, J. A., Foidart, A., Balthazart, J., and Vandesande, F. (1992). Immunocytochemistry and immunoblotting of avian prolactins using polyclonal and monoclonal antibodies toward a synthetic fragment of chicken prolactin. *Gen. Comp. Endocrinol.* **85,** 346–357.

Book, C. M., Millam, J. R., Guinan, M. J., and Kitchell, R. L. (1991). Brood patch innervation and its role in the onset of incubation in the turkey hen. *Physiol. Behav.* **50,** 281–285.

Brisbin, I. L. (1969). Bioenergetics of the breeding cycle of the ring dove. *Auk* **86,** 54–74.

Buntin, J. D. (1977). Stimulus requirements for squab-induced crop sac growth and nest occupation in ring doves (*Streptopelia risoria*). *J. Comp. Physiol. Psychol.* **91,** 17–28.

Buntin, J. D. (1979). Prolactin release in parent ring doves after brief exposure to their young. *J. Endocrinol.* **82,** 127–130.

Buntin, J. D. (1989). Time course and response specificity of prolactin-induced hyperphagia in ring doves. *Physiol. Behav.* **45,** 903–909.

Buntin, J. D., and Figge, G. R. (1988). Prolactin and growth hormone stimulate food intake in ring doves. *Pharmacol. Biochem. Behav.* **31,** 533–540.

Buntin, J. D., and Ruzycki, E. (1987). Characterization of prolactin binding sites in the brain of the ring doves (*Streptopelia risoria*). *Gen. Comp. Endocrinol.* **65,** 243–253.

Buntin, J. D., and Tesch, D. (1985). Effects of intracranial prolactin administration on maintenance of incubation readiness, ingestive behavior, and gonadal condition in ring doves. *Horm. Behav.* **19,** 188–203.

Buntin, J. D., and Walsh, R. J. (1988). In vivo autoradiographic analysis of prolactin binding sites in brain and choroid plexus of the domestic ring dove. *Cell Tissue Res.* **251,** 105–109.

Buntin, J. D., Cheng M.-F., and Hansen, E. W. (1977). Effect of parental feeding activity on squab-induced crop sac growth in ring doves (*Streptopelia risoria*). *Horm. Behav.* **8,** 297–309.

Buntin, J. D., Lea, R. W., and Figge, G. R. (1988). Reductions in plasma LH concentration and testicular weight in ring doves following intracranial injection of prolactin or growth hormone. *J. Endocrinol.* **118,** 33–40.

Buntin, J. D., Becker, G. M., and Ruzycki, E. (1991). Facilitation of parental behavior in ring doves by systemic or intracranial injections of prolactin. *Horm. Behav.* **25,** 424–444.

Buntin J. D., Ruzycki, E., and Witebsky, J. (1993). Prolactin receptors in dove brain: Autoradiographic analysis of binding characteristics in discrete brain regions and accessibility to blood-borne prolactin. *Neuroendocrinology* **57,** 738–750.

Burke, W. H., and Dennison, P. T. (1980). Prolactin and luteinizing hormone levels in female turkeys (*Meleagris gallopavo*) during a photoinduced reproductive cycle and broodiness. *Gen. Comp. Endocrinol.* **41**, 92–100.

Burley, N. (1980). Clutch overlap and clutch size: Alternative and complementary reproductive tactics. *Am. Nat.* **115**, 223–246.

Burrows, W. H., and Byerly, T. C. (1938). The effect of certain groups of environmental factors upon the expression of broodiness. *Poult. Sci.* **17**, 324–330.

Cain, J. R., Snodgrass, J. D., and Gore, H. G. (1978). Induced broodiness and imprinting in wild turkeys. *Poult. Sci.*, **57**, 1122–1123.

Campbell, R. R., Etches, R. J., and Leatherland, J. F. (1981). Seasonal changes in plasma prolactin concentration and carcass lipid levels in the lesser snow goose (*Anser caerulescens caerulescens*). *Comp. Biochem. Physiol.* **68**, 653–657.

Camper, P. M., and Burke, W. H. (1977). The effects of prolactin on the gonadotropin induced rise in serum estradiol and progesterone of the laying turkey. *Gen. Comp. Endocrinol.* **32**, 72–77.

Cheng, M.-F. (1975). Induction of incubation behavior in male ring doves (*Streptopelia risoria*): A behavioral analysis. *J. Reprod. Fertil.* **42**, 267–276.

Cheng, M.-F. (1992). For whom does the female dove coo? A case for the role of vocal self-stimulation. *Anim. Behav.* **43**, 1035–1044.

Cheng, M.-F., and Burke, W. H. (1983). Serum prolactin levels and crop-sac development in ring doves during a breeding cycle. *Horm. Behav.* **17**, 54–65.

Cheng, M.-F., and Silver, R. (1975). Estrogen-progesterone regulation of nest building and incubation behavior in ovariectomized ring doves (*Streptopelia risoria*). *J. Comp. Physiol. Psychol.* **88**, 256–263.

Cherel, Y., Mauget ,R., Lacroix, A., and Gilles, J. (1994). Seasonal and fasting-related changes in circulating gonadal steroids and prolactin in king penguins, *Aptenodytes patagonicus*. *Physiol. Zool.* **76**, 1154–1173.

Collias, N. E. (1950). Hormones and behavior with special reference to birds and the mechanisms of hormone action. *In* "A Symposium on Steroid Hormones" (E. S. Gordon, ed.), pp. 277–329. Univ. Wisconsin Press (Business Press, Inc.), Lancaster, PA.

Crispens, C. G., Jr. (1956). Prolactin: An evaluation of its use in ring neck pheasant propagation. *J. Wildl. Manage.* **20**, 453–455.

Crispens, C. G., Jr. (1957). Use of prolactin to induce broodiness in two wild turkeys. *J. Wildl. Manage.* **21**, 462.

Dawson, A., and Goldsmith, A. R. (1982). Prolactin and gonadotropin secretion in wild starlings (*Sturnus vulgaris*) during the annual cycle and in relation to nesting, incubation, and rearing young. *Gen. Comp. Endocrinol.* **48**, 213–221.

Dawson, A., and Goldsmith, A. R. (1983). Plasma prolactin and gonadotrophins during gonadal development and the onset of photorefractoriness in male and female starlings (*Sturnus vulgaris*). *J. Endocrinol.* **97**, 253–260.

deGreef, W. J., Ooms, M. P., Vreeburg, J. T. M., and Weber, R. F. A. (1995). Plasma levels of luteinizing hormone during hyperprolactinemia: Response to central administration of antagonists of corticotropin-releasing factor. *Neuroendocrinology* **61**, 19–26.

Deviche, P., and Buntin, J. D. (1992). Hypothalamic localization of prolactin and opioid binding sites in a male migratory songbird. *Soc. Neurosci. Abstr.* **18**, 1164.

DeVito, W. J., Avakian, C., Stone, S., and Ace, C. I. (1992). Estradiol increases prolactin synthesis and prolactin messenger ribonucleic acid in selected brain regions in the hypophysectomized female rat. *Endocrinology* **131**, 2154–2160.

Dittami, J. (1981). Seasonal changes in the behavior and plasma titers of various hormones in barheaded geese, *Anser indicus*. *Z. Tierpsychol.* **55**, 289–324.

Dufty, A. M., Jr., Goldsmith, A. R., and Wingfield, J. C. (1987). Prolactin secretion in a brood parasite: The brown-headed cowbird, *Molothrus ater. J. Zool. (London)* **212,** 669–675.
Dutt, A., Kaplitt, M. G., Kow, L.-M., and Pfaff, D. W. (1994). Prolactin, central nervous system and behavior: A critical review. *Neuroendocrinology* **59,** 413–419.
Ebling, F. J. P., Goldsmith, A. R., and Follett, B. K. (1982). Plasma prolactin and luteinizing hormone during photoperiodically-induced testicular growth and regression in starlings (*Sturnus vulgaris*). *Gen. Comp. Endocrinol.* **48,** 485–490.
Eisner, E. (1969). The effect of hormone treatment upon the duration of incubation in the Bengalese finch. *Behaviour* **33,** 262–276.
El Halawani, M. E., Burke, W. H., and Dennison, P. T. (1980a). Effect of nest deprivation on serum prolactin level in nesting female turkeys. *Biol. Reprod.* **23,** 113–123.
El Halawani, M. E., Burke, W. H., and Dennison, P. T. (1980b). Effects of p-chlorophenylalanine on the rise in serum prolactin associated with nesting in broody turkeys. *Biol. Reprod.* **23,** 815–819.
El Halawani, M. E., Silsby, J. L., Fehrer, S. C., and Behnke, E. J. (1983). Effects of estrogen and progesterone on serum prolactin and luteinizing hormone levels in ovariectomized turkeys (*Meleagris gallopavo*). *Gen. Comp. Endocrinol.* **52,** 67–78.
El Halawani, M. E., Silsby, J. L., Behnke, E. J., and Fehrer, S. C. (1986). Hormonal induction of incubation in ovariectomized female turkeys (*Meleagris gallopavo*). *Biol. Reprod.* **35,** 59–67.
El Halawani, M. E., Silsby, J. L., and Mauro, L. J. (1990). Vasoactive intestinal peptide is a hypothalamic prolactin-releasing neuropeptide in the turkey (*Meleagris gallopavo*). *Gen. Comp. Endocrinol.* **78,** 66–73.
El Halawani, M. E., Silsby, J. L., Rozenboim, I., and Pitts, G. R. (1995). Increased egg production by active immunization against vasoactive intestinal peptide in the turkey (*Meleagris gallopavo*). *Biol. Reprod.* **52,** 179–183.
Etches, R. J., Garbutt, A., and Middleton, A. L. (1979). Plasma concentrations of prolactin during egg laying and incubation in the ruffed grouse (*Bonasa umbellus*). *Can. J. Zool.* **57,** 1624–1627.
Fechner, J. H., Jr., and Buntin, J. D. (1989). Localization of prolactin binding sites in ring dove brain by quantitative autoradiography. *Brain Res.* **487,** 245–254.
Feder, H. H., Storey, A., Goodwin, D., Reboulleau, C., and Silver, R. (1977). Testosterone and "5α-dihydrotestoterone" levels in periheral plasma of male and female ring doves (*Streptopelia risoria*). *Biol. Reprod.* **16,** 666–667.
Fivizzani, A. J., and Oring, L. W. (1986). Plasma steroid hormones in relation to behavioral sex role reversal in the spotted sandpiper, *Actitis macularia. Biol. Reprod.* **35,** 1195–1201.
Fivizzani, A. J., El Halawani, M. E., Ottinger, M. A., and Buntin, J. D. (1993). Autoradiographic localization of CNS prolactin receptors and hypothalamic LHRH in the sex-role reversed Wilson's phalarope. *Soc. Neurosci. Abstr.* **19,** 819.
Fleming, A. S., and Rosenblatt, J. S. (1974). Maternal behavior in the virgin and lactating rat. *J. Comp. Physiol. Psychol.* **86,** 957–972.
Friedman, M. C., and Lehrman, D. S. (1968). Physiological conditions for the stimulation of prolactin secretion by external stimuli in the male ring dove. *Anim. Behav.* **16,** 233–237.
Goldsmith, A. R. (1982a). Plasma concentrations of prolactin during incubation and parental feeding throughout repeated breeding cycles in canaries (*Serinus canarius*). *J. Endocrinol.* **94,** 51–59.
Goldsmith, A. R. (1982b). The Australian black swan (*Cygnus atratus*): Prolactin and gonadotrophin secretion during breeding including incubation. *Gen. Comp. Endocrinol.* **46,** 458–462.

Goldsmith, A. R. (1983). Prolactin in avian reproductive cycles. *In* "Hormones and Behavior in Higher Vertebrates" (J. Balthazart, E. Pröve, and R. Gilles, eds.), pp. 375–387. Springer-Verlag, Berlin.

Goldsmith, A. R. (1991). Prolactin and avian reproductive strategies. *Acta Congr. Int. Ornithol. XX* **4,** 2063–2071.

Goldsmith, A. R., and Williams, D. M. (1980). Incubation in mallards (*Anas platyrhynchos*): Changes in plasma levels of prolactin and luteinizing hormone. *J. Endocrinol.* **86,** 371–379.

Goldsmith, A. R., Edwards, C., Koprucu, M., and Silver, R. (1981). Concentrations of prolactin and luteinizing hormone in plasma of doves in relation to incubation and development of the crop gland. *J. Endocrinol.* **90,** 437–443.

Goldsmith, A. R., Burke, S., and Prosser, J. M. (1984). Inverse changes in plasma prolactin and LH concentrations in female canaries after deprivation and reinitiation of incubation. *J. Endocrinol.* **103,** 251–256.

Gratto-Trevor, C. L., Oring, L. W., Fivizzani, A. J., El Halawani, M. E., and Cooke, F. (1990). The role of prolactin in parental care in a monogamous and a polyandrous shorebird. *Auk* **107,** 718–729.

Gray, J. M., Yarian, D., and Ramenofsky, M. (1990). Corticosterone, foraging behavior, and metabolism in dark-eyed juncos, *Junco hyemalis. Gen. Comp. Endocrinol.* **79,** 375–384.

Hall, M. R. (1986). Plasma concentrations of prolactin during the breeding cycle in the Cape gannet (*Sula capensis*): A foot incubator. *Gen. Comp. Endocrinol.* **64,** 112–121.

Hall, M. R. (1987). External stimuli affecting incubation behavior and prolactin secretion in the duck (*Anas platyrhynchos*). *Horm. Behav.* **21,** 269–287.

Hall, M. R., and Goldsmith, A. R. (1983). Factors affecting prolactin secretion during breeding and incubation in the domestic duck (*Anas platyrhynchos*). *Gen. Comp. Endocrinol.* **49,** 270–276.

Hansen, B. L., and Hansen, G. N. (1982). Immunocytochemical demonstration of somatotropin-like and prolactin-like activity in the brain of *Calamoichthys calabaricus* (Actinopterygii). *Cell Tissue Res.* **222,** 615–627.

Hansen, E. W. (1966). Squab-induced crop growth in ring dove foster parents. *J. Comp. Physiol. Psychol.* **62,** 120–122.

Hansen, E. W. (1971). Squab-induced crop growth in experienced and inexperienced ring dove (*Streptopelia risoria*) foster parents. *J. Comp. Physiol. Psychol.* **77,** 375–381.

Harlan, R. E., Shivers, B. D., Fox, S. R., Kaplove, K. A., Schachter, B. S., and Pfaff, D. W. (1989). Distribution and partial characterization of immunoreactive prolactin in the rat brain. *Neuroendocrinology* **49,** 7–22.

Harwood, D., and Vowles, D. M. (1967). Defensive behavior and the after effects of brain stimulation in the ring dove (*Streptopelia risoria*). *Neuropsychologia* **5,** 345–366.

Hector, J. A. L., and Goldsmith, A. R. (1985). The role of prolactin during incubation: Comparative studies of three *Diomedea* albatrosses. *Gen. Comp. Endocrinol.* **60,** 236–243.

Hegner, R. E., and Wingfield, J. C. (1987). Effects of experimental manipulation of testosterone levels on parental investment and breeding success in male house sparrows. *Auk* **104,** 462–469.

Hiatt, E., Goldsmith, A. R., and Farner, D. S. (1987). Plasma levels of prolactin and gonadotropins during the reproductive cycle of white crowned sparrows (*Zonotrichia leucophrys*). *Auk* **104,** 208–217.

Hnasko, R. M., and Buntin, J. D. (1993). Functional mapping of neural sites mediating prolactin-induced hyperphagia in doves. *Brain Res.* **623,** 257–266.

Hohn, E. O. (1971). Attempted hormonal induction of brood patches and broodiness in ducks. *Auk* **88,** 674–676.

Hohn, E. O. (1981). Failure to induce incubation behavior with estradiol and prolactin and hormonal induction of brood patches in Japanese quail (*Coturnix coturnix japonica*). *Gen. Comp. Endocrinol.* **44,** 396–399.

Horseman, N. D., and Buntin, J. D. (1995). Regulation of pigeon crop milk secretion and parental behaviors by prolactin. *Annu. Rev. Nutr.* (in press).

Hutchison, R. E. (1975). Effects of ovarian steroids and prolactin on the sequential development of nesting behavior in female budgerigars. *J. Endocrinol.* **67,** 29–39.

Janik, D. S., and Buntin, J. D. (1985). Behavioral and physiological effects of prolactin in incubating ring doves. *J. Endocrinol.* **105,** 201–209.

Jones, R. E. (1971). The incubation patch of birds. *Biol. Rev.* **46,** 315–329.

Juss, T. S., and Goldsmith, A. R. (1992). Intracerebroventricular prolactin is potently gonado-inhibitory but does not induce photorefractoriness. Presented at V International Symposium on Avian Endocrinology, Edinburgh, U.K., September 13–17, 1992. AFRC Institute of Animal Physiology and Genetics, Roslin, Midlothian, U.K.

Kendeigh, S. C. (1952). Parental care and its evolution in birds. *Ill. Biol. Monogr.* **22,** 1–356.

Kent, J. P. (1992). Maternal aggression and inter-individual distance in the broody hen (*Gallus gallus*). *Behav. Proc.* **27,** 37–44.

Ketterson, E. D., Nolan, V., Jr., and Wolf, L. (1990). Effect of sex, stage of reproduction, season, and mate removal on prolactin in dark-eyed juncos. *Condor* **92,** 922–930.

Ketterson, E. D., Nolan, V., Jr., Wolf, L., and Ziegenfus, C. (1992). Testosterone and avian life histories: Effects of experimentally elevated testosterone on behavior and correlates of fitness in the dark-eyed junco (*Junco hyemalis*). *Am. Nat.* **140,** 980–999.

King, B. M. (1991). Ventromedial hypothalamic obesity: A reexamination of the irritative hypothesis. *Neurosci. Biobehav. Rev.* **15,** 341–347.

Klinghammer, E., and Hess, E. H. (1964). Parental feeding in ring doves (*Streptopelia risoria*): Innate or learned? *Z. Tierpsychol.* **21,** 338–347.

Komisaruk, B. R. (1967). Effects of local brain implants of progesterone on reproductive behavior in ring doves. *J. Comp. Physiol. Psychol.* **64,** 219–224.

Korenbrot, C. C., Schomberg, D. W., and Erickson, C. J. (1974). Radioimmunoassay of plasma estradiol during the breeding cycle of ring doves (*Streptopelia risoria*). *Endocrinology* **94,** 1126–1132.

Kuenzel, W. J. (1994). Central neuroanatomical systems involved in the regulation of food intake in birds and mammals. *J. Nutr.* **124,** 1355S–1370S.

Lea, R. W. (1987). Prolactin and avain incubation: A comparison between Galliformes and Columbiformes. *Sitta* **1,** 117–141.

Lea, R. W., and Sharp, P. J. (1982). Plasma prolactin concentrations in broody turkeys: Lack of agreement between homologous chicken and turkey prolactin radioimmunoassays. *Br. Poult. Sci.* **23,** 451–459.

Lea, R. W., and Sharp, P. J. (1989). Concentrations of plasma prolactin and luteinizing hormone following nest deprivation and renesting in ring doves (*Streptopelia risoria*). *Horm. Behav.* **23,** 279–289.

Lea, R. W., and Sharp, P. J. (1991). Effects of presence of squabs upon plasma concentrations of prolactin and LH and length of time of incubation in ringdoves on "extended" incubatory patterns. *Horm. Behav.* **25,** 275–282.

Lea, R. W., and Vowles, D. M. (1985). The control of prolactin secretion and nest defense in the ring dove (*Streptopelia risoria*). *Ital. Boll. Zool.* **52,** 323–329.

Lea, R. W., and Vowles, D. M. (1986). Vasoactive intestinal polypeptide stimulates prolactin release in vivo in the ring dove (*Streptopelia risoria*). *Experientia* **42,** 420–422.

Lea, R. W., Dods, A. S. M., Sharp, P. J., and Chadwick, A. (1981). The possible role of prolactin in the regulation of nesting behavior and the secretion of luteinizing hormone in broody bantams. *J. Endocrinol.* **91,** 89–97.

Lea, R. W., Sharp, P. J., and Chadwick, A. (1982). Daily variations in the concentrations of plasma prolactin in broody bantams. *Gen. Comp. Endocrinol.* **48,** 275–284.

Lea, R. W., Vowels, D. M., and Dick, H. R. (1986). Factors affecting prolactin secretion during the breeding cycle of the ring dove (*Streptopelia risoria*) and its possible role in incubation. *J. Endocrinol.* **110,** 447–458.

Lea, R. W., Talbot, R. T., and Sharp, P. J. (1991). Passive immunization against chicken vasoactive intestinal polypeptide suppresses plasma prolactin and crop sac development in incubating ring doves. *Horm. Behav.* **25,** 283–294.

Lea, R. W., Klandorf, H., Harvey, S., and Hall, T. R. (1992). Thyroid and adrenal function in the ring dove (*Streptopelia risoria*) during food deprivation and a breeding cycle. *Gen. Comp. Endocrinol.* **86,** 138–146.

Leboucher, G., Richard-Yris, M.-A., Williams, J. A., and Chadwick, A. (1990). Incubation and maternal behaviour in domestic hens: Influence of the presence of chicks on circulating luteinising hormone, prolactin, and oestradiol and on behaviour. *Br. Poult. Sci.* **31,** 851–862.

Leboucher, G., Richard-Yris, M.-A. Guémené, D., and Chadwick, A. (1993). Respective effects of chicks and nest on behavior and hormonal concentrations of incubating domestic hens. *Physiol. Behav.* **54,** 135–140.

Lehrman, D. S. (1955). The physiological basis of parental feeding behavior in the ring dove (*Streptopelia risoria*). *Behavior* **7,** 241–286.

Lehrman, D. S. (1958). Effect of female sex hormones on incubation behavior in the ring dove (*Streptopelia risoria*). *J. Comp. Physiol. Psychol.* **51,** 142–145.

Lehrman, D. S., and Brody, P. N. (1961). Does prolactin induce incubation behavior in the ring dove? *J. Endocrinol.* **22,** 269–275.

Lehrman, D. S., and Brody, P. N. (1964). Effect of prolactin on established incubation behavior in the ring dove. *J. Comp. Physiol. Psychol.* **57,** 161–165.

Lehrman, D. S., Brody, P. N., and Wortis, R. P. (1961). The presence of the mate and of nesting material as stimuli for the development of incubation behavior and for gonadotrophin secretion in the ring dove (*Streptopelia risoria*). *Endocrinology* **68,** 507–516.

Leinhart, R. (1927). Contribution a l'étude de l'incubation. *C. R. Soc. Biol. (Paris)* **97,** 1296–1297.

Li, C., Kelly, P. A., and Buntin, J. D. (1995). Inhibitory effects of anti-prolactin receptor antibodies on prolactin binding in brain and prolactin-induced feeding behavior in ring doves. *Neuroendocrinology* **61,** 125–135.

Lott, D. F., and Comerford, S. (1968). Hormonal initiation of parental behavior in inexperienced ring doves. *Z. Tierpsychol.* **25,** 71–75.

Macnamee, M. C., Sharp, P. J., Lea, R. W., Sterling, R. J., and Harvey, S. (1986). Evidence that vasoactive intestinal polypeptide is a physiological prolactin-releasing factor in the bantam hen. *Gen. Comp. Endocrinol.* **62,** 470–478.

March, J. B., Sharp, P. J., Wilson, P. W., and Sang, H. M. (1994). Effect of active immunization against recombinant-derived chicken prolactin fusion protein on the onset of broodiness and photoinduced egg laying in bantam hens. *J. Reprod. Fertil.* **101,** 227–233.

Maxon, S. J., and Oring, L. W. (1980). Breeding season time and energy budgets of the polyandrous spotted sandpiper. *Behavior* **74,** 200–263.

Medway, L. (1961). Domestic pigeons: The stimulus provided by the egg in the nest. *J. Endocrinol.* **23,** 9–18.

Michel, G. F., and Moore, C. L. (1985). Contribution of nesting experience to progesterone-induced incubation in ring doves (*Streptopelia risoria*). *J. Comp. Physiol. Psychol.* **99,** 259–266.

Michel, G. F., and Moore, C. L. (1986). Contributions of reproductive experience to observation-maintained crop growth and incubation in male and female ring doves. *Anim. Behav.* **34,** 790–796.

Miller, R. A., and Riddle, O. (1943). Ability of adrenal cortical hormones, prolactin and thyroxin to sustain weight of body and viscera of hypophysectomized pigeons. *Endocrinology* **32,** 463–474.

Mondloch, C. J., and Timberlake, W. (1991). The effect of parental food supply on parental feeding and squab growth in pigeons, *Columba livia. Ethology* **88,** 236–248.

Moore, C. L. (1976). Experiential and hormonal conditions affect squab-egg choice in ring doves (*Streptopelia risoria*). *J. Comp. Physiol. Psychol.* **90,** 583–589.

Moore, M. C. (1982). Hormonal response of free-living male white-crowned sparrows to experimental manipulation of female sexual behavior. *Horm. Behav.* **16,** 323–329.

Moore, M. C. (1983). Effect of female displays on the endocrine physiology and behavior of male white-crowned sparrows, *Zonotrichia leucophrys. J. Zool. (London)* **199,** 137–148.

Muccioli, G., Bellussi, G., Ghe, C., Pagnini, G., and DiCarlo, R. (1988). Regional distribution and species variation of prolactin binding sites in the brain. *Gen. Comp. Endocrinol.* **69,** 399–405.

Muccioli, G., Guardabassi, A., and Pattono, P. (1990). Biochemical study of prolactin binding sites in *Xenopus laevis* brain and choroid plexus. *J. Exp. Zool.* **253,** 311–318.

Myers, S. A., Millam, J. R., and El Halawani, M. E., (1989). Plasma LH and prolactin levels during the reproductive cycle of the cockatiel (*Nymphicus hollandicus*). *Gen. Comp. Endocrinol.* **73,** 85–91.

Nalbandov, A. V., and Card, L. E. (1945). Endocrine identification of the broody genotype of cocks. *J. Hered.* **36,** 251–258.

Numan, M. (1994). Maternal behavior. *In* "The Physiology of Reproduction," 2nd ed. (E. Knobil and J. D. Neill, eds.), pp. 221–302. Raven Press, New York.

Opel, H. (1971). Induction of incubation behavior in the hen by brain implants of prolactin. *Poult. Sci.* **50,** 1613.

Opel, H., and Proudman, J. (1980). Failure of mammalian prolactin to induce incubation behavior in chickens and turkeys. *Poult. Sci.* **59,** 2250–2258.

Opel, H., and Proudman, J. A. (1988). Effects of poults on plasma concentrations of prolactin in turkey hens incubating without eggs or a nest. *Br. Poult. Sci.* **29,** 791–800.

Opel, H., and Proudman, J. A. (1989). Plasma prolactin levels in incubating turkey hens during pipping of the eggs and after introduction of poults into the nest. *Biol. Reprod.* **40,** 981–987.

Oring, L. W., and Fivizzani, A. J. (1991). Reproductive endocrinology of sex-role reversal. *Acta Congr. Int. Ornithol. XX* **4,** 2072–2080.

Oring, L. W., Fivizzani, A. J., and El Halawani, M. E. (1986a). Changes in plasma prolactin associated with laying and hatch in the spotted sandpiper. *Auk* **103,** 820–822.

Oring, L. W., Fivizzani, A. J., El Halawani, M. E., and Goldsmith, A. R. (1986b). Seasonal changes in prolactin and luteinizing hormone in the polyandrous spotted sandpiper, *Actitis macularia. Gen. Comp. Endocrinol.* **62,** 394–403.

Oring, L. W., Fivizzani, A. J., Colwell, M. A., and El Halawani, M. E. (1988). Hormonal changes associated with natural and manipulated incubation in the sex-role reversed Wilson's phalarope. *Gen. Comp. Endocrinol.* **72,** 247–256.

Oring, L. W., Fivizzani, A. J., and El Halawani, M. E. (1989). Testosterone-induced inhibition of incubation in the spotted sandpiper (*Actitis macularia*). *Horm. Behav.* **23,** 412–423.

Patel, M. D. (1936). The physiology of the formation of "pigeon's milk." *Physiol. Zool.* **9,** 129–152.

Pedersen, H. C., (1989). Effects of exogenous prolactin on parental behavior in free-living female willow ptarmigan *Lagopus l. lagopus. Anim. Behav.* **38,** 926–934.

Phillips, R. E. (1964). "Wildness" in the mallard duck: Effects of brain lesions and stimulation on "escape behavior" and reproduction. *J. Comp. Neurol.* **122,** 139–156.

Phillips, R. E., and Youngren, O. M. (1971). Brain stimulation and species-typical behavior: Activities evoked by electrical stimulation of the brains of chickens (*Gallus gallus*). *Anim. Behav.* **19,** 757–779.

Pietras, R. J., and Wenzel, B. M. (1974). Effects of androgens on body weight, feeding and courtship behavior in the pigeon. *Horm. Behav.* **5,** 289–302.

Pinxten, R., Eens, M., and Verheyen, R. F. (1993). Male and female nest attendance during incubation in the facultatively polygynous European starling. *Ardea* **81,** 125–133.

Pitts, G. R., Youngren, O. M., Silsby, J. L., Rozenboim, I., Chaiseha, Y., Phillips, R. E., Foster, D. N., and El Halawani, M. E. (1994). Role of vasoactive intestinal peptide in the control of prolactin-induced turkey incubation behavior. II. Chronic infusion of vasoactive intestinal peptide. *Biol. Reprod.* **50,** 1350–1356.

Posner, B. I., van Houten, M., Patel, B., and Walsh, R. J. (1983). Characterization of lactogen binding sites in choroid plexus. *Exp. Brain Res.* **49,** 300–306.

Ramos, C., and Silver, R. (1992). Gonadal hormones determine sex differences in timing of incubation by doves. *Horm. Behav.* **26,** 586–601.

Ramsey, S. M., Goldsmith, A. R., and Silver, R. (1985). Stimulus requirements for prolactin and LH secretion in incubating ring doves. *Gen. Comp. Endocrinol.* **59,** 246–256.

Richard-Yris, M.-A., and Leboucher, G. (1987a). Effects of exposure to chicks on maternal behavior in domestic chickens. *Bird Behav.* **7,** 31–36.

Richard-Yris, M.-A., and Leboucher, G. (1987b). Induction and maintenance of maternal behavior in the domestic hen: Influence of testosterone and oestradiol treatments. *Ethology* **75,** 337–347.

Richard-Yris, M.-A., Garnier, D. H., and Leboucher, G. (1983). Induction of maternal behavior and some hormonal and physiological correlates in the domestic hen. *Horm. Behav.* **17,** 345–355.

Richard-Yris, M.-A. Leboucher, G., Chadwick, A., and Garnier, D. H. (1987a). Induction of maternal behavior in incubating and non-incubating hens: Influence of hormones. *Physiol. Behav.* **40,** 193–199.

Richard-Yris, M.-A., Leboucher, G., Williams, J., and Garnier, D. H. (1987b). Influence of food restriction and of the presence of chicks on the reproductive system of the domestic hen. *Br. Poult. Sci.* **28,** 251–260.

Richard-Yris, M.-A., Chadwick, A., Guémené, D., Grillou-Schuelke, H., and Leboucher, G. (1995). Influence of the presence of chicks on the ability to resume incubation behavior in domestic hens (*Gallus domesticus*). *Horm. Behav.* **29,** 425–441.

Riddle, O., Bates, R. W., and Lahr, E. L. (1935). Prolactin induces broodiness in fowl. *Am. J. Physiol.* **111,** 352–360.

Rosenblatt, J. S. (1967). Nonhormonal basis of maternal behavior in the rat. *Science (Washington, D. C.)* **156,** 1512–1514.

Rosenblatt, J. S. (1992). Hormone-behavior relations in the regulation of parental behavior. *In* "Behavioral Endocrinology" (J. B. Becker, S. M. Breedlove, and D. Crews, eds.), pp. 219–260. MIT Press, Cambridge, MA.

Saeki, Y., and Tanabe, Y. (1955). Changes in prolactin content of fowl pituitary during broody periods and some experiments on the induction of broodiness. *Poult. Sci.* **34,** 909–919.

Schoech, S. J., Mumme, R. L., and Wingfield, J. C. (1996). Prolactin and helping behavior in the cooperatively breeding Florida scrub jay (*Aphelocoma c. coerulescens*). *Anim. Behav.* (in press).

Schooley, J. P., Riddle, O., and Bates, R. W. (1941). Replacement therapy in hypophysectomized juvenile pigeons. *Am. J. Anat.* **69,** 124–154.

Seiler, H. W., Gahr, M., Goldsmith, A. R., and Güttinger, H.-R. (1992). Prolactin and gonadal steroids during the reproductive cycle of the Bengalese finch (*Lonchura striata var. domestica,* Estrildidae), a nonseasonal breeder with biparental care. *Gen. Comp. Endocrinol.* **88,** 83–90.

Selander, R. K., and Kuich, L. L. (1960). Hormonal control and development of the incubation patch in icterids, with notes on behavior of cowbirds. *Condor* **65,** 73–90.

Selander, R. K., and Yang, S. Y. (1966). Behavioral responses of brown-headed cowbirds to nests and eggs. *Auk* **83,** 207–232.

Sharp, P. J., Scanes, C. G., Williams, J. B., Harvey, S., and Chadwick, A. (1979). Variations in concentrations of prolactin, luteinizing hormone, growth hormone, and progesterone in the plasma of broody bantams. *J. Endocrinol.* **80,** 51–57.

Sharp, P. J., Sterling, R. J., and Pedersen, H.-C. (1986). An assessment of the roles of prolactin and ovarian hormones in the maintenance of low plasma LH levels in broody bantams (*Gallus domesticus*). *J. Endocrinol.* **108** Suppl., Abst. No. 93.

Sharp, P. J., Macnamee, M. C., Sterling, R. J., Lea, R. W., and Petersen, H. C. (1988). Relationships between prolactin, luteinizing hormones and broody behavior in bantam hens. *J. Endocrinol.* **118,** 279–286.

Sharp, P. J., Sterling, R. J., Talbot, R. T., and Huskisson, N. S. (1989). The role of hypothalamic vasoactive intestinal polypeptide in the maintenance of prolactin secretion in incubating bantam hens: Observations using passive immunization, radioimmunoassay and immunocytochemistry. *J. Endocrinol.* **122,** 5–13.

Sherry, D. F. (1981). Parental care and the development of thermoregulation in red junglefowl. *Behavior* **76,** 250–279.

Shivers, B. D., Harlan, R. E., and Pfaff, D. W. (1989). A subset of neurons containing immunoreactive prolactin is a target for estrogen regulation of gene expression in rat hypothalamus. *Neuroendocrinology* **49,** 23–27.

Silver, R., and Buntin, J. D. (1973). Role of adrenal hormones in incubation behavior of male ring doves (*Streptopelia risoria*). *J. Comp. Physiol. Psychol.* **84,** 453–463.

Silver, R., and Feder, H. H. (1973). Role of gonadal hormones in incubation behavior of male ring doves (*Streptopelia risoria*) *J. Comp. Physiol. Psychol.* **84,** 464–471.

Silver, R., Feder, H. H., and Lehrman, D. S. (1973). Situational and hormonal determinants of courtship, aggressive and incubation behavior in male ring doves (*Streptopelia risoria*) *Horm. Behav.* **4,** 163–172.

Silver, R., Reboulleau, C., Lehrman, D. S., and Feder, H. H. (1974). Radioimmunoassay of plasma progesterone during the reproductive cycle of male and female ring doves. (*Streptopelia risoria*). *Endocrinology* **94,** 1547–1554.

Silver, R., Andrews, H., and Ball, G. F. (1985). Parental care in an ecological perspective: A quantitative analysis of avian subfamilies. *Am. Zool.* **25,** 823–840.

Silverin, B. (1980). Effects of long-acting testosterone treatment on free-living pied flycatchers, *Ficedula hypoleuca,* during the breeding period. *Anim. Behav.* **28,** 906–912.

Silverin, B. (1981). Reproductive effort, as expressed in body and organ weights, in the pied flycatcher. *Ornis Scand.* **12,** 133–139.

Silverin, B. (1982). Endocrine correlates of brood size in adult pied flycatchers, *Ficedula hypoleuca. Gen. Comp. Endocrinol.* **47,** 18–23.

Silverin, B. (1986). Corticosterone-binding proteins and behavioral effects of high plasma levels of corticosterone during the breeding period in the pied flycatcher. *Gen. Comp. Endocrinol.* **64,** 67–74.

Silverin, B. (1990). Testosterone and corticosterone and their relation to territorial and parental behavior in the pied flycatcher. *In* "Hormones, Brain and Behavior in Vertebrates. 2. Behavioral Activation in Males and Females-Social Interaction and Reproductive Endocrinology" (J. Balthazart, ed.), pp. 129–142. S. Karger, Basel.

Silverin, B. (1991a). Annual changes in plasma levels of LH, and prolactin in free-living female great tits (*Parus major*). *Gen. Comp. Endocrinol.* **83,** 425–431.

Silverin, B. (1991b). Behavioral, hormonal, and morphological responses of free-living male pied flycatchers to estradiol treatment of their mates. *Horm. Behav.* **25,** 38–56.

Silverin, B., and Goldsmith, A. R. (1983). Reproductive endocrinology of free-living pied flycatchers (*Ficedula hypoleuca*): Prolactin and FSH secretion in relation to incubation and clutch size. *J. Zool. (London)* **200,** 119–130.

Silverin, B., and Goldsmith, A. R. (1984). The effects of modifying incubation on prolactin secretion in free-living pied flycatchers. *Gen. Comp. Endocrinol.* **55,** 239–244.

Silverin, B., and Goldsmith, A. R. (1990). Plasma prolactin concentrations in breeding pied flycatchers (*Ficedula hypoleuca*) with an experimentally prolonged brooding period. *Horm. Behav.* **24,** 104–113.

Slawski, B. A., and Buntin, J. D. (1995). Preoptic area lesions disrupt prolactin-induced parental feeding behavior in ring doves. *Horm. Behav.* **29,** 248–266.

Stern, J. M. (1974). Estrogen facilitation of progesterone-induced incubation behavior in castrated male ring doves. *J. Comp. Physiol. Psychol.* **87,** 332–337.

Stern, J. M., and Ehrman, D. S. (1969). Role of testosterone in progesterone-induced incubation behavior in male ring doves (*Streptopelia risoria*). *J. Endocrinol.* **44,** 13–22.

ten Cate, C., Lea, R. W., Ballintijn, M. R., and Sharp, P. J. (1993). Brood size affects behavior, interclutch interval, LH levels, and weight in ring dove (*Streptopelia risoria*) breeding pairs. *Horm. Behav.* **27,** 539–550.

van Tienhoven, A. (1958). Effect of progesterone on established broodiness and egg production of turkeys. *Poult. Sci.* **37,** 428–433.

Vleck, C. M., and Brown, J. L. (1992). Testosterone and prolactin in the "cooperatively-breeding" Mexican jay. Presented at the V International Symposium on Avian Endocrinology, Edinburgh, U.K. September 13–17, 1992. AFRC Institute of Animal Physiology and Genetics, Roslin, Midlothian, U.K.

Vleck, C. M., and Dobrott, S. J. (1993). Testosterone, antiandrogen, and alloparental behavior in bobwhite quail foster fathers. *Horm. Behav.* **27,** 92–107.

Vleck, C. M., Mays, N. A., Dawson, J. W., and Goldsmith, A. R. (1991). Hormonal correlates of parental and helping behavior in cooperatively breeding Harris' hawks (*Parabuteo unicinctus*). *Auk* **108,** 638–648.

Vowles, D. M., and Beazley, L. (1974). The neural substrate of emotional behavior in birds. *In* "Birds: Brain and Behavior" (M. Schein and D. Goodman, eds.), p. 221–258. Academic Press, New York.

Vowles, D. M., and Harwood, D. (1966). The effect of exogenous hormones on aggressive and defensive behavior in the ring dove (*Streptopelia risoria*). *J. Endocrinol.* **36,** 35–51.

Vowles, D. M., and Lea, R. W. (1986). External factors affecting the duration of broody behavior in the ring dove (*Streptopelia risoria*). *Horm. Behav.* **20,** 249–262.

Walsh, R. J., Slaby, F. J., and Posner, B. I. (1987). A receptor-mediated mechanism for the transport of prolactin from blood to cerebrospinal fluid. *Endocrinology* **120,** 1846–1850.

Williams, T. D., and Sharp, P. J. (1993). Plasma prolactin during the breeding season in adult and immature macaroni (*Eudyptes chrysolophus*) and gentoo (*Pygoscelis papus*) penguins. *Gen. Comp. Endocrinol.* **92,** 339–346.

Wingfield, J. C., and Goldsmith, A. R. (1990). Plasma levels of prolactin and gonadal steroids in relation to multiple-brooding and renesting in free-living populations of the song sparrow, *Melospiza melodia*. *Horm. Behav.* **24,** 89–103.

Wingfield, J. C., Ronchi, E., Goldsmith, A. R., and Marler, C. (1989). Interactions of sex steroid hormones and prolactin in male and female song sparrows, *Melospiza melodia. Physiol. Zool.* **62,** 11–24.

Wingfield, J. C., Hegner, R. E., Dufty, A. M., Jr., and Ball, G. F. (1990). The "challenge hypothesis": Theoretical implications for patterns of testosterone secretion, mating systems, and breeding strategies. *Am. Nat.* **136,** 829–846.

Wood-Gush, D. G. M., and Gilbert, A. B. 1973. Some hormones involved in the nesting behavior of hens. *Anim. Behav.* **21,** 99–103.

Wright, G. M. (1986). Immunocytochemical demonstration of growth hormone, prolactin, and somatostatin-like immunoreactivities in the brain of larval, young adult, and upstream migrant adult sea lamprey, *Petromyzon marinus. Cell Tissue Res.* **246,** 23–31.

Youngren, O. M., El Halawani, M. E., Phillips, R. E., and Silsby, J. L. (1989). Effects of preoptic and hypothalamic lesions in female turkeys during a photoinduced reproductive cycle. *Biol. Reprod.* **41,** 610–617.

Youngren, O. M., El Halawani, M. E., Silsby, J. L., and Phillips, R. E. (1991). Intracranial prolactin perfusion induces incubation behavior in turkey hens. *Biol. Reprod.* **44,** 425–431.

Zadworny, D., Shimada, K., Ishida, H., Sumi, C., and Sato, K. (1988). Changes in plasma levels of prolactin and estradiol, nutrient intake, and time spent nesting during the incubation phase of broodiness in the Chabo hen (Japanese bantam). *Gen. Comp. Endocrinol.* **71,** 406–412.

# Biochemical Basis of Parental Behavior in the Rat

ROBERT S. BRIDGES

DEPARTMENT OF COMPARATIVE MEDICINE
TUFTS UNIVERSITY SCHOOL OF VETERINARY MEDICINE
NORTH GRAFTON, MASSACHUSETTS 01536

The focus of this article concerns the involvement of biochemical factors in the regulation of parental behavior in the *Rattus norvegicus*. Specifically, research on the roles of various hormones and neurochemical agents will be reviewed within the context of normal development in female and, to a lesser extent, in male rats. The effects of parental experience on the biochemical regulation of this evolutionarily critical behavior will then be discussed. Finally, a working model for the biochemical regulation of parental care is presented, suggesting possible future avenues of research.

## I. Behavioral Responses of Parental Rats

Parturient rats display a set of behavioral responses over the course of lactation that help to ensure the growth and survival of their offspring. These behaviors can be divided into those that are pup directed and others that are not (see Table I). Specific pup-directed responses exhibited by mother rats over the course of lactation include retrieval of the young to the nest site, grouping the young together, crouching over the pups to provide warmth and the opportunity to suckle, anogenital licking to stimulate micturition and defecation, and tactile stimulation. Non-pup-directed maternal responses include nest building, placentophagia, and defense of the young. Each of these maternal responses are quantifiable and have been used in numerous experimental studies to examine the proximate mechanisms underlying the expression of maternal behavior.

## II. Parental Behavior — A Developmental Perspective

Whereas the basic capacity to display maternal or parental (refers to either sex) care in the rat is present from early development prior to puberty

TABLE I
Behavioral Responses in Parental Rats

| Pup-directed behaviors | Non-pup–directed behaviors |
|---|---|
| Retrieval | Nest building |
| Grouping | Placentophagia |
| Crouching | Defense of the young |
| Anogenital licking | |
| Tactile stimulation–contact | |

throughout adulthood, varying amounts and kinds of sensory stimuli are required to elicit parental care as a function of the animal's physiological and experiential state. The early studies of Rosenblatt (1967) and Cosnier and Couturier (1966) established that behaviorally naive adult female and male rats would show parental responses if kept in constant contact with young pups (approximately 3–10 days of age) for about 1 week, confirming an earlier report by Wiesner and Sheard (1933). The latencies to display retrieval, grouping, and crouching responses ranged from 5 to 8 days of age for both sexes, indicating that the neural substrate necessary to engage in parental behavior was present in both sexes and that the shortened latencies to display parental care at parturition were apparently the result of the physiological conditions associated with pregnancy. The average latency to display parental behavior in female rats during development is depicted in Fig. 1. Interestingly, the latency to display parental care is quite short during the postweaning–prepubertal period in both female and male rats with response latencies equal to 1–2 days. By 30 days of age the latency increases to 5–7 days, a response latency generally retained until the female becomes pregnant. Once pregnant the latency remains elevated at 5–6 days until just prior to parturition when rats exhibit a spontaneous onset of maternal care when presented with foster young (Slotnick et al., 1973; Rosenblatt and Siegel, 1975; Mayer and Rosenblatt, 1984). Female rats that have given birth and raised a litter continue to display short-latency maternal behavior when presented with young weeks after weaning or removal of the young (Bridges, 1975). This retention of maternal responsiveness appears to depend on mother–young interactions during the immediate postpartum period, but it is not dependent upon establishment of nursing (Bridges, 1975; see Fleming, et al., this volume).

Defense of the young, which is often referred to as maternal aggression, appears to occur at two stages. First, during pregnancy dams display heightened levels of pregnancy-induced aggression (Mayer and Rosenblatt, 1984). Second, parturient and lactating females display high levels of postpartum

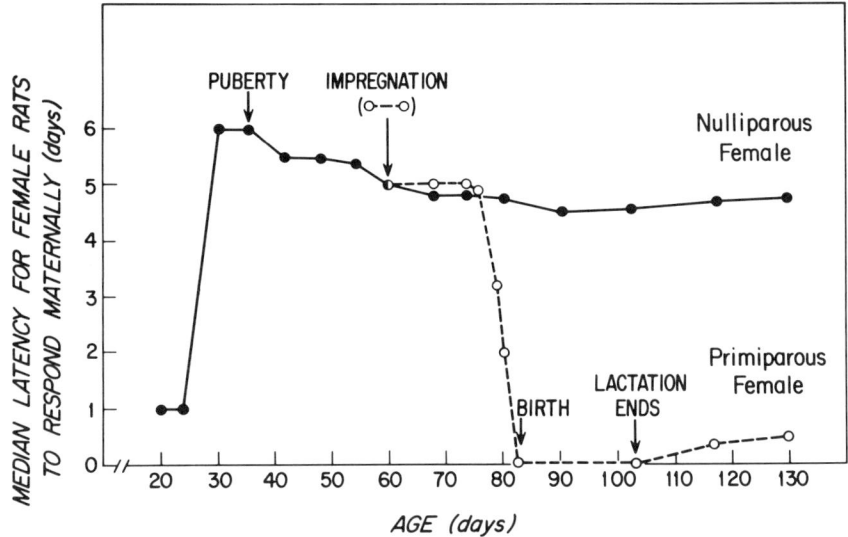

Fig. 1. A developmental profile of maternal responsiveness in female rats. Values represent the median number of days of pup exposure necessary to induce maternal behavior, that is, retrieval, grouping, and crouching, during a 1-h test. (From Mammalian Parenting: *Biochemical, Neurobiological, and Behavioral Determinants,* edited by Norman A. Krasnegor and Robert S. Bridges. Copyright © 1990 by Oxford University Press, Inc. Reprinted by permission.)

aggression. Postpartum maternal aggression is highest at parturition and remains elevated through the first 2 weeks of lactation, declining after Day 14 (Erskine *et al.,* 1978; Mayer *et al.,* 1987). These forms of defensive aggression are typically evoked by an intruder male or female.

## III. Physiological Underpinnings of Parental Behavior

### A. Role of Hormones in Parental Behavior

Endogenous shifts in secretory patterns of hormones and neurochemicals are thought to underlie the shifts in incidences and intensities of parental behavior in female as well as male rats. A substantive body of research has explored the role of hormones and various neurotransmitters in the regulation of parental care. That female rats show pronounced shifts in latencies to display maternal behavior at the end of gestation initially led investigators to explore the relationships between the physiological changes of pregnancy and the expression of maternal behavior. The early studies of Moltz *et al.* (1970) and Zarrow *et al.* (1971) were the first to clearly demonstrate a role for hormones in the induction of maternal behavior. Their studies suggested that the spontaneous onset of maternal behavior in the newly parturient rat was stimulated by a pregnancy-like hormone cocktail consisting of estradiol ($E_2$), progesterone (P), and prolactin (PRL). These three hormones have subsequently been the subject of the majority of studies on the hormonal control of maternal behavior. The specific involvements of these hormones are presented as follows.

*1. Estradiol*

Both Moltz's and Zarrow's studies indicated that $E_2$ was important for the hormonal stimulation of maternal behavior. In these studies injections of P and PRL without $E_2$ failed to stimulate maternal care in the ovariectomzied, nulliparous rat. Strong evidence that emphasized the importance of $E_2$ in stimulating short-latency maternal behavior was provided by the studies of Rosenblatt and Siegel (Rosenblatt and Siegel, 1975; Siegel and Rosenblatt, 1975). Using a hysterectomy–ovariectomy (HO) model in which both the uterus and the ovaries are removed to eliminate potential sites of sex steroid uptake and production, respectively, it was shown that virgin rats injected with estradiol benzoate (EB) displayed full maternal behavior (defined as retrieval, grouping, and crouching within the standard 1-h test session) within 1–2 days. In contrast, vehicle-treated females became maternal after about 5 days of pup exposure (Siegel and Rosenblatt, 1975). Moreover, when behaviorally inexperienced, pregnant rats were hysterectomized on Day 16 of gestation and were first tested for maternal behavior 48 h later, a high percentage (>60%) displayed maternal behavior after only 24 h of exposure to foster pups (Rosenblatt and Siegel, 1975). This high level of responsiveness appears to have resulted from the prolonged hormonal priming associated with pregnancy followed by the decline in P levels and the relatively unopposed $E_2$ exposure in the hysterectomized female (Rosenblatt and Siegel, 1975; Bridges *et al.*, 1978; see Fig. 2 for the normal hormone patterns during pregnancy in rats). Maternal behavior, likewise, rapidly appears in estrogen-treated rats whose pregnancies are surgically terminated by a combination of HO on Day 16 of pregnancy (Rosenblatt and Siegel, 1975). HO on Day 16 of pregnancy results in a rapid and demonstrable decline in $E_2$ and P, which are produced by the ovaries, and in placental hormones; pregnancy followed by HO appears to prime the female to the stimulatory effects of estrogen. It is worth noting that $E_2$ appears to be a key hormone in the induction of maternal behavior. Essentially all reported hormone (P, PRL) as well as peptide (oxytocin, cholecystokinin) stimulatory effects on maternal behavior *require* some exposure to $E_2$. However, it is unknown whether estrogen's actions occur directly on the brain independent of P and PRL to stimulate maternal

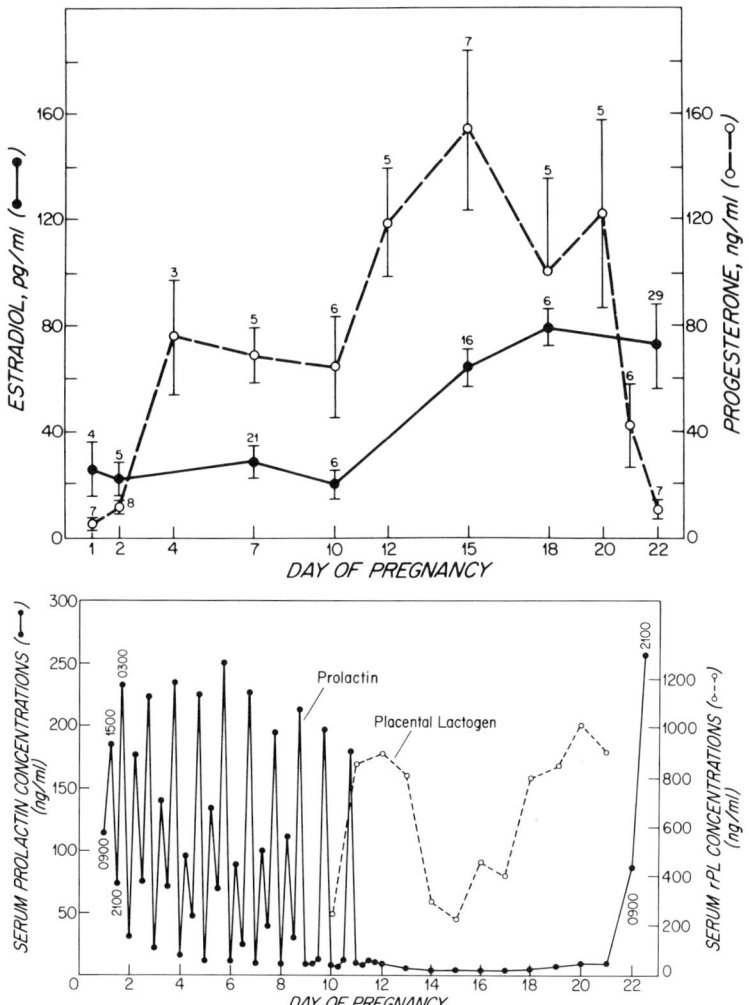

FIG. 2. Profiles of steroid (upper panel) and lactogenic (lower panel) hormone concentrations during pregnancy in rats. (From Mammalian Parenting: *Biochemical, Neurobiological, and Behavioral Determinants,* edited by Norman A. Krasnegor and Robert S. Bridges. Copyright © 1990 by Oxford University Press, Inc. Reprinted by permission.)

behavior under pregnancy conditions or whether estrogen has a permissive action, that is, it allows P and/or PRL to stimulate the neural processes mediating the expression of parental care. It is possible that $E_2$ exerts both direct and permissive actions.

The studies of Numan *et al.* (1977) and Fahrbach and Pfaff (1986) demonstrated that one site of $E_2$ action in the stimulation of maternal behavior is the medial preoptic area (MPOA). This region of the brain is crucial for the expression of both the induction of maternal behavior at the end of pregnancy (Cohn and Gerall, 1989) and in the maintenance of maternal care during lactation in rats (Numan, 1990). Application of $E_2$ to the MPOA of 16-day pregnant, hysterectomized–ovariectomized rats stimulates an almost immediate onset of maternal behavior in these females (Numan *et al.*, 1977). Rosenblatt and his colleagues have quantified nuclear estrogen receptors in various neural regions during pregnancy and after parturition, and have found that estrogen receptor (ER) concentrations in the MPOA are increased during pregnancy and postpartum (Rosenblatt *et al.*, 1994). Injections of EB into Day-16 pregnant rats that are HO also produced transient elevations in ER concentrations. Wagner and Morrell (1995) reported that mRNA for the ER increased in the ventromedial nucleus (VMN) of the hypothalamus just prior to parturition. The actions of $E_2$ in the VMN in controlling maternal behavior are unestablished, although perturbation of this region by lowering cannulas into the VMN of steroid-primed virgins appears to stimulate the induction of maternal care (Bridges and Mann, 1994). Thus, while one mode of estrogen action appears to be through a receptor mechanism, the cellular and neurochemical actions of estrogen in regulating maternal behavior have not been fully characterized.

## 2. *Progesterone*

Progesterone is secreted in large amounts by the corpora lutea of the ovaries during most of gestation in the rat and is essential for pregnancy maintenance. P serum levels first increase on Day 3 of pregnancy and reach an initial plateau at midpregnancy, then increase further during the second half of the 22-day gestation period before declining precipitously about 24–48 h prior to parturition (Morishige *et al.*, 1974; Bridges, 1984; see Fig. 2). Two roles for P in the stimulation of maternal behavior have been proposed. First, the long-term exposure to P together with $E_2$ during gestation helps to prime the female to respond maternally at parturition. Latencies of female rats to respond maternally to foster young progressively decrease as gestation proceeds (Bridges *et al.*, 1977) and as a function of the duration of steroid priming (Bridges, 1984). However, this increased responsivity is only revealed when pregnancy is terminated by procedures such as HO and is first statistically apparent around midgestation (Rosenblatt, 1969; Bridges *et al.*, 1977). P's second important function is to regulate the timing of maternal behavior at the end of pregnancy. If one maintains high circulating P levels after HO on Day 17 of pregnancy, the rapid onset of maternal behavior induced by P withdrawal is prevented (Bridges *et al.*,

1978). Similarly, P blocks the rapid expression of maternal behavior in estrogen-treated, nulliparous rats (Doerr et al., 1981). Numan explored the possible neural sites of P's inhibitory action using a pregnancy-termination model (Day 16 HO), but was unable to localize a specific site of P's action (Numan, 1978). He postulated that P's inhibitory action may occur at multiple dependent sites such that application of P to one area was insufficient to inhibit maternal behavior. The interactions between P with $E_2$ merit further study at the neural level, since these two steroids act synergistically to stimulate the onset of maternal behavior. Little is also known about neural P receptors during pregnancy and whether P's actions on maternal behavior are receptor-mediated as is suggested by work in mice (Wang et al., 1995) in which treatment of mice during late pregnancy with RU-486, a P receptor antagonist, increased retrieval onset latencies postpartum.

3. *Prolactin and Lactogenic Hormones*

Early research by Riddle et al. (1935) first suggested a role for PRL in maternal behavior in the rat. However, subsequent efforts to replicate this effect were unsuccessful (Lott and Fuchs, 1962; Beach and Wilson, 1963; Baum, 1978). During the past decade, strong evidence supporting a role for PRL in the induction of maternal behavior has emerged. By employing an experimental model that used a combination of steroid treatment and PRL administration, a stimulatory role for PRL in the induction of maternal behavior has been established (Bridges et al., 1985; Bridges and Ronsheim, 1990). In our initial studies it was found that adult, hypophysectomized, nulliparous female rats failed to respond to the stimulatory effects of sequential exposure to P (Days 1–11) and $E_2$ (Day 11 to the completion of testing), whereas females with intact pituitary glands became maternal in 1–2 days (Bridges et al., 1985). Replacement of PRL in steroid-treated, hypophysectomized animals by either giving subjects ectopic pituitary grafts under the renal capsule that results in the secretion of PRL or injecting ovine (o) PRL produced a the rapid onset of maternal behavior in PRL-exposed animals.

In a second series of studies a nonhypophysectomized model was developed to address the possible concern about the overall health and responsivity of hypophysectomized rats as a behavioral model. This second model ultilized ovariectomized, steroid-treated nulliparous rat in which endogenous PRL was suppressed by twice daily injections of bromocriptine (CB-154), a dopamine agonist (Bridges and Ronsheim, 1990). Suppression of the $E_2$-induced rise in PRL from treatment Day 11 (a day before behavior testing began) onward delayed the expression of maternal behavior. Maternal latencies averaged about 4–5 days for steroid-primed rats treated with CB-154, while controls responded in 1–2 days. When steroid-primed rats

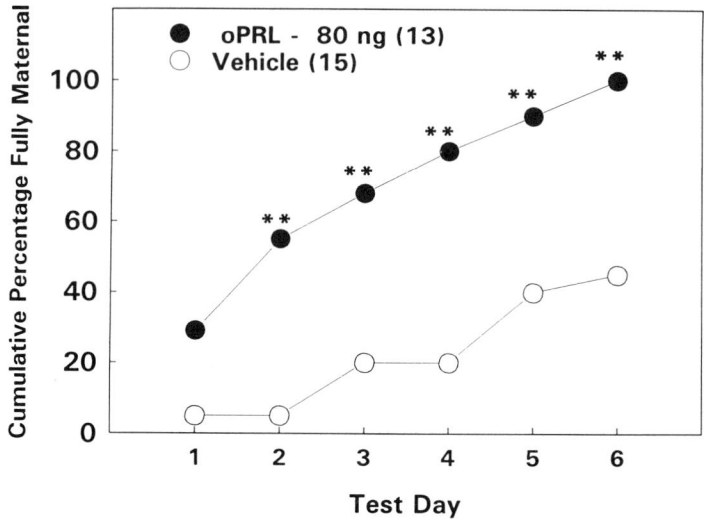

FIG. 3. Maternal behavior after bilateral infusions of oPRL (40 ng/side) into the MPOA of steroid-primed, bromocriptine-treated, virgin rats. **$p < .01$ versus controls. Animals were scored as fully maternal, if they retrieved all three foster test pups, grouped them together, and crouched over them within the daily 1-h test session. (Taken with permission from Bridges et al., 1990.)

were treated with both CB-154 plus exogenous PRL, a rapid onset of maternal behavior occurred, again demonstrating a stimulatory role for PRL (Bridges and Ronsheim, 1990).

One site of PRL's behavioral action is the central nervous system (CNS). Intracerebroventricular (ICV) infusions of oPRL or bilateral infusions of oPRL into the MPOA of steorid-primed, bromocriptine-treated, nulliparous rats stimulates short-latency maternal behavior (Bridges et al., 1990; see Fig. 3). The central actions of PRL in stimulating the onset of maternal behavior in adult female rats appear to be steroid-dependent. ICV infusions of oPRL into nonsteroid-primed animals fails to affect the rate of onset of maternal behavior (Bridges et al., 1990). Similarly, infusions of rat (r) PRL into the MPOA of ovariectomized, nulliparous rats only stimulates maternal behavior in rats also treated sequentially with P and $E_2$; rPRL MPOA infusions are ineffective in nonsteroid-treated females and in animals primed only with either P or $E_2$ (Bridges et al., 1995). While it is possible that PRL would stimulate the onset of maternal behavior in females given higher doses of P or $E_2$, at this time the most effective hormone regimen includes the combination of steroids plus PRL.

PRL also appears to help stimulate the rapid expression of parental behavior in juvenile rats, although PRL's presence may not always be essential (Stern, 1987). Reducing circulating levels of PRL in 24- to 30-day-old juvenile male rats by injecting CB-154 results in longer response latencies, while concurrent injections with oPRL attenuate the actions of CB-154 (Kinsley and Bridges, 1988b). These data indicate that juvenile animals are sensitive to the influence of humoral factors on parental-like behaviors. Koranyi *et al.* (1976) found that injections of maternal plasma into juvenile rats stimulated short-latency parental care. Likewise, Brunelli *et al.* (1987) reported that juveniles were sensitive to the behavioral-inducing properties of maternal blood, although handling the juvenile rats may have potentiated their parental-like responses. One possible factor in the maternal plasma that, when injected into juveniles, may stimulate parental behavior is PRL, a hormone secreted in large amounts by nursing mothers (Terkel *et al.*, 1972). During early development and under these conditions, PRL may act selectively to stimulate retrieval behavior (Loundes and Bridges, 1986). The elegant early transfusional studies of Terkel found that virgin rats that were parabiotically connected to rats during late pregnancy and parturition would retrieve pups soon after parturition in the donor females (Terkel and Rosenblatt, 1972). Again, the precise factor responsible for producing these behavioral changes has not been identified. However, PRL as well as $E_2$, which are elevated during this reproductive period, are likely candidates that may contribute to these changes in parental care.

The actions of PRL on maternal behavior in rats are shared by other lactogenic molecules, including ovine growth hormone (oGH), human placental lactogen (hPL), and rat placental lactogens (rPLs) (Bridges and Millard, 1988; Bridges *et al.*, 1994; Bridges and Freemark, 1995). When oGH is injected daily subcutaneously into steroid-primed, hypophysectomized rats, maternal behavior is stimulated. Likewise, bilateral infusions of hPL, rPL-I or rPL-II into the MPOA of steroid-primed, bromocriptine-treated, nulliparous rats result in a rapid onset of maternal behavior when compared with vehicle-infused controls (Bridges and Freemark, 1995; Bridges *et al.*, 1994). Latencies of experimental animals ranged from 1 to 2 days, while those of controls averaged about 6 days. These findings indicate that the normal rapid onset of maternal behavior at parturition is brought about by the prolonged exposure of the female to P, $E_2$, and lactogenic hormones during gestation. As shown in Fig. 2, pregnant rats are exposed to high titers of PRL during the first half of gestation and during the final day prior to parturition, whereas from midpregnancy onward the female is exposed to high levels of placental lactogens. These lactogenic hormones appear to gain access to the brain by binding to lactogenic receptors on cells of the choroid plexus (Walsh *et al.*, 1987; Pihoker *et al.*, 1993), which transport these molecules or their derivatives into the cerebrospinal fluid (CSF). Once in the CSF, these hormones can diffuse into neuropil, such as the MPOA, which is adjacent to the ventricles.

Biochemical measurements of messenger RNA for the PRL receptor in rat brains indicate that the expression of the long, but not short, form of the PRL receptor is increased on Day 7 of pregnancy relative to its expression in nonpregnant rats (Sugiyama et al., 1994). Expression of mRNA for the long form of the PRL receptor continues to increase during midpregnancy, reaching peak levels during the second half of gestation and continuing to remain high throughout most of lactation (Sugiyama et al., 1994). PRL itself, P, and estrogen (through its stimulation of PRL) are all able to stimulate expression of the long form of the rat PRL receptor and likely contribute to the increased expression of mRNA for the PRL receptor during pregnancy and lactation. Whereas the sites of increased expression of mRNA for the PRL receptor within the brain have not been identified, it appears that one site of increased expression is within the choroid plexus. Whether similar changes in receptor expression occur in the MPOA, which mediate PRL's central stimulation of maternal behavior, remain to be determined. It would be worthwhile to know, for example, whether placental lactogens bind to the long form of the PRL receptor when they stimulate the induction of maternal behavior.

The relative potencies of PRL and placental lactogens appears to be similar, although further studies are needed to clarify this issue. Preliminary studies indicate that both rPRL and rPL-I are equally effective when infused into the MPOA of steroid-primed rats; varying the number of infusions from one to five doses at the beginning of behavioral testing does not differentiate between the stimulatory potencies of rPRL and rPL-I (Bridges et al., 1995). It is not known, however, whether these two hormones have equal access to the CSF. During late pregnancy, rPL-II may have greater access to the CSF and the brain, since immunoreactive PRL is not detectable in the CSF just prior to parturition even though serum PRL levels are elevated (Rubin and Bridges, 1989). At this same time lactogenic activity in the CSF is high as measured in the $Nb_2$ lymphoma cell bioassay; the activity is neutralized with antibodies to rPL-II, but not affected by antibodies to rPRL (Bridges et al., 1994). Thus, while the behavioral potencies of PRL and rPLs appear comparable when infused directly into the MPOA, under physiological conditions when these lactogenic hormones may compete for the same lactogenic receptors on epithelial cells of the choroid plexus to gain access to the brain (Walsh et al., 1987), rPLs may actually have greater access to the brain and play a more critical role in stimulating the onset of maternal behavior. Communication between the developing conceptus through its secretion of placental lactogenic hormones provides an efficient mechanism to help prime the expectant mother to become maternal.

B. PEPTIDERGIC INVOLVEMENT IN PARENTAL BEHAVIOR

A series of peptides have been shown to affect parental care in female rats. The neuropeptide, oxytocin (OXY), and the gut peptide, cholecystokinin (CCK), stimulate the onset of parental behavior, while corticotropin-releasing factor (CRF) disrupts the establishment of maternal care. In addition, $\beta$-endorphin ($\beta$E) and endogenous opioids appear to have dual roles in the induction and maintenance of parental care. In general, the actions of these peptides seem to be more acute, and turn the behavior on or off and increase or decrease the response tendencies within a restricted time frame. This is in contrast to the actions of the hormones that act over a much longer period of time, for example, days versus hours, to prime the maternal brain to respond positively to young.

*1. Oxytocin*

Oxytocin appears to play an important role in acutely stimulating the onset of maternal behavior at parturition. Pedersen and colleagues first reported that central administration of OXY stimulated a rapid onset of maternal behavior in nulliparous rats (Pedersen and Prange, 1979). ICV infusions of OXY facilitated maternal care in estrogen-primed females, but were ineffective in nonsteroid-treated animals. The stimulatory actions of OXY were also demonstrated in EB-treated, Day-16 pregnant HO rats. As shown in Fig. 4, central infusions of anti-OXY or the oxytocin antagonist $d(CH_2)_5$-8-ornithine-vasotocin delay the rapid onset of maternal behavior (Fahrbach *et al.*, 1985). The actions of OXY in the rat appear to be dependent upon the test environment and the olfactory status of the female. Fahrbach found that OXY stimulated the onset of maternal behavior in animals tested in a novel cage, but was not effective when the subject was tested in its home cage (Fahrbach *et al.*, 1986). OXY's action was also demonstrated to be most effective in oflactory-impaired rats. ICV infusions of OXY-stimulated maternal behavior in rats made anosmic with zinc sulfate, but were not effective in EB-treated, control females (Wamboldt and Insel, 1987).

Sites of OXY's general stimulatory action include the paraventricular nucleus (PVN) of the hypothalamus, the MPOA, and the ventral tegmental area (VTA) in nulliparous animals (Insel and Harbaugh, 1989; Pedersen *et al.*, 1994). The PVN appears to be only a site of OXY stimulation of maternal behavior. Lesions of the PVN prior to parturition disrupt the establishment of full maternal care (Insel and Harbaugh, but fail to disrupt

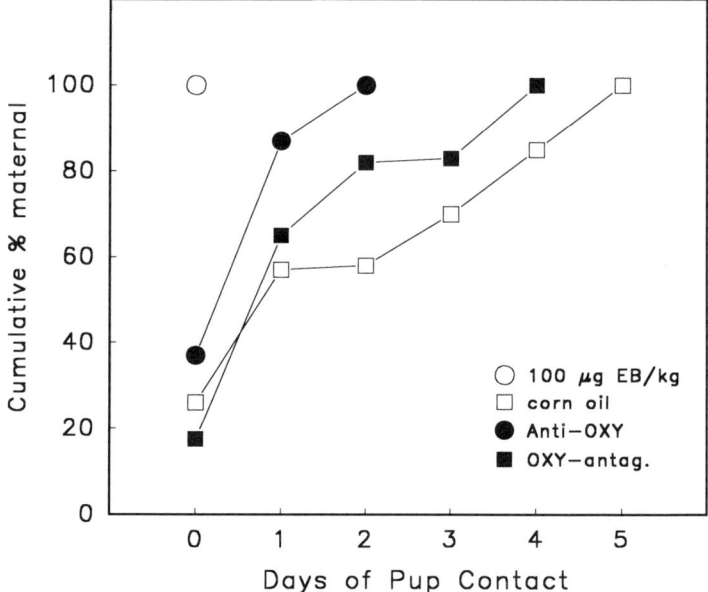

FIG. 4. Delaying effect of ICV infusions of anti-OXY agents on the onset of maternal behavior. Rats were hysterectomized and ovariectomized on Day 16 of pregnancy and first tested for maternal behavior 48 h later. Experimental animals received EB alone, EB plus anti-OXY, EB plus an OXY antagonist, or corn oil. EB and oil were given at the time of surgery, while the anti-OXY agents were given repeatedly from the beginning to completion of testing. Animals were tested with three foster pups daily. A maternal score was assigned if a female displayed four of the following behaviors: licked pups, carried pups, grouped pups together, crouched over the pups, and built a maternal nest. (Adapted from Fahrbach *et al.*, *Neuroendocrinology* **40**, 529. Copyright © 1985 by S. Karger AG, Basel. Reprinted with permission)

ongoing maternal responsiveness (Numan and Corodimas, 1985). Another region of the CNS that appears to modulate OXY stimulation of maternal behavior is the bed nucleus of the stria terminalis (BNST). OXY receptor concentrations increase in the BNST prepartum (Insel, 1990) and bilateral knife cuts that sever fibers passing into and out of this region block active maternal care in lactating rats (Terkel *et al.*, 1979). This region is an attractive candidate for OXY involvement in maternal behavior, since it receives input from the amygdala, a structure shown to regulate olfactory input associated with the establishment of maternal behavior (Fleming *et al.*, 1980).

The dependence of OXY's action on estrogen is a relatively unexplored phenomenon. $E_2$ and P appear to modulate OXY receptor concentrations in specific neural loci, including the ventromedial hypothalamus (VMH; Schumacher et al., 1989). While this region of the CNS is intimately involved in the hormonal control of feminine sexual behavior in rats, only recently has evidence emerged that the VMH may also be part of the neural network underlying the control of maternal behavior. Perturbation of this region in steroid-primed female rats by lowering cannulas into the region potentiates a rapid onset of maternal behavior (Bridges et al., 1993). Interestingly, OXY receptor concentrations increase in the VMH at parturition (Insel, 1990). Whether the VMH is a site of OXY-steroid interaction in stimulating the spontaneous onset of maternal care at parturition merits study.

Finally, studies in rats with prior maternal experience have shown that the vaginal and cervical stimulation associated with delivery help potentiate the expression of maternal behavior, possibly by central activation of oxytocinergic systems (Yeo and Keverne, 1986). Thus, it appears that the sensory input received by the female during parturition helps to acutely tune the maternal responses of the mother to the newborn young.

## 2. Cholecystokinin

The gut octapeptide CCK also appears to exert a mild stimulatory action on maternal care in rats. Linden et al. (1989) reported that systemic infusions of CCK into EB-treated nulliparous adult Wistar rats induced a more rapid onset of maternal behavior in experimental rats than in vehicle controls during a 4-h test period. CCK was ineffective in the absence of EB exposure. This action of CCK may be strain specific, since systemic infusions of CCK did not stimulate a more rapid onset of maternal behavior in EB-treated, nulliparous Sprague-Dawley animals (Mann et al., 1995). CCK also appears to modulate ongoing maternal behavior in lactating rats. Subcutaneous injections of proglumide, a CCK antagonist, increased retrieval latencies and decreased crouching responses on Days 5–6 of lactation (Mann et al., 1995). Moreover, CCK treatment prevents or attenuates the inhibitory actions of βE on maternal behavior in lactating rats. Concurrent infusions of CCK with βE into either the lateral ventricle or the MPOA antagonizes the disruptive effects of βE on ongoing maternal responsiveness (Felicio et al., 1991; Mann et al., 1995; see Fig. 5). The physiological changes in the CCK system during pregnancy and lactation have received minimal attention. This type of information, for example, CCK release during gestation and lactation, would be useful in order to place the CCK findings associated with the regulation of maternal behavior into a physiological perspective.

## 3. Corticotropin-Releasing Factor

CRF is a neuropeptide that stimulates adrenocorticotropin (ACTH) release from the anterior pituitary gland. CRF functions in part by mediating

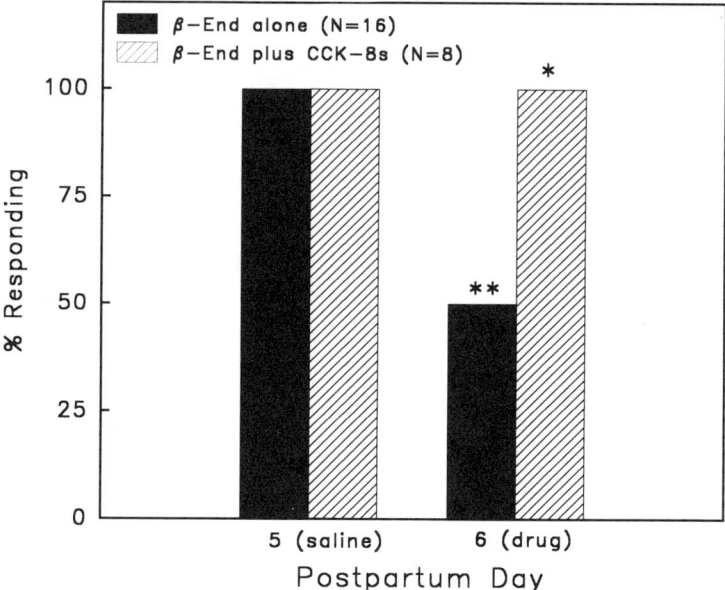

FIG. 5. Prevention of central opioid disruption of maternal behavior by CCK. Shown is the percentage of lactating rats displaying full maternal responsiveness (retrieval, grouping, and crouching over six pups within the 1-h test) after MPOA infusions of saline (Day 5), βE (Day 6), or βE plus CCK-8s. *$p < .05$ versus βE plus CCK-8s animals; **$p < .05$ versus Day-5 saline-treated responses. (Taken with permission from Mann et al., *Hormones and Behavior* **29,** 399. Copyright © 1995 by Academic Press, Inc.)

the neuroendocrine axis during an organism's stress response. In an initial study, Pedersen et al. (1991) reported that ICV infusions of CRF into nulliparous rats resulted in increased incidences of pup killing and lower incidences of maternal care. Further work on this peptide is needed in order to access whether this CRF effect is dose dependent or absolute and what the normative CRF release patterns are during the peripartum period. Increased CRF secretion during this period would be expected, given the involvement of ACTH in lactation. It is possible that severe stress during this peripartum period might produce abnormally elevated patterns of CRF release that, in turn, may disrupt parental care and induce pup killing. Again, the role of this peptide and its interactions with the hormones of pregnancy and important peptides and neurotransmitters merits attention.

### 4. β-Endorphin

The endogenous opioid peptide, βE, appears to have dual roles in regulating the expression of maternal behavior. βE may help stimulate the establishment of maternal behavior during the peripartum period, while in lactating rats it may inhibit maternal care during the suckling bout. Moreover, the sensitivity of the female to opioid regulation of maternal behavior quite interestingly appears to shift as a function of parity.

Our initial studies using Day-17 pregnancy-terminated (HO) rats showed that daily injections of morphine prior to testing inhibited the establishment of maternal behavior, while concurrent injections with naloxone prevented morphine's action (Bridges and Grimm, 1982). Morphine treatment during lactation also blocked ongoing maternal behavior (Grimm and Bridges, 1983; Kinsley and Bridges, 1988a; Mann *et al.*, 1991), apparently by altering the female's olfactory attraction to the young (Kinsley and Bridges, 1990). However, subsequent studies attempting to stimulate the onset of maternal behavior during late pregnancy with opiate antagonists have proven unsuccessful (R. S. Bridges *et al.*, unpublished data). This has led us to question whether the opiate blockade of the onset of maternal behavior was pharmacological.

In fact, the studies of Mayer *et al.* (1985) and Thompson and Kristal (1992) indicate that opioids may be stimulatory to the induction of maternal behavior. Injections of naloxone, an opiate antagonist, during late pregnancy interfered with placentophagia at parturition (Mayer *et al.*, 1985). Thompson and Kristal (1992) reported that infusions of morphine into the VTA of the midbrain stimulated the onset of maternal behavior in nulliparous rats; latencies decreased from about 8 days to 4 days in morphine-treated rats. The concept that opioids may enhance the establishment of affiliative behavior is in agreement with Panksepp's earlier work in mice (Panksepp *et al.*, 1980). More recently, we examined the role of opioids in the regulation of parental behavior in juvenile rats. Consistent with the idea that opioids may stimulate aspects of maternal behavior, chronic treatment with the opiate antagonist naltrexone delayed the expression of parental care in both female and male juvenile rats first tested for parental care on Day 30 of age after 9 days of naltrexone injections (Zaias *et al.*, 1996; see Fig. 6). These findings support the hypothesis that opioids may facilitate the establishment of parental care. The existing physiological data regarding endogenous opioid systems during pregnancy and lactation can also be interpreted to support a stimulatory role for opioids around the time of parturition. First, $\mu$-opiate receptor densities in the MPOA ($\mu$ receptors can bind $\beta E$) increase during pregnancy, reaching peak levels around the time of parturition (Dondi *et al.*, 1991; Hammer *et al.*, 1992). Pain thresholds, as assessed by the tail-jump response, also increase through parturition (Gintzler, 1980), while $\beta E$ concentrations in the MPOA likewise increase during gestation, declining prior to parturition (Bridges and Ronsheim, 1987). Further studies on $\beta E$ release and mRNA for proopiomelanocortin

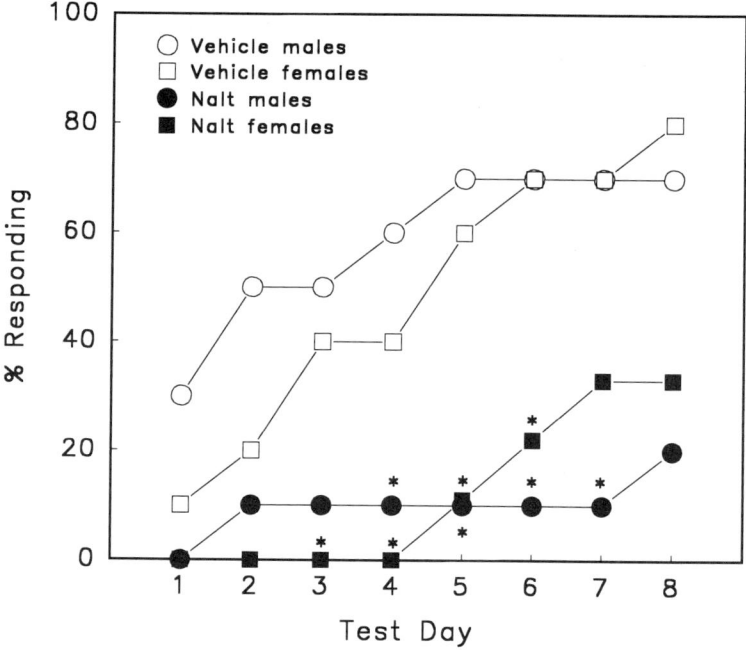

FIG. 6. Naltrexone inhibition of parental behavior in juvenile rats. Cumulative percentage of male and female juvenile rats displaying full parental behavior (retrieval, grouping, and crouching over three foster pups) during a 1-h test session. Rats were injected with naltrexone twice daily beginning at 21 days of age and were tested for parental behavior for 8 days from 30 to 37 days of age. $*p < .05$ versus same-sex, vehicle-injected controls. (Reprinted by permission of the publisher from Zaias et al., Pharmacology, Biochemistry and Behavior **53**, 990. Copyright © 1996 by Elsevier Science Inc.)

(POMC), the precursor for βE, during the prepartum and postpartum periods should provide additional valuable information relevant to opioid involvement in the establishment of maternal behavior, that is, whether opioids are stimulatory or inhibitory at this time.

The role of opioids in the maintenance of ongoing maternal behavior has been studied to a greater extent. During the lactational period, opioids that bind to the μ-opiate receptor interfere with maternal behavior by blocking retrieval and grouping responses (Kinsley and Bridges, 1988a; Mann et al., 1991; Mann and Bridges, 1992), possibly by decreasing the olfactory attraction of the pups (Kinsley and Bridges, 1990). Infusions of βE or the specific μ-receptor agonist, D-Ala$^2$, MePhe$^4$, Gly-ol$^5$-enkephalin (DAMGO), into the lateral ventricle of lactating, primiparous dams interfere with these elements of maternal behavior for 1–3 h (Mann et al., 1991).

Since suckling is known to stimulate βE release (Selmanoff and Gregerson, 1986), one interpretation of the behavioral findings is that suckling stimuli from the young cause an endogenous release of βE in the MPOA, which functions to quiet the female during the nursing bout. Only after βE release declines or possibly body temperature increases (Leon et al., 1978) does the dam leave the nest and reinitiate active maternal care. Nursing may begin again when the young become hungry and/or the intramammary pressure increases to the point where the dam initiates a nursing bout to relieve pressure. While these findings indicate that opioids may be involved in controlling various aspects of maternal behavior, additional research is needed to clarify the specific roles of endogenous opioids both in the establishment of maternal care and in the maintenance phase of this behavior.

C. BIOGENIC AMINES AND PARENTAL BEHAVIOR

The involvement of the classical neurotransmitters, dopamine (DA), and norepinephrine (NE), and serotonin (5-HT), have received limited attention in biological studies of parental behavior in rats and mammals in general. While some efforts have been made to explore the roles of dopamine and norepinephrine in parental care, studies of serotonin's involvement are quite limited.

*1. Dopamine*

Researchers have examined dopaminergic involvement in both the onset and maintenance of parental care. Using lesion and pharmacological approaches, it has been established that certain aspects of maternal behavior in lactating dams are affected by dopaminergic function. Szechtman et al. (1977) first suggested that the dopaminergic system may mediate maternal behavior in female rats. They reported that tail-pinch, a procedure thought to activate the nigrostriatal dopaminergic system, shortened retrieval latencies in virgin rats. The neuroanatomical studies of Numan and Nagle (1983) provided support for DA's involvement in maternal behavior in the rat. They found that small electrolytic lesions of the substantia nigra, a rich source of dopaminergic neurons, disrupt retrieval behavior but not nursing behavior. Likewise, Hansen *et al.* (1991) showed that infusions of the neurotoxin 6-hydroxydopamine (6-OHDA) into the ventral striatum, a procedure that depletes mesolimbic dopamine, produced retrieval deficits in lactating rats. Interestingly, this deficit could be partially overcome by separating the dams from their young for an extended period prior to testing (Hansen, 1994). Hansen (1994) has suggested that separation of the young from the OHDA-lesioned dam increases maternal motivation, thereby overcoming some of the lesion effects. Using a pharmacological approach, Giordano and colleagues found that systemic injections of the DA antagonist, haloperidol, also disrupted retrieval behavior, an effect that was reversed by concurrent treatment with the DA-agonist, apomorphine (Giordano et al., 1990). Similar haloperidol effects were found by Stern and Taylor (1991), who reported that haloperidol treatment interfered with pup retrieval and licking in lactating rats. Using a pregnancy-termination model, Rodriguez-Sierra and Rosenblatt (1977) found that apomorphine did not interfere with the onset of maternal behavior. Clarke-Hall et al. have reported that the dopamine $D_2$ receptor appears to mediate dopaminergic actions on maternal behavior. Infusions of a 19 base $D_2$ receptor antisense S-oligodeoxynucleotide into the lateral ventricle of lactating rats disrupted nest building and retrieval and nursing behaviors (Clarke-Hall et al., 1995). These findings are consistent with a possible stimulatory role for DA in maternal care. In a related experiment, Bridges found that when maternal nulliparous rats were given ectopic pituitary grafts that secrete large amounts of PRL, retrieval latencies decreased (Bridges, 1990). Since elevated PRL levels stimulate DA turnover (Ben-Jonathan, 1985), then it is possible that the endocrine actions of PRL in the maternal virgin may act through the stimulation of DA release. This suggested action requires verification. Moreover, whether the actions of DA on parental responsiveness are specific to maternal care or increase activity and motivation, in general, is unknown. Studies that explore DA's mode of action, like those of the hormones and neuropeptides, are needed to clarify how these systems act and interact.

## 2. Norepinephrine

The involvement of the noradrenergic system in parental behavior in the rat has received limited attention. The initial work of Rosenberg et al. (1977) suggested that NE had a role in the induction, but not the maintenance, of maternal behavior. Primiparous rats were infused ICV with 6-OHDA either 2 days prior to delivery or on Day 4 of lactation. The parental behavior of the animals infused prepartum was disrupted by the 6-OHDA infusions, while that of the lactating rats was unaffected. Together with the findings of Moltz et al. (1975), showing that NE turnover increased at parturition in the hypothalamus, it was proposed that hypothalamic NE may be involved in the induction of maternal care. Other work, however, suggests that noradrenergic involvement in the induction of maternal behavior more likely involves nonhypothalamic structures. Steele et al. (1979) reported that electrolytic lesions of the dorsal bundles, which contain numerous noradrenergic fibers that project to the hippocampus and cortex during early pregnancy, resulted in deficits in retrieval behavior, nursing, and nest quality. These lesions also reduced NE content in the hippocampus and cortex by 70% and 68%, respectively, but failed to alter hypothalamic NE content. Using a deafferentation procedure, Bridges *et al.* (1982) found that bilateral transection of both dorsal and ventral ascending noradrenergic bundles at the level of the midbrain during midpregnancy also interfered with the female's nest quality, but did not eliminate the establishment of other maternal behaviors. This surgical approach resulted in a 70% depletion of hypothalamic NE. Therefore, it appears that NE regulation of the onset of maternal behavior is reflected primarily in poorer nest building and quality, possibly through the depletion of NE within the hippocampus. While NE appears to play a stimulatory role in the onset maternal behavior, a careful re-examination of NE's role in parental behavior in the rat is warranted in order to define more thoroughly the precise role of NE and identify its sites and modes of action.

*3. Serotonin*

Serotonin's role in parental behavior in the rat has not been defined. Neurotoxic lesions of the median raphe serotoninergic nuclei have been reported to disrupt maternal behavior in lactating rats (Barofsky *et al.*, 1983). Lesions performed on Day 1 of lactation resulted in deficits in maternal behavior on Day 2, but normal maternal care returned when lesioned rats were tested on Day 5 of lactation. Thus, the effects of these neurotoxic lesions were temporary and may have been due to acute surgical stress. It would be worthwhile to determine whether the behavioral involvement of 5-HT is substantial and how the surgery affected sensory, motivational, and motor performance in these animals. Interestingly, suckling-induced PRL release was not affected by the same median raphe lesions (Barofsky *et al.*, 1983).

In summary, the roles of the classic neurotransmitters, DA, NE, and 5-HT, in the regulation of parental behavior in rats have only received limited research attention and as a result are not well understood. DA has been studied most extensively and appears to modulate mechanisms underlying retrieval and perhaps maternal motivation. NE also appears to be involved in the induction of some aspects of maternal behavior, most notably nest building.

IV. NEUROCHEMICAL ELEMENTS AND EXPERIENTIAL ASPECTS OF PARENTAL CARE

The dependency and involvement of hormones, peptides, and biogenic amines in parental behavior changes as a function of the reproductive experience of the rat. Females that have given birth previously appear to retain the ability to display short-latency maternal care regardless of hormonal status. Specifically, rats undergoing a second pregnancy respond maternally within a day during midgestation. In contrast, females at the same stage of their initial pregnancy require 5–7 days before maternal care emerges (Bridges, 1978). Hypophysectomy after parturition, likewise, appears to have minimal effects on ongoing maternal care (Erskine *et al.*, 1980). Hormones, therefore, appear to prepare the female to respond maternally at the end of the initial pregnancy after which experiential and neurochemical factors are sufficient to regulate expression of most aspects of this behavior (Rosenblatt, 1990).

The establishment of the long-term retention of maternal care requires only a short period of maternal–young interactions postpartum (Bridges, 1975; also see Fleming *et al.*, this volume). If primiparous rats are allowed to interact with their young throughout parturition prior to pup removal, these females will display rapid onsets of maternal care toward foster young 4 weeks later. In contrast, if each pup is removed immediately after its birth, preventing maternal care from being expressed, response latencies remain prolonged a month later (Bridges, 1975). Fleming *et al.* (1990) has shown that the development of this maternal memory can be blocked by treatment with the protein synthesis inhibitor, cycloheximide, indicating that the establishment of maternal "memory" involves some form of genomic activation. Numan reported increased expression of the protooncogene c-fos in the MPOA of maternal virgins (Numan and Numan, 1994). c-fos appears to be a marker of neuronal gene transcription and may be associated with neuronal activity involved in controlling the expression of maternal care. It would be of interest to determine whether c-fos activity increases or is modified in the MPOA and other neuronal populations when "maternal memory" is activated. For example, is there a specific region of the CNS where this memory is stored and reflected in alterations in c-fos activity? It would also be valuable to know whether c-fos activity increases in males showing parental behavior, since the biochemical and neurochemical control of paternal behavior in rats are virtually unexplored.

The role of peptides in regulating previously established maternal behavior has recently gained more interest. One important factor that influences the re-establishment of parental care is the effect of repeated reproductive experience on maternal care. In rats, as in many other mammals, maternal behavior improves with maternal experience. That is, multiparous females are "better" mothers than are primiparous females. Interestingly, it appears that the neural sensitivity to opioids shifts as a function of parity. Specifically, multiparous lactating rats are less sensitive to the behavioral actions of systemic morphine as well as to centrally administered $\beta E$ (Kinsley and

Bridges, 1988a; Mann and Bridges, 1992). For example, bilateral MPOA infusions of βE block ongoing maternal behavior in primiparous rats much more effectively than in age-matched, multiparous rats. Furthermore, increases in opiate receptor densities in the MPOA are found in multiparous rats (Bridges and Hammer, 1992). Studies by Pedersen suggest that OXY may modulate certain aspects of the re-emergence of maternal care (Pedersen et al., 1995). The interactions of the opiate systems with OXY and other neurochemical systems needs study. For example, since opioids can alter DA turnover (decrease) in the CNS, how might the dopaminergic system be modified as a function of parity? What effects might parity have on peptidergic control of parental behavior?

Together, these series of studies indicate that whereas hormonal regulation of maternal care lessens after the initial parturition, central biochemical modulation of the behavior continues, even though the sensitivity of the systems change. Finally, one remaining proposal merits consideration regarding the role of hormones, and specifically lactogenic hormones, that is, PRL and PLs, in ongoing maternal behavior. Is it possible that the control of maternal behavior after parturition may shift from a systemic endocrine to some form of central endocrine-like regulation? Although unproven, the expression of ongoing maternal care may be influenced after birth more by central lactogenic elements. It is conceivable that central lactogen-producing neurons and receptors (Paut-Pagano et al., 1993; Chui et al., 1992) become upregulated once maternal behavior is established. This possibility requires investigation.

### V. Overview: A Model for the Biochemical Regulation of Parental Care

A working model that illustrates the biochemical factors which influence the expression of parental behavior in rats is shown in Fig. 7. A series of hormones, neuropeptides, and transmitters affect the induction and maintenance of care in rats. Those factors that have the greatest influence on parental care in rats are estrogens and lactogenic hormones (PRL, PLs). It is well established that the stimulatory actions of PRL, P, and OXY, as well as CCK, are all estrogen dependent. $E_2$ appears to act together with these substances to stimulate maternal care. The actions of $E_2$, likewise, appear to be mediated in part through its stimulation of PRL, since the ability of $E_2$ to stimulate maternal behavior is reduced or eliminated in hypophysectomized females (Bridges et al., 1985) and in steroid-treated rats in which PRL is suppressed with bromocriptine (Bridges and Ronsheim, 1990). The estrogen dependence of OXY for its action may be mediated

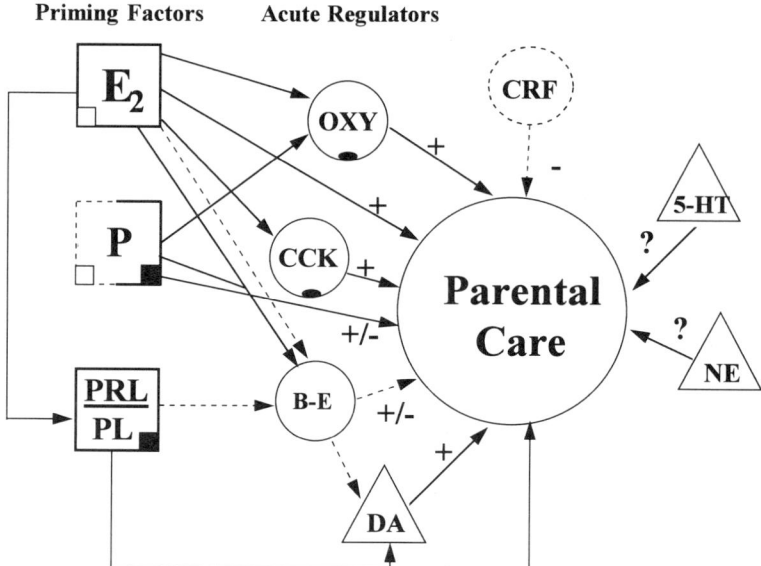

FIG. 7. Schematic diagram illustrating the biochemical factors shown to regulate parental care in rats. Relationships among factors are noted by arrows. Solid lines represent stimulatory actions, while dashed lines represent inhibitory effects of these molecules. Note that P has both stimulatory and inhibitory effects. □ = hormones; △ = peptides; ○ = biogenic amines. A filled-in square or oval within a symbol denotes that the action of that factor is estrogen dependent, while an empty square within a symbol denotes that the action of the factor is PRL or lactogen dependent. $E_2$ = estradiol; P = progesterone; PRL = prolactin; PL = placental lactogen; OXY = oxytocin; CCK = cholecystokinin; B-E = $\beta$-endorphin; DA = dopamine; CRF = corticotropin-releasing factor; 5-HT = serotonin; NE = norepinephrine.

under physiological conditions through steroidal modulation of OXY receptors (Insel, 1990; Kremarik *et al.*, 1995). Likewise, the possible involvement of $\beta$E and endogenous opioids in maternal care at parturition may be under steroidal influence. $\beta$E concentrations in the brain increase during gestation, and $E_2$ and P act together to increase both $\beta$E and $\mu$-opioid receptor densities (Hammer and Bridges, 1987; Hammer *et al.*, 1992). The involvement of the dopaminergic system in parental behavior, in particular, appears to be a promising avenue of inquiry. The regulation of dopaminergic systems by PRL and PL is well established, as are the inhibitory actions of opioids on DA release. The effects of reproductive experience on these systems are intriguing and should provide new and useful information for our understanding of biochemical underpinnings to learning.

One final consideration that needs to be addressed in future studies is how these various molecules affect sensory, integrative, and motor functions. For example, do the hormones of pregnancy alter the stimulatory characteristics of the newborn so that the mother finds the young attractive, leading to the initiation of maternal care? Likewise, how the birthing process and the associated stimulation of the female's reproductive tract affects central neurochemical events, that is, OXY, $\beta$E release, associated with the induction of maternal behavior is an important avenue to investigate. Moreover, examination of the factors or molecules that may increase parental motivation and/or activity to enhance the likelihood of positive maternal care will also help identify biological events that contribute to the expression of the maternal state. Such studies should help us to understand which biological events produce normal as well as poor parenting, and might contribute to clinical conditions, such as postpartum depression.

Whereas significant progress has been made during the past two decades in identifying factors that influence parental care, one challenge for the future is to understand more about the interactions of these substances within a physiological and developmental context, which aspects of neural functions are regulated by these molecules, and to what extent that the biochemical processes underlying the control of parental behavior in rats extend to females and males of other vertebrate and mammalian species.

### Acknowledgments

I would like to thank Dr. Phyllis Mann, Jodi Sturgis, and Brenda Henriquez for their assistance and comments in the preparation of this manuscript. The research on the biochemical regulation of parental behavior in our laboratory has been supported by PHS Grants from the NIH (HD19789) and NIDA (DA04291), and a grant from the Whitehall Foundation, Inc. (D92-09).

### References

Barofsky, A.-L., Taylor, J., Tizabi, Y., Kumar, R., and Jones-Quartey, K. (1983). Specific neurotoxin lesions of median raphe serotoninergic neurons disrupt maternal behavior in the lactating rat. *Endocrinology* **113**, 1884–1893.

Baum, M. (1978). Failure of pituitary transplants to facilitate the onset of maternal behavior in ovariectomized virgin rats. *Physiol. Behav.* **20**, 87–90.

Beach, F. A., and Wilson, J. R. (1963). Effects of prolactin, progesterone and estrogen on reactions of nonpregnant rats to foster young. *Psychol. Rep.* **13**, 231–239.

Ben-Jonathan, N. (1985). Dopamine: A prolactin-inhibiting hormone. *Endocrinol. Rev.* **6**, 564–589.

Bridges, R. S. (1975). Long-term effects of pregnancy and parturition upon maternal responsiveness in the rat. *Physiol. Behav.* **14**, 245–249.

Bridges, R. S. (1978). Retention of rapid onset of maternal behavior during pregnancy in primiparous rats. *Behav. Biol.* **24**, 113–117.

Bridges, R. S. (1984). A quantitative analysis of the roles of dosage, sequence and duration of estradiol and progesterone exposure in the regulation of maternal behavior in the rat. *Endocrinology* **114**, 930–940.

Bridges, R. S. (1990). Endocrine regulation of parental behavior in rodents. *In* "Mammalian Parenting: Biochemical, Neurobiological, and Behavioral Determinants" (N. A. Krasnegor and R. S. Bridges, eds.), pp. 93–117. Oxford Univ. Press, New York.

Bridges, R. S., and Freemark, M. (1995). Human placental lactogen and rat prolactin infusions into the medial preoptic area stimulate maternal behavior in steroid-primed, nulliparous female rats. *Horm. Behav.* **29**, 216–226.

Bridges, R. S., and Grimm, C. T. (1982). Reversal of morphine disruption of maternal behavior by concurrent treatment with the opiate antagonist naloxone. *Science (Washington, D. C.)* **218**, 166–168.

Bridges, R. S., and Hammer, R. P., Jr. (1992). Parity-associated alterations in medial preoptic opiate receptors in female rats. *Brain Res.* **578**, 269–274.

Bridges, R. S., and Mann, P. E. (1994). Prolactin-brain interactions in the induction of maternal behavior in rats. *Psychoneuroendocrinology* **19**, 611–622.

Bridges, R. S., and Millard, W. J. (1988). Growth hormone is secreted by ectopic pituitary grafts and stimulates maternal behavior in rats. *Horm. Behav.* **22**, 194–206.

Bridges, R. S., and Ronsheim, P. M. (1987). Immunoreactive beta endorphin concentrations in brain and plasma during pregnancy in rats: Possible modulation by estradiol and progesterone. *Neuroendocrinology* **45**, 381–388.

Bridges, R. S., and Ronsheim, P. M. (1990). Prolactin (PRL) regulation of maternal behavior in rats: Bromocriptine treatment delays and PRL promotes the rapid onset of behavior. *Endocrinology* **126**, 837–848.

Bridges, R. S., Feder, H. H., and Rosenblatt, J. S. (1977). Induction of maternal behaviors in primigravid rats by ovariectomy, hysterectomy or ovariectomy plus hysterectomy: Effect of length of gestation. *Horm. Behav.* **9**, 156–169.

Bridges, R. S., Rosenblatt, J. S., and Feder, H. H. (1978). Serum progesterone concentrations and maternal beahvior in rats after pregnancy termination: Behavioral stimulation following progesterone withdrawal and inhibition by progesterone maintenance. *Endocrinology* **102**, 258–267.

Bridges, R. S., Clifton, D. K., and Sawyer, C. H. (1982). Postpartum luteinizing hormone release and maternal behavior in the rat after late-gestational depletion of hypothalamic norepinephrine. *Neuroendocrinology* **34**, 286–291.

Bridges, R. S., DiBiase, R., Loundes, D. D., and Doherty, P. C. (1985). Prolactin stimulation of maternal behavior in female rats. *Science (Washington, D. C.)* **227**, 782–784.

Bridges, R. S., Numan, M., Ronsheim, P. M., Mann, P. E., and Lupini, C. E. (1990). Central prolactin infusions stimulate maternal behavior in steroid-treated, nulliparous female rats. *Proc. Natl. Acad. Sci. USA* **87(20)**, 8003–8007.

Bridges, R. S., Okimoto, L., Pellerin, L. J., and Mann, P. E. (1993). Infusions of saline into the ventromedial hypothalamus of steroid-primed female rats stimulate a rapid onset of maternal care. *Soc. Neurosci. Abstr.* **241.19**, 587.

Bridges, R. S., Robertson, M. C., Shiu, R. P. C., Friesen, H. G., Stuer, A. M., and Mann, P. E. (1994). Rat placental lactogens: Access to the brain during pregnancy and central stimulation of maternal behavior in rats. *Soc. Neurosci. Abstr.* **591.3**, 1447.

Bridges, R. S., Sturgis, J. D., Robertson, M. C., Shiu, R. P. C., Stuer, A. M., and Mann, P. E. (1995). Steroidal dependence and potencies of rat prolactin and rat placental lactogen-I in maternal behavior. *Soc. Neurosci. Abstr.* **190.2**, 464.

Brunelli, S. A., Shindledecker, R. D., and Hofer, M. A. (1987). Behavioral responses of juvenile rats (*Rattus norvegicus*) to neonates after infusion of maternal blood plasma. *J. Comp. Psychol.* **101**, 47–59.

Chui, S., Koos, R. D., and Wise, P. M. (1992). Detection of prolactin receptor (PRL-R) mRNA in the rat hypothalamus and pituitary gland. *Endocrinology* **130,** 1747–1749.

Clarke-Hall, Y. M., Rosenblatt, J. S., and Creese, I. (1995). A role for dopamine $D_2$ receptors in maternal behavior in lactating rats. *Soc. Neurosci. Abstr.* **149.7,** 364.

Cohn, J., and Gerall, A. A. (1989). Pre- and postpubertal medial preoptic area lesions and maternal behavior in the rat. *Physiol. Behav.* **46,** 333–336.

Cosnier, J., and Couturier, C. (1966). Comportement maternal provoqué chez les rattes adultes castrees. *C. R. Seances Soc. Biol. Ses Fil.* **160,** 789–791.

Doerr, H. K., Siegel, H. I., and Rosenblatt, J. S. (1981). Effects of progesterone withdrawal and estrogen on maternal behavior in nulliparous rats. *Behav. Neural Biol.* **32,** 35–44.

Dondi, D., Maggi, R., Panerai, A. E., Piva, F., and Limonta, P. (1991). Hypothalamic opiatergic tone during pregnancy, parturition and lactation in the rat. *Neuroendocrinology* **53,** 460–466.

Erskine, M. S., Barfield, R. J., and Goldman, B. D. (1978). Intraspecific fighting during late pregnancy and lactation in rats and effects of litter removal. *Behav. Biol.* **23,** 206–218.

Erskine, M. S., Barfield, R. J., and Goldman, B. D. (1980). Postpartum aggression in rats: I. Effects of hypophysectomy. *J. Comp. Physiol. Psychol.* **94,** 484–494.

Fahrbach, S. E., and Pfaff, D. W. (1986). Effect of preoptic region implants of dilute estradiol on the maternal behavior of ovariectomized, nulliparous rats. *Horm. Behav.* **20,** 354–363.

Fahrbach, S. E., Morrell, J. I., and Pfaff, D. W. (1985). Possible role for endogenous oxytocin in estrogen-facilitated maternal behavior in rats. *Neuroendocrinology* **40,** 526–532.

Fahrbach, S. E., Morrell, J. I., and Pfaff, D. W. (1986). Effect of varying the duration of pretest cage habituation on oxytocin induction of short-latency maternal behavior. *Physiol. Behav.* **37,** 135–139.

Felicio, L. F., Mann, P. E., and Bridges, R. S. (1991). Intracerebroventricular cholecystokinin infusions block beta-endorphin-induced disruption of maternal behavior. *Pharmacol. Biochem. Behav.* **39,** 201–204.

Fleming, A. S., Vaccarino, F., and Luebke, C. (1980). Amygdaloid inhibition of maternal behavior in the nulliparous female rat. *Physiol. Behav.* **25,** 731–743.

Fleming, A. S., Cheung, U. S., and Barry, M. (1990). Cycloheximide blocks the retention of maternal experience in postpartum rats. *Behav. Neural Biol.* **53,** 64–71.

Gintzler, A. R. (1980). Endorphine-mediated increases in pain threshold during pregnancy. *Science (Washington, D. C.)* **210,** 193–195.

Giordano, A. L., Johnson, A. E., and Rosenblatt, J. S. (1990). Haloperidol-induced disruption of retrieval behavior and reversal with apomorphine in lactating rats. *Physiol. Behav.* **48,** 211–214.

Grimm, C. T., and Bridges, R. S. (1983). Opiate regulation of maternal behavior in the rat. *Pharmacol. Biochem. Behav.* **19,** 609–616.

Hammer, R. P., Jr., and Bridges, R. S. (1987). Preoptic area opioids and opiate receptors increase during pregnancy and decrease during lactation. *Brain Res.* **420,** 48–56.

Hammer, R. P., Jr., Mateo, A. R., and Bridges, R. S. (1992). Hormonal regulation of medial preoptic $\mu$-opiate receptor density before and after parturition. *Neuroendocrinology* **56,** 38–45.

Hansen, S. (1994). Maternal behavior of female rats with 6-OHDA lesions in the ventral striatum: Characterization of the pup retrieval deficit. *Physiol. Behav.* **55,** 615–620.

Hansen, S., Harthon, C., Wallin, E., Lofberg, L., and Svensson, K. (1991). The effects of 6-OHDA-induced dopamine depletions in the ventral or dorsal striatum on maternal and sexual behavior in the female rat. *Pharmacol. Biochem. Behav.* **39,** 71–77.

Insel, T. R. (1990). Regional changes in brain oxytocin receptors post-partum: Time-course and relationship to maternal behaviour. *J. Neuroendocrinol.* **2,** 539–545.

Insel, T. R., and Harbaugh, C. R. (1989). Lesions of the hypothalamic paraventricular nucleus disrupt the initiation of maternal behavior. *Physiol. Behav.* **45,** 1033–1041.

Kinsley, C. H., and Bridges, R. S. (1988a). Parity-associated reductions in behavioral sensitivity to opiates. *Biol. Reprod.* **39,** 270–278.

Kinsley, C. H., and Bridges, R. S. (1988b). Prolactin modulation of the maternal-like behavior displayed by juvenile rats. *Horm. Behav.* **22,** 49–65.

Kinsley, C. H., and Bridges, R. S. (1990). Morphine treatment and reproductive condition alter olfactory preferences for pup and adult odors in female rats. *Dev. Psychobiol.* **23,** 331–347.

Koranyi, L., Lissak, K., Tamasy, V., and Kamaras, L. (1976). Behavioral and electrophysiological attempts to elucidate central nervous system mechanisms responsible for maternal behavior. *Arch. Sex. Behav.* **5,** 503–510.

Kremarik, P., Freund-Mercier, M. J., and Stoeckel, M. E. (1995). Estrogen-sensitive oxytocin binding sites are differently regulated by progesterone in the telencephalon and the hypothalamus of the rat. *J. Neuroendocrinol.* **7,** 281–289.

Leon, M., Crosberry, P. G., and Smith, G. K. (1978). Thermal control of mother-young contact in rats. *Physiol. Behav.* **21,** 793–811.

Lindén, A., Uvnäs-Moberg, K., Eneroth, P., and Södersten, P. (1989). Stimulation of maternal behaviour in rats with cholecystokinin octapeptide. *J. Neuroendocrinol.* **1,** 389–392.

Lott, D., and Fuchs, S. S. (1962). Failure to induce retrieving by sensitization or the injection of prolactin. *J. Comp. Physiol. Psychol.* **65,** 111–113.

Loundes, D. D., and Bridges, R. S. (1986). Length of prolactin priming differentially affects maternal behavior in female rats. *Biol. Reprod.* **34,** 495–501.

Mann, P. E., and Bridges, R. S. (1992). Neural and endocrine sensitivities to opioids decline as a function of multiparity in rats. *Brain Res.* **580,** 241–248.

Mann, P. E., Kinsley, C. H., and Bridges, R. S. (1991). Opioid receptor subtype involvement in maternal behavior in lactating rats. *Neuroendocrinology* **53,** 487–492.

Mann, P. E., Felicio, L. F., and Bridges, R. S. (1995). Investigation into the role of cholecystokinin (CCK) in the induction and maintenance of maternal behavior in rats. *Horm. Behav.* **29,** 392–406.

Mayer, A. D., and Rosenblatt, J. S. (1984). Prepartum changes in maternal responsiveness and nest defense in *Rattus norvegicus*. *J. Comp. Psychol.* **98,** 177–188.

Mayer, A. D., Faris, P. L., Komisaruk, B. R., and Rosenblatt, J. S. (1985). Opiate antagonism reduces placentophagia and pup cleaning by parturient rats. *Pharmacol. Biochem. Behav.* **22,** 1035–1044.

Mayer, A. D., Reisbick, S., Siegel, H. I., and Rosenblatt, J. S. (1987). Maternal aggression in rats: Changes over pregnancy and lactation in a Sprague-Dawley strain. *Aggressive Behav.* **13,** 29–43.

Moltz, H., Lubin, M., Leon, M., and Numan, M. (1970). Hormonal induction of maternal behavior in the ovariectomized nulliparous rat. *Physiol. Behav.* **5,** 1373–1377.

Moltz, H., Rowland, D., Steele, M., and Halaris, A. (1975). Hypothalamic norepinephrine concentration and metabolism during pregnancy and lactation in the rat. *Neuroendocrinology* **19,** 252–259.

Morishige, W. K., Pepe, G. J., and Rothchild, I. (1974). Serum luteinizing hormone, prolactin and progesterone levels during pregnancy in the rat. *Endocrinology* **92,** 1527–1530.

Numan, M. (1978). Progesterone inhibition of maternal behavior in the rat. *Horm. Behav.* **11,** 209–231.

Numan, M. (1990). Neural control of maternal behavior. *In* "Mammalian Parenting: Biochemical, Neurobiological, and Behavioral Determinants" (N. A. Krasnegor and R. S. Bridges, eds.), pp. 231–259. Oxford Univ. Press, New York.

Numan, M., and Corodimas, K. P. (1985). The effects of paraventricular hypothalamic lesions on maternal behavior in rats. *Physiol. Behav.* **35,** 417–425.

Numan, M., and Nagle, D. S. (1983). Preoptic area and substantia nigra interact in the control of maternal behavior in the rat. *Behav. Neurosci.* **97,** 120–139.

Numan, M., and Numan, M. J. (1994). Expression of fos-like immunoreactivity in the preoptic area of maternally behaving virgin and postpartum rats. *Behav. Neurosci.* **108,** 379–394.

Numan, M., Rosenblatt, J. S., and Komisaruk, B. R. (1977). Medial preoptic area and onset of maternal behavior in the rat. *J. Comp. Physiol. Psychol.* **91,** 146–164.

Panksepp, J., Herman, B. H., Villberg, T., Bishop, P., and DeEskinazi, F. G. (1980). Endogenous opioids and social behavior. *Neurosci. Biobehav. Rev.* **4,** 473–487.

Paut-Pagano, L., Roky, R., Valatx, J.-L., Kitahama, K., and Jouvet, M. (1993). Anatomical distribution of prolactin-like immunoreactivity in the rat brain. *Neuroendocrinology* **58,** 682–695.

Pedersen, C. A., and Prange, A. J., Jr. (1979). Induction of maternal behavior in virgin rats after intracerebroventricular administration of oxytocin. *Proc. Natl. Acad. Sci. USA* **76,** 6661–6665.

Pedersen, C. A., Caldwell, J. D., McGuire, M., and Evans, D. L. (1991). Corticotropin-releasing hormone inhibits maternal behavior and induces pup-killing. *Life Sci.* **48,** 1537–1546.

Pedersen, C. A., Caldwell, J. D., Walker, C., Ayers, G., and Mason, G. A. (1994). Oxytocin activates the postpartum onset of rat maternal behavior in the ventral tegmental and medial preoptic areas. *Behav. Neurosci.* **108,** 1163–1171.

Pedersen, C. A., Johns, J. M., Musiol, I., Perez-Delgado, M., Ayers, G., Faggin, B., and Caldwell, J. D. (1995). Interfering with somatosensory stimulation from pups sensitizes experienced, postpartum mothers to oxytocin antagonist inhibition of maternal behavior. *Behav. Neurosci.* **109,** 980–990.

Pihoker, C., Robertson, M. C., and Freemark, M. (1993). Rat placental lactogen-I binds to the choroid plexus and hypothalamus of the pregnant rat. *J. Endocrinol.* **139,** 235–242.

Riddle, O., Lahr, E. L., and Bates, R. W. (1935). Maternal behavior induced in virgin rats by prolactin. *Proc. Soc. Exp. Biol. Med.* **32,** 730–734.

Rodriguez-Sierra, H. F., and Rosenblatt, J. S. (1977). Does prolactin play a role in estrogen-induced maternal behavior in rats: Apomorphine reduction of prolactin release. *Horm. Behav.* **9,** 1–7.

Rosenberg, P., Halaris, A., and Moltz, H. (1977). Effects of central norepinephrine depletion on the initiation and maintenance of maternal behavior in the rat. *Pharmacol. Biochem. Behav.* **6,** 21–24.

Rosenblatt, J. S. (1967). Nonhormonal basis of maternal behavior in the rat. *Science (Washington, D. C.)* **156,** 1512–1513.

Rosenblatt, J. S. (1969). The development of maternal responsiveness in the rat. *Am. J. Orthopsychiatry* **39,** 36–56.

Rosenblatt, J. S. (1990). Landmarks in physiological study of maternal behavior with special reference to the rat. *In* "Mammalian Parenting: Biochemical, Neurobiological, and Behavioral Determinants" (N. A. Krasnegor and R. S. Bridges, eds.), pp. 40–60. Oxford Univ. Press, New York.

Rosenblatt, J. S., and Siegel, H. I. (1975). Hysterectomy-induced maternal behavior during pregnancy in the rat. *J. Comp. Physiol. Psychol.* **89,** 685–700.

Rosenblatt, J. S., Wagner, C. K., and Morrell, J. I. (1994). Hormonal priming and triggering of maternal behavior in the rat with special reference to the relations between estrogen receptor binding and ER mRNA in specific brain regions. *Psychoneuroendocrinology* **19,** 543–552.

Rubin, B. S., and Bridges, R. S. (1989). Immunoreactive prolactin in the CSF of ovariectomized, estrogen-treated and lactating rats as determined by push-pull perfusion of the lateral ventricles. *J. Neuroendocrinol.* **1,** 345–349.

Schumacher, M., Coirini, H., Frankfurt, M., and McEwen, B. S. (1989). Localized actions of progesterone in hypothalamus involve oxytocin. *Neurobiology* **86,** 6798–6801.

Selmanoff, M., and Gregerson, K. A. (1986). Suckling-induced prolactin release is suppressed by naloxone and stimulated by $\beta$-endorphin. *Neuroendocrinology* **42,** 255–259.

Siegel, H. I., and Rosenblatt, J. S. (1975). Estrogen-induced maternal behavior in hysterectomized-ovariectomized virgin rats. *Physiol. Behav.* **14,** 465–471.

Slotnick, B. M., Carpenter, M. L., and Fusco, R. (1973). Initiation of maternal behavior in pregnant nulliparous rats. *Horm. Behav.* **4,** 53–59.

Steele, M. K., Rowland, D., and Moltz, H. (1979). Initiation of maternal behavior in the rat: Possible involvement of limbic norepinephrine. *Pharmacol. Biochem. Behav.* **11,** 123–130.

Stern, J. M. (1987). Pubertal decline in maternal responsiveness in Long-Evans rats: Maturational influences. *Physiol. Behav.* **41,** 93–98.

Stern, J. M., and Taylor, L. A. (1991). Haloperidol inhibits maternal retrieval and licking, but enhances nursing behavior and litter weight gains in lactating rats. *J. Neuroendocrinol.* **3,** 591–596.

Sugiyama, T., Minoura, H., Kawabe, N., Tanaka, M., and Nakashima, K. (1994). Preferential expression of long form prolactin receptor mRNA in the rat brain during the oestrous cycle, pregnancy and lactation: Hormones involved in its gene expression. *J. Endocrinol.* **141,** 325–333.

Szechtman, H., Siegel, H. I., Rosenblatt, J. S., and Komisaruk, B. R. (1977). Tail-pinch facilitates onset of maternal behavior in rats. *Physiol. Behav.* **19,** 807–809.

Terkel, J., and Rosenblatt, J. S. (1972). Humoral factors underlying maternal behavior at parturition. *J. Comp. Physiol. Psychol.* **80,** 365–371.

Terkel, J., Blake, C. A., and Sawyer, C. H. (1972). Serum prolactin levels in lactating rats after suckling or exposure to ether. *Endocrinology* **91,** 49–53.

Terkel, J., Bridges, R. S., and Sawyer, C. H. (1979). Effects of transecting lateral neural connections of the medial preoptic area on maternal behavior in the rat: Nest building, pup retrieval, and prolactin secretion. *Brain Res.* **169,** 369–380.

Thompson, A. C., and Kristal, M. B. (1992). Opioids in the ventral tegmental area facilitate the onset of maternal behavior in the rat. *Soc. Neurosci. Abstr.* **18,** 280.13, 659.

Wagner, C. K., and Morrell, J. I. (1995). In situ analysis of estrogen receptor mRNA expression in the brain of female rats during pregnancy. *Mol. Brain Res.* **33,** 127–135.

Walsh, R. J., Slaby, F. J., and Posner, B. I. (1987). A receptor-mediated mechanism for the transport of prolactin from blood to cerebrospinal fluid. *Endocrinology* **120,** 1846–1850.

Wamboldt, M. Z., and Insel, T. R. (1987). The ability of oxytocin to induce short latency maternal behavior is dependent on peripheral anosmia. *Behav. Neurosci.* **101,** 439–441.

Wang, M.-W., Crombie, D. L., Hayes, J. S., and Heap, R. B. (1995). Aberrant maternal behaviour in mice treated with a progesterone receptor antagonist during pregnancy. *J. Endocrinol.* **145,** 371–377.

Wiesner, B. P., and Sheard, N. M. (1933). "Maternal Behaviour in the Rat." Oliver and Boyd, Edinburgh, Great Britain.

Yeo, J. A. G., and Keverne, E. B. (1986). The importance of vaginal-cervical stimulation for maternal behavior in the rat. *Physiol. Behav.* **37,** 23–26.

Zaias, J., Okimoto, L., Trivedi, A., Mann, P. E., and Bridges, R. S. (1996). Inhibitory effects of naltrexone on the induction of parental behavior in juvenile rats. *Pharmacol. Biochem. Behav.* **53,** 987–993.

Zarrow, M. X., Gandelman, R., and Denenberg, V. H. (1971). Prolactin: Is it an essential hormone for maternal behavior in the mammal? *Horm. Behav.* **2,** 343–354.

# Somatosensation and Maternal Care in Norway Rats

JUDITH M. STERN

DEPARTMENT OF PSYCHOLOGY
RUTGERS UNIVERSITY
NEW BRUNSWICK, NEW JERSEY 08903

## I. Introduction

"Why does a ring dove sit on its eggs?" asked the charismatic man before a captivated gathering in the kitchen of my Manhattan apartment during a party 30 years ago. The speaker was my mentor, the late Danny Lehrman. He proceeded to enthusiastically describe Tinbergen's four questions about evolution, adaptive significance, and developmental and proximate causation. We worked together on hormones and incubation behavior in male ring doves (Stern and Lehrman, 1969). Later his question—transposed to "Why does a rat nurture her young?"—is still the inspiration of my longtime research program on maternal behavior of rats. The particular "whys" I focus on concern proximate and developmental mechanisms, but I address the other issues in my research designs and writing.

The Norway rat (*Rattus norvegicus*) is among the most-studied creatures, at least in the laboratories of the world. Knowledge about the natural environment of rats (e.g., Calhoun, 1963) enables investigators to devise ethologically meaningful situations in which to study rat behavior in artificial settings. The vast amount known about the anatomy and physiology (e.g., Paxinos, 1995) as well as the behavior of rats facilitates sophisticated analyses of how a given behavior is mediated.

The maternal behavior of rats provides an excellent model of mammalian maternal behavior because the number of activities subsumed under this complex category is extensive, the mother spends a great deal of time with her litter, and the behavior has been well studied for over 60 years (e.g., Wiesner and Sheard, 1933; Rosenblatt and Lehrman, 1963). Although virgin females and males may be induced to display aspects of maternal behavior (Section II,B), only females of this species provide parental care in nature and the full repertoire requires that maternal behavior be integrated with the physiology of pregnancy, parturition, and lactation. Therefore, this article focuses on the maternal behavior of postpartum rats.

The starting point for understanding a given behavior is a detailed description of its sensory regulation and motoric elements, that is, sensorimotor analysis. Basic components of parental care in rats—indeed, in all species—involve physical contact between the caregiver and objects (nest building), the offspring (licking, retrieval, nursing), and an intruder (biting). To varying extents, depending on the behavior, distal stimuli play a role in arousal by and orientation toward young (Stern, 1989, 1990), but receipt of appropriate tactile stimuli appears to be essential to actual execution in rats, and this may be largely true of mammals in general. Since the source of tactile stimuli for nurturant activities is the young, unraveling the role of somatosensation in the regulation of maternal behavior clarifies the very nature of the mother–young interaction, including changes as the young develop. Because hormones often act by changing responsiveness to stimuli, understanding how the most critical stimuli mediate maternal behavior will inevitably clarify the role of hormones as well (Bridges, this volume). Further, because memories are largely stored as sensory experiences, understanding somatosensory regulation sheds light on experiential consequences of maternal behavior (Section VI; Fleming *et al.,* this volume).

What follows next is an overview of somatosensation and maternal behavior: the major components of maternal care in rats; the neurobiology of the somatosensory system; the relationships between somatosensory reflexes, behavioral sequences, and motivated behavior; research strategies; and the role of physical interaction with pups on maternal behavior in rats. The bulk of the article is then devoted to a detailed analysis of trigeminal and ventral trunk somatosensory contributions to the postpartum display of nurturant activities toward young (Sections III and IV) and defense of the young (Section V). The article concludes with discussions of neurobiological consequences and implications of the findings on somatosensory regulation of maternal behavior to date, including cortical plasticity, gene activation, and neural circuits (Sections VI). Because of space limitations, early work is sometimes summarized briefly, often without citations; a more extensive review provides detailed references (Stern, 1989).

## II. Somatosensation and Maternal Behavior

### A. Major Components of Maternal Care in Rats

Maternal behavior in rats consists of the *motorically inactive nursing behavior,* which involves a series of postural adjustments and limb immobility, and *motorically active behaviors,* involving mouth, paws, and body movements. Behaviors that typically precede and contribute to the success of dam-initiated nursing, including nest repair, retrieval, mouthing, and licking, are herein termed *pronurturant.*

In nature, *maternal nest building*—providing shelter and warmth—consists of digging an underground maternal burrow, carrying a variety of materials to it by mouth, and then manipulating these with both mouth and paws. *Retrieval*—ensuring that pups are available for provision of warmth and nutriment—is typically accomplished by locating a pup displaced from the nest; briefly sniffing it; grasping it gently by the pup's dorsal trunk skin with the incisors; carrying it to another location, typically a nest site; and depositing it there. Repeated retrievals, usually without interruption, result in the *grouping* of the pups in the nest. Excursions continue one or more times before the dam settles down in the nest with her gathered young. During the onset of maternal behavior, retrieval typically takes place a short distance from the nest, the female stretching forward to bring the pup closer (i.e., *mouthing*); only later does this complex behavior become integrated with ambulation—often quite variable—within the female's environment. *Licking*—ensuring cleaning, elicitation of elimination, behavioral activation, and stimulation of growth—usually occurs during long bouts (≥30 sec). While the pup is held in the paws, the repetitive tongue licking and head movements of the dam proceed from head to trunk to anogenital region. *Maternal aggression*—ensuring protection from an intruder, conspecific or allospecific—includes bouts of *biting* and *attacking.*

Dam-initiated *nursing behavior* occurs when the dam hovers over her litter gathered in the nest, usually while licking them (Fig. 1A); this enables the pups to begin suckling, thereby stimulating her quiescence (i.e., inhibition of locomotion) and postural adjustments that culminate in the *upright crouching posture,* herein termed *kyphosis*[1] (Section IV). This posture is characterized by rigid limb support, with limbs splayed if the litter mass is large, bilateral symmetry of the trunk and limbs, and *ventroflexion,* which involves all or most of the spinal cord and results in a dorsal arch (Fig. 1B). Upon further stimulation by the pups, the dorsal arch becomes more pronounced and the head is depressed (Fig. 1C). This posture provides room under the female's body for the pups so that they are not squashed while they are suckling. The *supine (passive) nursing posture,* in which the dam lies on her side while the pups suckle (catlike), is more common in the second postpartum week when the then-mobile pups (frequently) suckle their recumbent dam (Stern and Levine, 1972).

---

[1] *Kyphosis* (n.) [Modern L.; Gr. kyphøsis, from kyphos, humpbacked] is a medical term meaning abnormal curvature of the spine resulting in a hump. Therefore, a *kyphotic* (adj.) posture in a rat is one with a pronounced dorsal arch resulting from ventroflexion. The opposite spinal curvature is known as *lordosis,* a medical term adopted to describe the posture of sexual receptivity of female rats (and other animals). I am grateful to Emeritus Professor Anna Benjamin of Rutgers University for finding this appropriate term.

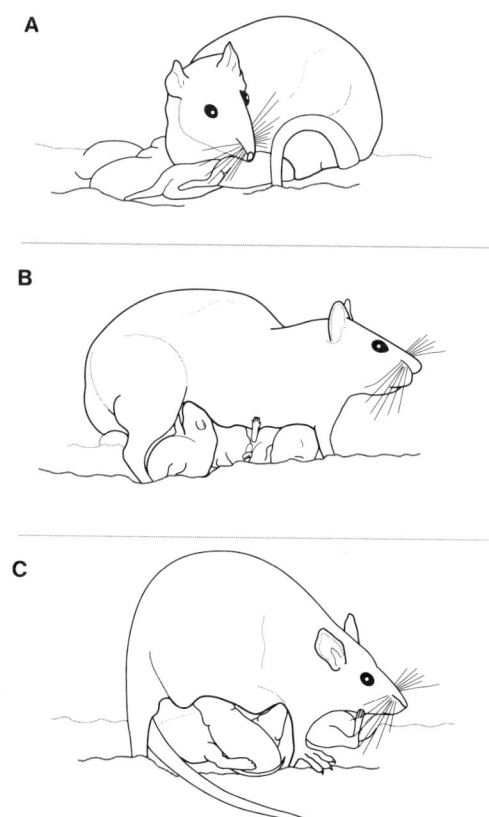

FIG. 1. Sketches of a lactating rat dam and her litter illustrating the different stages in the development of nursing behavior. (A) Hovering over. The dam stands over all or some of the gathered young while engaged in another activity, in this case licking one of the pups. Other activities consistent with this stage include nuzzling and mouthing pups, self-grooming, and nest repair. The dam continues to show head or limb movements, or both, while hovering over the pups, but the pups can gain access to her ventrum. (B) Low crouch. In response to sufficient ventrum stimulation from pups the dam becomes quiescent, that is, limb movements cease and head movements cease or are infrequent; her back is either fairly flat or, as depicted here, slightly arched, and not all of her legs are extended. (C) High crouch. In response to additional ventral stimulation from pups, while they are rooting or after all or most have become attached to a nipple, the dam's ventrum is raised (*ventroflexion*), resulting in a pronounced dorsal arch. In addition, all legs are extended rigidly, typically in a splayed position to accommodate the litter mass. (Reprinted from *Physiology & Behavior,* Volume 47, J. M. Stern and S. K. Johnson, Ventral somatosensory determinants of nursing behavior in Norway rats. I. Effects of variations in the quality and quantity of pup stimuli, pp. 993–1011, Copyright 1990, with kind permission from Elsevier Science Ltd, The Boulevard, Langford Lane, Kidlington 0X5 1GB, UK.)

B. SOMATOSENSORY SYSTEM

*1. Neurobiology of the Somatosensory System*

The interested reader is urged to consult other sources to supplement this necessarily brief overview (e.g., Tracey and Waite, 1995; Waite and Tracey, 1995). There is a rich array of *receptors* in the skin, mucous membranes, muscles, joints, and viscera of mammals that are responsive to touch, pressure, vibration, temperature, and pain. These receptors may be specialized in their anatomy and function or they may be nonspecialized, unmyelinated free nerve endings, responsive to mechanical or chemical stimuli that may or may not be painful. Large collections of receptors and afferent fibers are located in the orofacial region, which includes both perioral (furry buccal pads, lips, rhinarium, and vibrissae) and oral (gums, tongue, teeth) structures innervated by sensory branches of the trigeminal (cranial V) nerve (Table IA). On the ventral trunk—ventrum—of females, the special structures are the nipples, innervated by cutaneous nerves.

The innervation of the face and the trunk is mediated by cell bodies of primary afferents that are located in the trigeminal (gasserian) ganglion and the chain of dorsal root ganglia, the myelinated fibers of which enter the brain stem and spinal cord, respectively, via the trigeminal and dorsal sensory roots. Fine discrimination and localization of tactile stimulation on the body surface is conveyed by the trigeminal and dorsal column-medial lemniscal pathways, which decussate in the brain stem, synapse in the ventrobasal and posterior thalamus, then project to the somatosensory cortex. Also in the trunk, the lateral columns of the spinal cord convey pain, temperature, and diffuse touch, including suckling (Section III), sensations. Afferents from nipples are carried by the spinocervicothalamic pathways in the dorsolateral columns, which ascend ipsilaterally, synapse in the lateral cervical nucleus in the first two segments of the spinal cord, and then cross in the lower brain stem, projecting to a portion of the ventral posterolateral nucleus of the thalamus (Geisler *et al.*, 1979).

The trigeminal system also includes a variety of nonlemniscal projections to "nontrigeminal" structures such as dorsal horn, cerebellum, superior colliculus, reticular formation (Zeigler *et al.*, 1985), and midbrain periaqueductal gray (Shipley *et al.*, 1991). Given connections of these structures with effector mechanisms, they probably contribute to the sensorimotor functions of the trigeminal system. For example, in the cerebellar cortex the representation of the body is organized as a group of patches, each of which contains a contiguous array of disjunctive but functionally related receptive fields, such as for perioral structures and paws. This "fractured somatopy" is probably significant for species-specific oromotor sequences involving two or more body parts (Shambes *et al.*, 1978), for exam-

TABLE I
(A) Trigeminal Sensory Nerves and (B) Effects on Maternal Behavior and Maternal Aggression of Various Denervations

| A. Trigeminal division and nerve branch (V) | Areas of innervation |
|---|---|
| Opthalmic | |
|   1. Frontal nerve | Upper eyelid, skin of forehead |
| Maxillary | |
|   2. Zygomatic nerve | Skin of side of head and cheek |
|   3. Infraorbital nerve | Skin of nose, muzzle, upper lip; mouth mucous membrane |
|   4. Superior alveolar (dental) nerve (anterior and posterior) | Upper incisor and three molar teeth; nasal and cheek mucous membranes; gums; lining of maxillary sinus |
|   5. Sphenopalatine nerve | Soft and hard palate, cheeks and nasal mucous membranes |
| Mandibular | |
|   6. Inferior alveolar (dental) nerve | |
|     a. Mental nerve | Skin of chin; mucous membrane of lower lip |
|     b. Mylohyoid nerve | Skin of lower jaw |
|     c. Inferior dental nerve | Molar teeth |
|     d. Incisor nerve | Lower incisor tooth |
|   7. Lingual nerve | Anterior two-thirds of tongue, gums, mouth mucous membrane, sublingual glands |

| B. Study | Nerves cut | Time after surgery | Major findings |
|---|---|---|---|
| a | 1,2,3,6a,b | Unspecified | Slight difficulties in grasping young during retrieval |
| b | 3 | 12–24 h | Large decrease in percentage retrieving for 12–18 h |
| c-1 | 3 | 2–24 h | Large decrease in percentage retrieving, licking, and fighting, <12 h |
| c-2 | 3,6a | 2 h to 3 days | As in c-1, but recovery delayed and then retrieval duration prolonged |
| c-2 | 6a | 2 h to 3 days | Essentially normal |
| d | 3 | 3–28 h | Parturition behaviors severely impaired; postpartum retrieval erratic |
| e | 3,6 | 1–3 days | Retrieval virtually absent; alternate strategies used by some females |

a Beach and Jaynes, 1956a, p. 114.
b Kenyon et al., 1983.
c Stern and Kolunie, 1991.
d Stern, 1996a.
e Stern, 1990.

ple, licking a pup while holding it in the paws. The somatotopic representation of the superior colliculus is dominated by the vibrissae and is closely related to retinotopic projections, suggesting a role in orienting behavior for trigeminocollicular inputs. Reticular formation nuclei contribute to general arousal and serve premotor functions such as facilitation of patterned movements and oromotor reflexes, including their state-dependent modulation.

Somatotopic organization—characterized by topographic layout, differential innervation densities of different body regions, and submodality specificity—is found in all levels of the somatosensory system. In the primary somatosensory cortex (Section V,A) this representation is dominated by the oral cavity, the vibrissae, and the paws, regions with dense innervations. In contrast, the very large trunk has a small representation. In parallel, the size of a cutaneous receptive field, stimulation of which excites a given neuron in the somatosensory system, is inversely related to innervation density.

## 2. Somatosensory Reflexes and Species-Specific Behavior

Analogy is a profound source of insight.
—Stephen Jay Gould, 1980

Reflexes—somatic, autonomic, and neuroendocrine—are integral building blocks of motivated behaviors. The somatosensory stimulation of behavioral reflexes often requires motivational augmentation of responses to the incentive that may be mediated by appropriate hormones. Receipt of relevant somatosensory stimuli also provides continued interest in the incentive once approached and sensory feedback to guide subsequent actions. Often there are complex chains of reflexes such that a failure to elicit the first one prevents the expression of an entire behavioral sequence. However, one hallmark of motivated behavior is the ability to achieve a goal via alternative means, which is possible in some cases.

Many species-specific behaviors in mammals are carried out largely or in part with the mouth, and most of these are elicited by and directed to an external stimulus. These behaviors either do not occur or occur in an aberrant manner following desensitization or denervation of various trigeminal sensory nerves (Table I, see Stern, 1989). Neonatal rat pups are unable to attach to a nipple and suck following bilateral infraorbital denervation (IO-x). There is a moderate disruption of mouse killing after infraorbital denervation in adult rats, but a severe inhibition of biting attack after denervation of both infraorbital (IO) and inferior alveolar (IA) nerves in cats, a difference probably due to the more extensive denervation in cats, rather than to the species. In adult rats, while IO-x produces relatively slight and transient deficits in ingestion, there is a prolonged and severe aphagia and adipsia after lesions to the sphenopalatine, superior alveolar (SA), and IA nerves. In contrast, self-grooming is essentially intact following bilateral denervation of the IA, SA, IO, and lingual nerves, apparently because the jaw and tongue movements involved in this behavior are elicited by extraoral stimuli.

The reflexive and motivational role of trigeminal input in ingestive behavior in rats, extensively explored by Zeigler, Jacquin, and their coworkers (e.g., Zeigler et al., 1985), provides a model for my own work on the trigeminal nerve and maternal behavior (Section III). A film analysis of feeding reveals that the mouth-opening phase of ingestive behavior does not occur during approach to the food or water source, and therefore this specific aspect of feeding is not controlled by distal stimuli (vision, olfaction). Rather, it occurs after perioral contact, including repetitive palpations of vibrissae during "whisking" cycles, nosing, and mouthing, contact that engages trigeminal stimulation of the *jaw-opening reflex*. When trigeminal stimulation is reduced by selective sensory lesions, the contact phase is greatly prolonged, culminating in either jaw opening or an aborted attempt to grasp the food. When trigeminal orosensory-denervated and oromotor-denervated rats were compared on various measures of ingestive motivation, such as operant responding for food reinforcement, the former were motivationally impaired, that is, they quickly lost interest in operant responding for food, whereas the latter were merely motorically impaired in their execution of relevant oromotor responses.

The reflexive and motivational role of cutaneous nerves of the trunk in lordosis, the posture of female sexual receptivity, extensively explored in rats by Pfaff and his coworkers (Pfaff et al., 1994), provides a model for my work on ventrum somatosensory-induced nursing behavior (Section IV). Given that full lordosis takes over 300 msec to achieve and depends on several component stimuli followed by several component responses, it is "best conceived of as a chain of component reflexes in sequence" (Pfaff and Modianos, 1985, p. 475). Reductions in the effective cutaneous stimuli to the dorsolateral trunk result in a decreased intensity or likelihood of lordosis. Estradiol, acting in the hypothalamus and midbrain periaqueductal gray, as well as peripherally in the perigenital region, stimulates the antecedent, proceptive behaviors of hopping and darting, and facilitates reflex responsiveness to the relevant cutaneous stimuli from the mounting male.

*3. Research Strategies*

Lengthy, continuous observations, aided by tools such as computerized event recorders and videotape, are critical to obtain accurate descriptions of mother–offspring interactions. To ensure that a full sequence of maternal behavior, including a nursing bout, occurs during an observation, a lengthy (e.g., 4 h) dam–litter separation is imposed, during which the pups are kept at approximately nest temperature (34°C) in a humid incubator (unless otherwise indicated). In most such situations, litters are expressed of urine and feces before weighing and return to a dam so that the measure of subsequent litter weight gain is more accurate. If an experimental procedure prevents a dam from retrieving, pups are placed in the nest or older, mobile pups are used.

Manipulations of the sensory system include modifications of the pup stimuli presented or of the dam's receipt of them. The latter includes selective blockade of sensory nerves by local anesthesia or by surgical sectioning. Local anesthesia is easy to perform and is reversible, but it may not be possible to confine the blockade to particular sensory branches. Recall Novocain injections you may have received by your dentist into the gum surrounding a single molar tooth. Personally, I usually prefer to forego the temporary pain relief because the anesthetic effect extends unilaterally to my cheek, tongue, lip, and nose, and lasts longer than the dental procedure. In rats, injecting lidocaine subcutaneously into the mystacial pads to anesthetize the IO nerve is accompanied by blockade of the facial motor nerve as well, which results in vibrissae paralysis. Surgical denervation varies in difficulty, is relatively permanent (since regrowth of fibers is unlikely during the length of most experiments), and is highly selective, so that sensory, and not motor, branches can be cut. In addition to these strategies, a variety of central nervous system manipulations are used to trace the relationship between somatosensation and maternal behavior; these will be described along with the relevant studies.

C. Effects of Physical Interaction with Pups on Maternal Behavior

The heightened maternal responsiveness of the parturient rat requires interaction with pups for the *establishment of maternal behavior* and, because the action of pregnancy hormones wanes with time postpartum, such interaction is also essential for the *maintenance of maternal behavior* (Rosenblatt and Lehrman, 1963). Further, even a brief experience with pups during birth promotes *long-term retention of maternal responsiveness,* which is shown by its short-latency re-establishment once pups are available again, up to 30 days later (Stern, 1989). By reducing perioral somatosensations (with mystacial pad anesthesia) or ventral trunk somatosensations (with a full spandex jacket), or both, during a 1-h interaction with pups 36 h after cesarean delivery, Morgan *et al.* (1992) showed that at least one of these sources of somatosensation is required for an abbreviated latency to onset of maternal behavior 8 days later compared with rats without such interaction. Maternal behavior experience also leads the female to prefer the characteristic odor and size of her pups; renders her relatively invulnerable to various endocrine, sensory, or neural insults (Stern, 1989); and enhances her retrieval efficiency (Section III). Possible mechanisms for the encoding of maternal memory are considered elsewhere (Section VI; Fleming *et al.*, this volume).

*Pup-induced maternal behavior (sensitization)* occurs in nonhormonally treated female and male rats after several days of cohabitation with pups (Wiesner and Sheard, 1933; Rosenblatt, 1967); however, compared with the maternal behavior of postpartum lactating or hormone-induced maternal virgin rats, there are limitations in its intensity and repertoire, and there are strain and sex differences in the likelihood and speed of becoming maternal (Stern, 1989). Whereas nontactile stimuli from pups are either irrelevant or inhibitory to sensitization, tactile interaction with pups during this cohabitation is essential. Thus, continuous exposure to distal stimuli from pups does not reduce the subsequent latency of sensitization (Stern, 1983), peripheral anosmia hastens it (Stern, 1989), and deafening or blinding have no effect (Stern, 1996c). In contrast, increased contact hastens it (Terkel and Rosenblatt, 1971; Stern, 1983). Twenty-four-hour videotape recordings of virgin females housed with pups reveals a gradual change from avoidance to increasing tolerance of pup contact, with mouthing and licking of individual pups preceding retrieval and hovering over the grouped pups (Stern, 1996c). Sensitization may occur in nature when older pups remain in the maternal burrow and interact with younger siblings, and it may relate to how maternal behavior is maintained postpartum (Stern and Rogers, 1988).

## III. Trigeminal Somatosensation and Maternal Nurturance

As we have seen (Section II,C), physical interaction with pups, inevitably involving the dam's snout, is required for the elicitation, establishment, maintenance, retention, and induction of maternal behavior. The role of trigeminal sensations in the regulation of maternal behavior is reviewed in this section. Also, because recovery of function occurs with partial trigeminal desensitizations, the possible bases for this phenomenon will be explored.

### A. Maternal Behavior Depends on Perioral Stimuli

#### 1. Retrieval

The skin of pups is by far the most potent stimulus for eliciting retrieval. Beach and Jaynes (1956a) found a substantial inhibition of retrieval by intact lactating dams in response to altered pup *tactile* characteristics, that is, pups coated with petroleum jelly, petroleum jelly plus powdered pumice, or collodion (which forms a tough, elastic film). Compared with the retrieval of normal pups, that of texturally altered pups was more likely to be incomplete and slow, and live pups were often attacked. Because the impairments in retrieval were greater with textural alterations of pups than with a coating of the strong-smelling oil of lavender, they were probably not due to the females' reactions to a strange odor. Others later found that a rubber toy the size of a newborn rat is not retrieved readily by maternal rats (Stern, 1989).

Beach and Jaynes (1956a) also assessed the effects of altered pup *thermal* characteristics on retrieval. Frozen (stiff, dead) pups, when carried at all, were treated like small pieces of meat and were usually brought to places other than the nest, where most were then eaten. Similarly, Stern and Johnson (1990) showed that disruptions in maternal behavior in response to cold, live pups are related to the degree of chilling. In response to the coldest pups, with a mean skin temperature of 7°C at reunion, aberrancies included prolonged sniffing and retrieval; the latter also was delayed in onset and was often incomplete or with grouping in more than one site. These changes in maternal behavior may be mediated by activation of cold receptors in the face, which fire maximally at 10–30°C (Heinz et al., 1990), possibly a signal for the presence of food.

In lactating rats, bilateral anesthesia or denervation of the IO nerves prevents or profoundly impairs retrieval of pups (Kenyon et al., 1981, 1983; Stern and Kolunie, 1989, 1991). This impairment is not due to dysfunction of the masseter (jaw closing) muscle, vibrissae paralysis, akinesia, anosmia, or systemic effects of lidocaine (Kenyon et al., 1981, 1983). Grasping a pup, like grasping a food pellet (Zeigler et al., 1984, 1985), is dependent upon elicitation of the *jaw-opening* reflex, stimulated predominantly by trigeminal perioral afferents (Stern and Kolunie, 1989). Whereas disruption of olfaction, taste, or jaw proprioception do not interfere with jaw opening, extensive trigeminal orosensory denervation does (Zeigler et al., 1984). In these denervated rats, self-grooming occurs in response to extraoral stimuli (Berridge and Fentress, 1986), showing that jaw-opening and tongue muscles are not impaired.

By assessing retrieval on Day 3 postpartum in a specially designed apparatus and with an event recorder, we were able to analyze temporally its component parts: the time spent in the dark compartment, approaching the pup (after emergence from the dark compartment), contacting the pups by sniffing or nosing them, and carrying the pups (Fig. 2; Stern and Kolunie, 1989). Control dams received injections of saline into the mystacial pads (SMP) for comparison with dams rendered anaptic (without feeling) in the upper, rostral snout following similar injections of lidocaine (LMP), which

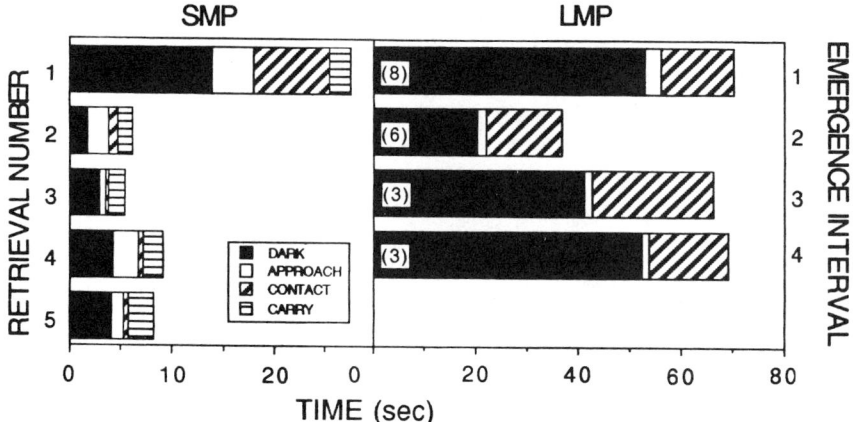
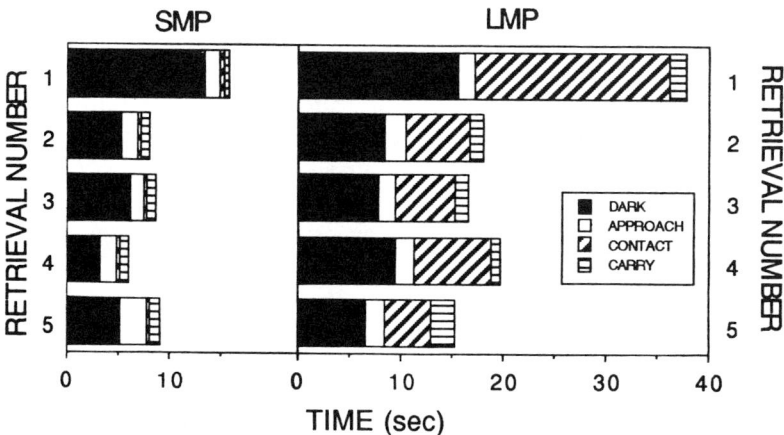

FIG. 2. Retrieval on Day 3 postpartum as a function of perioral anesthesia and pretreatment retrieval testing experience on Day 2. On Day 3 postpartum dams were anesthetized with methoxyflurane, a short-acting inhalant, and then each mystacial pad was injected subcutaneously with 0.2 ml of either saline or 2% lidocaine. (A) No pretreatment retrieval test. Analysis of the major components of (left) retrieval and (right) emergence intervals 30 min after mystacial pad injections of saline (SMP, left) or lidocaine (LMP, right; $n = 8$ per group) into time in the *dark* compartment prior to emergence, time from emergence into the light compartment to contact with pups(s) (*approach*), time in *contact* with pups, and time to *carry* a pup into the dark compartment. Each component represents the group mean. Note that on the abscissa, 10 sec on the right = 5 sec on the left. The numbers in parentheses to the left of the bars for LMP dams are the numbers that emerged and made contact with pups in each interval. (B) Pretreatment retrieval test (Day 2). Analysis of the major components of retrieval intervals for SMP (left, $n = 10$) and LMP (right, $n = 8$) dams that retrieved on Day 3 postpartum. (From Stern and Kolunie, 1989 (figure legend modified); reprinted with permission from Academic Press.)

blocks the IO nerve. Anaptia is assessed by failure to respond to a tactile stimulus, such as a pencil, brought into contact with the mystacial pads and by failure in tactile placing when the rat is held by the tail and lowered head first, resulting in nose rather than paws touching the surface first. During anaptia, rats tend to compensate for reduction of vibrissae sensations by keeping their head down and *nosing,* that is, placing the rhinarium in direct contact with, and sometimes under, objects of interest.

Initial retrieval by SMP dams took almost three times longer than subsequent retrievals (Fig. 2A, left), a rapid increase in efficiency noted earlier by Beach and Jaynes (1956b). Control dams made sniffing contact with a pup before picking it up, for about 7 sec initially and for $\leq 0.5$ sec subsequently. Anaptic dams, by contrast, did not retrieve, despite contact with pups of $\geq 10$ sec. In fact, following their aborted retrieval attempts, LMP dams began abandoning these attempts altogether by failing to emerge into the compartment containing the pups. Home-cage retrieval also was not elicited in LMP dams, with one exception (Stern and Kolunie, 1989).

Despite these dramatic findings, the failure to elicit retrieval following IO blockade or denervation is neither absolute nor permanent. During perioral anesthesia retrieval may occur, especially with increased experience with this task, either during repeated tests under this condition (Kenyon *et al.,* 1983) or, as illustrated in Fig. 2B, after pretreatment retrieval experience in the same situation (Stern and Kolunie, 1989). Among controls, retrieval of the first pup in the test situation not only increases retrieval efficiency within the test (Fig. 2A, left), but the increased efficiency is retained at least until the next day (Fig. 2B, left). Among anaptic dams, the most marked effect of pretreatment retrieval experience is that 80% did retrieve on the subsequent test versus 0% without such prior experience (Fig. 2B versus 2A, right). Among anaptic retrievers, there was 24 times as much sniffing contact with a pup before each retrieval compared with controls (Fig. 2B, right versus left), similar to the prolonged contact with a food pellet before jaw opening after extensive trigeminal orosensory denervation (Zeigler *et al.,* 1985).

Similarly, although retrieval is not elicited shortly after bilateral IO denervation, there is substantial recovery 12–24 h later (Kenyon *et al.,* 1983; Stern and Kolunie, 1991; Stern, 1996b). Thus, other nerves, including other trigeminal branches, contribute to the elicitation of retrieval. Given the mild and transient hypophagia when trigeminal orosensory denervation is confined to the IO nerves (Zeigler *et al.,* 1985), it is not surprising that the effect of this procedure on maternal behavior is also transient. Indeed, Beach and Jaynes (1956a) carried out an extensive cutaneous denervation of the snout of five lactating, albino rat mothers and found their retrieval performance to be only slightly impaired (Table IB,a). However, the operation-testing interval in this study was not specified; assuming that it was at least a day or so, the operation most likely caused a profound impairment initially, followed by substantial recovery, with the residue being the long-lasting need for prolonged contact with a pup in order to elicit jaw opening. When the maxillary rostral snout cutaneous denervation due to IO denervation of rats on Day 3 postpartum was extended to the mandible by severing the mental nerves (IO/M-x), the effects on retrieval were prolonged and exacerbated, although mental denervation alone had little or no effect (Stern and Kolunie, 1991; Table IB,c-2). Thus, on Days 2 and 3 postsurgery, duration of retrieval was two to three times longer in IO/M-x dams than in sham-operated or unoperated controls. When both the IO and IA nerves (Table IA) were cut, normal retrieval did not occur up to a week postsurgery, but this was accompanied by ingestive behavior deficits as well (Table IB,e). Finally, according to Lashley (1938), Beach found that cutting the trigeminal sensory root interfered with the retrieval of pups, but also with finding and grasping food in the mouth.

Taken together, the previous findings suggest that the only way to produce a long-lasting deficit in retrieval via surgical trigeminal orosensory deprivation requires a denervation, including intraoral afferents, that causes aphagia and adipsia as well. It is unlikely that selectivity of deficits to particular classes of stimuli can be achieved with this approach. With less extensive trigeminal orosensory denervations, recovery of function is probably mediated largely by recruitment of remaining trigeminal afferents. In addition, there may be conditioned associations with distal or other somatosensory stimuli (e.g., from paws) acquired rapidly during performance (cf. Stern, 1990). The rapidity with which retrieval efficiency is achieved suggests that the initial experience may lead to a removal of inhibitory control.

## 2. Licking of Pups

Except for nursing, licking of pups is the most regularly recurring and time consuming of maternal activities. After a 4-h separation from their litter, rat mothers typically lick the pups for 15–30% of a 30-min observation, even when the pups are manually expressed of urine and feces by the experimenter before the reunion. In our initial experiments with perioral anesthesia, we found that when retrieval was inhibited, so was licking (Stern and Kolunie, 1989). Since licking of pups is typically delayed until after the completion of retrieval and may be heightened after a long dam–litter separation, a more valuable assessment of this behavior occurs following long dam–litter separations (4–6 h), extended observations (15–30 min), and placement of pups in the nest before reunion (Stern and Johnson, 1989;

Table II). In a 15-min observation on Day 2 or Days 12–14 postpartum, whereas control dams licked their pups for about one-third of the time, LMP dams, subjected to mystacial pad anesthesia, did not lick at all (with the exception of two dams later postpartum that licked relatively briefly) (Table II,a,b).

The effects of bilateral IO denervation on licking of pups also paralleled its effects on retrieval during a 5-min retrieval test, that is, 0% of the subjects displayed pup licking at 2 h after surgery, with substantial recovery in percentage licking at 12 and 24 h (Stern and Kolunie, 1991). However, in a 30-min test 24 h after IO denervation on Day 5 postpartum, there was a one-third reduction in licking duration and a close to 50% reduction in duration of licking bouts compared with sham-operated controls, regardless

TABLE II
EFFECTS OF ALTERATIONS OF DAM'S PERIORAL SOMATOSENSORY AFFERENTS ON LICKING OF PUPS AND NURSING BEHAVIOR[a,b]

| Study | Day | Treatment (Tx) | Test Time Post-Tx | Min | Licking Dams (%) | Duration[c](s) | Crouching (%) |
|---|---|---|---|---|---|---|---|
| a | 2 | Saline mystacial pads | 30 min | 15 | 100.0** | 253 ± 56 | 100.0** |
|   |   | Lidocaine mystacial pads |   |   | 0.0 | — | 12.5 |
| b | 12–14 | Saline mystacial pads | 30 min | 15 | 100.0** | 311 ± 50 | 88.9 |
|   |   | Lidocaine mystacial pads |   |   | 22.2 | 129 ± 4 | 55.6 |
| c | 5 | SH-4 | 24 h | 30 | 100.0 | 648 ± 50 | 77.7 |
|   |   | SH-24 |   |   | 88.9 | 540 ± 103 | 77.7 |
|   |   | IO-4 |   |   | 100.0 | 360 ± 55† | 87.5 |
|   |   | IO-24 |   |   | 87.5 | 406 ± 76† | 50.0 |
| d | 1 | No tongue injections (NT) | 30 min | 30 | 100.0 | 336 ± 71*** | 100.0 |
|   |   | Saline tongue (ST) |   |   | 100.0 | 153 ± 35 | 88.9 |
|   |   | Lidocaine tongue (LT) |   |   | 88.9 | 64 ± 17 | 88.9 |
| e | 2 | Control sutured mouth | 90 min | 30 | 100.0 | 416 ± 82* | 100.0 |
|   |   | Sutured mouth |   |   | 100.0 | 186 ± 27 | 100.0 |
| f | 2 | Control muzzle | 90 min | 30 | 100.0 | 326 ± 55 | 66.7** |
|   |   | Full muzzle |   |   | — | — | 0.0 |
| g | 14–16 | No muzzle | 90 min | 30 | 100.0 | 422 ± 81 | 100.0 |
|   |   | Control muzzle |   |   | 100.0 | 221 ± 67 | 80.0 |
|   |   | Full muzzle |   |   | — | — | 60.0 |

[a] $N$s = 8–10 per group. With the exception of Study c, all dams experienced short-acting general anesthesia at the time of treatment, that is, 30 or 90 min before the reunion with pups. In Study c, an operation was carried out 24 h earlier; SH = sham-infraorbital denervation; IO = infraorbital denervation; 4, 24 = dam–litter separation of 4 or 24 h prior to test. *$p <$ .05, **$p <$ .01, ***$p <$ .001 (NT > ST > LT).

[b] Studies a,b,d–g: Stern and Johnson, 1989; Study c: Stern, 1996c (foster pups used).

[c] Responders only; Study e, sutured-mouth dams showed pseudo-licking movements of head and snout.

of whether or not the dams interacted with pups in the interval between surgery and testing (Table II,c; Table IV). It is not yet known whether this reduction in pup licking after IO denervation is permanent, or whether it generalizes to other licking. This deficiency suggests a role for IO sensations in maintaining rhythmical tongue movements, which are mediated by the

TABLE III
Effects of Altering Dams' Ventral Trunk Somatosensations on Nursing and Pup Licking[a]

| | | | | | Effect on kyphosis | Effect on licking |
|---|---|---|---|---|---|---|
| Study | Day | Min | Manipulation | Ss (%) | Latency (versus Controls) | (versus Controls) |
| Qualitative and quantitative alteration of pups[b] | | | | | | |
| a | 2 | 15 | Dead, warm | 0.0* | | 1.2× |
| b | 2 | 30 | 20 min of chilling | 0.0** | | 1.6×* |
| | | | 30 min of chilling | 0.0** | | 1.2× |
| c | 7 | 20 | Anesthetized mystacial pads | 0.0+ | | 1.3× |
| d | 14 | 20 | Anesthetized mystacial pads | 0.0** | | 0.7×* |
| e | 14 | 30 | Mouth sutured closed | 0.0** | | 1.7×** |
| b' | 2 | 30 | 10 min of chilling | 57.1 | 2.4×** | 1.4×† |
| f | 4 | 30 | 1 pup | 12.5* | 1.5×** | ~1.0 |
| | | | 2 pups | 37.5* | 1.5×** | ~1.0 |
| | | | 4 pups | 75.0 | 1.5×** | 1.2× |
| g | 7 | 30 | 1–2 awake: 7–6 anesthetized | 25.0* | 2.1×** | 1.4×* |
| h | 7 | 30 | Warm (4 h at 39°C) | 89.0 | 1.5×† | 1.6×** |
| i | 13 | 30 | Warm (4 h at 39°C) | 100.0 | 3.0×** | 1.7×** |
| Alteration of dam's ventral trunk afferents[c] | | | | | | |
| j | 4–5 | 30 | Thelectomy (nipple removal) | 37.5* | 1.7×† | 1.3×† |
| k | 6–8 | 30 | Anesthetized ventrum | 40.0* | 1.6×* | 2.0×** |

[a] Ns = 6–12 per group. Observations occurred after a dam–litter separation of 4 h. Control litters included eight pups kept at nest temperature before reunion (~34°C). Intrusive control procedures to pups included saline injected into mystacial pads (c,d) and stiches to the upper lip only while under general anesthesia (e). Intrusive control procedures to dams included 16 injections of 0.1 ml saline into the ventrum while under general anesthesia (k). Latency includes responders only. 100% of dams in each group licked their pups. In 30-min observations with control pups untreated except for incubation at 34°C, dams licked between 6.8 and 9 min on average, 100% crouched and they began to crouch consistently after 7.1–11 min on average. †$p \leq .10$, *$p \leq .05$, **$p \leq .01$.

[b] a–g: Stern and Johnson, 1990; h,i: Stern and Lonstein, 1996.

[c] j,k: Stern et al., 1992.

hypoglossal nerve (Travers, 1995). Following combined IO and mental denervations, licking was absent in a 5-min test at 2 and 12 h, but was not assessed thereafter (Stern and Kolunie, 1991). Because self-grooming is not inhibited after trigeminal blockade or denervation, the failure to lick pups cannot be attributed to impairments in mouth opening or tongue movements. Although other nonmaternal behaviors involving the mouth were not tested after cutaneous trigeminal denervation, a transient hypophagia and hypodypsia following IO lesions (Zeigler *et al.*, 1985) was reflected in a larger dam weight loss in the first day after surgery in trigeminal orosensory-denervated dams compared with sham-operated controls. However, retrieval deficits continued even after dams subjected to combined IO and mental denervation gained the most weight, between 1 and 2 days postsurgery.

We next asked whether licking is affected as well by *intraoral stimuli* from the tongue. Whereas injections into the tongue, of saline and especially of lidocaine, reduced the duration of pup licking (by 54 and 81%, respectively), it did not prevent it (Table II,d). Most surprisingly, when the mouth was sutured closed (temporarily), thereby preventing tongue contact completely, all dams displayed pseudo-licking (Table II,e). Although pseudo-licking movements were reduced in duration compared with real licking (by 55%), they appeared remarkably similar to real licking in that the dam held the pup in her paws and moved her head rhythmically along the pup's body. However, when perioral contact with pups was prevented completely on Day 2 or Days 14–16 postpartum with a full muzzle—which permitted females to see, hear, and smell the young, but prevented snout contact with them; see Stern and Johnson, 1989. Fig. 2—pseudo-licking movements of pups were not observed (Table II,f,g), although attempts at self-grooming of the head and face continued.

As a consequence of absent or delayed crouching (Section IV,B,2), there is usually an increase in pup licking, significant in most cases (Table III). Study d is an exception, with a significant decrease in pup licking, explained by the tendency of 14-day-old pups with anesthetized mystacial pads to wander away from the nest. With one- or two-pup litters, licking duration is the same as toward eight-pup litters, but it is directed entirely at a small number of pups. These results suggest a predominant somatosensory control of this behavior because chemosensory stimuli from the skin and anogenital secretions would decrease over time consequent to the dam's thorough licking.

### 3. *Dam-Initiated Nursing Behavior: Dependence upon Preceding Perioral Feedback*

Probably the most important discovery in the series of experiments reviewed in this section is that dam-initiated nursing behavior, which does not involve the dam's mouth, is dependent upon perioral feedback from antecedent dam behaviors such as licking. When dams were tested briefly for retrieval in their home cage, we noted that the usually appearing crouching over the pups did not occur in dams given mystacial pad injections of lidocaine until retrieval and grouping of pups had recovered, correlated with the wearing off of the local anesthesia (Stern and Kolunie, 1989). Because this behavior, like licking, usually occurs over pups gathered in the nest, we assessed this behavior after various perioral manipulations, with pups placed in the nest before returning the dam to the home cage (Stern and Johnson, 1989). The results are summarized in Table II. In brief, when the pups were 2 days old and largely immobile, if licking did not occur (due to perioral anesthesia or muzzling), nursing behavior did not occur, with one exception (Table II,a,f). (The exceptional LMP dam nosed her pups for twice as long as the other dams in her group). If licking occurred, though reduced in duration due to diminution of perioral (Table II,c) or intraoral (Table II,d,e) somatosensations, nursing behavior occurred as well.

Figure 3 summarizes the sensory regulation of maternal behavior from reunion with young pups to the establishment of a nursing bout. The role of distal stimuli in attracting the mother to the litter has been reviewed elsewhere (Stern, 1989, 1990). In order for *young* pups to gain access to a nipple, the dam must hover over them for a sufficient length of time while displaying pronurturant behaviors, that is, licking, mouthing, or nuzzling, interspersed with self-grooming and nest repair activities. If maternal oral attentions are not expressed, prolonged hovering over pups does not occur and, consequently, dam-initiated nursing behavior is not possible. The role of pups in providing adequate ventral trunk somatosensory stimulation to induce kyphosis, as well as the ability of older pups to bypass their dam's perioral attentions and stimulate her nursing behavior directly (Table II,b,g), are addressed in Section IV.

Also not illustrated in Fig. 3 is a striking motivational difference between the effects of mystacial pad anesthesia and muzzling. After briefly sniffing or nosing the pups, LMP dams remained inactive in a recumbent posture away from the pups. In contrast, muzzled dams did not abandon attempts to make perioral contact with the pups; indeed, these attempts continued unabated throughout the test, resulting in the scattering, and occasional injury, of the pups. This difference suggests that, in conjunction with distal stimuli from pups, IO nerve sensations (blocked during perioral anesthesia) contribute to the motivation to make snout contact with the pups even when actual contact is prevented (with the muzzle). We quantified this motivation as the duration of time spent pushing pups with the muzzle and pawing them; these behaviors are greatly diminished by treatment with haloperidol,

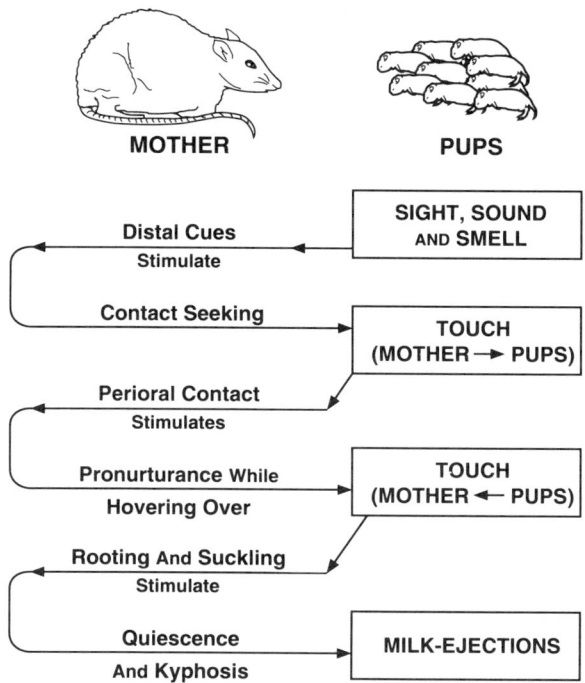

FIG. 3. The sequence of mother–litter interactions that culminate in a nursing bout in Norway rats. Although licking of pups takes up a substantial amount of time before crouching onset, snout, but not actual tongue, contact is required for hovering over pups. Older pups can directly provoke crouching by burrowing under their dam's ventrum, thereby bypassing the stage of dam's perioral contact while hovering over the litter. If pups are unable to suckle, the dam does not become quiescent and continues a variety of behaviors, including licking, self-grooming, nest repair, and excursions from the nest. (Modified from Stern, 1989, with permission from Academic Press.)

a dopamine receptor blocker, at dosages too low (0.05 or 0.10 mg/kg) to prevent dams with control muzzles to retrieve and lick their pups (Keer and Stern, 1996a; cf. Stern and Taylor, 1991, and Section III,B,3).

*4. Parturition and the Establishment of Maternal Behavior*

When a baby is born a mother is also born.

—*Ashley Montagu, 1986*

The high reliability of maternal care at birth is a marvel of mammalian evolution, one that is probably overdetermined by a variety of mediating mechanisms. For example, are the crucial maternal oral activities that make normal parturition possible in rats triggered by internal stimuli from fetuses as they pass through the cervix and vagina (cf. Stern, 1996a), olfactory and gustatory stimuli from birth fluids (Holloway et al., 1980), trigeminal somatosensory stimuli from contact with the neonates, or a combination of these? To assess the role of IO nerves, bilateral IO-x was carried out on Gestation Day 22, between 3 and 28 h prior to a dam's first birth, and births recorded with a low-light sensitive camera and time-lapse videorecorder (Stern, 1996a).

Parturition behavior was normal in sham-operated controls, whereas it was severely disrupted in all seven IO-x dams. Impairments included absent, abbreviated, or delayed perioral behaviors—nest building, participation in delivery, placentophagia, and licking and retrieval of pups—accompanied by increased general activity and delayed onset of nursing. The dam's usual role of facilitating delivery with teeth and tongue is revealed by astonishing instances of a pup half out of its mother's vagina while the dam moved about the cage in an agitated state for up to 40 min or an expelled pup being dragged about because it was not detached from its retained umbilical cord and placenta. Not surprisingly, pup mortality was low in control litters and seven to eight times higher in the IO-x group, both by the end of parturition (2 versus 14%) and cumulatively by postpartum day (PD) 3 (4 versus 35%). Because maternal licking is important for the onset of respiratory function, mortality was probably due to anoxia at birth. Although postpartum retrieval occurred in all but one IO-x dam, it was impaired relative to controls on PD 1–3 in terms of latency and completion. A surgery-birth interval of <7 versus ≥22 h was associated additionally only with inadequate litter growth, probably due to the poor condition of the pups from the birth experience because several foster pups grew at a normal rate.

Trigeminal IO sensations appear to be necessary for the normal onset of maternal behavior during parturition, with remaining trigeminal orosensory and extraoral cervical and vaginal sensations contributing to retained oral functions. Since marked reductions in duration of pup licking lasts at least 24 h after postpartum IO-x (Table II,c)—and perhaps much longer—deficiencies in cleaning pups of their fetal membranes are not surprising. The impairment in retrieval lasted longer than that after postpartum IO-x, but testing with healthy foster litters in all subjects is required to assess this possible difference. Previous maternal experience may not ameliorate these effects as suggested by a similar pattern of deficits in one rat subjected to IO-x 43 h before her second delivery compared with normal parturition in a sham-operated biparous control.

B. Recovery of Function

Behavioral recovery after neural damage may be due to one or more of the following processes: functional neural regeneration, substitution of a new behavior with the same function, or compensatory mediation of the original behavior by other nerves (Goldberger, 1980). Given the time course of recovery of maternal behavior and the retention of anaptia, neural regeneration—which takes at least several weeks—is not a plausible explanation (see Stern and Kolunie, 1991). We have already seen that several trigeminal orosensory branches contribute to retrieval. Behavioral substitution, the role of experience with pups on recovery of function, and sensorimotor responsiveness will be discussed in this section.

*1. Behavioral Substitution*

When normal retrieval was not elicited due to trigeminal orosensory blockade or denervation, some dams demonstrated the retention of their maternal motivation by behavioral substitution. Three dams with pretreatment retrieval experience displayed ineffective retrieval attempts prior to their first or second successful retrieval during perioral anesthesia, which led to some movement of a pup with the mouth (grasped and then dropped), chin (rolled), snout (pushed), or paws (pulled) (Stern and Kolunie, 1989). Following prepartum IO denervation of primiparous rats, some postpartum tests with thwarted retrieval culminated with the dam using her snout and paws to gather nest materials and pups in the new location (Stern, 1996a). Similarly, three dams subjected to both IO and IA denervations (including one primiparous rat operated on prepartum) successfully gathered pups and nest material by pulling and rolling these with their paws. These behaviors are reminiscent of alternative ingestion strategies observed by Jacquin and Zeigler (1983) in their orosensory-denervated rats, such as scooping, shoveling, and postural adjustments to the water spout.

Somatosensations from snout regions not deafferented or desensitized and from the paws—body parts activated during the elicitation and performance of pup licking, mouthing, or retrieval—may help mediate these alternate behavioral strategies. The adjacent topographic representation of snout and paws in the somatosensory cortex and the functional representation of these body parts called *fractured somatotopy* in the cerebellar cortex (Section II,B,1) provide anatomical and neurophysiological bases for using a relevant body part to achieve a goal when the body part typically used is not stimulatable.

*2. Role of Interaction with Pups after Trigeminal Denervation*

We have seen that the experience of retrieval causes a rapid improvement in its performance in control rats, enables retrieval to occur during anaptia due to acute IO anesthesia, and may contribute to the reliable execution of retrieval upon recovery from IO denervation (prepartum versus postpartum effects) (Section III,A,1). Accordingly, the question arises as to whether physical interaction with pups between postpartum IO denervation and testing 24 h later contributes to the recovery of retrieval and other aspects of maternal behavior. This was evaluated by observing dams' maternal behavior for 30 min on PD 6, 24 h after IO denervation or sham operation. Females were separated from their litter either at the time of surgery, that is, 24 h before the test, or 4 h before the test. All dams were tested with six foster pups the same age as their own pups, which were separated from their dam 4 h before the test and were kept in an incubator at 34°C in the interim (Table IV; Stern, 1996b).

Long separation from pups reduced retrieval likelihood and increased its latency, regardless of the dam's surgical condition. These findings may shed light on the decline of maternal behavior with time postpartum and following longer periods of separation from pups (Rosenblatt and Lehrman, 1963). IO denervation resulted in a marked reduction in duration of pup licking regardless of pup-separation interval, as discussed earlier (Section III,A,2; Table II,c; Table IV).

Separation from pups for 24 h after IO denervation did not prevent recovery of retrieval, but it resulted in greatly increased pup-sniffing latencies and facial self-grooming durations, effects not found in the IO-x dams

TABLE IV
Effects of Infraorbital Denervation (IO-x) or Sham Surgery (SH) on Maternal Behavior[a,b]

| Factor(s) | Significant changes in behavior | Comparison | p |
| --- | --- | --- | --- |
| 24-h SEP | ↓ pups retrieved in 10 min | 60 ± 0.0 (4 h) versus 4.5 ± 0.6 (24 h) | <.03 |
| | ↑ retrieval latency (RL), all dams | 24 ± 6.5 (4 h) versus 196 ± 59 (24 h) | <.01 |
| | ↑ RL, responders only | RL ≥ 20 sec: 3/17 (4 h) versus 7/13 (24 h) | <.05 |
| IO-x | ↓ (35%) pup-licking duration | 594 ± 57 (SH) versus 357 ± 52 (IO-x) | <.01 |
| | ↓ (~50%) pup-licking bout duration | 52 ± 8 (SH) versus 27 ± 3 (IO-x) | <.02 |
| 24-h SEP × IO-x | ↑ (≥3.5×) sniff-pup latency | 10 ± 2 (IO-4) versus 38 ± 10 (IO-24) | <.02 |
| | ↑ (≥5×) facial grooming duration | 62 ± 15 (IO-4) versus 351 ± 95 (IO-24) | <.01 |
| 24-h SEP; IO-x | ↑ retrieval duration | 2–4× versus SH-4 | ≤.01 |
| 24-h SEP × IO-x | ↑ pup contact in retrieval sequence | 3–20× versus SH-4 | ≤.01 |

[a] Stern, 1996b; surgery on Day 5 postpartum. All latencies and durations are in seconds.
[b] During 30 min, after 24 h and a separation (SEP) interval from pups of 4 or 24 h.

separated for only 4 h. The delay in pup investigation may be due to deficiencies in arousal or orientation due to IO-x that is compensated by interaction with pups between operation and testing (see Section III,B,3). The increased self-grooming is probably not due to altered sensation in the snout because it did not occur in IO-4 dams; therefore, it may occur as a displacement activity (e.g., Cohen and Price, 1979) in response to pup stimuli when maternal responses are abnormal, and it may provide self-arousal.

Both 24-h separation and IO denervation resulted in prolonged retrieval duration, the retrieval sequence being interrupted by other pup contact, such as licking, mouthing, or nosing; this was exacerbated by the interaction of these factors. Indeed, whereas normal retrieval is characterized by an inhibition of extraneous activity until after grouping of pups is completed, the abnormal retrieval seen in this experiment is not. This may be because longer contact with pups than normal is needed following trigeminal deprivation to stimulate each retrieval. Alternatively, if stimuli received while carrying a pup suppress extraneous activity, then such stimuli received following or during reduced trigeminal orosensations from pups may be insufficient for this suppression. Physical interaction with pups may contribute to recovery of function after IO denervation by recruitment of other trigeminal nerves. These findings are possibly relevant to other lesion-induced deficits in this behavior, as well as to the decline of maternal behavior with time postpartum.

## 3. Orientation, Arousal, and Dopamine-Dependent Sensorimotor Responsiveness

When rat dams subjected to perioral anesthesia are not stimulated to engage in the typical maternal sequence, they become inactive (Stern and Kolunie, 1989; Stern and Johnson, 1989). To determine whether this lethargy is specific to the lack of maternal oral activities, we tested the effects of perioral anesthesia on female sexual behavior, a behavior with no obvious oral component (Stahlbaum *et al.,* 1989). To our surprise, proceptivity (hopping and darting) was essentially absent, although receptivity (lordosis) was normal. Thus, even behaviors that do not appear to be elicited by trigeminal stimuli may nonetheless be dependent on them. Possibly, because proceptive movements are orientation specific, they may rely upon trigeminal sensations, in particular those arising from the movement of whiskers. Further, the acute loss of IO sensations may lead to decreased arousal. Reduction of activity in trigeminal afferents to the superior colliculus and reticular formation provide anatomical bases for difficulty with orientation and decreased arousal.

An alternate, but not mutually exclusive, explanation is that lethargy consequent to perioral desensitization is related to a time-limited deficiency in dopaminergic activation, and therefore alterations in sensorimotor responsiveness. There are functional and anatomical links between the trigeminal somatosensory and the nigrostriatal dopamine systems (e.g., Huston *et al.,* 1988). For both female sexual behavior (e.g., Caggiula *et al.,* 1979) and maternal behavior (Stern and Taylor, 1991), inhibition of dopamine pathways blocks active behaviors—proceptivity and pronurturance, respectively—while enhancing immobile postures elicited by tactile stimulation of the trunk—lordosis and kyphosis, respectively.

## C. Summary

Trigeminal orosensations contribute to maternal behavior in various ways. Tactile and thermal characteristics of pup skin are the most salient stimuli for eliciting retrieval. The elicitation of activities that involve the mouth, such as licking and retrieval, is eliminated or impaired when cutaneous somatosensation of the snout is reduced by local anesthesia or surgical denervation. The duration and extent of the impairment in retrieval is related to the number of trigeminal orosensory branches denervated. Retrieval by anaptic dams—during perioral anesthesia after pretreatment retrieval experience or after sufficient time following trigeminal orosensory denervations—requires prolonged contact with a pup before it is picked up. By 24 h after postpartum IO denervation, there is recovery in the percentage of subjects displaying retrieval and licking. However, retrieval is greatly prolonged due to interruptions in the sequence with increased pup contact, an effect exacerbated by absence of interaction with pups between surgery and testing. Further, there is a marked reduction in pup licking duration, regardless of whether or not interaction with pups occurs between operation and testing. Following prepartum IO denervation, impairments in facilitation of delivery with teeth and tongue, and in rapid, thorough cleaning of pups leads to increased pup mortality. During snout anaptia, alternate retrieval strategies include use of a related body part, such as paws or chin, to gather pups. Recovery of retrieval with the mouth may depend upon adaptation to the loss of nontrigeminal projections that influence orientation, arousal, and dopamine-mediated effects on sensorimotor responsiveness. When pronurturant activities such as licking are not elicited during anaptia, nursing of young pups also does not occur because hovering over pups—from which position pups gain access to nipples—is dependent upon trigeminal perioral sensory feedback. Reduced or absent tongue feedback results in reduced duration of licking or pseudo-licking movements, respectively, but does not prevent nursing. When snout contact with pups is physically prevented in the presence of normal perioral sensations, dams are highly motivated to make such contact; this contact-seeking behavior is blocked by low dosages of a dopamine receptor antagonist.

Like all reflexes, when there is a reduction of relevant trigeminal somatosensory stimuli that elicit a maternal behavior reflex, the behavior either does not occur or it is elicited after greatly increased stimulation, perhaps via recruitment of remaining trigeminal and other somatosensory nerves. Physical interaction with pups may alter the secretion of a neuromodulator with varying outcome such as promotion of retrieval efficiency. Finally, trigeminal perioral and oral sensations also have a motivational role, such as the dependence of a critical behavioral sequence—dam-initiated nursing behavior—on these sensations or behaviors elicited by them.

## IV. Ventral Trunk Somatosensation and Nursing Behavior

During an experiment on the perioral somatosensory determinants of licking and nursing behaviors (Stern and Johnson, 1989), I watched a dam with anesthetized mystacial pads lying recumbent away from her pups after briefly sniffing them. Her hungry, 2-week-old pups eventually surrounded her, began burrowing under and rooting on her ventrum, finally gaining access to her nipples. In response, the immobile, disinterested though tolerant, mother had a series of trunk twitches that elevated her ventrum and arched her spine convexly; gradually her posture was transformed from lying flat to the characteristic upright crouch of nursing. In this memorable moment, I realized that the upright crouching nursing posture, like lordosis of female sexual receptivity, is reflexive in nature, and I envisioned a series of experiments to evaluate this insight. This insight, coupled with an abundance of experimental verification reviewed as follows, have rendered obsolete the then-prevailing belief that suckling is irrelevant to nursing behavior in rats (e.g., Moltz *et al.,* 1967).

### A. Suckling- and Nonsuckling-Induced Nursing Behavior

#### 1. *Ventrum Stimulation from Pups*

When the dam hovers over her gathered young, the pups gain access to her ventrum, which stimulates a cascade of *neonatal reflexes* (Brake *et al.,* 1988; Lorenz, 1992; Shair and Hofer, 1992). Pups crawl under objects (*thigmotaxis*), ideally the mother's inviting (warm, furry, odor laden) ventrum, turn their head toward cheek stimulation (*rooting*), use olfactory and tactile cues to locate a nipple, and attach to a nipple (*jaw opening* and *jaw closing*). Because the nipple is moistened with pups' saliva, a vacuum seal is formed, facilitating sustained attachment. Lengthy *sucking,* accompanied at times by forelimb padding, leads eventually to the dam's milk ejection. Individual rat pups suck every few seconds, asynchronously usually and synchronously when milk is released. Upon receipt of milk, there is a frenzy of activity. Pups *swallow* and display a whole-body stretch, with intensified forelimb pressure; this stretch response is used as a behavioral assay for the dam's milk letdown. Detachment from nipples may occur during the excitement accompanying receipt of milk followed by nipple switching. Both rhythmic (nutritive) and arhythmic (nonnutritive) types of sucking have been identified with electromyographic recordings. The frequency of rhythmic sucking, in particular, varies as a function of sucking and nutritional deprivation (Section V,C,2). In sum, rat dams receive frequent, intensive, and varied stimulation of the glabrous (hairless) nipples and surrounding hairy skin of the ventrum from the mouth and paws of the suckling pups.

## 2. Alterations of Litters or Nipples

Dams without perioral feedback from contacting their young pups do not hover over them and therefore cannot be stimulated to adopt the crouching posture (Section III,A,3; Table II,a,f). In contrast, when dams are tested with pups *older* than 1 week—when pups often initiate nursing (Rosenblatt and Lehrman, 1963)—the dam's perioral contact with them can be bypassed, and the pups are able to elicit crouching from their otherwise unwilling, recumbent dam (Table II,b,g). Dams continue to initiate nursing bouts with older pups and these are dependent upon preceding perioral feedback, as they are with young pups.

When untreated dams are reunited, after a 4-h separation, with pups rendered incapable of suckling, at least for the duration of the observation, the dam retrieves, licks, and hovers over them normally, but kyphosis does not occur, regardless of the time postpartum (Day 2–14) (Table III,a–e) (Stern and Johnson, 1990). Effective alterations to the pups include recent sacrifice, acute anaptia (via mystacial pad anesthesia), sutured mouth, or 20–30 min of chilling, which renders the nearly poikilothermic pups inactive (Table III,a–e). If the pups are initially impaired in sucking, but may recover before the end of the observation (Table III,b',h,i), or there are an insufficient number of pups capable of suckling (Table III,f,g), there may be a reduction in the percentage of dams crouching and there is a marked delay in onset of kyphosis by dams that display this posture. Alterations that produced these effects are 4 h at 39°C (Stern and Lonstein, 1996), 10 min of chilling, and a reduced litter mass ($\leq$ four pups) or with the litter mass constant at eight pups but anesthetizing all but one or two of them (Stern and Johnson, 1990). Single pups, which are unlikely to maintain lactation before Day 12 (Leigh and Hofer, 1973), are especially unlikely to stimulate their dam's immobility.

A complementary demonstration of the role of cutaneous sensations from suckling in nursing behavior is to alter the dam's ventrum and provide her with a full, unmanipulated litter. This was accomplished by removing the nipples (thelectomy) or anesthetizing the nipples and surrounding skin (Table III,j,k; Stern *et al.,* 1992). (Ventrum and nipple anesthesia were verified by tests of cutaneous sensitivity and by the lack of litter weight gains up to 60 min post-reunion, indicating failure of the dam to respond to suckling with milk-ejection reflexes.) Despite the difference in the sucking ability of the pups in these two conditions, the effect on nursing behavior was comparable: fewer than 50% of the mothers crouched in 30 min, those that crouched did so after a long latency, sustained it briefly, and none showed the high crouch posture (Fig. 1C). However, dams capable of being suckled were likely to lie *prone* over the pups, that is, lying flat, with little or no leg support, a posture not seen in the thelectomized dams.

When crouching immobility is delayed or prevented, pronurturant maternal activities increase in duration, especially licking (Table III) but also pawing, mouthing, and nuzzling of pups and nest repair; these activities are interspersed with increases in the dam's self-grooming, repositioning, rearing, and excursions from the nest (see Stern and Johnson, 1990, Fig. 12).

Thus, suckling stimulates the mother's full nursing behavior, not vice versa. The stimuli from the suckled nipples summate spatially and temporally to effect both behavioral and neuroendocrine changes. When the dam is not stimulated to become quiescent by the suckling of her young, she remains active.

3. *Nonsuckling Ventrum Stimulation, Dopamine Blockade, and Induction of Nursing Posture in Nonmaternal Rats*

If suckling is needed to induce kyphosis in lactating rats, then can sensitized virgin females, with undeveloped nipples, and males, with no nipples, show this posture? In lactating rats, systemic treatment with the dopamine (DA) receptor blocker haloperidol inhibits or reduces pronurturant, motorically active maternal behaviors such as retrieval, licking, and nest building, but enhances quiescent nursing behavior (Stern and Taylor, 1991), findings that complement earlier ones on the inhibition of female sexual proceptivity but enhancement of lordosis (e.g., Caggiula *et al.,* 1979; Hansen *et al.,* 1981). Thus suckling—well known to inhibit tuberoinfundibular DA, thereby releasing prolactin from tonic inhibition—may promote quiescence and facilitate kyphosis at least in part by inhibiting the release of extrahypothalamic DA. Indeed, because the posture induced by a cataleptic dosage of haloperidol subserves "the maintenance of stable static equilibrium . . . at the expense of phasic locomotor reactions" (De Ryck *et al.,* 1980), it provides a precursor for kyphosis.

Nonmaternal rats—virgin females and males without hormone treatment—are typically intolerant of remaining in physical contact with pups and may even attack them. In contrast, when such rats were treated with a cataleptic dosage of haloperidol, they readily adopted the nursing posture in response to ventral trunk stimulation provided by a litter of eight hungry pups (Stern, 1991). Given the supersensitivity to trunk stimulation and the postural changes induced by haloperidol, even nonsuckling tactile and pressure stimulation evokes the reflexive responses—stretch, withdrawal, and bilateral leg extension—that culminate in the kyphotic posture. However, when placed over anesthetized pups, haloperidol-treated rats merely remained prone over them, revealing that neither hair deflection nor the feel of an inert litter mass are adequate stimuli to induce kyphosis. Nonlactating females scored higher than males in terms of likelihood and intensity of crouching in both intact and castrated conditions. This sex difference, comparable with findings on sensitization (Stern, 1989, Table 5), may be due to an androgen-induced reduction in ventrum sensitivity.

These results cast doubt on whether sensitized rats, in the absence of appropriate hormone or DA-antagonism treatment, are capable of demonstrating the quiescent, upright nursing posture typical of lactating rats. Also, the number of pups used to induce and test maternal responsiveness in virgin rats (usually two to four) tends to be insufficient to elicit normal nursing behavior even in lactating rats (Table III,f,g; Stern and Johnson, 1990). Sensitized rats do display hovering over the gathered litter, a criterion adopted by Reisbick *et al.* (1975) in their comparison of the maternal behavior of virgin and lactating rats.

A potentially valuable deduction from these results (Stern, 1991) is that appropriate hormones induce maternal behavior *centrally* by promoting preference for snout contact with pups, which elicits motorically active, pronurturant maternal behaviors such as retrieval and licking, and *peripherally* via nipple development, which permits suckling-induced nursing behavior to occur once the dam and litter are in close proximity.

### B. Spinal Pathway of Kyphosis

We next investigated which part of the spinal cord conveys the signals from the nipples and surrounding ventrum that culminates in the stimulation of the reflexive nursing posture. Suckling elicits well-known neuroendocrine reflexes principally resulting in the secretions of prolactin and oxytocin. Extensive lesions to the lateral but not the dorsal or ventral columns of the spinal cord eliminated lactation in most rats rearing a litter (Eayrs and

Baddeley, 1956) and oxytocin-dependent milk ejections in all urethane-anesthetized rats (Fukuoka *et al.,* 1984); details of nursing behavior were not included in the former report. We hypothesized that the crouching posture would be eliminated, along with lactation, following lateral column lesions (Stern *et al.,* 1993).

On Day 6 ± 1 postpartum, dams were subjected to a bilateral section of either the dorsal (DC-x) or dorsolateral (LC-x) columns of the spinal cord, or a control procedure (CON). CON and DC-x dams continued to lactate and display normal maternal behavior. In contrast, LC-x dams stopped lactating, due to a loss of both suckling-induced oxytocin and prolactin secretion. Soon after the operation, LC-x dams were largely lethargic, including a loss of maternal behavior. However, after a few days licking of pups returned in all LC-x dams and retrieval recovered in six of eight, but the upright crouching posture of nursing did not recover for the duration of testing, 1 week postoperation. The dams assumed a *hunched posture,* typical of rats with no pups present, even when surrounded by pups suckling or attempting to suckle. Less often LC-x dams lay prone over pups, when a sufficient number of the pups managed to burrow under their ventrum. Thus, the postural reflexive responses to the suckling of pups were lost along with the neuroendocrine reflexes. The LC-x dams did remain quiescent in proximity to the pups, as did the ventrum-anesthetized dams with suckling young, perhaps in response to remaining ventrum sensations carried by the dorsal columns or simply to the quiescence of the young. Because the typical nursing postures occur in bromocriptine-treated postpartum dams (Protomastro and Stern, 1994), cessation of suckling-induced neuroendocrine secretions cannot be responsible for elimination of nursing behavior in LC-x dams. Also, because LC-x dams recovered other physiological and behavioral functions, their maternal behavior deficit is specific to kyphosis.

The present results indicate that kyphosis, like lordosis, is organized at a supraspinal level (Kow *et al.,* 1977). Unlike lordosis, which is activated by somatosensory stimulation to the dorsal trunk that is carried by the ventrolateral columns and results in dorsoflexion, kyphosis is activated by somatosensory stimulation to the ventral trunk that is carried by the dorsolateral columns and results in ventroflexion. It is likely that spinal-cord–lateral-cervical-nucleus–brain-stem connections mediate kyphosis (Section VI,B).

## C. Termination of Nursing Bouts

Given that suckling stimulation is necessary for the induction and maintenance of the dam's nursing behavior, then it follows that the cessation or diminution of suckling stimulation should herald the termination of a nursing bout. There may be, in addition, suckling-induced physiological changes in the dam that contribute to bout termination. However, one particular physiological change—a rise in the dam's body temperature during nesting bouts—has received the most attention as an important basis for bout terminations (Croskerry et al., 1978; Leon et al., 1978, 1985). This section provides a brief critical analysis of the thermoregulatory theory and offers an alternate view, based on the role of pup stimulation and physiological processes related to lactation (cf. Lorenz, 1992).

*1. Thermal Limitation of Mother–Young Contact Revisited*

The theory assumes that the chronic increase in maternal temperature due to lactation renders Norway rat dams vulnerable to an acute increase in temperature during the nest bout, especially after the second week postpartum when the litter mass is large. Further, this increased temperature "cannot be diminished by autonomic thermoregulatory mechanisms alone" but "rather, the behavior of bout termination halts the rise in maternal temperature" (Leon et al., 1978). Considerations of mother–infant behavior and thermal physiology led to the conclusion that this theory is irrelevant to the termination of *nursing* behavior and is unsupportable in general.

First, the evidence obtained to support the theory is based on nest bouts, not actual nursing behavior. A small nest box (11- × 14-cm floor area) was used—to which the pups were confined—and the dam's occupancy of it was measured by automatic electromechanical recording. Ironically, by restricting mother–infant contact to a space too small for the dam to stretch out, the apparatus designed to measure this contact changed its pattern, and probably its thermal dynamics as well, by preventing all pup-initiated nursing away from the nest and all supine nursing (Stern and Lonstein, 1996), which occur increasingly as the pups grow older (Rosenblatt and Lehrman, 1963; Stern and Levine, 1972). Nesting time was presumed to be a measure of actual nursing, but there were no direct behavioral observations or reports of litter weight gains. The theory was formulated when pups were viewed as largely passive behavioral participants in the mother–infant nursing interaction. Thus, effects on nest-bout duration consequent to extreme changes in the room or nest-floor temperature were viewed as due to changes only in the dam (Leon et al., 1978). In contrast, we have shown that extreme temperatures prevent the pups from suckling and, thereby, from stimulating their dam's quiescent nursing posture (Section IV,A,3). A nesting bout, therefore, is not necessarily synonymous with a nursing bout that includes actual suckling. This caution is especially true when behavior is altered due to thermal, sensory, or endocrine manipulations of the dam, pups, or both.

Second, none of the assumptions about the lactating dam's vulnerability to dangerous levels of hyperthermia have held up. It is highly unlikely that small increases in dams' temperature during many nursing bouts *compel* them to leave the nest because their temperature rises much more (up to 1°C) when active away from the nest and pups than at anytime during or at the end of nest bouts (Kittrell and Satinoff, 1987). Dams actually choose a *warm* location during intervals between nest bouts (Jans and Leon, 1983) and they are usually active, not inactive, at this time (Kittrell and Satinoff, 1987). Neither lowering the dam's temperature with sodium salicylate (Bates *et al.*, 1985) nor increasing it by treatment with morphine + naloxone (Azzara *et al.*, 1993) alters the duration of either nest bouts or interbout intervals. In addition, the previously untested assumption that the large litter mass in the second week postpartum is a heat *source* for the dam was found to be untrue, that is, heat always flows from dam to pups, not the reverse (Dr. Jeffrey Alberts, unpublished observations, 1990; see Stern and Lonstein, 1996).

## 2. Alterations in Suckling Stimuli and Nursing-Bout Terminations

In the course of a long nursing bout (e.g., ≥20 min) that includes milk ejections, the pups become sated. Although nipple attachment per se is not a good indicator of hunger, probably because suckling is its own reward, there are qualitative and quantitative changes in suckling (Section IV,A,1) that probably affect the dam. The degree of nutritional deprivation of the pups and their receipt of milk affects their state of arousal, the persistence and type of suckling, and the likelihood of nipple shifting (Brake *et al.*, 1988); collectively these factors can account for individual differences in pup weight gain despite equivalent amounts of milk available at each nipple (Lorenz, 1992). A decrease in the number of pups suckling, and the intensity of suckling of those still attached, may result in a cumulative diminution of the suckling stimulus, thereby freeing the dam from suckling-induced quiescence and terminating the nursing bout. Indeed, rat dams become increasingly aroused after several milk ejections—in response to behavioral changes in the pups and perhaps to their own internal changes as well—and terminate nursing most often within a minute after a milk ejection (J. M. Stern, unpublished observations). To date, we have tested the effects on nursing behavior of presumed alterations in pup suckling by inhibiting milk flow to pups.

Milk secretion was reduced or inhibited entirely by treating lactating dams for several days with bromocriptine—a DA receptor antagonist that inhibits prolactin secretion, and thereby synthesis of milk—but suckling still occurred on the well-developed nipples (Protomastro and Stern, 1994). During a 1-h posttreatment observation on Day 12, the nursing bout was significantly longer in lactation-suppressed than control dams, suggesting that something related to milk delivery contributes to bout termination. Further, maternal interactions with 9- to 10- and 13- to 14-day-old litters over 24 h were assessed from time-lapse videorecordings. When the dam was lactating, the pups were likely to remain in the nest after the termination of a nursing bout and the dam initiated most nest bouts. In contrast, when the dam was no longer lactating, the hungry pups were likely to wander away from the nest after the nursing bout and most nursing bouts were initiated in this way. Therefore, activity changes due to satiety or hunger are true of mobile rat pups as well as adults; these alterations, in turn, affect the nursing pattern of their mother.

D. SUMMARY

The rat dam's transformation from frenetic activity, mostly directed toward her young, to quiescence and the sustained adoption of kyphosis—the upright, dorsally arched crouching posture—occurs in response to effective ventral somatosensory stimuli from pups, especially suckling. Pups gain access to the dam's ventrum either when the dam hovers over the gathered litter while engaged in pronurturant activities, such as licking, or when they become mobile enough to locate their recumbent dam away from the nest and burrow under her ventrum. In either case, pups provide a rich source of tactile stimulation that includes pressure, from their head while crawling and rooting, from their paws while pushing against the ventrum, and mostly from their mouth during suckling. Pups must be active, at or near nest temperature, capable of suckling, and sufficient in number to elicit normal nursing behavior. Dams must have nipples from which they feel suckling sensations for normal nursing behavior to be displayed. Suckling stimulation is summated spatially and temporally for the elicitation not only of well-known neuroendocrine reflexes—release of prolactin and oxytocin—but for kyphotic nursing as well. However, nonmaternal rats with undeveloped nipples or no nipples treated with a cataleptic dosage of the DA receptor blocker haloperidol can be induced to display kyphosis when placed over a litter of rooting pups. This supports the hypothesis that suckling inhibits extrahypothalamic DA, which, in turn, facilitates nursing behavior by enhancing somatosensory responsiveness to trunk stimulation and by stimulating a posture that is a precursor of kyphosis. Suckling sensations are carried by the dorsolateral columns of the spinal cord, bilateral lesions of which in the lower cervical cord eliminate lactation and kyphotic nursing; control procedures or dorsal column lesions did not impair maternal behavior or lactation. The effective lesions are specific to nursing behavior because licking of pups recovered in all and retrieval in most dams, and all permitted suckling. Nursing bout termination, like its initiation and maintenance, is probably related to a decrease in intensity or occurrence of suckling stimulation. One change previously proposed to cause rat dams to terminate nest bouts, a rise in the dam's temperature while nursing, was rejected on the basis of behavioral and physiological evidence.

## V. Somatosensation and Maternal Aggression

One of the paradoxes of maternal behavior in rats, and other mammals, is that the potential for vicious aggression toward a strange intruder coexists with gentle nurturant activities toward offspring. Clearly, stimuli received from the intruder must be sufficiently distinct from those received from offspring to signal the dramatic, and selective, alteration in the female's demeanor. Because the offending intruder can be a male or female conspecific, or even an allospecific individual, the relevant distinguishing cues may be unfamiliar "nonpup."

Maternal aggression in rats begins late postpartum, reaches a peak early postpartum, and persists at high levels for about another week. Guarding the entry of the underground maternal burrow may be the most effective way for the mother to ward off an intruder and the most common in nature, but in laboratory studies of maternal aggression the intruder is typically placed in the territory of the dam, usually with the litter present, and it is not provided an escape route. In this situation, the resident female approaches the intruder, *sniffing* it extensively (for about 1 min before first aggressive act), often *aggressively grooms* (investigates intruder's head with nose, mouth, and paws), and then begins *biting* and *attacking* (sudden lunge, grabbing with paws, rolling, pinning, kicking, and biting). This sequence is repeated frequently, interspersed with various postures, while the unfortunate intruder often freezes. Once the initial bite or attack has occurred, the dam is in an aggressive mood—indicated by posture, appearance of her fur, and behavior—until the offending intruder is removed. The reader is referred to other sources for more detailed reviews (e.g., Svare, 1981; Rosenblatt *et al.*, 1994).

### A. Elicitation by Perioral Stimuli

Neither acute deprivation of the sight nor the sound of the intruder affects the occurrence or the normal intensity of maternal aggression in rats (Kolunie *et al.*, 1994). Although volatile odors may play some role (e.g., Ferreira and Hansen, 1986), neither peripherally induced anosmia nor vomeronasal-organ removal impairs the expression of this behavior during early lactation (Kolunie and Stern, 1995; Mayer and Rosenblatt, 1993). In contrast, biting and attacking the intruder, like retrieval and licking of pups, are unlikely during perioral anesthesia (Kolunie and Stern, 1990) or for several hours after cutaneous trigeminal orosensory denervations (Stern and Kolunie, 1991). When biting and attacking do occur after reduction of trigeminal perioral sensory stimulation, it is infrequent and brief, and it is elicited after a threefold increase in frequency and duration of sniffing the intruder, analogous to the stimulus requirement for retrieval under similar circumstances (Section III,A,1). In the absence of oral aggressive acts during perioral anesthesia, the transformation to an aggressive mood does not occur and nonoral aggressive acts, such as kicking, are not displayed. Indeed, the male intruders paired with these dams are more likely to explore the cage and less likely to freeze than males paired with control dams.

Fighting experience on PD 1 during control treatment did not increase the likelihood of fighting on Day 5 in females during acute anaptia. This is in contrast to the beneficial effects of pretreatment retrieval experience on retrieval during acute anaptia (Section III,A,1; Stern and Kolunie, 1989), perhaps because aspects of retrieval are well practiced in the home cage, whereas fighting an intruder is not (Kolunie and Stern, 1990). In male rats the absence of biting during perioral anesthesia could be largely counteracted by *repeated* pretreatment experience with shock-induced fighting or mouse killing (reviewed by Kolunie and Stern, 1990).

B. Maintenance by Ventral Trunk Stimuli

Maternal aggression toward intruders, like maternal behavior toward pups, is stimulated by the hormones of pregnancy (Rosenblatt *et al.,* 1994). Similarly, whereas the maintenance of maternal responsiveness postpartum depends on continual physical interaction with pups, this is true of maternal aggression as well.

When Norway rats are completely separated from their litters, maternal aggression remains high for 1 h, is reduced after 4–5 h, and is eliminated by 24 h (Stern and Kolunie, 1993). During a separation of 4 h, the odor but not the sight or sound of pups contributes to the maintenance of maternal aggression (Ferreira and Hansen, 1986). Indeed, it makes most sense, functionally and mechanistically, that the propensity to display maternal aggression is brought about by a relatively long-lasting neurobiological change resulting from the actual nursing of the offspring. Yet postpartum rats without nipples show normal levels of maternal aggression (Mayer *et al.,* 1987; Stern and Kolunie, 1993), unlike the case for early postpartum thelectomized house mice (Svare, 1981). However, acute anethesia of the nipples and surrounding skin of dams eliminates or greatly reduces maternal aggression (Stern and Kolunie, 1993). These findings suggest a greater sensitivity of rats than mice to nonnipple ventrum somatosensations, perhaps because of the large size difference between these two rodent species.

Thus, ventral somatosensory stimuli from pups normally received during suckling—responsible for neuroendocrine secretions that maintain lactation and the full pattern of normal nursing behavior (Section V,A,B)—seem to be responsible as well for maintaining the mother's willingness to fight for the protection of her offspring. Each consequence of suckling and suckling-related afferents requires activation of separate sites in the brain. Lesions to a midbrain site, the peripeduncular nucleus, diminishes both milk secretion and maternal aggression in some circumstances (Hansen and Ferreira, 1986), but may diminish maternal aggression and not lactation in others (Rosenblatt et al., 1994). The peripeduncular nucleus may be in a region where the neuroendocrine and behavioral consequences of ventral trunk somatosensory afferents become dissociated, a view suggested by the diffuse suckling-stimulated pathways in the brain (Wakerley et al., 1994).

C. SUMMARY

Investigation of an intruder placed in a home cage of a lactating rat is followed swiftly by biting and attacking. Maternal aggression occurs during selective deprivation of seeing, hearing, or smelling, but it is not likely to be elicited during snout anaptia due to perioral anethesia or shortly after trigeminal orosensory denervations, even after prolonged sniffing. When biting is not elicited, fighting does not occur, including kicking, a component that does not involve the mouth; this finding reveals the motivational role of sufficient trigeminal sensations. Pretreatment fighting experience does not ameliorate the effects of trigeminal afferent deprivation. Recovery of fighting occurs within a day or so after trigeminal cutaneous sensory denervations and is then elicited after prolonged sniffing of the intruder. The maintenance of maternal aggression occurs via a functional link to lactation, demonstrated by its loss 24 h following separation from the litter or following acute anesthesia of the nipples and surrounding ventrum. This mechanism ensures that a potentially life-threatening behavior to the dam—litter defense—does not occur if the young are no longer viable. The peripeduncular nucleus may be a site that receives suckling stimuli and promotes maternal aggression.

VI. NEUROBIOLOGICAL CONSEQUENCES AND IMPLICATIONS

Maternal behavior is a complex category that includes a variety of components, each with different sensorimotor mechanisms and functions. Elucidation of the central neural control of maternal behavior must take these components into account (Stern, 1989, 1990). Accordingly, a main approach used in my research is the detailed behavioral analysis of the components of maternal behavior, coupled with selective somatosensory alterations. This work has yielded new insights into the relationship among these components, including behavioral sequences and state change. Before attempting to integrate these findings along with possible neural bases (Section VI,C), the results of two other approaches used recently to shed light on the interface between somatosensations, maternal behavior, and neural functioning are described. Neurophysiological mapping of the trunk representation in the primary somatosensory cortex (S-I) of lactating and nonlactating rats revealed striking changes in response to ventral trunk somatosensations received while nursing (Section IV,A). Immunocytochemical mapping of the activity of a gene that is turned on by a variety of sensory stimuli revealed the involvement in nursing behavior of a site not previously implicated in maternal behavior, the caudal periaqueductal gray (Section VI,B).

A. CORTICAL PLASICITY: NURSING BEHAVIOR CHANGES THE BRAIN

We have seen that somatosensory stimulation from the snout and the nipple-bearing trunk plays a critical role in all aspects of maternal behavior. Further, experience with pups facilitates subsequent responses to them. Subcortical sites contain the neuroendocrine and behavioral reflex circuits that make the expression of maternal behaviors possible, but the neocortex has also been implicated (Beach, 1937; Stern, 1989, pp. 166–168). The neocortex is dynamic, changing in response to afferent inputs (e.g., Merzenich *et al.*, 1988). This neocortical plasticity contributes to current and long-term experiential influences (e.g., Merzenich *et al.*, 1990), which may include those affecting maternal behavior.

Many studies show that representational changes in S-I in adult mammals in response to new tactual experiences require extensive training that in cludes attention to the task. Owl monkeys (Aoutus trivergatus), for example, are trained to perform a task for about 1 hour per day for many weeks, which results in greatly increased tactile stimulation to a finger, and the increase in this finger's S-I representation is determined by mapping before and after the training (Merzenich *et al.*, 1988). The ventrum stimulation received by a lactating rat, however, occurs naturally, without training, for long periods each day over several weeks postpartum. Although nursing behavior is largely reflexive in nature, it is reinitiated many times per day (e.g., Leon *et al.*, 1978), at which times the dam is alert and attentive to her pups.

There is a topographical, functional representation of the body in rats in S-I, which is dominated by the face, especially the whiskers, whereas a very small area represents the much larger trunk (Section II,A; Chapin and Lin, 1984). We (Xerri *et al.,* 1994) mapped in detail the representation of the trunk, especially the ventrum, in young, adult female Long Evans rats, 10 primiparous postpartum dams—6 lactating (Days 6–19) and 4 nonlactating (Days 12–18) that had pups removed on the day of birth—and 3 virgins. The S-I representation of the ventral trunk skin was about 1.6 times larger in maternal, lactating than in nonmaternal rats, both virgin and postpartum nonlactating controls (Fig. 4A,B). The greatest representational increase, about twofold, was on the nipple-bearing skin in the anterior and posterior regions, whereas the change in the S-I map for the middle third of the ventrum, which has no nipples, was not significant. Corresponding to this increased representation, the sizes of individual receptive fields (RFs) on the ventrum in lactating rats was about one-third that of control females, and again, this difference was magnified in the nipple-bearing skin (Fig. 5A,B). The glabrous nipple and areolar skin were weakly represented in S-I. Because the major inputs to S-I are from the dorsal columns, and it is the lateral columns that convey suckling information (Section IV,B), perhaps the nipples are represented elsewhere. Preliminary findings suggest that these receptive field changes occur as early as PD 6 and are reversible at 2.5 months following weaning.

Rats with even large lesions of the *neocortex* are capable of showing all the components of maternal behavior and of raising a litter, unless the repeated tests are too disruptive (Beach, 1937), but there are deficits in performance that may be related to poor integration of voluntary and reflexive activities (Stern, 1989). However, the effects of specific neocortical lesions on detailed aspects of maternal behavior have not yet been investigated. Further, the findings on somatosensory cortex plasticity during lactation (Xerri *et al.,* 1994) could provide a new impetus for studying the role of cortex in the regulation of maternal behavior in rats. There are five implications of these findings:

1. There may be many other examples of cortical plasticity than the small part of the S-I we studied in response to changing tactile and other sensory stimuli received in the course of maternal behavior. For example, the already extensive, finely grained representation of the face may change in response to the trigeminal stimulation received from pups during lactation.

2. The cortical changes may, in turn, affect the ongoing behavior by increasing sensitivity to or discrimination of appropriate afferents.

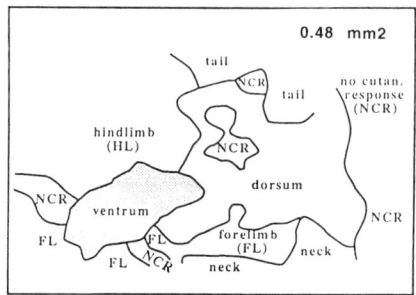
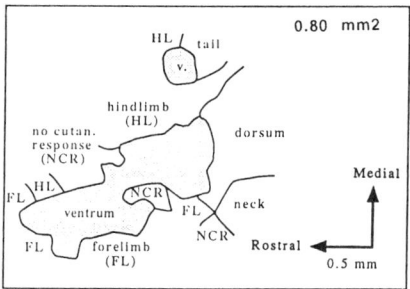
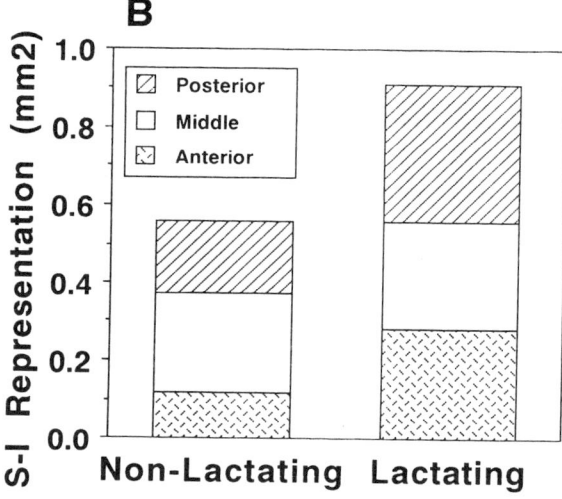

FIG. 4. (A) Cartoon reconstructions of two representative experiments in which skin surfaces of the trunk were mapped in the middle cortical layers of the primary somatosensory cortex (S-I) in postpartum primiparous nonlactating control (left) and lactating (right) rats. These reconstructions were based on 101 and 85 microelectrode penetration sites, respectively, at which receptive fields (RFs) were defined for neurons or small clusters of neurons. Cortical zones bordering the ventra (shaded) represent the skin of the forelimb (FL), hindlimb (HL), neck, tail, and dorsolateral trunk (dorsum). In regions labeled NCR (no cutaneous response), neurons were not excited by light tactile stimulation, as is typical for many S-I sectors adjacent to the trunk representation (e.g., Chapin and Linn, 1984). The area of the ventrum representation is shown in the top right corner of each figure. (B) Mean areas of S-I representation of the anterior, middle, and posterior, and total ventrum surfaces for nonlactating and lactating rats ($n = 4$ each). Group differences in the nipple-bearing anterior and posterior subdivision areas, but not in the nippleless middle subdivision, are statistically significant. Virgins (not shown) are comparable to the nonlactating controls in terms of cortical ventrum maps. (Modified from Xerri et al., 1994; reprinted with permission from Oxford University Press, The Journal of Neuroscience.)

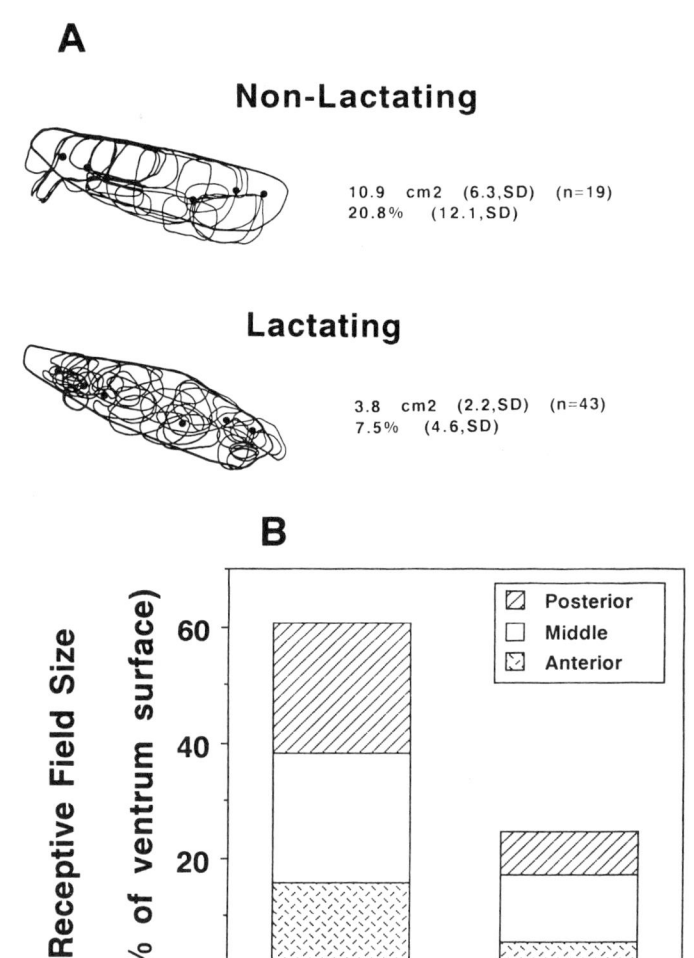

FIG. 5. (A) All receptive fields (RFs) defined in the primary somatosensory cortex (S-I) ventrum representational zone in representative nonlactating and lactating rats, both at 16 days postpartum. The heavy dark lines outline the ventrum skin, small solid circles mark the location of the nipples, and thinner lines outline cortically recorded cutaneous RFs. The means and SDs of measured and normalized (with reference to total ventrum skin area) RF sizes are shown for each case at the right. (B) Mean RF areas, normalized with reference to total ventrum skin area, centered over the anterior, middle, and posterior thirds of the ventrum, for nonlactating ($n = 4$) and lactating ($n = 5$) rats. Each area differed significantly by group. For nonlactating rats, the middle and posterior RFs do not differ significantly and both are significantly larger than anterior RFs; for lactating rats, the middle RFs are significantly larger than the anterior and posterior RFs, which do not differ from each other. Virgins (not shown) are comparable to the nonlactating controls in terms of ventrum RF sizes. (Modified from Xerri et al., 1994; reprinted with permission from Oxford University Press, *The Journal of Neuroscience.*)

3. Because memories are largely sensory in nature, the cortical changes may contribute to the mediation of experiential influences, including retention of short-latency onset of maternal responsiveness over a span of several weeks.

4. The maternal behavior of other species may also induce changes in cortical representations, the details of which are species specific. For example, males and females of the monogomous New World primate, *Callicebus* (titi monkey), share parental responsibilities; the male generally carries the infant except during nursing bouts, and he does so on his back (Mendoza and Mason, 1986). Thus, in this species the dorsum of the male may show a greater cortical representational change than the ventrum of the female.

5. Cortical plasticity in both sensory and motor regions probably occurs in the course of other species-specific behaviors when certain stimuli and movements are experienced to a much greater extent than ever before.

B. ACTIVATION OF NEURAL SITES DURING INTERACTION WITH PUPS

As part of the primary genomic response to external stimulation, transcriptional activity of the immediate early gene, *c-fos*, is triggered in many cell types by a wide variety of excitors (Morgan and Curran, 1991), such as stimulation of primary sensory afferents (Chinapen *et al.*, 1992; Hunt *et al.*, 1987). Consequently, measures of increased *c-fos* activity have been used to identify specific areas in the brain that are stimulated after exposure to, or behavioral interaction with, reproductively relevant stimuli. Studies using immunocytochemical visualization of Fos, the protein product of *c-fos*, have confirmed the well-known involvement of the medial preoptic area (MPOA) in the mediation of maternal behavior in rats by showing increased Fos compared with controls during the expression of this behavior in maternal virgins (Numan and Numan, 1994), in parturient dams (Fleming *et al.*, 1994), and in lactating dams reunited with pups after separation of close to 3 days (Numan and Numan, 1994, 1995).

In our Fos study (Lonstein *et al.*, 1995, 1996; also see latter for additional references), dams were separated from their pups for 48 h beginning on PD 5, then were given a 1-h period with no stimulation, distal stimulation from pups (pups in a wire-mesh box) or its control (empty box), nonsuckling interaction with pups (pups incapable of suckling due to mystacial pad anesthesia), or full interaction with pups, including suckling. Interactions with the stimulus were recorded continuously and litter weight changes were obtained. Fos labeling was subsequently quantified in 25 brain sites.

Distal stimuli from pups—olfactory (volatile), auditory, and visual—did not differentially activate Fos in any site, confirming and extending previous reports (Fleming and Walsh, 1994; Numan and Numan, 1995). Nonsuckled dams had significantly more Fos than dams in all other groups, including suckled dams, only in the anterior cortical amygdala; this currently puzzling finding may reflect longer attention to olfactory stimuli or frustration due to lack of suckling stimulation.

Physical interaction with pups, with or without suckling, elicited high levels of active maternal behaviors and similarly high Fos expression compared with other conditions in sites previously shown to impair or eliminate retrieval after damage (Numan, 1994), that is, MPOA, nucleus accumbens, lateral septum, and lateral habenula, in sites activated by stimuli from pups, that is, somatosensory cortex (both face and trunk representations), and in sites not previously implicated at all with maternal behavior, that is, paraventricular thalamic nucleus and rostral midbrain periaqueductal gray (PAG). Thus, these known sites may not play a role in nursing behavior, while the newly identified sites may play a hitherto unknown role in the mediation of motorically active maternal behaviors.

The most exciting finding of this study is that suckled dams had significantly more Fos than nonsuckled dams in the caudal PAG, mostly in the lateral and ventrolateral sections. Given the known neuroanatomy and functions of the PAG (Depaulis and Bandler, 1991), this structure may mediate the quiescent nursing posture. For example, this region receives strong innervation from the cervical enlargement of the spinal cord—where afferents from suckling stimulation are thought to synapse—and projects extensively to a wide area within the medulla—where there are control mechanisms of both quiescence and postural changes (Klemm, 1990). Indeed, following electrolytic lesions of the lateral or ventrolateral part of the caudal PAG either pre- or postpartum, there is a selective impairment of the kyphotic nursing posture (Lonstein *et al.*, 1996).

C. BEHAVIORAL ELEMENTS AND SEQUENCES, STATE CHANGES, AND NEURAL PATHWAYS

The complex category of maternal behavior includes simpler components combined by a chain of stimuli and responses into sequences that may be accompanied by state changes. A summary follows of what was gleaned about several aspects of maternal behavior in rats from the research reviewed previously, along with an integrative sketch of possible underlying neural pathways.

*Licking* is the first maternal behavior expressed toward pups, occurring as each neonate emerges from the vaginal canal; it is also perhaps the simplest element, and it occupies more of the dam's time than any other active behavior. Although licking during parturition is probably initiated by extraoral stimuli related to the emerging fetuses (cf. Lecci *et al.*, 1994), it is elicited and maintained normally by perioral trigeminal stimuli (Section III,A,2). Even once licking recovers 24 h after postpartum IO denervation, there is a marked reduction in its duration (Section III,B,2), which contributes to impaired parturition behavior following prepartum IO denervation.

Licking of pups undoubtedly shares many neurobiological mechanisms with licking in any context, regulated by an oscillator in the hypoglossal nucleus (Weisenfeld *et al.*, 1977). This nucleus contains tongue motoneurons, expresses Fos during parturition (Luckman *et al.*, 1994), and is influenced by many rostral brain projections (Travers, 1995). The reduction in likelihood or duration of pup licking during DA receptor antagonism (Stern and Taylor, 1991) may be due to effects on the striatum (Pisa, 1988), resulting in lingual dystonia (Fowler and Mortell, 1992). However, the presence of pseudo-licking, that is, pup-licking movements without tongue protrusion when the mouth is sutured closed—reminiscent of facial grooming movements in mice with amputated forelimbs (Fentress, 1973)—indicates that pup licking, unlike licking to drink, is a fixed action pattern with greater motoric complexity that includes holding a pup in the paws, rhythmic head movements, and postural adjustments. Reduction or even elimination of tongue feedback does not prevent pup licking or alter its form, but it does reduce its duration. A small litter receives as much licking as a full litter because this behavior is maintained by trigeminal feedback and it is inhibited by suckling-induced quiescence, not by the depletion of chemosensory stimuli, however attractive (see Stern, 1989).

*Retrieval* emerges during parturition from *mouthing* when the dam simply stretches to pick up a pup and move it closer to her. Retrieval combines locomotion with trigeminal orosensory reflexes such as jaw opening and grasping; this "loop" is then repeated, usually without interruption, until completion of pup grouping, typically at the nest site in the laboratory paradigm. Retrieval shares many elements with nest building and hoarding of food, that is, mouth carrying, nest deposition, and repetitions.

Retrieval is blocked or impaired by bilateral damage to the MPOA—where estradiol acts to induce short-latency onset of maternal behavior—or its connections, or to one of several sites interconnected with the MPOA, for example, ventral tegmental area (VTA), lateral habenula (LHb), and nucleus accumbens (NAcc) (Numan, 1994). Therefore, a dopaminergic circuit (VTA–LHb–NAcc) exists that affects retrieval, consonant with the inhibition of retrieval following systemic treatment with haloperidol (Giordano et al., 1990; Stern and Taylor, 1991). Indeed, DA is released in the NAcc during the display of active maternal behaviors (Hansen et al., 1993) and microinfusion of cis-flupenthixol, a DA receptor antagonist, into the NAcc dose-dependently blocks retrieval and reduces licking duration (Keer et al., 1994). Trigeminal afferents might interact with MPOA in various ways (Numan and Numan, 1991). The nucleus of the solitary tract has reciprocal connections with the trigeminal system and projects to the MPOA. Integration of afferents in somatosensory cortex from the trigeminal system and the MPOA via the lateral-preoptic-area–substantia-innominata region (Simerly and Swanson, 1988) may affect oromotor functions. Also, MPOA efferents important for maternal behavior terminate in various brain stem sites, including the principal sensory nucleus of the trigeminal, the spinal trigeminal nucleus (Numan and Numan, 1991), and the midbrain PAG (Shipley et al., 1991).

What accounts for the typical efficiency of the retrieval sequence, in which extraneous activity and pup contact is inhibited until after the completion of grouping (see Stern, 1996b)? The combination of IO denervation and 24-h separation from pups results in a pronounced breakdown in this efficiency characterized by long retrieval durations interrupted with such pup contact (Section III,B,2). Trigeminal afferents stimulated by the initial retrieval may provoke the secretion of a neuromodulatory peptide transmitter that is long acting, that is, at least several seconds for short carrying distances and longer for transport to a new nest site. A possible site of action is the lateral septum, lesions of which result in disorganized retrieval (Fleisher and Slotnick, 1978).

The *nursing* behavior characteristics described herein, that is, quiescence and postural adjustments to teat-seeking offspring, is generalizable to the guinea pig (Hennessy and Jenkins, 1994), as well as lesser bushbaby, pig, pony, rabbit, and sheep (see Stern and Johnson, 1990). Dam-initiated nursing behavior in rats occurs as the culmination of a sequence that begins with the dam's return to the nest and display of pronurturant, oral attentions toward the pups. The behavioral link between the trigeminal perioral requirement for hovering over pups and the subsequent adoption of the kyphotic posture in response to suckling may be paralleled by a mechanistic link because there is a projection from the brainstem trigeminal complex to the afferent milk-ejection pathway in the mesencephalic lateral tegmentum (Dubois-Dauphin et al., 1985). Given the overlap in the spinal pathway of afferents from nipples with respect to neuroendocrine secretions and the nursing posture (Section IV,B), this projection may be relevant to nursing behavior as well as to neuroendocrine secretions.

The state change during kyphosis, supported as well by the relationship between slow-wave sleep and milk ejections (e.g., Voloschin and Tramezzani, 1979), may be mediated by suckling-induced inhibition of extrahypothalamic DA and activation, probably via the midbrain PAG (Section VI,B), of brainstem mechanisms that inhibit movement (Klemm, 1990). The behavioral and stimulus parallels between lordosis (Pfaff and Modianos, 1985; Pfaff *et al.*, 1994) and kyphosis, the upright crouching posture during nursing (Stern, 1990), have neuroanatomical parallels as well in terms of spinal pathways (Section IV,B), hypothalamic or MPOA facilitation, and midbrain completion of the reflex loop. Our recent Fos and electrolytic lesion results suggest that the midbrain PAG may be the site of completion of reflex pathways for kyphosis, as it is for lordosis, a site with spinal cord, trigeminal, MPOA, and limbic system afferents and projections to medullary sites that affect posture and inhibition of movement (Bandler and DePaulis, 1991).

Although dam-initiated nursing behavior is extremely different from *maternal aggression* (Section V), they are related in several ways. Both are part of a behavior sequence that is initiated by a trigeminally elicited response: in the absence of maternal oral behaviors, hovering over pups, a precursor of kyphosis, does not occur; in the absence of biting, no fighting occurs, not even with another body part. Both involve a state change that is probably mediated in part by altered activity of a neurotransmitter: decreased DA during nursing (discussed previously) and decreased serotonin during maternal aggression (Olivier *et al.*, 1989) are possibilities. Finally, the maintenance of the propensity to display maternal aggression is linked to continued nursing behavior and lactation.

## VII. General Conclusions and Summary

Maternal behavior in rats consists of components involving physical interaction between the dam and pups (licking, mouthing, retrieval, nursing), objects (nest building), or an intruder (maternal aggression). Physical interaction with pups is crucial for the establishment, maintenance, long-term retention, and induction of maternal behavior. Behavioral reflexes are critical elements of maternal behavior and these are elicited by appropriate somatosensory stimuli. During reduced trigeminal orosensory stimulation, jaw opening in response to pups or an intruder does not occur or occurs only after prolonged stimulation; the severity of these effects depends on the extent of the loss and, in the case of retrieval, relevant experience. Most dramatically, IO denervation carried out within a day prior to term results in marked deficiencies in oral facilitation of delivery and efficient pup cleaning, leading to increased pup mortality. During reduced ventral trunk somatosensory stimulation, especially from suckling, nursing behavior is impaired. Pups must be capable of suckling and sufficient in number to stimulate their dam's nursing. Suckling inhibits the dam's display of motorically active, trigeminally dependent, pronurturant activities by promoting quiescence and the reflexive, sustained kyphotic nursing posture; this posture, resulting from a concatenation of reflexes, is characterized by bilateral symmetry, ventroflexion, and rigid leg support. A reduction in suckling, rather than a rise in the dam's temperature, probably accounts for termination of most nursing bouts.

Somatosensory stimuli promote several motivational, as well as eliciting, effects:

1. Behavioral chains are truncated in the absence of the initiating stimuli and responses. Thus, when maternal oral behaviors are not elicited during snout anaptia, hovering over the young pups does not occur, so nursing cannot be elicited. When biting of an intruder is not elicited during snout anaptia, no fighting occurs, including behaviors not involving the mouth, such as kicking.

2. Somatosensory stimuli from pups cause changes in state of varying duration, possibly mediated by a neurochemical change. Whereas trigeminal stimulation from pups elicits behavioral and dopaminergic activation, suckling stimulation elicits quiescence and dopaminergic inhibition, probably in extrahypothalamic systems. Retrieval efficiency and aggressive mood may result from altered secretion of a neuromodulator in response to the initiating stimulus.

3. Continual attempts to make snout contact with pups when it is blocked by a muzzle and behavioral substitution when retrieval is thwarted during snout anaptia, for example, gathering pups with paws, are indications of high maternal motivation, probably mediated largely by available somatosensations.

Various links between sensorimotor analyses and central neural pathways of maternal behavior are described. Neurophysiological mapping of the primary somatosensory cortex revealed cortical plasticity indicative of increased sensitivity in the ventrum representation of lactating rats that may affect ongoing behavior or mediate certain maternal experiences that last several weeks. Immunocytochemical visualization of Fos, a gene product expressed in response to a variety of sensory stimuli, confirms a role for several sites in maternal oral activities (MPOA, lateral septum, and two DA target sites: LHb and NAcc), and suggests others, as well as the involvement of the caudal PAG in the kyphotic nursing posture. Retrieval must involve integration of hormone responsiveness, trigeminal and other sensory afferents, communication with a dopaminergic pathway, and motor mechanisms; the MPOA may be such a site, and the PAG a hitherto unexplored link. A kyphosis circuit—analogous to the lordosis circuit—includes the dorsolateral columns of the spinal cord and the caudal PAG, which receives trigeminal, spinal cord, and MPOA afferents, and projects to the medulla where inhibition of movement and postural alterations are mediated.

### Acknowledgments

The research from my laboratory reviewed in this article was supported largely by Grant MH-40459 from NIMH, with additional support from Busch Biomedical Research Grants from Rutgers University. I thank the editors, Dr. Jay S. Rosenblatt and Dr. Charles T. Snowden, and Mr. Joseph Lonstein for their helpful comments on the manuscript.

### References

Azzara, A., Lonstein, J. L., and Stern, J. M. (1993). Duration of mother-young contact of rats is *not* limited by a rise in maternal temperature. *Abstr., Int. Soc. Develop. Psychobiol.,* Alexandria, VA, Nov. 4–7.

Bandler, R., and Depaulis, A. (1991). Midbrain periqueductal gray control of defensive behavior in the cat and the rat. *In* "The Midbrain Periaqueductal Gray Matter: Functional, Anatomical, and Neurochemical Organization" (A. Depaulis and R. Bandler, eds.), pp. 175–198. Plenum, New York.

Bates, A., Adels, L., and Leon, M. (1985). Thermal control of maternal contact bouts: The interbout interval. *Physiol. Behav.* **34,** 835–837.

Beach, F. A. (1937). The neural basis of innate behavior. I. Effects of cortical lesions upon the maternal behavior pattern in the rat. *J. Comp. Psychol.* **24,** 393–436.

Beach, F. A., and Jaynes, J. (1956a). Studies on maternal retrieving in rats. III. Sensory cues involved in the lactating female's response to her young. *Behaviour* **10,** 104–125.

Beach, F. A., and Jaynes, J. (1956b). Studies on maternal retrieving in rats. II. Effects of practice and previous parturitions. *Am. Nat.* **90,** 803–809.

Berridge, K. C., and Fentress, J. C. (1986). Contexual control of trigeminal sensorimotor function. *J. Neurosci.* **6,** 325–330.

Brake, S. C., Shair, H. N., and Hofer, M. A. (1988). Exploiting the nursing niche: The infant's sucking and feeding in the context of the mother-infant interaction. *In* "Handbook of Behavioral Neurobiology: Developmental Psychobiology and Behavioral Ecology," Vol. 9 (E. M. Blass, ed.), pp. 347–388. Plenum Press, New York.

Caggiula, A. R., Herndon, J. J., Jr., Scanlon, R., Greenstone, D., Bradshaw, W., and Sharp, D. (1979). Dissociation of active from immobility components of sexual behavior in female rats by central 6-hydroxydopamine: Implications for CA involvement in sexual behavior and sensorimotor responsiveness. *Brain Res.* **172,** 505–520.

Calhoun, J. B. (1963). "The Ecology and Sociology of Norway Rats." U. S. Public Health Service Publ. no. 1008. U. S. Gov. Print. Office, Washington, D. C.

Chinapen, S., Swann, J. M., Steinman, J. L., and Komisaruk, B. R. (1992). Expression of *c-fos* protein in lumbosacral spinal cord in response to vaginocervical stimulation in rats. *Neurosci. Lett.* **145,** 93–96.

Chapin, J., and Lin, C.-S. (1984). Mapping of the body representation in the SI cortex of anesthetized and awake rats. *J. Comp. Neurol.* **229,** 199–213.

Cohen, J. A., and Price, E. O. (1979). Grooming in the Norway rat: Displacement activity or "boundary-shift"? *Behav. Neural. Biol.* **26,** 177–188.

Croskerry, P. G., Smith, G. K., and Leon, M. (1978). Thermoregulation and the maternal behaviour of the rat. *Nature (London)* **273,** 299–300.

Depaulis, A., and Bandler, R. (1991). "The Midbrain Periaqueductal Gray Matter: Functional, Anatomical and Neurochemical Organization." NATO ASI Ser. A, Life Sciences, Vol. 213. Plenum Press, New York.

De Ryck, M., Schallert, T., and Teitelbaum, P. (1980). Morphine versus haloperidol catalepsy in the rat: A behavioral analysis of postural support mechanisms. *Brain Res.* **201,** 143–172.

Dubois-Dauphin, M., Armstrong, W. E., Tribollet, E., and Dreifuss, J. J. (1985). Somatosensory systems and the milk-ejection reflex in the rat. II. Lesions of the mesencephalic lateral tegmentum disrupt the reflex and damage mesencephalic somatosensory connections. *Neuroscience* **15,** 1111–1129.

Eayrs, J. T., and Baddeley, R. M. (1956). Neural pathways in lactation. *J. Anat.* **90,** 161–171.

Fentress, J. C. (1973). Development of grooming in mice with amputated forelimbs. *Science (Washington, D. C.)* **179,** 704–705.

Ferreira, A., and Hansen, S. (1986). Sensory control of maternal aggression in *Rattus norvegicus*. *J. Comp. Psychol.* **100,** 173–177.

Fleischer, S., and Slotnick, B. M. (1978). Disruption of maternal behavior in rats with lesions of the septal area. *Physiol. Behav.* **21,** 189–200.

Fleming, A. S., Suh, J. S., Korsmit, M., and Rusak, B. (1994). Activation of Fos-like immunoreactivity in the medial preoptic area and limbic structures by maternal and social interactions in rats. *Behav. Neurosci.,* **108,** 724–734.

Fowler, S. C., and Mortell, C. (1992). Low doses of haloperidol interfere with rat tongue extensions during licking: A quantitative analysis. *Behav. Neurosci.* **106,** 386–395.

Fukuoka, T., Negoro, H., Honda, K., Higuchi, T., and Nishida, E. (1984). Spinal pathway of the milk-ejection reflex. *Biol. Reprod.* **30,** 74–81.

Giesler, G. J., Urca, G., Cannon, J. T., and Liebeskind, J. C. (1979). Response properties of neurons of the lateral cervical nucleus in the rat. *J. Comp. Neurol.* **186,** 65–78.

Giordano, A. L., Johnson, A. E., and Rosenblatt, J. S. (1990). Haloperidol-induced disruption of retrieval behavior and reversal with apomorphine in lactating rats. *Physiol. Behav.* **48,** 211–214.

Goldberger, M. E. (1980). Motor recovery after lesions. *Trends Neurosci.* **3,** 288–291.

Gould, S. J. (1980). "The Panda's Thumb: More Reflections in Natural History." Norton, New York.

Hansen, S., Bergvall, A., and Nyiredi, S. (1993). Interaction with pups enhances dopamine release in the ventral striatum of maternal rats: A microdialysis study. *Pharmacol. Biochem. Behav.* **45,** 673–676.

Hansen, S., and Ferreira, A. (1986). Food intake, aggression and fear behavior in the mother rat: Control by neural systems concerned with milk ejection and maternal behavior. *Behav. Neurosci.* **100,** 64–70.

Hansen, S., Stanfield, E. J., and Everitt, B. J. (1981). The effects of lesions of lateral tegmental noradrenergic neurons on components of sexual behaviour and pseudopregnancy in female rats. *Neuroscience* **6,** 1105–1117.

Heinz, M., Shafer, K., and Braun, H. A. (1990). Analysis of facial cold receptor activity in phenomena the rat. *Brain Res.* **521**, 289–295.

Hennessy, M. B., and Jenkins, R. (1994). A descriptive analysis of nursing behavior in the guinea pig (*Cavia porcellus*). *J. Comp. Psychol.* **108**, 23–28.

Holloway, W. R., Dollinger, M. J., and Denenberg, V. H. (1980). Parturition in the rat: Description and assessment. *In* "Maternal Influences and Early Behavior" (R. W. Bell and W. P. Smotherman, eds.), pp. 1–26. SP Med. Sci. Books, New York.

Hunt, S. P., Pini, A., and Evan, G. (1987). Induction of c-fos-like protein in spinal cord neurons following sensory stimulation. *Nature* (*London*) **328**, 632–634.

Huston, J. P., Steiner, H., Schwarting, R. K., and Morgan, S. (1988). Parallels in behavioral and neural plasticity induced by unilateral vibrissae removal and unilateral lesion of the substantia nigra. *In* "Postlesion Neural Plasticity" (H. Flohr, ed.), pp. 537–551. Springer-Verlag, Berlin.

Jacquin, M., and Zeigler, H. P. (1983). Trigeminal orosensation and ingestive behavior in the rat. *Behav. Neurosci.* **97**, 62–97.

Jans, J., and Leon, M. (1983). Determinants of mother-young contact in Norway rats. *Physiol. Behav.* **30**, 959–1003.

Keer, S. E., and Stern, J. M. (1996). Haloperidol disrupts maternal motivation in lactating Long-Evans rats. *Conference on Reproductive Behavior*, June 15–19.

Keer, S., Protomastro, M., and Stern, J. M. (1994). Dopamine antagonists and maternal behavior: D1 vs. D2 receptor subtypes and sites of action. Conference on Reproductive Behavior, Lehigh Univ., Bethlehem, PA, June 11–14.

Kenyon, P., Cronin, P., and Keeble, S. (1981). Disruption of maternal retrieving by perioral anesthesia. *Physiol. Behav.* **27**, 313–321.

Kenyon, P., Cronin, P., and Keeble, S. (1983). Role of the infraorbital nerve in retrieving behavior in lactating rats. *Behav. Neurosci.* **97**, 255–269.

Kittrell, E. M., and Satinoff, E. (1987). Diurnal rhythms of body temperature, drinking and activity over reproductive cycles. *Physiol. Behav.* **42**, 477–484.

Klemm, W. R. (1990). Behavioral inhibition. *In* "Brainstem Mechanisms of Behavior" (W. R. Klemm and R. B. Vertes, eds.), pp. 497–533. Wiley, New York.

Kolunie, J. M., and Stern, J. M. (1990). Maternal aggression: Disruption by perioral anesthesia in lactating Norway rats. *J. Comp. Psychol.* **104**, 352–360.

Kolunie, J. M., and Stern, J. M. (1995). Maternal aggression in rats: Effects of olfactory bulbectomy, $ZnSO_4$-induced anosmia, and vomeronasal organ removal. *Horm. Behav.* **29**, 492–518.

Kolunie, J. M., Stern, J. M., and Barfield, R. J. (1994). Maternal aggression in rats: Effects of visual or auditory deprivation of the mother and dyadic pattern of ultrasonic vocalizations. *Behav. Neural Biol.* **62**, 41–49.

Kow, L.-M., Montgomery, M. O., and Pfaff, D. W. (1977). Effects of spinal cord transections on lordosis reflex in female rats. *Brain Res.* **123**, 75–88.

Lashley, K. (1938). Experimental analysis of instinctive behavior. *Psychol. Rev.* **45**, 445–471.

Lecci, A., Giulian, S., Lazzeri, M., Benaim, G., Turini, D., and Maggi, C. A. (1994). The behavioral response induced by intravesical instillation of capsaicin in rats is mediated by pudendal urethral sensory fibers. *Life Sci.* **55**, 429–436.

Leigh, H., and Hofer, M. A. (1973). Behavioral and physiological effects of littermate removal on the remaining single pup and mother during the pre-weaning period in rats. *Psychosom. Med.* **35**, 497–508.

Leon, M., Adels, L., and Coopersmith, R. (1985). Thermal limitation of mother-young contact in Norway rats. *Dev. Psychobiol.* **18**, 85–105.

Leon, M., Croskerry, P. G., and Smith, G. K. (1978). Thermal control of mother-young contact in rats. *Physiol. Behav.* **21**, 793–811.

Lonstein, J. S., Simmons, D. A., and Stern, J. M. (1996). Periaqueductal gray lesions, pre- or postpartum, specifically impair kyphotic nursing behavior but facilitate maternal aggression in Long-Evans rats. *Society for Neuroscience,* **22,** in press.

Lonstein, J. S., Swann, J. M., and Stern, J. M. (1995). C-fos expression in the brain of lactating Long-Evans rats after mother-young interaction with or without suckling. *Soc. Neurosci.* **21,** 465.

Lonstein, J. S., Swann, J. M., and Stern, J. M. (1996). Maternal behavior stimulates *c-fos* activity in the brain of lactating rats. I. Effects of suckling, non-suckling contact, and distal stimulation from pups. In preparation.

Lorenz, D. N. (1992). Suckling physiology and behavior of rats: An integrated theory of ingestion and satiety. *Prog. Psychobiol. Physiol. Psychol.* **15,** 1–83.

Luckman, S. M., Dyball, R. E. J., and Leng, G. (1994). Induction of c-fos expression in hypothalamic magnocellular neurons requires synaptic activation and not simply increased spike activity. *J. Neurosci.* **14,** 4825–4830.

Mayer, A. D., Carter, L., Jorge, E. A., Mota, M. J., Tannu, S. M., and Rosenblatt, J. S. (1987). Mammary stimulation and maternal aggression in rodents: Thelectomy fails to reduce pre- or postpartum aggression in rats. *Horm. Behav.* **21,** 501–510.

Mayer, A. D., and Rosenblatt, J. S. (1993). Contributions of olfaction to maternal aggression in laboratory rats (*Rattus norvegicus*): Effects of peripheral deafferentation of the primary olfactory system. *J. Comp. Psychol.* **107,** 12–24.

Mendoza, S. P., and Mason, W. A. (1986). Parental division of labour and differentiation of attachments in a monogomous primate (*Callicebus moloch*). *Anim. Behav.* **34,** 1336–1347.

Merzenich, M. M., Recanzone, G., Jenkins, W. M., Allard, T., and Nudo, R. J. (1988). Cortical representational plasticity. *In* "Neurobiology of Neocortex" (P. Rakic, and W. Singer, eds.), pp. 42–67. Wiley, New York.

Merzenich, M. M., Recanzone, G., Jenkins, W. M., and Grajski, K. A. (1990). Adaptive mechanisms in cortical networks underlying cortical contributions to learning and nondeclarative memory. *Cold Spring Harbor Symp. Quant. Biol.* **55,** 873–887.

Moltz, H., Geller, D., and Levin, R. (1967). Maternal behavior in the totally mammectomized rat. *J. Comp. Physiol. Psychol.* **61,** 383–387.

Montagu, A. (1986). "Touching: The Human Significance of the Skin," 3rd ed. Harper & Row, New York.

Morgan, H. D., Fleming, A. S., and Stern, J. M. (1992). Somatosensory control of the onset and retention of maternal responsiveness in primiparous Sprague-Dawley rats. *Physiol. Behav.* **251,** 549–556.

Morgan, J. I., and Curran, T. (1991). Stimulus-transcription coupling in the nervous system: Involvement of the inducible proto-oncogenes Fos and Jun. *Annu. Rev. Neurosci.* **14,** 421–451.

Numan, M. (1994). Maternal behavior. *In* "Physiology of Reproduction," Vol. 2, 2nd ed. (E. Knobil and J. D. Neill, eds.), pp. 221–302. Raven Press, New York.

Numan, M., and Numan, M. J. (1991). Preoptic-brainstem connections and maternal behavior in rats. *Behav. Neurosci.* **105,** 1013–1029.

Numan, M., and Numan, M. J. (1994). Expression of fos-like immunoreactivity in the preoptic area of maternally behaving virgin and postpartum rats. *Behav. Neurosci.* **108,** 379–394.

Numan, M., and Numan, M. J. (1995). Importance of pup-related sensory inputs and maternal performance for the expression of Fos-like immunoreactivity in the preoptic area and ventral bed nucleus of the stria terminalis of postpartum rats. *Behav. Neurosci.* **109,** 135–149.

Olivier, B., Mos, J., and van Oorschot, R. (1989). Etho-experimental studies of similarities and differences in male and female agonistic behaviour. *In* "Ethoexperimental Approaches to the Study of Behavior" (R. Blanchard, P. F. Brain, C. Blanchard, and S. Parmigiani eds.), pp. 494–507. NATO ASI Ser. Ser. D, Vol. 48. Kluwer Academic Publ., Boston.

Paxinos, G. (ed.) (1995). "The Rat Nervous System," 2nd ed. Plenum, New York.

Pfaff, D. W., and Modianos, D. (1985). Neural mechanisms of female reproductive behavior. In "Handbook of Behavioral Neurobiology: Reproduction" (N. Adler, D. W. Pfaff, and R. Goy, eds.), pp. 423–493. Plenum, New York.

Pfaff, D. W., Schwartz-Giblin, S., McCarthy, M., and Kow, L-M. (1994). Cellular and molecular mechanisms of female reproductive behavior. In "Physiology of Reproduction" (E. Knobil and J. D. Neill, eds.), pp. 107–220. Raven Press, New York.

Pisa, M. (1988). Motor functions of the striatum in the rat: Critical role of the lateral region in tongue and forelimb reaching. Neuroscience 24, 453–463.

Protomastro, M., and Stern, J. M. (1995). Nursing behavior in lactation-suppressed postpartum Long-Evans rats during and after treatment with bromocriptine. Abstr. Develop. Psychobiol., 28, 193.

Reisbick, S., Rosenblatt, J. S., and Mayer, A. D. (1975). Decline of maternal behavior in the virgin and lactating rat. J. Comp. Psychol. 89, 722–732.

Rosenblatt, J. S. (1967). Nonhormonal basis of maternal behavior in rats. Science (Washington, D. C.) 156, 1512–1514.

Rosenblatt, J. S., Factor, E. M., and Mayer, A. D. (1994). Relationship between maternal aggression and maternal care in the rat. Aggressive Behav. 20, 243–255.

Rosenblatt, J. S., and Lehrman, D. S. (1963). Maternal behavior of the laboratory rat. In "Maternal Behavior in Mammals" (H. Rheingold, ed.), pp. 8–57. Wiley, New York.

Shair, H. N., and Hofer, M. A. (1992). Sleep-wake states, sucking, and nursing patterns in young rats. Prog. Psychobiol. Physiol. Psychol. 15, 141–229.

Shambes, G. M., Gibson, J. M., and Welker, W. I. (1978). Fractured somatotopy in granule cell tactile areas of rat cerebellar hemispheres revealed by micromapping. Brain Behav. Evol. 15, 94–140.

Shipley, M. T., Ennis, M., Rizvi, T. A., and Behbehani, M. M. (1991). Topographical specificity of forebrain inputs to the midbrain periaqueductal gray: Evidence for discrete longitudinally organized input columns. In "The Midbrain Periaqueductal Gray Matter Functional, Anatomical, and Neurochemical Organization" (A. Depaulis and R. Bandler, eds.), pp. 417–448. Plenum, New York.

Simerly, R. B., and Swanson, L. W. (1988). Projections of the medial preoptic nucleus: A Phaseolus vulgaris leucoagglutinin anterograde tract-tracing study in the rat. J. Comp. Neurol. 270, 209–242.

Stahlbaum, C., Matochik, J., White, N., Barfield, R. J., and Stern, J. M. (1989). Perioral anesthesia disrupts proceptive but not receptive behavior patterns in female rats. Conference on Reproductive Behavior. Saratoga Springs, NY, June 11–15.

Stern, J. M. (1983). Maternal behavior priming in virgin and Caesarean-delivered Long-Evans rats: Effects of brief contact or continuous exteroceptive pup stimulation. Physiol. Behav. 31, 757–763.

Stern, J. M. (1989). Maternal behavior: Sensory, hormonal, and neural determinants. In "Psychoendocrinology" (F. R. Brush and S. Levine, eds.), pp. 105–226. Academic Press, New York.

Stern, J. M. (1990). Multisensory regulation of maternal behavior and masculine sexual behavior: A revised view. Neurosci. Biobehav. Revs. 14, 183–200.

Stern, J. M. (1991). Nursing posture is induced in haloperidol-treated maternally-naive female and male rats in response to ventrum stimulation from active pups. Horm. Behav. 25, 504–517.

Stern, J. M. (1996a). Trigeminal lesions and maternal behavior in Norway rats: II. Disruption of parturition. *Physiol. Behav.*, **60**, in press.

Stern, J. M. (1996b). Trigeminal lesions and maternal behavior in Norway rats. III. Experience with pups affects. *Dev. Psychobiol.*, **30**, in press.

Stern, J. M. (1996c). Offspring-induced nurturance: Animal-human parallels. *In* "Growing Points in Developmental Psychobiology," proceedings of a conference in honor of Dr. Seymour Levine, Tuscon, Arizona, October 8–10, 1995. *Develop. Psychobiol.*, **30**, in press.

Stern, J. M., and Johnson, S. K. (1989). Perioral somatosensory determinants of nursing behavior in Norway rats. *J. Comp. Psychol.* **103**, 269–280.

Stern, J. M., and Johnson, S. K. (1990). Ventral somatosensory determinants of nursing behavior in Norway rats. I. Effects of variations in the quality and quantity of pup stimuli. *Physiol. Behav.* **47**, 993–1011.

Stern, J. M., and Kolunie, J. M. (1989). Perioral anesthesia disrupts maternal behavior during early lactation in Long-Evans rats. *Behav. Neural Biol.* **52**, 20–38.

Stern, J. M., and Kolunie, J. M. (1991). Trigeminal lesions and maternal behavior in rats. I. Effects of cutaneous rostral snout denervation on maintenance of nurturance and maternal aggression. *Behav. Neurosci.* **105**, 984–997.

Stern, J. M., and Kolunie, J. M. (1993). Maternal aggression of rats is impaired by cutaneous anesthesia of ventral trunk, but not by nipple removal. *Physiol. Behav.* **54**, 861–868.

Stern, J. M., and Lehrman, D. S. (1969). Role of testosterone in progesterone-induced incubation behaviour in male ring doves (*Streptopelia risoria*). *J. Endocrinol.* **44**, 13–22.

Stern, J. M., and Levine, S. (1972). Pituitary-adrenal activity in the post-partum rat in the absence of suckling stimulation. *Horm. Behav.* **3**, 237–246.

Stern, J. M., and Lonstein, J. S. (1996). Nursing behavior in rats is impaired in a small nestbox and with hyperthermic pups. *Dev. Psychobiol.* **29**(2), 101–122.

Stern, J. M., and Rogers, L. (1988). Experience with younger siblings facilitates maternal responsiveness in pubertal Norway rats. *Dev. Psychobiol.* **21**, 575–589.

Stern, J. M., and Taylor, L. A. (1991). Haloperidol inhibits maternal retrieval and licking, but facilitates nursing behavior and milk ejection in lactating rats. *J. Neuroendocrin.* **3**, 591–596.

Stern, J. M., Dix, L., Bellomo, C., and Thramann, C. (1992). Ventral trunk somatosensory determinants of nursing behavior in Norway rats. II. Role of nipple and surrounding sensations. *Psychobiology* **20**, 71–80.

Stern, J. M., Yu, Y. L., and Crockett, D. C. (1993). Spinal pathway mediating suckling-induced nursing behavior and neuroendocrine reflexes. *Soc. Neurosci.* **19**, 1610.

Svare, B. R. (1981). Maternal aggression in mammals. *In* "Parental Care in Mammals" (D. J. Gubernick and P. H. Klopfer, eds.), pp. 179–210. Plenum Press, New York.

Terkel, J., and Rosenblatt, J. S. (1971). Aspects of nonhormonal maternal behavior in the rat. *Horm. Behav.* **2**, 161–171.

Tracey, D. J., and White, P. M. E. (1995). Somatosensory system. *In* "The Rat Nervous System" (G. Paxinos, ed.), pp. 689–704.

Travers, J. B. (1995). Oromotor nuclei. *In* "The Rat Nervous System" (G. Paxinos, ed.), pp. 239–255.

Voloschin, L. M., and Tramezzani, J. H. (1979). Milk ejection reflex linked to slow wave sleep in nursing rats. *Endocrinology* **105**, 1201–1207.

Wakerley, J. B., Clarke, G., and Summerlee, A. J. S. (1994). Milk ejection and its control. *In* "The Physiology of Reproduction," Vol. 2, 2nd ed. (E. Knobil and J. D. Neill, eds.), pp. 1131–1177. Raven Press, New York.

Waite, P. M. E., and Tracey, D. J. (1995). Trigeminal sensory system. *In* "The Rat Nervous System" (G. Paxinos, ed.), pp. 705–724.

Wiesenfeld, Z., Halpern, B. P., and Tapper, D. N. (1977). Licking behavior: Evidence of hypoglossal oscillator. *Science (Washington, D. C.)* **196,** 1122–1124.

Wiesner, B. P., and Sheard, N. M. (1933). "Maternal Behaviour in the Rat." Oliver and Boyd, London.

Xerri, C., Stern, J. M., and Merzenich, M. M. (1994). Alterations of the cortical representation of the rat ventrum induced by nursing behavior. *J. Neurosci.* **14,** 1710–1721.

Zeigler, H. P., Jacquin, M. F., and Miller, M. G. (1985). Trigeminal orosensation and ingestive behavior in the rat. *Prog. Psychobiol. Physiol. Psychol.* **11,** 63–196.

Zeigler, H. P., Semba, K., Egger, M. D., and Jacquin, M. F. (1984). Trigeminal sensorimotor mechanisms and eating in the rat. *Brain Res.* **308,** 149–154.

# Experiential Factors in Postpartum Regulation of Maternal Care

ALISON S. FLEMING, HYWEL D. MORGAN, AND CAROLYN WALSH

DEPARTMENT OF PSYCHOLOGY
ERINDALE COLLEGE
UNIVERSITY OF TORONTO
MISSISSAUGA, ONTARIO

## I. INTRODUCTION

In this article we focus primarily on mechanisms that underlie experiences acquired when a new mammalian mother first interacts with her young, and the effects of this experience on her later responses to these same young and to subsequent offspring. Although emphasis will be given to our own and related work based on rat studies, we also discuss examples from other species, where relevant.

We first discuss ways in which experiences acquired by the mother reinforce and, in some instances, supplant the "motivational" effects of the maternal hormones and thereby increase the probability of the offspring's survival, and hence, the mother's reproductive fitness. These experiences usually enhance mothers' nurturant behavior but, if negative, may have the opposite effect (Harlow, 1963).

We then concentrate on mechanism; we first discuss experiences acquired when a mother rat first interacts with her young and the role of sensory factors in this plasticity. We consider the role of (1) the parturitional hormones, estradiol, progesterone, prolactin, and oxytocin; and (2) stimulation provided by the young, primarily through the smell and touch senses. We explore ways in which the postpartum experience impacts on mechanisms regulating later maternal behavior and lactational processes. We then consider the extent to which maternal learning shares properties with learning within other behavioral contexts in terms of its temporal and behavioral parameters, as well as in terms of its physiology. We also consider the involvement in maternal learning of proteins and neurotransmitters known to be implicated in other appetitive learning, focusing primarily on the monoamines, dopamine and norepinephrine. We end by discussing the neuroanatomy of maternal experience based on lesion, stimulation, and immunohistochemical studies. The neuroanatomy of interest includes the medial preoptic area, important for the expression of maternal behavior, the olfactory–amygdala circuit involved in processes underlying olfactory perception and emotional behavior, and cortical and limbic structures involved in reinforcement and the formation of memories.

These analyses are based on the use of a variety of convergent physiological methodologies. These include the following: (1) brain lesions, where localized damage to specific parts of the brain are produced; (2) psychopharmacologic techniques in which chemical compounds are injected into the brain, either mimicking (agonists) or antagonizing (antagonists) the effects of naturally occurring neurochemicals and neurotransmitters; or (3) c-fos immunohistochemistry, which involves the staining of a specific brain protein that is expressed by specialized genes within cells that are activated. By use of such a marker for brain activation, one can trace cell groups in the brain that are activated by a particular behavior (like maternal behavior) or process (like learning).

## II. Types of Experience

When a young animal is being cared for by its mother, it is gaining experience that becomes incorporated into its own mothering (Harlow, 1963; Suomi, 1990; Kraemer, 1992; Main et al., 1985; Fairbanks, 1989). In rhesus macaques, for instance, variations in the kind of mothering infants receive result in variations in the infant's cognitive style; emotional reactivity; neuroendocrine function; and, in consequence, adult social and parental behavior. When a juvenile animal interacts with its younger siblings or peers, it is gaining relevant experience, through play-mothering and, possibly, through the formation of emotional attachment to peers (Pryce, 1993). When, as nonreproductive adults, females help care for an infant that is not their own, they become experienced with infant cues and they practice behaviors which they will themselves be exhibiting with their own offspring (Pryce, 1993; Terkel and Rosenblatt, 1971; Fleming and Rosenblatt, 1974a; Keverne, 1995; among rats, such prepartum contact reduces their natural withdrawal tendencies (Fleming and Luebke, 1981). When new mothers initially interact with their own young, the experience gained familiarizes them with the context (Bridges, 1975, 1977; Orpen and Fleming, 1987; Fleming and Sarker, 1990) and promotes individual or litter recognition (Holmes, 1990; Keverne and Kendrick, 1990). These experiences, gained under the influence of the parturitional hormones, influence their subsequent responding to these same offspring as well as to subsequent offspring

(Cohen and Bridges, 1981; Fleming and Sarker, 1990; Ruppenthal *et al.*, 1976).

While these different varieties of experience influence the quality of maternal behavior, they do so by very different behavioral mechanisms—appropriate to the species, the particular stage in development, and reproductive state during which the experience is obtained. Thus, experience can influence social–emotional processes and/or cognitions as well as provide the opportunity to become familiar with infant cues and to practice the component behaviors.

### III. Functional Adaptiveness of Experience in the Service of "Good" Mothering

Variations in maternal responsiveness within a population result in variations in infant survival and, hence, in mothers' reproductive fitness. Mothers who are not adequately attentive to their offspring and who are not motivated to nurse them or protect them from harm are likely to lose those young before the time of natural weaning. It is true that there are circumstances when neglect of, or aggression towards, present young may in fact enhance fitness and likelihood of future young. Examples would be cannibalism of young at times of reduced resources (hamsters, Schneider and Wade, 1989, 1991; Miceli and Malsbury, 1982) or reduced investment when young are sick, damaged, or otherwise at severe risk (Elwood and Kennedy, 1990). In general, however, enhanced fitness depends on "good" mothering.

A corollary of this statement is that factors that promote better mothering also have adaptive value; this applies to an appropriate hormonal state and appropriate life experience which both directly and indirectly influence a mother's response to her offspring. Among marmosets, for instance, the early prepubertal experience of carrying infants (allocaring) results in mothers who are more likely to show sensitive and adequate mothering when they give birth to their first young. In the absence of some previous experience caring for nonrelated young, first-time marmoset mothers will often neglect or desert their first offspring (Pryce, 1993).

Moreover, experiences rearing one set of offspring will enhance mothering of subsequent offspring. There are many examples of multiparous mothers who have reared one set of young showing more efficient, less expensive, and more effective mothering. In mice, pup cues that are initially ineffective in eliciting maternal behavior in first-time mothers, come to be effective in multiparous animals (Noirot, 1972). On the other side, multiparous sheep mothers (ewes) are less disrupted than are primiparous mothers by alterations in the stimulus properties of the young (Kendrick *et al.*, 1991). They are also more sensitive than first-time mothers to the hormones and to vaginocervical stimulation that normally activate maternal responsiveness (Kendrick *et al.*, 1991; Keverne, 1995). Multiparous mothers are also better able to maintain their mothering in the face of nutritional, environmental, or social stressors. Moreover, they are less disturbed than primiparous mothers by a variety of experimental manipulations including cesarean section (Moltz *et al.*, 1966), endocrine manipulations (Moltz and Wiener, 1966a,b; Moltz *et al.*, 1969a), morphine administration (Bridges, 1990; Kinsley and Bridges, 1988), and brain lesions (Fleming and Rosenblatt, 1974b; Franz *et al.*, 1986; Numan, 1988; Schlein *et al.*, 1972) that could disrupt maternal behavior.

The value of experience in promoting appropriate mothering is most clear, perhaps, in situations where the mother comes to recognize individual offspring and responds selectively and exclusively to them (Holmes, 1990; Poindron and Lévy, 1990). For example, sheep and goats are nomadic, moving across pastures in herds and give birth to precocial young who soon after birth join the herd. Because there is a potential for young of different, possibly unrelated, ewes to become confused with one another, it is advantageous that the mother recognize her own young and invest primarily in that lamb. Such selective recognition develops during a brief period postpartum and is based on exposure during the postpartum sensitive period to the scent surrounding the offspring (Poindron and Lévy, 1990). The familiarity rule applies more generally to many mammalian species that show individual kin recognition (Holmes, 1990).

Maternal experience not only influences subsequent behavior, but it also influences subsequent lactation, which has an impact on infant viability. Thus, in rats and other species as well, the processes of lactation and milk letdown depend on suckling-induced release of hormones (Grosvenor *et al.*, 1990). However, after some experience with suckling young (14 or so days in the primiparous mother), distal olfactory and visual cues from the young can also elicit the release of prolactin and the maintenance of milk production. Initially, these distal pup cues must be located underneath the female to produce a conditioned release of hormones; as lactation proceeds, however, exteroceptive cues can be located at a greater distance from the female (Grosvenor *et al.*, 1990). As a result, the growing young, who are becoming increasingly mobile, do not need to suckle continuously or remain close by the female to obtain adequate nutrition; instead, milk is available to them when they reattach and they can nurse for brief intense bouts appropriate to their developmental age. Interestingly, in multiparous animals the conditioned release of hormones occurs a week earlier than first seen in primiparous mothers (on Day 7 rather than Day 14) (Grosvenor et al., 1990).

As these examples illustrate, experiences caring for young have long-term effects that impact on the mother's effectiveness as a caregiver and, hence, her reproductive fitness. That they influence the quality and intensity of mothering is without question; how they do so constitutes the focus of the remainder of this article.

## IV. The Physiology of Maternal Behavior

The work described as follows focuses primarily on our own work using laboratory rats. While use of these highly inbred strains facilitates the experimental analysis of physiological mechanisms, it must be acknowledged that the laboratory setting does not expose the animal to the usual vagaries of the social and inanimate environment that would occur in a natural setting and that may interfere with (or prime) mothering in the inexperienced mother.

In the absence of these influences, the first-time maternally inexperienced mother rat is very maternal to offspring as soon as they emerge from the birth canal. They retrieve pups to a nest site; engage in intensive licking of the pups, especially of their anogenital regions; and they adopt a crouch-nursing posture over the pups. The enhanced responsiveness at birth of the litter depends on the action of a sequence of hormonal changes that characterize the end of pregnancy and parturition (Rosenblatt, 1990; Bridges, 1990; Insel, 1990). Included among these are elevated levels of progesterone followed by their decline and a rise in circulating estradiol, prolactin, and oxytocin. Although the hormonally primed primiparous mother rat shows all the components of maternal behavior on first exposure to pups, as mothers gain experience they become motorically more efficient; they are more likely to pick pups up at the nape of the neck (rather than by a limb or tail), thereby facilitating their transport, and they show more rapid retrieval behavior (Carlier and Noirot, 1965; Fleming and Rosenblatt, 1974a).

Once mothers have interacted with pups, the parturitional hormones undergo a decline, and after 5–6 days, no longer contribute to maternal responding (Orpen et al., 1987). Although lactational hormones, including the glucocorticoids, prolactin and oxytocin, may well contribute to the patterning of the mother's nursing behavior, maternal responsiveness will be maintained following the decline of these hormones (Leon et al., 1990). Thus, it is assumed that maintenance of the behavior is based on stimulation provided by the litter and the ongoing experience of interacting with the young (Stern, 1983; Jakubowski and Terkel, 1986; Orpen and Fleming, 1987). The importance of processes of learning and memory to this stage of the mother–young relationship is evident.

A. THE EFFECTS OF MATERNAL EXPERIENCE

The effects of experience with pups on subsequent maternal responsiveness can be demonstrated by removing pups from the female at various points after parturition and testing for maternal behavior to foster pups after a period of mother–litter separation (Rosenblatt and Lehrman, 1963; Bridges, 1975, 1977; Cohen and Bridges, 1981; Siegel and Greenwald, 1978; Jakubowski and Terkel, 1986). Using this strategy, we have found that if females are prevented from interacting with pups at parturition (by removing pups as they emerge from the birth canal or by cesarean delivery), dams gradually lose their maternal responsiveness over the next few days. When tested with foster pups on Day 10 postpartum, they do not retrieve, lick, or crouch over the young; like virgin animals first presented with foster pups, they tend to withdraw from them. However, if pups are left continuously with dams, they show the same pattern of sensitization as shown by the virgin animal, responding maternally to pups only after a period of 8 or more days of continuous exposure to pups (Orpen and Fleming, 1987). This process of becoming maternal in the absence of hormonal changes and in response to continuous contact with foster pups has been variously called *maternal induction* or *sensitization* (Rosenblatt, 1967; Fleming and Rosenblatt, 1974a).

If new mothers are permitted to interact with young for a brief period (a few hours or less) after parturition, before removal of the litter (exposure phase), they respond very rapidly to foster pups in later maternal induction tests (retention phase), showing the full repertoire of maternal behaviors within a few days (Bridges, 1975, 1977; Orpen and Fleming, 1987). As shown in Fig. 1, how rapidly dams respond to pups at retention testing is directly related to the duration of the initial experience; it is also related inversely to the interval between experience and test (Cohen and Bridges, 1981; Fleming and Sarker, 1990). It is also highly dependent on the rat strain being tested (Jakubowski and Terkel, 1986; Bridges, 1975, 1977; Cohen and Bridges, 1981). In one study we found that if new mothers were permitted as little as 1/2 h of maternal interactions with a litter of pups on Day 1 postpartum, they showed latencies of 1–2 days to the reinduction of maternal behavior 10 days later (Orpen and Fleming, 1987).

B. ROLE OF HORMONES

Although the parturitional hormones are not necessary for the maternal experience to produce a long-term effect on maternal behavior, as can be

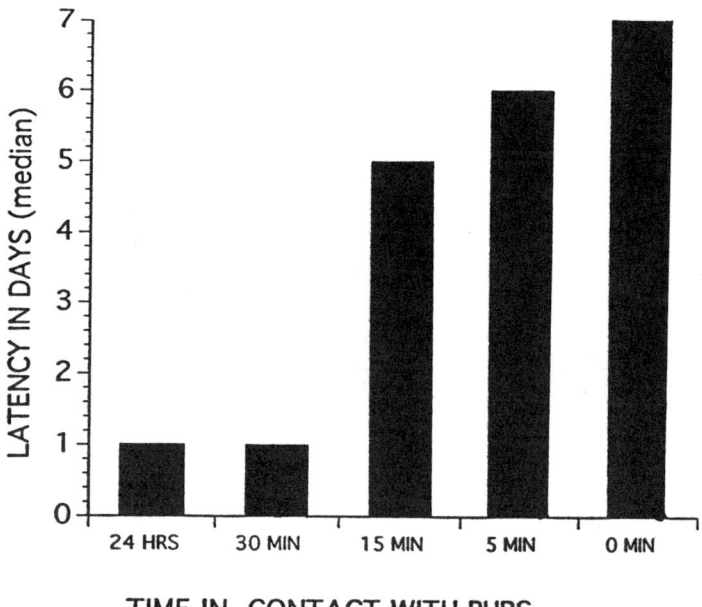

FIG. 1. Latency (in days) to become maternal in tests 10 days after exposure to pups as a function of time in contact with pups. Numbers at bottom represent time in contact with pups on day 1 post-partum. (Based on Orpen and Fleming, 1987.)

seen in Fig. 2, the postpartum state does enhance the effectiveness of that experience. Maternal experiences acquired by the postpartum mother are better retained than are experiences acquired by sensitization procedures in the nonhormonally primed virgin animal (Fleming and Sarker, 1990). Maternal responsiveness on re-exposure to pups is further enhanced if retention testing also occurs during a period of hormonal priming during a subsequent pregnancy (Fleming and Sarker, 1990). These effects of reproductive state on maternal responsiveness can be mimicked in virgin animals injected with parturitional hormones (Fleming and Sarker, 1990).

These data suggest that although the parturitional hormones are not necessary for a maternal experience to be acquired and retained, they do enhance the effectiveness of that experience at both the acquisition and retrieval stages.

There are several ways hormones could act to promote the robust experience effects. They could increase the salience of the unconditioned olfactory or somatosensory cues perceived during proximal mother–litter interactions. They could strengthen the association between these unconditioned stimuli and distal olfactory, and other cues that become associated with the

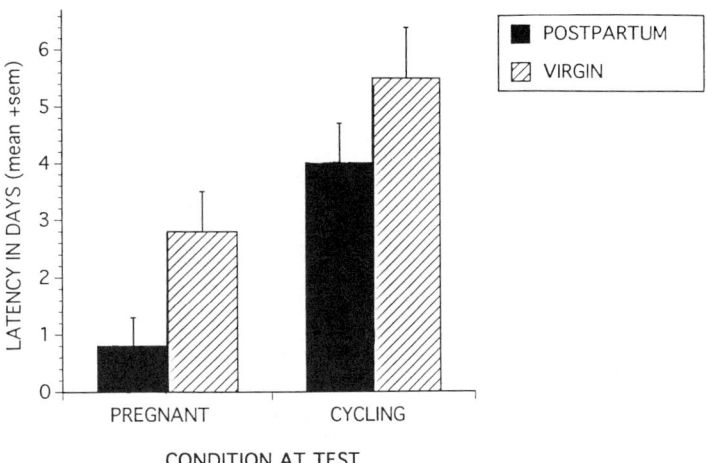

FIG. 2. Latency (in days) to become maternal in tests 30 days after exposure to pups as a function of reproductive state at exposure and at test. (Based on Fleming and Sarker, 1990.)

pups. They could produce internal cues that themselves act as conditioned stimuli, a mechanism that could explain the apparent state dependency of the hormone–experience interactions described previously. Research has not yet identified which of these hormone mechanisms is important.

The next section analyzes in greater detail the nature of the experience being acquired, emphasizing the role of sensory factors.

C. ROLE OF SENSORY INPUT

1. *Maternal Responsiveness*

Precisely what the dam is acquiring or "learning" when she gains a maternal experience is not well understood. We know that the actual experience of parturition is not itself necessary because experience effects occur also if dams are cesarean sectioned prior to interaction with pups (Cohen and Bridges, 1981; Orpen and Fleming, 1987). Moreover, parturition itself, in the absence of experience, does not result in long-term responsiveness (Bridges, 1975, 1977; Fleming and Sarker, 1990; Suomi, 1990). We also know that a robust long-term retention of responsiveness requires that the mother physically interact with pups during the postpartum period. Usually the effective experience involves chemosensory and somatosensory stimulation that occurs in the context of the initiated behavior, during nosing, licking, and nursing of pups (Stern *et al.*, 1992; Stern *et al.*, 1990; Stern and

Johnson, 1989; Stern and Kolunie, 1989). If dams are physically separated from their litters and are then continuously exposed to pups under a perforated clear floor, mothers lose their responsiveness (Jakubowski and Terkel, 1986; Stern, 1983; Orpen and Fleming, 1987). The fact that under some circumstances and in some highly responsive dams a pre-exposure to distal pup cues may facilitate the subsequent expression of maternal behavior suggests that the quality of the required experience depends in part on the animals baseline level of responsiveness (Orpen and Fleming, 1987).

These data suggest either that the relevant experience involves the actual execution of the maternal behaviors and hence involves some form of motor learning and/or that while interacting proximally with pups, the dam is multiply stimulated by more proximal tactual, chemosensory, and, possibly, thermal cues, and these are the important features of the maternal experience. That is, that the experience involves some form of perceptual learning. Alternately, the experience may involve some form of conditioned response where, for instance, one set of distal pup cues (odors) come to substitute for another (e.g., tactile) (Rosenblatt, 1983).

In order to determine whether somatosensory input from the ventral and perioral regions are necessary for an enduring maternal experience, new mother rats underwent manipulations that reduced somatosensory input to the perioral and ventral trunk regions during a 1-h period of interaction with pups and were tested for their maternal behavior towards foster pups 10 days later. As shown in Fig. 3, reduction of somatosensory input to either region alone did not eliminate the experienced-based short maternal onset; however, reduction of tactile input to both regions did. While the desensitized animals were fully maternal at the time of the exposure, and engaged in both pup licking and crouching, at test they responded with long latencies like inexperienced animals and with latencies that were significantly longer than shown by the experienced sham control group (Morgan et al., 1992).

We interpret these results to mean that an important unconditioned stimulus for the maternal learning includes stimulation derived through the skin surface when the animal interacts with the litter. Interestingly, suckling stimulation per se appears not to be necessary for an experience effect since thelectomized (with nipples removed) animals who care for foster young, but are not suckled by them, show both maintenance of maternal behavior (Stern, 1989) and a rapid reinduction after separation from pups (C. Walsh and A. S. Fleming, unpublished data).

As indicated earlier, the chemosensory systems are also used when an animal interacts with pups, although in rats (unlike mice and many other mammals) such input seems not to be necessary either for the initial expression of maternal behavior (Schlein et al., 1972; Fleming and Rosenblatt,

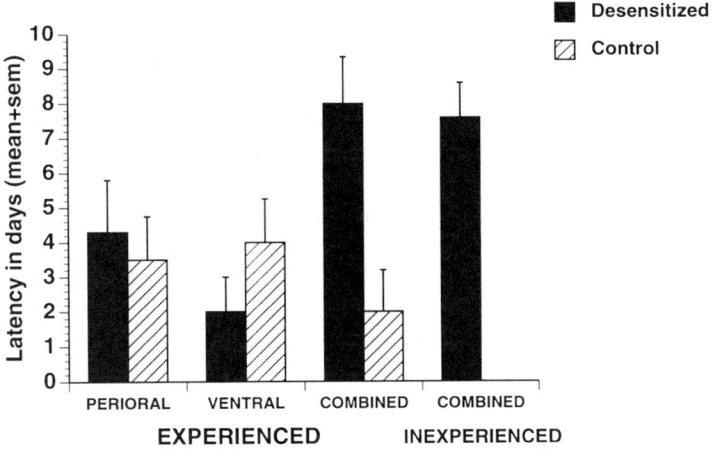

FIG. 3. Latency (in days) to become maternal in maternally experienced: (1) perioral groups (experienced/lidocaine (desen) and experienced/saline (control)); (2) ventral-trunk groups (experienced/full jacket (desen) and experienced/control jacket (control)); (3) combined groups (experienced/lidocaine/full jacket (desen) and experienced/saline/control jacket (control)), and in maternally inexperienced combined; and (4) perioral or ventral-trunk groups (inexperienced/controls). (Based on Morgan et al., 1992.)

1974a) or for its long-term retention (Fleming et al., 1992). Thus, while new mothers are attracted to odors associated with pups (Fleming et al., 1989), primary olfactory or vomeronasal denervations do not substantially affect the pattern of behaviors exhibited, although denervated animals do show some reductions in pup licking (Brouette-Lahlou et al., 1992). Similarly, chemosensory denervations prior to a maternal experience in a new mother rat does not reduce the long-term effectiveness of the experience (Fleming et al., 1992). That said, there is considerable evidence showing that when dams interact with pups, they do learn about pup-associated odors and come to respond selectively to them at a later point in time (Bauer, 1983; Malenfant et al., 1991). If new mothers are exposed to pups that have been scented with an artificial odorant, in maternal tests 10 days later they respond more quickly to pups scented with the same odorant than to ones differently scented; they also come to prefer the pup-associated odorant to an equally familiar odorant, not previously paired with pups (Malenfant et al., 1991).

Through a rather clever manipulation, Noirot (1970, 1972) was able to show that exposing virgin female mice, which are already highly responsive, to auditory and olfactory stimuli (a live pup in a perforated metal box) increased maternal responding to a dead pup, a suboptimal stimulus. Moreover, exposure to different sensory modalities seems to augment different components of the behavioral response repertoire. Thus, female mice exposed to pup vocalizations spent more time nest building, while those exposed to pup odors spent more time licking pups, suggesting the existence of preset sensory–motor associations that are modality dependent. These studies show that in some highly responsive species (e.g., mice), proximal somatosensory cues may not be essential for an experience effect. Interestingly, maternal behavior and recognition of young in mice is more dependent on intact chemosensory function than it is in rats (Calamandrei *et al.*, 1992).

## 2. Individual Recognition

In contrast to these rodents, the role of chemosensory cues is more pronounced and specific among species that are able to recognize individual young and respond selectively to them (Poindron and Lévy, 1990; Romeyer *et al.*, 1994). As described by Lévy *et al.*, (this volume) in sheep, for instance, mothers develop a hormone-dependent attraction to uterine odors that were aversive to the animal before (Lévy *et al.*, 1983). Eliminating this odorant by washing the lamb is more disruptive to primiparous than to multiparous ewes (Poindron and Lévy, 1990). They also exhibit an increased willingness to accept and nurse all young; in this respect, sheep and rats show a similar pattern of hormone-dependent change in maternal responsiveness with parturition. However, unlike rats, which will nurse all young, once ewes have gained experience with one lamb, they rapidly develop an aversion to alien young, which they butt away. This individual recognition is based on experience with the individual odors of the lambs within the context of both hormonal and vaginal–cervical stimulation, and occurs during a very brief 10- to 20-min postpartum period (Poindron and Lévy, 1990).

## 3. Pup Reinforcement

The issue of which modalities are important to the maternal experience effect in rats has been addressed in a somewhat different way by exploring what pup characteristics become reinforcing to the mother as a result of an interactive experience. We investigated the effects of a variety of sensory denervations on the ability of mothers to form a preference for a previously neutral environment that had been paired with pups. New mother rats were tested in a conditioned-place preference paradigm consisting of two chambers differing in their visual and tactile characteristics. Animals were placed into one chamber along with a litter of rat pups on Days 2 and 4 postpartum and into the other without pups on the alternate days. On the fifth day animals were placed into the center compartment separating the two (now empty) chambers and were observed for where they spent their time. A significant proportion of new mothers came to prefer the side that had been associated with pups. Nonmaternal virgins showed no such preference. These effects were based on a full functional olfactory or somatosensory system. As shown in Fig. 4, peripheral anesthetization with lidocaine of either the ventral trunk region or the perioral region, or peripherally induced hyposmia by application of zinc sulfate into the nasal epithelium prior to the pup-exposure phases eliminated the development of a conditioned place preference.

Collectively, these studies show that maternal learning is a robust phenomenon that is based on activation of both chemosensory and somatosensory systems during mother–litter interactions. Although the new mother seems primed to respond to certain cues over others by the action of hormones, the primary effect of the maternal hormones is to activate maternal behavior. Once preset maternal behavior has been exhibited, general mechanisms of learning, reinforcement, and memory are used to further consolidate experiences acquired during mother–litter interactions. These experiences include the activation (and hence modification) in the mother of both chemosensory and somatosensory systems. The next section considers ways in which these sensory and psychological experiences are encoded by in the brain. The role of the monoamines is considered first, followed by a discussion of the relevant neural structures.

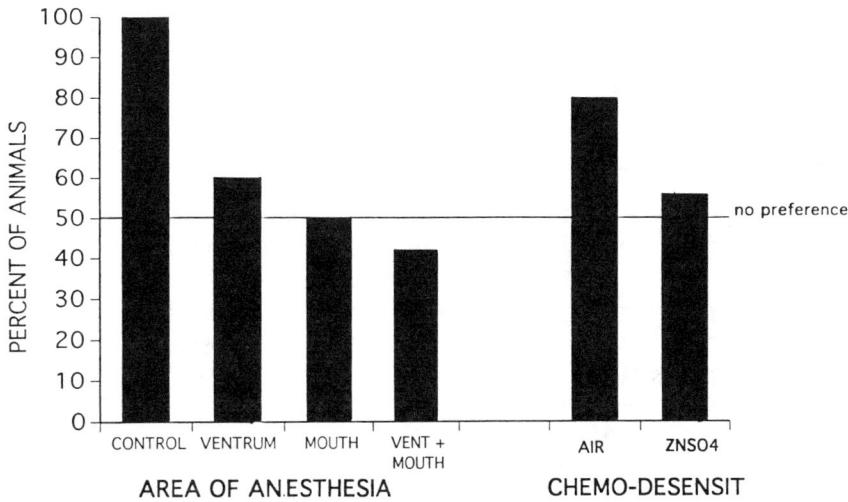

FIG. 4. Percent of animals showing a conditioned place preference after somatosensory or olfactory desensitizations. (Based on Magnusson and Fleming, 1995.)

## D. NEUROCHEMISTRY OF MATERNAL EXPERIENCE

### 1. Maternal Responsiveness and Individual Recognition

Because the maternal experience effect reflects a long-term change in behavior, similar to other forms of learning, there must occur within the brain mediating structural (i.e., synaptic) or functional (i.e., neurotransmitter, receptor) changes. As one approach to understanding underlying neurochemistry, Fleming et al. (1990) asked whether maternal memory requires the synthesis of proteins in the brain in the same way as do other more traditional forms of memory (Davis and Squire, 1984). In their investigations of the maternal experience effect, Fleming et al. (1990) and Malenfant et al. (1991) found that, if a drug that inhibits protein synthesis (cycloheximide) was injected into dams immediately after a 1-h exposure to pups, the long-term retention of maternal behavior was blocked and test animals behaved like inexperienced virgin animals. If, however, cycloheximide was injected 24 h after the experience when the "consolidation" of memory is complete, or immediately before the experience at the time of acquisition, the effect of the maternal experience was retained (see Fig. 5).

These results suggest that, as with other forms of memory, protein synthesis is necessary for acquisition of the experience. The additional observation that protein synthesis inhibitors administered at the time of consolidation do not block the reactivation of maternal behavior on Day 4 postpartum,

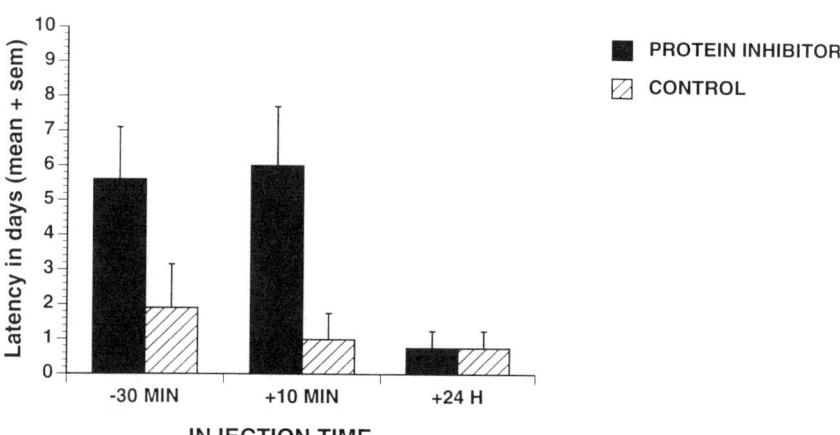

FIG. 5. Latency (in days) to become maternal in tests 10 days after exposure to pups in groups injected with CYX or SAL 30 min before (−30 min), 10 min after (+10 min), or 24 h after (+24 h) a 2-h maternal experience. (Based on Fleming et al., 1990.)

when hormonal effects are still in evidence, indicates that they are not having direct negatively reinforcing effects nor are they affecting the actual expression of the behavior. Moreover, the consolidation of a specifically olfactory experience acquired during the exposure phase was also blocked by drugs that inhibit protein synthesis (Malenfant *et al.*, 1991).

It is not clear how these synthesis-blocking drugs interfere with the consolidation of the maternal experience. These protein synthesis inhibitors block an array of proteins and thus might be altering, for example, levels of structural proteins, enzymes, or neurotransmitters necessary for the consolidation process (Davis and Squire, 1984; Nelson and Alkon, 1989). One interesting possibility is the noradrenergic system that we know is activated at the time of parturition (Rosenberg *et al.*, 1976) and has been implicated in the initial expression of maternal behavior (Rosenberg *et al.*, 1976, 1977; Steele *et al.*, 1979).

In fact, the noradrenergic system is demonstrably involved in the consolidation of a wide variety of learning tasks (Decker *et al.*, 1990; McGaugh *et al.*, 1988) including olfactory memories in adult rats (Rosser and Keverne, 1985) and the acquisition of young rat pup's conditioned preference for odors that had been previously paired with tactile stimulation of their doral surfaces (as in maternal licking) (Wilson and Sullivan, 1994; Rangel *et al.*, 1994; Woo and Leon, 1994). The noradrenergic system is involved in a variety of other functions that may contribute to its learning and memory effects. For instance, it has been implicated in processes of reinforcement, attention, and stress (Glavin, 1985; Oades, 1985; Iversen, 1984) and undergoes a change with early manipulations of the social environment (Kraemer, 1992).

The importance of norepinephrine for the experience effects within the maternal context is further suggested by norepinephrine involvement in the expression of maternal behavior (Rosenberg *et al.*, 1976, 1977; Steele *et al.*, 1979) and its elevated levels in the hypothalamus at the time of parturition (Rosenberg *et al.*, 1976). However, most importantly, the noradrenergic system is clearly involved in the development of both the ewe's individual recognition of her lamb and the ability of mice mothers to recognize their young (Calamandrei *et al.*, 1992); a recognition that is based on olfactory cues and noradrenergic innervation of the olfactory bulbs by nerve terminals originating within the midbrain locus coeruleus (Lévy *et al.*, 1990; Pissionnier *et al.*, 1985). Moreover, there is now evidence that release of norepinephrine, along with glutamate and GABA, in the olfactory bulbs in response to parturition is higher in multiparous than in primiparous ewes (Lévy *et al.*, this volume).

Based on these and other observations, Moffat *et al.* (1993) injected drugs that mimic or block the effects of norepinephrine (noradrenergic agonists and antagonists, respectively) into dams immediately after a brief maternal experience and tested them 10 days later for the retention of maternal behavior. By using this pharmacologic technique we hoped to establish whether the neurotransmitter, norepinephrine, is indeed involved in the consolidation of a maternal experience. As shown in Fig. 6, animals receiving the adrenergic blocker (propranolol) exhibited reduced responsiveness, whereas those receiving the agonist (isoproterenol) exhibited elevated responsiveness during subsequent induction tests.

Although these data indicate that the noradrenergic system may, indeed, be involved in the consolidation of the maternal experience effect, the fact that latencies among animals injected with antagonists were not as long as found for virgin or inexperienced animals suggests again that more than one neurochemical system is probably involved. What these are is not clear. However, likely candidates include other neuropeptides like oxytocin and the endorphins that are present at parturition, influence maternal behavior, and have been shown in other contexts to influence learning and/or memory formation (Martinez and Kesner, 1991).

Another neurotransmitter that is potentially important for maternal experience is the catecholamine, dopamine. This neurotransmitter is biochemically similar to norepinephrine, but is released from different nerve terminals in different brain systems. Of particular interest in this context is the dopaminergic mesolimbic system that originates in the midbrain ventral

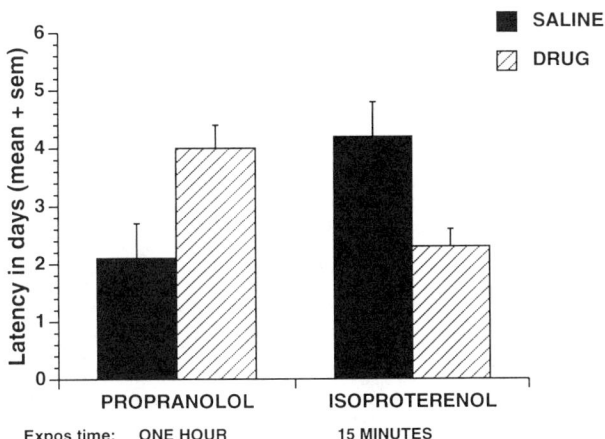

FIG. 6. Effects of propranolol and of isoproterenol on latency (in days) to become maternal in tests 10 days after exposure to pups. Groups tested include experienced (1 h) propranolol and saline groups, and experienced (15 min) isoproterenol and saline groups. (Based on Moffat et al., 1993.)

tegmental area and terminates in the nucleus accumbens. This neural system and neurotransmitter have been implicated in processes of reinforcement within several behavioral contexts including feeding, sex, and a variety of drug addictions (Bechera *et al.,* 1992; Wise and Rompre, 1989). Because rat pups become more reinforcing to the mother rat through interactive experience with them, it seemed likely that dopamine would play a role. Fleming *et al.* (1994a) explored the effects of blocking the dopamine system on the development of a conditioned place preference, using pups as reinforcing stimuli. Normal dams readily prefer to spend time in the chamber of the apparatus that has become associated with pups. However, when a dopamine antagonist is administered systemically, the conditioned place preference did not develop, even though dams responded very maternally to pups during the pup–chamber pairing.

It seems that for mothers to exhibit a high level of maternal responsiveness postpartum, under the influence of hormones, the pups do not need to be strongly reinforcing. However, for pups to elicit responsiveness when hormones are no longer in evidence, they must become more highly reinforcing, through the dam's prior interactive experience with them. The enhanced salience of pup cues and associative processes that are activated during an interaction is probably mediated by brain norepinephrine released during parturition as well as in response to stimulation; the reinforcing effect of the pups derived through the interaction seems to be mediated by a general dopamine reward system. How these two systems work together is still unclear, although there are extensive neural connections between the dopamine reward mesolimbic system and amygdaloid sites believed to be involved in stimulus evaluation that are rich in noradrenergic innervation (Everitt and Robbins, 1992). The next section explores the relevant neuroanatomy more directly.

E. Neuroanatomy of Maternal Experience

*1. Maternal Responsiveness*

We know very little about how the brain mediates the acquisition, consolidation, and retention of a maternal experience. We know considerably more about the neural control of the maternal behavior itself. Research by Numan and his colleagues (Numan, 1994) shows the importance of the medial preoptic area (MPOA) and its brain stem projections (Numan, 1994). Lesions to MPOA or knife cuts to its midbrain efferents eliminate maternal behavior in hormonally primed postpartum animals (Numan, 1990, 1994; Numan and Numan, 1991), while a single daily electrical stimulation of the MPOA facilitates the behavior in the virgin animal (Morgan *et al.,* in prep.). Other limbic sites feeding into the MPOA have also been implicated in aspects of the behavior, including the amygdala (Fleming *et al.*, 1980; Numan *et al.*, 1993), the hippocampus (Kimble *et al.*, 1967; Terlecki and Sainsbury, 1978), the septum (Fleischer and Slotnick, 1978), and the cortex (Beach, 1937; Slotnick, 1967; Stamm, 1955).

Each of the major neural sites within the maternal circuitry contributes to the expression of maternal behavior by exerting different effects (Fleming and Walsh, 1994). For instance, the amygdala, which receives extensive multimodal input from both the chemosensory and somatosensory systems and which also contains estradiol-binding neurons (Pfaff and Keiner, 1973), is believed to be involved in the evaluation of stimulus salience and in the mediation of affective states (Davis, 1992; Everitt and Robbins, 1992; LeDoux, 1992). Hence, the medial and cortical nuclei of this structure may well be activated (and modified) at the time of parturition by hormones organizing the new mother's attraction to novel pup odors and reducing the animal's natural neophobic responses to new stimuli. Thus, hormones may function to open sensory pathways, giving them greater control of the effector systems. In contrast, hippocampal involvement in the expression of maternal behavior is most likely to organize the individual behaviors sequentially, whereas MPOA nuclei are probably part of the final effector system (Numan, 1994). The MPOA receives extensive input from these other limbic sites (Numan, 1994). Which of these brain systems are involved in the enhanced maternal responding of the experienced animal and by activating what behavioral changes is not known. Although maternal experience involves many of the same processes as involved in learning and memory within other behavioral contexts, the brain systems involved appear not to be entirely identical.

Different laboratories have addressed this issue, from quite different perspectives. For instance, work in Modney's laboratory on the supraoptic nucleus shows that, with suckling experience in the lactating animal or with sensitization experience during maternal inductions in virgins, there occur changes in the structural and functional properties of supraoptic neurons responsible for suckling-induced oxytocin release that is necessary for the milk letdown reflex and, under some conditions, the expression of maternal behavior (Modney and Hatton, 1990). Modney and Hatton (1990) found two kinds of change associated with the end of pregnancy and the initiation of nursing. One involves the formation of new (double) synapses on the cell body and dendrites of the supraoptic nucleus, a group of cells in the brain that releases the milk letdown hormone, oxytocin, during nursing; the other involves the increase in electrical coupling among these same oxytocinergic supraoptic neurons. These changes very likely contribute to the synchronized activity in these oxytocin cells prior to milk ejection. Of particular interest in this context are the additional findings that some of these modifications also appear in virgin animals induced to become maternal through long-term exposure to pups but who did not receive suckling stimulation. Instead, there is evidence that olfactory and/or other chemosensory stimulation derived through licking and retrieving of pups may induce these changes in the virgin (Modney and Hatton, 1990, p. 318; Modney *et al.,* 1987). The time course of these effects in the nursing animal may be relevant; some synaptic changes are reversible, but apparently others are more permanent. Such long-term structural changes could provide a possible basis for the findings that, once dams have had some nursing experience, distal olfactory and other exteroceptive pup cues come to elicit the conditioned release of oxytocin in the absence of suckling (Grosvenor *et al.,* 1990).

Using quite a different approach, in an attempt in our laboratory to locate structures relevant to the acquisition and/or retention of a maternal memory or experience, we systematically infused the protein synthesis inhibitor, cycloheximide, or lesioned sites within the limbic system known to be implicated in the formation of memories within other behavioral contexts, including the dorsal hippocampus (Cohen, 1989; Otto and Eichenbaum, 1992; Winocur and Moscovitch, 1990), the prefrontal cortex (Kolb, 1984; Winocur, 1991), and the central and basolateral nuclei of the amygdala (LeDoux, 1992; Everitt and Robbins, 1992; Gaffan, 1992; Kesner, 1992; Otto and Eichenbaum, 1992), as well as amygdaloid sites that process sensory information and are responsive to hormones (e.g., medial and cortical amygdaloid nuclei). Cycloheximide was infused immediately after the maternal experience to block consolidation, while lesions were administered both before and after a 1-h maternal experience. In no case were we able to eliminate or reduce the long-term effectiveness of the maternal experience. At test 10 days later, all infused groups and all lesioned groups showed the same enhanced responsiveness shown by experienced sham groups (see Fig. 7A and 8) (Fleming and Lee, in prep). Cycloheximide animals were also tested in a two-arm maze to determine whether blocking protein synthesis indeed blocks consolidation of other experiences. As shown in Fig. 7B, these infusions did block the consolidation and, hence, retention of the maze learning. The maternal results are shown in Fig. 7A (cycloheximide) and Fig. 8 (lesion).

These results indicate that either maternal learning is mediated by different systems than are other types of learning or that there is considerable redundancy in the control mechanisms. Alternately, because of the complexity of the pup cues and of the task, multiple systems may be activated. In fact, the following immunohistochemical studies suggest that this may indeed be the case, leading to the prediction that for lesions to effectively eliminate the effects of a maternal experience, several contributing systems must be simultaneously disrupted.

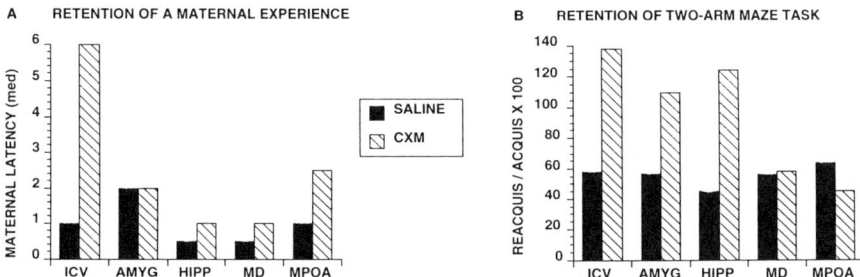

FIG. 7. (A) Latency (in days) to become maternal in tests 10 days after exposure to pups in groups receiving infusions of cycloheximide or saline immediately after pup exposure; infusions given either intracerebroventricular (ICV) or into the amygdala (AMYG), hippocampus (HIPP), mediodorsal thalamus (MD), or medial preoptic area (MPOA). (B) Retention of performance on a two-arm maze task in animals receiving infusions of cycloheximide or saline immediately after pup exposure; infusions given either ICV or into the AMYG, HIPP, MD, or MPOA.

To explore these possibilities more fully, in a set of studies we used Fos immunohistochemistry as a way of determining which brain sites undergo change during the acquisition of a maternal experience and at the time of retention testing. Fos is a protein produced by a class of immediate-early

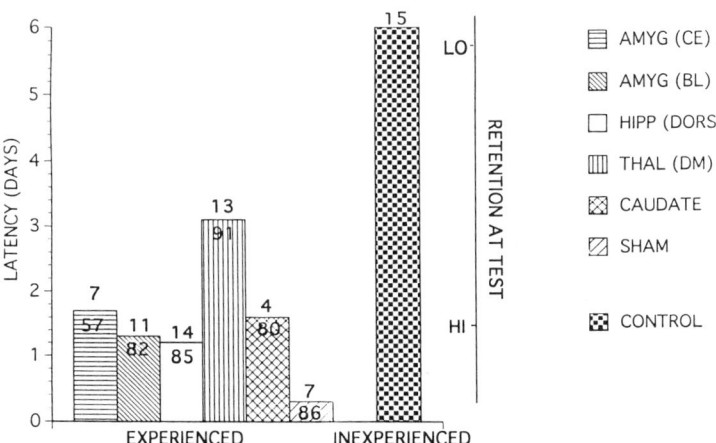

FIG. 8. Latency (in days) to become maternal in tests 10 days after exposure to pups in animals receiving lesions in the amygdala (AMYG) (central (CE) or basolateral nuclei (BL)), dorsal hippocampus (HIPP(DORS)), dorsomedial thalamus (THAL (DM)), caudate nucleus (CAUDATE), or SHAM sites. Numbers at top of histograms represent group size (n); numbers within histograms reflect percent of animals becoming maternal.

genes known as c-fos oncogenes, present in brain neurons and activated by a variety of stimulus situations and behavioral states (Sagar *et al.,* 1988). When produced in conjunction with other proto-oncogene proteins, Fos protein "turns on" other genes within the cell that may lead to long-term structural or functional changes in the brain (Morgan and Curran, 1991). By exploring which cell groups within the brain express the Fos protein in response to stimulation or during ongoing behavior, one can trace the neural circuitry relevant to the behavior (e.g., Baum and Everitt, 1992; Erskine and Rowe, 1992; Pfaus *et al.,* 1992).

Using Fos immunohistochemistry as a tool to delineate a functional neuroanatomy, we have been able to confirm that the MPOA is indeed activated when an animal expresses maternal behavior; Fos density in this region is elevated after a 1-h interaction with newborn pups in both postpartum and sensitized virgin animals (Calamandrei and Keverne, 1994; Fleming *et al.,* 1994b; Numan and Numan, 1994). Other sites that were changed by pup interactions include the primary olfactory (piriform) and somatosensory (parietal) cortices, sensory-receiving areas in the amygdala (medial and cortical amygdaloid nuclei) and cortical regions implicated in associative processes (e.g., basolateral and central amygdaloid nuclei, cingulate cortex) and amygdala and the other reinforcing sites (e.g., nucleus accumbens) (Fleming *et al.,* 1994a; Fleming and Walsh, 1994).

Because the acquisition of a maternal experience involves remodeling of anatomical systems involved with chemosensory and somatosensory processing, we investigated the role of input through these modalities in the pup-induced activation of the MPOA and amygdaloid nuclei (Walsh, Fleming & Magnusson, 1996). Prior to a 1-h postpartum interaction with pups, new mothers in different groups were treated with the following: (1) zinc sulfate, to produce anosmia peripherally; (2) were administered a local topical anesthetic on the perioral (mouth) or ventral trunk regions, which anesthetized the skin surfaces, or (3) received combinations of these treatments. These manipulations resulted in an interesting pattern of Fos-lir in the brain (see Fig. 9).

Surprisingly, there were no decrements in Fos-lir in the MPOA of any of the desensitized groups. In contrast, in animals sustaining chemosensory desensitizations, there were Fos-lir decrements in the medial and cortical amygdaloid nuclei, to which the main and accessory olfactory tracts project. Of greatest interest, however, was the observation of a summative effect of these denervations in the basolateral nucleus—a region believed to receive multimodal input and to mediate the formation of associations. Animals sustaining manipulations of both somatosensory and chemosensory systems showed the greatest reduction in basolateral Fos-lir.

Based on these data we have some idea of which brain regions are activated during the expression of maternal behavior and during the acquisi-

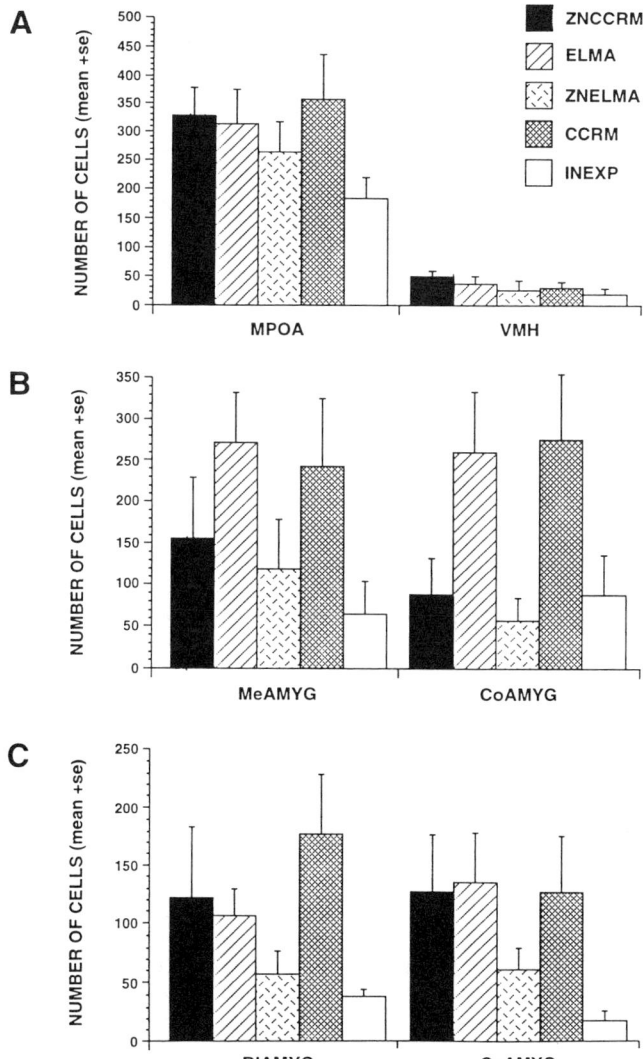

FIG. 9. Number of cells showing Fos-lir in (A) the MPOA and ventromedial hypothalamus (VMH), (B) the medial (MeAMYG) and cortical (CoAMYG) amygdala, and (C) the basolateral (BlAMYG) and central (CeAMYG) amygdala. Groups include (1) experienced dams with pups ZNCCRM, zinc sulfate (to produce hyposmia) and control cream; (2) EMLA, Emla cream (to produce local anesthetic effect); (3) ZNEMLA, zinc sulfate and Emla cream; (4) CCRM, control cream; and (5) INEXP, inexperienced animals. (Based on Walsh et al., (1996).)

tion of a maternal experience. What we do not know is whether these brain regions are involved in the processes associated with the storage or retrieval of a maternal memory.

To investigate this issue, we explored the effects on the distribution of Fos-lir of an interaction with pups (including all the conditioned and unconditioned pup stimuli) or exposure to pup-associated cues (hence, only conditioned stimuli (Fleming & Korsmit, in press); Plexiglas box next to pups) in postpartum animals that had either received an earlier interactive experience with pups or had not. In this study a 2-h exposure period was used, and both the experience phase (Day 1) and the retention phase (Day 5) occurred during the first 5 postpartum days, when hormonal effects are still present.

In this study we found that three brain regions differed as a function of earlier maternal experience. In comparison with inexperienced animals, animals who were experienced and who interacted with pups at the time of retention testing showed higher concentrations of Fos-lir in the MPOA, the basolateral amygdaloid nucleus, and the parietal cortex. Animals who did not interact with pups at retention testing showed no experience effect.

In subsequent studies we extended the first study by providing groups with a 4-h postpartum interactive experience (with PUPS next to a perforated BOX in a new CAGE) or no experience and testing for Fos-lir in response to either distal pup cues (PUPS in the BOX in the CAGE), BOX and CAGE alone, or no cues (HOME) 10 days after the initial experience, when hormonal effects are no longer present. These studies differed from the first in the exposure-test interval (4 versus 9 days), and the hormonal status at test (postpartum versus cycling). The third study differed from the second by adding an additional artificial odor cue to the pup complex at the time of the initial 4-h exposure and at test. This was done to enhance the conditionable cues associated with the pups and to permit use of identical odors (i.e., same artificial scent) at exposure and test.

Results of both studies were essentially identical to one another in showing the same pattern of elevated Fos-lir in the MPOA and basolateral nucleus in the experienced over the inexperienced animals, especially in groups that were re-exposed to pups in a Plexiglas box (PUPS/BOX/CAGE) at the time of retention testing. Groups receiving no stimulation (HOME) at retention testing showed no experience effects, whereas those exposed only to BOX and CAGE cues showed intermediate differences. In Study 2, there was one additional experience effect where experienced animals showed lower Fos-lir in the prefrontal cortex than did inexperienced animals, in response to all the pup cues and in response to the distal, pup-associated cues. These results are shown in Fig. 10 and 11.

These data are consistent with our behavioral data showing that for maternal experience to be encoded, some interaction with pups must occur.

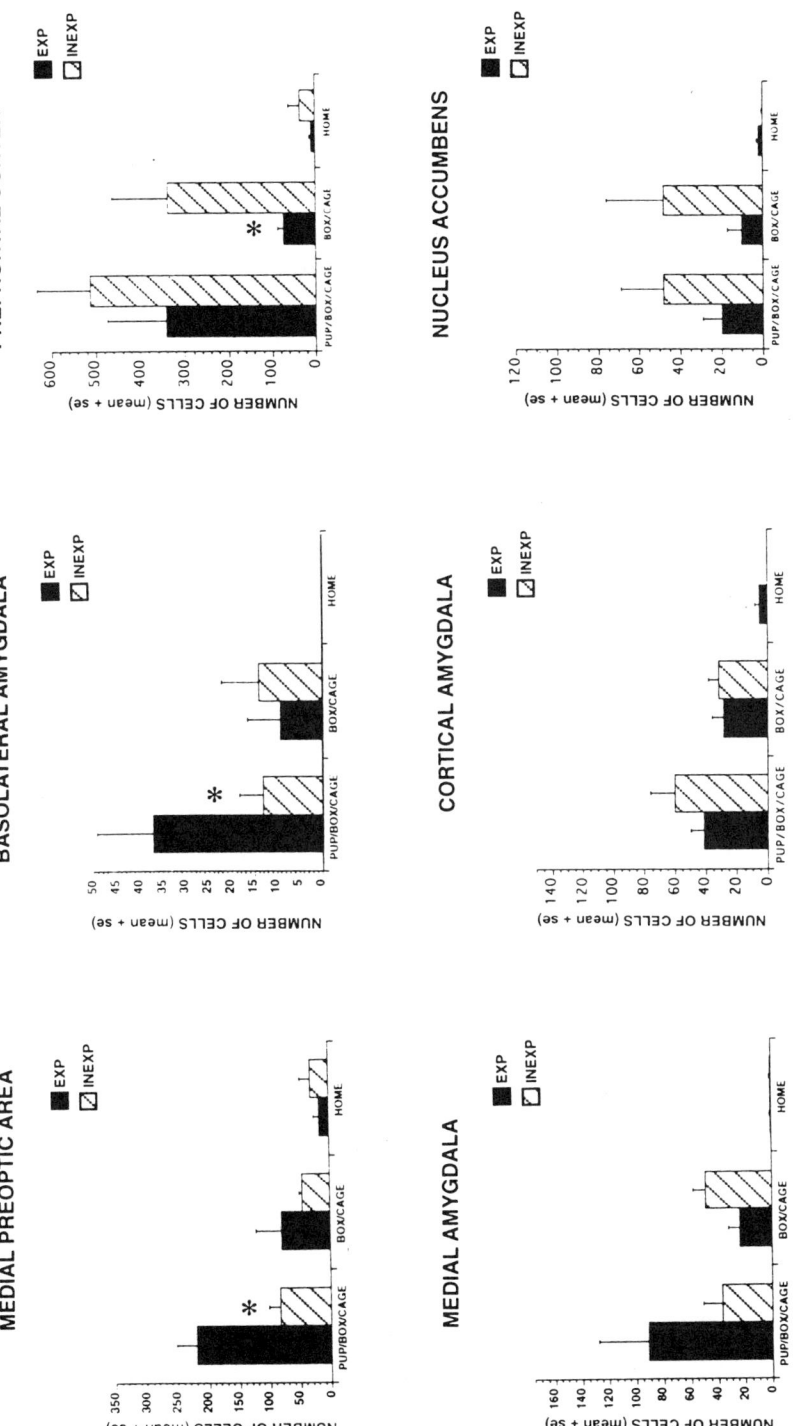

FIG. 10. Number of cells showing Fos-lir in medial preoptic area (MPA), basolateral amygdala, prefrontal cortex, medial and cortical amygdala, nucleus accumbens, lateral habenula, parietal cortex (P2), and piriform cortex. Groups include experienced (EXP) and inexperienced (INEXP) animals with PUPS, BOX, CAGE, etc. (EXP/PUPS/BOX/CAGE, EXP/BOX/CAGE, and EXP/HOME). (From Fleming and Korsmit, in press). For description of conditions, see text.

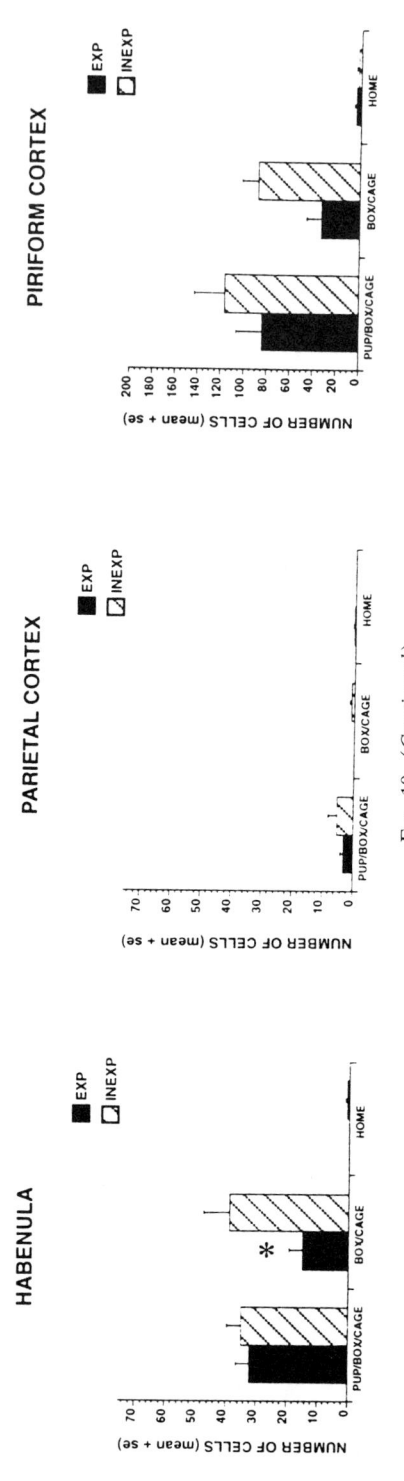

FIG. 10. (*Continued*)

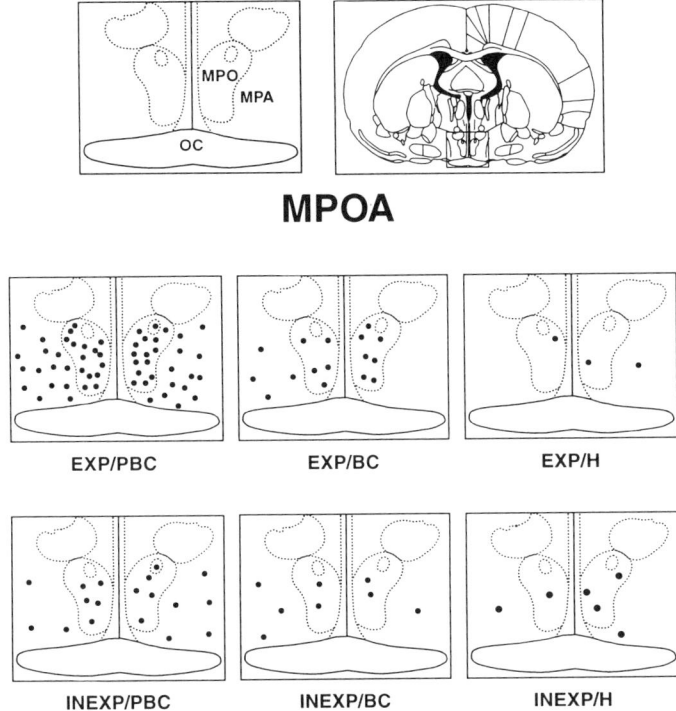

FIG. 11. Schematics of Fos-labeled cells in the MPOA and amygdala in experienced (EXP) and inexperienced (INEXP) groups exposed on Day 10 postpartum of PUPS/BOX/CAGE (PBC), BOX/CAGE (BC), OR HOME (H).

In general, being suckled by pups is not necessary, although crouching over pups and licking them may be. The experience of parturition and/or stimulation by the parturitional hormones is not, however, adequate to effect a maternal experience. Simple exposure to parturition and the parturitional hormones is not adequate.

The fact that elevated Fos-lir was observed in the MPOA of experienced animals in response to distal pup cues (PUPS/BOX/CAGE) or pup-associated cues (BOX/CAGE) in animals that were not actually interacting with pups was intriguing. We interpret these results to mean that through an association with the unconditioned (somatosensory?) stimuli present during physical interactions with pups, distal pup cues become conditioned stimuli that can then activate "the final common path" for the expression of maternal behavior. These results lead us to question whether the MPOA is part of the effector system for the expression of maternal behavior, as we have always assumed, or part of the motivational system.

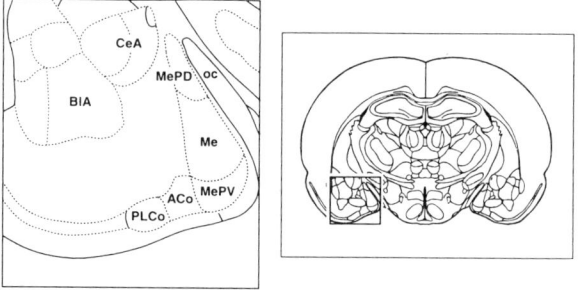

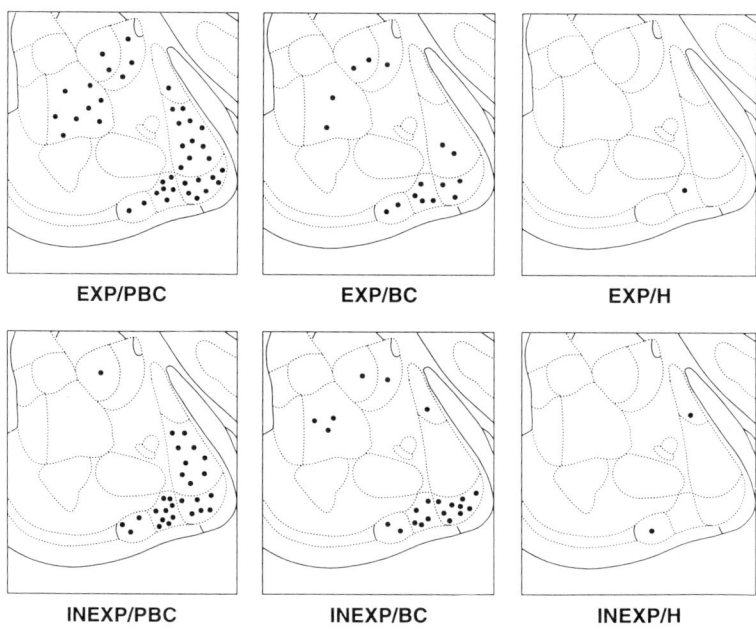

Fig. 11. (*Continued*)

Which MPOA cells undergo change with experience is not known. Whether it is cells that process the sensory input, or those that mediate the motor output or cells that respond to the internal hormonal environment is, of course, not known. There is precedence for all these possibilities. For instance, Bridges and his colleagues (Bridges and Hammer, 1992; Mann *et al.*, 1990) report that in comparison to primiparous mothers, multiparous animals exhibit a reduced suckling-induced release of prolactin that may mediate the elevated opiate receptor densities in the MPOA of these animals. Also, two abstracts appeared showing enhanced activation of MPOA as a result of previous sexual experience in sexually active male rodents (Fernandez-Fewell and Meredith, 1994; Lumley and Hull, 1994).

Although experience-dependent changes in the MPOA constitute the most robust findings of these studies, Fos-lir in the basolateral nucleus of the amygdala also showed experience effects on re-exposure to pup-associated stimuli; interestingly, the greatest effect was seen when an initially neutral stimulus (the box) was paired with pups, rather than pups themselves. Based on the work of Everitt and Robbins (1992), showing a role for this nucleus in the formation of associations between neutral stimuli and biologically relevant reinforcers, we were surprised not to find an even greater effect of experience on Fos-lir in the basolateral nucleus than we found. However, because there was quite a high level of Fos-lir in this nucleus during the initial interactions with pups, when presumably, the experience was first obtained, it is possible that the basolateral nucleus is primarily implicated in the formation of the memory, rather than in its long-term storage or retrieval.

The third site exhibiting experience effects is the parietal cortex, but only when mothers actually interacted with pups at the time of retention testing and tests took place on Day 4 postpartum when the effects of the parturitional hormones were still in evidence. In contrast to experience-dependent changes in the MPOA and amygdala, in the parietal cortex no experience effect was found in this region if, at the time of retention testing, animals were simply exposed to distal and associated pup cues. These somatosensory effects are interesting in light of work in the monkey showing that somatosensory experience can produce a reorganization of cells in the somatosensory cortex in the adult animal, indicating surprising plasticity in the system. There is, interestingly, evidence of projections from the parietal cortex to the MPOA, providing a route by which somatosensory experience could affect the MPOA and a related behavioral change, (Simerly and Swanson, 1988). A final site showing experience effects is the prefrontal cortex. In this case, however, a reduction, rather than an enhancement, of Fos-lir was produced by the additional experience. We interpret these results to mean that the medial prefrontal region is involved in the processing of novelty and the processes underlying habituation to initially novel stimuli (for review of prefrontal function, see Kolb, 1984). This form of learning clearly occurs when an animal is exposed to pups and may well be reflected in reduced prefrontal activation.

2. *Individual Recognition between Mother and Young*

Although it is likely that the enhanced maternal responsiveness that occurs with experience nursing and rearing a previous lamb is based on processes that are similar in sheep and rats, experiential processes that underlie the development of the ewe's recognition of individual lambs are probably quite different (Poindron and Lévy, 1990). In fact, unlike the situation with rats where we know that multiple brain systems are involved in the experience effect, in sheep recognition is based on quite specialized changes within the olfactory bulb itself (Kendrick et al., 1992). However, as indicated earlier, there is evidence that parturition-induced activation of centrifugal noradrenergic inputs to the olfactory bulbs, although not specific to own lamb's odors, is necessary for the formation and, possibly, recall of recognition memory (Kendrick et al., 1992) (see Lévy et al., this volume).

Interestingly, these changes are similar also to changes in olfactory processing that occurs when neonatal and young rats become conditioned to the maternal odor within the nest (Coopersmith and Leon, 1986; Sullivan and Leon, 1986; reviewed by Wilson and Sullivan, 1994). Because the experience effects in rat pups are reminiscent of the experience effects in the dam, and may well be their precursors, we describe their physiology in some detail. Sullivan and Hall (1988) found that if rat pups are exposed to artificial scents in association with dorsal tactile stimulation, simulating mother's licking behavior, they undergo a change in function and structure of the olfactory bulbs that is specific to the learned-about odor. These changes underlie both behavioral activation and the preference the young animal develops in response to either maternal odors or artificial odors to which they have become conditioned (Sullivan and Hall, 1988). Such conditioning may well underlie the subsequent development of their food preferences and of preferences for social stimuli.

If a conditioned odor association is formed during the first 2 postpartum weeks, specific glomerular cells within the pups' olfactory bulbs show enhanced focal uptake of 2-deoxyglucose (2-DG) in response to presentation of the conditioned odor but not in response to nonconditioned odors, indicating that these cells are being activated by the conditioned odor. Moreover, extensive pairing of odor and tactile input over the first 18 days of life results in more juxtaglomerular neurons in the affected region of the bulb, which may account for the enhanced 2-DG activation. Moreover, in contrast to olfactory cells elsewhere in the bulb that show excitation to odors in single units, the tufted-mitral cells within the vicinity of the specific glomeruli show a suppression in single unit activity (see Wilson and Sullivan, 1994). The region that undergoes change as a result of this early learning seems to be restricted to the midlateral juxtaglomerular portion of the olfactory bulbs, the region that also receives norepinephrine terminals from the locus coeruleus (Johnson et al., 1994; Woo and Leon, 1994).

Thus, in the rat pup the learning that occurs in response to odors that have become paired with the mother's behavior (or simulations thereof) that produces behavioral activation is organized within the olfactory bulbs. Of interest from the point of view of maternal learning are the additional observations that other limbic, primarily amygdaloid, sites are involved in the learning that underlies the development of a preference for one odorant over another. This is not surprising because activation of the olfactory bulbs affects the amygdala through direct lateral olfactory tract projections from the main and accessory olfactory bulbs into the medial and cortical regions of the amygdala (Scalia and Winans, 1975). These limbic sites, in turn, project to the hypothalamus (Scott and Chapin, 1975; Amaral *et al.,* 1992). In the rat pup, lesions of the cortical amygdala eliminate a CS-induced preference for (and approach towards) the learned odor but do not eliminate behavioral activation produced by this same odor (Sullivan and Wilson, 1993); thus, the amygdala is not implicated in the formation of the conditioned response so much as in the the salience or hedonic quality that develops as a result of the conditioning. In light of these results it would be interesting to know whether specific pairing of ventral stimulation of the mother and exposure to pup-related cues during the postpartum period would produce specific and localized changes in the mothers olfactory bulbs.

## V. Conclusion

Maternal behavior in the rat is a highly buffered system with multiple regulatory mechanisms. Although clearly species characteristic and influenced by hormones at the time of parturition, maternal behavior nevertheless depends on an array of environmental and experiential processes for its reliable expression. Mechanisms underlying these experiential influences have been the focus of this article. The analysis has raised several interesting questions. The first concerns the type of learning a maternal experience represents. It seems that several types of learning are involved. Although some stimulus–response (S-R) conditioning may be involved, in general there is little specialization and evidence that new responses develop; some may, however, become more efficient; for instance, with experience dams become more successful at transporting pups from one location to another and do so more quickly. In general, through experience the animal seems to be learning about the eliciting stimuli; familiarity with pup odors results in a reduction in the animal's natural neophobic responses to these cues, which permits the expression (and disinhibition?) of maternal responding to these cues. In addition, stimulus–stimulus (S-S) perceptual associative processes are activated and the animal comes to respond to cues or conditioned stimuli that have been associated with the unconditioned cues. Thus, pup-associated stimuli that are intrinsic to the pups (their visual characteristics) or are part of the surrounding environment (nest site) come to be reinforcing to the dam when they are present during a mother–litter interaction; she then approaches these cues to initiate interactions with young. Although these forms of plasticity reflect general processes, in some cases, as with recognition learning in sheep, the conditioning looks "prepared" in the sense that the mother learns to recognize the young's odor cues very rapidly and during a highly circumscribed (sensitive) postpartum period. Exposure to these same cues at other times or exposure to nonodor cues associated with lambs will not result in a selective recognition. Even in the rat the parturitional hormones have the effect of increasing the animal's plasticity and ability to be modified by experience—an effect that argues for some degree of "preparedness" in the maternal learning by rat dams. This conclusion is consistent with evidence of experience-induced brain changes in hormone receptor densities (Koch, 1990; Pederson et al., 1994) and neurotransmitters (Lévy et al., 1993).

A second question raised by this article concerns how maternal experience (or experiences) are processed by the brain. Does the brain encode them by general mechanisms or are there mechanisms dedicated to processing information within this specialized functional context? Both conclusions seem to be the case.

In the rat, the involvement of protein synthesis and of the noradrenergic system in the consolidation of the maternal experience and of dopamine in pup-mediated reinforcement suggests the existence of general processes. These neurochemicals are involved in many forms of learning. However, the absence of an effect on maternal experience of individual lesions, known to effectively block learning within other behavioral contexts, suggests either that the "correct" sites or systems have not yet been identified for the maternal context or that the substrates underlying these processes are not entirely localized in any one brain site. Our c-fos studies suggest the existence of both specialized systems specific to maternal or perhaps reproductive contexts (e.g., MPOA) and systems with more distributed effects (e.g., basolateral amygdala–somatosensory cortex). Similar conclusions are suggested by studies of the infant's learning about the mother (see Sullivan and Hall, 1988; Wilson and Sullivan, 1994).

How such distributed processing might work is not clear. It could be that these processes are distributed in having redundant (or shared?) representations and hence lesions or damage to any one site has no effect, or, alternately, that it results in only a slightly degraded (but nondetectable) representation of the experience (Rose, 1973). In this case, one might expect a lesion in one site to produce some change in the quality of the behaviors expressed at retention testing. Unfortunately, this level of detailed analysis of behavior has not been done. Another possibility is that different aspects of experience acquisition and/or storage depend on different functions and hence different brain systems. In this case, one might imagine that damage to a site that mediates stimulus salience in one modality (e.g., olfactory) could be compensated for by a region that processes salience within another modality. This leads to the prediction that behaviors based on one modality (e.g., licking) would be compensated for by behaviors depending on another (e.g., crouching). Alternately, damage to a system that mediates dopaminergic reinforcement systems could result in recruitment of alternate reinforcement systems (e.g., opioid systems), demonstrable proposition. Choosing among these different hypotheses requires tests that make very specific predictions and that can sensitively reveal the deficits produced.

There is much to learn about the processes and mechanisms underlying maternal experience and its effects on the dam's behavior. On a behavioral level, we still need to understand how experience affects the animal's emotional state, its perceptual capabilities, its strategies for compensating for sensory, and other deficits. We need to better understand the different kinds of learning that are activated when animals gain a maternal experience and how these different types of learning are expressed behaviorally within the maternal context. We also have a lot to learn about the underlying physiological mechanisms, the relevant neurochemistries and brain circuits, and how these interact with one another. Knowing which specific neural sites are activated during the experience or at the time of representation of the pups permits more detailed analyses through such powerful techniques as microdialysis, of the involvement of the different neurochemicals in the acquisition, and the consolidation and the retrieval of a maternal experience.

Although we have much to learn about the control of maternal experience, our c-fos studies provide some interesting clues. The multiple types of learning recruited during a mother–litter interaction may be reflected in activation (and hence, remodeling?) of different neural systems. Thus, sensitization to pup cues and the habituation to their withdrawal-eliciting properties on their repeated presentation may be reflected at the time of acquisition in a change in cortical and medial amygdaloid nuclei. A change in the activation properties of these nuclei produces long-term changes in hypothalamic mechanisms. In contrast, true associative learning involving the pairing of conditioned with unconditioned stimuli may depend on amygdaloid function (e.g., in basolateral and central nuclei) both at acquisition and at retention when the CSs are present. Finally, the enhanced reinforcing effects of pups through the dam's interactive experience with them recruits mesolimbic activity. All of these changes surely act together to ensure that the mother will be adequately motivated to nurture offspring when hormones are no longer present or when extraneous influences threaten the viability of the young and, hence, the mother's reproductive fitness.

## References

Amaral, D. G., Price, J. L., Pitkänen, A., and Carmichael, S. T. (1992). Anatomical organization of the primate amygdaloid complex. *In* "The Amygdala: Neurobiological Aspects of Emotion, Memory and Mental Dysfunction" (J. Aggleton, ed.), pp. 1–66. Wiley-Liss, Inc., New York.

Bauer, J. H. (1983). Effects of maternal state on the responsiveness to nest odors of hooded rats. *Physiol. Behav.* **30**, 229–232.

Baum, M. J., and Everitt, B. J. (1992). Increased expression of c-fos in the medial preoptic area after mating in male rats: Role of afferent inputs from the medicil amygdala and mid brain central tegmental field. *Neuroscience* **50**, 627–646.

Beach, F. A. (1937). The neural basis of innate behavior: I. Effects of cortical lesions upon the maternal behavior pattern in the rat. *J. Comp. Physiol. Psychol.* **24**, 393–436.

Bechara, A., Harrington, F., Nader, K., and van der Kooy, D. (1992). Neurobiology of motivation: Double dissociation of two motivational mechanisms mediating opiate reward in drug-naive versus drug-dependent animals. *Behav. Neurosci.* **106**, 798–807.

Bridges, R. S. (1975). Long-term effects of pregnancy and parturition upon maternal responsiveness in the rat. *Physiol. Behav.* **14**, 245–249.

Bridges, R. S. (1977). Parturition: Its role in the long-term retention of maternal behavior in the rat. *Physiol. Behav.* **18**, 487–490.

Bridges, R. S. (1990). Endocrine regulation of parental behavior in rodents. *In* "Mammalian Parenting: Biochemical, Neurobiological, and Behavioral Determinants" (N. A. Krasnegor and R. S. Bridges, eds.), pp. 93–117. Oxford Univ. Press, New York.

Bridges, R. S., and Hammer, R. P. (1992). Parity-associated alterations of medial preoptic opiate receptors in female rats. *Brain Res.* **578**, 269–274.

Brouette-Lahlou, I., Vernet-Maury, E., Godinot, and Chanel, J. (1992). Vomeronasal organ sustains pups' anogenital licking in primiparous rats. *In* "Chemical Signals in Vertebrates VI" (R. L. Doty and D. MüOler-Schwarze, eds.), pp. 551–555. Plenum Press, New York.

Calamandrei, G., and Keverne, E. G. (1994). Differential expression of Fos protein in the brain of female mice dependent on pup sensory cues and maternal experience. *Behav. Neurosci.* **108**, 113–120.

Calamandrei, G., Wilkinson, L. S., and Keverne, E. G. (1992). Olfactory recognition of infants in laboratory mice: Role of noradrenergic mechanisms. *Physiol. Behav.* **52**, 901–907.

Carlier, C., and Noirot, E. (1965). Effects of previous experience on maternal retrieving by rats. *Anim. Behav.* **13**, 423–426.

Cohen, J., and Bridges, R. S. (1981). Retention of maternal behavior in nulliparous and primiparous rats: Effects of duration of previous maternal experience. *J. Comp. Physiol. Psychol.* **95**, 450–459.

Cohen, N. (1989). Further studies of hippocampal representation during odor discrimination learning. *Behav. Neurosci.* **103**, 1207–1216.

Coopersmith, R., and Leon, M. (1986). Enhanced response by adult rats to odors experienced early in life. *Brain Res.* **371**, 400–403.

Davis, M. (1992). The role of the amygdala in conditioned fear. *In* "The Amygdala: Neurobiological Aspects of Emotion, Memory, and Mental Dysfunction" (J. Aggleton, ed.), pp. 255–306. Wiley-Liss, New York.

Davis, H. P., and Squire, L. R. (1984). Protein synthesis and memory: A review. *Psychol. Bull.* **96**, 518–559.

Decker, M. W., Gill, T. M., and McGaugh, J. L. (1990). Concurrent muscarinic and b-adrenergic blockade in rats impairs place-learning in a water maze and retention of inhibitory avoidance. *Brain Res.* **513**, 81–85.

Elwood, R. W., and Kennedy, H. F. (1990). The relationship between infanticide and pregnancy block in mice. *Behav. Neural Biol.* **53**, 277–283.

Erskine, M. S., and Rowe, D. W. (1992). C-fos proto-oncogene activity increases in the rat amygdala and preoptic area of the female rat after cervical stimulation: Role of afferent input via the pelvic nerve. *Soc. Neurosci. Abstr.* **18**, 891.

Everitt, B. J., and Robbins, T. W. (1992). Amygdala-ventral striatal interactions and reward-related processes. *In* "The Amygdala: Neurobiological Aspects of Emotion, Memory, and Mental Dysfunction" (J. Aggleton, ed.), pp. 401–429. Wiley-Liss, New York.

Fairbanks, L. A. (1989). Early experience and cross-generational continuity of mother-infant contact in vervet monkeys. *Dev. Psychobiol.* **22**, 669–681.

Fernandez-Fewell, G., and Meredith, M. (1994). Fos patterns in male hamster vomeronasal pathway: Pheromone stimulation and effect of experience. *Soc. Neurosci. Abstr.* **501.9**, 242.

Fleischer, S., and Slotnick, B. M. (1978). Disruption of maternal behavior in rats with lesions of the septal area. *Physiol. Behav.* **21**, 189–200.

Fleming, A. S., and Luebke, C. (1981). Timidity prevents the nulliparous female from being a good mother. *Physiol. Behav.* **27**, 863–868.

Fleming, A. S., and Rosenblatt, J. (1974a). Maternal behavior in the virgin and lactating rat. *J. Comp. Physiol. Psychol.* **86**, 957–972.

Fleming, A. S., and Rosenblatt, J. (1974b). Olfactory regulation of maternal behavior in rats: I. Effects of olfactory bulb removal in experienced and inexperienced lactating and cycling females. *J. Comp. Physiol. Psychol.* **86**, 221–232.

Fleming, A. S., and Sarker, J. (1990). Experience-hormone interactions and maternal behavior in rats. *Physiol. Behav.* **47**, 1165–1173.

Fleming, A. S., and Walsh, C. J. (1994). Neuropsychology of maternal behavior in the rat: c-fos expression during mother-litter interactions. *Psychoneuroendocrinology* **19**, 429–443.

Fleming, A. S., Vaccarino, F., and Leubke, C. (1980). Amygdaloid inhibition of maternal behavior in the nulliparous female rat. *Physiol. Behav.* **25**, 731–743.

Fleming, A. S., Cheung, U. S., Myhal, N., and Kessler, Z. (1989). Effects of maternal hormones on 'timidity' and attraction to pup-related odors in females. *Physiol. Behav.* **46**, 449–453.

Fleming, A. S., Cheung, U. S., and Barry, M. (1990). Cyclohexamide blocks the retention of maternal experience in postpartum rats. *Behav. Neural Biol.* **53**, 64–73.

Fleming, A. S., Garvagh, K., and Sarker, J. (1992). Effects of transections to the vomeronasal nerves or to the main olfactory bulbs on the initiation and long–term retention of maternal behavior in primiparous rats. *Behav. Neural Biol.* **57**, 177–188.

Fleming, A. S., Korsmit, M., and Deller, M. (1994a). Rat pups are potent reinforcers to the maternal animal: Effects of experience, parity, hormones and dopamine function. *Psychobiology* **22**, 44–53.

Fleming, A. S., Suh, E. J., Korsmit, M., and Rusak, B. (1994b). Activation of Fos-like immunoreactivity in MPOA and limbic structures by maternal interactions and social interactions in rats. *Behav. Neurosci.* **108**(4), 724–734.

Fleming, A. S., and Lee, A. (in prep). Effects of lesions of the brain on the acquisition and retention of a maternal experience.

Fleming, A. S., and Korsmid, M. (in press). Plasticity in the maternal circuit: Effects of maternal experience on Fos-lir in hypothalamic, limbic sites and cortical structures in postpartem rats. *Behavioral Neuroscience.*

Franz, J. R., Leo, R. J., Steuer, M. A., and Kristal, M. B. (1986). Effects of hypothalamic knife cuts and experience on maternal behavior in the rat. *Physiol. Behav.* **38**, 629–640.

Gaffan, D. (1992). Amygdala and the memory of reward. *In* "The Amygdala: Neurobiological Aspects of Emotion, Memory, and Mental Dysfunction" (J. Aggleton, ed.) pp. 471–483. Wiley-Liss, New York.

Glavin, G. B. (1985). Stress and brain noradrenaline. *Neurosci. Biobehav. Rev.* **9,** 233–243.

Grosvenor, C. E., Shah, G. V., and Crowley, W. R. (1990). Role of neurogenic stimuli and milk production in the regulation of prolactin secretion during lactation. *In* "Mammalian Parenting: Biochemical, Neurobiological, and Behavioral Determinants" (N. A. Krasnegor and R. S. Bridges, eds.), pp. 324–342. Oxford Univ. Press, New York.

Harlow, H. F. (1963). The maternal affectional system of rhesus monkeys. *In* "Maternal Behavior in Mammals" (H. L. Rheingold, ed.), pp. 254–281. John Wiley & Sons, New York.

Holmes, W. G. (1990). Parent-offspring recognition in mammals: Approximate and ultimate perspective. *In* "Mammalian Parenting: Biochemical, Neurobiological, and Behavioral Determinants" (N. A. Krasnegor and R. S. Bridges, eds.), pp. 441–460. Oxford Univ. Press, New York.

Insel, T. (1990). Oxytocin and maternal behavior. *In* "Mammalian Parenting: Biochemical, Neurobiological, and Behavioral Determinants" (N. A. Krasnegor and R. S. Bridges, eds.), pp. 260–280. Oxford Univ. Press, New York.

Iversen, S. (1984). Cortical monoamines and behavior. *In* "Monoamine Innervation of the Cerebral Cortex" (L. Descarries and H. H. Jasper, eds.), pp. 321–349. Alas: R. Liss, New York.

Jakubowski, M., and Terkel, J. (1986). Establishment and maintenance of maternal responsiveness in postpartum Wistar rats. *Anim. Behav.* **34,** 256–262.

Johnson, B. A., Woo, C. C., and Leon, M. (1994). Early olfactory preference training increases the Fos-like response of juxta glomfruian cells in the rat olfactory bulb. *Soc. Neurosci. Abstr.* **143.13,** 65.

Kendrick, K. M., Lévy, F., and Keverne, E. B. (1991). Importance of vaginocervical stimulation for the formation of maternal bonding in primiparous and multiparous parturient ewes. *Physiol. Behav.* **50,** 595–600.

Kendrick, K. M., Lévy, F., and Keverne, E. B. (1992). Changes in the sensory processing of olfactory signals induced by birth in sheep. *Science* (Washington, D. C.)**256,** 833–836.

Kesner, R. P. (1992). Learning and memory in rats with an emphasis on the role of the amygdala. *In* "The Amygdala: Neurobiological Aspects of Emotion, Memory, and Mental Dysfunction" (J. Aggleton, ed.), pp. 379–400. Wiley-Liss, New York.

Keverne, E. B. (1995). Neurochemical changes accompanying the reproductive process: Their significance for maternal care in primates and other mammals. *In* "Motherhood in Human and Nonhuman Primates: Biosocial Determinants" (C. R. Pryce, R. D. Martin, and D. Skuse, eds.), pp. 69–77. S. Karger, Basel.

Keverne, E. B., and Kendrick, K. M. (1990). Neurochemical changes accompanying parturition and their significance for maternal behavior. *In* "Mammalian Parenting: Biochemical, Neurobiological and Behavioral Determinants" (N. A. Krasnegor and R. S. Bridges, eds.), pp. 281–304. Oxford Univ. Press, New York.

Kimble, D. P., Rogers, L., and Hendrickson, C. W. (1967). Hippocampal lesions disrupt maternal, not sexual behavior in the albino rat. *J. Comp. Physiol. Psychol.* **63,** 401–405.

Kinsley, C. H., and Bridges, R. S. (1988). Parity-associated reductions in behavioral sensitivity to opiates. *Biol. Reprod.* **39,** 270–278.

Koch, M. (1990). Effects of treatment with estradiol and parental experience on the number and distribution of estrogen-binding neurons in the ovariectomized mouse brain. *Neuroendocrinology* **51,** 505–514.

Kolb, B. (1984). Functions of the frontal cortex of the rat: A comparative review. *Brain Res. Rev.* **8,** 65–98.

Kraemer, G. W. (1992). A psychobiological theory of attachment. *Behav. Brain Sci.* **15**, 493–541.
LeDoux, J. E. (1992). Emotion and the Amygdala. *In* "The Amygdala: Neurobiological Aspects of Emotion, Memory, and Mental Dysfunction" (J. Aggleton, ed.), pp. 339–351. Wiley-Liss, New York.
Leon, M., Coopersmith, R., Beasley, L., and Sullivan, R. M. (1990). Thermal aspects of parenting. *In* "Mammalian Parenting: Biochemical, Neurobiological, and Behavioral Determinants" (N. A. Krasnegor and R. S. Bridges, eds.), pp. 400–415. Oxford Univ. Press, New York.
Lévy, F., Gervais, R., Kinderman, U., and Orgeur, P. (1990). Importance of b-noradrenergic receptors in the olfactory bulb of sheep for recognition of lambs. *Behav. Neurosci.* **104**, 464–469.
Lévy, F., Guevara-Guzman, R., Hinton, M. R., Kendrick, K. M., and Keverne, E. B. (1993). Effects of parturition and maternal experience on noradrenaline and acetylcholine release in the olfactory bulb of sheep. *Behav. Neurosci.* **107**(4), 662–668.
Lévy, F., Poindron, P., and Le Neindre, P. (1983). Attraction and repulsion by amniotic fluids and their olfactory control in the ewe around parturition. *Physiol. Behav.* **31**, 687–692.
Lumley, L. A., and Hull, E. M. (1994). Sexual experience increases copulation induced Fos-Like Immunoreactivity in the MPOA. *Soc. Neurosci. Abstr.* **16**, 210.
Main, M., Kaplan, N., and Cassidy, J. (1985). Security in infant childhood and adulthood—A move to the level of representation. *In* "Growing Points of Attachment Theory and Research" (I. Bretherton and E. Waters, eds.), Vol. 50, pp. 1–2, Ser. no. 209. Monographs of the Society for Research in Child Development.
Malenfant, S. A., Barry, M., and Fleming, A. S. (1991). Effects of cycloheximide on the retention of olfactory learning and maternal experience effects in postpartum rats. *Physiol. Behav.* **49**, 289–294.
Mann, P. E., Lupini, C. E., Ronsheim, P. M., and Bridges, R. S. (1990). Reproductive experience alters opiode-mediated maternal behavior and prolactin secretion in lactating rats. *Soc. Neurosci. Abstr.*, **16**, 210.
Martinez, J. L., Jr., and Kesner, R. P. (1991). Pharmacology and biochemistry: memory: Drugs and hormones. *In* "Learning and Memory: A Biological View" (J. L. Martinez, Jr., and R. P. Kesner eds.) pp. 127–163. Academic Press, San Diego.
McGaugh, J. L., Introini-Collison, I. B., and Nagahara, A. H. (1988). Memory-enhancing effects of post-training naloxone: Involvement of b-noradrenergic influences in the amygdaloid complex. *Brain Res.* **446**, 37–49.
Miceli, M. O., and Malsbury, C. W. (1982). Availability of a food hoard facilitates maternal behavior in Virgin female hamsters. *Physiol. Behav.* **28**, 855–856.
Modney, B. K., and Hatton, G. I. (1990). Motherhood modifies magnocellular neuronal relationships in functionally meaningful ways. *In* "Mammalian Parenting: Biochemical, Neurobiological, and Behavioral Determinants" (N. A. Krasnegor and R. S. Bridges, eds.) pp. 305–323. Oxford Univ. Press, New York.
Modney, B. K., Yang, Q. Z., and Hatton, G. I. (1987). Pup-induced maternal behavior in virgin rats is associated with increased frequency of dye coupling among supraoptic nucleus (SON) neurons. *Soc. Neurosci. Abstr.* **13**, 1593.
Moffat, S. T., Suh, E. J., and Fleming, A. S. (1993). Noradrenergic involvement in the consolidation of maternal experience in postpartum rats. *Physiol. Behav.* **53**, 805–811.
Moltz, H. and Wiener, E. (1966a). Effects of ovariectomy on maternal behavior of primiparous and multiparous rats. *J. Comp. Physiol. Psychol.* **61**, 455–460.
Moltz, H. and Wiener, E. (1966b). Ovariectomy: Effects on the maternal behavior of the primiparous and multiparous rat. *J. Comp. Physiol. Psychol.* **62**, 383–387.

Moltz, H., Robbins, D., and Parks, M. (1966). Caesarian delivery and the maternal behavior of primiparous and multiparous rats. *J. Comp. Physiol. Psychol.* **61,** 455–460.

Moltz, H., Levin, R., and Leon, M. (1969a). Differential effects of progesterone on the maternal behavior of primiparous and multiparous rats. *J. Comp. Physiol. Psychol.* **67,** 36–50.

Morgan, J. I., and Curran, T. (1991). Stimulus-transcription coupling in the nervous system: Involvement of the inducible protoconcogenes fos and jun. *Annu. Rev. Neurosci.* **14,** 421–451.

Morgan, H. D., Fleming, A. S., and Stern, J. M. (1992). Somatosensory control of the onset and retention of maternal responsiveness in primiparous Sprague-Dawley rats. *Physiol. Behav.* **51,** 541–555.

Nelson, T. J., and Alkon, D. L. (1989). Specific protein changes during memory acquisition and storage. *BioEssays* **10,** 75–79.

Noirot, E. (1970). Selective priming of maternal responses by auditory and olfactory cures from mouse pups. *Dev. Psychobiol.* **2,** 273–276.

Noirot, E. (1972). The onset of maternal behavior in rats, hamsters, and mice. *In* "Advances in the Study of Behavior" Vol. 4 (D. S. Lehrman, R. A. Hinde, and E. Shaw, eds.), pp. 107–140. Academic Press, New York.

Numan, M. (1988). Maternal behavior. *In* "Physiology of Reproduction" Vol. 12 (E. Knobil and J. Neill eds.), pp. 1569–1645. Raven Press, New York.

Numan, M. (1990). Neural control of maternal behavior. *In* "Mammalian Parenting: Biochemical, Neurobiological, and Behavioral Determinants" N. S. Krasnegor and R. S. Bridges (eds.), pp. 231–259. Oxford Univ. Press, New York.

Numan, M. (1994). Maternal behavior. *In* "Physiology of Reproduction" 2nd ed. (E. Knobil and J. Neill, eds.), pp. 221–302. Raven Press, New York.

Numan, M., and Numan, M. J. (1991). Preoptic-brainstem connections and maternal behavior in rats. *Behav. Neurosci.* **104,** 1013–1029.

Numan, M., and Numan, M. J. (1994). Expression of Fos-like-immunoreactivity in the preoptic region of maternally behaving virgin and postpartum rats. *Behav. Neurosci.* **108,** 379–394.

Numan, M., Corodimas, K., Numan, M., Factor, E., and Piers, W. (1988). Axon-sparing lesions of the preoptic region and substantia innominata disrupt maternal behavior in rats. *Behav. Neurosci.* **102,** 381–396.

Numan, M., Numan, M. J., and English, J. B. (1993). Excitotoxic amino acid injections into the medial amygdala facilitate maternal behavior in virgin female rats. *Horm. Behav.* **27,** 56–81.

Oades, R. D. (1985). The role of adrenaline in tuning and dopamine in switching behavior signals in the CNS. *Neurosci. Biobehav. Rev.* **9,** 261–282.

Orpen, B. G., and Fleming, A. S. (1987). Experience with pups sustains maternal responding in postpartum rats. *Physiol. Behav.* **40,** 47–54.

Orpen, B. G., Furman, N., Wong, P. Y., and Fleming, A. S. (1987). Hormonal influences on the duration of postpartum maternal responsiveness in the rat. *Physiol. Behav.* **40,** 307–315.

Otto, T., and Eichenbaum, H. (1992). Olfactory learning and memory in the rat. A "model system" for studies of neurobiology of memory. *In* "Science of Olfaction" (M. J. Serby and K. L. Chobar, eds.), pp. 213–244. Springer–Verlag New York.

Pederson, C. A., Johns, J. M., Faggin, B. M., Avers, G., and Caldwell, J. D. (1994). Proximal separation from pups re-establishes oxytocin control of maternal behavior in experienced rat mothers. *Soc. Neurosci. Abstr.* **441.9,** 1069.

Pfaff, D. W., and Keiner, M. (1973). Atlas of estradiol-concentrating cells in the central nervous system of the female rat. *J. Comp. Neurol.* **151,** 121–158.

Pfaus, J. G., Kleopoulos, S. P., Mobbs, C. V., Gibbs, R. B., and Pfaff, D. W. (1992). Fos and jun expression in the female rat forebrain following hormone treatment and sexual stimulation. *Soc. Neurosci. Abstr.* **18,** 892.

Pissonnier, D., Theiry, J. C., Fabre-Nys, P., Poindron, P., and Keverne, E. B. (1985). The importance of olfactory bulb noradrenalin for maternal recognition in sheep. *Physiol. Behav.* **35**, 361–363.

Poindron, P., and Lévy, F. (1990). Physiological, sensory and experiential determinants of maternal behavior in sheep. *In* "Mammalian Parenting: Biochemical, Neurobiological, and Behavioral Determinants" (N. A. Krasnegor and R. S. Bridges, eds.), pp. 133–157. Oxford Univ. Press, New York.

Pryce, C. R. (1993). The regulation of maternal behavior in marmosets and tamarins. *Behav. Proc.* **30**, 201–224.

Rangel, S., Leon, J. M., and Leon, M. (1994). Early odor preference training increases olfactory bulb norepinephrine. *Soc. Neurosci. Abstr.* **24**, 65.

Romeyer, A., Poindron, P., and Orgeur, P. (1994). Olfaction mediates the establishment of selective bonding in goats. *Physiol. Behav.* **56(4)**, 693–700.

Rose, S. (1973). "The Conscious Brain." Alfred A. Knopf Inc., New York.

Rosenberg, P., Leidahl, L., Halaris, A., and Moltz, H. (1976). Changes in the metabolism of hypothalamic norepinephrine associated with the onset of maternal behavior in the nulliparous rat. *Pharmacol. Biochem. Behav.* **4**, 647–649.

Rosenberg, P., Halaris, A., and Moltz, H. (1977). Effects of central norepinephrine depletion on the initiation and maintenance of maternal behavior in the rat. *Pharmacol. Biochem. Behav.* **6**, 21–24.

Rosenblatt, J. S. (1967). Nonhormonal basis of maternal behavior in the rat. *Science* **156**, 1512–1514.

Rosenblatt, J. S. (1983). Olfaction mediates developmental transition in the altricial newborn of selected species of mammals. *Develop. Psychobiol.* **16**, 347–475.

Rosenblatt, J. S. (1990). Landmarks in the physiological study of maternal behavior with special reference to the rat. *In* "Mammalian Parenting: Biochemical, Neurobiological, and Behavioral Determinants" (N. A. Krasnegor and R. S. Bridges eds.), pp. 40–60. Oxford Univ. Press, New York.

Rosenblatt, J. S., and Lehrman, D. S. (1963). Maternal behavior in the laboratory rat. *In* "Maternal Behavior in Mammals" (H. L. Rheingold, ed.), pp. 8–57. John Wiley & Sons, New York.

Rosser, A. E., and Keverne, E. B. (1985). The importance of central noradrenergic neurons in the formation of an olfactory bulb memory in the prevention of pregnancy block. *Neuroscience* **15**, 1141–1147.

Ruppenthal, G. C., Arling, G. L., Harlow, H. F., Sackett, G. P., and Suomi, S. J. (1976). A 10-year perspective of motherless-mother behavior. *J. Abnorm. Psychol.* **85**, 341–349.

Sagar, S. M., Sharp, F. R. and Curran, T. (1988). Expression of *c-fos* protein in brain: Metabolic mapping at the cellular level. *Science* (Washington, D. C.) **240**, 1328–1330.

Scalia, F., and Winans, S. S. (1975). The differential projection of the olfactory bulb and accessory olfactory bulb in mammals. *J. Comp. Neurol.* **161**, 31–56.

Schneider, J. E., and Wade, G. N. (1989). Effects of maternal diet, body weight and body composition on infanticide in Syrian hamsters. *Physiol. Behav.* **46**, 815–821.

Schneider, J. E., and Wade, G. N. (1991). Effects of ambient temperature and body fat content on maternal litter reduction in Syrian hamsters. *Physiol. Behav.* **49**, 135–139.

Schlein, P. A., Zarrow, M. X., Cohen, H. A., Denenberg, V. H., and Johnson, N. P. (1972). The differential effect of anosmia on maternal behavior in the virgin and primiparous rat. *J. Reprod. Fertil.* **30**, 139–142.

Scott, J. W., and Chapin, B. R. (1975). Origin of olfactory projections to lateral hypothalamus and nuclei geminz of the rat. *Brain Res.* **88**, 64–88.

Siegel, H. S., and Greenwald, G. S. (1978). Effects of mother-litter separation on later maternal responsiveness in the hamster. *Physiol. Behav.* **21**, 147–149.

Simerly, R. B., and Swanson, L. W. (1988). Projections of the medial preoptic nucleus: A *Phaseolus vulgaris* Leucoagglutinin anterograde tract-tracing study in the rat. *J. Comp. Neurol.* **270**, 209–242.

Slotnick, B. M. (1967). Disturbances of maternal behavior in the rat following lesions of the cingulate cortex. *Behaviour* **29**, 204–236.

Stamm, J. S. (1955). The function of the median cerebral cortex in maternal behavior in rats. *J. Comp. Physiol. Psychol.* **48**, 347–356.

Steele, M., Rowland, D., and Moltz, H. (1979). Initiation of maternal behavior in the rat: Possible involvement of limbic norepinephrine. *Pharmacol. Biochem. Behav.* **11**, 123–130.

Stern, J. M. (1989). Maternal behavior: Sensory, hormonal, and neural determinants. *In* "Psychoendocrinology" (F. R. Brush and S. Levine, eds.), pp. 105–226. Academic Press, New York.

Stern, J. M. (1983). Maternal behavior priming in virgin and Caesarean-delivered Long-Evans rats: Effects of brief contact or continuous exteroceptive pup stimulation. *Physiol. Behav.* **31**, 757–763.

Stern, J. M., and Johnson, S. K. (1989). Perioral somatosensory determinants of nursing behavior in Norway rats, *Rattus norvegicus*. *J. Comp. Psychol.*, **103**, 269–280.

Stern, J. M., and Kolunie, J. M. (1989). Perioral anesthesia disrupts maternal behavior during early lactation in Long-Evans rats. *Behavior. Neural Biol.* **52**, 20–38.

Stern, J. M., Dix, L., Pointek, C., and Thramann, C. T. (1990). Ventral somatosensory determinants of nursing behavior in rats: Effects of nipple loss or anesthesia. *Soc. Neurosci. Abstr.* **16**, 600.

Stern, J. M., Dix, L., Bellomo, C., and Thramann, C. T. (1992). Ventral trunk somatosensory determinants of nursing behavior in Norway rats: Role of nipple and surrounding sensations. *Psychobiology* **20**, 71–80.

Sullivan, R. M., and Hall, W. G. (1988). Reinforcers in infancy: Classical conditioning using stroking or intra-oral infusions of milk as a UCS. *Develop. Psychobiol.* **21**, 215–223.

Sullivan, R. M., and Leon, M. (1986). Early olfactory learning induces an enhanced olfactory bulb response in young rats. *Develop. Brain Res.* **27**, 278–282.

Sullivan, R. M., and Wilson, D. A. (1993). The role of the amygdala complex in early olfactory associative learning. *Behav. Neurosci.* **107**, 254–263.

Suomi, S. J. (1990). The role of tactile contact in rhesus monkey social development. *In* "Touch: The Foundation of Experience" (K. E. Barnard and T. B. Brazellton eds.), pp. 129–164. Internal Univ. Press, Inc., Madison, CN.

Terkel, J., and Rosenblatt, J. S. (1971). Aspects of nonhormonal maternal behavior in the rat. *Horm. Behav.* **2**, 161–171.

Terlecki, L. J., and Sainsbury, R. S. (1978). Effects of fimbria lesions on maternal behavior of the rat. *Physiol. Behav.* **21**, 89–97.

Walsh, C., Fleming, A. S., Lee, A., and Magnusson, J. (1996). Effects of olfactory and somatosensory desensitization on Fos-like immunoreactivity in the brains of pup-exposed postpartum rats. *Behavioral Neuroscience* **110**(1), 1–20.

Wilson, D. A., and Sullivan, R. M. (1994). Review: Neurobiology of associative learning in the neonate: Early olfactory learning. *Behav. Neural Biol.* **61**, 1–18.

Winocur, G. (1991). Functional dissociation of the hippocampus and prefrontal cortex in learning and memory. *Psychobiology* **19**, 11–20.

Winocur, G., and Moscovitch, M. (1990). Hippocampal and prefrontal cortex contributions to learning and memory: Analysis of lesion and aging effects on maze learning in rats. *Behav. Neurosci.* **104**, 544–551.

Wise, R. A., and Rompre, P. P. (1989). Brain dopamine and reward. *Annu. Rev. Psychol.* **40**, 191–225.

Woo, C. C., and Leon, M. (1994). Beta adrenergic reception density in the main olfactory bulb is altered by early olfactory experience. *Soc. Neurosci. Abstr.* **143.12**, 65.

# Maternal Behavior in Rabbits
## A Historical and Multidisciplinary Perspective

GABRIELA GONZÁLEZ-MARISCAL

CENTRO DE INVESTIGACIÓN EN REPRODUCCIÓN ANIMAL
CINVESTAV-UNIVERSIDAD
AUTÓNOMA DE TLAXCALA
TLAXCALA, MEXICO

JAY S. ROSENBLATT

INSTITUTE OF ANIMAL BEHAVIOR
RUTGERS UNIVERSITY
NEWARK, NEW JERSEY 07102

## I. INTRODUCTION: WHY ARE RABBITS INTERESTING TO STUDY?

The image of an ever-present mother providing continuous care to the young is commonly considered the pattern of maternal care in mammals. Indeed, in rodents and ungulates (by far the most studied mammalian orders) mothers display, across most of lactation, an intense interaction with their young: they may lick their anogenital region, drink their urine, nurse them frequently, retrieve pups that have strayed away from the nest, emit distress vocalizations if separated from the litter, and show aggression towards intruders approaching the nest (for review, see Numan, 1994). By contrast, lagomorphs (i.e., rabbits, hares, and pikas) have adopted an "absentee" mothering system in which the mother–young interaction has been restricted to a minimum. Nonetheless, as warranted by their worldwide distribution, this mothering system has allowed lagomorphs to multiply and colonize a variety of habitats.

As other mammals, lagomorph mothers prepare a nest where they will deliver the young. Shortly before parturition, snowshoe (*Lepus americanus;* Rongstad and Tester, 1971) and European (*Lepus europaeus;* Broekhuizen and Maaskamp, 1980) hares—that give birth to precocious young—gather grass from the field, place it on the surface of the gound in a secluded place selected for parturition, and shape the grass into a nest. The young remain inside the nest for the first days of life, after which they begin to disperse and initiate the ingestion of solid food. Yet, across the whole of lactation, the young will find their way back to the nest for the daily nursing episode. Pikas (*Ochotona princeps;* Whitworth, 1984), in the field or the laboratory, also construct a nest (above ground) with the surrounding vegetation, where nursing takes place. In contrast to hares and pikas, the maternal nest of swamp (*Sylvilagus aquaticus;* Sorensen *et al.,* 1972) and European (*Oryctolagus cuniculus;* Denenberg *et al.,* 1963; Deutsch, 1957; González-Mariscal *et al.,* 1994b; Ross *et al.,* 1956; Sawin *et al.,* 1960; Zarrow *et al.,* 1961, 1962, 1963) rabbits is a much more elaborate structure. Nest building in these species begins some days before parturition as females dig several shallow basins (swamp rabbits) or true underground burrows (European rabbits). Straw or grass is then uprooted from the field, collected, carried into the basins or burrows, and shaped into so-called "straw nests" (Denenberg *et al.,* 1963; Sawin and Crary, 1953; Sorensen *et al.,* 1972; Verga *et al.,* 1978). As parturition approaches, the mother selects one of the straw nests as the one that will hold the young. Within this structure, preparturient females culminate the construction of the "maternal nest." As their tightly rooted hair becomes loose, doe rabbits pluck it with their mouth and incorporate it into the straw nest to form a compact "hairy" structure—the maternal nest (Sawin and Crary, 1953; Sawin *et al.,* 1960). To our knowledge, rabbits are the only mammals that use their body hair for constructing a maternal nest.

Few studies in the field or the laboratory have documented the maternal care of lagomorphs other than European rabbits. However, from these works a common feature emerges as characteristic of the "absentee" mothering system: the mother–young interaction occurs only during nursing, which takes place once a day for a very brief period—3–5 min in European rabbits (Cross, 1952; Drewett *et al.,* 1982; González-Mariscal *et al.,* 1994b; Lincoln, 1974; Verga *et al.,* 1978; Zarrow *et al.,* 1965a), 5–10 min in snowshoe hares (Rongstad and Tester, 1971), 10–20 min in swamp rabbits (Sorensen *et al.,* 1972). In European hares, the duration of the nursing bout decreases from around 4 to 6 min at the beginning of lactation to 1 to 2 min as weaning approaches (Broekhuizen and Maaskamp, 1980). Pikas present an intermediate situation, nursing the litter for 10 min every 2 h during the first week of lactation, thereafter decreasing both the duration and the number of nursing bouts per day (Whitworth, 1984).

Despite the short duration of the nursing episode, rabbit pups, who are born altricial (i.e., hairless, with closed eyelids, and unable to regulate their body temperature), can locate the mother's nipples and suckle enough milk to sustain them for the next 24 h. This remarkably efficient way of nursing is achieved through the operation of refined mechanisms in the mother and the young. Hudson and Distel have shown (1983, 1984) that lactating rabbit does emit from their ventrum an olfactory signal (so-called *nipple pheromone*) that is perceived by the main olfactory system of the young (Hudson and Distel, 1986). The perception of this olfactory signal triggers in the pups a stereotyped *searching* behavior on the mother's ventrum that allows them to locate the maternal nipples within seconds, grasp them, and suckle. Whether such a remarkable mechanism exists in other lagomorphs to optimize nursing remains to be determined.

Another surprising aspect of the maternal care in lagomorphs is that, aside from its short duration, nursing is displayed with circadian periodicity (Broekhuizen and Maaskamp, 1980; Jilge, 1993, 1995; Rongstad and Tester, 1971). The obvious fact that organisms (mother and young) that are separated for most of the day need to reunite for a brief period, at a particular time of day, presents the formidable challenge of coordinating the activity and spatiotemporal location of such organisms across the whole of lactation. The mechanisms that operate to achieve the remarkably efficient operation of this system are unknown. However, behavioral observations in hares and European rabbits indicate that both mother and young engage in specific activities that facilitate their encounter once every 24 h. One of these activities is the location of the maternal nest. As described earlier, lagomorph mothers leave the nest after each nursing bout and must, therefore, find their way back every day. Similarly, the young hares (albeit not the young rabbits) leave the maternal nest a few days after birth, meeting their mother (and each other) only for nursing at the birth site (Broekhuizen and Maaskamp, 1980; Rongstad and Tester, 1971). Though young rabbits are not faced with the problem of locating the maternal nest, Hudson and Distel (1982) have shown that the behavior of pups in the nest is tuned to the mother's circadian nursing visits. Shortly before the mother's entrance, pups display much activity, uncovering themselves from the nest material and appearing highly agitated. Thus, pups are fully awake by the time the mother enters the nest and can immediately initiate nursing, guided only by the emission of the nipple pheromone from her ventrum. Jilge (1993, 1995) has shown, under carefully controlled laboratory conditions, that the mother rabbit's nursing visits occur with circadian periodicity, during the dark phase, with a few minutes of delay across successive days. He has suggested that the synchronization of pup and mother activity may be achieved through the operation of a "food-entrained" oscillator in the young. Whatever the nature of the mechanisms involved, it seems common to the maternal behavior of lagomorphs their capacity to synchronize specific activities in mother and young, at a particular time of day, that facilitate their encounter and allow nursing.

Aside from the obvious function of providing nutrients for the pups, nursing in rabbits acts as a means for transferring chemical information from mother to young (Bilkó et al., 1994; Hudson and Altbäcker, 1992). Milk apparently contains traces of the food ingested by the mother and, as pups drink it, they receive information on food selection that is crucial for their survival after weaning. Aside from providing milk, during each nursing bout the mother rabbit deposits a mean of four fecal pellets before leaving the nest (Hudson and Altbäcker, 1992). These pellets seem to contain not only bacteria from her intestinal flora but also traces of the mother's meal. As they grow older and consume less milk, the young nibble on these pellets and may, therefore, continue receiving chemical information from their mother. Whether this means of transferring chemical information from mother to litter also operates in other lagomorphs is unknown. It is tempting to speculate that this mechanism may be even more relevant in hares, whose precocial young face the challenge of selecting solid food a few days after birth.

The unique characteristics of maternal behavior in rabbits present a multitude of questions regarding the control and integration of the diverse components of maternal behavior and, at the same time, provide a model for testing the generality of current notions about the participation of hormones and sensory information in the triggering and maintenance of maternal care in mammals (Bridges, 1990; Poindron et al., 1988; Pryce, 1992; Rosenblatt and Siegel, 1981). In this work we will review past and current studies that have explored particular aspects of rabbit maternal behavior and we shall try to integrate such data into a comprehensive unit that incorporates ecological, endocrine, somatosensory, and experiential factors in the regulation of rabbit maternal behavior.

## II. Studying Rabbit Maternal Behavior in the Laboratory: What Can We Measure?

The group of Zarrow, Denenberg, and their colleagues was the first to study the behavior of mother rabbits in the laboratory. Based on observations of a large number of European rabbits from different strains, these investigators identified specific behavioral patterns displayed by mother rabbits that could be quantified under laboratory conditions (see as follows). As in nature, the behavior of mother rabbits before parturition was directed towards nest building, while her postpartum actions were oriented towards the pups and human "intruders." To quantify nest building, the group of Zarrow and Denenberg determined: (1) when (relative to the day of parturition) females began to collect material for constructing the straw nest (Denenberg et al., 1958, 1959); (2) how many excelsior wads (provided for nest building) were shredded and shaped as a nest (Denenberg et al.,

1963); and (3) what was the quality of the maternal nest, through a scale that assessed its shape and the presence of straw and hair (Ross et al., 1956). At parturition, mother rabbits—like other mammals (Kristal, 1991)—lick their pups as they are being born, ingest the emerging placentas, and engage in nursing. Therefore, the display of cannibalism, the scattering of pups outside the nest, and the lack of nursing were considered as indicators of a deficient maternal behavior (Denenberg et al., 1958, 1959). To quantify maternal aggression, Denenberg et al. (1958) assessed the behavior of mother rabbits towards humans that approached the maternal nest. By using all these measures, Zarrow, Denenberg, and their colleagues determined that the time of occurrence, incidence, and intensity of the previous behaviors varied among the different strains (Denenberg et al., 1959; Ross et al., 1956, 1963; Zarrow et al., 1965b) and with successive parturitions (Denenberg et al., 1958). Moreover, these investigators found significant correlations among some of the parameters measured. For instance, earlier time of nest building correlated with better nest quality and this, in turn, with a high percentage of young nursed at parturition and an absence of scattering and cannibalism (Denenberg et al., 1958, 1959). By contrast, maternal aggression was not correlated with any of the previous parameters (Denenberg et al., 1959). Based on these findings, Denenberg et al. (1959) proposed the existence of a "package" of maternal care that would include early and good quality nest building, large percentage of liveborn young suckled at parturition, and a lack of scattering and cannibalism.

The research of Zarrow, Denenberg, and collaborators laid the groundwork for investigating rabbit maternal behavior under laboratory conditions. However, from those studies we cannot fully assess how the various components of maternal behavior change across gestation, after parturition, and in relation to each other. Moreover, other aspects of rabbit maternal behavior (e.g., digging of a burrow) and of the pregnant doe's physiology (e.g., food ingestion, emission of nipple pheromone) were not investigated in this pioneering work. Therefore, we performed a study aimed at describing and quantifying a variety of somatic events and motor patterns displayed by multiparous New Zealand white rabbits under laboratory conditions (González-Mariscal et al., 1994b).

As described earlier, maternal behavior in rabbits is initiated before parturition; during the last third of the 30-day pregnancy, females begin to excavate an underground burrow where they will construct the maternal nest. To measure this behavior in the laboratory, we introduced into the cage of each female a wooden box with a round opening on one side where the female could construct the maternal nest. We then cut a piece of compressed cardboard to fit exactly into the floor of the nest box, weighed it, and left it inside for 24 h. We then removed the cardboard and weighed it again. The decrease in the cardboard's weight indicated the amount of substrate removed by the female in 1 day. Figure 1 (upper panel) shows that digging was the first motor pattern to emerge, increasing steadily from midpregnancy up to Gestation Day 25. The next stage in the nest-building process is the shaping of the straw nest with vegetation collected by the female. In the laboratory, we measured this behavior by placing in the cage of each female (but outside the nest box) a container with 100 g of straw. We removed and weighed the straw the female had introduced into the nest box 24 h later. As shown in Fig. 1, straw carrying began as digging declined (around 3 days before parturition), was maximally expressed on prepartum Day 1, and declined thereafter. During the last stage of nest building, we quantified how many females had incorporated tufts of their loose hair into the previously built straw nest. We found that hair pulling

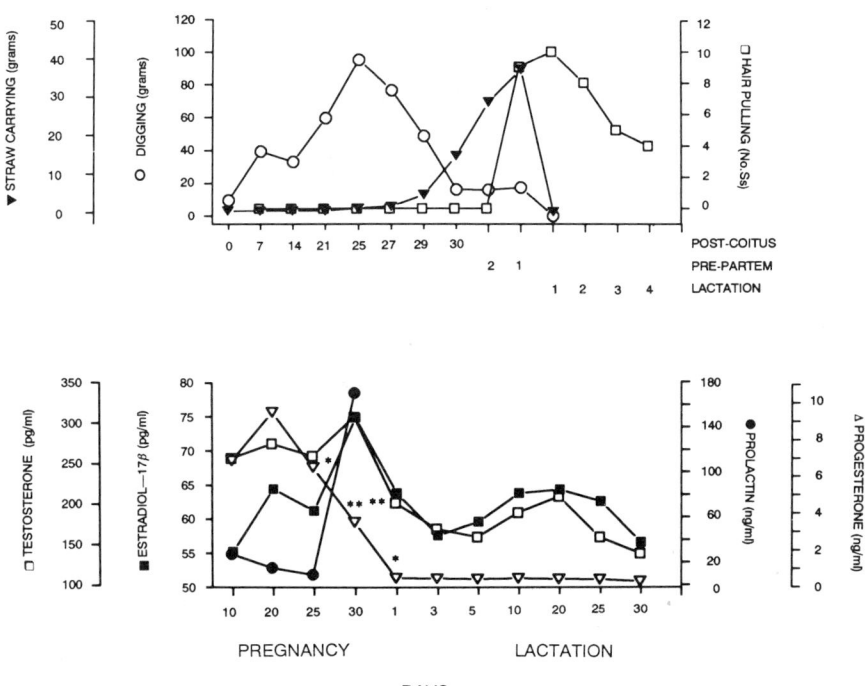

FIG. 1. (Upper panel) Variations in digging, straw carrying, and hair pulling in New Zealand white does across pregnancy and 4 days of lactation. (Lower panel) Concomitant changes in the plasma concentration of estradiol-17$\beta$, testosterone, progesterone, and prolactin. (From González-Mariscal et al., 1994b, with the kind permission of Pergamon Press.)

was initiated on Day 1 prepartum and that it persisted in some does for up to 4 days after parturition (Fig. 1).

Though not a component of maternal behavior, Yaschine and Beyer (1969) had reported that food and water intake decrease markedly as parturition approaches. We quantified food ingestion across pregnancy by providing females with 400 g of pellets and determining the weight of the food remaining in the container 24 h later. As shown in Fig. 2A, we confirmed that a marked decrease in food intake occurs on the 3 days preceding

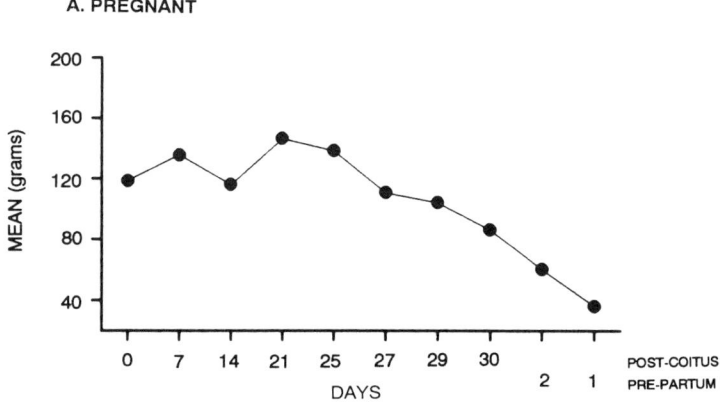

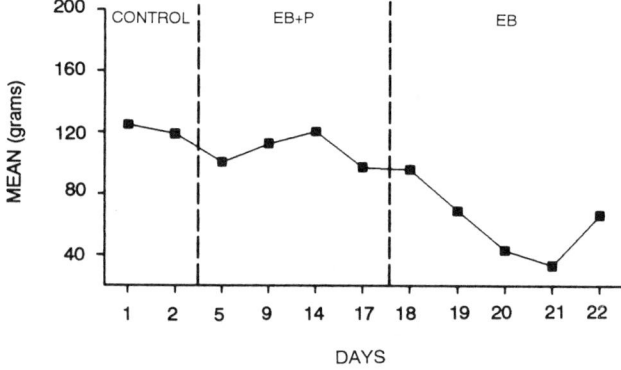

FIG. 2. Variations in food intake observed in (A) pregnant New Zealand white does across gestation and (B) ovariectomized females of the same strain, treated with estradiol benzoate (EB; 5 µg/day) and progesterone (P; 10 mg/day). (A, modified from González-Mariscal et al., 1994a).

parturition. The functional significance of this transitory and reversible decline in food intake is unknown, but several mammals have been reported to show a similar phenomenon before parturition (Ota, 1981; Penzhorn and Meintjes, 1972; Shirley, 1984; Weiss, 1991). Perhaps it is advantageous to reduce intestinal transit as parturition approaches because the released oxytocin may contract not only the uterine myometrium to expulse the pups (Fuchs and Dawood, 1980) but also the intestinal smooth muscle and, thus, provoke an unwanted defecation.

As mentioned earlier, maternal care after parturition is restricted in rabbits to a single nursing bout per day. To determine the duration of the nursing episode, we removed the litter after birth (usually four to eight pups per female), we reduced litter size to five young per mother, and we returned these pups to the maternal nest 24 h later. We determined when the mother entered the nest box, how long she remained inside it, and the amount of milk ingested by the pups. On perceiving the pups in the nest, the mother immediately entered the nest box, adopted a crouching posture over the litter, and allowed the young to search for her nipples and suckle. Nursing was terminated as the mother abruptly left the nest box. In agreement with previous data (Cross, 1952; Drewett *et al.*, 1982; Lincoln, 1974; Zarrow *et al.*, 1965a), we found that, from Days 2 to 30 of lactation, the duration of the nursing bout was remarkably constant and short—199 ± 7 sec (mean ± standard error). However, on the first day of lactation, the mother remained inside the nest box for a longer time—from 215 up to 700 sec. This observation does not necessarily indicate a longer duration of suckling. Rather, it may reflect that, as other mammals (Rosenblatt and Siegel, 1981; Vince, 1993), mother rabbits perceive during the first nursing interaction a variety of sensory cues from the litter that may be crucial for the maintenance of an adequate maternal care across lactation (see as follows).

The remarkable effectiveness of the mother rabbit's nursing strategy was further substantiated when we evaluated the milk intake and the weight gain of the pups across lactation (González-Mariscal *et al.*, 1994b). Simply by weighing the young before and after the nursing bout we confirmed earlier reports (Cowie, 1969; Drewett *et al.*, 1982; Zarrow *et al.*, 1965a) that, despite the constant duration of the nursing bout, milk intake increased steadily, from around 8 g per pup on Day 1 of lactation to a maximum of 32 g per pup on Day 19 of lactation. Interestingly, pup weight increased linearly from about 50 g per pup at birth to 385 g per pup at weaning on Day 30 of lactation, despite a gradual reduction in milk intake after Day 19 of lactation. Cowie (1969) has suggested that this may ensue from the abrupt increase in the fat and protein content of milk after Day 20 of lactation and the greater metabolic efficiency of the growing young.

Unlike other mammals, maternal care in rabbits is terminated at weaning. Interestingly, the duration of lactation is not rigidly fixed, though it usually lasts around 30 days (Cowie, 1969), it is modulated by whether the mother is pregnant or not. Female rabbits show a clear postpartum estrus after parturition (Beyer and Rivaud, 1969; Brambell, 1944; Foxcroft and Hasnain, 1973) and, indeed, males in the wild have been reported to wait outside the maternal burrow to mate with the female as soon as she exits (Mykytowycz and Rowley, 1958). Hudson and Altbäcker (1992) reported that lactating females wean their young 3–5 days earlier when they are pregnant than when they are not. This early weaning may prevent the pregnant female from prematurely entering into labor as a consequence of the release of oxytocin provoked by the suckling litter (Fuchs *et al.*, 1984).

As discussed in several articles of this book, in many mammalian and nonmammalian species, males participate in the care of infants. In rabbits, Mykytowycz (1972) reported that females are very aggressive towards juveniles (i.e., lactating young who have grown enough to occasionally leave the maternal burrow during the day) that are not their own. When this occurs, males are protective of such juveniles, standing between the aggressive female and the young. Whether the male displays this behavior towards all juveniles or just towards his own progeny is unknown.

In the following sections we shall see that the coordinated expression of many of the somatic and motor events that characterize the maternal behavior of rabbits is regulated by hormonal factors and sensory stimuli.

### III. Participation of Estradiol, Progesterone, and Prolactin in the Initiation of Maternal Behavior

Female rabbits show a pattern of circulating levels of estradiol, progesterone, and prolactin similar to that observed in other mammals across pregnancy and lactation (Browning *et al.*, 1980; Challis *et al.*, 1973; Rosenblatt and Siegel, 1981). Figure 1 (lower panel) shows the plasma concentrations of estradiol-17$\beta$, progesterone, and testosterone, determined in multiparous New Zealand white rabbits across pregnancy and lactation (González-Mariscal *et al.*, 1994b). Estradiol remains at around 60 pg/ml across most of pregnancy (though a tendency to rise is apparent shortly before parturition) and lactation. By contrast, progesterone attains maximal levels of around 9 ng/ml on pregnancy Day 20 and steadily declines thereafter, negligible levels of this hormone being observed across lactation. Testosterone remains at around 270 pg/ml across most of pregnancy but declines from 308 pg/ml on pregnancy Day 30 to 202 pg/ml on Day 1 of lactation. Prolactin rises abruptly, from the negligible levels detected across most of pregnancy, to reach 100 ng/ml on pregnancy Day 30 (McNeilly and Friessen, 1978).

If we compare both panels in Fig. 1, it becomes apparent that particular motor patterns and somatic events are expressed by pregnant does under specific hormonal conditions. Thus, digging was maximally expressed when circulating levels of both estradiol and progesterone were high. By contrast, straw carrying and the decline in food intake (Fig. 2A) occurred as progesterone declined, prolactin increased, and estradiol levels remained high. One to two days later, hair loosening–pulling was expressed, also under low-progesterone–high-prolactin conditions, coinciding with a decline in testosterone levels. Overall, these data suggest that the precise temporal expression of specific motor patterns and somatic events associated with maternal behavior in rabbits largely depends on the presence and subsequent decline of progesterone, on a background of estradiol, and an abrupt increase in prolactin at the end of pregnancy.

The participation of steroid hormones and prolactin in the regulation of maternal behavior in mammals has been studied mainly in rodents (for reviews, see Bridges, 1990; Numan, 1994). These studies have used two main experimental strategies: (1) endocrinectomies, that is, the removal of endocrine organs that produce specific hormones (e.g., ovaries, pituitary, placenta); and (2) selective replacement of the hormone under study (usually with injections or by placing beneath the skin pellets containing the hormone). The group of Zarrow and Denenberg, who initiated the studies on the endocrine control of maternal behavior in rabbits (Farooq *et al.*, 1963), used similar experimental strategies. To assess the role of estradiol and progesterone, these investigators removed the ovaries of pregnant females on different days of pregnancy. They found that ovariectomy performed on Days 15–17 of pregnancy (but not earlier) led to maternal nest building in 56–100% of subjects (Zarrow *et al.*, 1962). This finding would suggest that ovarian hormones are required during the first half of pregnancy but that, past this time, the presence of some ovarian hormone(s) is inhibiting the display of maternal nest building. The reverse experiment consisted of giving ovariectomized (nonpregnant) females a synthetic estrogen (stylbestrol, 10 $\mu$g/kg) combined with progesterone (1 or 2 mg/day) for 14 days and injecting prolactin (200 IU/day) for 5–7 days after giving the steroids. This treatment induced the construction of a straw nest in 65% of ovariectomized females but only 23% of them built a maternal (hairy) nest (Zarrow *et al.*, 1961). When pups and their placentas were removed through a cesarean section performed between Days 20 to 22 of pregnancy, 59–97% of subjects built a maternal nest (Zarrow *et al.*, 1961). Because the removal of placentas through a cesarean section eliminates a source of progesterone; provokes the release of estrogen; and this, in turn, induces the secretion of prolactin from the pituitary (reviewed in Neill, 1988; Numan, 1994), the previous results would suggest that maternal nest building is facilitated by the removal of progesterone, a rise in estrogen, and the release of prolactin. Indeed, the injection of progesterone (4 mg/day) from Days 28 to 35 of pregnancy inhibited maternal nest building and delayed parturition (Zarrow et al., 1961). A role of prolactin in the facilitation of nest building is further supported by three additional observations:

1. When a rise in estrogen was provoked earlier than usual by giving estradiol benzoate (10 µg/day) from Days 20 to 22 of pregnancy, nest building was advanced by 4–5 days (Zarrow et al., 1963).

2. When ergocornine (0.5 or 1.0 mg/kg), which blocks the release of prolactin (Martin and Bateson, 1982; Meltzer et al., 1982; Mena et al., 1982; Taylor and Peaker, 1975), was given from pregnancy Day 26 until parturition, the incidence of maternal nest building was reduced. This effect of ergocornine was counteracted by injecting ovine prolactin (8 mg/day; Anderson et al., 1971; Zarrow et al., 1971).

3. Removing the pituitary (thereby eliminating the source of endogenous prolactin) prevented the activation of maternal nest building normally induced in ovariectomized does by the combined administration of estradiol benzoate and progesterone (Anderson et al., 1971; Zarrow et al., 1971).

These studies strongly support the participation of estradiol, progesterone, and prolactin in the activation of nest building in rabbits. However, as stated in Section I, we cannot assess from such work which of the several components of maternal nest building are regulated by specific hormones. Moreover, the endocrine regulation of other aspects of maternal behavior (e.g., emission of nipple pheromone, decline in food ingestion) was not explored in the studies of the group of Zarrow and Denenberg. Therefore, to expand our understanding of the endocrine control of maternal behavior, we assessed in ovariectomized New Zealand white rabbits the effect of administering estradiol benzoate, progesterone, and prolactin as follows.

The combined administration of estradiol benzoate (5 µg/day) plus progesterone (10 mg/day) for 14 days effectively stimulated digging in ovariectomized does (Fig. 3). Straw carrying was initiated after digging, following progesterone withdrawal and continuation of estradiol benzoate, as in pregnant does (Fig. 3). Hair pulling was also stimulated after progesterone withdrawal but this behavior occurred, at most, in 40% of treated subjects (Fig. 3). As in pregnant does, food intake in ovariectomized estradiol-benzoate-primed females steadily declined after progesterone withdrawal despite the continued action of estradiol benzoate (Fig. 2B). On the fourth day after progesterone removal, pellet consumption was reduced to a third

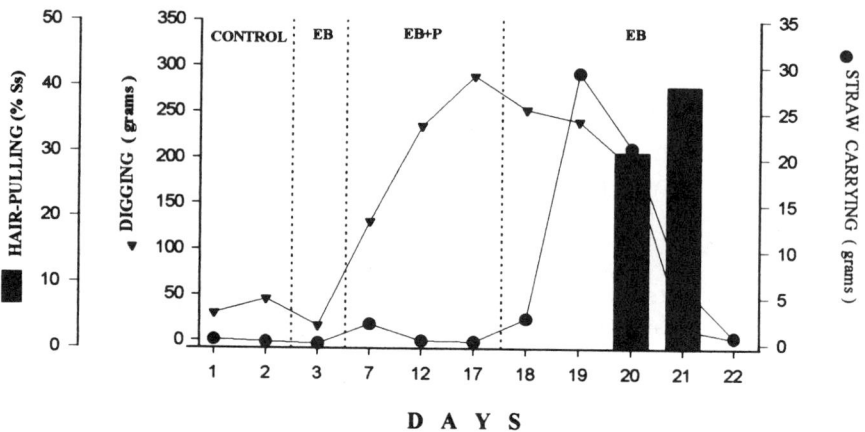

FIG. 3. Activation of digging, straw carrying, and hair pulling in ovariectomized New Zealand white does by the administration of estradiol benzoate (EB; 5 µg/day) and progesterone (P; 10 mg/day).

of the levels observed under the combined estradiolbenzoate–progesterone treatment.

Aside from the previously described behaviors, Hudson and Distel (1983, 1984) have shown that pregnant females emit the nipple pheromone from their ventrum as do lactating does. Though the significance of this event, in the absence of pups, remains to be determined, we assessed if the administration of estradiol and progesterone could mimic in ovariectomized does this other aspect of the pregnant doe's physiology. To quantify the emission of nipple pheromone, we used the method described by Hudson and Distel (1983, 1984). This bioassay relies on the behavior displayed by rabbit pups (1–10 days old) when placed on the ventrum of a female that is emitting the nipple pheromone. They show very distinctive and rapid head movements, they quickly (less than 10 sec) locate her nipples, and try to suckle. The percentage of pups that show such a response (i.e., the percentage of positive trials) indicates the level of nipple pheromone emission. We found that the daily administration of estradiol benzoate (1 or 10 µg/day) plus progesterone (10 mg/day) stimulated the emission of nipple pheromone in ovariectomized New Zealand white does to levels comparable to those observed in pregnant females (Fig. 4, upper panel; Hudson et al., 1990).

The results presented so far indicate that estradiol and progesterone regulate specific motor patterns and somatic events observed in mother rabbits before parturition. However, as described earlier, several pieces of evidence suggest that other hormones (namely, prolactin) may be also

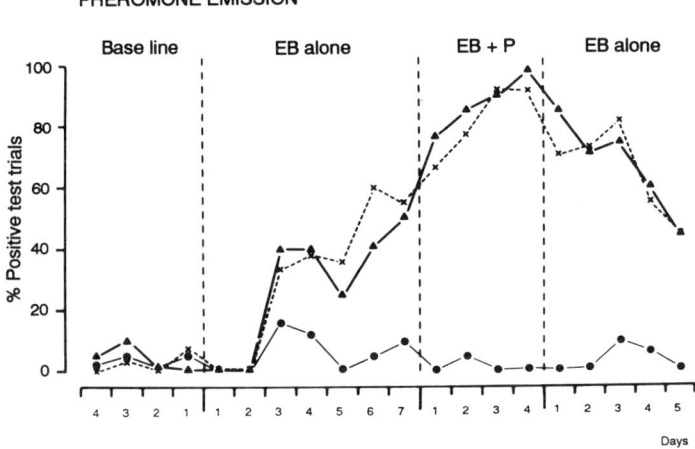
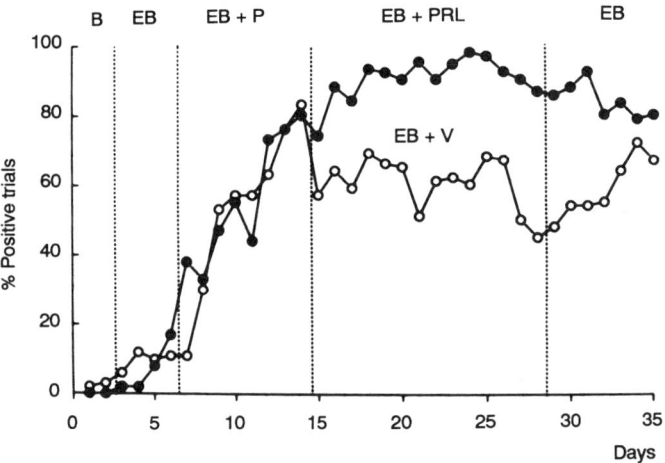

FIG. 4. Stimulation of nipple pheromone emission in ovariectomized New Zealand white does by the administration of (upper panel) estradiol benzoate (EB; 1 µg/day, crosses, or 10 µg/day, filled triangles) combined with progesterone (P; 10 mg/day). Filled circles show lack of nipple pheromone emission in ovariectomized, untreated does. (From Hudson et al., 1990, with the kind permission of Academic Press.) (Lower panel) EB (0.5 µg/day) + P (1 mg/day) followed by EB + ovine prolactin (PRL; 1.5 mg/day). Note decline in nipple pheromone emission after progesterone withdrawal in group that received vehicle (V) instead of prolactin. B = base line levels. (From González-Mariscal et al., 1994a, with the kind permission of Biology of Reproduction.)

involved. Therefore, to further substantiate a role of prolactin in the expression of rabbit maternal behavior, we blocked its release from the pituitary by administering bromocriptine to ovariectomized does treated with estradiol benzoate plus progesterone. As expected, bromocriptine blocked the estrogen-stimulated release of prolactin (data not shown), but it did not modify the incidence of digging (Fig. 5A). This behavior occurred in five of eight treated females with a similar temporal pattern as the one previously observed in subjects treated only with estradiol benzoate plus progesterone

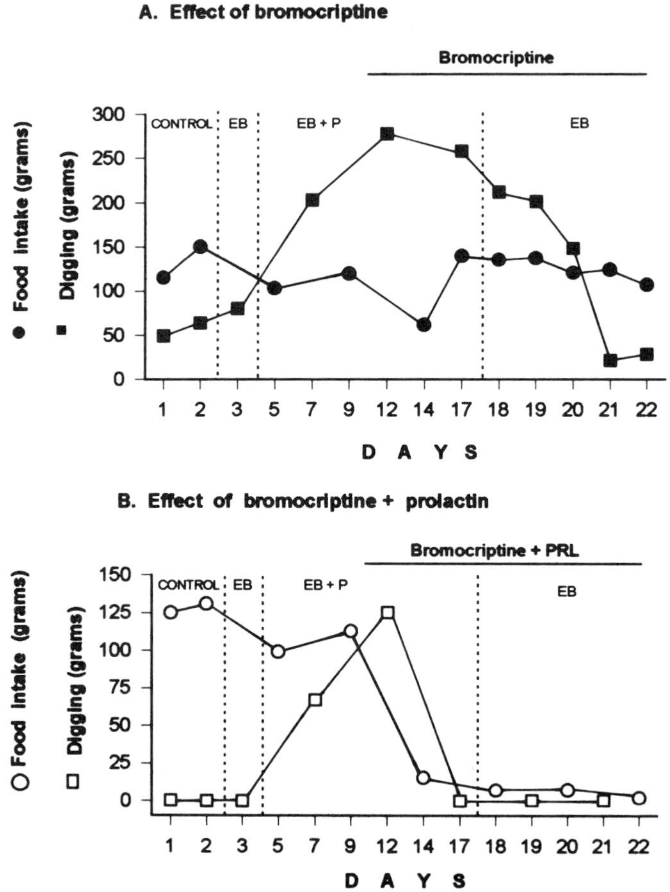

FIG. 5. (A) Effect of bromocriptine (1 or 3 mg/kg per day) on digging and food intake in ovariectomized New Zealand white does treated with estradiol benzoate + progesterone as in Fig. 3. (B) Effect of ovine prolactin (3 mg/day) on digging and food intake in ovariectomized estradiolbenzoate–progesterone-treated does given bromocriptine (1 mg/kg per day).

(see Fig. 3). By contrast, straw carrying and hair pulling, which normally follow digging, were totally abolished in bromocriptine-treated subjects. None of the nine treated does displayed these behaviors. Food intake was depressed on the fourth day after initiating the injection of bromocriptine (Fig. 5A). This effect, which was no longer evident 3 days later, had not been observed in females treated only with estradiol benzoate plus progesterone (see Fig. 2B). By contrast, the decline in food intake normally observed in such females after progesterone withdrawal (Fig. 2B) did not occur in bromocriptine-treated does (Fig. 5A). Because the agent used to prevent prolactin release (bromocriptine) is a potent dopaminergic agonist (Meltzer *et al.,* 1982), we cannot discard the possibility that some of the effects observed may have been due to the participation of dopamine. Therefore, to distinguish between prolactin-mediated and dopamine-mediated effects, we assessed if the administration of ovine prolactin could reverse the actions of bromocriptine in ovariectomized does treated with estradiol benzoate plus progesterone. As shown in Fig. 5B, the administration of ovine prolactin (3 mg/day) to bromocriptine-treated does induced an abrupt decline in both digging and food intake. Ovine prolactin did not restore the expression of straw carrying or hair pulling, abolished by bromocriptine in females treated with estradiol benzoate plus progesterone (data not shown). These data suggest a complex role of prolactin in the regulation of specific aspects of maternal behavior in the rabbit. Contrary to data obtained in rats (Moore *et al.,* 1986), prolactin seems to *inhibit* food intake in female rabbits exposed to estradiol and progesterone. By contrast, prolactin is apparently *stimulatory* to straw carrying and hair pulling because (1) both behaviors are expressed when plasma levels of prolactin are high and progesterone has declined (i.e., shortly before parturition in pregnant does and after progesterone withdrawal in ovariectomized does primed with estradiol benzoate plus progesterone), and (2) neither behavior occurs when the release of endogenous prolactin (induced by estradiol benzoate) is prevented by bromocriptine. Nonetheless, the fact that ovine prolactin did not restore straw carrying or hair pulling in ovariectomized does given estradiol benzoate, progesterone, and bromocriptine disagrees with the previous suggestion. Though the dose of ovine prolactin used may have been insufficient for restoring the previous behaviors (yet, it effectively suppressed food intake and digging), it is also possible that the expression of straw carrying and hair pulling (possibly stimulated by prolactin) was counteracted by the activation of dopaminergic receptors by bromocriptine. Future experiments should assess the participation of dopamine in straw carrying and hair pulling.

The role of prolactin in the control of digging seems to fit within a broader multihormonal regulation of this behavior. As shown in Fig. 1, digging gradually increased across gestation, along with a rise in progesterone plasma levels, and decreased shortly before parturition as progesterone declined. Similarly, digging increased in ovariectomized estradiolbenzoate-primed does as the treatment with progesterone progressed and gradually declined after progesterone withdrawal (Fig. 3). These data support the idea that digging is stimulated by progesterone (on a background of estrogen) and coincide with our observation that estrous does (exposed only to follicular estrogen but lacking progesterone) display small amounts of digging (González-Mariscal et al., 1994b). Prolactin, by contrast, seems to curtail digging because in preparturient does prolactin release (a consequence of a rise in estrogen at the end of gestation) coincides with a decline in digging (Fig. 1, upper and lower panels). In ovariectomized females primed with estradiol benzoate plus progesterone, prolactin plasma levels are high (30–50 ng/ml; unpublished results) across the duration of the combined steroid treatment. Yet, digging occurred steadily, declining only after progesterone withdrawal (Fig. 3). This observation would seemingly contradict the proposition that prolactin is inhibitory to digging. However, as discussed previously, if progesterone stimulates digging, the inhibitory action of prolactin may not become evident before progesterone withdrawal. Indeed, the capacity of progesterone to counteract the action of prolactin has been documented in some target organs (reviewed in Tucker, 1994). In summary, we believe that prolactin participates in the regulation of digging as part of an "accelerator–brake" system (represented by progesterone and prolactin, respectively) that regulates the expression of this behavior.

To assess the role of prolactin in the emission of nipple pheromone observed in lactating does, we first treated ovariectomized females with estradiol benzoate (0.5 $\mu$g/day) plus progesterone (10 mg/day) to mimic the endocrine conditions of pregnancy normally preceding lactation. As in our previous work (Hudson et al., 1990), this treatment induced maximal levels of nipple pheromone emission, which declined gradually following progesterone withdrawal (Fig. 4, lower panel). However, if ovine prolactin (1.5 mg/day) was given after removing progesterone, nipple pheromone emission did not decline; it was maintained at the highest levels, observed in lactating does (Fig. 4, lower panel; González-Mariscal et al., 1994a). These results indicate that the emission of nipple pheromone is regulated by the same multihormonal complex that regulates the behaviors characteristic of mother rabbits.

### IV. Maintaining Maternal Behavior after Parturition: The Importance of the Interaction between Mother and Young

As discussed in the previous section, in rabbits—as in other mammals (Bridges, 1990; Numan, 1994; Rosenblatt and Siegel, 1981)—the endocrine profile that prevailed during pregnancy begins to change as parturition approaches. This observation has led to the question of how maternal behavior is regulated after parturition. Evidence has accumulated to substantiate the view that, at least in rodents and ungulates, the continuous, unfailing, expression of maternal care postpartum relies heavily on the continuous perception of a variety of stimuli (olfactory, acoustic, visual, tactile) provided by the young (Orpen and Fleming, 1987; Poindron et al., 1988; Rosenblatt and Siegel, 1981; Vince, 1993). Rosenblatt and Siegel (1981) have proposed that this *nonhormonal phase,* which becomes more consolidated as contact with the young increases, is preceded by a *transition phase* around the time of parturition. During this period the *hormonal phase,* that triggered the onset of maternal behavior across gestation, progressively wanes while the somatosensory input from the young gradually takes control over the expression of maternal behavior. It must be recalled, however, that unlike rodents and ungulates, rabbits and other lagomorphs spend little time with their young after parturition (Broekhuizen and Maaskamp, 1980; Rongstad and Tester, 1971; Sorensen et al., 1972; Whitworth, 1984). Therefore, rabbits offer the extreme model for testing the generality of the previous model in the regulation of maternal behavior postpartum.

Early studies by the groups of Beyer (Tindal et al., 1963; Beyer and Mena, 1965) and Findlay (Findlay and Roth, 1970) reported that anesthetizing mother rabbits during the nursing bout across 7 days starting at parturition prevented the subsequent resumption of nursing. Anesthetization during nursing for an equivalent period in late lactation did not provoke such an effect. These studies, more oriented towards the neuroendocrine regulation of nursing, scarcely documented the consequences of anesthetizing the mother rabbit on her subsequent behavior towards the young. Therefore, by using the same methodology, we evaluated the impact of deprivation of contact between mother and young during early lactation or midlactation on the display of maternal behavior in multiparous New Zealand white does.

Anesthesia during early lactation (Days 1–7) or midlactation (Days 11–17) provoked some alterations in the resumption of nursing on subsequent days. These alterations consisted of a refusal to nurse or an increased latency to initiate nursing (Fig. 6), a decreased incidence of licking–sniffing of pups (Fig. 7), a higher occurrence of startle–rejection reactions upon perceiving the pups (Fig. 7), and an increased number of entrances into the nest box that did not culminate in nursing (Fig. 8). These alterations in maternal behavior were not a consequence of the anesthetic per se because control subjects, anesthetized (during Days 1–7 of lactation) 8 h *after* they had displayed normal nursing, continued showing the usual pattern of maternal behavior after the anesthetic was discontinued (data not

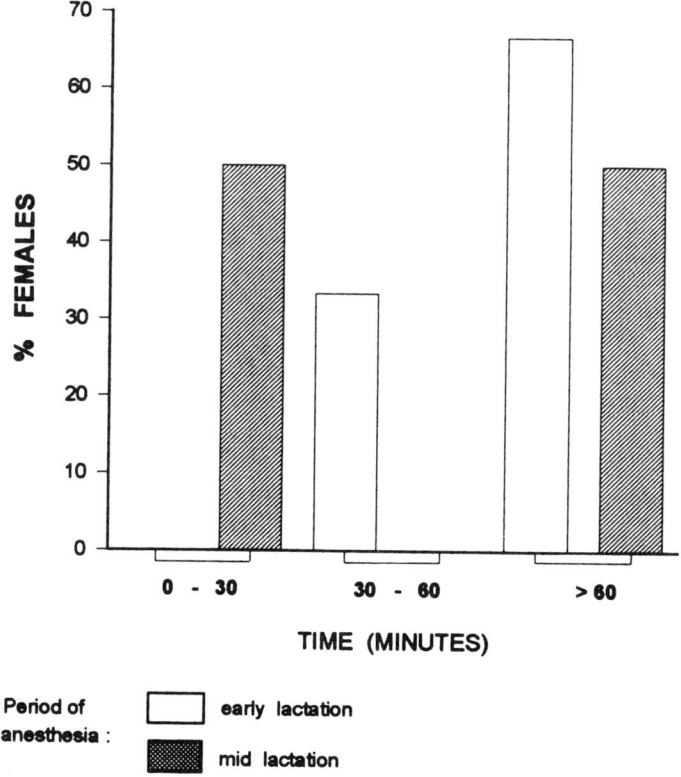

FIG. 6. Effect of anesthetizing New Zealand white mothers during the nursing bout across early lactation (Days 1–7; white bars) or midlactation (Days 11–17; shaded bars) on the latency to initiate nursing (determined on the next day after anesthesia withdrawal). Only three of eight mothers in the early lactation group engaged in nursing, while all six females in the midlactation group did.

shown). Though the previously described alterations in maternal behavior were evident in both experimental groups and their intensity was greater in the group anesthetized during early lactation, maternal behavior was *not* abolished by the experimental procedure. Mother rabbits were still clearly attracted to the pups and tried to be in contact with them. It must be recalled, however, that because we did not prevent the mother–young contact at parturition, the mother rabbit may have received at this time a variety of stimuli that may have aided in the resumption of maternal interest after anesthesia withdrawal. In summary, the deprivation of somatosensory input from the offspring shortly after parturition produces a greater deficit in nursing than a similar deprivation in midlactation. After more than

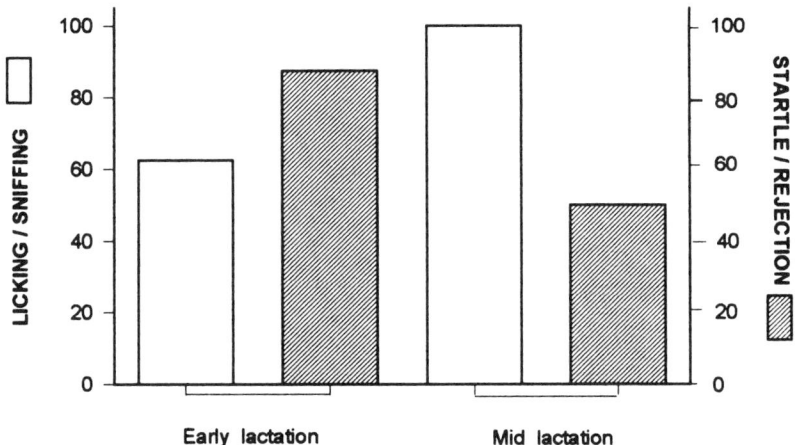

FIG. 7. Effect of anesthetizing New Zealand white mothers during the nursing bout across early lactation (Days 1–7) or midlactation (Days 11–17) on the percentage displaying pup-directed behaviors after anesthesia withdrawal.

a week of contact between mother and young, a transitory absence of somatosensory input from the litter produces some disturbances of nursing but does not prevent it. However, even under these conditions, maternal behavior does not show the "fine-tuning" characteristic of mother rabbits during midlactation, that is, an immediate entrance into the nest box, adoption of a crouching posture that facilitates nipple search by the pups, and a lack of rejection or fright reactions upon perceiving the litter. Our data, therefore, allow us to conclude the following: (1) as in other mammals, the mother–litter interaction in rabbits plays a crucial role for maintaining maternal behavior postpartum; (2) during the first days postpartum (transition phase?), maternal behavior is more liable to being disrupted; and (3) with the experience gained across successive lactation days, maternal behavior is consolidated and is less liable in response to deprivation of offspring stimulation.

## V. Recapitulation and Discussion

The data presented here show the rich variety of motor patterns and somatic events displayed by mother rabbits that can be readily quantified in the laboratory. As in other mammals, estradiol and progesterone present

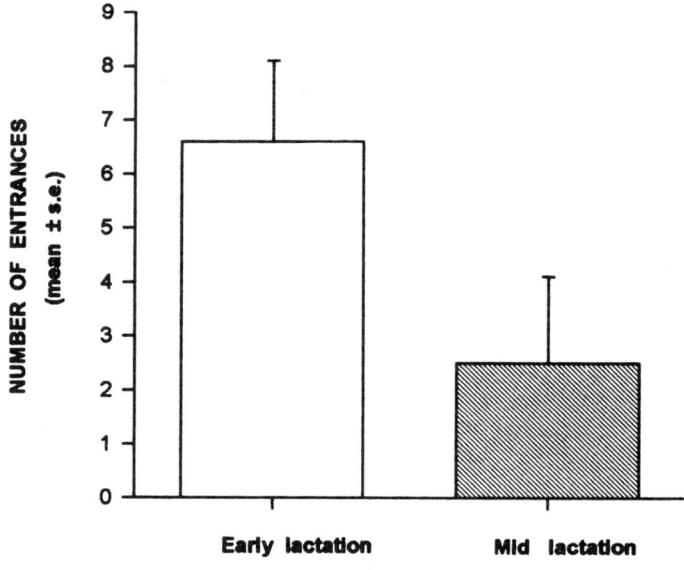

FIG. 8. Effect of anesthetizing New Zealand white mothers during the nursing bout across early lactation (Days 1–7) or midlactation (Days 11–17) on the number of entrances into the nest that did not culminate in nursing (the day after anesthesia withdrawal).

during pregnancy play a crucial role in regulating the precise expression of such events. Though also present at high concentrations during pregnancy and declining before parturition, the role of testosterone in the regulation of specific aspects of maternal behavior remains to be explored. We proposed (González-Mariscal et al., 1994b) that, as in male baldness (Mauvais-Jarvis, 1977; Schweikert and Wilson, 1974; Verhoeven et al., 1974), testosterone could participate, through its local conversion to 5α-dihydrotestosterone, in regulating hair loosening from the ventral skin. The local transformation of testosterone into 5α-dihydrotestosterone (achieved through the action of the enzyme 5α-reductase) would be regulated by progesterone, a well-known inhibitor of 5α-reductase (Cassidenti et al., 1991; Mauvais-Jarvis et al., 1974; Metcalf et al., 1989; Rasmusson et al., 1986). As parturition approaches and progesterone levels drop, the enzyme would become disinhibited, 5α-dihydrotestosterone would be generated *in situ*, it would act on the hair follicle, and thus promote hair loss.

Prolactin may also participate in modulating specific aspects of maternal behavior (e.g., straw carrying, hair pulling, food intake) in the immediate prepartum period. As discussed in Section III, levels of this hormone increase dramatically shortly before parturition (McNeilly and Friessen, 1978) and preventing the release of endogenous prolactin in pregnant does through hypophysectomy or by giving bromocriptine to ovariectomized females primed with estradiol benzoate plus progesterone (Anderson *et al.*, 1971; Zarrow *et al.*, 1971) antagonizes the expression of maternal behavior. Moreover, we have found (unpublished results) that the estradiol benzoate plus progesterone treatment that effectively stimulates maternal behavior in ovariectomized does provokes an increase in plasma prolactin. Furthermore, prolactin binding sites have been reported in the rabbit hypothalamus (Di Carlo and Muccioli, 1981a,b; Walsh *et al.*, 1990) and placental lactogen binding sites were detected in the choroid plexus of preparturient rabbits (Mangurian *et al.*, 1994).

Several works have documented the participation of oxytocin in the expression of maternal behavior in rats (Pedersen *et al.*, 1992; Fahrbach *et al.*, 1985; Van Leengood *et al.*, 1987) and sheep (Keverne and Kendrick, 1992). In rabbits, a hypothalamic oxytocinergic system has been described in detail (Schimchowitsch *et al.*, 1989) but the participation of oxytocin in the regulation of rabbit maternal behavior has remained unexplored. We investigated the immunoreactivity to oxytocin in the hypothalamus of estrous, late pregnant (Day 29) and postpartum (Day 1) rabbits (Caba *et al.*, 1996). A significant increase in the somal area was found in the supraoptic nucleus, paraventricular nucleus, and lateral hypothalamic area of postpartum females compared with estrous and late pregnant does. An increased number of oxytocin-immunoreactive cell bodies was found in the paraventricular nucleus of postpartum (Day 1) females (compared with late pregnant and estrous females) and, to a lesser extent, also in the lateral hypothalamic area (compared with estrous does). Because the paraventricular nucleus is the source of most extrahypothalamic oxytocinergic pathways (Sofroniew, 1983) and these, in turn, may participate in the control of maternal behavior in rodents (Pedersen *et al.*, 1992; Insel, 1992; Insel and Shapiro, 1992), our results agree with the idea that oxytocin may regulate specific aspects of rabbit maternal behavior. Future studies should assess this possibility.

The hormonal dependency of the various parameters that characterize rabbit maternal behavior does not necessarily indicate that they are independent of environmental modulations. We tested the importance of hair availability on the construction of the maternal nest by preparturient multiparous females (González-Mariscal and Cuamatzi, 1995). We found that does, who had been shaved from their ventrum and inner thighs, readily collected a variety of hair types (their own, male, or synthetic) from a container and incorporated this material into the straw nest previously built inside the nest box. Hair collecting was not dependent on the absence of their own hair because unshaved preparturient females also displayed this behavior. Surprisingly, when hair was provided earlier, in the absence of straw, hair collecting was initiated sooner, that is, at a time when females normally collect straw. It is tempting to speculate that straw carrying and the shaping of a straw nest normally precede hair pulling and the construction of a hairy nest simply because these events depend on the previous loosening of the mother's hair. If this limitation is bypassed and an alternative source of hair is provided, the attractiveness towards hair and the construction of a hairy nest is revealed much earlier. Do these data indicate that the hormones of pregnancy build up a "package" of behavioral patterns (digging, straw carrying, shaping of a straw nest, hair pulling, building of a hairy nest) aimed at the construction of the maternal nest but that the actual expression of such patterns is modulated by the available elements in the environment?

The maintenance of rabbit maternal behavior after parturition relies heavily on a very limited somatosensory stimulation received from the young during nursing across early lactation. The deprivation of mother–young contact for the first 7 days of lactation was enough to alter the display of maternal behavior. This paradigm, however, did not allow us to determine what is the minimum lack of contact that can provoke this effect. We have assessed the impact of removing pups (and placentas) at parturition on the subsequent display of maternal behavior. Preliminary observations have revealed that maternal behavior is normal on the first 1 or 2 days postpartum. However, a gradual deterioration of maternal behavior is observed thereafter. Females display "abnormal" behaviors similar to those shown by mothers anesthetized during early lactation, namely, an increased latency to initiate nursing, several entrances into the nest box that do not led to nursing, startle–rejection reactions upon perceiving the litter, and, in some cases, a refusal to nurse. Are these data indicative that in rabbits, as in other mammals (Poindron *et al.*, 1988; Pryce, 1992; Rosenblatt and Siegel, 1981; Vince, 1993), the mother–young contact at parturition is crucial for the onset of a mechanism that will maintain maternal behavior postpartum?

In summary, research on the behavior of the mother rabbit has revealed that some hormones (namely, estradiol, progesterone, prolactin, and possibly oxytocin) known to regulate the maternal behavior of other mammals also operate in the rabbit. However, the ways in which these hormones direct the complex pattern of maternal behavior displayed by the rabbit have only begun to be unveiled. The interaction between mother and young, that allows the unfailing expression of maternal behavior after parturition in many mammals, has been restricted to a minimum in the rabbit. Nonetheless, this limited contact, initiated at parturition and continued across the earliest days of lactation, provides the means for transferring the control of maternal behavior from a mechanism operating throughout pregnancy to another one acting postpartum. Future studies should attempt to integrate the relative participation of hormonal factors, environmental influences, somatosensory input, and experiential elements in the final control of the expression of maternal behavior in rabbits.

## VI. Summary

The unique characteristics of maternal behavior in rabbits are described. From midgestation until parturition mother rabbits construct a maternal nest that will hold the young. Nest building begins with the digging of an underground burrow, continues with the collection of straw and construction of a straw nest inside the burrow, and culminates as the mother plucks hair from her ventrum and thighs and incorporates it into the straw nest. Under laboratory conditions, digging behavior appeared earliest (6–8 days prepartum), its decline preceding the onset of straw carrying (1–3 days prepartum). Hair pulling occurred from prepartum Day 1 to postpartum Day 4. Food intake decreased on prepartum Days 2 and 1. Mother rabbits nurse only once a day for around 3 min. Yet, during this brief time, pups suckle enough milk to sustain them for 24 h. The blind pups can find the mother's nipples guided by and olfactory cue (so-called nipple pheromone) emitted from her ventrum.

Steroids and prolactin regulated specific aspects of rabbit maternal behavior. Digging occurs in pregnant does under high levels of estradiol and progesterone, and is stimulated in ovariectomized females by the combined administration of estradiol benzoate and progesterone. Straw carrying is expressed as progesterone declines in pregnant does (on a background of estrogen) or after withdrawing progesterone in ovariectomized females treated with estradiol benzoate plus progesterone. Hair pulling begins and food intake decreases as progesterone declines while estradiol and prolactin rise shortly before parturition. Bromocriptine (prolactin release blocker) abolished straw carrying and hair pulling in ovariectomized, estradiol benzoate plus progesterone-treated does, but the addition of ovine prolactin did not restore these behaviors. The decline in food intake, normally following progesterone withdrawal, was counteracted by bromocriptine and reinstated even more sharply when prolactin was given. Though digging in ovariectomized does treated with estradiol benzoate plus progesterone was not modified by bromocriptine, the addition of prolactin abruptly reduced this behavior. These data suggest a complex role of prolactin in regulating specific aspects of rabbit maternal behavior. While prolactin apparently inhibits food intake, the possible stimulation of straw carrying and hair pulling should be confirmed by using higher doses of prolactin and by assessing the role of dopamine in these behaviors. Prolactin may regulate digging within an accelerator–brake system, represented by progesterone and prolactin, respectively.

Nipple pheromone emission is stimulated across gestation by estradiol and progesterone; high levels are maintained after partutition and across lactation, despite the drop in progesterone, by the action of prolactin on a background of estradiol.

Following parturition, the expression of maternal behavior relies heavily on the interaction with the litter. Anesthetizing mother rabbits during the daily nursing bout for 7 days (and thus preventing conscious contact with the litter) altered (but dit not abolish) maternal behavior on the days following anesthesia withdrawal. This effect was more intense when anesthesia was performed during early lactation (Days 1–7) than during midlactation (Days 11–17). Data suggest that the mother–litter interaction initiated at parturition and continued across early lactation allows the consolidation of maternal behavior postpartum.

### References

Anderson, C. O., Zarrow, M. X., Fuller G. B., and Denenberg, V. H. (1971). Pituitary involvement in maternal nest-building in the rabbit. *Horm. Behav.* **2,** 183–189.

Beyer, C., and Mena, F. (1965). Effect of ovariectomy and barbiturate administration on lactation in the cat and the rabbit. *Bol. Inst. Esud. Med. Biol. Univ. Nac. Auton. Mex.* **23,** 89–99.

Beyer, C., and Rivaud, N. (1969). Sexual behavior in pregnant and lactating domestic rabbits. *Physiol. Behav.* **4,** 753–757.

Bilkó, A., Altbäcker, V., and Hudson, R. (1994). Transmission of food preference in the rabbit: The means of information transfer. *Physiol. Behav.* **56,** 907–912.

Brambell, F. (1944). The reproduction of the wild rabbit *Oryctolagus cuniculus* (L.). *Proc. Zool. Soc. London* **114,** 1–45.

Bridges, R. S. (1990). Endocrine regulation of parental behavior in rodents. *In* "Mammalian Parenting: Biochemical, Neurobiological, and Behavioral Determinants" (N. A. Krasnegor and R. S. Bridges, eds), pp. 93–117. Oxford Univ. Press, New York.

Broekhuizen, S., and Maaskamp, F. (1980). Behaviour of does and leverets of the European hare (*Lepus europaeus*) whilst nursing. *J. Zool.* (*London*) 191, 487–501.

Browning, J. Y., Keyes, P. L., and Wolf, R. C. (1980). Comparison of serum progesterone, 20-alpha-dihydroprogesterone, and estradiol-17$\beta$ in pregnant and pseudopregnant rabbits: Evidence for post-implantation recognition of pregnancy. *Biol. Rep.* **23,** 1014–1019.

Caba, M., Beyer, C., González-Mariscal, G., Jiménez, M. A., and Silver, R. (1996). Oxytocin and vasopressin immunoreactivity in rabbit hypothalamus during estrus, late pregnancy, and parturition. *Brain Res.* in press.

Cassidenti, D. L., Paulson, R. J., Serafini, P., Sanczyk, F. Z., and Lobo, R. A. (1991). Effects of sex steroids on skin 5-alpha-reductase activity in vitro. *Obstet. Gynecol.* **78,** 103–107.

Challis, J. R. G., Davies, I. J., and Ryan, K. J. (1973). The concentration of progesterone, estrone and estradiol-17β in the plasma of pregnant rabbits. *Endocrinology* **93,** 971–976.

Cowie, A. T. (1969). Variations in the yield and composiiton of the milk during lactation in the rabbit and the galactopoietic effect of prolactin. *J. Endocrinol.* **44,** 437–450.

Cross, B. A. (1952). Nursing behavior and the milk ejection reflex in rabbits. *J. Endocrinol.* **8,** 13–14.

Denenberg, V. H., Sawin, P. B., Frommer, G. P., and Ross, S. (1958). Genetic, physiological and behavioral background of reproduction in the rabbit: IV. An analysis of maternal behavior at successive parturitions. *Behaviour* **13,** 131–142.

Denenberg, V. H., Petropolus, S. F., Sawin, P. B., and Ross, S. (1959). Genetic, physiological and behavioral background of reproduction in the rabbit: VI. Maternal behavior with reference to scattered and cannibalized newborn and mortality. *Behaviour* **15,** 71–76.

Denenberg, V. H., Huff, R. L., Ross, S., Sawin, P. B., and Zarrow, M. X. (1963). Maternal behaviour in the rabbit: The quantification of nest building. *Anim. Behav.* **11,** 494–499.

Denenberg, V. H., Zarrow, M. X., Kalberer, W. D., and Farooq, A. (1963). Maternal behaviour in the rabbit: Effects of environmental variation. *Nature (London)* **197,** 161–162.

Deutsch, J. A. (1957). Nest-building behaviour of domestic rabbits under semi–natural conditions. *Br. J. Anim. Behav.* **5,** 53–54.

Di Carlo, R., and Muccioli, G. (1981a). Presence of specific prolactin binding sites in the rabbit hypothalamus. *Life Sci.* **28,** 2299–2307.

Di Carlo, R., and Muccioli, G. (1981b). Changes in prolactin binding sites in the rabbit hypothalamus induced by physiological and pharmacological variations of prolactin serum levels. *Brain Res.* **230,** 445–450.

Drewett, R. F., Kendrick, K. M., Sanders, D. J., and Trew, A. M. (1982). A quantitative analysis of the feeding behavior of suckling rabbits. *Dev. Psychobiol.* **15,** 25–32.

Fahrbach, S. E., Morrell, J. I., and Pfaff, D. W. (1985). Possible role for endogenous oxytocin in oestrogen facilitated maternal behaviour in rats. *Neuroendocrinology* **40,** 526–532.

Farooq, A., Denenberg, V. H., Ross, S., Sawin, P. B., and Zarrow, M. X. (1963). Maternal behavior in the rabbit: Endocrine factors involved in hair loosening. *Am. J. Physiol.* **204,** 271.

Findlay, A. L. R., and Roth, L. L. (1970). Long-term dissociation of nursing behavior and the condition of the mammary gland in the rabbit. *J. Comp. Physiol. Psychol.* **72,** 341–344.

Foxcroft, G. R., and Hasnain, H. (1973). Effects of suckling and time to mating after parturition on reproduction in the domestic rabbit. *J. Reprod. Fertil.* **33,** 367–377.

Fuchs, A. R., and Dawood, M. Y. (1980). Oxytocin release and uterine activation during parturition in rabbits. *Endocrinology* **107,** 1117–1126.

Fuchs, A. R., Cubile, L., Dawood, M. Y., and Jorgensen, F. S. (1984). Release of oxytocin and prolactin by suckling rabbits throughout lactation. *Endocrinology* **114,** 462–469.

González-Mariscal, G., and Cuamatzi, E. (1995). Does hair-plucking for maternal nest-building promote maternal care in the rabbit? *Dev. Psychobiol.* **28,** 185(abstr. 26).

González-Mariscal, G., Chirino, R., and Hudson, R. (1994a). Prolactin stimulates emission of nipple pheromone in ovariectomized New Zealand white rabbits. *Biol. Rep.* **50,** 373–376.

González-Mariscal, G., Díaz-Sánchez, V., Melo, A. I., Beyer, C., and Rosenblatt, J. S. (1994b). Maternal behavior in New Zealand white rabbits: Quantification of somatic events, motor patterns, and steroid plasma levels. *Physiol. Behav.* **55,** 1081–1089.

Hudson, R., and Altbäcker, V. (1992). Development of feeding and food preference in the European rabbit: Environmental and maturational determinants. *In* "Ontogeny and Social Transmission of Food Preferences in Mammals: Basic and Applied Research" (B. G. Galef, M. Mainardi, and P. Valsecchi, eds.), pp. 125–145. Harwood Academic Publ., London.

Hudson, R., and Distel, H. (1982). The pattern of behaviour of rabbit pups in the nest. *Behaviour* **79**, 255–271.

Hudson, R., and Distel, H. (1983). Nipple location by newborn rabbits: Behavioural evidence for pheromonal guidance. *Behaviour* **85**, 260–275.

Hudson, R., and Distel, H. (1984). Nipple-search pheromone in rabbits: Dependence on season and reproductive state. *J. Comp. Physiol.*, Sect. [A] **155**, 13–17.

Hudson, R., and Distel, H. (1986). Pheromonal release of suckling in rabbits does not depend on the vomeronasal organ. *Physiol. Behav.* **37**, 123–129.

Hudson, R., González-Mariscal, G., and Beyer, C. (1990). Chin-marking behavior, sexual receptivity, and pheromone emission in steroid-treated, ovariectomized rabbits. *Horm. Behav.* **24**, 1–13.

Insel, T. R. (1992). Oxytocin: A neuropeptide for affiliation. Evidence from behavioral, receptor autoradiographic and comparative studies. *Psychoneuroendocrinology* **17**, 3–36.

Insel, T. R., and Shapiro, L. E. (1992). Oxytocin receptors and maternal behavior. *In* "Oxytocin in Maternal, Sexual, and Social Behaviors" (C. A. Pedersen, J. D. Caldwell, G. F. Jirikowski, and T. R. Insel, eds.), pp. 122–141. Ann. N.Y. Acad. Sci., New York.

Jilge, B. (1993). The ontogeny of circadian rhythms in the rabbit. *J. Biol. Rhythms* **8**, 247–260.

Jilge, B. (1995). Ontogeny of the rabbit's circadian rhythms without an external zeitgeber. *Physiol. Behav.* **58**, 131–140.

Keverne, E. B., and Kendrick, K. M. (1992). Oxytocin facilitation of maternal behavior in sheep. *In* "Oxytocin in Maternal, Sexual, and Social Behaviors" (C. A. Pedersen, J. D. Caldwell, G. F. Jirikowsky, and T. R. Insel, eds.), pp. 83–101. Ann. N.Y. Acad. Sci., New York.

Kristal, M. B. (1991). Enhancement of opioid-mediated analgesia: A solution to the enigma of placentophagia. *Neurosci. Biobehav. Rev.* **15**, 425–435.

Lincoln, D. W. (1974). Suckling: A time-constant in the nursing behavior of the rabbit. *Physiol. Behav.* **13**, 711–714.

Mangurian, L., Lewis, R., and Walsh, R. J. (1994). Placental lactogen binding sites in the pregnant rabbit choroid plexus. *J. Anat.* **184**, 425–428.

Martin, P., and Bateson, P. (1982). The lactation-blocking drug bromocriptine and its application to studies of weaning and behavioral development. *Dev. Psychobiol.* **15**, 139–157.

Mauvais-Jarvis, P. (1977). Androgen metabolism in human skin: Mechanisms of control. *In* "Androgens and Antiandrogens" (L. Martini and M. Motta, eds.), pp. 229–245. Raven Press, New York.

Mauvais-Jarvis, P., Kuttenn, F., and Baudot, B. (1974). Inhibition of testosterone conversion to dihydrotestosterone in men treated percutaneously by progesterone. *J. Clin. Endocrinol. Metab.* **38**, 142–147.

McNeilly, A. S., and Friessen, H. G. (1978). Prolactin during pregnancy and lactation in the rabbit. *Endocrinology* **102**, 1548–1554.

Meltzer, H. Y., Gudelsky, G. A., Simonovic, M., and Fang, V. S. (1982). Effect of dopamine agonists and antagonists on prolactin and growth hormone secretion. *Psychopharmacology* **28**, 200–218.

Mena, F., Martínez-Escalera, G., Aguayo, D., Clapp, C., and Grosvenor, E. (1982). Latency and duration of the effects of bromocriptine and prolactin on milk secretion in lactating rabbits. *J. Endocrinol.* **94**, 389–395.

Metcalf, B. W., Levy, M. A., and Holt, D. A. (1989). Inhibitors of steroid 5-alpha-reductase in benign prostatic hyperplasia, male pattern baldness and acne. *Trends Pharmacol. Sci.* **10**, 491–495.

Moore, B. J., Gerardo-Gettens, T., Horowitz, B. A., and Stern, J. S. (1986). Hyperprolactinemia stimulates food intake in the female rat. *Brain Res. Bull.* **17**, 563–569.

Mykytowycz, R. (1972). Aggressive and protective behavior of adult rabbits, *Oryctolagus cuniculus* (L.), towards juveniles. *Behaviour* **43**, 97–120.
Mykytowycz, R., and Rowley, I. (1958). Continuous observations of the activity of the wild rabbit, *Oryctolagus cuniculus* (L.), during 24-hr periods. *CSIRO Wildl. Res.* **3**, 26–31.
Neill, J. D. (1988). Prolactin secretion and its control. *In* "The Physiology of Reproduction," Vol. 2 (E. Knobil and J. D. Neill, eds.), pp. 1379–1390. Raven Press, New York.
Numan, M. (1994). Maternal behavior. *In* "The Physiology of Reproduction," Vol. 2 (E. Knobil and J. D. Neill, eds.), pp. 221–302. Raven Press, New York.
Orpen, G., and Fleming, A. S. (1987). Experience with pups sustains maternal responding in postpartum rats. *Physiol. Behav.* **40**, 47–54.
Ota, K. (1981). Food consumption by pregnant and lactating animals. *Jpn. J. Anim. Reprod.* **27**, 19–30.
Pedersen, C. A., Caldwell, J. D., Peterson, G., Walker, C. H., and Mason, G. A. (1992). Oxytocin activation of maternal behavior in the rat. *In* "Oxytocin in Maternal, Sexual, and Social Behaviors" (C. A. Pedersen, J. D. Caldwell, G. F. Jirikowsky, and T. R. Insel, eds.), pp. 58–69. Ann. N.Y. Acad. Sci., New York.
Penzhorn, E. J., and Meintjes, J. P. (1972). Influence of pregnancy and lactation on the voluntary feed intake of Afrikaner heifers and cows. *Agroanimalia* **4**, 83–92.
Poindron, P., Lévy, F., and Krehbiel, D. (1988). Genital, olfactory and endocrine interactions in the development of maternal behaviour in the parturient ewe. *Psychoneuroendocrinology* **13**, 99–125.
Pryce, C. R. (1992). A comparative systems model of the regulation of maternal motivation in mammals. *Anim. Behav.* **43**, 417–441.
Rasmusson, G. H. (1986). Biochemistry and pharmacology of 5-alpha-reductase inhibitors. *In* "Pharmacology and Clinical Uses of Inhibitors of Hormone Secretion and Action" (B. J. A. Furr and A. E. Wakeling, eds.), pp. 308–325. Baillère-Tindall, London.
Rongstad, O. J., and Tester, J. R. (1971). Behavior and maternal relations of young snowshoe hares. *J. Wildl. Manag.* **35**, 338–346.
Rosenblatt, J. S., and Siegel, H. I. (1981). Factors governing the onset and maintenance of maternal behavior among non-primate mammals: The role of hormonal and non-hormonal factors. *In* "Parental Care in Mammals" (D. J. Gubernick and P. H. Klopfer, eds.), pp. 13–76. Plenum Press, New York.
Ross, S., Denenberg, V. H., Sawin, P. B., and Meyer, P. (1956). Changes in nest building behaviour in multiparous rabbits. *Br. J. Anim. Behav.* **4**, 69–74.
Ross, S., Zarrow, M. X., Sawin, P. B., Denenberg, V. H., and Blumenfield, M. (1963). Maternal behaviour in the rabbit under semi-natural conditions. *Anim. Behav.* **11**, 283–285.
Sawin, P. B., and Crary, D. D. (1953). Genetic and physiological background of reproduction in the rabbit. 1. Some racial differences in the pattern of maternal behavior. *Behaviour* **6**, 128–146.
Sawin, P. G., Denenberg, V. H., Ross, S., Hafter, E., and Zarrow, M. X. (1960). Maternal behaviour in the rabbit: Hair loosening during gestation. *Am. J. Physiol.* **198**, 1099–1102.
Schimchowitsch, S., Moreau, S. C., Laurent, F., and Stoeckel, M. E. (1989). Distribution and morphometric characteristics of oxytocin- and vasopressin-immunoreactive neurons in the rabbit hypothalamus. *J. Comp. Neurol.* **285**, 304–324.
Schweikert, H. U., and Wilson, J. D. (1974). Regulation of human hair growth by steroid hormones. I. Testosterone metabolism in isolated hairs. *J. Clin. Endocrinol. Metab.* **38**, 811–819.

Shirley, B. (1984). The food intake of rats during pregnancy and lactation. *Lab. Anim. Sci.* **34,** 169–172.

Sofroniew, M. V. (1983). Morphology of vasopressin and oxytocin neurons and their central and vascular projections. *Prog. Brain Res.* **60,** 101–114.

Sorensen, M. F., Rogers, J. P., and Baskett, T. S. (1972). Parental behavior in swamp rabbits. *J. Mammal.* **53,** 840–849.

Taylor, J. C., and Peaker, M. (1975). Effects of bromocriptine on milk secretion in the rabbit. *J. Endocrinol.* **67,** 313–314.

Tindal, J. S., Beyer, C., and Mena, F. (1963). Milk ejection reflex and maintenance of lactation in the rabbit. *Endocrinology* **72,** 720–724.

Tucker, H. A. (1994). Lactation and its hormonal control. *In* "The Physiology of Reproduction," Vol. 2 (E. Knobil and J. D. Neill, eds.), pp. 1065–1098. Raven Press, New York.

Van Leengood, E., Kerker, E., and Swanson, H. H. (1987). Inhibition of post-partum maternal behavior in the rat by injecting an oxytocin antagonist into the cerebral ventricles. *J. Endocrinol.* **112,** 275–282.

Verga, M., Dell'Orto, V., and Carenzi, C. (1978). A general review and survey of maternal behaviour in the rabbit. *Appl. Anim. Ethol.* **4,** 235–252.

Verhoeven, G., Lamberigts, G., and DeMoor, P. (1974). Nucleus associated steroid 5-alpha-reductase activity and androgenic responsiveness. A study in various organs and brain regions of rats. *J. Steroid Biochem.* **5,** 93–100.

Vince, M. A. (1993). Newborn lambs and their dams: The interaction that leads to sucking. *Adv. Study Behav.* **22,** 239–268.

Walsh, R. J., Mangurian, L. P., and Posner, B. I. (1990). The distribution of lactogen receptors in the mammalian hypothalamus: An in vitro autoradiographic analysis of the rabbit and rat. *Brain Res.* **530,** 1–11.

Weiss, V. (1991). Feeding of mink during pregnancy. *Dan. Pelsdyravl* **54,** 41–42.

Whitworth, M. R. (1984). Maternal care and behavioural development in pikas *Ochotona princeps. Anim. Behav.* **32,** 743–752.

Yaschine, T., and Beyer, C. (1969). Ingestión acuosa y alimenticia en la coneja durante las diversas fases del ciclo reproductor. *Acta Physiol. Latinoam.* **16,** 88–89.

Zarrow, M. X., Sawin, P. B., Ross, S., Denenberg, V. H., Crary, D., Wilson, E. D., and Farooq, A. (1961). Maternal behaviour in the rabbit: Evidence for an endocrine basis of maternal nest building and additional data on maternal nest building in the dutch-belted race. *J. Reprod. Fertil.* **2,** 152–162.

Zarrow, M. X., Farooq, A., and Denenberg, V. H. (1962). Maternal behavior in the rabbit: Critical period for nest-building following castration during pregnancy. *Proc. Soc. Exp. Biol. Med.* **111,** 537–538.

Zarrow, M. X., Farooq, A., Denenberg, V. H., Sawin, P. B., and Ross, S. (1963). Maternal behaviour in the rabbit: Endocrine control of maternal nest-building. *J. Reprod. Fertil.* **6,** 375–383.

Zarrow, M. X., Denenberg, V. H., and Anderson, C. (1965a). Rabbit: Frequency of suckling in the pup. *Science (Washington, D. C.)* **150,** 1835–1836.

Zarrow, M. X., Denenberg, V. H., and Kalberer, W. D. (1965b). Strain differences in the endocrine basis of maternal nest-building in the rabbit. *J. Reprod. Fertil.* **10,** 397–401.

Zarrow, M. X., Gandelman, R., and Denenberg, V. H. (1971). Prolactin: Is it an essential hormone for maternal behavior in the mammal? *Horm. Behav.* **2,** 343–354.

# Parental Behavior in Voles

ZUOXIN WANG AND THOMAS R. INSEL

DEPARTMENT OF PSYCHIATRY AND BEHAVIORAL SCIENCES
EMORY UNIVERSITY
ATLANTA, GEORGIA 30322

## I. INTRODUCTION

Voles belong to the genus *Microtus* within the family Arvicolidae or subfamily Arvicolinae (Anderson, 1985). The microtine rodents are an ideal group of rodents for comparative studies because many of these species, in spite of their close taxonomic relationship, show profound differences in reproductive biology and social organization. For the purposes of this volume, voles are informative not only for the insights they have provided on male and female parental care but also for studies on the neurobiologic basis of parental behaviors.

In this article, we will summarize the available literature on parental behavior in voles. An attempt is also made to examine the interactions of the environment with the behavior, as well as to explore the possible neural mechanisms underlying parental care in voles. We will focus on monogamous prairie (*Microtus ochrogaster*) and pine (*M. pinetorum*) voles and promiscuous meadow (*M. pennsylvanicus*) and montane (*M. montanus*) voles because most of the work has been done on these species.

## II. PARENTAL BEHAVIOR AND SOCIAL ORGANIZATION

Rodents show substantial interspecific variation in their social organization. They may be solitary, monogamous, polygamous, or promiscuous, as reflected in their different mating strategies and patterns of social behaviors (Elwood, 1983). Parental care shows a curious relationship to social organization. In general, although not always, biparental care is associated with monogamy, while sole maternal care is the most frequent pattern associated with promiscuity (Kleiman, 1977; Dewsbury, 1987). This pattern of association between parental care and social organization is evident in microtine rodents whether one relies on field– or laboratory–based data.

In the field studies using live trapping and radiotelemetry, meadow and montane voles appear promiscuous in that they do not exhibit nest sharing or pair bonding during the breeding season (Getz, 1972; Jannett, 1978, 1980, 1982; Madison, 1978, 1980; Webster and Brooks, 1981). The basic social unit is one of mother–young, and the mother is presumably the only provider of parental care. In contrast, prairie and pine voles appear to be monogamous in that females and males share a nest and home range throughout the year (Carter and Getz, 1993; FitzGerald and Madison, 1983; Getz and Hofmann, 1986). In prairie voles, if the resident breeding male or female is removed, the remaining adult does not accept a new mate 80% of the time (Getz et al., 1993). In these monogamous voles, the basic social unit consists of an extended family of parents and offspring, and both mothers and fathers may share the responsibilities of rearing young.

These field observations are supported by behavioral studies in the laboratory (Fig. 1). For example, after mating, male and female prairie voles show a partner preference for the mate in a social choice test, as one would predict for a monogamous species (Dewsbury, 1987; Shapiro and Dewsbury, 1990; Williams et al., 1992; Winslow et al., 1993). In the postpartum period, males and females of monogamous prairie and pine voles share the nest, and both parents exhibit parental behavior towards their offspring (Gruder-Adams and Getz, 1985; McGuire and Novak, 1984; Oliveras and Novak, 1986; Wilson, 1982). In contrast, promiscuous montane voles do not show a partner preference after mating; males mate randomly either with the

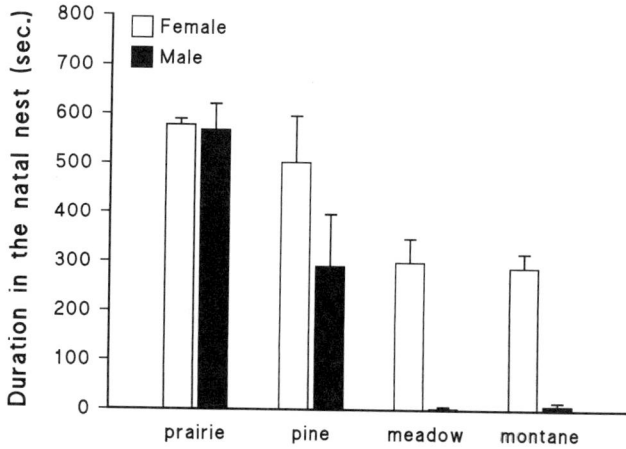

FIG. 1. Species differences in the mean duration that voles spend in the natal nest over 15-min observation period. Bars = mean ± SEM. (Adapted from McGuire and Novak, 1984; 1986.)

familiar partner or with strangers (Dewsbury, 1987; Shapiro and Dewsbury, 1990). After birth of young, male and female montane and meadow voles do not share the nest, and females provide all of the parental care (Gruder-Adams and Getz, 1985; McGuire and Novak, 1984, 1986; Oliveras and Novak, 1986; Wilson, 1982). Table I summarizes microtine species that have been studied for their social organization and parental behavior.

### III. Patterns of Parental Behavior

#### A. Maternal Behavior

Parental behavior in rodents can be divided into indirect and direct forms (Kleiman, 1977). Indirect parental behavior includes nest building and maintenance as well as protection from predators or conspecifics, while direct parental behavior involves responsiveness to the young. From the day of parturition onward, female voles (both monogamous and promiscuous species) spend significant amounts of time in the nest caring for the young until weaning. Females exhibit all the patterns of direct parental behavior that are observed in other rodent species, such as nursing, huddling over, grooming, retrieving, and side-by-side contact with the litter. The time that mothers spend in direct parental care decreases across days of litter development (McGuire and Novak, 1984; Wilson, 1982). As suggested previously, monogamous prairie and pine vole mothers spend significantly more time exhibiting direct parental care throughout the lactation period than promiscuous meadow voles (McGuire and Novak, 1984).

In addition to direct parental behavior, female voles, irrespective of species, display rodent-typical patterns of indirect parental behavior, such as food hoarding, nest building, and runway construction. These behaviors, however, neither change during the development of young nor differ between species that show differences in their social organization (McGuire and Novak, 1984).

#### B. Paternal Behavior

Paternal behavior has been studied in several species of rodents under the laboratory condition. From a theoretical perspective, the association between monogamy and paternal care has been attributed to the increased reproductive advantage for a male with a single mate to ensure that "his" offspring survive. Consistent with this notion, male prairie voles share the nest with their mates and exhibit all the patterns of parental behavior except nursing throughout the entire gestation period. These males huddle over, contact, groom, and retrieve young. They also contribute to the nest–

TABLE I
Social Organization and Parental Behavior in Microtine Rodents

| Rodent | Scientific name | Social organization | Parental care | References |
|---|---|---|---|---|
| Prairie vole | *M. ochrogaster* | Monogamous–extended family | Biparental | Getz and Carter, 1980; Getz *et al.*, 1981; Gruder-Adams and Getz, 1985; McGuire and Novak, 1984; Oliveras and Novak, 1986; Thomas and Birney, 1979 |
| Pine vole | *M. pinetorum* | Monogamous–extended family | Biparental | FitzGerald and Madison, 1983; McGuire and Novak, 1984; Oliveras and Novak, 1986 |
| Meadow vole | *M. pennsylvanicus* | Promiscuous | Maternal | Gruder-Adams and Getz, 1985; Madison, 1980; McGuire and Novak, 1984; Oliveras and Novak, 1986 (see also Wilson, 1982; Storey and Snow, 1987) |
| Montane vole | *M. montanus* | Promiscuous | Maternal | Jannett, 1980, 1982; McGuire and Novak, 1986 (see also Hartung and Dewsbury, 1979) |
| California vole | *M. californicus* | Monogamous–polygynous | Biparental | Hartung and Dewsbury, 1979; Hatfield, 1935; Lidicker, 1980; Ostfeld, 1986 |
| Taiga vole | *M. xanthognathus* | Polygynous | Maternal | Wolff, 1980; Wolff and Lidicker, 1981 |
| Water vole | *M. richardsoni* | Promiscuous | Maternal | Ludwig, 1981 |
| Townsend vole | *M. townsendii* | Polygynous–monogamous | ? | Lambin and Krebs, 1991 |
| Tundra vole | *M. oeconomus* | Polygynous | ? | Lambin *et al.*, 1992 |
| Field vole | *M. agrestis* | Polygynous | Maternal | Myllymaki, 1977 |

runway building and food hoarding (Gruder-Adams and Getz, 1985; Oliveras and Novak, 1986; Thomas and Birney, 1979; Wilson, 1982). After the birth of younger offspring, male parents continue to care for the juvenile offspring and play an important role in their behavioral development (Wang and Novak, 1992, 1994a). Although monogamous male and female parents exhibit similar patterns of parental behavior, males huddle over and groom the litter less frequently than their female partners, whereas no differences are found in their indirect parental behavior (Solomon, 1993).

In contrast to monogamous species, promiscuous meadows and montane vole males neither share the nest with the female nor display paternal behavior towards their young (Fig. 1; Gruder-Adams and Getz, 1985; McGuire and Novak, 1986; Oliveras and Novak, 1986; Wang and Novak, 1992). In fact, meadow vole males usually withdraw from pups they encounter in the runway (Oliveras and Novak, 1986).

C. ALLOPARENTAL BEHAVIOR

In monogamous species such as prairie voles, most surviving juveniles remain in the natal nest beyond weaning and contribute to the care of subsequent litters—a behavior that has been called *alloparenting* (Getz *et al.*, 1987; Gruder-Adams and Getz, 1985; Solomon, 1991; Thomas and Birney, 1979; Wang and Novak, 1992, 1994a). Young male and female prairie voles remaining within the family group are usually reproductively suppressed (Batzli *et al.*, 1977; Getz *et al.*, 1983; Richmond and Conaway, 1969). Juvenile's pup caring behavior in prairie voles qualitatively resembles paternal care, with grooming, huddling over, contacting, and retrieving younger siblings (Solomon, 1991; Wang and Novak, 1992, 1994a). Quantitatively, the time that juveniles invest in alloparental care is similar to the time males devote to paternal care (Wang and Novak, 1992, 1994a). Remarkably, the time that a given juvenile spends in the natal nest exhibiting alloparental behavior tends to be correlated with the time the juvenile's father spent in paternal care (Wang and Novak, 1994a). These similarities in the behavioral patterns and time investment suggest that juvenile alloparental behavior in monogamous prairie voles resembles paternal behavior. Juveniles may reduce the workload of parents and contribute to the development of younger siblings at the expense of delaying or inhibiting their own reproduction. Juvenile helpers may also derive benefits from this strategy, and both indirect and direct fitness may increase by rearing siblings (Solomon, 1991; Wang, 1991). Continued association with the family group may provide prolonged parental protection and aid in the acquisition of experience in rearing young (Kleiman, 1977), which may make them more effective breeders when they start to produce their own litter (Emlen, 1984; Leonard *et al.*, 1989). Indeed, prairie voles that had juvenile experience with younger siblings display higher levels of parental behavior in their first reproductive effort, and their litters are left alone in the nest less often (see as follows). These litters develop more rapidly, and have a larger size at weaning than those of the prairie voles without juvenile experience with younger siblings (Wang, 1991).

### D. Onset of Parental Behavior

In addition to the juvenile prairie vole's display of alloparental behavior, more than 50% of sexually naive adult male prairie voles also show parental responsiveness when they are exposed to a 3- to 5-day-old conspecific pup (Bamshad *et al.*, 1994). After mating and cohabitation with a female for 3 days, male prairie voles show an increased level of paternal responsiveness, which reaches a plateau around Day 13 of the female's gestation and remains at the same high level after birth of the young (Bamshad *et al.*, 1994). The parental responsiveness of sexually naive female prairie voles is still not clear. Males of promiscuous meadow voles following 3 days of mating and cohabitation with a female ignore or show little parental responsiveness towards a conspecific pup. Female meadow voles following 3 days with a male either attack pups or show little parental responsiveness (see as follows).

## IV. Species Differences in the Rate of Litter Development

The rates of litter development differ between vole species that show differences in their social organization and parental care. Generally, promiscuous litters develop more rapidly, eating solid food earlier than monogamous litters (McGuire and Novak, 1984, 1986). The last days of nipple attachment for meadow and montane voles are 13 and 14 days, while for prairie and pine voles suckling persists until 20 and 21 days, respectively (McGuire and Novak, 1984, 1986). In a study describing species differences in the patterns of nipple attachment, Salo *et al.* (1994) tested the strength of the offspring to remain attached to the nipples of their mothers. The females were led via a nylon leader from one end of the cage to the other while the distance that each of the offspring became detached from the female's nipples was recorded. Monogamous prairie and pine vole pups remained attached to the mother's nipple for the entire length of the cage during the test, while promiscuous montane and meadow vole pups remained attached for significantly shorter distances than the former two species (Fig. 2). In a study of temporal development of the brain, monogamous voles that display biparental care showed a slow rate of overall brain growth and cerebral growth, and a later peak of enzyme activity than promiscuous voles that display sole maternal care (Gutierrez *et al.*, 1989). Patterns of incisor growth (incisor-length–body-weight ratio) during development are also different among species. Prairie and pine

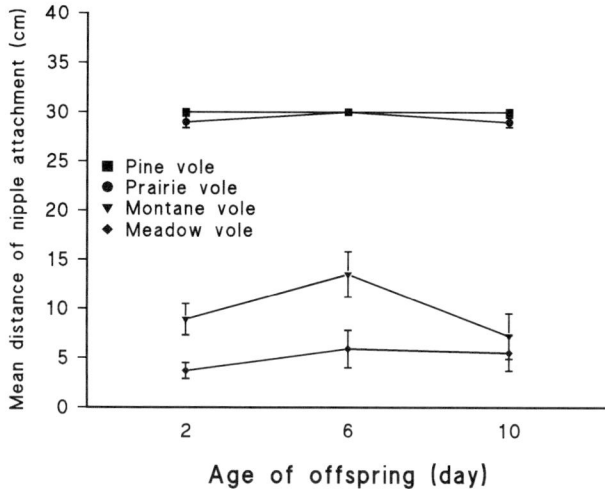

FIG. 2. Distances that offspring traveled before detaching from the nipples of females as a function of species and age of offspring. Points indicate mean ± SEM. (Adapted from Salo et al., 1994.)

vole pups have higher incisor ratios on Days 2, 4, and 6 than meadow and montane vole pups. The latter two species, however, had accelerated incisor growth as Day 10 and weaning approached (Salo et al., 1994). Meadow and montane vole pups are weaned at approximately 2 weeks of age when their mothers abandon the nest and construct a new nest, while prairie and pine vole pups remain in the nest with both parents and younger siblings (McGuire and Novak, 1984, 1986; Wang and Novak, 1992).

Species differ not only in the rate of growth but in the size of their litters, placing different behavioral and metabolic demands on their parents. The average litter sizes in the field are 6.0 and 4.9 for montane and meadow voles, and 2.1 and 3.5 for pine and prairie voles, respectively (Nadeau, 1985). In the laboratory, the litter sizes are 6.1 and 5.1 for montane and meadow voles, and 2.3 and 3.6 for pine and prairie voles, respectively (Dewsbury, 1981).

Such variations in behavioral and physical development of litters among microtine rodents are consistent with the patterns of life-history strategies originally described by MacArthur and Wilson (1967). Promiscuous species such as meadow voles usually live in unstable habitats, show extreme population fluctuations, and breed rapidly. Therefore, selection in this species favors large numbers of offspring with relatively low investment and rapid rates of growth, a pattern originally termed *r selection.* In contrast, monogamous species such as prairie voles usually live in a stable but limited resource habitats. They are long lived, mature slowly, and have extensive parental care, a pattern previously called *K selection*. The pattern is not perfect however, as both species undergo population fluctuations and habitat stability varies regionally within each species.

Is the species-typical pattern of parental care regulating infant development or does the rate of infant development dictate the pattern of parental care? At lease some evidence supports the influence of parental care on litter development. In monogamous prairie voles, litters are active outside of the nest and eat solid food significantly earlier when both parents provide care than when only mothers provide care in the absence of fathers (Wang and Novak, 1992). In the condition involving both parents, fathers are almost always in the natal nest when mothers are away. Because the presence of fathers significantly reduces the time that the litter is left alone in the nest (Wang and Novak, 1992, 1994a), the litters could use conserved energy for rapid growth and development. Nest attendance by fathers, which reduces exposure of offspring to the ambient temperature, has also been correlated with enhanced rates of litter development in other monogamous rodent species, such as Mongolian gerbils (*Meriones unguiculatus*, Elwood and Broom, 1978; Gerling and Yahr, 1979; Ostermeyer and Elwood, 1984), California mice (*Peromyscus californicus;* Dudley, 1974), and southern grasshopper mice (*Onychomys torridus;* Schultz, 1979).

Thermal regulation, however, is not the only possible explanation for rapid litter development in the biparental condition. Olfactory stimuli from the fathers could also play some role inasmuch as olfactory stimulation can facilitate or retard development and growth generally (Drickamer, 1986). That thermal regulation may not be the best explanation is suggested by the fact that prairie vole litters reared by mothers with juveniles showed slowed behavioral development. They were observed to eat solid food and to move out of the nest significantly later than the litters reared by both parents, although both litters received similar amounts of physical contact (Wang and Novak, 1992, 1994a). Thus, olfactory stimulation from the monogamous fathers may also have strong effects on growth and development of litters other than keeping them warm. This speculation, however, needs to be further studied.

In contrast to monogamous voles, promiscuous fathers in a seminatural environment usually do not exhibit paternal behavior towards their young, and thus do not influence litter growth and development directly. However, the presence of fathers may alter maternal behavior that, in turn, influences litter development. In the presence of fathers, mothers spend more time in the nest, show higher levels of maternal behavior, and paradoxically produce litters that develop more slowly. Higher maternal scores under this condition are associated with mother's active defense of the nest against male's entering (Wang and Novak, 1992). Therefore, delayed maturation of litters may have been the result of several factors, including but not limiting to maternal stress and alterations in the length of nursing bouts. In another study, however, meadow vole litters gained more weight when both parents were present than when only mothers were present (Storey and Snow, 1987).

## V. Environmental Factors Regulating Parental Behavior

The presence of absence of one parent can influence the behavior of the remaining parent. In monogamous grasshopper mice, removal of the male decreases maternal care (Duvall *et al.,* 1982). In California mice, presence of mothers maintain paternal responsiveness, but fathers have no such influence on maternal care (Gubernick and Alberts, 1989). The effects of the social environment on parental behavior have been studied in monogamous prairie voles. No significant differences are found in the duration of mother–young contact between the father-present and father-absent groups (Wilson, 1982). However, if juveniles are present, both mothers and fathers spend less time in the nest and more time foraging (Getz and Carter, 1980; Wang and Novak, 1992). Curiously, prairie vole fathers appear to influence juvenile's alloparental behavior inasmuch as juveniles spend more time in the nest displaying alloparental behavior when housed with both parents than when housed with mothers only. The father's effect on behavior appears stronger in male than in female juveniles (Wang and Novak, 1994a).

Parental behavior is also influenced by the early postnatal social environment. Meadow voles neonatally cross-fostered to and reared by prairie vole parents receive more parental contact than meadow voles in-fostered to and reared by meadow vole parents. In their first reproductive effort, cross-fostered meadow vole pairs showed nest cohabitation after birth of the young, and males frequently huddled over and groomed pups. In contrast, in-fostered meadow vole pairs nest separately and males rarely enter the natal nest. Furthermore, cross-fostered females also show more huddling over, nursing, and pup grooming than in-fostered females (Fig. 3; McGuire, 1988). These data suggest that the early environment influences development of subsequent parental care in voles. This hypothesis is, indeed, supported by another study. Prairie voles that had experience and "practices" alloparental behavior with younger siblings as juveniles exhibited higher levels of parental behavior in their first reproduction than the prairie voles without such juvenile experience with younger siblings (Fig. 4; Wang, 1991). Such alterations in parental behavior could be explained by a "learning hypothesis" (Hrdy, 1976), that is, rearing siblings provides opportunities for juveniles to practice and improve their parental skills.

Of course, parental behavior can also be influenced by earlier experience rearing one's own offspring. Litters reared by multiparous parents are left

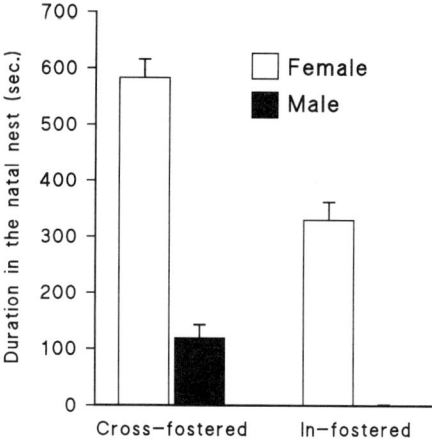

FIG. 3. Mean duration in the natal nest in meadow vole parents that were either neonatally cross-fostered to and reared by prairie vole parents, or in fostered to and reared by meadow vole parents. Bars = mean ± SEM. (Adapted from McGuire, 1988.)

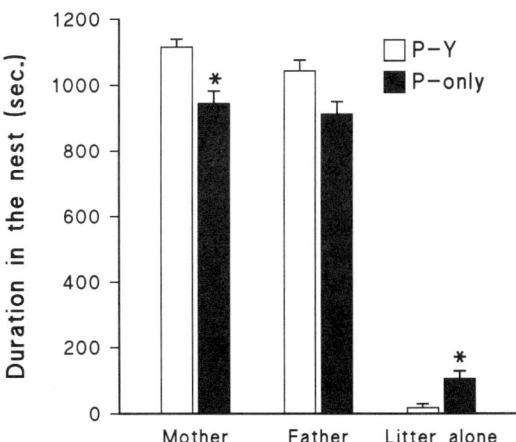

FIG. 4. Behavioral effects of exposure to younger siblings as juveniles. Prairie voles observed over 20-min interval with their first litters (in home cage) had been exposed to one of two rearing conditions: P-Y (exposure to parents and younger siblings as juveniles) and P-only (exposure to parents without younger siblings as juveniles). Litter alone represents time that litters were not in contact with either parent. Bars = mean ± SEM;* $p < .05$. (Adapted from Wang, 1991.)

alone in the nest less often, develop more rapidly, have a higher survival rate, and are larger at weaning than litters reared by primiparous parents (Wang and Novak, 1994b). These data suggest that just as juveniles acquire or improve their parental skills by interacting with younger siblings, so might parents enhance their parental skills by repeatedly rearing their own offspring. That parity affects subsequent parental behavior and litter development has also been reported in other rodent species (Carlier and Noirot, 1965; Myers and Master, 1983; Swanson and Campbell, 1979).

Parental behavior could also be influenced by the nonsocial environment. In the field, for example, meadow vole males usually do not share the nest with the female during the breeding season, and do not display paternal behavior in the seminatural environment (Madison, 1978, 1980; Oliveras and Novak, 1986). In the laboratory condition, however, meadow vole parents are found to share the nest, although males exhibit paternal behavior at a level which is significantly lower than that of the monogamous prairie vole (Hartung and Dewsbury, 1979; Wilson, 1982). The limited space and lack of cover may alter male–female tolerance, so that the male can enter the maternal nest and socialize with the young (Hartung and Dewsbury, 1979; Wilson, 1982).

## VI. Neurobiological Basis of Parental Behavior

Parental behavior involves changes in a great number of functions, including but not limited to sensory processing, memory, motivation, and hormonal responsiveness. Although it is unlikely that a single brain area or a single neurotransmitter will subserve such a diverse array of functions, there have been some intriguing observations that have emerged from a classical comparative approach, correlating some features of brain anatomy or brain chemistry with the presence or absence of parental behavior. Some of the most instructive comparisons have involved the closely related microtine species that exhibit marked differences in parental behavior.

### A. Neuroanatomic Basis of Vole Parental Behavior

One approach to defining brain areas involved in parental care has utilized c-fos expression to map neuronal activation. c-fos is the protein product of the "immediate early gene" *fos* that is rapidly expressed in neurons during activation. By staining for the c-fos protein immunocytochemically, one can obtain important information concerning regional changes in neuronal activity (Hoffman *et al.*, 1993). After 3 h of exposure to a conspecific pup, male and female prairie voles show an increase in c-fos staining in the accessory olfactory bulb, lateral septum, the medial preoptic area

(MPA), and the bed nucleus of the stria terminalis (BST), suggesting an involvement of these brain area in parental behavior in voles (Kirkpatrick *et al.,* 1994b). A separate study compared parental responsiveness and associated c-fos expression in monogamous prairie voles and promiscuous meadow voles (Z. X. Wang, L. Zhou, T. R. Insel, and G. J. DeVries, unpublished data). During a 10-min exposure to a conspecific pup, prairie voles showed a higher level of parental behavior in terms of grooming, huddling over, and contacting conspecific pups than meadow voles (Fig. 5a). Overall, prairie voles had a higher level of c-fos expression in the MPA, the BST, the medial amygdaloid nucleus (MA), and the lateral septum than meadow voles. As with the earlier study in prairie voles, pup exposure significantly induced c-fos expression in the MPA, the BST, and the MA; and a similar trend was also found in the lateral septum, although the difference in this region did not reach statistical significance (Fig. 5b). The induction of c-fos expression was equivalent in both male and female prairie voles. Most important, pup exposure did not induce any significant changes in c-fos staining in any of these four brain areas in meadow voles (Fig. 5c). These data suggest that (1) these brain areas are activated by pup exposure only in the species that responds with biparental behavior, and (2) these regional differences in neural activity may subserve pup responsiveness in both male and female prairie voles.

The MPA, the BST, and the MA form a circuit that has been previously implicated in processing socially relevant information. The MPA has, in particular, been shown to be critical in the mediation of rat maternal behavior (Numan, 1990). This region includes a sexually dimorphic nucleus in the rat. It may be significant therefore, that prairie voles, which are biparental, do not exhibit sexual dimorphism in this or a second hypothalamic nucleus as measured by volume and cell density (Shapiro *et al.,* 1991). Montane voles show the sexual dimorphisms that have been previously reported in rats (Shapiro *et al.,* 1991). Another study in prairie voles has implicated the same region of the amygdala that shows the c-fos activation in paternal behavior. Male prairie voles that receive lesions of the rostral aspects of the medial nucleus of the amygdala show significant deficits in paternal responsiveness, although a range of other social and nonsocial behaviors are unaffected (Kirkpatrick *et al.,* 1994a). Taken together, these studies suggest that the MA–BST–MPA circuit and possibly the lateral septum are important for processing pup stimuli and, by extension, parental care. We still know very little about how any of these regions function and the extent to which sensory processing or motivation or motor outputs are guided by each of these regions.

B. NEUROCHEMICAL BASIS OF VOLE PARENTAL BEHAVIOR

Two neuropeptides, vasopressin (AVP) and oxytocin (OT), have been implicated in vole parental behavior. These are closely related peptides,

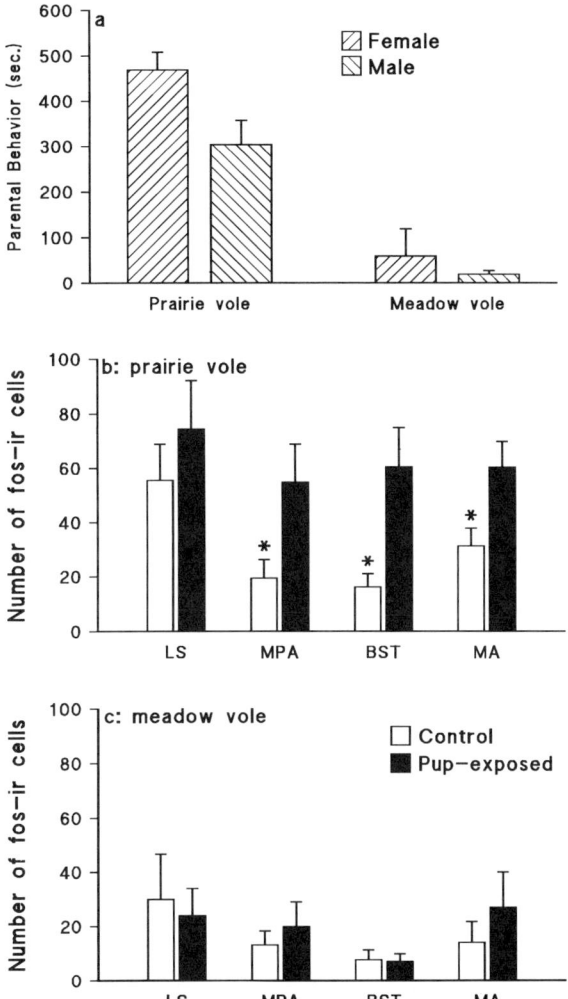

FIG. 5. Effects of 10-min pup exposure on (a) parental behavior, and c-fos immunoreactive staining in brains of (b) prairie voles and (c) meadow voles. Regions showing increased c-fos staining in the prairie vole but not the meadow vole include medial preoptic area (MPA), the bed nucleus of the stria terminalis (BST), and medial amygdaloid nucleus (MA). In the lateral septum (LS), there was a species difference in c-fos staining, but the increase in prairie voles failed to reach statistical significance after pup exposure. Bars = mean ± SEM;* $p < .05$.

each is composed of nine amino acids and synthesized in the same parts of the hypothalamus. These particular peptide hormones are found exclusively in mammals, although other nine amino acid peptides are found in other vertebrates and appear to influence reproductive behaviors. Traditionally AVP and OT have been called "neurohypophyseal hormones" because though synthesized in the hypothalamus, they are released from the posterior pituitary (neurohypophysis). In the past two decades, we have learned that these hormones are also released within the brain where they function as neurotransmitters influencing cognition and behavior. It is intriguing to consider that the neurohypophyseal roles of these hormones, which include labor and milk ejection (OT), may have a matching role in the brain related to the central organization of parentl behavior.

*1. Vasopressin*

There are profound differences in central AVP receptor binding, immunoreactive staining (AVP-ir), AVP mRNA expression, and changes in central AVP activity during reproduction between vole species that show different patterns of social organization. For instance, an autoradiographic study indicates that monogamous and promiscuous voles show virtually a nonoverlapping pattern of AVP receptor distribution in the brain (Insel *et al.*, 1994). Prairie voles have a higher density of AVP receptors labeled by $^{125}$I-sarc-AVP in the granule cell layer of olfactory bulb, diagonal band, the BST, amygdala, and thalamus than montane voles. On the other hand, montane voles have a higher density of AVP receptors in the lateral septum, lateral habenula, and central gray than prairie voles (Fig. 6). A similar discrepancy in AVP receptor distribution has been found in monogamous pine voles and promiscuous meadow voles (Insel *et al.*, 1994), but these four vole species do not differ in the distribution of opiate or benzodiazepine receptors (Insel and Shapiro, 1992).

---

FIG. 6. Vasopressin receptors in two monogamous and nonmonogamous voles. Brightfield images of $^{125}$I-sarc-AVP binding to 20-μm coronal sections through forebrain and midbrain of prairie vole (A, C, E, G, and I) and montane vole (B, D, F, H, and J). The plates represent anatomically matched sections comparing vasopressin receptors in the two species. Receptor-rich areas appear as dark regions on the sections. Regions identified by counterstains include the following: ac, anterior commissure; BNST, bed nucleus of the stria terminalis; Ce, central nucleus of amygdala; Cg, cingulate cortex; CG, central gray; CM, centromedial thalamus; DB, diagonal band; DG, dentate gyrus; H, lateral habenula nucleus; LD, laterodorsal thalamus; LP, lateroposterior thalamus; LS, lateral septum; MG, medial geniculate; Po, posterior thalamus; PV, paraventricular thalamus; SC, superior colliculus; VS, ventral subiculum; VT, ventral posterolateral thalamus; VTA, ventral tegmental area. Bar = 1.0 mm. (Adapted from Insel *et al.*, 1994.)

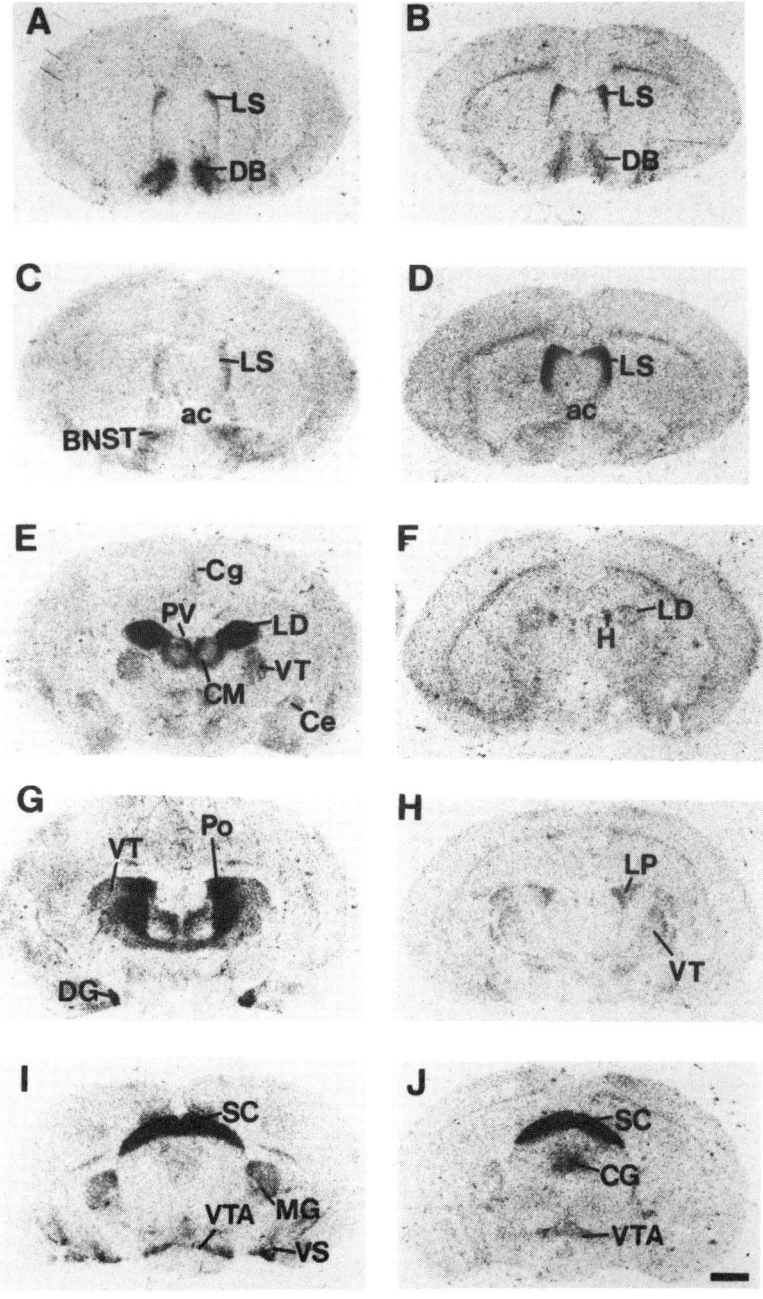

In both monogamous and nonmonogamous voles, males have more AVP-ir or AVP mRNA labeled cells in the BST and the MA, and a higher density of AVP-ir fibers in the lateral septum and lateral habenular nucleus than females (Bamshad et al., 1993, 1994; Wang et al., 1994b)—a gender difference found in other species of rodents (Bittman et al., 1991; Crenshaw et al., 1992; Hermes et al., 1990; Van Leeuwen et al., 1985). A species difference also exists, in which male prairie voles have a higher density of AVP-ir fibers in the lateral septum than male meadow voles, whereas no differences are found between the females (Wang, 1995). After 3 days of cohabitation with a female or after becoming parental, male prairie voles have a reduced density of AVP-ir fibers in the lateral septum (Fig. 7a) and a higher level of AVP mRNA expression in the BST (Fig. 7b) relative to their sexually naive counterparts (Bamshad et al., 1993, 1994; Wang et al., 1994b). These data suggest increased septal AVP release because the lateral septum receives AVP-ir projections from AVP-producing cells in the BST (Bamshad et al., 1993; DeVries and Buijs, 1983). Such changes in AVP activity associated with reproduction are not found in male meadow voles nor are they apparent in females of either species (Bamshad et al., 1993, 1994; Wang et al., 1994b).

These data suggest that AVP activity during reproduction differs between species that show different patterns of social organization, but they do not address the role of AVP in parental care. More direct evidence that AVP may be specifically critical for paternal behavior in prairie voles comes from studies that begin with the observation that mating–cohabitation with a female induces an increase in selective affiliation and paternal behavior in male prairie voles (Bamshad et al., 1994; Winslow et al., 1993). Infusion

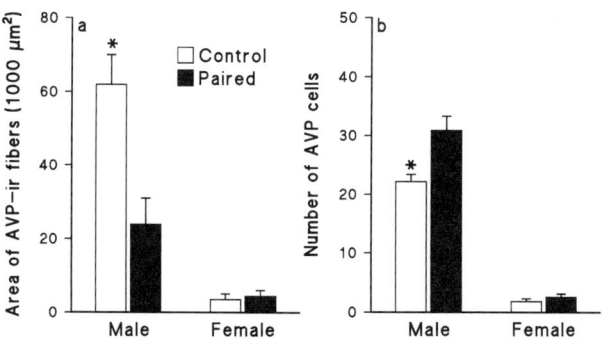

FIG. 7. Effects of 3 days mating and cohabitation between a male and a female on (a) the denisty of AVP-ir fibers in the lateral septum and (b) the number of AVP mRNA labeled cells in the bed nucleus of the stria terminalis in male and female prairie voles. Bars = mean ± SEM;* $p < .05$. (Adapted from Bamshad et al., 1994 and Wang et al., 1994b.)

of AVP into the lateral ventricle in sexually naive male prairie voles enhances their affiliative behavior, whereas infusion of the AVP antagonist, $d(CH_2)_5[Tyr(Me)]AVP$, reduces mating-induced partner preferences and mate guarding (Winslow et al., 1993). Injecting AVP into the lateral septum (the region with increased AVP-ir and c-fos expression in prairie vole males) increases male parental responsiveness, while injecting an AVP antagonist reduces this behavior (Fig. 8, Wang et al., 1994a). Cental AVP pathways have been previously implicated in parental behavior in other rodents. Long-Evans rats, for example, display superior parental behavior in comparison to the AVP-deficient mutant Brattleboro rats (Wideman and Murphy, 1990). Injections of AVP into the lateral ventricle induces persistent parental behavior in female rats (Pedersen et al., 1982). These studies suggest that AVP is involved in the process by which mating changes subsequent parental and other social behaviors in monogamous prairie voles. However, it is still not clear whether central AVP plays a role in maintenance of parental behavior in male prairie voles, or in the initiation and maintenance of parental behavior in female prairie voles. In addition, whether central AVP is involved in regulating social behaviors in promiscuous voles is still unknown.

## 2. Oxytocin

Several studies in rats have implicated OT in maternal behavior (Insel, 1990). In the rat, maternal behavior is influenced by an increase in estrogen

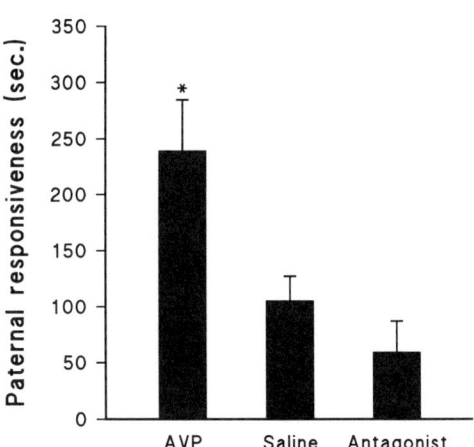

FIG. 8. Differences in paternal responsiveness in male prairie voles that received septal injections of AVP (0.1 ng), saline, or the AVP $V_{1a}$ antagonist (1 ng), respectively. Bars = mean ± SEM;* $p < .05$. (Adapted from Wang et al., 1994a.)

at the end of gestation. As OT receptors in specific regions of the rat brain are dependent on circulating estrogen levels, one hypothesis is that OT is released with parturition and the steroid-induced increase in brain receptors amplified the central response to this hormone.

In the vole, the scenario appears quite different. OT receptors are found in the vole brain, but in different regions from the pattern observed in the rat brain. Comparing species of vole, prairie and pine voles show patterns that are different from the patterns in montane and meadow voles (Fig. 9; Insel

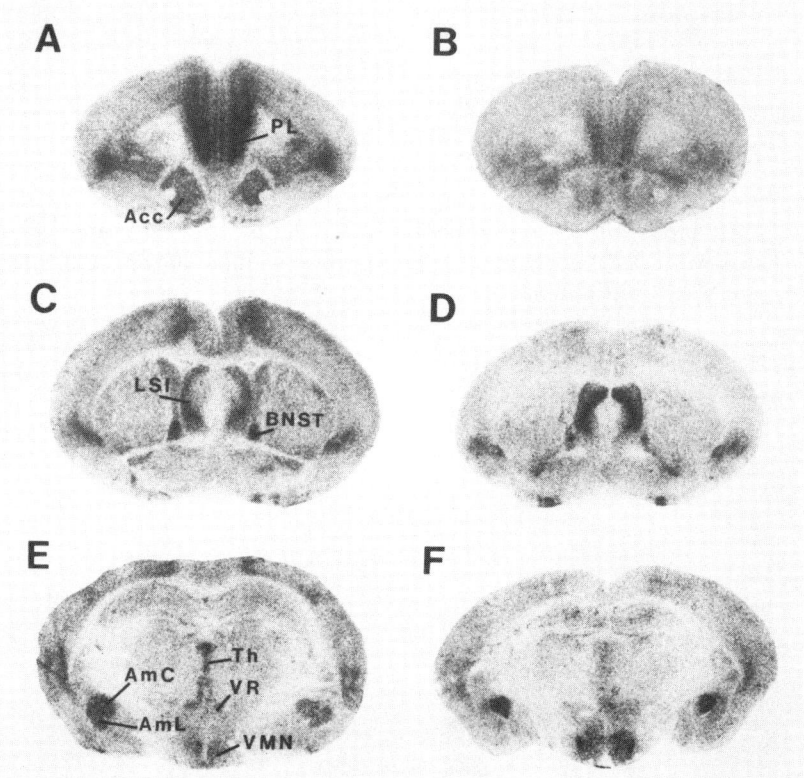

FIG. 9. Oxytocin receptors in monogamous and nonmonogamous voles. Brightfield images of oxytocin receptor binding to 20-$\mu$m coronal sections of prairie (A,C,E) and montane (B,D,F) vole brains. The plates represent anatomically matched sections comparing oxytocin receptors in the two species. Binding to nucleus accumbens (Acc), prelimbic cortex (PL), the bed nucleus of the stria terminalis (BNST), the midline thalamus (Th), ventral reuniens (VR), and the lateral amygdala (AmL) is greater in prairie voles, whereas binding to the lateral septum (LSI) and ventromedial nucleus of the hypothalamus (VMN) is greater in montane voles. Both species show high levels of binding to the central nucleus of the amygdala (AmC). (Adapted from Insel and Shapiro, 1992.)

and Shapiro, 1992). This degree of species difference in receptors is similar to what we observed with AVP receptors, but is unusual in neurobiology. What makes the OT receptor difference particularly intriguing is the observation that the receptor distribution in the montane vole female changes at parturition. With the onset of maternal care, OT receptors in the amygdala increase in the montane female to resemble the pattern observed in the prairie vole (Fig. 10). The mechanism for this regional induction is not yet clear, although studies in the prairie vole have demonstrated that, unlike the rat, OT receptors in this species are not dependent on estrogen.

Injections of OT and an OT antagonist directly into the brain indicate that this peptide is important for the selective affiliation observed in the female

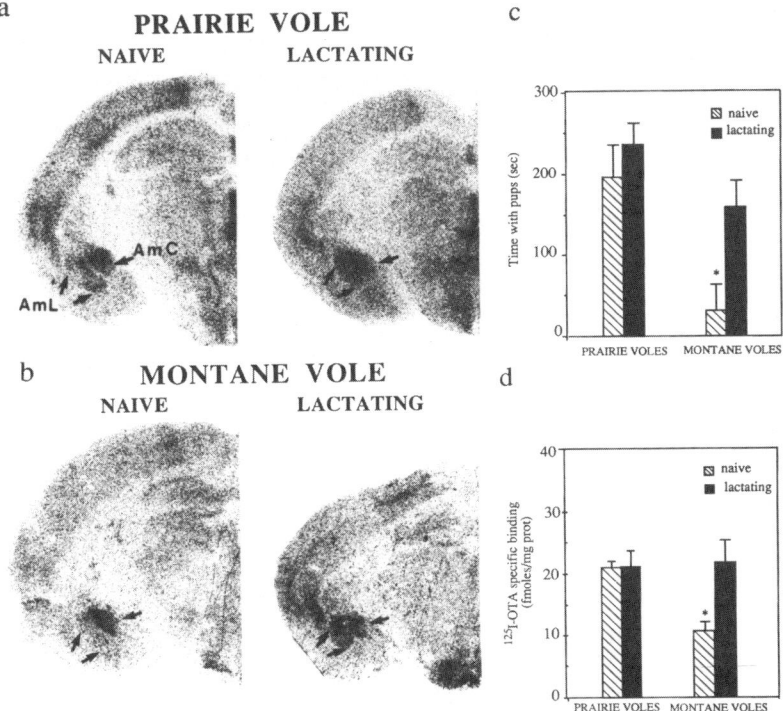

FIG. 10. Effects of parturition on OT receptor binding in the amygdala and on parental behavior in sexually naive and lactating females comparing prairie and montane voles. There are no differences in oxytocin receptor binding in (a) amygdala and (c) parental behavior in sexually naive and lactating female prairie voles. By contrast, lactating female montane voles have a higher density of OT receptor binding in the lateral aspects of the amygdala (b and d) and display a higher level of parental behavior (c) than their sexually naive counterparts. AmC, central nucleus of the amygdala; AmL, lateral amygdala;* $p < .05$.

prairie vole following mating (Williams *et al.*, 1994). OT appears to have no effect on the analogous behaviors in the male (Winslow *et al.*, 1993) and AVP, which is critical in the male prairie vole, appears ineffective in the female (Insel and Hulihan, 1995). Although OT appears critical for affiliative behavior in the female prairie vole, its importance for parental behavior in either monogamous or nonmonogamous voles has yet to be determined.

## VII. Summary

The microtine rodents are an important group for studies of social behavior. Monogamous species with biparental care have provided useful models for studying the role of the father and juveniles in the care and development of pups. Although we know relatively little about the behavioral mechanisms that regulate either form of parental behavior, the presence of closely related species without paternal behavior or alloparenting provides a promising tool for future research. For instance, the observation that cross-fostering meadow vole pups to prairie vole parents increases subsequent parental behavior in the mature meadow voles suggests the importance of early experience for acquiring patterns of parental care. At the same time, the varying rates and patterns of development in different vole species indicates the critical importance of species-typical ontogeny for understanding the species differences in parental care.

Voles have also provided excellent model systems for studying the neurobiological basis of parental care. Although this work is still very recent, the data from several sources suggest that specific brain areas— the MPA, the amygdala, and the lateral septum—are important for various aspects of affiliation including parental care. Two neuropeptide hormones, AVP and OT, with pathways in these brain regions appear to influence pair bonding and parental care in prairie voles. AVP in the lateral septum appears especially important for paternal care; OT (possibly in the amygdala) may influence maternal behavior. Exactly how and where these neuropeptides affect behavior remains to be defined, as does their roles in the nonmonogamous vole species.

It has been said that the best experiments are those that nature has performed for us. The microtine rodents represent an elegant experiment in natural selection with closely related species developing dichotomous patterns of social organization. The challenge for us is to understand the methods and the results of this experiment; to determine how these animals were crafted to exhibit such contrasts in behavior and to recognize the rules that governed this process.

**Acknowledgments**

This article was written while Z. X. W. was supported by grant MH54368-01 from NIMH.

## References

Anderson, S. (1985). Taxonomy and systematics. In "Biology of New World *Microtus*" (R. H. Tamarin, ed.), pp. 52–83. Am. Soc. of Mammalogists, Lawrence, KS.

Bamshad, M., Novak, M. A., and DeVries, G. J. (1993). Sex and species differences in the vasopressin innervation of sexually naive and parental prairie voles, *Microtus ochrogaster* and meadow voles, *Microtus pennsylvanicus. J. Neuroendocrinol.* **5,** 247–255.

Bamshad, M., Novak, M. A., and DeVries, G. J. (1994). Cohabitation alters vasopressin innervation and paternal behavior in prairie voles, *Microtus ochrogaster. Physiol. Behav.* **56,** 751–758.

Batzli, G. O., Getz, L. L., and Hurley, S. S. (1977). Suppression of growth and reproduction of microtine rodents by social factors. *J. Mammal.* **58,** 583–591.

Bittman, E. L., Bartness, T. J., Goldman, B. D., and DeVries, G. J. (1991). Suprachiasmatic and paraventricular control of photoperiodism in Siberian hamsters. *Am. J. Physol.* **260,** R90–R101.

Carlier, C., and Noirot, E. (1965). Effects of previous experience on maternal retrieving by rats. *Anim. Behav.* **13,** 423–426.

Carter, C. S., and Getz, L. L. (1993). Monogamy and the prairie vole. *Sci. Am.* **268,** 100–106.

Crenshaw, G. J., DeVries, G. J., and Yahr, P. (1992). Vasopressin innervation of sexually dimorphic structures of the gerbil forebrain under various hormonal conditions. *J. Comp. Neurol.* **322,** 589–598.

DeVries, G. J., and Buijs, R. M. (1983). The origin of the vasopressinergic and oxytocinergic innervation of the rat brain with special reference to the lateral septum. *Brain Res.* **273,** 307–317.

Dewsbury, D. A. (1981). An exercise in the prediction of manogamy in the field from laboratory data on 42 species of muroid rodents. *The Biologist* **63,** 138–162.

Dewsbury, D. A. (1987). The comparative psychology of monogamy. *Nebr. Symp. Motiv.* **35,** 1–50.

Drickamer, L. C. (1986). Urinary chemosignals that influence puberty in mice; Ecological and evolutionary considerations. In "Chemical Signals in Vertebrates" (D. Duvall, D. Muller-Schwarze, and R. M. Silverstein, ed.), pp. 441–455. Plenum, New York.

Dudley, D. (1974). Contributions of paternal care to the growth and development of the young in *Peromyscus californicus. Behav. Biol.* **11,** 155–166.

Duvall, D., Scudder, K. M., Southwick, C. H., and Schultz, N. J. (1982). Paternal urine elicits increased maternal care in grasshopper mice. *Behav. Neural Biol.* **34,** 221–225.

Elwood, R. W. (1983). Paternal care in rodents. In "Parental Behavior of Rodents" (R. W. Elwood, ed.), pp. 235–257. John Wiley & Sons, New York.

Elwood, R. W., and Broom, D. M. (1978). The influence of litter size and parental behavior on the development of Mongolian gerbil pups. *Anim. Behav.* **26,** 438–454.

Emlen, S. T. (1984). Cooperative breeding in birds and mammals. In "Perspectives in Ornithology" (J. R. Krebs and N. B. Davies, eds.), pp. 305–339. Cambridge Univ. Press, Cambridge.

FitzGerald, R. W., and Madison, D. M. (1983). Social organization of a free-ranging population of pine voles, *Microtus pinetorum. Behav. Ecol. Sociobiol.* **13,** 183–187.

Gerling, S., and Yahr, P. (1979). Effect of the male parent on pup survival in Mongolian gerbils. *Anim. Behav.,* **27,** 310–311.

Getz, L. L. (1972). Social structure and aggressive behavior in a population of *Microtus pennsylvanicus. J. Mammal.* **53,** 310–317.

Getz, L. L., and Carter, C. S. (1980). Social organization in *Microtus ochrogaster* populations. *The Biologist* **60,** 134–146.

Getz, L. L., and Hofmann, J. E. (1986). Social organization in free living prairie voles, *Microtus ochrogaster. Behav. Ecol. Sociobiol.* **18,** 275–282.

Getz, L. L., Carter, C. S., and Gavish, L. (1981). The mating system of the prairie vole, *Microtus ochrogaster:* Field and laboratory evidence of pair-bonding. *Behav. Ecol. Sociobiol.* **8,** 189–194.

Getz, L. L., Dluzen, D., and McDermott, J. (1983). Suppression of reproductive maturation in male-stimulated virgin female *Microtus* by a female urinary chemosignal. *Behav. Proc.* **8,** 189–195.

Getz, L. L., Hofmann, J. E., and Carter, C. S. (1987). Mating system and population fluctuations of the prairie vole, *Microtus ochrogaster. Am. Zool.* **27,** 909–920.

Getz, L. L., McGuire, B., Pizzuto, T., Hofmann, J. E., and Frase, B. (1993). Social organization of the prairie vole (*Microtus ochrogaster*). *J. Mammal.* **74,** 44–58.

Gruder-Adams, S., and Getz, L. L. 1985. Comparison of the mating system and paternal behavior in *Microtus ochrogaster* and *Microtus pennsylvanicus. J. Mammal.* **66,** 165–167.

Gubernick, D. J., and Alberts, J. R. (1989). Postpartum maintenance of paternal behaviour in the biparental California mouse, *Peromyscus californicus. Anim. Behav.* **37,** 656–664.

Gutierrez, P. J., Meyer, J. S., and Novak, M. A. (1989). Comparison of postnatal brain development in meadow voles (*Microtus pennsylvanicus*) and pine voles (*Microtus pinetorum*). *J. Mammal.* **70,** 292–299.

Hartung, T. G., and Dewsbury, D. A. (1979). Paternal behavior in six species of muroid rodents. *Behav. Neural Biol.* **26,** 446–478.

Hatfield, D. M. (1935). A natural history study of *Microtus californicus. J. Mammal.* **16,** 261–271.

Hermes, J. L. H. J., Buijs, R. M., Masson-Pevet, M., and Pevet, P. (1990). Seasonal changes in vasopressin in the brain of the garden dormouse (*Eliomys quercinusl*). *J. Comp. Neurol.* **293,** 340–346.

Hoffman, G. E., Smith, M. S., and Verbalis, J. G. (1993). c-fos and related immediate early gene products as markers of activity in neuroendocrine systems. *Front. Neuroendocrinol.* **14,** 173–213.

Hrdy, S. B. (1976). Care and exploitation of nonhuman primate infants by conspecifics other than the mother. *In* "Advances in the Study of Behavior," Vol. 6 (J. S. Rosenblatt, R. A. Hinde, E. Shaw, and C. Beer, eds.), pp. 101–158. Academic Press, New York, 284.

Insel, T. R. (1990). Oxytocin and maternal behavior. *In* "Mammalian Parenting" (N. A. Drasnegor and R. S. Bridges, eds.), pp. 260–280. Oxford Univ. Press, New York.

Insel, T. R., and Hulihan, T. J. (1995). A gender specific mechanism for pair bonding: Oxytocin and partner preference formation in monogamous voles. *Behav. Neurosci.* **109,** 782–789.

Insel, T. R., and Shapiro, L. E. (1992). Oxytocin receptor distribution reflects social organization in monogamous and polygamous voles. *Proc. Natl. Acad. Sci., U. S. A.* **89,** 5981–5985.

Insel, T. R., Wang, Z. X., and Ferris, C. F. (1994). Patterns of brain vasopressin receptor distribution associated with social organization in microtine rodents. *J. Neurosci.* **14,** 5381–5392.

Jannett, F. J., Jr. (1978). The density-dependent formation of extended maternal families of the montane vole, *Microtus montanus nanus. Behav. Ecol. Sociobiol.* **3,** 245–263.

Jannett, F. J., Jr. (1980). Social dynamics of the montane vole, *Microtus montanus,* as a paradigm. *The Biologist* **62,** 3–19.

Jannett, F. J., Jr. (1982). Nesting patterns of adult vole, *Microtus montanus,* in field populations. *J. Mammal.* **63,** 495–498.

Kirkpatrick, B., Carter, C. S., Newman, S. W., and Insel, T. R. (1994a). Axon-sparing lesions of the medial nucleus of the amygdala decrease affiliative behaviors in the prairie vole (*Microtus ochrogaster*): Behavioral and anatomical specificity. *Behav. Neurosci.* **108,** 501–513.

Kirkpatrick, B., Kim, J. W., and Insel, T. R. (1994b). Limbic system fos expression associated with paternal behavior. *Brain Res.* **658,** 112–118.

Kleiman, D. F. (1977). Monogamy in mammals. *Q. Rev. Biol.* **52,** 36–69.

Lambin, X., and Krebs, C. J. (1991). Spatial organization and mating system of *Microtus townsendii*. *Behav. Ecol. Sociobiol.* **28,** 353–363.

Lambin, X., Krebs, C. J., and Scott, B. (1992). Spacing system of the tundra vole (*Microtus oeconomus*) during the breeding season in Canada's western arctic. *Can. J. Zool.* **70,** 2068–2072.

Leonard, M. L., Horn, A. G., and Eden, S. F. (1989). Does juvenile helping enhance breeder reproductive success? *Behav. Ecol. Sociobiol.* **25,** 357–361.

Lidicker, W. Z., Jr. (1980). The social biology of the California vole. *The Biologist* **62,** 46–55.

Ludwig, D. R. (1981). The population biology and life history of the water vole (*Microtus richardsoni*). Ph.D. dissertation, Univ. of Calgary, Calgary, Alberta.

MacArthur, R. H., and Wilson, E. O. (1967). "The Theory of Island Biogeography" (Monographs in population Biology no. 1). Princeton Univ. Press, Princeton, NJ.

Madison, D. M. (1978). Movement indications of reproductive events among female meadow voles as revealed by radiotelemetry. *J. Mammal.* **59,** 835–843.

Madison, D. M. (1980). An integrated view of the social biology of *Microtus pennsylvanicus*. *The Biologist* **62,** 20–30.

McGuire, B. (1988). Effects of cross-fostering on parental behavior of meadow voles (*Microtus pennsylvanicus*). *J. Mammal.* **69,** 332–341.

McGuire, B., and Novak, M. (1984). A comparison of maternal behaviour in the meadow vole (*Microtus pennsylvanicus*), prairie vole (*M. ochrogaster*), and pine vole (*M. pinetorum*). *Anim. Behav.* **32,** 1132–1141.

McGuire, B., and Novak, M. (1986). Parental care and its relation to social organization in the montane vole. *J. Mammal.* **67,** 305–311.

Myers, P., and Master, L. L. (1983). Reproduction by *Peromyscus maniculatus:* Size and compromise. *J. Mammal.* **64,** 1–18.

Myllymaki, A. (1977). Intraspecific competition and home range dynamics in the field vole *Microtus agrestis*. *Oikos* **29,** 553–569.

Nadeau, J. H. (1985). Ontogeny. *In* "Biology of New World *Microtus* (R. H. Tamarin, ed.), pp. 254–285. Am. Soc. Mammalogists, Lawrence, KS.

Numan, M. (1990). Neural control of maternal behavior. *In* "Mammalian Parenting" (N. A. Drasnegor and R. S. Bridges, eds.), pp. 231–259. Oxford Univ. Press, New York.

Oliveras, D., and Novak, M. (1986). A comparison of paternal behavior in the meadow vole, *Microtus pennsylvanicus*, the pine vole, *Microtus pinetorum*, and prairie vole, *Microtus ochrogaster*. *Anim. Behav.* **34,** 519–526.

Ostermeyer, M. C., and Elwood, R. W. (1984). Helpers (?) at the nest in the Mongolian gerbil, *Meriones unguiculatus*. *Behavior* **91,** 61–77.

Ostfeld, R. S. (1986). Territoriality and mating system of California voles. *J. Anim. Ecol.* **55,** 691–706.

Pedersen, C. A., Asche, J. A., Monroe, Y. L., and Prange, A. J. (1982). Oxytocin induces maternal behavior in virgin female rats. *Science (Washington, D. C.)* **216,** 648–649.

Richmond, M., and Conaway, C. H. (1969). Management, breeding, and reproductive performance of the vole, *Microtus ochrogaster*, in a laboratory colony. *Lab. Anim. Care* **19,** 80–87.

Salo, A. L., Shapiro, L. E., and Dewsbury, D. A. (1994). Comparisons of nipple attachment and incisor growth among four species of voles (*Microtus*). *Dev. Psychobiol.* **27,** 317–330.

Schultz, N. J. (1979). Parental role and behavior of the southern grasshopper mouse, *Onychomys torridus*. Ph.D. thesis, Johns Hopkins Univ., Baltimore.

Shapiro, L. E., and Dewsbury, D. A. (1990). Differences in affilative behavior, pair bonding, and vaginal cytology in two species of vole (*Microtus ochrogaster*) and (*M. montanus*). *J. Comp. Psychol.* **104,** 268–274.

Shapiro, L. E., Leonard, C. M., Sessions, C. E., Dewsbury, D. A., and Insel, T. R. (1991). Comparative neuroanatomy of the sexually dimorphic hypothalamus in monogamous and polygamous voles. *Brian Res.* **541,** 232–240.

Solomon, N. G. (1991). Current indirect fitness benefits associated with philopatry in juvenile prairie voles. *Behav. Ecol. Sociobiol.* **29,** 277–282.
Solomon, N. G. (1993). Comparison of parental behavior in male and female prairie voles (*Microtus ochrogaster*). *Can. J. Zool.* **71,** 434–437.
Storey, A. E., and Snow, D. T. (1987). Male identity and enclosure size affect paternal attendance of meadow vole, *Microtus pennsylvanicus*. *Anim. Behav.* **35,** 411–419.
Swanson, L. J., and Campbell, C. S. (1979). Maternal behavior in the primiparous and multiparous golden hamster. *Z. Tierpsychol.* **50,** 96–104.
Thomas, J. A., and Birney, E. C. (1979). Parental care and mating system of the prairie vole, *Microtus ochrogaster*. *Behav. Ecol. Sociobiol.* **5,** 171–186.
Van Leeuwen, F. W., Caffe, A. R., and DeVries, G. J. (1985). Vasopressin cells in the bed nucleus of the stria terminalis of the rats: Sex differences and the influence of androgens. *Brain Res.* **325,** 391–394.
Wang, Z. X. (1991). Effects of social environment and experience on parental care, behavioral development, and reproductive success of prairie voles (*Microtus ochrogaster*). Ph.D. dissertation, Univ. of Massachusetts, Amherst, MA.
Wang, Z. X. (1995). Species differences in the vasopressin immunoreactive pathways in the bed nucleus of the stria terminalis and medial amygdaloid nucleus in prairie voles (*Microtus ochrogaster*) and meadow voles (*Microtus pennsylvanicus*). *Behav. Neurosci.,* **109,** 305–311.
Wang, Z. X., and Novak, M. A. (1992). Influence of the social environment on parental behavior and pup development of meadow voles (*Microtus pennsylvanicus*) and prairie voles (*M. ochrogaster*). *J. Comp. Psychol.* **106,** 163–171.
Wang, Z. X., and Novak, M. A. (1994a). Alloparental care and the influence of father presence on juvenile prairie voles, *Microtus ochrogaster*. *Anim. Behav.* **47,** 281–288.
Wang, Z. X., and Novak, M. A. (1994b). Parental care and litter development in primiparous and multiparous prairie voles (*Microtus ochrogaster*). *J. Mammal.* **75,** 18–23.
Wang, Z. X., Ferris, C. F., and DeVries, G. J. (1994a). The role of septal vasopressin innervation in paternal behavior in prairie voles (*Microtus ochrogaster*). *Proc. Natl. Acad. Sci., U. S. A.* **91,** 400–404.
Wang, Z. X., Smith, W., Major, D. E., and DeVries, G. J. (1994b). Sex and species differences in the effects of cohabitation on vasopressin messenger RNA expression in the bed nucleus of the stria terminalis in prairie voles (*Microtus ochrogaster*) and meadow voles (*Microtus pennsylvanicus*). *Brain Res.* **650,** 212–218.
Webster, A. B., and Brooks, R. J. (1981). Social behavior of *Microtus pennsylvanicus* in relation to seasonal changes in demography. *J. Mammal.* **62,** 738–751.
Wideman, C. H., and Murphy, H. M. (1990). Vasopressin, maternal behavior and pup wellbeing. *Curr. Psychol. Res. Rev.* **9,** 285–295.
Williams, J. R., Carter, C. S., and Insel, T. R. (1992). Partner preference development in female prairie voles (*Microtus ochrogaster*) is facilitated by mating or the central infusion of oxytocin. *In* "Oxytocin in Maternal, Sexual and Social Behavior," Vol. 652 (C. A. Pedersen, J. D. Caldwell, G. F. Jirikowski, and T. R. Insel, eds.), pp. 487–489. Ann. N. Y. Acad. Sci., New York.
Williams, J. R., Insel, T. R., Harbaugh, C. R., and Carter, C. S. (1994). Oxytocin administered centrally facilitates formation of a partner preference in female prairie voles (*Microtus ochrogaster*). *J. Neuroendocrinol.* **6,** 247–50.
Wilson, S. C. (1982). Parent-young contact in prairie and meadow voles. *J. Mammal.* **63,** 300–305.
Winslow, J. T., Hastings, N., Carter, C. S., Harbaugh, C. R., and Insel, T. R. (1993). A role for central vasopressin in pair bonding in mongamous prairie voles. *Nature (London)* **365,** 545–548.
Wolff, J. O. (1980). Social organization of the taiga vole (*Microtus xanthognathus*). *The Biologist* **62,** 34–45.
Wolff, J. O., and Lidicker, W. Z., Jr. (1981). Communal winter nesting and food sharing in taiga voles. *Behav. Ecol. Sociobiol.* **9,** 237–240.

# Physiological, Sensory, and Experiential Factors of Parental Care in Sheep

F. Lévy and R. H. Porter

LABORATOIRE DE COMPORTEMENT ANIMAL
INRA/CNRS, URA 1291
NOUZILLY, FRANCE

K. M. Kendrick

THE BABRAHAM INSTITUTE
BABRAHAM, CAMBRIDGE, ENGLAND

E. B. Keverne

SUB DEPARTMENT OF ANIMAL BEHAVIOUR
MADINGLEY, CAMBRIDGE, ENGLAND

A. Romeyer

UNIVERSIDAD AUTONOMA DE TLAXCALA
CENTRO DE INVESTIGACION EN REPRODUCCION ANIMAL
CINVESTAV, TLAXCALA, MEXICO

## I. Introduction

Maternal behavior presents a wide variety of patterns among the mammals as the numerous contributions of this book demonstrate. Its manifestation depends mainly on the degree of precocity of the young at birth, but also on the social structure of the species and their ecology.

Sheep are seasonal breeders and lambing typically occurs in early spring, during a relatively short period of time, thereby affording the young a maximal period of growth and development. Amongst wild sheep, as well as domestic sheep living in free-ranging conditions, adult females and their suckling young form flocks whose members share a common home range (Grubb and Jewell, 1966; Geist, 1971). These matriarchal flocks may also include yearling offspring and mothers of adult ewes. These social groups are constantly on the move in search of food and it is vital that young are well advanced to follow their mothers. Indeed newborn lambs are highly precocial with their senses as well as their capacities for thermoregulation and motor control. Thus, the demands of maternal care for precocial mammals could be somewhat different from altricial mammals.

Because of the co-occurrence of own and alien young in the same flock, nursing ewes potentially risk having their limited maternal resources usurped by lambs that are not their offspring. Indiscriminate acceptance of any lambs that approach the udder and attempt to suck might result in the ewe having insufficient milk for her own young and therefore have drastic negative consequences for her reproductive success. The strategy that has evolved for guarding against costly misdirected maternal investment is for ewes to become rapidly familiar with the salient phenotypic traits of their own neonates (i.e., learned individual recognition). Subsequently, mothers develop discriminative maternal care with those young alone, allowing them to suck while rejecting any alien young that may approach the udder. In this respect, *the establishment of a selective bond* between the ewe and her lamb within the first hour after parturition represents one of the essential characteristics of maternal behavior in this mammal. This characteristic is different from *maternal responsiveness,* which represents the interest towards any newborn lamb and occurs immediately at birth. Both aspects characterize maternal behavior in sheep. Thus, sheep offer the possibility of studying, in the adult, the mechanisms of an attachment that is renewed at each parturition.

Interest in sheep is further enhanced by the fact that the onset of maternal behavior is strictly dependent *on the process of parturition* by contrast to what is observed in rodents in which maternal behavior can be elicited by the presence of young (see articles of Bridges and Stern, this volume). Nonpregnant ewes do not display maternal behavior spontaneously and only within the last 2 h preceding birth do pregnant ewes show interest towards lambs (Arnold and Morgan, 1975). A series of studies have shown that parturition is the physiological key factor for triggering a cascade of neurobiological mechanisms that allows maternal care. This event induces a neural maturation process, at least in the olfactory structures, which persists and may facilitate the emergence of maternal behavior during subsequent parturitions (Kendrick *et al.,* 1992d; Keverne *et al.,* 1993; Lévy *et al.,* 1993).

Another intriguing characteristic of sheep maternal behavior is the primary importance of *olfactory cues* provided by the newborn. In fact, olfaction is involved both in the attractiveness of any neonate for the parturient ewe and in the individual recognition of the lamb that permits the formation of the exclusive bond. Consequently, sheep provide the opportunity to address the question of the underlying bases of lambs' individual odor and, in particular, the respective participation of genetic and environmental influences. Furthermore, the olfactory discrimination of the lamb by the mother involves mechanisms for learning this biological odor. This leads us to the view shared with other groups of researchers (Holley, 1991; Gervais *et al.*, 1988, 1990; Keverne and de la Riva, 1982) that the mammalian olfactory bulb, the first relay in the processing of olfactory information is able to show plastic changes during learning and this structure has to be integrated in forebrain circuits underlying olfactory memory.

The following article will focus on the immediate factors that control the onset of maternal behavior (maternal responsiveness and maternal selectivity) at parturition in sheep and, to a lesser extent, in goats. Section II concerns the determination of maternal responsiveness towards any newborn lamb. We will examine the respective roles of both physiological and sensory factors, mainly olfactory, and their close interactions that facilitate the emergence of maternal acceptance at parturition. Section III of this article concerns the development of bonding with the neonate and will consider the basis of individual's olfactory signature and the neural mechanisms involved in learning this signature.

## II. The Control of Maternal Responsiveness to the Neonate

In domestic and feral breeds of sheep and goats, as parturition approaches, the female shows a shift of behavior from a gregarious tendency to a tendency to seek isolation from the flock (Alexander *et al.*, 1990; Arnold and Dudzinski, 1978; Lécrivain and Janeau, 1987; Lickliter, 1984; O'Brien, 1984). This preliminary step is important because it allows the mother to give birth and become familiarized with her neonate without being disturbed by other conspecifics or predators. At this time, females show a strong attraction to *amniotic fluid*. The ground where the amniotic sac has ruptured is sniffed and licked, and birth usually takes place at this same time. Immediately after expulsion of the fetus, the mother licks her newborn without interruption until the young is cleaned of the amniotic fluid and the placental membranes that cover it at birth. This behavior is associated with the emission of low-pitched bleats (produced only by maternal females) and pawing, which stimulates the activity of the neonate. If the young is removed, the mother responds with high-pitched bleats. As soon as the lamb can stand, it searches for the udder and the mother responds by arching her body, thus facilitating access to the udder. Sniffing and licking of the lamb's anogenital region by the female help to push it in the direction of the udder. Suckling usually occurs within the first 2 h after parturition in the parallel inverse stance that is very common among ungulates (Poindron, 1974). From these behavioral observations, a series of criteria has been defined to assess the acceptance and the rejection of the newborn by a female. Licking behavior, emission of maternal bleats, udder acceptance, and suckling are indicative of maternal responsiveness, whereas aggressive behavior, emission of high bleats, and udder refusal are indicative of maternal rejection. Criteria used to categorize a ewe as maternal are varied, from the use of one criterion such as licking (Krehbiel et al., 1987), suckling (Poindron et al., 1980), to a combination of criteria such as licking and suckling (Kendrick et al., 1987) or suckling and the absence of aggressive behavior (Romeyer et al., 1993b). This lack of homogeneity sometimes renders the comparison of the results of different studies difficult.

Full maternal responsiveness is strictly related to the event of parturition, but fades after a few hours in the absence of a newborn lamb (Poindron and Le Neindre, 1980). Thus, 4 h of separation beginning at birth induced disturbances of maternal responsiveness in half of the mothers and after 12 h of deprivation, 75% of them rejected the lamb (Poindron et al., 1979). On the other hand, 4 or 24 h of immediate postpartum contact was sufficient to maintain maternal responsiveness for the following 24-h period of separation (Poindron et al., 1979; Lévy et al., 1991). Similar results have been reported in goats (Lickliter, 1982). Thus, maximal maternal receptivity to the neonate is limited to a particular period starting at parturition and lasting approximately 12 h. This period, which is related to the ability of a mother to care for any young, has been called the *sensitive period.* This characteristic emphasizes the importance of parturition for the onset of maternal responsiveness.

### A. Physiological Changes during Parturition

As in many other mammals, parturition in sheep and goats is characterized by a rapid increase in the ratio of estradiol to progesterone. In sheep, 2–4 days before parturition the concentration of plasma progesterone falls, while estradiol rises 1 day before parturition to reach a peak at parturition and returns to basal levels within the first 4 h after parturition (Challis, 1971; Chamley et al., 1973; Shipka and Ford, 1991). While similar changes have been observed in goats, there is no large acute prepartum increase in estradiol of the great magnitude known for sheep (Currie et al., 1988). As a result of the shift in balance in steroid exposure from progesterone to estradiol dominance, serum prolactin levels increase dramatically around 1 day before parturition in sheep (Chamley et al., 1973), during the last 3 days in goats (Currie et al., 1988), and are maintained at a high level after parturition because of suckling stimulation (Davis *et al.,* 1971; Chamley *et al.,* 1973). Levels of prostaglandins also rise successively for the last 3 days before parturition and peak at parturition in both sheep and goats (Currie *et al.,* 1988; Fredriksson, 1985).

Contrary to the other hormones, the plasma release of oxytocin is strictly associated with the expulsion of the fetus caused by the genital stimulation. In both species, oxytocin concentrations show significant elevation only at birth and up to 15 min postpartum due to the Fergusson reflex (Currie *et al.,* 1988; Kendrick *et al.,* 1991a). Plasma concentrations of a closely related peptide, arginine-vasopressin, show similar, although smaller, changes in sheep (Kendrick *et al.,* 1991a), whereas in goats no rise during parturition has been reported (McNeilly *et al.,* 1972). Nevertheless, the physiological events cited previously concern the peripheral peptide levels and, because of the existence of the blood-brain barrier, it is not clear if similar changes occur at the central level during parturition. A series of studies, beginning in 1986, related to the measurement of hormones and neurotransmitters in the brain have improved our knowledge of the physiological determinism of maternal responsiveness.

Clearly, parturition induces a large number of neurochemical changes in various brain regions. One of the major alterations concerns oxytocin, which increases in the cerebrospinal fluid (CSF), the olfactory bulbs (OB), the bed nucleus of the stria terminalis (BNST), the septum, the medial preoptic area (MPOA), the paraventricular nucleus of the hypothalamus (PVN), and the substantia nigra (SN; Da Costa *et al.,* 1993, 1994; Kendrick *et al.,* 1986, 1988a,b, 1992c; Lévy *et al.,* 1995a). This is associated with the increase of oxytocin immunoreactivity and mRNA level in OB, BNST, MPOA, the supraoptic nucleus (SON), and PVN (Broad *et al.,* 1993). Therefore, at the time of parturition, there is a dramatic increase of oxytocin synthesis, storage, and release. On the other hand, arginine-vasopressin levels show changes only in the CSF but not in the OB, suggesting that this peptide could only play a minor role in maternal responsiveness. As for classical neurotransmitters, the ascending noradrenergic system shows a general pattern of activation that parallels the oxytocinergic system because increase in noradrenaline release is observed in the same brain regions (Da Costa *et al.,* 1993, 1994; Kendrick *et al.,* 1986, 1988a, 1992c; Lévy *et al.,* 1993). As for dopamine, there is an increased release during parturition in OB, S, MPOA, PVN, and SN but not in CSF and BNST. Release of gamma-aminobutyric acid and glutamate also increases in OB, S, BNST, MPOA, and PVN (Da Costa *et al.,* 1993, 1994; Kendrick *et al.,* 1986, 1988a, 1992c; Keverne *et al.,* 1993). Interestingly, some of these neurochemical events occur in brain regions that are activated at parturition. Changes in the expression of the immediate early gene c-fos, a tool for mapping the activation of neural systems, have been observed at parturition in the OB, S, MPOA, PVN, and SON (Da Costa *et al.*, 1993).

Thus, at parturition in areas of the brain not directly connected with one another, such as the SN and OB or the septum and the BNST, the release of neurotransmitters and hormones are similar. This suggests that maternal responsiveness is determined by simultaneous release of a large number of substances in diverse brain regions and therefore one can predict that the physiological control of maternal responsiveness would be multifactorial. Moreover, among these numerous factors, some of them could be supernumerary and then lead to an overdetermination of maternal responsiveness that is not surprising in the view of the complexity of this behavior and its importance for the survival of the young. On the other hand, the literature on maternal responsiveness, at least in rodents, points out the fact that only one factor could stimulate the behavior (Stern, 1989). However, this apparent contradiction could be understood if one thinks firstly that a hormone or a transmitter could trigger a multiplicity of events through the possible interaction between hormonal and transmittor systems. For instance, studies in sheep have demonstrated an influence of oxytocin on noradrenaline release in the MPOA and the OB (Kendrick *et al.*, 1992b; Lévy *et al.*, 1995a). Secondly, previous maternal experience could have set up the network of the neurochemical interactions in such a way that only one factor of this network could subsequently be sufficient to induce maternal responsiveness. In this context, we have observed in the OB an influence of maternal experience on the interactions between oxytocin and noradrenaline (Lévy *et al.*, 1995a). Nevertheless, one cannot forget that these studies correlate physiological events with not only the emergence of maternal responsiveness per se but also with the onset of parturition and lactation. Therefore only some of these physiological events could be specific for the induction of maternal responsiveness.

## B. Neuroendocrine Control of Maternal Responsiveness

In sheep, the earliest studies on the physiological control of maternal behavior investigated the possibility of hormonally inducing maternal responsiveness in nonpregnant ewes (Le Neindre *et al.*, 1979). A short-term treatment (1 week) or a long-term treatment (4 weeks) of progesterone following estradiol at high doses were effective but only in 50% of the ewes that had been exposed to lambs for 1 hour. This treatment also induced abnormal behavior such as permanent estrus and malelike sexual behavior. Further investigations have revealed evidence for facilitation of maternal behavior by sex steroids. First, the spontaneous occurrence of maternal care during the reproductive cycle is limited to times of high levels of estrogen, that is, at estrus and the last days of gestation (Poindron and Le Neindre, 1980). Second, only a single injection of estradiol at high doses (25 mg) was able to induce maternal responsiveness in nongestant ewes. Third, when lambing was induced by estradiol benzoate, which leads to high plasmatic levels of estradiol, the sensitive period was lengthened from 4 to 24 h (Poindron and Le Neindre, 1980). In addition, relationships have been found between prepartum endogenous levels of plasma estradiol on the 2-3 days prepartum and some aspects of maternal behavior (Shipka and Ford, 1991). On the other hand, studies in another breed of sheep have reported that a short treatment (12 days) of progesterone following low doses of estradiol (200 $\mu$g) or a long-term treatment with both progesterone and estradiol induced only a few items of maternal responsiveness (Kendrick and Keverne, 1991; Kendrick et al., 1992a). The respective roles for both steroids are not clear. High levels of progesterone during gestation did not correspond with maternal responsiveness. On the other hand, high doses of progesterone induced maternal responsiveness in nongestant ewes but not as effectively as estradiol (Poindron and Le Neindre, 1980). Finally, although steroid treatment did not induce maternal responsiveness, progesterone facilitated the action of estradiol in reducing aggressiveness towards the lamb (Kendrick and Keverne, 1991). In goats, attempts to stimulate maternal responsiveness with estradiol or combined progesterone and estradiol treatment have failed so far (Rosenblatt and Siegel, 1981; K. M. Kendrick, unpublished observations).

Thus, although inconsistencies in these results could be due to the conditions of testing and/or the criteria chosen to define maternal responsiveness, only high doses of steroid have a notable effect on maternal responsiveness. Moreover, this effect was not specific and not all the components were induced. Therefore, it appears that ovarian steroids are not capable by themselves of stimulating maternal care, but have only a *priming* effect allowing other factors to *trigger* the behavior.

The synchrony between parturition and the various physiological changes cited previously have led us to suspect that the actual process of expulsion of the fetus is of primary importance. Indeed, we now have several lines of evidence indicating that vaginocervical stimulation occurring at that time is the starting point of a cascade of events that trigger the onset of maternal responsiveness. In nonpregnant ewes primed with progesterone and estradiol, the induction of maternal responsiveness in response to a neonate was obtained in 80% of the cases when ewes received 5 min of vaginocervical stimulation. By contrast, only 20% of females responded maternally without this stimulation (Keverne et al., 1983). These results have been further confirmed in two other breeds of sheep, with low doses of estradiol and even in ewes at estrus (Kendrick and Keverne, 1991; Kendrick et al., 1992a;

Poindron *et al.*, 1989). This stimulation not only increased the number of maternal ewes but induced the full complement of maternal responsiveness, licking, maternal bleats, acceptance at the udder, and reduction in aggressive behavior. Vaginocervical stimulation stimulated even the attraction towards amniotic fluid, which is a characteristic trait of maternal ewes at parturition (see Section II,C). On the other hand, vaginocervical stimulation without a pretreatment of steroids was ineffective, which confirms the importance of steroid priming action (Kendrick and Keverne, 1991; Poindron *et al.*, 1989). This facilitatory action of vaginocervical stimulation was also established in parturient ewes. Five minutes of vaginocervical stimulation 1 h after parturition reinduced maternal responses, especially licking towards a newborn lamb (Keverne *et al.*, 1983). Vaginocervical stimulation was even effective at 24 h postpartum (Kendrick *et al.*, 1991b). On the contrary, if stimulation of the genital tract was prevented at parturition by peridural anesthesia, maternal behavior was greatly disturbed (Krehbiel *et al.*, 1987).

What are the neural mechanisms by which vaginocervical stimulation initiates the onset of maternal responsiveness? The finding that CSF levels of oxytocin rise at parturition and following artificial vaginocervical stimulation (Kendrick *et al.*, 1986) led to the hypothesis of the involvement of this peptide. This was first demonstrated in nonpregnant ewes in which intracerebroventricular injection of either 10 or 20 $\mu$g of oxytocin, after a steroid pretreatment, induces maternal responses within a minute (Kendrick *et al.*, 1987). This finding was then replicated and it was shown that 50 $\mu$g was the more effective dose (Keverne and Kendrick, 1991). Interestingly, oxytocin, like vaginocervical stimulation, was ineffective when given without estradiol priming. In parturient females, inhibiting genital stimulation feedback with peridural anesthetic prevented central oxytocin release. Maternal care was absent in such ewes but was restored by intracerebroventricular infusions of 20 $\mu$g of oxytocin (Lévy *et al.*, 1992). The sites of action of oxytocin are now gradually becoming known. Among the numerous sites of oxytocin release during parturition, the PVN and the MPOA appear to be responsible for the induction of maternal responsiveness in nonpregnant steroid primed ewes. Using retrodialysis, infusions of oxytocin in the PVN induced full maternal responsiveness, whereas in the MPOA it only reduced rejection behavior towards lambs but without affecting the acceptance behavior (Da Costa *et al.*, 1994; Kendrick *et al.*, 1992c). Oxytocin could act then at different sites to exert its effects on the different components of maternal responsiveness.

However, in all the experiments, oxytocin administration was not as effective as parturition itself in inducing maternal responsiveness. This, and the fact that parturition induces changes in the release of many neurotransmitters in various brain regions suggest that other physiological factors might be involved in the onset of maternal responsiveness.

A series of experiments indicates that opioids potentiate the effects of vaginocervical stimulation to promote maternal responsiveness. Intracerebroventricular infusions of an opioid receptor blockade, naltrexone, prevented the induction of maternal responsiveness in estrogen primed ewes following vaginocervical stimulation (Kendrick and Keverne, 1989). Furthermore, intracerebroventricular infusions of morphine, given together with 5 min of vaginocervical stimulation, induced maternal responses similar to that of the postparturient ewe (Kendrick and Keverne, 1991). The facilitatory effect of opiates have been confirmed in parturient ewes. Intravenous injection of naltrexone blocked the display of maternal responsiveness (Caba et al., 1995). Opiates could play a modulatory role rather than a primary one. The effect of naltrexone at parturition was not dramatic and long lasting as when blocking vaginocervical stimulation by peridural anesthesia. Intracerebroventricular infusions of morphine without vaginocervical stimulation were also ineffective in nongestant ewes. The influence of opiates could be partly due to a modulation of oxytocin release. Preproenkephalin mRNA expression increased in parallel with that of oxytocin in the PVN at parturition (Broad et al., 1993). Morphine alone did not affect CSF levels of oxytocin but potentiated it after vaginocervical stimulation (Kendrick and Keverne, 1991). However, this opioid action could also be direct because even high doses of intracerebroventricular oxytocin did not induce maternal responses similar to those of a parturient ewe. To further study this point, it will be of interest to show first that changes of central concentrations of opiates occur at parturition when maternal responsiveness is induced. Second, to test the hypothesis of a direct action of opiates, it could be interesting to investigate if intracerebroventricular infusion of morphine could restore maternal responsiveness in peridural anesthetized ewes.

Among the possible factors involved in the determination of maternal responsiveness, corticotrophin-releasing factor was found to play a similar role to morphine (Keverne and Kendrick, 1991). Corticotropin-releasing factor potentiated the effects of vaginocervical stimulation in increasing acceptance behaviors and reducing the rejection behaviors towards the newborn but had little effect when given alone. As for morphine, the action of corticotropin-releasing factor is thought to consist of potentiating the oxytocin release. The coexistence of corticotropin-releasing factor in oxytocin neurons in the PVN and the increase of corticotropin-releasing factor mRNA expression in the PVN of sheep at parturition strengthens the possible relationship between these two peptides (Pretel and Piekut, 1990; Broad et al., 1995).

As we have reported in the previous section, there is a general activation of the noradrenergic system at parturition in areas of the brain that seem to be important for maternal behavior such as the MPOA or the PVN. However, intracerebroventricular infusion of the alpha-adrenergic receptor blocker, phentolamine, or the beta-adrenergic antagonist, timolol, did not inhibit maternal responsiveness following vaginocervical stimulation in nonpregnant ewes (Kendrick and Keverne, 1989). This failure could be due to the use of an inappropriate dose or antagonist. Also, it will be of interest to block the action of noradrenaline directly in neural sites in which the release of noradrenaline has been observed such as the MPOA and the PVN. Another possibility is that the noradrenaline is not directly involved in the onset of maternal responsiveness, but is preferentially associated with the hypothalamic mechanisms of parturition and lactation. Nevertheless, the activation of the noradrenaline system in the OB has been found to be of great importance for the regulation of attention to olfactory cues implicated in the mother's recognition of her offspring (see Section III,D).

While the evidence for prolactin involvement in the onset of maternal responsiveness in the rat is strong (Bridges, 1990), attempts to show an action of this hormone in ewes have failed so far. In parturient ewes, injection of dibromoergocriptine to block prolactin release did not prevent the onset of maternal responsiveness (Poindron and Le Neindre, 1980). In nongestant ewes primed with steroids, intracerebroventricular infusions of 50 or 100 $\mu$g of prolactin did not induce maternal responsiveness (F. Lévy, unpublished observations). This discrepancy could reflect an important difference in the physiological control of maternal responsiveness between rodents and ungulates.

Maternal experience greatly influences the action of the physiological factors identified. In inexperienced nongestant ewes, neither steroid priming nor vaginocervical stimulation induced acceptance of the lamb (Le Neindre et al., 1979; Keverne and Kendrick, 1991). Also, estrogen-primed nulliparous ewes were unresponsive to other artificial means of inducing maternal responsiveness, either with oxytocin, opiates, or corticotropin-releasing factor (Keverne and Kendrick, 1991). Furthermore, preventing vaginocervical stimulation at parturition by means of a peridural anesthesia had more profound effects on the onset of maternal responsiveness in primiparous females (Krehbiel et al., 1987). It seems possible that one of the effects of maternal experience is to increase the brain's sensitivity to the feedback action of sex steroids, probably in terms of their ability to upregulate the oxytocin receptor. This is suggested by evidence showing that maternal responsiveness can be induced in nulliparous ewes by vaginocervical stimulation if they receive long-term priming with progesterone and estradiol (Kendrick et al., 1992a).

C. SENSORY CONTROL OF MATERNAL RESPONSIVENESS

The initial phase of maternal responsiveness controlled by physiological factors is not sufficient for the normal development of maternal behavior. The absence of the newborn lamb at parturition rapidly induces a loss of maternal responsiveness (Poindron and Le Neindre, 1980). Thus, the sensory cues provided by the lamb allow the maintenance of maternal responsiveness beyond the sensitive period. Among the possible sensory modalities involved, a series of studies have shown that olfaction is of primary importance in sheep.

First, to investigate the kind of sensory cues involved, lambs were separated from their mother at birth for 8–12 h, so that only some cues were available (Poindron and Le Neindre, 1980; Poindron et al., 1988). After this period of separation, mothers were tested for their maternal responsiveness. Deprivation of tactile stimulation of licking or suckling by placing the lamb in a double-walled mesh cage led to minor disturbances of maternal responsiveness. By contrast, suppression of olfactory cues by placing the lamb in an airtight transparent box (thereby eliminating olfactory cues) led to a significant decrease of maternal responsiveness. When only olfactory cues were available, the rate of acceptance was significantly greater than when only the visual cues were provided. Thus, it is only the perception of olfactory cues that allows ewes to remain maternal beyond the sensitive period. However, from these data it is not clear if olfactory cues are of importance for the maintenance of maternal responsiveness beyond the sensitive period and/or for the transition between the phase of internal control and the phase of sensory control occurring at parturition and responsible for the onset of maternal responsiveness. To clarify this point, studies were undertaken to observe, at parturition, the effects of deprivation of either olfaction or olfactory cues on the onset of maternal behavior.

In an experiment, the respective roles of the main and accessory olfactory systems in the development of maternal responsiveness were investigated in parturient ewes. Disruption of the main olfactory system by irrigation of the nasal cavities with zinc sulfate before parturition delayed the onset of maternal responsiveness. Females spent less time licking their neonates, emitted fewer maternal bleats, and emitted more protest bleats (Lévy et al., 1995b). On the other hand, females with only lesion of the vomeronasal organ showed little disturbances of maternal care. These results underline the importance of the main olfactory system in the establishment of maternal responsiveness at parturition. The accessory olfactory system appears to be not necessary, whereas in rat deafferentation of this system plays some role in the induction of maternal responsiveness (Fleming et al., 1979). The next step was to identify the type of olfactory cues meaningful for the mother. A report showing that the newborn lamb is more attractive than an older lamb (Poindron and Le Neindre, 1980) led us to postulate that amniotic fluid on the neonate could be responsible for this attractiveness.

Initially, ewes were tested for their preferences for food that was either treated with amniotic fluid or water (Lévy et al., 1983). Females were repelled by amniotic fluid throughout their estrous cycle and gestation, but for a few hours following parturition they became attracted to amniotic fluid. These results were further confirmed by Vince et al. (1985), showing that parturient ewes were more attracted to a model lamb containing amniotic fluid than to one without amniotic fluid. Such a rapid change of preference occurring at parturition did not exist in anosmic females that were neither repelled nor clearly attracted to amniotic fluid (Lévy et al., 1983). In addition, lesions of the vomeronasal organ were without effect on the attraction–repulsion responses to amniotic fluid (Lévy et al., 1995b). To test the hypothesis that attraction to amniotic fluid is a necessary step for the development of maternal responsiveness, amniotic fluid was removed from the newborn lamb of primiparous and multiparous ewes (Lévy and Poindron, 1987). Washing the neonate, with either soap or just water, greatly reduced licking behavior and, in primiparous ewes, this treatment abolished acceptance at the udder and increased aggressive behavior. Amniotic fluid is also sufficient by itself to induce maternal responsiveness in a context wherein females typically reject young. Parturient experienced ewes were tested with 1-day-old lambs whose coats were either treated with amniotic fluid or water (Lévy and Poindron, 1984). This treatment led to a significant increase in the rate of acceptance of the lambs in comparison with the control condition. Similar results were found in a study in which fostering of alien lambs onto parturient ewes was facilitated by covering the lambs in jackets soaked in amniotic fluid (Basiouni and Gonyou, 1988). The origin of amniotic fluid (from the tested ewe or from an alien mother) had no reliable effect on maternal acceptance, which suggests that amniotic fluid would only contain cues responsible for general attractiveness but not for individual recognition (see Section III,A).

Overall, aside from contributing to the maintenance of maternal responsiveness beyond the sensitive period, olfactory cues provided by amniotic fluid are necessary to ensure appropriate maternal behavior at parturition, especially by naive ewes. Experienced females are able to compensate for the loss of olfactory information by relying on other cues associated with newborns. On the other hand, in experienced mothers amniotic fluid is sufficient to induce maternal care of lambs bearing that substance. Therefore, as stated by Schneirla et al. (1963), amniotic fluid appears to be "a potent organizer of maternal behavior" in focusing the attention of the dam from the licking of her body to the licking of the neonate.

How the changes in the olfactory reaction to amniotic fluid occurring at parturition can be explained? The simultaneous occurrence of attraction to amniotic fluid and the sensitive period was an indication that the same factors controlled maternal olfactory functioning and maternal behavior. Indeed various experiments indicate that vaginocervical stimulation facilitates attraction to amniotic fluid as it does for maternal responsiveness (Lévy *et al.,* 1990b). Deprivation of vaginocervical stimulation feedback to the brain by peridural anesthesia in parturient ewes induced a loss of preference for amniotic fluid. Furthermore, attraction to amniotic fluid, once it has disappeared in postparturient ewes, can be restored by artificial vaginocervical stimulation up to 4 h postpartum. This is also in agreement with the fact that, in estrous ewes, vaginocervical stimulation diminished the aversion to amniotic fluid of the nongestant ewe (Rubianes, 1992b). The effect of vaginocervical stimulation appears to be mediated in part by oxytocin. Intracerebroventricular infusion of oxytocin in peridural anesthetized ewes restored the preference for amniotic fluid (Lévy *et al.,* 1990b). Therefore, this, together with the studies showing, in the OB, an increase of oxytocin release at parturition and a facilitation of release of noradrenaline and acetylcholine by oxytocin (Kendrick *et al.,* 1988a,b; Lévy *et al.,* 1995a), suggests that oxytocin could modulate olfactory processing at the level of OB, for instance, by increasing the response of mitral cells to maternal odors. Electrophysiological recordings as well as anatomical studies to localize the sites of action of oxytocin would be necessary to further understand such a humoral regulation of the OB. In addition, the underlying mechanisms mediating the change of preference for amniotic fluid could also involve other physiological factors such as the opiates. Indeed, in sheep, they are implicated in the induction of maternal responsiveness and, in rodents, it was demonstrated that morphine modified the olfactory preference of the mother towards the pups (Kinsley and Bridges, 1990).

## III. The Control of Maternal Selectivity to the Neonate

During the phase of maternal responsiveness, ewes and does progressively learn the characteristics of their offspring, so that after a few hours they are able to recognize them and only to accept them at the udder. The purpose of the next section is to show the primary role of olfactory cues and the mechanisms by which these cues are learned.

### A. Olfactory Mediation of Lamb Recognition

Although olfactory cues may be less salient than the lamb's visual appearance and voice for distal recognition (Lindsay and Fletcher, 1968; Poindron and Carrick, 1976; Walser *et al.*, 1981), numerous observational and experimental studies consistently indicate that the sense of smell plays a primary role in ewes' selective acceptance of lambs for nursing (Poindron *et al.*, 1993b; Porter and Lévy, 1995).

The coordinated behavior of mother and young during the recurring feeding bouts facilitates olfactory investigation and identification of the lamb (Poindron, 1974). Like the young of other ungulate species, lambs usually adopt a parallel inverse position (relative to that of the mother) when suckling. As the lamb approaches the udder, it moves past the ewe's head, enabling her to smell the trunk and anogenital region. This latter area of the lamb continues to be sniffed frequently as nursing proceeds.

When ewes or does suffer experimentally induced olfactory deficits prior to parturition, via bulbectomy (Bouissou, 1968; Baldwin and Shillito, 1974), sectioning the olfactory nerve (Morgan *et al.*, 1975), or irrigating the olfactory mucosa with zinc sulfate solution (Lévy *et al.*, 1995b; Poindron, 1976; Romeyer *et al.*, 1994), they subsequently show no evidence of recognizing their neonate, but accept alien young as well as their own. It should be pointed out that the anatomically distinct accessory olfactory system does not appear to be implicated in lamb recognition, since severing the nerves of the accessory olfactory system does not prevent ewes to be selective (Lévy *et al.*, 1995b). In related experiments, the degree of mother–offspring contact during the first 12 h following birth was systematically manipulated (Poindron and Le Neindre, 1980). Ewes that had been exposed to their lamb confined in a double-walled mesh cage that prevented physical contact nonetheless developed a selective bond with that lamb. In contrast, few ewes subsequently accepted their lamb if it had been kept in an airtight transparent box (thereby eliminating olfactory cues) during the exposure period. The pre-eminence of olfactory cues for maternal recognition and acceptance at the udder is further demonstrated by ewes that were rendered anosmic 2 weeks after parturition (Poindron and Le Neindre, 1980; see also Alexander and Stevens, 1981). Whereas treated ewes discriminated between their own and alien young at a distance, in the majority of cases, those same mothers displayed disrupted nursing behavior ranging from rejection of their lamb through indiscriminate acceptance of their own and alien young.

There have been few investigations of either the chemical nature or the source (site of production or emission) of the odors by which ewes distinguish between familiar and unfamiliar lambs. As mentioned previously, ewes tend to focus on the lamb's anogenital region for identification at close quarters, but odor cues from other areas of the body surface may also be sufficient for maternal recognition (Alexander, 1978). Alexander and Stevens (1982a) approached the question of the source of lambs' individual odors by testing selective ewes for their responses to anesthetized alien lambs treated with various odor samples. Ewes displayed more interest in the odor cues taken from their own lamb compared with comparable samples from alien young for the following regions: a small amount of wool clipped from the rump, trunk, or head–neck; a swab that had been rubbed over the stimulus lamb's rump area. Urine elicited only a weak preference for the ewes' own lamb. With feces samples, however, there was no evidence of ewes recognizing the odor of their offspring. In a similar manner, alien young were not accepted after they had been scrubbed and then smeared with fresh feces taken from the lamb of the subject ewes (Alexander and Stevens, 1981). It was concluded that neither feces nor glands in the digestive tract are the source of the lamb's olfactory signature. Odors associated with amniotic fluid likewise appear not to be implicated in selective maternal acceptance. Ewes that already became bonded to their familiar offspring rejected alien young coated with amniotic fluid, regardless of whether that substance came from their own or an alien lamb (Porter *et al.*, 1994).

Arbitrarily selected scents, such as thyme and lavender, may not be as salient as the lamb's natural biological odor for mediating learned maternal recognition. In tests of maternal selectivity conducted several hours postpartum, ewes rejected equal numbers of alien lambs anointed with a novel scent and those bearing a familiar artificial odor that had been applied to their offspring at birth (Porter *et al.*, 1994). None of these ewes rejected their own (odorized) lamb, however. It was concluded that artificial odorants do not impede learning the lamb's natural olfactory phenotype, nor do they substitute effectively for that odor. Moreover, ewes may be predisposed to learn a limited range of biological odors associated with conspecific neonates. Alternatively, ewes in the previous experiment may have recognized unique mixtures of odors (mosaics) composed of the lamb's individual biological signature plus the artificial scent. The artificial odorant would thus constitute a component of the mosaic signature rather than serve as that recognizable phenotype per se.

B. Underlying Bases of Lambs' Individual Odor Signatures

Demonstrating that lambs can be discriminated by the odor of their wool (or by chemical samples collected from other bodily sources) does not address the question of the underlying basis of those individually distinct signatures. That is, recognizable olfactory signatures could reflect genetically influenced metabolic and biochemical processes, be acquired from the lamb's early environment (possible beginning prenatally), or arise from interactions between these two ultimate bases.

## 1. Genotypic Influences

Because of their unique genetic relationship, twins are a valuable "tool" for investigating genotypic influences on the development of specific phenotypic characteristics. Twin births are common in various domestic breeds of sheep, however, it has been estimated that more than 99% of naturally occurring twins are dizygotic (Rowson and Moor, 1964; Skjervold, 1979). Poindron et al. (1980) reported that immediately after the birth of their first-born twin, Prealpes-du-sud ewes will indiscriminately accept alien neonates as well as their own offspring. Second-born twins, in contrast, were treated differently than alien newborns (when the first twin had remained with the mother after birth). In subsequent experiments, recently parturient ewes more readily accepted their familiar lamb than its (second born) twin that had been removed at birth and maintained in isolation prior to the test session (Porter et al., 1991). Nevertheless, the separated twins were treated more positively than were the alien lambs. It was tentatively concluded that the genetic variability among the (presumably) dizygotic twins may have been sufficient for the development of individually distinctive odor profiles. Even though the signatures of dizygotic twins are not identical, they may be more similar than those of unrelated lambs. Accordingly, ewes may have discriminated between their isolated (or second born) twin and alien young because of a phenotypic resemblance between the former lamb and the familiar first-born twin.

These data are consistent with the hypothesis that individual olfactory signatures may reflect the lamb's genotype, but they could also be accounted for by environmental influences alone. For example, discrimination between familiar versus isolated dizygotic twins might be based upon chemical cues that the former lamb acquired postnatally from its mother (while sucking or when licked by the ewe). To assess further the underlying bases of lambs' recognizable odors, ewes were tested for their responses to both dizygotic twins and monozygotic twins created by bisecting fertilized ova and implanting the genetically identical embryos in the same recipient ewe (Romeyer et al., 1993b). As soon as the first twin was born, it was immediately placed into a double-wall wire-grid cage that prevented the ewe from licking or nursing that lamb, but the ewe could still gain access to its odor. The second-born twin was removed from the mother's pen at birth and isolated. Approximately 4–5 h later, isolated monozygotic twins were better accepted than alien lambs, but this was not the case for the isolated dizygotic twins. Furthermore, (caged) familiar and isolated monozygotic twins did not elicit differential responses from their mothers, but the corresponding dizygotic twins were discriminated. It thus appears that the odor signatures of monozygotic twins may have been more similar than those of the dizygotic twins, enabling the mothers to discriminate more effectively between twins of the latter category. Because the monozygotic twins were not the biological offspring of the ewes that bore them, recognition of the separated lamb by those same females could not have been mediated by genetically influenced cues shared between that lamb and the subject ewe. Rather, once the mother became familiar with the odor of the caged lamb left with her after parturition, she may have discerned a resemblance in its twin that had been removed, and therefore treated it differently than an alien—a process termed *phenotype matching* (Holmes and Sherman, 1982) or *indirect familiarization* (Porter and Blaustein, 1989). The positive relationship that appears to exist between the resemblance of lambs' signatures and their degree of genetic relatedness supports the hypothesis that those recognizable phenotypes are genetically influenced, as shown in mice and rats (Beauchamp *et al.*, 1990; Brown *et al.*, 1990).

## 2. Environmental Influences

In the previously discussed experiment with monozygotic twins, the likelihood of differential contamination of the familiar and isolated twins with odorous substances acquired from the early postnatal environment was reduced by confining the ewe's first-born lamb in a grid cage that precluded direct physical contact between mother and neonate. When ewes in an additional condition were allowed to interact freely (i.e., complete contact) with their first-born lamb, they subsequently discriminated between that lamb and its monozygotic twin that had been removed at birth (Romeyer *et al.*, 1993b). Because these twins were genetically identical, discernible differences in their odor signatures must have had a nongenetic base. Plausible candidates for the odors that enabled ewes to distinguish between their monozygotic twins include substances that the familiar lamb came into contact with while interacting with its mother (e.g., maternal milk, colostrum, saliva, urine).

A possible role of acquired maternal labels in offspring recognition in ungulates was originally proposed in a series of articles by Gubernick (1980, 1981; Gubernick *et al.*, 1979) concerning the behavior of goats. Other data indicate, however, that such maternal labels may not be necessary for either the acceptance of familiar kids or rejection of aliens (Romeyer *et al.*, 1993a), and a similar conclusion follows from relevant studies with sheep. As previously mentioned, ewes developed a selective bond with their newborn offspring, even though direct physical contact was prevented, providing that they otherwise had access to the lamb's salient odor (Poindron and Le Neindre, 1980; Romeyer *et al.*, 1993a). The odor signature by which the lambs were discriminated could not have been transferred to them postnatally from their mother. Selective ewes have also been observed to reject alien young regardless of whether those lambs had been isolated after birth or housed with (and presumably labeled by) their own mother (Porter et al., 1991; see also Lévy et al., 1991).

Maternal labels may not be necessary for ewes to discriminate between their own versus alien young, but if such cues are present, they might be incorporated into the lamb's recognizable odor phenotype. It will be recalled that ewes discriminated between their monozygotic twins after having full contact with one of those lambs, but not when mother–young physical contact had been eliminated during the exposure session (Romeyer et al., 1993b). The recognizable odor profile of a lamb that interacted freely with its mother might reflect a complex mosaic of chemical by-products of bodily processes (genetically mediated and otherwise) and acquired odors (including cues transferred from the mother, and the result of bodily microfauna activity). In contrast, the odor profile of the isolated lamb would have included the same genetically mediated component as its twin, but the postnatal maternal contribution was missing. Such differential labeling was presumably sufficient for the ewes to discern an olfactory difference between their two monozygotic twins. According to this model, when the first-born twin was confined to a mesh cage in the mother's pen, neither that lamb nor its isolated monozygotic twin had the opportunity to acquire maternal labels. Thus, their (genetically influenced) odor signatures remained indistinguishable.

## C. Mechanisms Mediating Discriminative Responses to Own versus Alien Lambs

What is the decision rule that determines whether the familiar lamb will be accepted or rejected? In theory, maternal acceptance could be dependent upon the presence of the familiar odor, the absence of foreign scents, or the joint occurrence of both of these criteria. Data relevant to this question can be found in studies in which ewes have been tested with: (1) their familiar lamb after its characteristic odor has been adulterated by the addition of a foreign scent, or (2) stimulus lambs that lack a salient component of the familiar odor mosaic of the ewes' offspring (Porter et al., 1995).

### 1. Adding a Foreign Scent

Artificial scents that were presumed to mask natural olfactory signatures have often been used in attempts to foster lambs. The effectiveness of this strategy has been assessed in a series of experiments by Alexander and Stevens. For example, subject ewes' own and alien lambs were anesthetized and placed behind a mesh barrier, and their rumps anointed with substances such as household solvents, carnivore anal secretions, fecal odors (i.e., skatole and mercaptoethanol), and plant extracts (Alexander and Stevens, 1982b). Despite the application of the strong odorants, ewes displayed more interest in their own lambs than the aliens—indicating that the lambs' unique odor signatures were still discernible. In related experiments, stimulus lambs were coated more thoroughly with foreign scents. In the majority of the odorant conditions (e.g., detergent, vanilla, eucalyptus, wintergreen, butyric acid), ewes immediately accepted their 2- to 8-day-old odorized lambs, but rejected alien young anointed with the same scent (Alexander and Stevens, 1981, 1985a).

Additional attempts to disguise lambs by having them wear a body "stocking" or coat impregnated with the scent of an alien lamb also had no noticeable effect on individual recognition (Alexander and Stevens, 1985b; Martin *et al.*, 1987). In a similar manner, the (presumed) acquisition of foreign maternal labels did not affect lambs' treatment by their own mother. Ewes readily accepted their familiar offspring following a 6- to 12-h separation period, during which the lambs had sucked from an alien dam (Lévy *et al.*, 1991).

It can be concluded from the data reviewed in this section that once the ewe develops a selective bond with her offspring, application of a novel scent to that lamb (e.g., artificial odorant, alien lamb's odor, foreign maternal "label") has little effect on subsequent maternal acceptance. The rare exceptions involved thorough anointing with strong odorants that may have masked temporarily the natural olfactory signature. This conclusion agrees with that of Price *et al.* (1984), that is, the criterion for maternal acceptance is presence of the lamb's own odor rather than the absence of a foreign scent.

2. *Incomplete Odor Signatures*

Experimental manipulations that provide insights into ewes' responsiveness to incomplete versions of their familiar lamb's olfactory signature include degrading the odor of that same lamb, and the use of alien stimulus lambs that share components of the familiar offspring's odor mosaic. Washing the entire body surface with a detergent, followed by thorough rinsing with clean water, is a relatively simple method of diluting (partially eliminating) a lamb's whole-body odor. Such washing has been found to disrupt markedly ewes' recognition of their own familiar lambs (Alexander and Stevens, 1981; Alexander *et al.*, 1983, 1987). The previously discussed study in which ewes were tested with their isolated and familiar monozygotic twins is also relevent in this context (Romeyer *et al.*, 1993b). That is, fewer mothers accepted the lamb that had been removed and isolated at birth as compared with its monozygotic twin with which they had been allowed full postpartum contact. In this condition, the signature of the isolated lamb may have resembled that of the familiar twin (because of their identical genotypes) but nevertheless lacked a salient environmental component that enabled the two monozygotic twins to be discriminated.

Attempts to facilitate fostering of alien young by transferring odorized coats from ewes' familiar offspring have met with limited success. Rubianes (1992a) reported that only 1 of 13 mothers immediately accepted an alien lamb wearing a body stocking bearing the scent of her own offspring. Higher rates of initial acceptance of aliens (of different breeds) have been reported in related odor-transfer studies (Price et al., 1984; Alexander and Stevens, 1985b); but even in the most successful attempt, 45% of the alien lambs were rejected when first introduced to the subject ewes (Alexander et al., 1985). Eliminating the alien lamb's odor by thorough washing prior to the coat-transfer manipulation had no noticeable effect on acceptance rates (relative to that of unwashed aliens). Accordingly, ewes' treatment of alien lambs did not vary as a function of presence versus absence of a foreign olfactory signature. It appears that an odor stocking may not acquire the complete olfactory signature of the lamb that had worn it. Ewes may reject an alien lamb wearing their own offspring's coat because the alien does not possess all salient components of the familiar odor mosaic.

Artificial odorants have been employed in additional attempts to manipulate olfactory resemblance between own and alien lambs. Immediately after birth, lambs were treated with oil of thyme or lavender (Porter et al., 1994). During tests of maternal acceptance conducted several hours later, ewes rejected alien lambs treated with the familiar odorant (i.e., the one that had been applied to their own offspring) as frequently as aliens anointed with a novel scent. In contrast, not one of those same ewes rejected their own scented lamb following a brief separation period. Thus, the artificial odorant shared by the ewes' own offspring and alien young neither masked nor substituted effectively for the natural olfactory signatures of those lambs.

Similar results have been reported when analogous artificial-odor treatments were initiated after ewes had already established a selective bond with their own lamb. Alexander and Stevens (1985a) anointed familiar and alien lambs with the same odorant at 1-2 days postpartum, then returned them to their respective mothers for the next 24 h. Of the nine scents used, eight were not effective in inducing immediate acceptance of alien lambs treated with the same odor as the ewes' offspring. The sole exception was found when neat's-foot oil was used; however, the highest rate of successful fostering occurred when both the ewe's familiar lamb and the alien had been thoroughly dampened (Alexander et al., 1987). It was suggested that neat's-foot oil (a product derived from cattle skin) is uniquely effective in facilitating fostering because it may resemble lambs' natural odors and therefore more readily substitute for those signatures than would other arbitrarily selected scents.

It can be tentatively concluded from the consistent pattern of results discussed in this section that ewes discriminate between their own lamb's familiar scent and odor signatures that overlap with—but lack relevant features of—that recognizable phenotype (Fig. 1). Thus, full maternal acceptance appears to be dependent upon the presence of all of the salient components of the learned odor mosaic (i.e., products of genetically medi-

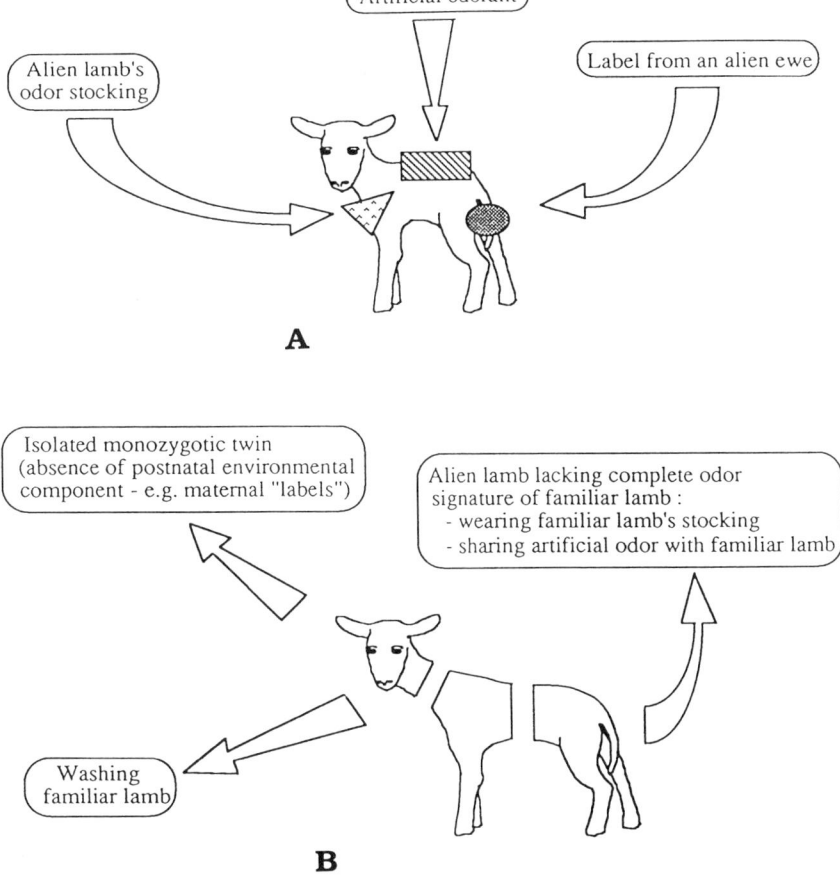

FIG. 1. Mosaic hypothesis of the individual odor signature of the lamb. (A) Addition of a foreign scent to the ewe's familiar lamb has little effect on maternal acceptance. (B) Absence or deletion of salient component of familiar odor signature reduces maternal acceptance.

ated biochemical and metabolic processes, as well as environmentally acquired cues). The addition of foreign odors to the familiar lamb does not disrupt maternal recognition, unless such odorants are sufficiently strong to interfere with the ewe's perception of her offspring's characteristic signature. Therefore, once the ewe establishes a selective bond with her neonate, she rejects alien young because they are not endowed with the complete (learned) odor signature of the familiar lamb, rather than because of the presence of their foreign scent per se.

D. NEUROBIOLOGICAL MECHANISMS OF LAMB ODOR RECOGNITION

A mother's interest in lambs and her ability to selectively recognize her own offspring depend not only on odor but also on parturition triggering her interest in these sensory cues (Keverne *et al.*, 1983). Neural signals resulting from stimulation of the vagina and cervix feed back to the brain to induce both maternal responsiveness and the recognition process, bringing about changes in the olfactory sensory processing system (Kendrick *et al.*, 1992d). The mechanism by which parturition brings about changes in the processing of olfactory signals involve the olfactory bulb, the first relay of olfactory information, and are influenced by maternal experience.

*1. Changes in the Sensory Processing of Olfactory Signals Induced by Birth in Sheep*

The way in which the brain handles olfactory sensory information is undeniably complex, even in relatively simple systems. One such simple system involves mate recognition in mice, which has much in common with mother–infant recognition in sheep (Brennan *et al.*, 1990). Both occur in critical periods following vaginocervical stimulation (mating in mice, parturition in sheep) and both depend on noradrenergic projections. In mice, the changes that occur at the first relay in the accessory olfactory bulb are both necessary and sufficient for recognition to take place. For these reasons, and because previous microdialysis studies had revealed significant changes in neurotransmitter release in the olfactory bulb at parturition in sheep, we decided to analyze systematically how olfactory processing is altered electrophysiologically and neurochemically to accommodate the ewes' behavioral requirement of recognizing their own lambs.

Electrophysiological recordings were made from mitral cells in the same conscious ewes before and after they gave birth (Kendrick *et al.*, 1992d). In recordings made during the last 2 months of pregnancy, none of these cells responded preferentially to lamb or amniotic fluid odors. Indeed, in only 11 cells (10%) were these odors capable of eliciting any significant change in firing rate. The majority of cells (72%) responded preferentially to food odors. Some 3 days after birth, there was a dramatic increase in the number of cells, from this region of the bulb, that now responded preferentially to lamb odors (60%). However, the majority of these cells that responded to lamb odors (70%) did not differentiate between the odor of the ewe's own lamb and that of an alien lamb, and were remarkably resistant to habituation. Nevertheless, a proportion of the cells (30%) did respond preferentially to the odor of the ewe's own lamb. A small proportion of cells was also found that responded preferentially to amniotic fluid odors (11%), while a large reduction was recorded in the number of cells that responded primarily to food odors. These results indicate that, although the odor of lambs have almost no influence on the activity of olfactory bulb neurons during the period before birth, when lambs have no behavioral attraction, they are a very potent olfactory stimulus in the period after birth, when the recognition of lamb odors has a very behavioral priority. Moreover, a proportion of cells respond differentially to the odor of the lamb with which the ewe has formed a selective bond.

The olfactory bulb is a relatively simple trilaminar structure, and its network comprises three basic neural types (Fig. 2; Shepherd, 1972). The mitral cells, which show this increased responsiveness to lamb odors after birth, receive and transmit olfactory signals, and their activity is modulated at their apical dendrites by periglomerular cells and at their lateral dendrites

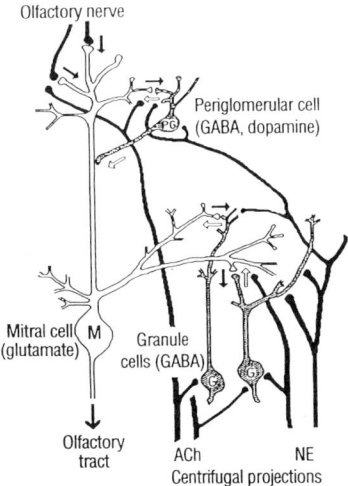

FIG. 2. The synaptic organization of the olfactory bulb. Solid arrows represent excitatory connections and hollow arrows represent inhibitory connections. GABA, gamma-aminobutyric acid; ACh, acetylcholine; NE, norepinephrine.

by granule cells. Intrinsic connections within this network contain both excitatory and inhibitory amino acid transmitters and dopamine (Mori, 1987). Transmissions among neurons in the network is further influenced by centrifugal projections from noradrenergic, cholinergic, and serotonergic neurons that lie deep in the brain (Shipley et al., 1985; Godfrey et al., 1980; Gervais et al., 1984). To further understand how the mitral cells increase their responsiveness to lamb odors, we used in vivo microdialysis to measure their effect on the release of acetylcholine, amino acid, and monoamine transmitters in the olfactory bulb before and after birth (Kendrick et al., 1992d). Before birth, lamb odors produced no changes in the neurotransmitter release that could be detected using microdialysis. After parturition, when ewes had established a selective bond with their lambs, the odors of these lambs, but not those of alien ones, increased the release of both the excitatory amino acid glutamate and the inhibitory gamma-aminobutyric acid. Release of another intrinsic transmitter, dopamine, was not influenced by lamb odors. These changes in glutamate and gamma-aminobutyric acid release occurred only during the first 5 min of exposure to the lamb odor, and the increase of gamma-aminobutyric acid after birth was significantly greater than that of glutamate. Basal release of gamma-aminobutyric acid and glutamate in the period after birth was also significantly higher than in the period before birth and, again, this change in gamma-aminobutyric acid release was significantly higher than that of glutamate. The release of acetylcholine and noradrenaline was influenced by birth, but the increase in their release on exposure to lamb odors did not distinguish between own and alien.

Because the gamma-aminobutyric-acid-containing granule cells are intrinsic bulbar neurons excited by mitral cells and provide feedback inhibition to the mitral cells by way of reciprocal dendrodendritic synapses, the proportionately higher release of gamma-aminobutyric acid compared with glutamate might be explained in terms of a changed efficacy of glutamate at these synapses after birth. Glutamate and gamma-aminobutyric acid release were correlated both before and after birth in all animals, but glutamate release was not always correlated with that of acetylcholine, noradrenaline, or serotonin. Although both glutamate and gamma-aminobutyric acid release were positively correlated before and after birth, the regression slopes for the two periods were significantly different. The overall increase in both glutamate and gamma-aminobutyric acid release in the period after birth is synonymous with more mitral cell activity in response to lamb odors, whereas the significant shift in the regression slope may be a result of an increased efficacy of glutamate in promoting gamma-aminobutyric acid release. Such enhancement of neurotransmitter release at the granule to mitral cell synapses after birth, and in response to the odor of the animal's own lamb, is consistent with the changed firing frequency of these neurons. This situation would then produce a bias in the network with respect to the odors that code for own lamb. This capacity of olfactory bulb circuits to respond to sensory information, contingent upon parturition, in such a way that plastic changes render the processing of olfactory information different on subsequent exposures to stimuli provides the basis for long-term changes that underlie recognition memory.

2. *Influence of Maternal Experience on Lamb Recognition*

The mechanisms by which vaginocervical stimulation, which mimics parturition, induces maternal recognition and behavior in multiparous ewes, are ineffective in nulliparous ewes. However, the same stimulation conducted only 6 h after parturition in ewes having given birth for the first time, now produces acceptance of an alien lamb with complete patterns of normal maternal behavior (Kendrick *et al.,* 1991b). It would seem, therefore, that the neural changes underlying maternal recognition have been produced within 6 h following parturition. These changes are long term and facilitate recognition of lambs at all subsequent births.

Olfactory recognition of lambs is therefore accomplished in two phases starting with the completely naive, nonparturient ewe that rejects all lambs. In the first few hours after parturition, ewes accept all lambs, but within 6 h they have become highly selective, recognizing and accepting only their own lamb. Ewes remain in the unselective phase if the olfactory bulbs are depleted of noradrenaline or if β-adrenergic receptors are blocked; they fail to make specific recognition and therefore accept any lamb (Pissonnier *et al.,* 1985; Lévy *et al.,* 1990a). Following the first birth experience, in all subsequent births the nonselective phase is much shorter and multiparous ewes rapidly develop selective recognition. Moreover, they no longer require the covering of amniotic fluids on their lamb, which was necessary for recognition at the first birth experience (Lévy and Poindron, 1987). Clearly changes in bulbar connection strengths established at the first birth are carried over into subsequent births, and enable the specialist or selective recognition phase to be accomplished faster. This selective recognition phase always requires some remodeling of synaptic connection strengths between intrinsic neurons, and also with their centrifugal noradrenergic and cholinergic connections. The effectiveness of the latter are further influenced by the peptides oxytocin and vasopressin released at birth and during suckling (Lévy *et al.,* 1995a).

The neurotransmitters released at parturition from intrinsic neurons within the olfactory bulb all show significant increases in ewes that have previously experienced olfactory recognition and selective bonding with lambs (Keverne *et al.,* 1993). This would suggest that enhanced activity across the reciprocal mitral granule-cell dendrodendritic synapses (glutamate–gamma-aminobutyric-acid) as well as increased activity in the olfactory glomeruli (dopamine–glutamate) are important for the recognition process. In ewes giving birth for the first time there is no measurable increase in glutamate or gamma-aminobutyric acid release in the olfactory bulb at parturition. However, dopamine release does show increases at the first parturition and basal levels are significantly higher in the olfactory bulb of ewes giving birth for the first time. Hence, parturient ewes experiencing their first olfactory exposure to lambs show enhanced inhibition by dopamine at glomerular synapses without any contingent changes at the reciprocal mitral granule-cell synapses. However, these become established within hours. Ewes with previous maternal experience have overall lower levels of dopamine, but do show increased activity at mitral granule-cell reciprocal synapses during the process of parturition itself.

Although the connection strengths of bulbar intrinsic neurons are essential to the selective recognition process, they cannot achieve this without the assistance of centrifugal inputs. These inputs, aided by the neurohormonal release of intracerebral oxytocin and vasopressin at birth, inform the intrinsic neurons that are responding to the lamb odor that this odor is special. They provide the signal for the laying down of a long-term memory trace. However, it would further appear that the centrifugal inputs are also brought into play during recall, because subsequent lamb odor exposure also results in their release (Kendrick et al., 1992a). This release requires experience, but is not in itself sufficient to distinguish own from alien lamb odors. Unlike the intrinsic transmitters, the release of noradrenaline and acetylcholine are indistinguishable between own and alien lamb odors. However, experience is also important for the release of noradrenaline and acetylcholine at birth itself, because only after the first birth experience can significant changes in acetylcholine transmitter be detected, while noradrenaline shows a fivefold increase at birth in experienced ewes (Lévy et al., 1993). In part these increases in classical transmitter release, dependent on experience, may be facilitated by the peptides oxytocin and vasopressin, both of which are released in significantly greater amounts in multiparous than in primiparous ewes at birth (Lévy et al., 1995a). However, it is not just the greater amount of these peptides that facilitate the release of noradrenaline and acetylcholine, because experience also influences the magnitude of the response of acetylcholine and noradrenaline release to controlled, dose-dependent infusions of these peptides directly into the olfactory bulb by retrodialysis.

The present studies have gone some way to revealing the neurochemical changes that underlie the different phases of olfactory recognition. Parturition induces massive increases in the release of noradrenaline and acetylcholine in the OB, a release that is enhanced by the peptides oxytocin and vasopressin, also recovered in high concentrations from the OB by microdialysis at parturition. Although the release of transmitters from centrifugal projections contingent on stimulation of mitral cells by lamb odors is essential for recognition, there appears to be no direct interaction of acetylcholine or noradrenaline on mitral cells. Nor does the release of bulbar acetylcholine and noradrenaline distinguish between familiar and strange lambs. Hence, any influence that these centrifugal transmitters exert requires the involvement of periglomerular and granule cell interneurons. The data reported here show that once learning has occurred, there is a decrease in periglomerular dopamine release, and resulting disinhibition at the glomerular level may well underlie the increased sensitivity to lamb odors in the early postpartum period. Selectivity in recognition is best explained by the changes that occur between mitral granule cells and their interaction with centrifugal afferents. Changes in the granule cells make them more sensitive to glutamate, which would enhance feedback inhibition on the mitral cell. However, because there is also an enhancement of inhibiting influences on the granule cell via noradrenergic innervation and the autoreceptor feedback inhibition, the overall outcome is a change in the oscillatory frequency of the mitral granule-cell unit, producing an increased frequency in mitral cell burst firing. This proposed mechanism is substantiated by the electrophysiological recordings following lamb recognition when more units respond to lamb odors, and some units differentiate between own and alien lambs with increased firing frequency. Of course, these changes in firing frequency of mitral cells only represents the first stage in the processing of information that underlies recognition memory. How the rest of the brain handles this information has to be the subject of future studies.

## IV. Conclusion

The data reviewed in this article are consistent with the pioneering conclusion of Rosenblatt and Lehrman (1963) that "The maternal behavior pattern, . . . , emerges from a set of interlocking causal relationships." These relationships involve physiological sensory and experiential factors that come into play for inducing both maternal responsiveness and selectivity (Fig. 3). Changes in steroid balance, together with the stimulation of the genital tract, appear to be the physiological starting point common to both phases. This stimulation increases oxytocin release in various brain regions and it seems that the PVN is the major site involved in maternal responsiveness. The action of this peptide modulated by different factors, such as opiates and corticotropin-releasing factor, could in part induce changes in

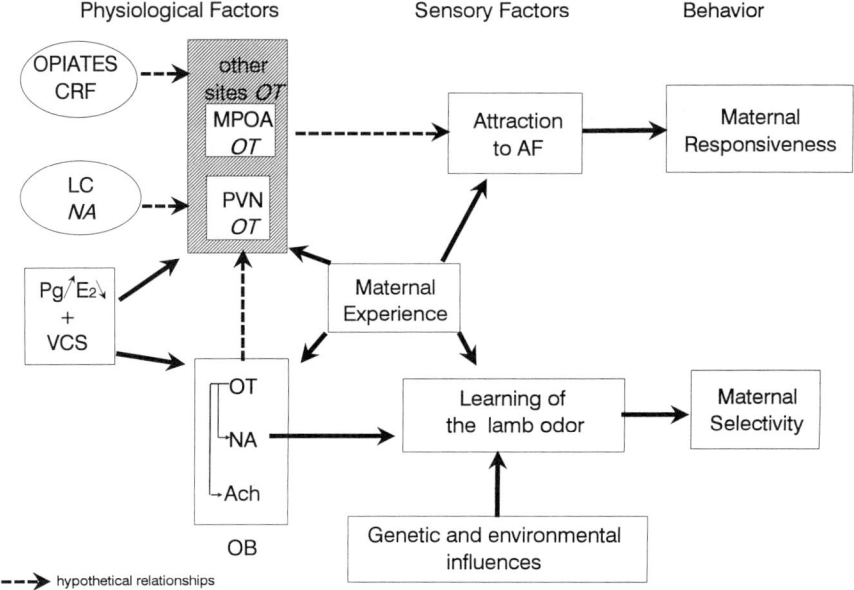

FIG. 3. Schema depicting the physiological and the sensory factors involved in the control of maternal responsiveness and selectivity in sheep. ACh, acetylcholine; AF, amniotic fluid; BO, olfactory bulb; CRF, corticotropin-releasing factor; E, estradiol; LC, locus coeruleus; MPOA, medial preoptic area; NA, noradrenaline; OB, olfactory bulb; Pg, progesterone; PVN, paraventricular nucleus of the hypothalamus; OT, oxytocin; VCS, vaginocervical stimulation.

preference towards amniotic fluid, thus allowing mother–young contact and an immediate onset of care. Steroids and vaginocervical stimulation are also responsible for neurochemical changes within the olfactory bulb that are a part of the learning mechanisms of the individual lamb odor. In particular, the release of noradrenaline enhanced by oxytocin appears to be essential for lamb recognition. Mothers learn all the salient components of the odor of their young, which are the results of genotypic and environmental influences. The functional significance of the development of selective maternal bonding within the first several hours following parturition becomes apparent when considering the natural history of sheep. Because ewes and their suckling lambs congregate in large flocks, and even the newborn are fully capable of independent locomotion, nursing mothers face the continual risk of an alien lamb gaining access to their limited milk supply and thereby depriving their own offspring of necessary nutrients. Accordingly, there is pressure for ewes to develop as quickly as possible the ability to discriminate between their own and alien young. In contrast, in those species that give birth to altricial offspring that are then confined to a nest, it is unlikely that alien young will attempt to suckle from the resident female. There is less urgency in this context for mothers to develop rapidly the ability to recognize their neonates. Maternal discrimination between own versus alien young may become biologically meaningful only when they emerge from their nests and offspring of different females begin to intermingle.

Obviously, this is an oversimplification. In particular, the neural circuitry involved in both aspects of maternal behavior has to be more carefully investigated. One can ask the question of the existence of one network that controls maternal behavior or rather the existence of several neural circuits that underline the expression of each component of maternal behavior as is believed to be the case in the rat (for review, see Stern, 1989). Apparently, in sheep the MPOA could participate in the inhibition of aggressive behavior towards lambs but not in the enhancement of acceptance behavior (Kendrick *et al.*, 1992c), whereas activation of the PVN is able to induce the full repertoire of maternal responsiveness (Da Costa *et al.*, 1996).

As for maternal selectivity, the olfactory bulb is certainly not the sole repository for olfactory recognition of the lamb. From the main olfactory bulb, two main pathways for olfactory memory have been identified (Wilson and Sullivan, 1994). One pathway involves the hippocampus via the entorhinal cortex and contributes to the representation of relations among odor so that they can use these representations in new contexts (Eichenbaum *et al.*, 1989). The other pathway involves the mediodorsal nucleus of the thalamus, which projects to the frontal cortex. This pathway is claimed to participate in complex olfactory learning set problems (Slotnick and Kaneko, 1981). Investigations of the respective role of these two pathways would further our understanding of the mechanisms by which lamb odor is memorized. However, anatomical studies of the relationships between the OB and the other parts of the brain need to be undertaken first.

Together with the assessment of a unique area of the brain that controls maternal behavior, it would also be naive to assume that one hormone or neurotransmitter mediates maternal behavior. For instance, oxytocin is certainly not "the" hormone of maternal behavior, and opiates and corticotropin-releasing factor appear to be implicated also. Activation of the noradrenaline system might participate not only in olfactory recognition but also in maternal responsiveness in acting at the level of the MPOA or the PVN. The participation in the processes of olfactory learning may include not only noradrenergic afferents of the OB but also other centrifugal systems. We have begun an investigation of the role of the cholinergic system and it appears that blocking the muscarinic receptors at parturition prevents the ewe from being selective (Richard *et al.,* 1994). Moreover, physiological studies have revealed that parturition induced numerous other changes in transmitters such as dopamine, gamma-aminobutyric acid, and glutamate (see Section II,A), but we have no precise idea concerning their role. Are they directly involved or do they act in modulating the release of other transmitters as oxytocin does on noradrenaline release in the OB?

Overall, it is clear that females that have undergone several parental cycles develop a greater ability to respond to the physiological factors and the presence of the young. In the naive animal, one physiological or sensory factor alone has no "meaning" in terms of the induction of maternal behavior. This meaning arises with the experience of pregnancy and parturition during which there is an association of all the internal and the environmental factors (i.e., sensory cues from the newborn). This association could set up a neural network that appears to be the underlying bases of what is called "maternal experience" so that experienced ewes become responsive to one or a few determinants. On the other hand, the absence of some factors in the normal situation of the emergence of maternal behavior is easily compensated by experienced mothers. This suggests that, at parturition, some remodeling occurs in the neural structures controlling maternal behavior. Such reorganization has already been demonstrated in the OB during learning of lamb's odor (see Section III,D). The nature of this restructuring is unknown so far. It could concern an increase in synaptic density or in the number of receptors. In mice, the number of cells containing E receptors increased in the limbic system after parental experience either in the female or in the male (Ehret *et al.,* 1993; Koch, 1990). There could also be a consequence of a change of neuronal–glial relationships allowing juxtaposition and association of neuronal elements that results in a better efficiency of transmitter release, as occurs at parturition in the oxytocinergic system of the rat (Theodosis and Poulain, 1987; Tweedle and Hatton, 1987).

This article has outlined our knowledge of the onset of maternal behavior, but little is known about the factors responsible for the maintenance and the breaking down of the bond between the female and her young. If the mother is first responsible for the maintenance of the mother–young bond during the first 4 weeks, after this period the ewe appears less interested in maintaining close proximity with the lambs (Hinch *et al.,* 1987, 1990). However, it must be stressed that, throughout all lactation and even thereafter, the mother and her offspring are most of the time more closely associated with one another than with any other member of the flock (Morgan and Arnold, 1974; Hinch *et al.,* 1987, 1990; Lawrence, 1990). Moreover, at 2 months postpartum, the mother still reacts strongly, although less than during the first weeks after birth, to the separation of the young (Poindron *et al.,* 1993a). Although at this stage the tendency for flocking behavior has already increased in the dam, she still remains undisturbed by social deprivation if her lamb is present in the test arena (Poindron *et al.*, 1993a). A study from Poindron and Le Neindre (1980) suggests that olfaction is also important for the maintenance of selectivity but extensive investigations examining the respective role of other sensory stimuli have not been carried out. As for the physiological factor, it was suggested that the tactile stimulation of sucking by the lamb and the level of the milk yield would be important for the maintenance of maternal behavior (Poindron and Le Neindre, 1980; Arnold *et al.*, 1979). However, in a preliminary study, Poindron *et al.* have observed that covering the udder for 1 week at 2 1/2 months after parturition does not alter the mother–young bond, a result similar to that found in cattle (Veissier *et al.*, 1989). Obviously, which factors and how they control the maintenance and breakdown of maternal–infant bonds are important issues and they offer a large field of further research.

## V. Summary

The present article focuses on the immediate factors that control the onset of maternal behavior at parturition in sheep and, to a lesser extent, in goats. Changes in steroid balance, together with the stimulation of the genital tract occurring at parturition, is the key factor for triggering a cascade of neurobiological mechanisms that allows maternal responsiveness and selectivity. Central release of oxytocin, mainly in the paraventricular nucleus of the hypothalamus, modulated by opiates and corticotropin-releasing factor, is involved in the induction of maternal responsiveness. These factors induce preference towards the odor of amniotic fluid, thus allowing mother–young contact and an immediate onset of care. Steroids and vaginocervical stimulation are also responsible for neurochemical and electrophysiological changes within the olfactory bulb that are a part of the learning mechanisms of the individual lamb odor, thus allowing the establishment of a selective bond between the ewe and her lamb. There is an increase in the number of mitral cells, the principal cells of the olfactory bulb, that respond to lamb odors, which is associated with increased release of glutamate and gamma-aminobutyric acid from the dendrodendritic synapses between the mitral and granule cells. The relation between the release of each transmitter after birth suggests an increased efficacy of glutamate-evoked-gamma-aminobutyric-acid release. Parturition is also accompanied by increased oxytocinergic, cholinergic, and noradrenergic neurotransmitter release that are essential for selective recognition of lambs. These increases in transmitter release depend on maternal experience in a way that greater amounts have been found in multiparous than in primiparous ewes.

Offspring recognition by ewes and selective maternal care are mediated by individually characteristic odors of the neonate that arise through a complex interaction of genetically influenced biochemical processes and nongenetic factors. Adding a foreign scent to a lamb has little effect on maternal recognition. On the other hand, recognition is disrupted if all of the salient components of the lamb's familiar odor signature are not available. Therefore, alien young are rejected because they lack the complete mosaic signature of the familiar lamb, and not because of the presence of foreign odors per se.

**References**

Alexander, G. (1978). Odour, and the recognition of lambs by Merino ewes. *Appl. Anim. Ethol.* **4**, 153–158.

Alexander, G., and Stevens, D. (1981). Recognition of washed lambs by Merino ewes. *Appl. Anim. Ethol.* **7**, 77–86.

Alexander, G., and Stevens, D. (1982a). Failure to mask lamb odour with odoriferous substances. *Appl. Anim. Ethol.* **8**, 253–260.

Alexander, G., and Stevens, D. (1982b). Odour cues to maternal recognition of lambs: An investigation of some possible sources. *Appl. Anim. Ethol.* **9**, 165–175.

Alexander, G., and Stevens, D. (1985a). Fostering in sheep. II. Use of hessian coats to foster an additional lamb on to ewes with single lambs. *Appl. Anim. Behav. Sci.* **14**, 335–344.

Alexander, G., and Stevens, D. (1985b). Fostering in sheep. III. Facilitation by the use of odorants. *Appl. Anim. Behav. Sci.* **14**, 345–354.

Alexander, G., Stevens, D., and Bradley, L. R. (1983). Washing lambs and confinement as aids to fostering. *Appl. Anim. Ethol.* **10**, 251–261.

Alexander, G., Stevens, D., and Bradley, L. R. (1985). Fostering in sheep. I. Facilitation by use of textile lamb coats. *Appl. Anim. Behav. Sci.* **14**, 315–334.

Alexander, G., Stevens, D., and Bradley, L. R. (1987). Fostering in sheep. V. Use of unguents to foster an additional lamb onto a ewe with a single lamb. *Appl. Anim. Behav. Sci.* **17**, 95–108.

Alexander, G., Stevens, D., and Bradley, L. R. (1990). Distribution of field birth-sites of lambing ewes. *Aust. J. Exp. Agric.* **30**, 759–767.

Arnold, G. W., and Dudzinski, M. L. (1978). Maternal behaviour. *In* "Ethology of Free-Ranging Domestic Animals" (G. W. Arnold and M. L. Dudzinski, eds.), pp. 137–165. Elsevier, Amsterdam.

Arnold, G. W., and Morgan, P. D. (1975). Behaviour of the ewe and lamb at lambing and its relationship to lamb mortality. *Appl. Anim. Ethol.* **2**, 24–46.

Baldwin, B. A., and Shillito, E. E. (1974). The effects of ablation of the olfactory bulbs on parturition and maternal behaviour in Soay sheep. *Anim. Behav.* **22**, 220–223.

Basiouni, G. F., and Gonyou, H. W. (1988). Use of birth fluids and cervical stimulation in lamb fostering. *J. Anim. Sci.* **66**, 872–879.

Beauchamp, G. K., Yamazaki, K., Duncan, H., Bard, J., and Boyse, E. A. (1990). Genetic determination of individual mouse odour. *In* "Chemical Signals in Vertebrates" (V. D. W. Macdonald, D. Muller-Schwarze, and S. E. Natynczuk, eds.), pp. 244–254. Oxford Univ. Press, Oxford.

Bouissou, M. F. (1968). Effet de l'ablation des bulbes olfactifs sur la reconnaissance du jeune par sa mere chez les ovins. *Rev. Comp. Anim.* **3,** 77–83.
Brennan, P., Kaba, H., and Keverne, E. B. (1990). Olfactory recognition: A simple memory system. *Science (Washington, D. C.)* **250,** 1223–1226.
Bridges, R. S. (1990). Endocrine regulation of parenting behaviour in rodents. *In* "Mammalian Parenting, Biochemical, Neurobiological and Behavioral Determinants" (N. A. Krasnegor and R. S. Bridges, eds.), pp. 93–117. Oxford Univ. Press, New York.
Broad, K. D., Kendrick, K. M., Sirinathsinghji, D. J. S., and Keverne, E. B., (1993). Changes in oxytocin immunoreactivity and mRNA expression in the sheep brain during pregnancy, parturition and lactation and in response to oestrogen and progesterone. *J. Neuroendocrinol.* **5,** 435–444.
Broad, K. D., Keverne, E. B., and Kendrick, K. M. (in press). Corticotrophin releasing factor expression in the sheep during pregnancy, parturition and lactation and following exogenous progesterone and oestrogen treatment. *Mol. Brain Res.*
Brown, R. E., Roser, B., and Singh, P. B. (1990). The MHC and individual odours in rats. *In* "Chemical Signals in Vertebrates" (V. D. W. Macdonald, D. Muller-Schwarze, and S. E. Natynczuk, eds.), pp. 228–243. Oxford Univ. Press, Oxford.
Caba, M., Poindron, P., Krehbiel, D. W., Lévy, F., Romeyer, A., and G. Venier (1995). Naltrexone delays the onset of maternal behavior in primiparous and multiparous parturient ewes. *Pharmacol. Biochem. Behav.* **52,** 743–748.
Challis, J. R. G. (1971). Sharp increase in free circulating oestrogen immediately before parturition. *Nature (London)* **229,** 208.
Chamley, W. A., Buckmaster, J., Cereni, M. E., Cumming, I. A., Goding, J. R., Obst, J. M., Williams, A., and Winfield, C. (1973). Changes in the levels of progesterone, corticosteroids, estrone, oestradiol-17β, luteinizing hormone and prolactin in the peripheral plasma of the ewe during late pregnancy and at parturition. *Biol. Reprod.* **9,** 30–35.
Currie, W. B., Gorewit, R. C., and Michel, F. J. (1988). Endocrine changes, with special emphasis on oestradiol-17β, prolactin and oxytocin, before and during labour and delivery in goats. *J. Reprod. Fert.* **82,** 299–308.
Da Costa, A., Broad, K., Guevara-Guzman, R., and Kendrick, K. M. (1993). C-Fos expression and neurotransmitter release in brain structures involved in the induction of maternal behaviour in sheep. *Soc. Neurosci. Abstr.* **19,** 1021.
Da Costa, A., Guevara-Guzman, R., Ohkura, S., Goode, J., and Kendrick, K. M. (1996). The role of the paraventricular nucleus in the control of maternal behaviour in sheep. *J. of Neuroendocrinology* **8,** 163–177.
Davis, S. L., Reichert, L. E., and Niswender, L. E. (1971). Serum levels of prolactin as measured by radioimmunoassay. *Biol. Reprod.* **4,** 145–147.
Ehret, G., Jürgens, A., and Koch, M. (1993). Oestrogen receptor occurrence in the male mouse brain: Modulation by paternal experience. *Neuroreport* **4,** 1247–1250.
Eichenbaum, H., Mathews, P., and Cohen, J. J. (1989). Further studies of hippocampal representation during odor discrimination learning. *Behav. Neurosci.* **103,** 1207–1216.
Fleming, A. S., Vaccarino, F., Tambosso, L., and Chee, P. (1979). Vomeronasal and olfactory system modulation of maternal behavior in the rat. *Science* (Washington, D. C.) **203,** 372–374.
Fredriksson, G. (1985). Release of PGF 2-alpha during parturition and postpartum period in the ewe. *Theriogenology* **24,** 331–335.
Geist, V. (1971). "Mountain Sheep." Univ. of Chicago Press, Chicago.
Gervais, R., Araneda, S., and Pujol, J. F. (1984). Effects of local 5,6–dihydroxytryptamine in the rat olfactory bulb responsiveness during wakefulness and sleep. *Electroencephalogr. Clin. Neurophysiol.* **57,** 462–472.

Gervais, R., Holley, A., and Keverne, E. B. (1988). The importance of central noradrenergic influences on the olfactory bulb in the processing of learned olfactory cues. *Chem. Senses* **13**, 3–12.

Gervais, R., Mouly, A. M., Elaagouby, A., and Lévy, F. (1990). Olfactory bulb plasticity. In "Chemosensory Information Processing. Information Processing of Chemical Sensory Stimuli in Biological and Artificial Systems" (D. Schild, ed.), pp. 191–200. Springer-Verlag, Berlin.

Godfrey, D. A., Ross, C. D., and Matschinsky, F. M. (1980). Distribution and derivation of cholinergic elements in the rat olfactory bulb. *Neuroscience* **5**, 273–292.

Grubb, P., and Jewell, P. A. (1966). Social grouping and home range in feral Soay sheep. *Symp. Zool. Soc. London* **18**, 179–210.

Gubernick, D. J. (1980). Maternal 'imprinting' or maternal 'labelling' in goats? *Anim. Behav.* **28**, 124–129.

Gubernick, D. J. (1981). Mechanisms of maternal 'labelling' in goats. *Anim. Behav.* **29**, 305–306.

Gubernick, D. J., Jones, K. C., and Klopfer, P. H. (1979). Maternal 'imprinting' in goats. *Anim. Behav.* **27**, 314–315.

Hinch, G. N., Lécrivain, E., Lynch, J. J., and Elwin, R. L. (1987). Changes in maternal-young associations with increasing age of lambs. *Appl. Anim. Behav. Sci.* **17**, 305–318.

Hinch, G. N., Lynch, J. J., Elwin, R. L., and Green, G. C. (1990). Long term association between Merino ewes and their offspring. *Appl. Anim. Behav. Sci.* **27**, 93–103.

Holley, A. (1991). Neural coding of olfactory information. In "Smell and Taste in Health and Disease" (T. V. Getchell, R. L. Doty, L. M. Bartoshuk, and J. B. Snow, Jr., eds.) pp. 329–343. Raven Press, New York.

Holmes, W. G., and Sherman, P. W. (1982). The ontogeny of kin recognition in two species of ground squirrels. *Am. Zool.* **22**, 491–517.

Kendrick, K. M., and Keverne, E. B. (1989). Effects of intracerebroventricular infusions of naltrexone and phentolamine on central and peripheral oxytocin release and on maternal behaviour induced by vaginocervical stimulation in the ewe. *Brain Res.* **505**, 329–332.

Kendrick, K. M., and Keverne, E. B. (1991). Importance of progesterone and estrogen priming for the induction of maternal behavior by vaginocervical stimulation in sheep: Effects of maternal experience. *Physiol. Behav.* **49**, 745–750.

Kendrick, K. M., Keverne, E. B., Baldwin, B. A., and Sharman, D. F. (1986). Cerebrospinal fluid levels of acetylcholinesterase, monoamines and oxytocin during labour, parturition, vagino-cervical stimulation, lamb separation and suckling in sheep. *Neuroendocrinology* **44**, 149–156.

Kendrick, K. M., Keverne, E. B., and Baldwin, B. A. (1987). Intracerebroventricular oxytocin stimulates maternal behaviour in sheep. *Neuroendocrinology* **46**, 56–61.

Kendrick, K. M., Keverne, E. B., Chapman, C., and Baldwin, B. A. (1988a). Microdialysis measurement of oxytocin, aspartate, gamma-aminobutyric acid and glutamate release from the olfactory bulb of the sheep during vaginocervical stimulation. *Brain Res.* **411**, 171–174.

Kendrick, K. M., Keverne, E. B., Chapman, C., and Baldwin, B. A. (1988b). Intracranial dialysis measurement of oxytocin, monoamine and uric acid release from the olfactory bulb and substantia nigra of sheep during parturition, suckling, separation from lambs and eating. *Brain Res.* **439**, 1–10.

Kendrick, K. M., Keverne, E. B., Hinton, M. R., and Goode, J. A. (1991a). Cerebrospinal fluid and plasma concentrations of oxytocin and vasopressin during parturition and vaginocervical stimulation in the sheep. *Brain Res. Bull.* **26**, 803–807.

Kendrick, K. M., Lévy, F., and Keverne, E. B. (1991b). Importance of vaginocervical stimulation for the formation of maternal bonding in primiparous and multiparous parturient ewes. *Physiol. Behav.* **50**, 595–600.

Kendrick, K. M., da Costa, A. P., Hinton, M. R., and Keverne, E. B. (1992a). A simple method for fostering lambs using anoestrous ewes with artificially induced lactation and maternal behaviour. *Appl. Anim. Behav. Sci.* **34,** 345–357.

Kendrick, K. M., Fabre-Nys, C., Blache, D., Goode, J. A., and Broad, K. D. (1992b). The role of oxytocin release in the mediobasal hypothalamus of the sheep in relation to female sexual receptivity. *J. Neuroendocrinol.* **4,** 1–9.

Kendrick, K. M., Keverne, E. B., Hinton, M. R., and Goode, J. A. (1992c). Oxytocin, amino acid and monoamine release in the region of the medial preoptic area and bed nucleus of the stria terminalis of the sheep during parturition and suckling. *Brain Res.* **569,** 199–209.

Kendrick, K. M., Lévy, F., and Keverne, E. B. (1992d). Changes in the sensory processing of olfactory signals induced by birth in sheep. *Science (Washington, D. C.)* **256,** 833–836.

Keverne, E. B., and de la Riva, C. (1982). Pheromones in mice: Reciprocal interactions between the nose and the brain. *Nature (London)* **296,** 148–150.

Keverne, E. B., and Kendrick, K. M. (1991). Morphine and corticotrophin-releasing factor potentiate maternal acceptance in multiparous ewes after vaginocervical stimulation. *Brain Res.* **540,** 55–62.

Keverne, E. B., Lévy, F., Poindron, P., and Lindsay, D. R. (1983). Vaginal stimulation: An important determinant of maternal bonding in sheep. *Science (Washington, D. C.)* **219,** 81–83.

Keverne, E. B., Lévy, F., Guevara-Guzman, R., and Kendrick, K. M. (1993). Influence of birth and maternal experience on olfactory bulb neurotransmitter release. *Neuroscience* **56,** 557–565.

Kinsley, C. H., and Bridges, R. S. (1990). Morphine treatment and reproductive condition alter olfactory preferences for pup and adult male odors in female rats. *Dev. Psychobiol.* **23,** 331–347.

Koch, M. (1990). Effects of treatment with estradiol and parental experience on the number and distribution of estrogen-binding neurons in the ovariectomized mouse brain. *Neuroendocrinology* **51,** 505–514.

Krehbiel, D., Poindron, P., Lévy, F., and Prud'Homme, M. J. (1987). Peridural anesthesia disturbs maternal behavior in primiparous and multiparous parturient ewes. *Physiol. Behav.* **40,** 463–472.

Lawrence, A. B. (1990). Mother-daughter bonds in sheep. *Anim. Behav.* **42,** 683–685.

Lécrivain, E., and Janeau, G. (1987). Comportement d'isolement et de recherche d'abri de brebis agnelant en plein air dans un système d'élevage à caractère extensif. *Biol. Behav.* **12,** 127–148.

Le Neindre, P., Poindron, P., and Delouis, C. (1979). Hormonal induction of maternal behavior in non-pregnant ewes. *Physiol. Behav.* **22,** 731–734.

Lévy, F., and Poindron, P. (1984). Influence du liquide amniotique sur la manifestation du comportement maternel chez la brebis parturiente. (Influence of amniotic fluid in the manifestation of maternal behaviour in parturient ewe). *Biol. Behav.* **9,** 65–88.

Lévy, F., and Poindron, P. (1987). The importance of amniotic fluids for the establishment of maternal behaviour in experienced and inexperienced ewes. *Anim. Behav.* **35,** 1188–1192.

Lévy, F., Poindron, P., and Le Neindre, P. (1983). Attraction and repulsion by amniotic fluids and their olfactory control in the ewe around parturition. *Physiol Behav.* **31,** 687–692.

Lévy, F., Gervais, R., Kindermann, U., Orgeur, P., and Piketty, V. (1990a). Importance of β-noradrenergic receptors in the olfactory bulb of sheep for recognition of lambs. *Behav. Neurosci.* **104,** 464–469.

Lévy, F., Keverne, E. B., Piketty, V., and Poindron, P. (1990b). Physiological determinism of olfactory attraction for amniotic fluids in sheep. *In* "Chemical Signals in Vertebrates 5"

(D. W. Macdonald, D. Müller-Schwarze, and S. E. Natynczuk, eds.), pp. 162–165. Oxford Univ. Press, Oxford.

Lévy, F., Gervais, R., Kindermann, U., Litterio, M., Poindron, P., and Porter, R. (1991). Effects of early post-partum separation on maintenance of maternal responsiveness and selectivity in parturient ewes. *Appl. Anim. Behav. Sci.* **31**, 101–110.

Lévy, F., Kendrick, K. M., Keverne, E. B., Piketty, V., and Poindron, P. (1992). Intracerebral oxytocin is important for the onset of maternal behavior in inexperienced ewes delivered under peridural anesthesia. *Behav. Neurosci.* **106**, 1–6.

Lévy, F., Guevara-Guzman, R., Hinton, M. R., Kendrick, K. M., and Keverne, E. B. (1993). Effects of parturition and maternal experience on noradrenaline and acetylcholine release in the olfactory bulb of sheep. *Behav. Neurosci.* **107**, 662–668.

Lévy, F., Kendrick, K. M., Goode, J. A., Guevara-Guzman, R., and Keverne, E. B. (1995a). Oxytocin and vasopressin release in the olfactory bulb of parturient ewes: Changes with maternal experience and effects on acetylcholine, gamma-aminobutyric acid, glutamate and noradrenaline release. *Brain Research* **669**, 197–206.

Lévy, F., Locatelli, A., Piketty, V., Tillet, Y., and Poindron, P. (1995b). Involvement of the main but not the accessory olfactory system in maternal behavior of primiparous and multiparous ewes. *Physiol. Behav.* **57**, 97–104.

Lickliter, R. E. (1982). Effects of a post-partum separation on maternal responsiveness in primiparous and multiparous domestic goats. *Appl. Anim. Ethol.* **8**, 537–542.

Lickliter, R. E. (1984). Mother-infant spatial relationships in domestic goats. *Appl. Anim. Ethol.* **13**, 93–100.

Lindsay, D. R., and Fletcher, I. C. (1968). Sensory involvement in the recognition of lambs by their dams. *Anim. Behav.* **16**, 415–417.

Martin, N. L., Price, E. O., Wallach, S. J. R., and Dally, M. R. (1987). Fostering lambs by odor transfer: The add-on experiment. *J. Anim. Sci.* **64**, 1378–1383.

McNeilly, A. S., Martin, M. J., Chard, T., and Hart, I. C. (1972). Simultaneous release of oxytocin and neurophysin during parturition in the goat. *J. Endocrinol.* **52**, 213–214.

Morgan, P. D., and Arnold, G. W. (1974). Behavioural relationships between Merino ewes and lambs during the four weeks after birth. *Anim. Prod.* **19**, 169–176.

Morgan, P. D., Boundy, C. A. P., Arnold, G. W., and Lindsay, D. (1975). The roles played by the senses of the ewe in the location and the recognition of lambs. *Appl. Anim. Ethol.* **1**, 139–159.

Mori, K. (1987). Membrane and synaptic properties of identified neurons in the olfactory bulb. *Prog. Neurobiol.* **29**, 275,

Nowak, R. (1990). Mother and sibling discrimination at a distance by three- to seven-day-old lambs. *Dev. Psychobiol.* **23**, 285–295.

O'Brien, P. H. (1984). Leavers and stayers: Maternal post-partum strategies in feral goats. *Appl. Anim. Behav. Sci.* **12**, 2233–2243.

Pissonnier, D., Thiery, J. C., Fabre-Nys, C., Poindron, P., Keverne, E. B. (1985). The importance of olfactory bulb noradrenaline for maternal recognition in sheep. *Physiol. Behav.* **35**, 361–364.

Poindron, P. (1974). Etude de la relation mere-jeune chez les brebis Ovis aries lors de l'allaitements. *C. R. Acad. Sci. Paris* **278**, 2691–2694.

Poindron, P. (1976a). Effets de la suppression de l'odorat, sans lesion des bulbes olfactifs, sur la selectivite du comportement maternel de la Brebis. *C. R. Acad. Sci. Paris* **282**, 489–491.

Poindron, P., and Carrick, M. J. (1976). Hearing recognition of the lamb by its mother. *Anim. Behav.* **24**, 600–602.

Poindron, P., and Le Neindre, P. (1980). Endocrine and sensory regulation of maternal behavior in the ewe. *Adv. Study Behav.* **11**, 75–119.

Poindron, P., Martin, G. B., and Hooley, R. D. (1979). Effects of lambing induction on the sensitive period for the establishment of maternal behaviour in sheep. *Physiol. Behav.* **23,** 1081–1087.

Poindron, P., Le Neindre, P., Raksanyi, I., Trillat, G., and Orgeur, P. (1980). Importance of the characteristics of the young for the manifestation and establishment of maternal behavior in sheep. *Reprod. Nutr. Dev.* **20,** 817–826.

Poindron, P. Lévy, F., and Krehbiel, D. (1988). Genital, olfactory, and endocrine interactions in the development of maternal behaviour in the parturient ewe. *Psychoneuroendocrinology* **13,**(1/2), 99–125.

Poindron, P., Rempel, N., Troyer, A., and Krehbiel, D. (1989). Genital stimulation facilitates maternal behavior in estrous ewes. *Horm. Behav.* **23,** 305–316.

Poindron, P., Caba, M., Gomora Arrati, P., Krehbiel, D., and Beyer, C. (1993a). Responses of maternal and non-maternal ewes to social and mother-young separation. *Behav. Proc.* **31,** 97–110.

Poindron, P., Nowak, R., Lévy, F., Porter, R. H., and Schaal, B. (1993b). Development of exclusive mother-young bonding in sheep and goats. *In* "Oxford Reviews of Reproductive Biology," Vol. 15 (S. R. Milligan, ed.), pp. 311–364. Oxford Univ. Press, Oxford.

Porter, R. H., and Blaustein, A. R. (1989). Mechanisms and ecological correlates of kin recognition. *Sci. Progr.* **73,** 53–66.

Porter, R. H., and Lévy, F. (1995). Olfactory mediation of mother-infant interactions in selected mammalian species. *In* "Biological Perspectives on Motivated and Cognitive Activities" (R. Wong, ed.), Ablex, Norwood, NJ, pp. 77–110.

Porter, R. H., Lévy, F., Poindron, P., Litterio, M., Schaal, B., and Beyer, C. (1991). Individual olfactory signatures as major determinants of early maternal discrimination in sheep. *Dev. Psychobiol.* **24,** 151–158.

Porter, R. H., Romeyer, A., Lévy, F., Krehbiel, D., and Nowak, R. (1994). Investigations of the nature of lambs' individual odour signatures. *Behav. Proc.* **31,** 301–308.

Porter, R. H., Lévy, F., Nowak, P., Orgeur, P., and Schaal, B. (in press). Lambs' individual odor signatures: Mosaic hypothesis. *In* "Chemical Signals in Vertebrates VII" (R. Apfelbach, ed.), Pergamon Press, Oxford.

Pretel, S., and Piekut, D. T. (1990). Coexistence of corticotrophin releasing factor peptide and oxytocin mRNA in the paraventricular nucleus. *Peptides* **11,** 621–624.

Price, E. O., Dunn, G. C., Talbot, J. A., and Dally, M. R. (1984). Fostering lambs by odor transfer: The substitution experiment. *J. Anim. Sci.* **59,** 301–307.

Richard, Ph., Meurisse, M., Ravel, N., and Lévy, F. (1994). Involvement of cholinergic system in recognition of lamb odor by parturient ewes. European Chernoreceptior Organization XI, pp. 89. Blous, France.

Romeyer, A., Porter, R. H., Lévy, F., Nowak, R., Orgeur, P., and Poindron, P. (1993a). Maternal labelling is not necessary for the establishment of discrimination between kids by recently parturient goats. *Anim. Behav.* **46,** 705–712.

Romeyer, A., Porter, R. H., Poindron, P., Orgeur, P., Chesne, P., and Poulin, N. (1993b). Recognition of dizygotic and monozygotic twin lambs by ewes. *Behaviour* **127,** 119–139.

Romeyer, A., Poindron, P., and Orgeur, P. (1994). Olfaction mediates the establishment of selective bonding in goats. *Physiol. Behav.* **56,** 693–700.

Rosenblatt, J. S., and Lehrman, D. S. (1963). Maternal behavior of the laboratory rat. *In* "Maternal Behavior in Mammals" (H. L. Rheingold, ed.), pp. 8–57. John Wiley & Sons, New York.

Rosenblatt, J. S., and Siegel, H. I. (1981). Factors governing the onset and maintenance of maternal behaviour among nonprimate mammals. *In* "Parental Care in Mammals" (D. J. Gubernick and P. H. Klopfer, eds.), pp. 13–76. Plenum Press, New York.

Rowson, L. E. A., and Moor, R. (1964). Occurrence and development of identical twins in sheep. *Nature (London)* **201,** 521–522.
Rubianes, E. (1992a). Will Corriedale ewes accept odor-transferred lambs? *Appl. Anim. Behav. Sci.* **35,** 91–95.
Rubianes, E. (1992b). Genital stimulation modifies behavior towards amniotic fluid in estrous ewes. *Appl. Anim. Behav. Sci.* **35,** 35–40.
Schneirla, T. C., Rosenblatt, J. S., and Tobach, E. (1963). Maternal behaviour in the cat. *In* "Maternal Behavior in Mammals" (H. L. Rheingold, ed.), pp. 122–168. John Wiley & Sons, New York.
Shepherd, G. M. (1972). Synaptic organization of the mammalian olfactory bulb. *Physiol. Rev.* **52,** 864–917.
Shipka, M. P., and Ford, S. P. (1991). Relationship of circulating estrogen and progesterone concentrations during late pregnancy and the onset phase of maternal behavior in the ewe. *Appl. Anim. Behav. Sci.* **31,** 91–99.
Shipley, M. T., Halloran, F. G., and de la Torre, J. (1985). Surprisingly rich projection from the locus coeruleus to the olfactory bulb in the rat. *Brain Res.* **329,** 294–299.
Skjervold, H. (1979). Causes of variation in sex ratio and sex combination in multiple births in sheep. *Livest. Prod. Sci.* **6,** 387–396.
Slotnick, B. M., and Kaneko, N. (1981). Role of mediodorsal thalamic nucleus in olfactory discrimination learning in rats. *Science (Washington, D. C.)* **214,** 91–92.
Stern, J. M. (1989). Maternal behavior: Sensory, hormonal, and neural determinants. *In* "Psychoendocrinology" (F. R. Brush and S. Levine, eds.), pp. 105–226. Academic Press, San Diego.
Theodosis, D. T., and Poulain, D. A. (1987). Oxytocin-secreting neurones: A physiological model for structural plasticity in the adult mammalian brain. *Trends in Neuroscience* **10,** 426–430.
Tweedle, C. D., and Hatton, G. I. (1987). Morphological adaptability at neurosecretory axonal endings on the neurovascular contact zone of the rat neurohypophysis. *Neuroscience* **20,** 241–246.
Veissier, I., Le Neindre, P., and Trillat, G. (1989). Adaptability of calves during weaning. *Biol. Behav.* **14,** 66–87.
Vince, M. A., Lynch, J. J., Mottershead, B., Green, G., and Elwin, R. (1985). Sensory factors involved in immediately postnatal ewe/lamb bonding. *Behaviour* **94,** 60–84.
Walser, E. E. S., Hague, P., and Walters, E. (1981). Vocal recognition of recorded lambs' voices by ewes of three breeds of sheep. *Behaviour* **78,** 261–272.
Wilson, D. A., and Sullivan, R. M. (1994). Neurobiology of associative learning in the neonate: Early olfactory learning. *Behav. Neural Biol.* **61,** 1–18.

# Socialization, Hormones, and the Regulation of Maternal Behavior in Nonhuman Simian Primates

Christopher R. Pryce*

ANTHROPOLOGY INSTITUTE
UNIVERSITY OF ZÜRICH
ZÜRICH, SWITZERLAND

## I. Introduction

In mammals, infant care is essential for infant survival, growth, and development. Only *maternal behavior* can promote all of these infant states, and then only if considerable energy and time are expended on maternal behavior by a female in the appropriate physiological state for lactation. The aim of this article is to synthesize the theory and evidence concerning the regulation of maternal behavior in mammals belonging to the order Primates. There are some 200 living species of primate, which can be grouped naturally into the prosimian lemurs, lorises, bush babies, and tarsiers (suborder Prosimii), and the simian monkeys, apes, and humans (suborder Anthropoidea). A simplified phylogenetic tree for living primates is provided in Fig. 1. The present synthesis is restricted to nonhuman simian primates, because the monkeys and apes have been studied more extensively in terms of their maternal behavior than have the prosimian primates (although see Klopfer and Boskoff, 1979; Shively and Mitchell, 1986a); there are unique sociocultural determinants of maternal behavior in humans (e.g., Levine *et al.*, 1994) that would render their inclusion here as inappropriate.

In essence, my aim here is to review the theory and evidence impinging on the following hypotheses of maternal behavior regulation in monkeys and apes: (1) preadult experience of other females' infants in the natal social group is sufficient for the demonstration of species-typical maternal behavior by a female toward her own infants; (2) preadult experience of other females' infants in the natal social group is essential for the demonstration of species-typical maternal behavior by a female toward her own infants; and (3) hormonal changes occurring during late pregnancy (prepar-

---

* Present address: Department of Anthropology, The University of Auckland, Private Bag 92019, Auckland, New Zealand.

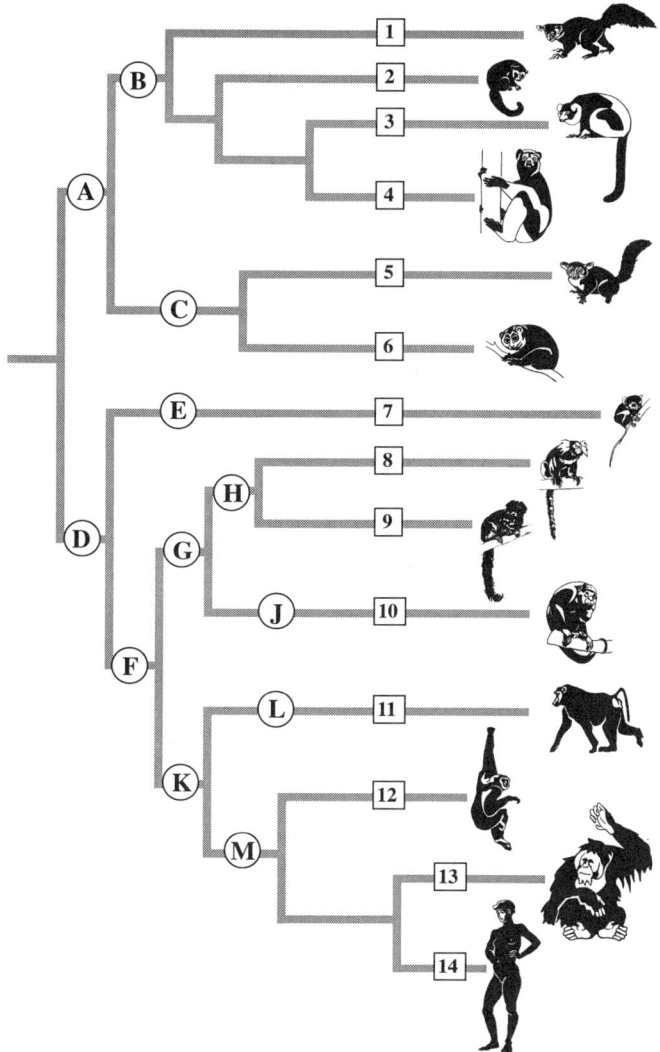

Fig. 1. Simplified phylogenetic tree for living primates. Numbers indicate individual groups: 1–7 = prosimians: 1 = aye-aye, 2 = dwarf lemur group, 3 = true lemurs, 4 = indri group, 5 = bush babies, 6 = lorisines, 7 = tarsiers; 8–10 = New World monkeys: 8 = marmosets and tamarins, 9 = Goeldi's monkey, 10 = cebid monkeys (e.g., squirrel monkey, titi monkey, owl monkey); 11 = cercopithecoid (Old World) monkeys (e.g., macaques, baboons, vervet monkeys, gelada, langurs, colobus monkeys); 12–13 = apes: 12 = lesser apes (gibbons and siamang), 13 = great apes (orang-utan, gorilla, chimpanzee, bonobo); 14 = humans. Twins–triplets, use of a maternal nest, infant transport by holding the dorsal skin between the teeth,

tum), and labor and parturition (peripartum), are essential for the demonstration of species-typical maternal behavior toward her own infants by a fully socialized female. Hypotheses 1 and 2 are referred to as follows as the preadult socialization hypotheses, and Hypothesis 3 as the neuroendocrine hypothesis of maternal behavior regulation in nonhuman simian primates. I have constructed these specific and explicit hypotheses in order to provide a clear rationale for this article. However, the cause–effect relationships contained therein have already been proposed by others. This is particularly true of the proposal, stated explicitly in Hypothesis 1, that preadult experience with infants is sufficient for the demonstration of maternal behavior in nonhuman simian primates (e.g., Coe, 1990, Keverne, 1995). This view is based on the large body of evidence describing the frequent infant-directed behavior of preadult females in primate social groups. It is, however, incompatible with the proposal that prepartum and peripartum neuroendocrine stimulation are essential components of maternal behavior regulation. This view, stated explicitly in Hypothesis 3 previously, has been proposed only rarely for primates (Rosenblatt, 1991; Pryce, 1992) and, accordingly, very few studies have attempted to test it in any primate species. This, in turn, is in direct contrast to the situation in nonprimate mammals, for which there is detailed evidence for neuroendocrine regulation of maternal behavior in two species—the laboratory rat (*Rattus norvegicus*) and domesticated sheep (*Ovis aries*)—which have been studied extensively in this respect.

A framework for the present synthesis is provided in Fig. 2. The approach taken is strongly influenced by the previous reviews of nonhuman primate maternal behavior, namely, those of Higley and Suomi (1986), Coe (1990), Nicolson (1991), Rosenblatt (1991), and Keverne (1995). Section II describes the "context" in which the regulation of maternal behavior occurs in monkeys and apes, by summarizing relevant aspects of their lifestyles, morphology, and reproductive life histories. Section III presents the evidence for and against the preadult socialization hypotheses (Fig. 2), and from both causal and functional perspectives. Section IV of this synthesis considers the neuroendocrine hypothesis. First, for comparative purposes, a resume of the evidence for the neuroendocrine regulation of maternal

---

FIG. 1. (*Continued*) occur in Groups 1, 2, 3, and 5. Twins occur in Group 8. Carriage of the infant by the mother and father occurs in Groups 8, 9, and in some species in 10. Otherwise, a single infant clinging to the mother's body is the typical pattern of simian primate care (groups 10–13). (From Martin, 1995. Reprinted with kind permission of the author and Karger Publishers.)

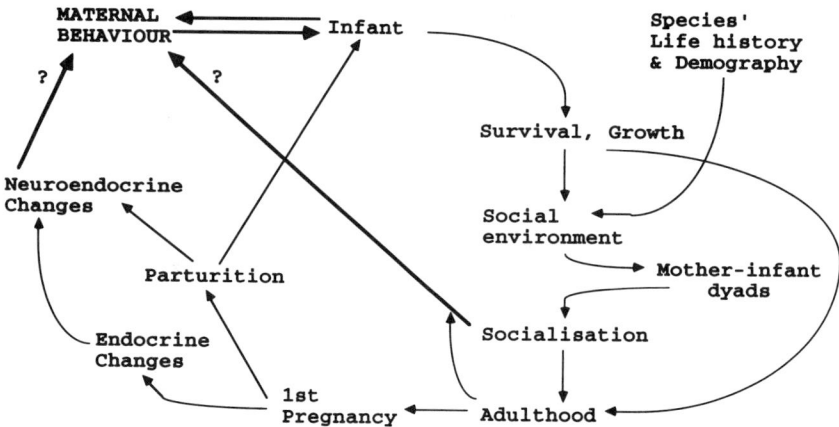

FIG. 2. Framework for the analysis of the regulation of maternal behavior in nonhuman simian primates. Question marks refer to the hypotheses that preadult socialization with mother–infant dyads is sufficient or essential for the demonstration of postpartum maternal behavior, and to the hypothesis that prepartum and peripartum neuroendocrine changes are essential for the demonstration of postpartum maternal behavior.

behavior in nonprimate mammals (i.e., the very evidence that prompted the hypothesis of neuroendocrine regulation of primate maternal behavior) is provided. Second, the hormonal changes that occur during pregnancy and at parturition in various monkeys and apes are described. Third, the evidence for and against the neuroendocrine regulation of maternal behavior in simian nonhuman primates (Fig. 2) is presented. Finally, Section V integrates the evidence for and against the socialization and neuroendocrine hypotheses, and, by logical extension of this evidence, proposes a model containing both social and neuroendocrine causal factors, as a basis for future research into the regulation of primate maternal behavior.

## II. The "Context" of Maternal Behavior Regulation

Existing estimates suggest that the evolution of primates from ancestral stocks of nonprimate mammals began about 80 million years ago (Martin, 1993). The proximate mechanisms that regulate maternal behavior in primates of modern aspect (Fig. 1) can be attributed to the following: (1) the mechanisms that were inherited from the ancestral stocks of nonprimate mammals; and (2) any additional selection pressures exerted during the course of primate evolution, by their ecological environments, lifestyles, morphological traits, and life-history traits. It is a parsimonious assumption that primate patterns of maternal behavior and the proximate regulation of this behavior have coevolved with the other aspects of primate biology listed in (2). The aim that follows is to provide an overview of this "coevolutionary context" of maternal behavior regulation. Detailed reviews of simian primate biology, from which the present overview is derived, are provided by Jolly (1985), Napier and Napier (1985), Smuts *et al.* (1987), Fleagle (1988), and Martin (1990).

A. LIFESTYLES

Referring to Fig. 1, all the New World monkeys are arboreal, whereas some Old World monkey genera are arboreal and others are terrestrial. Among the apes, the gibbons (*Hylobates* spp.), siamang (*Symphalangus syndactylus*), and orang-utan (*Pongo pygmaeus*) are arboreal, and the gorilla (*Gorilla gorilla*), chimpanzee (*Pan troglodytes*), and bonobo (*Pan paniscus*) are both arboreal and terrestrial. It is noteworthy that each of the major monkey and ape groups contains tree-living species that locomote by utilizing their ability to grasp and cling to branches with their forelimbs and hindlimbs; this adaptation is unique to primates and is very important in the context of their maternal behavior (Section II,C). With regard to activity patterns, all except one of the simian primates are diurnal, and parturition and the onset of maternal behavior occur after the onset of darkness, that is, in the inactive phase of the mother's diurnal rhythm (Jolly, 1972).

The term *primate society* (e.g., Hinde, 1976; Smuts *et al.,* 1987) refers to the social interactions, relationships, and structures that exist between group-living primate conspecifics. Primate groups are the product of interindividual attraction and aversion, and of life-history parameters, most notably, birth, death, and migration rates (Dunbar, 1987, 1988; Fig. 2). Species-typical group sizes are related to, and probably selected by, adult body size and ecological conditions such as predation and food distribution (Jolly, 1985; Smuts *et al.,* 1987); they range from 1 to more than 100 individuals. Typically, New World monkeys live in social groups of 4–20, and Old World monkeys in groups of 10–100. Among the apes, gibbons and siamangs live in groups of 2–4; orang-utans, with the exception of mothers and their immature offspring, are solitary; gorillas live in groups of 10–20; and chimpanzees in temporary parties of 2–10 within stable communities of 30–100. For any given species, the age, sex, and kinship composition of the typical social group is determined by the average rates of birth, death, and migration. Concerning migration out of the group, this normally occurs in young adulthood and is usually restricted to one sex while the other demonstrates philopatry (Pusey and Packer, 1987).

B. INFANT CHARACTERISTICS

Most monkeys and apes produce single infants, although twinning is typical of the New World marmosets and tamarins (Fig. 1). Compared with nonprimate mammals, monkey and ape infants are large relative to maternal body size and their brains make up a relatively large proportion of total infant body size (Martin, 1990). Accordingly, monkey and ape infants are relatively mature, or "precocial," at birth—their hair covering is dense, they are endohomeothermic, and their sensory and motor systems are well developed.

As in other animal groups, simian infants possess characteristics that differentiate them from older conspecifics, and that are unique in their ability to elicit maternal behavior (see Maestripieri and Call, this volume). In mammals, generally, infant-specific stimuli can take the form of odors and tastes, vocalizations, visual appearance in terms of body size and proportions and behavior, and the somatosensory stimuli of body contact and suckling. The evidence, although limited, indicates that each of these communication pathways is important in the elicitation of maternal behavior in nonhuman simian primates.

The newborn infant is covered in placental membranes, amniotic fluid, and secretions from the mother's reproductive tract; potentially, these are important sources of olfactory and gustatory communication between infant and mother. That this is indeed the case is suggested by the mother's intensive licking of the newborn infant, a behavior that constitutes an important component of the onset of maternal behavior in all monkeys and apes (Shively and Mitchell, 1986b; see as follows). Furthermore, there is one direct piece of evidence in support of this hypothesis. A common practice in some macaque (*Macaca* spp.; Fig. 1) laboratories is artificial delivery of infants by cesarean section. Infants delivered in this way acquire some "nonmacaque" odors; they also lack the maternal secretions with which vaginally delivered infants are normally covered. In rhesus (*M. mulatta*) and longtailed (*M. fasicularis*) macaques, cesarean-delivered infants do not elicit maternal behavior from their postoperative mothers unless they are smeared with the latters' prepartum vaginal secretions. If this procedure is performed, then females do demonstrate maternal behavior to their—perhaps familiar-smelling—infants (Lundblad and Hodgen, 1980).

Infant distress vocalization, or "crying," is a trait common to all primates (Newman, 1985; see Maestripieri and Call, this volume). Simian primate infants emit such calls when they are not in body contact with their mother or another familiar caregiver; indeed, they are already capable of crying immediately after birth if they are not able to cling to, or are rejected by, the postpartum mother. Infant distress calls are potent elicitors of maternal behavior (e.g., retrieval, carrying, see Maestripieri and Call, this volume) and of maternal arousal (Coe, 1990) in simian primates. With regard to infant crying as an elicitor of maternal arousal, or "maternal anxiety" (Pryce, 1992; Maestripieri, 1993a,b), studies in the squirrel monkey provide the key empirical evidence. Squirrel monkey mothers separated physically but not audibly from their infants demonstrate behavioral agitation (Kaplan, 1970) and increased cortisol output (Coe et al., 1978). Also, squirrel monkey mothers can discriminate between their own infants and another female's on the sole basis of its (familiar) vocalizations (Kaplan et al., 1978). Infants stop crying when they gain or regain clinging body contact with the mother, an event that presumably reduces arousal in (or is anxiolytic to) mothers.

Primates are specialized with respect to the visual nervous system (e.g., Allman, 1982), and it is possible that an infant monkey or ape's small body and relatively large head with foreshortened muzzle act as attractive or "cute" (Lorenz, 1971) elicitors of simian maternal behavior (see Coe, 1990). The limited empirical evidence supports this hypothesis. Female rhesus macaques were exposed simultaneously to an unfamiliar infant aged less than 1 month and an unfamiliar infant aged 12 months; the first exposure was visual only and this was followed by auditory, olfactory, and visual exposure. Multiparous and primiparous females, which were not caring for an infant of their own at the time they were tested, chose to look at and then to interact with the young–small infants for longer than the old–large infants (Sacket and Ruppenthal, 1974).

Turning to somatosensory stimulation, newborn monkeys and apes already possess the unique primate specialization of being able to grasp and cling with forelimbs and hindlimbs (see aforementioned). Even before they have emerged completely from the mother's reproductive tract, infants demonstrate reflex grasping with their hands until they grasp and cling to hair on the mother's leg or body. This sequence has been described for various primates, as reviewed by Rosenblatt (1991), and, in the usual case of the grasping newborn infant establishing clinging body contact with its mother, is followed by the reflexes of rooting and suckling that are common to all mammalian infants. Clinging, rooting, and suckling are demonstrated by infants only, and somatosensory stimulation must surely represent the most specialized form of infant–mother communication, when compared against the communication that occurs in simian social dyads generally.

C. Maternal Behavior Patterns

Nonhuman simian maternal behavior is stimulated exclusively by the stimuli described previously. It commences immediately after the expulsion of the infant and develops into a long-term dyadic discriminatory relationship, in which mothers are much more responsive to their own infant than to other infants in the social group, and infants are much more strongly attached to their own mother than to other mothers in the social group. Parturient maternal behavior is reviewed in detail by Shively and Mitchell (1986b). The parturient female may reach down to touch or pull the emerging infant and to support it as it grasps and initiates clinging contact. In order that the emerging infant can initiate clinging contact with the mother's body hair, it is essential that the mother remains stationary during and immediately after expulsion, so that she is in direct contact with or standing directly over the newborn infant. Immediately postpartum, intense licking and grooming of the infant alternate with licking and grooming of self. This activity is suspended temporarily as the placenta is expelled and consumed (e.g., common marmoset: Stevenson, 1976; Rothe, 1978; rhesus macaque: Tinklepaugh and Hartman, 1930; Teas *et al.,* 1981). This sequence of events differs between primiparous (first time) mothers, in which the physiological–psychological state of parturition and maternal behavior is a novel experience, and multiparous mothers, in which it is not. Primiparae are more generally active, spend more time engaged with the neonate and less time engaged with themselves, and have a longer latency before the neonate is in the ventroventral suckling position, compared with multiparae. An excellent study of these parity-related quantitative behavioral differences is provided by Kemps and Timmermans (1984) for the longtail macaque.

Postpartum patterns of maternal behavior, the formation of exclusive mother–offspring dyads, and the integration of changes in maternal behavior with infant development have been described in quite considerable detail in wild populations and/or captive colonies of various species of nonhuman simian primate. Some important and influential studies are those of wild savannah baboons (*Papio cynocephalus*: Altmann, 1980; Altmann and Samuels, 1992; Altman *et al.,* 1988); vervet monkeys (*Cercopithecus aethiops*), wild (Lancaster, 1972; Cheney *et al.,* 1988; Lee and Bowman, 1995) and captive (Fairbanks, 1990); wild gelada (*Theropithecus gelada:* Dunbar, 1988; Dunbar and Dunbar, 1988; Barrett *et al.,* 1995); rhesus macaques, wild (Johnson, 1986) and captive (Sackett *et al.,* 1967; Hinde and Simpson, 1975; Simpson and Simpson, 1986; Gomendio, 1995); wild hanuman langurs (*Presbytis entellus:* Hrdy, 1977); wild (Nishida, 1983; Goodall, 1986) and captive (Coe, 1990) chimpanzees; captive common marmosets (*Callithrix jacchus:* Box, 1977; Ingram, 1977; Locke-Haydon and Chalmers, 1983); captive cotton-top tamarins (*Saguinus oedipus:* Cleveland and Snowdon, 1984; Price, 1992a,b; Tardif, 1994; Tardif *et al.,* 1990); and captive squirrel monkeys (Rosenblum, 1968, 1972; Andrews *et al.,* 1993). The remainder of Section II is derived from the general picture that emerges from these important studies.

Monkey and ape maternal behavior is characterized by the infant clinging on to the mother's body hair with its hands and feet. The ability to grasp and cling with forelimbs and hindlimbs is adaptive for an arboreal lifestyle in the "fine-branch niche," which, the evidence suggests, was the scene of primate origins (Martin, 1990). If this is indeed why clinging evolved, then it served as a preadaptation for a major event in the evolution of simian maternal behavior; namely, mother–infant carriage replacing maternal nest building and transport of infants by holding their dorsal skin between the teeth (Fig. 1). Mother–infant carriage affords the infant transport, access to the nipples for suckling, thermoregulation, stimulation, secure attachment, a vantage point for observation of mother–environment interaction, opportunity for solid-food sharing, and protection. Mothers carry their infants dorsally and, particularly during nursing, ventrally. Ventral carrying without nursing occurs more often in terrestrial monkeys than in arboreal monkeys, and this is probably related to the navigational difficulties that arboreal species would experience with a ventrally clinging infant. Monkey and ape mothers nurse their infants frequently and in short bouts (Short, 1984). (The terminology used here is that of Hall *et al.,* 1988: mothers nurse and infants suckle and suck.)

In some New World monkeys reproductive males also demonstrate a substantial amount of infant carriage. In fact, this occurs in those species that demonstrate twinning and/or high infant growth rates and that are monomorphic (e.g., marmosets, tamarins, Goeldi's monkey (*Callimico*), owl monkey (*Aotus*), titi monkey (*Callicebus*); Fig. 1) (Tardif, 1994; Ross and MacLarnon, 1995). In some Old World monkeys infant carriage by adult females other than the mother occurs, and most notably in the langurs and the colobus monkeys (*Colobus* spp.) (Nicolson, 1987; Maestripieri, 1994). However, such "aunting behavior" by adult female Old World monkeys occurs to a lesser extent and is sometimes less-obviously affiliative than is paternal behavior in the monomorphic New World monkeys.

D. MATERNAL EXPENDITURE

Infant growth and development (Fig. 2) are slow and therefore extended processes in all simian primates compared with nonprimate mammals of equivalent adult body size (Payne and Wheeler, 1968; Harvey *et al.,* 1987). To give some examples of primate weaning ages (where weights refer to adult female body weights), the 0.3 to 0.4-kg captive common marmoset is weaned at Weeks 10–12 (Ingram, 1977), the 5 to 7-kg captive rhesus macaque is weaned at Weeks 60–70 (Hinde and Spencer-Booth, 1967;

Gomendio, 1989), and the 30 to 35-kg wild chimpanzee is weaned at Weeks 200–250 (Clark, 1977; Goodall, 1986). The slow process of simian infant growth and development is accompanied by a gradual reduction in the amount of time that a mother spends carrying her infant. However, reduction in maternal carriage does not equate to a reduction in maternal expenditure. Increased foraging and feeding to support increased lactation, grooming the infant, intervening in the infant's exchanges with other group members, and retrieving infants in situations of distress and uncertainty, are also examples of behavior that are essential components of motherhood and that promote infant survival, growth, and development (Gomendio, 1995; Lee and Bowman, 1995). Therefore, female simian primates expend considerable energy and time on each offspring, and expenditure on each offspring represents a quite substantial proportion of a female's lifetime reproductive potential. Variation in infant survival is likely to be the major contributor to variation in reproductive success between adult females (e.g., Cheney *et al.*, 1988), and such variation would exert high selection pressure on heritable components of maternal behavior regulation.

Social and ecological resources are limited in primate social groups, and life is based on "selfish" competition, agonism, and punishment (e.g., Clutton-Brock and Parker, 1995). Yet, at the birth of her infant, a female primate's behavior reveals a dramatic and unique shift in priorities. The needs of this infant cause the primate mother to rechannel her limited resources of energy and time into activity aimed at offspring benefit at maternal cost (Dunbar, 1988). Female nonhuman simian primates, like most female mammals: (1) demonstrate behavior conducive to infant survival, growth, and development; and (2) restrict this behavior to their own infants. The question now is, how, in proximate terms, is this strategy achieved?

### III. Preadult Socialization and Maternal Behavior

#### A. Infant-Directed Behavior of Preadult Females

Clearly then, the primary relationship of the primate infant is that experienced with the mother (Fig. 2). The classical maternal deprivation research instigated by H. F. Harlow has demonstrated that infant–mother attachment is the very *basis* of primate socialization, including the socialization of maternal behavior (e.g., Arling and Harlow, 1967; Ruppenthal *et al.*, 1976; Kraemer, 1992). However, consideration of the causal relationship that exists between simian primary attachment and maternal behavior is beyond the scope of this article, which is restricted to studies of the regulation of maternal behavior in *mother-reared* females (Fig. 2). A secure primary attachment with the mother promotes exploratory behavior in the developing infant, including interactions with and the formation of relationships with other members of the social group. The social environment comprises familiar conspecifics and interactions with certain of these are frequent and predictable. Social play is a major activity of preadults in social groups of monkeys and apes (Fagen, 1981; Chalmers, 1983; Lee, 1983a; Walters, 1987; Gandelman, 1992). Furthermore, because primate maturation is a slow process (Section II,D) and because primates remain in their natal social group until young adulthood at least (Section II,A), then juveniles and adolescents obtain social experience with the young infants born in their natal groups.

Juvenile and adolescent females observe and interact with mother–infant dyads, and observe and interact with young infants directly. Direct interactions between preadult females and infants include smelling, licking, touching, pulling, retrieving, carrying, holding, grooming, and playing. For example, among New World monkeys, captive common marmoset juveniles and adolescents carry each twin infant sibling aged 0–4 weeks for 10–15% of the time (Ingram, 1977; Locke-Haydon and Chalmers, 1983), and captive adolescent cotton-top tamarins carry each twin infant sibling aged 0–5 weeks for 15–20% of the time (Cleveland and Snowdon, 1984). Among Old World monkeys, juvenile females interact one to three times per hour during the first month with infants born in subsequent birth seasons, for example, wild savannah baboons (Walters, 1987), wild and captive vervet monkeys (Lancaster, 1972; Lee, 1983b; Fairbanks, 1990), and wild hanuman langurs (Hrdy, 1977). Among the apes, chimpanzee female adolescents and pre-reproductive (i.e., nulliparous) adults interact with infants for 2–3% of the time across the first 2 years of infant life (Nishida, 1983). The behavior patterns directed by preadult females at infants can elicit, on the one hand, infant crying, avoidance, returning to the mother, or retrieval by the mother; and, on the other hand, the infant clinging to, being held by, being groomed by, or playing with the preadult female (Lancaster, 1972; McKenna, 1981; Quiatt, 1979; Maestripieri, 1994). Several terms are in use to describe the apparent attempts at and the actual affiliative infant-directed behavior of preadult nonhuman simian females. A type of *infant handling* is probably the most parsimonious descriptor of such behavior (e.g., Nicolson, 1987; Maestripieri, 1994). *Play-mothering* is the term introduced by Lancaster (1971) to describe the holding, grooming, and carrying of other females' infants by juvenile female vervet monkeys, and has subsequently been applied to other Old World monkeys (e.g., Quiatt, 1979). In the New World marmosets and tamarins the amount of carrying demonstrated by juvenile and, in particular, adolescent females is substantial (see aforementioned) and elevates them to the status of "care-helpers" rather than infant handlers or play-mothers (e.g., Goldizen, 1987; Pryce, 1993).

On the basis of detailed observations of wild vervet monkeys, Lancaster proposed the learning-to-mother hypothesis as a causal and functional explanation for the affiliative behavior directed by preadult female nonhuman simians toward other females' young offspring (Lancaster, 1971). The central thrust of the hypothesis is that females will be socialized for future motherhood (Fig. 2) by repetitive preadult experience of the following specific chain of events: demonstrate nonmaternal-like behavior at infant (e.g., pulling, rough-and-tumble playing) → infant cries → loss of infant to mother → experience of punishment. According to this hypothesis, to avoid the punishment of infant loss, preadult females learn to restrict their playmothering to maternal-like behavior only, for example, holding, carrying.

Additional to direct interactions between preadult females and infants are the opportunities to observe and interact with mother–infant dyads. Depending on the cognitive abilities of preadult monkeys and apes, this type of social experience could be of major relevance to the socialization of maternal behavior (Fig. 2). Two types of social information exchange have been demonstrated in the animal kingdom. Imitative social learning is the cognitive ability to gain information about the performance and consequences of a *behavior* via visual exposure to a conspecific performing and being rewarded for that behavior (Heyes, 1993; Whiten and Ham, 1992). Nonimitative social learning is the cognitive ability to gain information about a *stimulus* via sensory exposure to a conspecific's interaction with that stimulus (Heyes, 1993). Both imitative and nonimitative social learning demand an ability to give meaning to information derived from conspecifics and both are processes via which the behavior of an "observer" individual can be regulated by the behavior of a "demonstrator" conspecific. A small number of studies have been carried out with monkeys aimed at identifying whether or not preadult experience with infants could involve, amongst other processes, acquisition of information about maternal behavior via imitative social learning, or about infants via nonimitative social learning. These studies are described here.

Dienske *et al.* (1980) studied captive female rhesus macaques that were reared by their mothers for about 4 months and then in physical isolation until adulthood. Although females were physically isolated, they did have visual and auditory access to conspecifics—one group ("observers") was exposed to breeding females with infants, while the other group ("nonobservers") was exposed to nonbreeding females. As adults, the study females were paired with an adult male until they conceived. Females that had not observed breeding females caring for infants did not permit their own first infants to gain body contact at parturition; rather, they avoided and neglected them so that their infants had to be removed. In contrast, observers demonstrated behavior towards their own first infants that was adequate for infant survival. However, observers did not demonstrate the species-typical maternal behavior to which they had been visually exposed; rather, they held the infant in an atypical position, for example, infants were held upside down, on a shoulder, on a leg, or by the head (Dienske *et al.,* 1980). This study provides evidence that female rhesus macaques are able to acquire information about infants but not about maternal behavior via visual and auditory exposure to mother–infant pairs and, furthermore, that nonimitative social learning can substitute for direct play-mothering experience. Within days such females were demonstrating species-typical maternal behavior, and this was probably acquired via trial-and-error learning of the relationship between ventral carrying and the cessation of infant crying.

In another study (Pryce, 1993) adolescent common marmosets without care-helping experience were isolated from their family groups and presented with their unfamiliar infant siblings in one-on-one tests. The effects of the following experiences were tested: *no* exposure to parent–infant dyads; *visual* exposure to parent–infant dyads; *full* physical access and exposure to parent–infant dyads. Adolescents with no exposure demonstrated carrying behavior because their crying infant siblings climbed on to them; however, average bouts of carrying were very short because adolescents responded by rubbing clinging infants against the cage wall or biting them, and infants responded to this by dismounting and resuming crying. Visual experience of parent–infant dyads did not alter this sequence of events, but full experience did. Adolescents with full physical access and exposure to parent–infant dyads demonstrated greater amounts of carrying behavior; they carried infant siblings for longer before using controlled-aggressive behaviors to dislodge them. This was already the case after only 12 h of full experience, and despite parents preventing the adolescents from retrieving and carrying their infant siblings during this time. Therefore, it would appear that during their full experience of parent–infant dyads in the typical setting of the social group, adolescent common marmosets acquire information about infant stimuli that regulates their behavior toward these stimuli, that is, nonimitative social learning. This might involve stimulus–stimulus conditioning during simultaneous exposure to unfamiliar infant stimuli and familiar (reinforcing) parental stimuli (e.g., smell infant odor–smell mother odor; groom infant–groom mother) (Pryce, 1993).

In the rhesus macaque study described previously (Dienske *et al.,* 1980), primiparous females that had not been exposed to mother–infant dyads, "considered a neonate as a strange, *fear-provoking* object which is not allowed to touch them and which must be removed, even with mutilating force if necessary" (Dienske *et al.,* 1980). In the common marmoset study, group-reared adolescents that had not been physically exposed to parent–infant dyads appeared to find an infant sibling clinging to their back as *aversive,* as demonstrated by their immediate attempts to rub the infant off or even to bite it, behavior that led to infants dismounting immediately (Pryce, 1993). This same behavior has also been observed in prereproductive cotton-top tamarins without care-helping experience (Cleveland and Snowdon, 1984). A study of the effects of play-mothering experience in rhesus macaques (Holman and Goy, 1981, 1995) provides further evidence of the motivational–emotional effects of such experience, as opposed to its effects on the ability to perform maternal-like behavior patterns per se. Study subjects were nulliparous adults that had been reared from birth either with mothers and/or peers only and were therefore deprived of play-mothering experience, or in large breeding groups where they were observed to obtain play-mothering experience. The experimental procedure consisted of daily one-on-one tests with 1 to 12-day-old unfamiliar infants. In both groups, ventral carrying was invariably initiated by the infant; furthermore, and also in both groups, the duration of each carrying bout was short because females always broke the infant's grip and invariably leapt away from the infant. However, females with play-mothering experience did demonstrate more ventral carrying than females without, and they also cuddled, approached, and touched the infants more. Females without play-mothering experience spent more time out of arm's reach of the infant and demonstrated more species-typical fear responses (e.g., lip smacking). Holman and Goy (1981, 1995) concluded that play-mothering experience attenuates the aversion and fear that are otherwise evoked in nulliparous adult rhesus macaques by unfamiliar infants.

The previous evidence indicates that, in simian nonhuman primates, preadult experience of mother–infant dyads and of infants directly, leads to the following: (1) nonimitative social learning that infants are not aversive and are not to be feared, and, although possibly to a lesser extent, (2) trial-and-error learning of maternal-like behavior per se. The preadult socialization hypotheses proposed in Section I contend that such preadult experience is: (1) sufficient for the demonstration of maternal behavior, and (2) essential (but not sufficient) for the demonstration of maternal behavior. Sections B and C that follow review the proximate and ultimate evidence, respectively, pertaining to these hypotheses.

B. PREADULT SOCIALIZATION OF MATERNAL BEHAVIOR AS A PROXIMATE MECHANISM

Figure 3 illustrates the hypothesis that preadult experience with infants is sufficient for the demonstration of maternal behavior. It proposes a

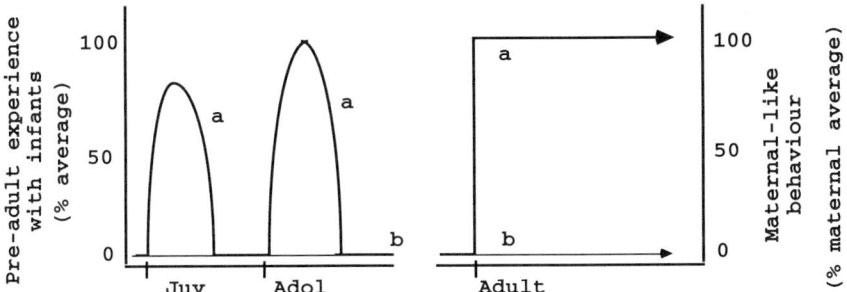

FIG. 3. Illustration and test of the hypothesis that preadult experience with mother–infant dyads is sufficient for the demonstration of maternal behavior by adult nonhuman simian primates. Female a obtains species-typical amounts of preadult experience with other females' infants during two birth seasons. On reaching adulthood and being presented with a young unfamiliar infant, she provides the infant with the same amount of carriage as a postpartum mother provides her own infant. Female b is deprived of preadult experience with other females' infants but has an otherwise species-typical socialization; she does not demonstrate maternal-like behavior.

female with species-typical preadult experience of infants reaching adulthood and being able *and* motivated to demonstrate maternal-like amounts of infant retrieval and carriage; only her inability to lactate prevents her from being a species-typical mother. A test of this hypothesis would require simulation of the conditions pertaining during the postpartum onset of maternal behavior. Preadult females would be allowed to obtain species-typical experience of infants. On reaching the species-typical age of first parturition (primiparity), but without ever becoming pregnant, females would be presented in a familiar environment with a young infant. The infant cries and the female retrieves and carries it just as a primiparous mother would. In comparison, females treated identically, but for the fact that their preadult social groups were manipulated so that they did not contain infants, do not retrieve or carry the crying infant (Fig. 3). This experiment and this result, I propose, would represent a proof of the hypothesis that preadult experience with other females' infants in the social group is sufficient for the demonstration of maternal behavior. To date, such an experiment has been performed with one primate only, and that is the study of rhesus macaques that was described previously in the context of motivational correlates of play-mothering experience (Holman and Goy, 1981, 1995). Nulliparous adult rhesus macaques with play-mothering experience *did not* demonstrate species-typical maternal retrieval or carriage with 1- to 12-day-old infants, in contradiction of the hypothesis illustrated in Fig. 3. However, it is possible that the scenario proposed previously and

Holman and Goy's experiment are not rigorous enough in their simulation of the conditions of postpartum maternal behavior. Postpartum mothers respond to newborn infants that are covered in amniotic fluid and vaginal secretions, and perhaps these stimuli are essential to the onset of maternal behavior (see Lundblad and Hodgen, 1980). Therefore, it is perhaps premature to conclude that preadult experience with infants is not sufficient for the demonstration of maternal behavior in the rhesus macaque or any other nonhuman simian primate.

Figure 4 illustrates the hypothesis that preadult experience with infants is essential for the demonstration of maternal behavior. It proposes, on the one hand, a female *without* preadult experience of infants reaching adulthood, conceiving, giving birth to but demonstrating no species-typical maternal behavior toward her own first infant, and, on the other hand, a female with species-typical preadult experience of infants reaching adulthood, conceiving, giving birth, and demonstrating species-typical maternal behavior toward her own first infant. The hypothesis has to be restricted to primiparous females because multiparous females might also have some maternal experience, that is, postpartum experience of maternal behavior, and this could compensate for the absence of preadult experience with

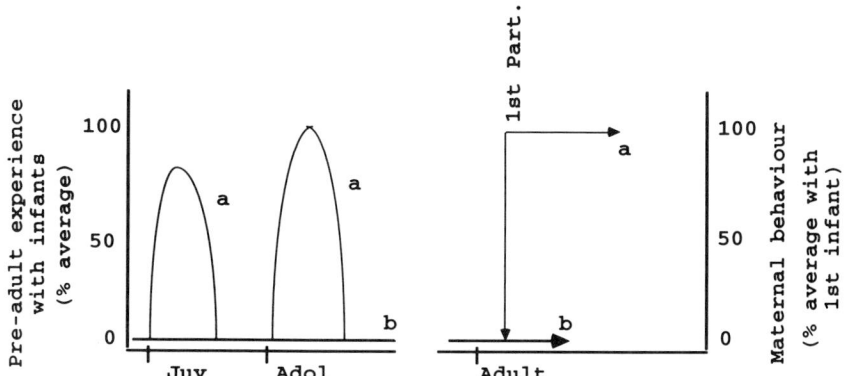

FIG. 4. Illustration and test of the hypothesis that preadult experience with mother–infant dyads is essential for the postpartum onset of maternal behavior by adult nonhuman simian primates. Female a obtains species-typical amounts of preadult experience with other females' infants during two birth seasons. On reaching adulthood, conceiving, and experiencing pregnancy and parturition, she demonstrates immediate postpartum onset of species-typical maternal behavior. Female b is deprived of preadult experience with other females' infants but has an otherwise species-typical socialization. On reaching adulthood, conceiving, and experiencing pregnancy and parturition, she does not demonstrate immediate postpartum onset of species-typical maternal behavior.

infants. The studies, described as follows, that are relevant to this hypothesis have been performed on captive rhesus macaques, vervet monkeys, common marmosets, and various species of tamarin.

Female rhesus macaques were reared by their mothers but did not have any opportunity to observe mother–infant dyads or to play-mother prior to adulthood (Gibber, 1986). After conceiving, these females were housed alone. Ten of 11 such females demonstrated maternal behavior with their first infants. Infant-directed behavior in the first hours after parturition was observed for some females; some demonstrated immediate ventral carrying, holding, and licking of the clinging infant, while others avoided the infant's grasps for 0.5–3 h before allowing the infant to cling and to gain ventral contact. One female avoided her infant continuously. These data contradict the hypothesis that preadult experience with infants is essential for the demonstration of maternal behavior in rhesus macaques. Female vervet monkeys were reared in social groups that approximated species-typical group composition and in which they therefore acquired play-mothering experience at the ages of 1 and 2 years (Fairbanks, 1990). When these females gave birth at aged 3–4 years, females with high amounts of play-mothering experience had higher reproductive success, measured in terms of infant survival over the first 3 months, than females with low play-mothering experience. Rather than representing a test of the hypothesis per se, this suggests that a high amount of play-mothering experience is important for the demonstration of species-typical maternal behavior in vervet monkeys. In retrospective studies of various marmoset and tamarin species, some females of each species did not obtain care-helping experience because their mothers did not reproduce again ("inexperienced"), while the other females did obtain care-helping experience as juveniles and/or adolescents with their infant siblings ("experienced"). As adults, all females were mated and conceived (at various ages) and the survival of their first (twin) infants was compared. In common marmosets 3 of 9 inexperienced females had surviving infants versus 12 of 12 experienced females (Tardif *et al.*, 1984; Pryce, 1993, data combined); in saddle-back tamarins (*Saguinus fuscicollis*) 3 of 16 inexperienced females had surviving infants versus 5 of 7 experienced females (Epple, 1978); in red-bellied tamarins (*S. labiatus*) 2 of 10 inexperienced females had surviving infants versus 4 of 14 experienced females (Pryce, 1988); in cotton-top tamarins 0 of 6 inexperienced females had surviving infants versus 4 of 13 experienced females (Tardif *et al.*, 1984; Snowdon *et al.*, 1985, data combined); in emperor tamarins (*S. imperator*) 2 of 5 inexperienced females had surviving infants versus 3 of 8 experienced females (Baker and Woods, 1992); and in golden lion tamarins (*Leontopithecus rosalia*) 1 of 6 inexperienced females had surviving infants versus 4 of 6 experienced females (Baker and Woods, 1992). Therefore, in none of these New World monkeys is care-helping essential for the demonstration of maternal behavior. However, in the common marmoset, saddle-back tamarin, and golden lion tamarin, care-helping experience was associated with an increase in infant survival, suggesting that this experience is very important in the regulation of primiparous maternal behavior in some, if not all, marmoset and tamarin species.

### C. Preadult Socialization of Maternal Behavior as an Adaptive Strategy

Therefore, the proximate evidence is that preadult experience with infants is important, rather than essential (Fig. 4), for the demonstration of species-typical maternal behavior in some simian species. That is, the effects of such a learning opportunity on maternal behavior and infant survival are quantitative rather than all-or-none. Even so, hypothetical "antisocial" females that lived alone between weaning and adulthood would leave fewer surviving offspring, on average, than do the real nonhuman simian females that do live in social groups and that do obtain preadult experience with infants. It would appear, therefore, that simian socialization, including socialization of maternal behavior, has evolved in the setting of the social group (Fig. 2). As such, social evolution in simian primates has been able to "assume" that females will have preadult opportunities to experience infants; behavioral propensities in some simian primates appear to have evolved in a direction that renders these opportunities important for the regulation of their maternal behavior.

The hypothesis that preadult experience with infants is sufficient for the demonstration of simian maternal behavior (Fig. 3) is often stated in the literature (see Section I), is not supported by the limited evidence available (see aforementioned), but remains to be tested rigorously (see aforementioned). However, it makes very little sense as a regulatory mechanism of *adaptive* maternal behavior. Monkeys and apes live in social groups, and the prediction that follows logically from the hypothesis is that any adult female with preadult experience of infants will be able and motivated to demonstrate maternal behavior with any available infant in its social group. The scene is barely imaginable. All adult females would continuously compete to retrieve and carry each infant; infants would sustain injuries from excessive interfemale competition and would become exhausted from insufficient time spent suckling on the lactating mother. In contrast to this hypothesis and its prediction, primate maternal behavior is discriminatory. Indeed, it is clearly the product of kin selection, where kin take the form of infant offspring (Maynard Smith, 1964). To be able to restrict maternal behavior to their own offspring, it is essential that nonhuman simian primates synchronize the onset of this behavior and of the ensuing mother–infant relationship to the birth of their own offspring. Synchronization of the onset of maternal behavior with birth of offspring has been described by Elwood (1994) as "temporal-based kinship recognition." For nonprimate mammals there is a substantial amount of empirical evidence demonstrating that such synchronization occurs and that it is mediated by changes in the central neuroendocrine state of the female during late pregnancy, labor, and parturition. For primates, in direct contrast, there are few answers at present to the question to which this synthesis now turns. How, in proximate terms, is the onset of maternal behavior and the mother–infant relationship synchronized with the birth of, and thereby restricted to, a female's own offspring?

## IV. Hormones and Maternal Behavior

The possibility that primate maternal behavior is regulated hormonally has received scant attention. This is despite the detailed evidence that has been accumulating for some 30 years now for the central and essential role of reproductive hormones in the regulation of maternal behavior in nonprimate mammals (see Rosenblatt, 1990; Numan, 1994). Via elegant experiments the endocrine regulation of maternal behavior as well as certain of the neural and neurochemical mediating pathways, have been demonstrated conclusively in two diverse nonprimate mammals, the rat (order Rodentia) and the sheep (order Artiodactyla). This evidence is summarized in Section IV,A. From the perspective of comparative biology, it would seem to be a logical progression if the evidence obtained for nonprimate mammals was applied to primates. One such application is the neuroendocrine hypothesis of primate maternal behavior, which states that prepartum and peripartum hormonal stimulation is essential for the demonstration of maternal behavior by fully socialized female monkeys and apes (see Section I and Fig. 2). The hypothesis, and the potential experimental tests of it, are illustrated in Fig. 5. The available evidence relating to this hypothesis is presented and discussed in Sections IV,B–E.

### A. Endocrine Regulation of Maternal Behavior in Nonprimate Mammals

Progesterone (P), estradiol-17$\beta$ (E2), prolactin (Prl), and oxytocin (OT) are hormones that are important in female mammals—nonprimate and primate—for implantation (P), maintenance of pregnancy (P, E2), development of the mammary glands (P, E2, Prl), onset of labor (P, E2 (nonprimates

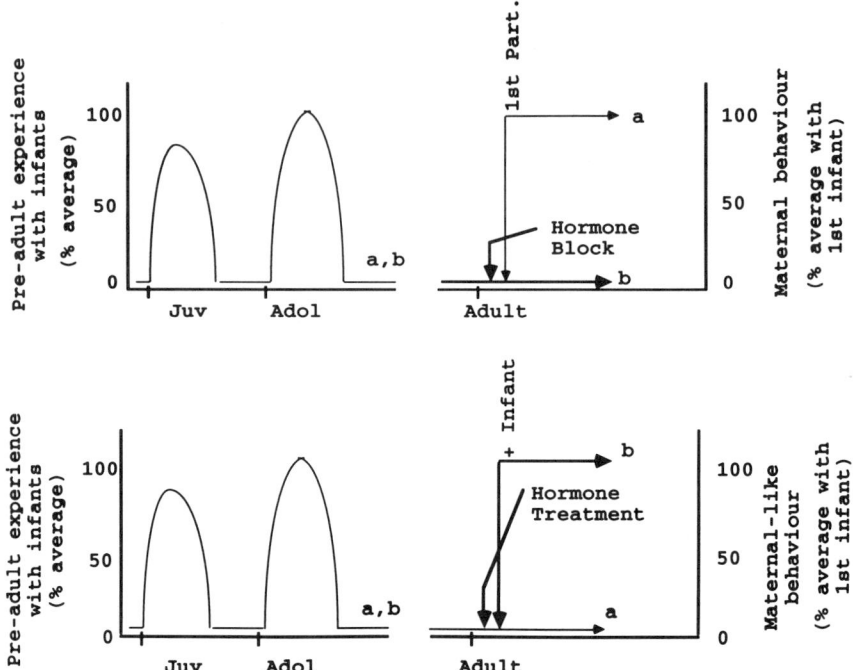

FIG. 5. Illustration and tests of the hypothesis that prepartum and peripartum neuroendocrine changes are essential for the postpartum onset of maternal behavior in fully socialized nonhuman simian primates. In (a), females a and b obtain species-typical amounts of preadult experience with other females' infants during two birth seasons. On reaching adulthood, they both conceive and experience the neuroendocrine changes of early-to-midpregnancy. Female a experiences typical prepartum and peripartum neuroendocrine changes and demonstrates immediate postpartum onset of maternal behavior. Female b is treated with hormone antagonists and does not experience typical prepartum and peripartum neuroendocrine changes, and does not demonstrate immediate postpartum onset of maternal behavior. In (b), females a and b obtain species-typical amounts of preadult experience with other females' infants during two birth seasons. On reaching adulthood, female a is presented with an infant and does not demonstrate the onset of maternal-like behavior; female b is treated with a hormone preparation that mimics prepartum and peripartum neuroendocrine changes and demonstrates the immediate onset of maternal-like behavior.

only)), parturition (OT), lactation (Prl, OT), and milk ejection (OT) (for a general introduction, see Johnson and Everitt, 1988). In nonprimate mammals the same four hormones are essential to the postpartum onset of maternal behavior. P and E2 are steroid hormones and are synthesized during pregnancy by the ovary alone or the ovary and the placenta, depending on the species (Heap and Flint, 1984). They are released into the maternal blood vascular system via which they can penetrate the blood-brain barrier and pass into the cerebrospinal fluid (CSF). Prl is a protein hormone and is produced in the anterior pituitary from where it is released into the circulation; although Prl can penetrate the blood-brain barrier, there is increasing evidence that it is also synthesized in the brain directly (DeVito, 1989; Emanuele *et al.,* 1992). OT is a nonapeptide that is synthesized in the hypothalamus of the brain, and released peripherally via the posterior pituitary and centrally via various neuronal projections (see as follows). There follows a summary of the essential relationship between these four hormones, their concentration changes during pregnancy and parturition, and the postpartum onset of maternal behavior. This evidence is derived from the major reviews of the endocrine regulation of rat maternal behavior, as provided by Rosenblatt and Siegel (1981); Stern (1989); Rosenblatt (1990); Bridges (1990); Bridges, this volume; Numan (1994); and of sheep maternal behavior, as provided by Poindron and Le Neindre (1980); Keverne (1988); Poindron *et al.* (1988); Keverne and Kendrick (1990); Poindron and Lévy (1990); Vince (1993); Kendrick (1994); Lévy *et al.,* this volume.

In the rat, the litter size is 12–18 and infants are altricial. Principal patterns of rat maternal behavior include building a maternal nest, retrieving pups to the nest by holding their dorsal skin between the incisors, licking pups, and adopting a nursing posture over pups (Rosenblatt and Lehrman, 1963, Fleming and Rosenblatt, 1974a). Certain of these patterns are also demonstrated by some prosimian primates (Fig. 1). Although, in nature, rat dams restrict provision of maternal behavior to their own offspring, it is an interesting and important phenomenon that females that are neither pregnant nor lactating can be induced to demonstrate maternal behavior in the laboratory. Nulliparous adult females (intact or ovariectomized) can be stimulated to demonstrate maternal behavior after 2–10 days of continuous exposure to consecutive litters of other females' 5- to 10-day-old pups (parous females with maternal exposure require shorter exposure) (Rosenblatt, 1990). Such induction, or sensitization, of maternal behavior in nulliparous adult females is preceded by frequent episodes of infant-directed approach-and-withdrawal behavior. This behavior suggests that nulliparae are in a conflicting composite motivational state in the presence of pup stimuli (Rosenblatt, 1990). Experiments have identified at least some of the components involved. Withdrawal is mediated in part by specific *aversion* to certain chemosignals emitted by pups (Fleming and Rosenblatt, 1974a,b,c,), and as part of a general *fearfulness* toward novel environments (i.e., *neophobia*) (Fleming and Leubke, 1981). These same emotional states have been invoked by various authors to explain the infant-directed behavior of nulliparous primates (Section III). Although the motivation underlying approach behavior has been less well studied, there is evidence to suggest that it is mediated in part by specific *attraction* to certain chemosignals emitted by pups (Fleming *et al.,* 1989). Approaching and investigating pups provides nulliparous females with the opportunity to habituate to their aversive and neophobic properties and to become more responsive to their attractive properties. Eventually the balance is shifted to the point where female–pup contact—and therefore pup–female sensory stimulation—becomes intense and maternal behavior ensues, that is, retrieval to the nest site, licking, and crouching over (Rosenblatt, 1990).

At parturition, the approach–withdrawal sequence is replaced by the rapid onset of maternal behavior. Primiparous rats demonstrate the onset of maternal behavior: (1) immediately after the cessation of parturition, and (2) can do so in the total absence of any prior experience with infants. This shift in responsiveness to pups develops during late pregnancy, peaks at parturition, is sustained for about 7 days postpartum, and subsides by 10 days (Orpen *et al.,* 1987), P, E2, Prl, and OT are essential to the composite process that changes the infant-directed approach–withdrawal behavior of nulliparous females into the maternal behavior of primiparous females. Pregnancy lasts 21–22 days in the rat; it is maintained by elevated P peripheral blood concentrations, which begin to decline rapidly at Days 19–21. E2 increases during pregnancy and is maximal at Days 18–22 (Bridges, 1990; Numan, 1994). Prl rises rapidly at Days 21–22 and high levels are maintained by suckling (Mena *et al.,* 1980). P receptors increase in the rat brain during pregnancy and are concentrated in the projection (stria terminalis) from the amygdala to the hypothalamus, the hypothalamus and medial preoptic area (MPOA) (Greenstein, 1986). Central E2 receptors are localized in the amygdala, stria terminalis, hypothalamus, MPOA and brain stem (Greenstein, 1986; Keverne, 1988); E2 receptors increase in the POA from midpregnancy onwards, and in the hypothalamus on pregnancy Day 22 (Giordano *et al.,* 1990). Exposure of nonpregnant, nulliparous females to the absolute (i.e., physiological) concentration profiles of P and E2 pertaining in the peripheral circulation during pregnancy, that is, high P and high E2 followed by P withdrawal, reduces the onset latency of their maternal behavior from 8 days to 1–2 days (Bridges, 1984). This stimulatory effect is mediated, at least partly, by E2 stimulation of Prl production (Bridges *et al.,* 1985; Bridges and Ronsheim, 1990), and is underlain by reduced fearfulness to the presence of pups, and reduced aversion and/or increased attraction by the odors of pups (Fleming *et al.,* 1989). OT-producing neurons in the hypothalamus have central projections to the olfactory bulb, amygdala, septum, and brain stem, while central OT receptors are localized in the olfactory bulb, amygdala, stria terminalis, and hypothalamus (Insel, 1990). OT release in the brain, at synaptic junctions specifically and into the CSF generally, is stimulated by parturition (Caldwell *et al.*, 1987), and OT binding by OT receptors (which increase in density prepartum) is a, if not the, rate-limiting factor in the postpartum onset of rat maternal behavior. Central administration of OT antagonist or antiserum delays the postpartum onset of rat maternal behavior (Fahrbach *et al.*, 1985; van Leengoed *et al.*, 1987; Insel, 1990).

In the sheep, the typical litter size is one to two and infants are precocial and capable of locomotion within minutes of birth. Maternal behavior comprises standing by and circling the lamb, licking the lamb, responding to the lamb's bleating with low-pitched gurgling bleats, standing stationary during suckling, facilitating suckling via arching the body and parting the hindlegs, and the maintenance of close proximity with the lamb (Collias, 1956; Lent, 1974; Vince, 1993). These behavior patterns are demonstrated: (1) only by ewes that have been exposed to a lamb within 1 day of parturition, (2) with this familiar lamb only, and (3) in the case of primiparous females, only if the lamb is covered with amniotic fluid during the initial exposure (for comparison see Lundblad and Hodgen, 1980, described in Section II). Otherwise, if these conditions are not met, nulliparous and parous ewes respond to all lambs with withdrawal and aggressive butting. The rapid onset of maternal behavior and establishment of an exclusive mother–infant bond are essential to adaptive (i.e., kin-selected) maternal behavior in species, such as the sheep, that live in large social groups, exhibit birth seasons, and produce precocial infants. Of course, these three features are also characteristic of many species of simian primates (Section II). The endocrine changes that occur during late pregnancy and parturition are essential to the rapid onset of sheep maternal behavior. P increases throughout the 145- to 150-day pregnancy until the final 5–10 days when it declines rapidly (Poindron and Lévy, 1990). E2 and Prl demonstrate massive prepartum increases from the low levels that characterize most of pregnancy, and peak at parturition (Poindron and Lévy, 1990). E2 declines postpartum and high Prl is maintained during lactation (McNeilly *et al.*, 1972). Suprapysiological doses of E2 can induce maternal behavior within 6–24 h in some multiparous females (Le Neindre *et al.*, 1979; Poindron and Lévy, 1990). Lower doses of P and E2 that simulate the prepartum profiles of these two steroid hormones reduce the withdrawal and butting behavior that nulliparae and multiparae direct at lambs but do not stimulate maternal behavior (Kendrick and Keverne, 1991). OT neurons project from the sheep hypothalamus to the olfactory bulb, stria terminalis and MPOA, and OT levels in the CSF are increased during parturition and nursing (Kendrick *et al.*, 1986, 1991, 1992). When OT is infused into the brain of E2-

primed, multiparous, nonpregnant ewes, they demonstrate an immediate, "postpartum-like" onset of maternal behavior (Kendrick et al., 1987).

To summarize, rat maternal behavior can be induced in nulliparous females via several days exposure to consecutive litters of young pups, after which approach–withdrawal behavior is replaced by approach–maternal behavior. Hormones are essential for the immediate onset of maternal behavior in postpartum females. The latency to the onset of rat maternal behavior is reduced to about 1 day if nulliparous females are exposed to the absolute physiological profiles of P and E2 that occur in pregnancy. At least part of the stimulatory effect of high E2 is due to increased Prl output. OT is essential to the immediate onset of maternal behavior. Rat maternal behavior is not restricted to specific pups, rather it is stimulated by any pups placed in the vicinity of the nest of a maternal dam. In sheep, in contrast to rats, maternal behavior cannot be induced by long-term exposure to lambs per se. Suprapsychological doses of E2 can stimulate maternal behavior within 1 day in some multiparous females. Exposure of nulliparous and multiparous females to doses of P and E2 that simulate prepartum profiles, that is, high P, followed by P withdrawal and high E2, reduces ewe–lamb avoidance and aggression but does not stimulate maternal behavior. Central OT treatment stimulates the immediate onset of sheep maternal behavior in E2-primed multiparous ewes. Sheep mothers form exclusive postpartum bonds with their lambs. Finally, the onset of maternal behavior is contemporaneous with reduced aversion (rat, sheep), reduced fear (rat), reduced aggression (sheep), and increased attraction (rat, sheep), relative to conspecific infant stimuli.

Before completing this section and returning to monkeys and apes, it is essential to consider one further aspect of the endocrine regulation of maternal behavior in nonprimate mammals. In terms of mechanism, there are two principal and interdependent effects of prepartum and peripartum hormones on maternal behavior: (1) the effects of hormones on female responsiveness to distal infant stimuli, that is, smell, sound, and sight, and on the appetitive behavior stimulated by these stimuli, for example, seeking, approaching; (2) the effects of hormones on responsiveness to proximate infant stimuli, that is, touch, and on the consummatory behavior stimulated by these stimuli, for example, nuzzling, licking, retrieval, hovering over, crouching over, arching. Appreciation of and differentiation between these two distinct processes are of fundamental importance to understanding the regulation of maternal behavior (e.g., see Stern, 1989; Fleming and Walsh, 1994; Numan, 1994). In experiments aimed at investigating hormonal effects, definitions of maternal responsiveness are generally based on consummatory responses. However, there are exceptions to this (e.g., Kendrick *et al.*, 1987; Fleming and Walsh, 1994) and what also needs to be emphasized is that: (1) consummatory maternal behavior can only occur after appetitive behavior has brought a female into proximate contact with the test infants, (2) some hormonal effects might be restricted to appetitive behavior and other hormonal effects to consummatory behavior. Furthermore, the distinction could turn out to be of paramount importance in the context of applying the evidence obtained from nonprimates to primates. Wallen (1990), in an outstanding comparative review, has demonstrated this last point in the context of mammalian sexual behavior. Whereas copulation in rodent females is completely dependent on ovarian hormones because they allow the release of lordosis by male somatosensory stimulation, in primate females the ability to copulate is completely emancipated from ovarian hormones. However, ovarian hormones do regulate sexually motivated appetitive behavior, for example, approaching males, not avoiding males, instrumental responding to gain access to males, in both rodent and primate females. As described in Section II,C, the principal component of postpartum onset of maternal behavior in simian primates comprises passive acceptance (as opposed to active avoidance) of the newborn infant's clinging contact. If hormones do regulate simian maternal behavior, the evidence for which is now considered, then it is possible that, as with simian sexual behavior, their effect is not on specific effector mechanisms of consummatory infant-directed behavior, rather on sensory and motivational mechanisms of appetitive infant-directed behavior.

B. Endocrinology of Pregnancy and Parturition

Amongst the nonhuman simian primates, the peripheral blood concentrations of P, E2, and Prl pertaining during pregnancy and lactation have been described for the common marmoset, the rhesus macaque, and the chimpanzee. In the common marmoset the gestation period is 20–21 weeks (Chambers and Hearn, 1979) (see Fig. 7 later). P is equivalent to luteal phase levels of the ovarian cycle for Weeks 1–12 after which it rises considerably until 1–2 weeks prior to parturition, when it begins to fall to prepregnancy levels (see Fig. 8 later). The common marmoset is the only primate for which the prepartum decline in P typical of nonprimate mammals (Rosenblatt and Siegel, 1981) has been demonstrated. E2 also begins to rise at Week 12 and continues to do so until parturition, when it is about 100 times greater than the highest pre-pregnancy level. P and E2 decrease to follicular phase levels at parturition, but it is important to note that common marmosets typically undergo an ovulation as early as Day 10 postpartum (Dixson and Lunn, 1987). Prl is slightly higher than during the luteal phase for Weeks 1–10 of pregnancy and then demonstrates a temporary rise between Weeks 10–14. It begins to rise again at Week 17 and this rise continues so that Prl peaks at parturition; this parturient level is maintained during full lactation (Moro *et al.*, 1995).

In the rhesus macaque the gestation period is 23–25 weeks. P peaks in Weeks 6–7 after which it is equivalent to luteal phase levels up to parturition. E2 increases gradually during pregnancy and begins to exceed ovarian cycle levels in Week 3 prepartum; E2 peaks in the final week of pregnancy (Hodgen *et al.*, 1972; Bosu *et al.*, 1973; Weiss *et al.*, 1976). In the prepartum period, Prl is maximal on the day of parturition and levels remain elevated during early lactation (Weiss *et al.*, 1976).

In the chimpanzee the gestation period is 32–34 weeks. P exceeds luteal phase levels throughout pregnancy; at parturition P is about 10 times higher than in the luteal phase. E2 increases gradually throughout pregnancy and is maximal and about 10 times higher than luteal phase levels, at parturition. Prl remains at luteal phase levels until Week 20, when it begins to increase gradually; it increases markedly in the week of parturition and remains elevated during early lactation (Reyes *et al.*, 1975).

Therefore, in the peripheral maternal circulation and in terms of general profiles, the common marmoset resembles the rat and sheep with respect to P, E2, and Prl profiles, and the rhesus macaque and chimpanzee resemble the rat and sheep with respect to E2 and Prl profiles. There is no systematic information on CSF levels of P, E2, Prl, or OT, between late pregnancy and the early postpartum period for any nonhuman simian (although see Takagi *et al.*, 1985 for OT during human pregnancy). Relative to the situation in rats and sheep, there is only a small amount of evidence for any primate that can be used to consider (not to mention test) the hypothesis that the previous prepartum and peripartum hormonal changes are essential to simian maternal behavior. The data that are available are considered as follows under three headings: relationship between female reproductive state and infant-directed behavior; relationship between reproductive hormones and infant-directed behavior; and effects of reproductive hormones on infant-directed behavior. Here, as with the hypothesis that preadult experience with infants is essential for maternal behavior, it is vital not to forget the potential confounding effects of female parity–maternal experience, and due attention is paid to these variables in the discussion that follows.

C. Relationship between Reproductive State and Infant-Directed Behavior

In Section III, in the context of socialization of simian maternal behavior, frequent reference was made to a group of studies conducted on play-mothering and maternal behavior in rhesus macaques (Holman and Goy, 1981, 1995; Gibber, 1986). These same studies also provide important information on the relationship between reproductive state and infant-directed behavior. Whereas nulliparous adult rhesus macaques with play-mothering experience do not demonstrate maternal-like behavior with unfamiliar infants (Section III), multiparous females do. Multiparous rhesus macaques that have not bred for at least 1 year, and irrespective of whether they are ovulatory, ovariectomized, or menopausal, demonstrate the immediate onset of maternal-like retreival and carrying when presented with unfamiliar 1- to 12-day-old infants (Holman and Goy, 1980, 1995). These data neither demonstrate that hormonal stimulation is nor that it is not essential for maternal behavior in rhesus macaques; rather, they demonstrate that hormonal stimulation is not essential in rhesus macaques that have already: (1) been exposed to pregnancy and parturition, and (2) obtained maternal experience with their first and subsequent infants. Interestingly, at 2 weeks prepartum, multiparous rhesus macaques are actually less interested in observing infants than are nonpregnant multiparae (Cross and Harlow, 1963), suggesting an inhibitory effect of the prepartum state on maternal motivation. In pigtail macaques (*Macaca nemestrina*), observed throughout pregnancy, interest in infants (touching, holding, carrying, and grooming other females' 0- to 12-week-old infants) increases during early and middle pregnancy but decreases just prior to parturition, further suggesting an inhibitory prepartum effect on macaque maternal movitation (Maestripieri and Wallen, 1995). In contrast, in another Old World monkey, the hanuman langur (Hrdy, 1977), as well as in the New World squirrel monkey (Rosenblum, 1972), adult females become more interested in other females' infants (e.g., retrieving, carrying) in the days preceding birth, suggesting a stimulatory prepartum effect. Perhaps the presence of clearly defined annual birth seasons in the macaques (Melnick and Pearl, 1987), with many females giving birth within a short space of time, has led to the evolution of prepartum maternal inhibition; elevated prepartum P might be responsible. Contrary to this hypothesis, however, is the fact that the squirrel monkey also has a pronounced annual birth season (Robinson and Janson, 1987). Nulliparous rhesus macaques without play-mothering experience did not demonstrate any change in infant-directed behavior in one-on-one tests conducted across their first pregnancy, right up to parturition (Gibber, 1986); that is, they remained avoidant of body contact (Section III). The same females (10 of 11 cases) demonstrated postpartum onset of maternal behavior with their own first infant (Gibber, 1986, Section III).

D. RELATIONSHIP BETWEEN REPRODUCTIVE HORMONES AND INFANT-DIRECTED BEHAVIOR

The first attempt to study the relationship between a specific hormone and maternal behavior in any nonhuman primate was a comparison of late-pregnancy urinary E2 profiles in red-bellied tamarins whose young infants either survived ("good mothers") or died ("poor mothers") (Pryce et al., 1988). Twelve adult male–female pairs of captive-born red-bellied tamarins were studied. From a total of 25 such pairs, animals were included in the study in the order that they became pregnant and gave birth, until each of four study groups (see as follows) contained three subject pairs. Ten of the 12 females were multiparous. Six of the females had obtained preadult care-helping experience with infant siblings and the other 6 had not (Section III). Because care-helping experience is associated with increased infant survival in multiparous red-bellied tamarins (Pryce et al., 1988), separate, experience-dependent definitions of good and poor mothering were constructed. For females with care-helping experience, good mothers had 2 surviving infants on Day 7 postpartum and poor mothers had only 1 or 0. For females without care-helping experience, good mothers had 2 or 1 surviving infants on Day 7 postpartum and poor mothers had 0. During the final 5 weeks of pregnancy in each study female, urine samples were collected on every second day. The total E2 level of urine samples was measured. New World monkeys retain unusually high amounts of conjugated (i.e., biologically inactive) estrogen in the circulation (e.g., Hodges et al., 1983), and it is possible that urinary total E2 reflected total blood E2 rather than biologically active E2 per se (Pryce et al., 1988; although see common marmoset study described as follows). Five females produced triplets, 6 produced twins, and 1 produced a singleton; all infants were alive at birth. Nine parturitions were observed as was the first 2 h of parturient and postpartum behavior; the other 3 cases were first observed at the beginning of the first day after parturition. On Day 7 postpartum, of the experienced females, 3 good mothers each had 2 surviving infants (including 1 primipara), and 2 poor mothers (including 1 primipara) had 1 surviving infant and the other had 0 surviving infants. Of the inexperienced females, 2 good mothers each had 2 surviving infants and the other had 1 surviving infant, and 3 poor mothers had 0 surviving infants. Of the 13 infants born to poor mothers, 11 died, and 10 of these had died within 12 h of birth. Infants of poor mothers were typically pushed off and avoided by the mother as they emerged and grasped but failed to initiate clinging contact. Two inexperienced poor mothers responded by leaping away from the infants as they were born followed by intense fear, expressed as bared-teeth screaming. While on the cage floor, newborn infants of poor mothers grasped reflexly with forelimb and hindlimbs, and vocalized frequently and loudly; but they were avoided and died, probably of hypothermia. Good mothers carried and licked their newborn clinging infants immediately and nursed them within 2 h.

In the six good mothers, urinary E2 levels were constant during the final 5 weeks of pregnancy. In the six poor mothers, urinary E2 levels decreased consistently as pregnancy advanced and in the final week of pregnancy urinary E2 levels were lower in poor mothers than in good mothers. Within the six experienced females, good mothers and poor mothers did not demonstrate consistent differences in their prepartum urinary E2 profiles (Fig. 6a). In the three experienced poor mothers, therefore, the cause of infant rejection immediately postpartum was not related to either the absence of preadult care-helping experience or distinctive prepartum E2 profiles. In the six inexperienced females, the three poor mothers had lower urinary E2 levels in the last week of pregnancy than did the three good mothers (Fig. 6b). One interpretation with which these data are consistent is that either a certain prepartum E2 profile or a certain prepartum E2 concentration is essential for the postpartum onset of maternal behavior in multiparous red-bellied tamarins that lack preadult care-helping experience. With regard to regulatory mechanisms, it is important to note that the three good inexperienced mothers had each acquired maternal experience during the rearing to weaning of at least one previous infant, as had two of the three poor mothers. Two of the poor mothers, including one with and one without maternal experience, demonstrated fear to and avoidance of their neonates immediately after expulsion. The third poor mother carried her single infant only rarely and although it was carried most of the time by the father it died on Day 3 postpartum, probably of starvation. It is possible that, as in rats (see aforementioned), high prepartum E2 reduces fear and aversion to infant odors in postpartum red-bellied tamarins. However, the study only demonstrated an E2–maternal-behavior association in such females and, in addition to this hormone-causes-behavior interpretation, several alternatives are possible. For example, it is important to consider the cause of the decrease in urinary (and presumably blood) E2 in the inexperienced poor mothers (Fig. 6b). In simians, E2 (and P) biosynthesis occurs in the placenta from early midpregnancy to parturition, and the simian placenta is dependent on mother *and* fetus for its supply of steroidal precursors (Albrecht and Pepe, 1990). Therefore, possible causes of the prepartum E2 decrease include any combination of small placentae, small fetuses, prolonged pregnancy, and maternal stress (e.g., see Dawood and Ratnam, 1974; Morishima *et al.*, 1979; Pratt and Lisk, 1991; Vermeulen *et al.*, 1982; Albrecht and Pepe, 1990). In this context it is interesting that, compared with the six good mothers, the six poor mothers received more aggression from and directed more submissive behavior at their male partners in the final 5 weeks of pregnancy, and had higher urinary cortisol levels in the first 2 weeks postpartum (Pryce *et al.*, 1991). Prepartum maternal stress might account for both the prepartum decrease in urinary E2

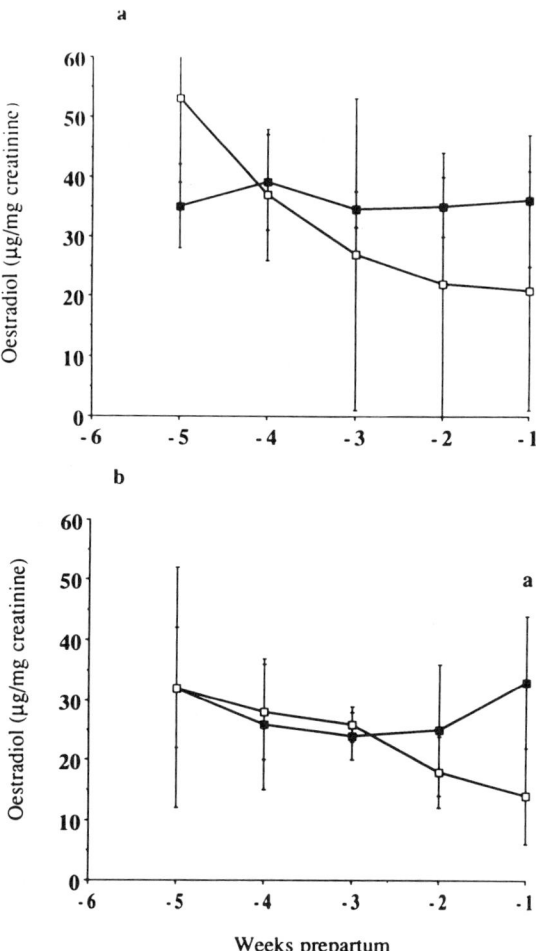

FIG. 6. Comparison of weekly prepartum changes in the mean urinary concentration of total estradiol in red-bellied tamarins defined as good mothers (■) and poor mothers (□): (a) mothers with preadult care-helping experience ($n = 3, 3$); (b) mothers without preadult care-helping experience ($n = 3, 3$). Values are expressed as the antilog of the mean of transformed data with 95% confidence limits. a = $p < 0.05$ versus poor mothers. (Figure reproduced from *Physiology and Behaviour* 44: 717–726, Copyright (1988), with permission from Pergamon Press Ltd., Headington Hill Hall, Oxford OX3 OBW, UK.)

and the absence of postpartum maternal behavior. To summarize, urinary E2 might be a predictor or a cause of poor maternal behavior in inexperienced, multiparous red-bellied tamarins, and only further studies can re solve this issue. Of course, reliable endocrine predictors of simian maternal care, even if without causal effects, would still be important in various contexts including, for example, captive primate conservation (e.g., Bahr, 1995).

One handicap in the design of the previous study is that it was performed primarily with multiparous females, that is, females with previous postpartum maternal experience. Reproductive hormones might be essential to the onset of maternal behavior, but only in females that are experiencing hormonal effects on responsiveness to infants for the first time, that is, primiparous females. If a primiparous female demonstrates maternal behavior with her first infant, then this maternal experience could reduce, even eliminate, the need for hormonal stimulation of maternal behavior with subsequent infants. That this could be the case for the rhesus macaque has been demonstrated by Holman and Goy (Section IV,C). A series of studies performed in the common marmoset overcomes this potential confound because it has used nulliparous-to-primiparous females. The study females were also experienced in the preadult care-helping that is typical of this species (Section III,A) and that is important in the regulation of its maternal behavior (Section III,B).

In one study, the hypothesis under examination was that variation in the postpartum maternal behavior of primiparous common marmosets is related to variation in their prepartum profiles of E2 and P (Pryce *et al.,* 1995). That is, rather than investigating the hypothesis that prepartum levels of E2 or P are *essential* for the onset of maternal behavior, this study aimed to determine whether prepartum levels of E2 or P are *important* for the demonstration of a high amount of maternal behavior in the first few days postpartum. As described previously (Section III,C; Fig. 1), marmoset and tamarin mothers *and* fathers provide infants with a substantial amount of care. Typically, a mother will carry and nurse her twin infants and then she will bite them in a restrained manner so that they cry; the father approaches, retrieves, licks, and carries the infants, then he bites them; the mother approaches, retrieves, licks, carries, nurses, etc. (Pryce, 1993). High amounts of maternal behavior in the first few days after parturition would minimize the need for interparental transfers of young infants and might also be important for efficient lactation. Extrapolating from the evidence for the rat and the sheep, high prepartum E2 levels combined with a marked prepartum P decline might provide optimal priming conditions for high postpartum maternal responsiveness.

Eight nulliparous common marmosets that had been paired with an adult male were studied prepartum and postpartum (Pryce *et al.,* 1995). All eight primiparous mothers reared two infants successfully and their maternal behavior was monitored during the first 7 days postpartum. Weekly levels of total E2 and total hydroxypregnanolone (HPO) were measured in urine samples collected at intervals of 2–3 days during the final 5 weeks of pregnancy. During this period, urinary total E2 predicted bioactive blood E2, and urinary total HPO predicted bioactive blood P. Urinary total E2 increased in the final 2 weeks of pregnancy and, for each female, this E2 level was expressed as a proportion of the E2 level in Weeks 5 to 3 prepartum ("E2 late–early increase"). Urinary total HPO decreased in the final week of pregnancy (see Section IV,B) and this HPO level was expressed as a proportion of the HPO level in Weeks 5 to 2 prepartum ("HPO late–early decrease"). Two measures of maternal motivation that are independent of paternal behavior (see aforementioned) were developed for use in this study: (1) retrieve-to-bite interval, that is, the time for which a mother carries before she bites her infant; (2) response latency to infant crying, that is, the time a mother takes to respond to an infant that cries in response to being bitten by its father. During Days 1–7 postpartum, the amount of time that mothers spent carrying their infants was predicted by the composite motivational score derived from these two measures. Furthermore, interfemale variation was marked, for example, average amount of time mothers spent carrying each infant varied from 88% to 29%.

Overall, neither the retrieve-to-bite interval nor the response latency to infant crying was predicted by the magnitude of the E2 late–early increase or the HPO late–early decrease. Despite the overall absence of any quantitative relationship between E2, HPO, and maternal motivation, some suggestive trends did emerge in this study. For example, the mother that demonstrated the largest prepartum increase in urinary total E2 and the second largest prepartum decrease in total HPO was the most responsive to infant crying and had the second longest retrieve-to-bite interval. The mother with the second largest prepartum increase in urinary total E2 was the second most responsive to infant crying and had the longest retrieve-to-bite interval. Conversely, the mother with the smallest decrease in urinary total HPO had the second shortest retrieve-to-bite interval, and the mother with the smallest prepartum E2 increase had the shortest retrieve-to-bite interval. That is, this study provides some limited evidence for the primiparous common marmoset with preadult experience with infants, that a prepartum increase in E2 and a prepartum decrease in P renders the mother highly sensitive to the crying of her infants. She responds rapidly to the infant crying elicited by the father's biting of the infant, and she carries her infant for a long time before biting it and eliciting infant crying herself. Perhaps, therefore, high levels of E2 and high levels of P, followed by marked P withdrawal in the prepartum period, serve to maximize maternal arousal, or maternal anxiety (see Section II,B), relative to infant vocalizations in the postpartum period. However, here, as in the red-bellied tamarin study described previously, caution needs to be exercised. This was a correlational and not an experimental study; the cause of interfemale variation in prepartum hormonal changes must also be considered with respect to its potential behavioral effects.

One feature of common marmoset maternal behavior that the previous study does demonstrate is the high responsiveness of primiparous mothers to infant crying. Infant crying can elicit rapid appetitive (e.g., approach crying infant) and consummatory (e.g., retrieve infant) maternal responses (Pryce *et al.*, 1995). The appetitive component of maternal responsiveness to infant crying was used as the basis of an operant experiment designed to examine the relationship between pregnancy blood profiles of E2 and P and maternal motivation in nulliparous common marmosets with preadult experience with infants (Pryce *et al.*, 1993). Artificial infant stimuli were produced with the intention of deploying them as invariant sensory reinforcers of operant behavior, that is, bar pressing. A visual stimulus consisted of an infant replica cast from a 2-day-old neonate, and an auditory stimulus consisted of a repeated 30-sec recorded sequence of infant crying. When a marmoset bar pressed in the operant chamber, it was exposed simultaneously to two events in the adjacent stimulus chamber: (1) switching on of a lamp and a 15-sec visual exposure to the infant replica, and (2) switching off of the playback and a 15-sec auditory time-out from the recorded infant crying. It was hypothesized that these two events would reinforce operant behavior and that this paradigm could be used to monitor E2- and P-related changes in responsiveness to artificial infant stimuli. Nonpregnant nulliparous females could indeed be trained to respond up to a stable and adequate level of baseline performance (Pryce *et al.*, 1993). Validation of the paradigm included a demonstration that bar pressing by postpartum primiparous mothers predicted the amount of time that they carried their own infants, their retrieve-to-bite intervals, and their latencies to respond to infant crying on the father.

The relationship between blood P and E2 and operant behavior was studied across pregnancy in four females; the results are summarized in Fig. 7a. The trained females were tested in blocks of five to eight 20-min operant sessions spread across the 140–143 days of pregnancy and Days 3–10 postpartum. Each female gave birth to and successfully reared twins. E2 increased 400-fold across pregnancy and was maximal in the final 10 days of pregnancy; P was maximal during Days $-25$ to $-11$ prepartum and then declined (see Section IV,B). Bar pressing increased as pregnancy advanced and peaked in the final 10 days of pregnancy, when E2 was maximal and P had declined. In the postpartum period, females were tested while not carrying their infants; their bar pressing was equivalent to levels

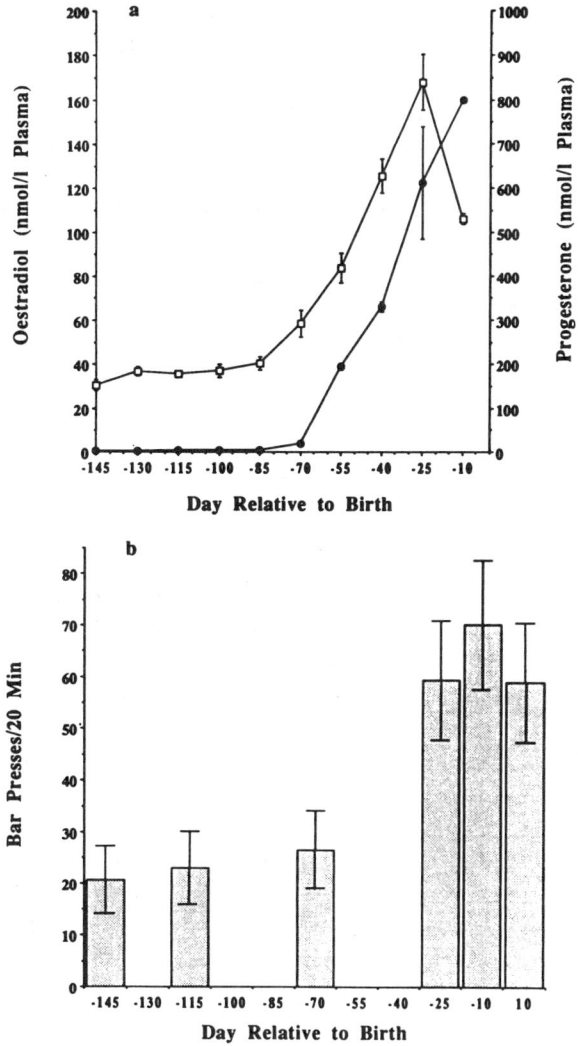

FIG. 7. (a) Changes in mean (± SEM) concentrations of plasma estradiol (●) and progesterone (□) across pregnancy in four primigravid common marmosets. (b) Changes in mean (± 95% comparison intervals) frequency of bar pressing for infant sensory reinforcement across pregnancy and postpartum in the same four females. In (b), values for bar-pressing frequencies were calculated as square-root means, and these have been retransformed. The 95% comparison intervals are for the minimum significant difference calculated *a posteriori*; values with nonoverlapping intervals are significantly different. (*Journal of Comparative Psychology* 107, 99–115, Copyright 1993 by the American Psychological Association. Adapted by permission of the publisher.)

pertaining in the final 10 days of pregnancy, despite the precipitous postpartum decline in E2.

This study demonstrated that the motivation to look at an infant replica and to terminate a playback of infant crying increases as blood levels of E2 and P increase in pregnant nulliparous common marmosets with preadult care-helping experience. There was no further increase in motivation associated with the prepartum decline in P; this is in contrast to what would be predicted from the rat evidence (Bridges, 1984; Section IV,A), but might have been the consequence of a "ceiling effect" of the reinforcement schedule at high response rates (Pryce et al., 1993). There was no postpartum decrease in motivation despite the decline in E2; the females reared their infants and it is possible that this maternal experience maintained responsiveness in the operant paradigm, just as infant stimuli maintain postpartum maternal behavior in the absence of hormonal priming in nonprimate mammals (Rosenblatt, 1990). By no means can this study be taken as evidence that E2 or P priming is essential to the onset of maternal behavior in nulliparous common marmosets. It does, however, suggest an important role for steroid priming in increasing peripartum responsiveness to visual and, particularly, auditory infant stimuli. This suggestion is supported by the results of an experimental study performed using the same operant paradigm, as described as follows.

E. EFFECTS OF REPRODUCTIVE HORMONES ON INFANT-DIRECTED BEHAVIOR

Three nulliparous, nonpregnant and anovulatory common marmosets were treated with a preparation of E2 and P that simulated the profiles of these steroids across the last 15 days of pregnancy (Fig. 8). The effects of this physiological treatment on operant responding for infant sensory reinforcement were tested (Pryce et al., 1993). In a 20-day control (basal) phase, operant test sessions were carried out in blocks of four on Days 3–6, 10–13, and 17–20, while females were injected daily with oil vehicle only, and their levels of E2 and P were about 1/100th of those pertaining in late pregnancy. Bar pressing demonstrated no change across the three periods of control testing, and an average bar-pressing rate was calculated for this phase ("basal," Fig. 9). The temporal relationship between steroid treatment and operant testing in the experimental phase was designed to detect specific effects of the P profile and E2 profile in late pregnancy. In his important study of the effects of physiological treatment with E2 and P on rat maternal behavior, Bridges (1984) demonstrated that the duration of P and E2 priming and the occurrence of P withdrawal were each an important factor in the regulation of maternal behavior onset. Based on this evidence, three E2–P treatments were identified for study: experimental

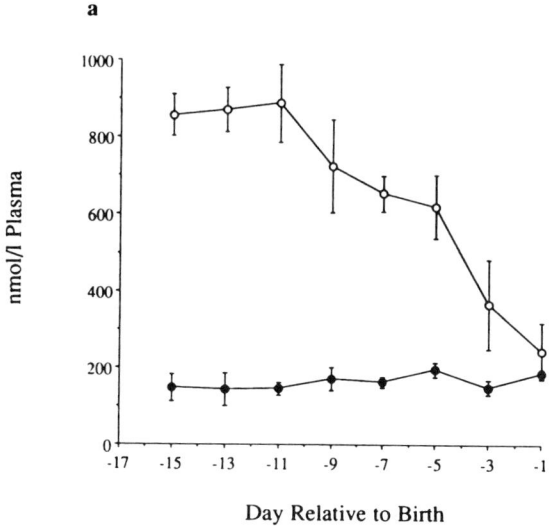
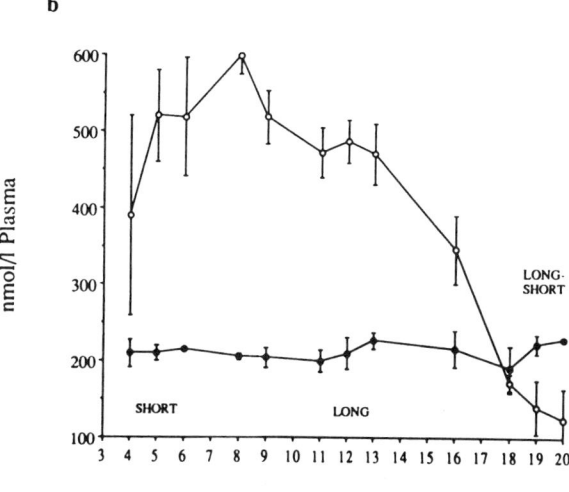

FIG. 8. Mean (± SEM) plasma concentrations of unconjugated estradiol (●) and progesterone (○) in (a) five primigravid common marmosets in the final 15 days of pregnancy, and (b) three nulliparae treated for 19 days with estradiol benzoate and progesterone. The labels in (b) describe the treatment conditions under which blocks of bar-pressing trials were carried out (see text for details). Females in (a) all demonstrated maternal behavior and reared their infants. (*Journal of Comparative Psychology* 107, 99–115, Copyright 1993 by the American Psychological Association. Adapted by permission of the publisher.)

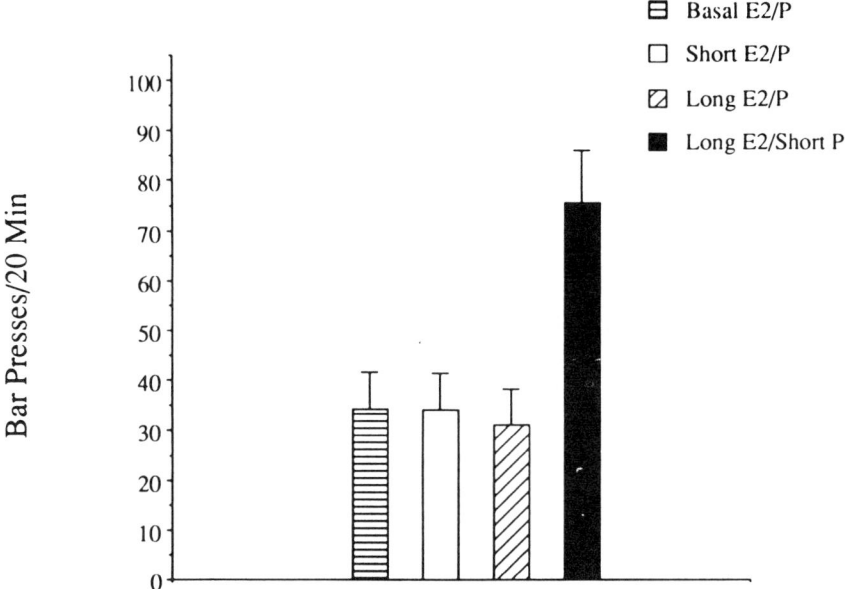

FIG. 9. Effect of treatment with estradiol and progesterone to simulate late pregnancy on frequency of bar pressing for infant sensory reinforcement. Values for bar-pressing frequencies were calculated as square-root means and these have been retransformed. The 95% comparison intervals are for the minimum significant difference (see legend, Fig. 7); treatment values with nonoverlapping intervals are significantly different. (*Journal of Comparative Psychology* 107, 99–115, Copyright 1993 by the American Psychological Association. Adapted by permission of the publisher.)

operant sessions were conducted after 3–6 days treatment with high E2 and high P ("short E2–P," Fig. 8b, 9); after 10–13 days treatment with high E2 and high P ("long E2–P"); and on Days 17–20, after 17–20 days treatment with high E2 and 2–5 days after P withdrawal ("long E2–short P"). Relating these treatment days to the prepartum period of pregnancy, Days 3–6 corresponded to Days −15 to −12 prepartum, Days 10–13 to Days −8 to −5 prepartum, and Days 17–20 to Day −1 prepartum (Fig. 8). P and E2 treatment resulted in consistent effects on the three marmosets' operant responding (Fig. 9). This effect was obtained with the long E2–short P sessions ("1 day prepartum"), when the frequency of bar pressing for infant sensory reinforcement increased 100% compared against the control and other experimental treatments. In summary, this study did not simulate the changes in E2 and P across the entire 140-day pregnancy in nonpregnant marmosets; that would be impracticable and unethical. However, it did simulate the physiological changes occurring in E2 and P levels in the maternal circulation during the final 15 days of pregnancy. Elevated E2 and P, followed by P withdrawal, stimulated maternal responsiveness to visual and, particularly, auditory infant stimuli. These operant studies, performed on the model of the nulliparous common marmoset with care-helping experience, provide some preliminary evidence for a role of reproductive steroids in the regulation of appetitive maternal behavior in nonhuman simian primates. The effects of E2 and P on maternal motivation might be direct, and they also might be indirect, for example, via stimulation of Prl synthesis or OT-receptor synthesis (see Section IV,A). Experimental studies, of the type depicted in Fig. 5, will be necessary to establish the role of neuroendocrine regulation of maternal behavior in nonhuman simian primates; to determine if it is essential and, if so, how it interacts with the important socialization processes described previously.

## V. A Socialization–Neuroendocrine Model

Clearly, there is a need for much more experimental work into the regulation of nonhuman simian material behavior if we are to understand the factors and mechanisms involved. However, before future experiments are designed and carried out, it is important to make full use of the evidence already available. A systems model, based on the available evidence, for the regulation of nonhuman simian postpartum maternal behavior by preadult socialization and prepartum and peripartum neuroendocrine changes is presented in Fig. 10. It is a motivational model, because of the many authors whose research has been cited in the previous synthesis, nearly all of them have invoked one or more motivations as the major variable mediating between a nonhuman simian female's sensory processing of infant stimuli and her behavioral responses to those stimuli. As such the model represents a departure from the generally accepted assumption, which is that preadult learning of maternal-like behavior patterns is the major variable in the regulation of simian maternal behavior (Section III). In the present model learning is certainly important, but its principal role is envisaged as being indirect, as a modulator of motivational processes.

The model proposes that the four major groups of infant stimuli, namely visual appearance (size, proportions, grasping behavior), audible crying, odor and taste, and tactile grasping and clinging (Section II,B), are processed by four motivational systems, namely attraction, anxiety, aversion, and neophobia (Fig. 10). Attraction processes infant visual appearance (Sackett and Ruppenthal, 1974) and infant odor and taste; anxiety processes infant crying (Coe et al., 1978); aversion processes infant odor and taste and infant grasping and clinging (Holman and Goy, 1981; Cleveland and

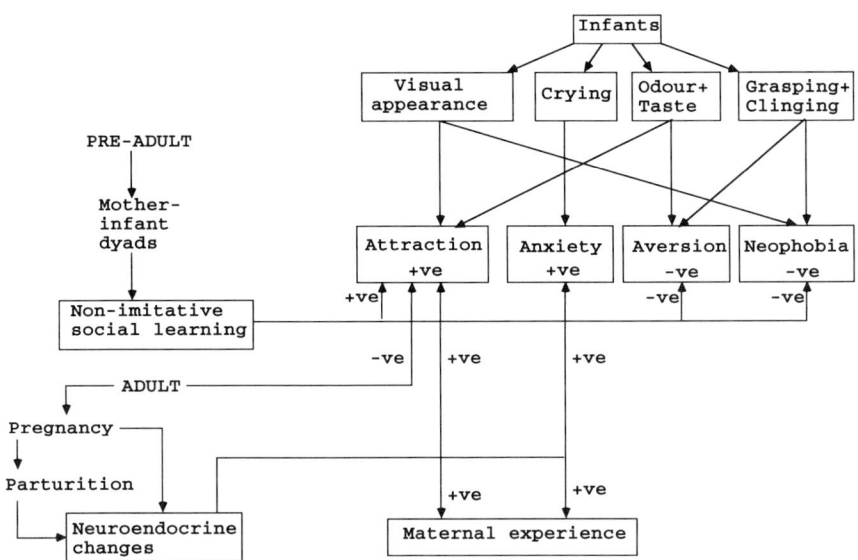

FIG. 10. A systems motivation model of preadult social regulation and prepartum and peripartum neuroendocrine regulation of maternal behavior in nonhuman simian primates. Attraction and anxiety increase (+ve) net maternal motivation/affect, aversion and neophobia reduce (−ve) net maternal motivation/affect. +ve denotes one variable increases the value of another variable in direction of arrow; −ve denotes one variable reduces the value of another variable. See text for details.

Snowdon, 1984; Pryce, 1993); and neophobia processes infant visual appearance and infant grasping and clinging (Dienske et al., 1980; Holman and Goy, 1981; Pryce, 1993). In the preadult female without experience of mother–infant dyads in her social group, the following motivational states would pertain if she is confronted with an infant: attraction by visual appearance and odor and taste is low, anxiety to crying is low, aversion to odor–taste and grasping–clinging is high, and fear of the novelty of visual appearance and grasping–clinging is high. The preadult female might investigate the infant but she avoids its attempts to grasp and cling, and therefore does not carry it (Pryce, 1993). Preadult experience with mother–infant dyads in the social group increases attraction by infant visual appearance and odor–taste, reduces aversion to infant odor–taste and grasping–clinging, and reduces fear of visual appearance and grasping–clinging (Holman and Goy, 1981; Pryce, 1993). It does this via the nonimitative social learning about these infant stimuli that occurs during interactions with familiar mother–infant dyads (Dienske et al., 1980; Pryce, 1993). The preadult female carries the infant as an integral part of its interactions with mother–

infant dyads. At the onset of adulthood, attraction by familiar mother–infant dyads is reduced and so, therefore, is attraction by infant visual appearance and odor–taste. The following motivational states would pertain if the nulliparous adult female with preadult experience of infants is confronted with an infant: attraction is low, anxiety is low, aversion is low, and neophobia is low (Holman and Goy, 1980, 1981). The adult female mates and conceives. In the late pregnant, nulliparous female, changes in vascular and therefore central concentrations of P, E2, and Prl cause the following: increased attraction by infant visual appearance and odor and taste, and increased anxiety to infant crying (Pryce et al., 1993). These shifts are compounded acutely by the increase in central OT that occurs at parturition. The following occur during the immediate postpartum period in the primiparous female with preadult experience of infants: attraction by infant visual appearance and odor–taste is high, anxiety to infant crying is high, aversion to infant odor–taste and grasping–clinging is low, and neophobia to infant visual appearance and grasping–clinging is low. This is the female's first experience of this combination of the four motivations; it is her first maternal experience. The postpartum primiparous female demonstrates maternal behavior and is reinforced because this behavior brings her into direct contact with her infant's attractive visual appearance and odor–taste, and because her infant does not emit anxiogenic crying (Coe et al., 1978) as long as it is being carried. With sufficient maternal experience, the adult female will set down a long-term memory of the high reinforcement value of (1) proximate contact with her infant's visual, olfactory, and gustatory stimuli; and (2) the cessation of her infant's crying.

## VI. Conclusions

Returning to the hypotheses that provided the rationale for this synthesis, the available evidence is consistent with the following conclusions:

1. Preadult experience of other females' infants in the natal social group is not sufficient for the demonstration of species-typical maternal behavior by nonhuman simian primates. The proximate evidence, limited as it is, is consistent with this negative conclusion and, furthermore, a functional consideration of the hypothesis indicates that a regulatory mechanism based solely on socialization would be maladaptive in any group living species with birth seasons.

2. Preadult experience of other females' infants in the natal social group is not essential but is important for the demonstration of species-typical postpartum maternal behavior in some nonhuman simian primates. Nonimitative social learning about infant stimuli during interactions with mother–infant dyads would appear to be one very important consequence of this

experience. Simians live in social groups and mature slowly in these groups, so that the probability of obtaining preadult experience with other females' infants is high. The socialization component of maternal behavior regulation in nonhuman simians appears to have evolved to a level where species-typical maternal responsiveness depends on a female having obtained preadult experience with other females' infants. This is in direct contrast to nonprimate mammals and is a primate specialization that is interesting from both causal and functional perspectives.

3. The neuroendocrine study of the regulation of simian maternal behavior is in its infancy, and it is too early to conclude on the hypothesis that prepartum and peripartum hormonal changes are essential for the demonstration of species-typical postpartum maternal behavior by a fully socialized simian female. That said, the adaptive (kin selected) need for a mechanism via which to synchronize the onset of maternal behavior with the birth of own offspring is as great in primates as it is in nonprimate mammals, and there is no *a priori* reason to assume that the neuroendocrine mechanism that fulfills this function in nonprimate mammals could not also do so in nonhuman simian primates. The available evidence is very preliminary but does provide support for neuroendocrine involvement in the regulation of maternal behavior in nonhuman simians.

The motivational (or emotional) state evoked in females by infants appears to be central to the regulation of maternal behavior in monkeys and apes. The motivational states of attraction, anxiety, aversion, and neophobia appear to be contributing to the composite motivational state that directs the behavior that simian females direct at infants. Socialization and hormones, the evidence suggests, regulate these four motivational states and thereby ensure the postpartum onset of a nurturant mother–infant relationship. This socialization–neuroendocrine hypothesis of maternal motivation is an interpretation of the available evidence and could form a basis for much-needed future investigation into the regulation of simian primate motherhood.

## VII. Summary

This article addresses the regulation of maternal behavior—behavior that promotes the survival, growth, and development of infant offspring when demonstrated by a lactating female—in nonhuman simian primates (i.e., New World monkeys and Old World monkeys and apes). Classical studies of maternal deprivation have demonstrated the importance of infant–mother attachment for species-typical behavioral development, including maternal behavior. However, the regulation of maternal behavior in *mother-reared* monkeys and apes has received very little experimental attention. Instead, it has been widely assumed and accepted that nonhuman simian maternal behavior is regulated by a female's preadult experience with other females' infants in the social group. This contrasts with the situation in nonprimate mammals, for which it has been demonstrated that maternal behavior is regulated by the neuroendocrine changes that occur late in pregnancy and during labor and parturition. Preadult monkeys and apes obtain experience with mother–infant dyads and with infants directly; for example, they investigate and carry other females' infants. As adults, females of most species carry other females' infants only rarely. In contrast, the cessation of parturition marks the onset of a relationship in which the adult female provides her own infant offspring with intensive and extensive maternal care. The present synthesis is based on three hypotheses of maternal behavior regulation in mother-reared monkeys and apes. Hypothesis 1 states that a female's preadult experience with other females' infants is sufficient for her to demonstrate species-typical maternal behavior with her own infants. Hypothesis 2 states that this experience is essential but not sufficient. Hypothesis 3 states that the neuroendocrine changes that occur late in pregnancy and during labor and parturition are essential for a female to demonstrate species-typical maternal behavior with her own infants; it is compatible with Hypothesis 2 and incompatible with Hypothesis 1. The majority of the evidence against which these hypotheses can be considered has been obtained in studies of captive rhesus macaques, squirrel monkeys, common marmosets, and various tamarin species. Preadult experience with other females' infants eliminates the aversion and fear that is otherwise demonstrated by inexperienced females toward infants. Preadult experience with other females' infants is not sufficient for the demonstration of maternal-like behavior by nulliparous adults. Preadult experience with other females' infants is important (if not essential) for the demonstration of maternal behavior by primiparous marmosets and tamarins, but is unimportant in primiparous rhesus macaques. Concerning Hypothesis 3, there is preliminary evidence that prepartum and peripartum neuroendocrine changes regulate simian maternal behavior. In multiparous red-bellied tamarins without preadult experience of other females' infants, maternal behavior is related to prepartum urinary estradiol levels; postpartum females with high prepartum estradiol carry and lick their infants and those with low prepartum estradiol reject them and are fearful of them. In nulliparous common marmosets with preadult experience of other females' infants, maternal motivation increases as pregnancy advances and in response to treatment with late-pregnancy levels of progesterone and estradiol. Rhesus macaques with maternal experience are attracted by the visual appearance of infants. Squirrel monkeys with maternal experience are made anxious by the crying of their infants. Based on the evidence, a motivational model of maternal behavior regulation in monkeys and apes is proposed, in which preadult experience with other females' infants reduces aversion to infant odor–taste and grasping–clinging, and reduces fear of the novelty of infant visual appearance and grasping–clinging; prepartum and peripartum neuroendocrine changes increase attraction by infant visual appearance and odor–taste, and increase anxiety to infant crying. This model provides a proximate explanation for an adaptive behavior, that is, how monkeys and apes living in social groups can restrict the demonstration of intensive and extensive patterns of maternal behavior to their own infants.

### Acknowledgments

I am very grateful to the Wellcome Trust, United Kingdom, and the National Science Foundation, Switzerland, for financial support, and to Chuck Snowdon and Jay Rosenblatt for constructive criticism of an earlier manuscript.

### References

Albrecht, E. D., and Pepe, G. J. (1990). Placental steroid hormone biosynthesis in primate pregnancy. *Endocrine Rev.* **11,** 124–150.
Allman, J. (1982). Reconstructing the evolution of the brain in primates through the use of comparative neurophysiological and neuroanatomical data. *In* "Primate Brain Evolution" (E. Armstrong and D. Falk, eds.), pp. 13–28. Plenum Press, New York.
Altmann, J. (1980). "Baboon Mothers and Infants." Harvard Univ. Press, Cambridge, MA.
Altmann, J., Hausfater, G., and Altmann, S. A. (1988). Determinants of reproductive success in savannah baboons, *Papio cynocephalus*. *In* "Reproductive Success" (T. H. Clutton-Brock, ed.), pp. 403–418. Univ. of Chicago Press, Chicago.
Altmann, J., and Samuels, A. (1992). Costs of maternal care: Infant-carrying in baboons. *Behav. Ecol. Sociobiol.* **29,** 391–398.
Andrews, M. W., Sunderland, G., and Rosenblum, L. A. (1993). Impact of foraging demands on conflict within mother-infant dyads. *In* "Primate Social Conflict" (W. A. Mason and S. P. Mendoza, eds.), pp. 229–251, State Univ. of New York Press, Albany, NY.
Arling, G. L., and Harlow, H. F. (1967). Effects of social deprivation on maternal behavior of rhesus monkeys. *J. Comp. Physiol. Psychol.* **64,** 371–377.
Bahr, N. I. (1995). Environmental factors and hormones: Their significance for maternal behavior in captive gorillas. *In* "Motherhood in Human and Nonhuman Primates: Biosocial Determinants" (C. R. Pryce, R. D. Martin, and D. Skuse, eds.), pp. 94–105. Karger, Basel.
Baker, A. J., and Woods, F. (1992). Reproduction of the emperor tamarin (*Saguinus imperator*) in captivity, with comparisons to cotton-top and golden lion tamarins. *Am. J. Primatol.* **26,** 1–10.
Barrett, L., Dunbar, R. I. M., and Dunbar, P. (1995). Mother-infant contact as contingent behaviour in gelada baboons. *Anim. Behav.* **49,** 805–810.

Bosu, W. T. K., Johansson, E. D. B., and Gemzell, C. (1973). Patterns of circulating oestrone, oestradiol-17B and progesterone during pregnancy in the rhesus monkey. *Acta Endocrinol.* **74,** 743–755.

Box, H. O. (1977). Quantitative data on the carrying of young captive monkeys (*Callithrix jacchus*) by other members of their family groups. *Primates* **18,** 475–484.

Bridges, R. S. (1984). A quantitative analysis of the roles of dosage, sequence, and duration of estradiol and progesterone exposure in the regulation of maternal behaviour in the rat. *Endocrinology* **114,** 930–940.

Bridges, R. S. (1990). Endocrine regulation of parental behavior in rodents. *In* "Mammalian Parenting: Biochemical, Neurobiological, and Behavioral Determinants" (N. A. Krasnegor and R. S. Bridges, eds.), pp. 93–117, Oxford Univ. Press, New York.

Bridges, R. S., DiBase, R., Loundes, D. D., and Doherty, P. C. (1985). Prolalctin stimulation of maternal behavior in female rats. *Science (Washington, D. C.)* **227,** 782–784.

Bridges, R. S., and Ronsheim, P. M. (1990). Prolactin (PRL) regulation of maternal behavior in rats: Bromocriptine treatment delays and PRL promotes the rapid onset of behavior. *Endocrinology* **126,** 837–848.

Caldwell, J. D., Greer, E. R., Johnson, M. F., Pederson, A. J., and Pederson, C. A. (1987). Oxytocin and vasopressin immunoreactivity in hypothalamic and extrahypothalamic sites in late pregnant and post-partum rats. *Neuroendocrinology* **46,** 39–47.

Chalmers, N. (1983). The development of social relationships. *In* "Animal Behaviour: Genes, Development and Learning," Vol. 3 (T. R. Halliday and P. J. B. Slater, eds.), pp. 82–113. Blackwell Scientific, Oxford.

Chambers, P. L., and Hearn, J. P. (1979). Peripheral plasma levels of progesterone, oestradiol-17$\beta$, oestrone, testosterone, androstenedione and chorionic gonadotrophin during pregnancy in the marmoset monkey, *Callithrix jacchus. J. Reprod. Fertil.* **56,** 23–32.

Cheney, D. L., Seyfarth, R. M., Andelman, S. J., and Lee, P. C. (1988). Reproductive success in vervet monkeys. *In* "Reproductive Success" (T. H. Clutton-Brock, ed.), pp. 384–402. Chicago University Press, Chicago.

Clark, C. B. (1977). A preliminary report on weaning among chimpanzees of the Gombe National Park, Tanzania. *In* "Primate Bio-social Development: Biological, Social, and Ecological Determinants" (S. Chevalier-Skolnikoff and F. E. Poirier, eds.), pp. 235–260. Garland Publ., New York.

Cleveland, J., and Snowdon, C. T. (1984). Social development during the first twenty weeks in the cotton-top tamarin (*Saguinus o. oedipus*). *Anim. Behav.* **32,** 432–444.

Clutton-Brock, T. H., and Parker, G. A. (1995). Punishment in animal societies. *Nature (London)* **373,** 209–216.

Coe, C. L. (1990). Psychobiology of maternal behavior in nonhuman primates. *In* "Mammalian Parenting: Biochemical, Neurobiological, and Behavioral Determinants" (N. A. Krasnegor and R. S. Bridges, eds.), pp. 157–183, Oxford Univ. Press, New York.

Coe, C. L., Mendoza, S. P., Smotherman, W. P., and Levine, S. (1978). Mother-infant attachment in the squirrel monkey: Adrenal response to separation. *Behav. Biol.* **22,** 256–263.

Collias, N. E. (1956). The analysis of socialization in sheep and goats. *Ecology* **37,** 228–239.

Cross, H. A., and Harlow, H. F. (1963). Observation of infant monkeys by female monkeys. *Percept. Motor Skills* **16,** 11–15.

Dawood, M. Y., and Ratnam, S. S. (1974). Serial estimations of serum unconjugated estradiol-17$\beta$ in high-risk pregnancies. *Obstet. Gynecol.* **44,** 201–207.

DeVito, W. J. (1989). Immunoreactive prolactin in the hypothalamus and cerebrospinal fluid of male and female rats. *Neuroendocrinology* **50,** 182–186.

Dienske, H., van Vreeswijk, W., and Koning, H. (1980). Adequate mothering by partially isolated rhesus monkeys after observation of maternal care. *J. Abnorm. Psychol.* **89,** 489–492.

Dixson, A. F., and Lunn, S. F. (1987). Post-partum changes in hormones and sexual behaviour in captive groups of marmosets (*Callithrix jacchus*). *Physiol. Behav.* **41,** 577–583.
Dunbar, R. I. M. (1987). Demography and reproduction. *In* "Primate Societies" (B. B. Smuts, D. L. Cheney, R. M. Seyfarth, R. W. Wrangham, and T. T. Struhsaker, eds.), pp. 240–249. Univ. of Chicago Press, Chicago.
Dunbar, R. I. M. (1988). "Primate Social Systems: Studies in Behavioral Adaptation." Crook Helm, London.
Dunbar, R. I. M., and Dunbar, P. (1988). Maternal time budgets of gelada baboons. *Anim. Behav.* **36,** 970–980.
Elwood, R. W. (1994). Temporal-based kinship recognition: A switch in time saves mine. *Behav. Proc.,* **33,** 15–24.
Emanuele, N. V., Jurgens, J. K., Halloran, M. M., Tentler, J. J., Lawrence, A. M., and Kelley, M. R. (1992). The rat prolactin gene is expressed in brain tissue: Detection of normal and alternatively spliced prolactin messenger RNA. *Mol. Endocrinol.* **6,** 35–42.
Epple, G. (1978). Reproductive and social behavior of marmosets and tamarins with special reference to captive breeding. *In* "Marmosets in Experimental Medicine" (N. Gengozian and F. Deinhardt, eds.), pp. 50–62. Karger, Basel.
Fagen, R. (1981). "Animal Play Behaviour." Oxford Univ. Press, Oxford.
Fahrbach, S. E., Morrell, J. I., and Pfaff, D. W. (1985). Possible role for endogenous oxytocin in estrogen-facilitated maternal behaviour in rats. *Neuroendocrinology* **40,** 526–532.
Fairbanks, L. A. (1990). Reciprocal benefits of allomothering for female vervet monkeys. *Anim. Behav.* **40,** 553–562.
Fleagle, J. (1988). "Primate Adaptation and Evolution." Academic Press, San Diego.
Fleming, A. S., and Leubke, C. (1981). Timidity prevents the virgin female rat from being a good mother: Emotionality differences between nulliparous and parturient females. *Physiol. Behav.* **27,** 863–868.
Fleming, A. S., and Rosenblatt, J. S. (1974a). Maternal behavior in the virgin and lactating rat. *J. Comp. Physiol. Psychol.* **86,** 957–972.
Fleming, A., and Rosenblatt, J. S. 1974b). Olfactory regulation of maternal behavior in rats: I. Effects of olfactory bulb removal in experienced and inexperienced lactating and cycling females. *J. Comp. Physiol. Psychol.* **86,** 221–232.
Fleming, A., and Rosenblatt, J. S. (1974c). Olfactory regulation of maternal behavior in rats: II. Effects of peripherally induced anosmia and lesions of the lateral olfactory tract in pup-induced virgins. *J. Comp. Physiol. Psychol.* **86,** 233–246.
Fleming, A. S., and Walsh, C. (1994). Neuropsychology of maternal behavior in the rat: *c-fos* expression during mother-litter interactions. *Psychoneuroendocrinology* **19,** 429–443.
Fleming, A. S., Cheung, U., Myhal, N., and Kessler, Z. (1989). Effects of maternal hormones on 'timidity' and attraction to pup-related odors in female rats. *Physiol. Behav.* **46,** 449–453.
Gandelman, R. (1992). "Psychobiology of Behavioral Development." Oxford Univ. Press, New York.
Gibber, J. R. (1986). Infant-directed behavior of rhesus monkeys during their first-pregnancy and parturition. *Folia Primatol.* **46,** 118–124.
Giordano, A. L., Ahdieh, H. B., Mayer, A. D., Siegel, H. I., and Rosenblatt, J. S. (1990). Cytosol and nuclear estrogen receptor binding in the preoptic area and hypothalamus of female rats during pregnancy and ovariectomized, nulliparous rats after steroid priming: Correlation with maternal behavior. *Horm. Behav.* **24,** 232–255.
Goldzion, A. W. (1987). Tamarins and marmosets: Communal care of offspring. *In* "Primate Societies" (B. B. Smuts, D. L. Cheney, R. M. Seyfarth, R. W. Wrangham, and T. T. Struhsaker, eds.), pp. 34–43. Univ. of Chicago Press, Chicago.

Gomendio, M. (1989). Suckling behaviour and fertility in rhesus macaques (*Macaca mulatta*). *J. Zool. (London)* **217,** 449–467.
Gomendio, M. (1995). Maternal styles in Old World primates: Their adaptive significance. In "Motherhood in Human and Nonhuman Primates: Biosocial Determinants" (C. R. Pryce, R. D. Martin, and D. Skuse, eds.), pp. 59–68. Karger, Basel.
Goodall, J. (1986). "The Chimpanzees of Gombe: Patterns of Behavior." Belknap Press of Harvard Univ. Press, Cambridge, MA.
Greenstein, B. D. (1986). Steroid hormone receptors in the brain. In "Neuroendocrinology" (S. L. Lightman and B. J. Everitt, eds.), pp. 32–48. Blackwell Scientific, Oxford.
Hall, W. G., Hudson, R., and Brake, S. C. (1988). Terminology for use in investigations of nursing and suckling. *Dev. Psychobiol.* **21,** 89–91.
Harvey, P. H., Martin, R. D., and Clutton-Brock, T. H. (1987). Life histories in comparative perspective. In "Primate Societies" (B. B. Smuts, D. L. Cheney, R. M. Seyfarth, R. W. Wrangham, and T. T. Struhsaker, eds.), pp. 181–196. Univ. of Chicago Press, Chicago.
Heap, R. B., and Flint, A. P. F. (1984). Pregnancy. In "Reproduction in Mammals, Hormonal Control of Reproduction," Vol. 3 (C. R. Austin and R. V. Short, eds.), pp. 153–194. Cambridge Univ. Press, Cambridge, UK.
Heyes, C. M. (1993). Imitation, culture and cognition. *Anim. Behav.* **46,** 999–1010.
Higley, J. D., and Suomi, S. J. (1986). Parental behavior in nonhuman primates. In "Parental Behavior" (W. Sluckin and M. Herbert, eds.), pp. 152–207. Blackwell, Oxford.
Hinde, R. A. (1976). Interactions, relationships and social structure. *Man* **11,** 1–17.
Hinde, R. A., and Simpson, M. J. A. (1975). Qualities of mother-infant relationships in monkeys. In "Parent-Infant Interactions" (CIBA Foundation Symp. 33), pp. 39–68. Wiley, Chichester, UK.
Hinde, R. A., and Spencer-Booth, Y. (1967). The behaviour of socially living rhesus monkeys in their first two and a half years. *Anim. Behav.* **15,** 169–196.
Hodgen, G. D., Dufau, M. L., Catt, K. J., and Tullner, W. W. (1972). Oestrogens, progesterone and chorionic gonadotrophin in pregnant rhesus monkeys. *Endocrinology* **91,** 892–900.
Hodges, J. K., Brand, H. M., Henderson, C., and Kelley, R. W. (1983). The levels of circulating and urinary oestrogens during pregnancy in the marmoset monkey (*Callithrix jacchus*). *J. Reprod. Fertil.* **67,** 73–82.
Holman, S. D., and Goy, R. W. (1980). Behavioral and mammary responses of adult female rhesus to strange infants. *Horm. Behav.* **14,** 348–357.
Holman, S. D., and Goy, R. W. (1981). Effects of prior experience with infants on behavior shown to unfamiliar infants by nulliparous rhesus monkeys. In "Primate Behaviour and Sociobiology" (A. B. Chiarelli, and R. S. Corruccini, eds.), pp. 72–74. Springer, Berlin.
Holman, S. D., and Goy, R. W. (1995). Experiental and hormonal correlates of care-giving in female rhesus macaques. In "Motherhood in Human and Nonhuman Primates: Biosocial Determinants" (C. R. Pryce, R. D. Martin, and D. Skuse, eds.), pp. 87–93. Karger, Basel.
Hrdy, S. B. (1977). "The Langurs of Abu." Harvard Univ. Press, Cambridge, MA.
Ingram, J. C. (1977). Interactions between parents and infants, and the development of independence in the common marmoset (*Callithrix jacchus*). *Anim. Behav.* **25,** 811–827.
Insel, T. R. (1990). Oxytocin and maternal behavior. In "Mammalian Parenting: Biochemical, Neurobiological, and Behavioral Determinants" (N. A. Krasnegor and R. S. Bridges, eds.), pp. 260–280. Oxford Univ. Press, New York.
Johnson, M. H., and Everitt, B. J. (1988). "Essential Reproduction," 3rd ed. Blackwell, Oxford.
Johnson, R. L. (1986). Mother-infant contact and maternal maintenance activities among free-ranging rhesus macaques. *Primates* **27,** 191–203.
Jolly, A. (1972). Hour of birth in primates and man. *Folia Primatol.* **18,** 108–121.

Jolly, A. (1985). "The Evolution of Primate Behaviour," 2nd ed. Macmillan Pub. Co., New York.
Kaplan, J. (1970). The effects of separation and reunion on the behavior of mother and infant squirrel monkeys. *Develop. Psychobiol.* **3,** 43–52.
Kaplan, J. N., Winship-Ball, A., and Sim, L. (1978). Maternal discrimination of infant vocalizations in squirrel monkeys. *Primates* **19,** 187–193.
Kemps, A., and Timmermans, P. (1984). Effects of social rearing conditions and partus experience on periparturitional behaviour in Java-macaques (*Macaca fascicularis*). *Behaviour* **88,** 200–214.
Kendrick, K. M. (1994). Neurobiological correlates of visual and olfactory recognition in sheep. *Behav. Proc.* **33,** 89–112.
Kendrick, K. M., and Keverne, E. B. (1991). Importance of progesterone and estrogen priming for the induction of maternal behavior by vaginocervical stimulation in sheep: Effects of maternal experience. *Physiol. Behav.* **49,** 745–750.
Kendrick, K. M., Keverne, E. B., Baldwin, B. A., and Sharman, D. F. (1986). Cerebrospinal fluid levels of acetylcholinesterase, monoamines and oxytocin during labour, parturition, vaginocervical stimulation, lamb separation and suckling in sheep. *Neuroendocrinology* **44,** 149–156.
Kendrick, K. M., Keverne, E. B., and Baldwin, B. A. (1987). Intracerebroventricular oxytocin stimulates maternal behaviour in the sheep. *Neuroendocrinology* **46,** 56–61.
Kendrick, K. M., Keverne, E. B., Hinton, M. R., and Goode, J. A. (1991). Cerebrospinal fluid and plasma concentrations of oxytocin and vasopressin during parturition and vaginocervical stimulation in the sheep. *Brain Res. Bull.* **26,** 803–807.
Kendrick, K. M., Keverne, E. B., Hinton, M. R., and Goode, J. A. (1992). Oxytocin, amino acid and monoamine release in the region of the medial preoptic area and bed nucleus of the stria terminalis of the sheep during parturition and suckling. *Brain Res.* **569,** 199–209.
Keverne, E. B. (1988). Central mechanisms underlying the neural and neuroendocrine determinants of maternal behaviour. *Psychoneuroendocrinology* **13,** 127–141.
Keverne, E. B. (1995). Neurochemical changes accompanying the reproductive process: Their significance for maternal care in primates and other mammals. *In* "Motherhood in Human and Nonhuman Primates: Biosocial Determinants" (C. R. Pryce, R. D. Martin, and D. Skuse, eds.), pp. 69–77. Karger, Basel.
Keverne, E. B., and Kendrick, K. M. (1990). Neurochemical changes accompanying parturition and their significance for maternal behavior. *In* "Mammalian Parenting: Biochemical, Neurobiological, and Behavioral Determinants" (N. A. Krasnegor, and R. S. Bridges, eds.), pp. 281–304. Oxford Univ. Press, New York.
Klopfer, P. H., and Boskoff, K. J. (1979). Maternal behavior in prosimians. *In* "The Study of Prosimian Behavior" (G. Doyle, and R. Martin, eds.), pp. 123–156. Academic Press, New York.
Kraemer, G. W. (1992). A psychobiological theory of attachment. *Behav. Brain Sci.* **15,** 493–541.
Lancaster, J. B. (1971). Play-mothering: The relations between juvenile females and young infants among free-ranging vervet monkeys (*Cercopithecus aethiops*). *Folia Primatol.* **15,** 161–182.
Lancaster, J. B. (1972). Play-mothering: The relations between juvenile females and young infants among free-ranging vervet monkeys. *In* "Primate Socialization" (F. E. Poirier, ed.), pp. 83–104, Random House, New York.
Le Neindre, P., Poindron, P., and Delouis, C. (1979). Hormonal induction of maternal behavior in non-pregnant ewes. *Physiol. Behav.* **22,** 731–734.

Lee, P. C. (1983a). Play as a means for developing relationships. *In* "Primate Social Relationships: An Integrated Approach" (R. A. Hinde, ed.), pp. 81–89. Blackwell, Oxford.

Lee, P. C. (1983b). Caretaking of infants and mother-infant relationships. *In* "Primate Social Relationships: An Integrated Approach" (R. A. Hinde, ed.), pp. 145–151. Blackwell Scientific, Oxford.

Lee, P. C., and Bowman, J. E. (1995). Influence of ecology and energetics on primate mothers and infants. *In* "Motherhood in Human and Nonhuman Primates: Biosocial Determinants" (C. R. Pryce, R. D. Martin, and D. Skuse, eds.), pp. 47–58. Karger, Basel.

Lent, P. C. (1974). Mother-infant relationships in ungulates. *In* "The Behaviour of Ungulates and its Relation to Management," Vol. 1, IUCN New Ser. No. 24 (V. Geist and F. Walther, eds.), pp. 14–55. International Union for the Conservation of Nature.

LeVine, R. A., Dixon, S., LeVine, S., Richman, A., Leiderman, P. H., Keefer, C. H., and Brazelton, T. B. (1994). "Child Care and Culture. Lessons from Africa." Cambridge Univ. Press, New York.

Lévy, F., and Poindron, P. (1987). The importance of amniotic fluids for the establishment of maternal behaviour in experienced and inexperienced ewes. *Anim. Behav.* **35,** 1188–1192.

Locke-Haydon, J., and Chalmers, N. R. (1983). The development of infant-caregiver relationships in captive common marmosets. *Int. J. Primatol.* **6,** 1–27.

Lorenz, K. (1971). "Studies in Animal and Human Behaviour." Methuen, London.

Lundbald, E. G., and Hodgen, G. D. (1980). Induction of maternal-infant bonding in rhesus and cynomolgous monkeys after cesarean delivery. *Lab. Anim. Sci.* **30,** 913.

Maestripieri, D. (1933a). Maternal anxiety in rhesus macaques (*Macaca mulatta*) I. Measurement of anxiety and identification of anxiety-eliciting situations. *Ethology* **95,** 19–31.

Maestripieri, D. (1993b). Maternal anxiety in rhesus macaques (*Macaca mulatta*) II. Emotional bases of individual differences in mothering style. *Ethology* **95,** 32–42.

Maestripieri, D. (1994). Social structure, infant handling, and mothering styles in group-living Old World monkeys. *Int. J. Primatol.* **15,** 531–553.

Maestripieri, D., and Wallen, K. (1995). Interest in infants varies with reproductive condition in group-living female pigtail macaques (*Macaca nemestrina*). *Physiol. Behav.* **57,** 353–358.

Martin, M. D. (1990). "Primate Origins and Evolution: A Phylogenetic Reconstruction." Chapman & Hall, London.

Martin, R. D. (1993). Primate origins: Plugging the gaps *Nature* (*London*) **363,** 223–234.

Martin, R. D. (1995). Phylogenetic aspects of primate reproduction: The context of advanced maternal care. *In* "Motherhood in Human and Nonhuman Primates: Biosocial Determinants." (C. R. Pryce, R. D. Martin, and D. Skuse, eds.), pp. 16–26, Karger, Basel.

Maynard Smith, J. (1964). Group selection and kin selection. *Nature* (*London*) **201,** 1145–1147.

McKenna, J. J. (1981). Primate infant caregiving behavior: Origins, consequences, and variability with emphasis on the common Indian langur monkey. *In* "Parental Care in Mammals" (D. J. Gubernick and P. H. Klopfer, eds.), pp. 389–416. Plenum Press, New York.

McNeilly, J. R., Moseley, S. R., and Lamming, G. E. (1972). Observations on the pattern of prolactin release during suckling in the ewe. *J. Reprod. Fertil.* **31,** 487–488.

Melnick, D. J., and Pearl, M. C. (1987). Cercopithecines in multimale groups: Genetic diversity and population structure. *In* "Primate Societies" (B. B. Smuts, D. L. Cheney, R. M. Seyfarth, R. W. Wrangham, and T. T. Struhsaker, eds.), pp. 121–134, Univ. of Chicago Press, Chicago.

Mena, F., Pacheco, P., Whitworth, N. S., and Grosvenor, C. E. (1980). Recent data concerning the secretion and function of oxytocin and prolactin during lactation in the rat and rabbit. *Front. Horm. Res.* **6,** 217–250.

Morishima, H. O., Yeh, M. N., and James, L. S. (1979). Reduced uterine blood flow and fetal hypoxemia with acute maternal stress: Experimental observation in the pregnant baboon. *Am. J. Obstet. Gynecol.* **134,** 270–275.

Moro, M., Torii, R., Koizumi, H., Inada, Y., Etoh, Y., Miyata, H., and Tamioka, Y. (1995). Serum levels of prolactin during the ovarian cycle, pregnancy, and lactation in the common marmoset (*Callithrix jacchus*). *Primates* **36,** 249–257.

Napier, J. R., and Napier, P. H. (1985). "The Natural History of the Primates." British Museum (Natural History), London.

Newman, J. D. (1985). The infant cry in primates: An evolutionary perspective. In "Infant Crying" (B. M. Lester and C. F. Z. Boukydis, eds.), pp. 307–323. Plenum Press, New York.

Nicolson, N. A. (1987). Infants, mothers and other females. In "Primate Societies" (B. B. Smuts, D. L. Cheney, R. M. Seyfarth, R. W. Wrangham, and T. T. Struhsaker, eds.), pp. 330–342. Univ. of Chicago Press, Chicago.

Nicolson, N. A. (1991). Maternal behavior in human and nonhuman primates. In "Understanding Behavior. What Primate Studies Tell Us About Human Behavior" (J. D. Loy and C. B. Peters, eds.), pp. 17–50. Oxford Univ. Press, New York.

Nishida, T. (1983). Alloparental behavior in wild chimpanzees of the Mahale mountains. *Folia Primatol.* **41,** 1–33.

Numan, M. (1994). Maternal behavior. In "The Physiology of Reproduction," 2nd ed. (E. Knobil and J. Neill, eds.), pp. 221–302. Raven Press, New York.

Orpen, B. G., Furman, N., Wong, P. Y., and Fleming, A. S. (1987). Hormonal influences on the duration of postpartum maternal responsiveness in the rat. *Physiol. Behav.* **40,** 307–315.

Payne, P. R., and Wheeler, E. F. (1968). Comparative nutrition in pregnancy and lactation. *Proc. Nutr. Soc.* **27,** 129–138.

Poindron, P., and Le Neindre, P. (1980). Endocrine and sensory regulation of maternal behaviour in the ewe. *Adv. Study Behav.* **11,** 76–119.

Poindron, P., and Lévy, F. (1990). Physiological, sensory, and experiential determinants of maternal behavior in sheep. In "Mammalian Parenting: Biochemical, Neurobiological, and Behavioral Determinants" (N. A. Krasnegor and R. S. Bridges, eds.), pp. 133–156. Oxford Univ. Press, New York.

Poindron, P., Lévy, F., and Krehbiel, D. (1988). Genital, olfactory, and endocrine interactions in the development of maternal behaviour in the parturient ewe. *Psychoneuroendocrinology* **13,** 99–125.

Pratt, N. C., and Lisk, R. D. (1991). Role of progesterone in mediating stress-related litter deficits in the golden hamster (*Mesocricetus auratus*). *J. Reprod. Fertil.* **92,** 139–146.

Price, E. C. (1992a). The benefits of helpers: Effects of group and litter size on infant care in tamarins. *Am. J. Primatol.* **26,** 179–190.

Price, E. C. (1992b). Contribution to infant care in captive cotton-top tamarins (*Saguinus oedipus*): The influence of age, sex, and reproductive status. *Int. J. Primatol.* **13,** 125–142.

Pryce, C. R. (1988). Endocrine and social correlates of tamarin behaviour and failure. Unpublished Ph.D. thesis, London.

Pryce, C. R. (1992). A comparative systems model of the regulation of maternal motivation in mammals. *Anim. Behav.* **43,** 417–442.

Pryce, C. R. (1993). The regulation of maternal behaviour in marmosets and tamarins. *Behav. Proc.* **30,** 201–224.

Pryce, C. R., Abbott, D. H., Hodges, J. K., and Martin, R. D. (1988). Maternal behavior is related to prepartum urinary estradiol levels in red-bellied tamarin monkeys. *Physiol. Behav.* **44,** 717–726.

Pryce, C. R., Abbott, D. H., Hodges, J. K., and Martin, R. D. (1991). Social and endocrine correlates of maternal behaviour and failure in a New World primate. *Appl. Anim. Behav. Sci.* **31,** 289–290.

Pryce, C. R., Döbeli, M., and Martin, R. D. (1993). Effects of sex steroids on maternal motivation in the common marmoset (*Callithrix jacchus*): Development and application of an operant system with maternal reinforcement. *J. Comp. Psychol.* **107,** 99–115.

Pryce, C. R., Mutschler, T., Döbeli, M., Nievergelt, C., and Martin, R. D. (1995). Prepartum sex steroid hormones and infant-directed behaviour in primiparous marmoset mothers (*Callithrix jacchus*). *In* "Motherhood in Human and Nonhuman Primates: Biosocial Determinants" (C. R. Pryce, R. D. Martin, and D. Skuse, eds.), pp. 78–86. Kargel, Basel.

Pusey, A. E., and Packer, C. (1987). Dispersal and philopatry. *In* "Primate Societies" (B. B. Smuts, D. L. Cheney, R. M. Seyfarth, R. W. Wrangham, and T. T. Struhsaker, eds.), pp. 250–266. Univ. of Chicago Press, Chicago.

Quiatt, D. (1979). Aunts and mothers: Adaptive implications of allomaternal behavior of nonhuman primates. *Am. Anthropol.* **81,** 310–319.

Reyes, F. I., Winter, J. S. D., Faiman, C., and Hobson, W. C. (1975). Serial serum levels of gonadotropins, prolactin and sex steroids in the nonpregnant and pregnant chimpanzee. *Endocrinology* **96,** 1447–1455.

Robinson, J. G., and Janson, C. H. (1987). Capuchins, squirrel monkeys, and atelines: Socioecological convergence with Old World primates. *In* "Primate Societies" (B. B. Smuts, D. L. Cheney, R. M. Seyfarth, R. W. Wrangham, and T. T. Struhsaker, eds.), pp. 69–82. Univ. of Chicago Press, Chicago.

Rosenblatt, J. S. (1990). Landmarks in the physiological study of maternal behavior with special reference to the rat. *In* "Mammalian Parenting: Biochemical, Neurobiological, and Behavioral Determinants" (N. A. Krasnegor and R. S. Bridges, eds.), pp. 40–60. Oxford Univ. Press, New York.

Rosenblatt, J. S. (1991). A psychobiological approach to maternal behaviour among primates. *In* "The Development and Integration of Behaviour" (P. Bateson, ed.), pp. 191–222. Cambridge Univ. Press, Cambridge.

Rosenblatt, J. S., and Lehrman, D. S. (1963). Maternal behaviour of the laboratory rat. *In* "Maternal Behaviour in Mammals" (H. L. Rheingold, ed.), pp. 8–57. John Wiley & Sons, New York.

Rosenblatt, J. S., and Siegel, H. I. (1981). Factors governing the onset and maintenance of maternal behaviour among nonprimate mammals. The role of hormonal and nonhormonal factors. *In* "Parental Care in Mammals" (D. J. Gubernick and P. H. Klopfer, eds.), pp. 13–76. Plenum Press, New York.

Rosenblum, L. A. (1968). Mother-infant relations and early behavioral development in the squirrel monkey. *In* "The Squirrel Monkey" (L. A. Rosenblum and R. W. Cooper, eds.), pp. 207–233. Academic Press, New York.

Rosenblum, L. A. (1972). Sex and age differences in response to infant squirrel monkeys. *Brain Behav. Evol.* **5,** 30–40.

Ross, C., and MacLarnon, A. (1995). Ecological and social correlates of maternal expenditure on infant growth in haplorhine primates. *In* "Motherhood in Human and Nonhuman Primates: Biosocial Determinants" (C. R. Pryce, R. D. Martin, and D. Skuse, eds.), pp. 37–46. Karger, Basel.

Rothe, H. (1978). Parturition and related behaviour in *Callithrix jacchus* (Ceboidea, Callitrichidae). *In* "The Biology and Conservation of the Callitrichidae" (D. Kleiman, ed.), pp. 193–206. Smithsonian Institution Press, Washington, D. C.

Ruppenthal, G. C., Arling, G. L., Harlow, H. F., Sackett, G. P., and Suomi, S. (1976). A 10-year perspective of motherless-mother behaviour. *J. Abnorm. Psychol.* **85,** 341–349.

Sackett, G. P., Griffin, G. A., Pratt, C. L., Joslyn, W. D., and Ruppenthal, G. C. (1967). Mother-infant and adult female choice behavior in rhesus monkeys after various rearing experiences. *J. Comp. Physiol. Psychol.* **63,** 376–381.

Sackett, G. P., and Ruppenthal, G. C. (1974). Some factor influencing the attraction of adult female macaque monkeys to neonates. *In* "The Effect of the Infant on the Caregiver" (M. L. Lewis and L. A. Rosenblum, eds.), pp. 163–186. Wiley, New York.

Shively, C., and Mitchell, G. (1986a). Perinatal behavior of prosimian primates. *In* "Comparative Primate Biology: Behavior, Conservation and Ecology," Vol. 2A (G. Mitchell and J. Erwin, eds.), pp. 217–243. Alan R. Liss, New York.
Shively, C., and Mitchell, G. (1986b). Perinatal behavior of anthropoid primates. *In* "Comparative Primate Biology: Behavior, Conservation and Ecology," Vol. 2A (G. Mitchell and J. Erwin, eds.), pp. 245–294. Alan R. Liss, New York.
Short, R. V. (1984). Breast feeding. *Sci. Am.* **250**, 23–29.
Simpson, M. J. A., and Simpson, A. E. (1986). The emergence and maintenance of interdyad differences in the mother-infant relationships of rhesus macaques: A correlational study. *Int. J. Primatol.* **7**, 379–399.
Smuts, B. B., Cheney, D. L., Seyfarth, R. M., Wrangham, R. W., and Struhsaker, T. T. (eds.) (1987). "Primates Societies." Univ. of Chicago Press, Chicago.
Snowdon, C. T., Savage, A., and McConnell, P. B. (1985). A breeding colony of cotton-top tamarins (*Saguinus oedipus*). *Lab. Anim. Sci.* **35**, 477–480.
Stern, J. M. (1989). Maternal behavior: Sensory, hormonal and neural determinants. *In* "Psychoendocrinology" (F. R. Brush and S. Levine, eds.), pp. 105–126. Academic Press, San Diego.
Stevenson, M. F. (1976). Birth and perinatal behaviour in family groups of the common marmoset (*Callithrix j. jacchus*) compared to other primates. *J. Hum. Evol.* **5**, 365–381.
Takagi, T., Tanizawa, O., Otsuki, Y., Sugita, N., Haruta, M., and Yamayi, K. (1985). Oxytocin in the cerebrospinal fluid and plasma of pregnant and nonpregnant subjects. *Horm. Metab. Res.* **17**, 308–310.
Tardif, S. D. (1994). Relative energetic cost of infant care in small-bodied neurotropical primates and its relation to infant-care patterns. *Am. J. Primatol.* **34**, 133–144.
Tardif, S. D., Carson, R. L., and Gangaware, B. L. (1990). Infant-care behavior of mothers and fathers in a communal-care primate, the cotton-top tamarin (*Saguinus oedipus*). *Am. J. Primatol.* **22**, 73–85.
Tardif, S. D., Richter, C. B., and Carson, R. L. (1984). Effects of sibling-rearing experience on future reproductive success in two species of Callitrichidae. *Am. J. Primatol.* **6**, 377–380.
Teas, J., Taylor, H. G., Richie, T. L., Shresthas, R. D., Turner, G. K., and Southwick, C. H. (1981). Parturition in rhesus monkeys (*Macaca mulatta*). *Primates* **22**, 580–586.
Tinklepaugh, O. L., and Hartman, C. G. (1930). Behavioral aspects of parturition in the monkey. *J. Comp. Psychol.* **11**, 63–98.
van Leengoed, E., Kerker, E., and Swanson, H. H. (1987). Inhibition of post-partum maternal behaviour in the rat by injecting an oxytocin antagonist into the cerebral ventricles. *J. Endocrinol.* **112**, 275–282.
Vermeulen, R. C. W., Kurver, P. H. J., Arts, N. F. T., Van Kessel, H., Wilson, G. R., and Klopper, A. (1982). The relationship between the surface area of the trophoblast and some placental products. *Placenta* **3**, 359–366.
Vince, M. A. (1993). Newborn lambs and their dams: The interaction that leads to sucking. *Adv. Study Behav.* **22**, 239–268.
Wallen, K. (1990). Desire and ability: Hormones and the regulation of female sexual behavior. *Neurosci. Biobehav. Rev.* **14**, 233–241.
Walters, J. R. (1987). Transition to adulthood. *In* "Primate Societies" (B. B. Smuts, D. L. Cheney, R. M. Seyfarth, R. W. Wrangham, and T. T. Struhsaker, eds.), pp. 358–369. Univ. of Chicago Press, Chicago.
Weiss, G., Butler, W. R., Hotchkiss, J., Dierschke, D. J., and Knobil, E. (1976). Periparturitional serum concentrations of prolactin, the gonadotropins, and the gonadal hormones in the rhesus monkey. *Proc. Soc. Exp. Biol. Med.* **151**, 113–116.
Whiten, A., and Ham, R. (1992). On the nature and evolution of imitation in the animal kingdom: Reappraisal of a century of research. *Adv. Study Behav.* **21**, 239–283.

# PART III

# FUNCTIONAL, ECOLOGICAL, AND ADAPTIVE ASPECTS OF PARENTAL CARE

# Field Studies of Parental Care in Birds
## New Data Focus Questions on Variation among Females[1]

PATRICIA ADAIR GOWATY

INSTITUTE OF ECOLOGY
UNIVERSITY OF GEORGIA
ATHENS, GEORGIA 30602-2602

## I. INTRODUCTION

The editors invited me to review parental care in birds, specifically focusing on field studies about adaptive significance. Therefore, in this article I review some current thinking—data, questions, and theories—about natural selection and parental care in birds. My review concentrates on seven areas of empirical effort and covers the last 20 or so years. Given that biparental care is the most common pattern in birds, past and even current field studies in avian parental care struck me as disproportionately focused on questions about male parental care—a trend to which my own studies have contributed. These field studies include: (1) descriptive studies of what *parental care* is and how it varies, though most of these seem to be attempts to explain variation in males rather than females; (2) experimental studies about the advantage of male parental care for females' fitnesses, usually done in an effort to understand the selective forces favoring social monogamy in birds; (3) descriptive studies evaluating the extent of uncertain maternity and paternity, which document the frequency of conspecific brood parasitism and extra-pair paternity using molecular genetic markers; (4) correlative studies that attempt to evaluate the relationship between paternal care and genetic paternity. Some studies focus more attention on females and include (5) studies about conflicts of interest between females and males cooperating in parental care and the resultant allocations of parental effort. (6) Even newer studies focus uniquely on females' maternal behavior as a function of the quality of their mates. Clutton-Brock's (1991) book was an excellent attempt at synthesis that suggested that the existing adaptive explanations for parental care in birds

[1] Note that scientific names for most species listed are in Table I. For species that do not appear in Table I, scientific names are in the text, after the first appearance of the species.

left something out and whatever it is that was/is left out is likely to be associated with our collective failure to come to a synthetic theory explaining parental care. The emerging foci on females has facilitated (7) the development of new adaptive explanations for parental care in birds, and suggest that what old theories left our were variations in females, variation in females' options, and opportunities for sexual conflict between males and females. Incorporation of ideas and data about variation in females is likely to change dramatically our current understanding of social behavior, including parenting behavior. I discuss new ideas throughout the article and discuss how attention to variation among females and the environments females are in may change future empirical studies.

I suspect that some of the things I have not done in this article will leave some readers frustrated. For example, I have intentionally not emphasized variation in males (though much discussion of males remains), because I thought it would be heuristically, theoretically, and empirically novel to put the spotlight on females. These female perspectives are more than just turning around of male perspectives; even so, in this article, I sometimes assume that variation in males is as uninformative as previous generations of scholars apparently thought variation in females. Here at the outset though, I want to make clear that I think one of the goals of social behavior research is to explore and explain sources of and effects of variation in *both* males and females, something that the new female perspectives and ideas about sexual conflict facilitate. However, I consider fully fleshed models explicitly incorporating simultaneous variation in both females and males work for the future and have included only modest attempts here. My focus on females represents my reactions to some of what has been left out about females so far; some of the implications of left-out variation-in-females occurred to me as I reviewed the literature. I suspect that inasmuch as some readers have already begun the task of incorporating "female perspectives" into their thinking, that my new ideas will provide opportunity for them to react as well and to continue our efforts to make social behavior research in birds more inclusive of both female and male interests.

## A. Biparental Care Is the Norm in Birds

Most descriptions of avian sociality begin with the classic facts that 90% of bird species exhibit biparental care and are socially monogamous (Lack, 1968). *Biparental care = social monogamy* is the norm among birds, but is relatively rare among vertebrates generally (see, e.g., Clutton-Brock, 1991). It is a pattern of sociality at least superficially similar to the most common mating patterns in humans, something that might account for the empirical attention paid to birds by students of the evolution of social behavior.

B. A Working Definition of Parental Care

*Parental care* is usually defined as any behavior by adults toward young that promotes the survival or quality of the offspring (e.g., Walters, 1984). Clutton-Brock (1991, p. 8) defined it as "Any form of parental behavior that appears likely to increase the fitness of a parent's offspring." This seems at first blush to be an operational definition of considerable utility (though further on I discuss an alternative definition with somewhat greater potential utility in studies exploring the evolution of parental care). Because these definitions limit parental care to behavior of parents to their offspring and because the literature on avian helpers-at-the-nest is huge (e.g., Stacey and Koenig, 1990; Brown, 1987), I have not discussed behavior towards offspring by "helpers" in this article. I have included cases of adult behavior towards young that are not theirs when it appears that the parental care is misdirected, as may be the case when young are from extra-pair paternity or conspecific nest parasitism.

## II. What Do Males Do? What Do Females Do?

A. Who Does What, How Much, and Which Came First?

Parental behavior of some sort occurs in all extant birds except for the 90 or so species of brood parasites that parasitize the parental behavior of individuals of other species. Despite the previous restrictive definition of parental care, what counts as parental care in birds has included feeding of the female by the male prior to and during egg laying and during incubation, territory defense, and nest building (Verner and Willson, 1969; Silver *et al.*, 1985). Incubation of eggs, brooding chicks, feeding nestlings and fledglings and escorting young to foraging sites or predator refugia, and protecting nestlings and fledglings from predators fit the definition of things parents do that promote survival or quality of their offspring (Verner and Willson, 1969; Silver *et al.*, 1985). In this article I have counted each of the previously mentioned as parental care.

Although biparental care seems to be a hallmark of the class Aves, uniparental care by either males or females occurs. Uniparental female care occurs in all species in which males display to females on leks, including cotingas, manakins, some hummingbirds, grouse, lyrebirds, bowerbirds, and birds of paradise; uniparental female care also occurs in pheasants, turkeys, and some flycatchers and Anseriformes. Uniparental male care occurs in several shorebirds as well as in kiwis, tinamous, cassowaries, and brush-turkeys. In the preceding list in all but the shorebirds, males have intromittent organs, a rare trait in birds shared only with Anseriformes. (The correlation between uniparental male care and the presence of intromittent organs is interesting in light of efforts to explain variation in social organizations as a function of males' efforts to control female reproductive capacities and females' efforts to resist males (Gowaty, 1992, 1996) in that creatures in which males have intromittent organs are theoretically more likely to have male-controlled copulations.) There are few studies of uniparental female care in birds, while a seemingly larger number of studies of uniparental male care exist (see citations in Clutton-Brock, 1991). In a small number of theoretically important species, biparental care seems to be a true partnership in which parental care is equitably shared between females and males and is essential for reproductive success of either parent (see Black, 1995 for some discussions of examples). Examples include species of penguins, Sphenisciformes; tropicbirds and pelicans, the Pelecaniformes; species of seabird, such as albatrosses, petrels, and storm petrels, the Procellariiformes. In many biparental species, particularly passeriformes, the rule seems to be that females do most parental care (Verner and Willson, 1969), but males in many species feed females during nest building and egg laying; and often feed nestlings and fledglings, so that the modal tendency in avian subfamilies is for the male to share parental duties with females (Silver *et al.*, 1985).

So far there is no definitive answer to the question of the phylogenetic origins of uniparental male, uniparental female, or biparental care. The answer to the question of which pattern is ancestral should guide theories about many other aspects of avian sociality, so "which came first?" is of no small importance. Some authors reason that the most widespread pattern is likely to be ancestral; these authors argue that biparental care is ancestral in birds (e.g., Silver *et al.,* 1985). McKitrick (1992) used an analysis of phylogenetic patterns and concluded that biparental incubation arose from no incubation by either parent; thus, she concluded that biparental care is the ancestral pattern in birds. Other authors note that uniparental male care is exhibited in the most primitive, extant avian taxa and argue that for birds uniparental male care is the ancestral condition (van Rhijn, 1984, 1985, 1990; Wesokowski, 1993) from which biparental and then uniparental female care evolved. If biparental care is ancestral and if uniparental female and uniparental male care evolved from biparental care, the logic of models of the evolution of behavior associated with parental care should reflect these assumptions, and I think that most of our theories do assume that biparental or uniparental female care was the ancestral pattern. If uniparental male care is ancestral, however, the structure of many of our current questions would seem to be backwards. For example, if uniparental male care is ancestral, would it not make more sense for avian behavioral ecologists to be asking questions about the evolution and function of female, rather than male, parental care? If biparental male care is ancestral, should we assume that biparental care evolved next and that uniparental female care is derived from this intermediate condition, or does it make more sense to assume that uniparental female care evolved from uniparental male care? I know of no definitive resolution to these questions, but I do want to suggest that conceptualizations of the evolution of co-occurring traits might be profitably thought about from the "turned around" perspective of ancestral uniparental male care in birds. I believe that such a turned-on-its-head perspective may facilitate novel and crucial predictions (in the sense of Platt, 1964). In fact, conceptual theorists of social behavior in birds might consider alternative scenarios originating at all three types of parental care systems: biparental, uniparental male, and uniparental female.

Increasingly, field studies are explicitly focused on how male parental care varies (see Section VI later). Verner and Willson (1969) have the latest review of the role of the male in parental care in passerines, specifically, North American passerines. They examined the distribution of males' activities as a function of sexual variation (plumage and size dimorphism and monomorphism). They point out that of 291 species of North American passerines, 190 and 101 were monomorphic or dimorphic, respectively. They concluded that males of most monogamous species assist with nest construction, but that monomorphic socially monogamous species are more likely to do so than dimorphic socially monogamous species; incubation by males is less common in dimorphic socially polygynous species than among monomorphic socially monogamous species; and, with rare exceptions, both dimorphic and monomorphic male passerines feed nestlings and fledglings. Their final conclusion was that the number of studies in which reliable details of the occurrence and allocation of parental duties between females and males was woefully small (Fraga (1992) is an exception to this rule).

My reading of the avian literature suggests that matters in this regard have changed little in the last 20 years; there are very few studies of who does what in nesting cycles except as an incidental effect of answering other questions arising from current theoretical perspectives. For example, in multiple studies of male parental care in eastern bluebirds, we observe that, on average, males feed nestlings as often as females feed nestlings (Gowaty, 1983; Gowaty and Droge, 1991; Gowaty *et al.*, Gowaty and Plissner). Our studies, like most studies of parental care, were done with other goals than describing differences and similarities in what females and males do, and I am currently wondering what sorts of things we may have missed given that our theoretical perspectives were based on the assumption that female parental care is the basic pattern upon which male parental care evolved.

An exception to studies that ask questions about who does what as a by-product of other investigations is Wright and Cuthill (1992) on starlings. They looked at how pairing patterns, covariance in male and female body size, male display, and phenological characters affected provisioning patterns of parents to nestlings. They found that larger females provisioned more, but that male provisioning rates were unrelated to song or any other attributes of males or their mates. Wright and Cuthill (1992) also handicapped one member of a pair by adding small lead weights to their tails. This manipulation decreased the ability of the handicapped partner to provision. They found that no matter which sex partner was handicapped, that the other partner partially compensated for the shortfall in handicapped-partner contributions. These are the first studies that I know of to look at how individuals cooperating over parental care actually achieve "cooperation."

I have the impression that most workers think that females do more parental care duties than males in birds; for example, I think this is an assumption that leads to language like "sex role reversal" to describe species in which males are the primary caregivers, as is the case in many shorebirds (e.g., Owens *et al.*, 1995). In contrast, male parental care is almost never described as "role reversal" in fishes in which uniparental male care is common. That females do most parental care is the point of view which informed the interesting analysis of Silver *et al.* (1985) that concluded that the best current statistical predictors of paternal behavior are mode of offspring development (more altricial young, more paternal care), mating system (monogamy = paternal care), habitat characteristics (diets of fish, insects, other invertebrates are associated with males escorting young), and clutch weight relative to female body weight (big investments in eggs by females are associated with males feeding females). What would a similar analysis of variation in female parental care gain us?

During my review of the parental care literature, I was struck with the possibility that our mammalian biases may have us on the wrong track, and that our explorations of avian sociality might be fast-tracked by a bit of skepticism in regard to these usual perceptions. For example, I am now wondering: What if female parental care in birds is derived? What if male passerines (the group I am most familiar with) have evolved from lineages in which male parental care was more prominent? What would our theories look like and what would our questions be, if we asked more specific questions about the origins of maternal care in birds, if we asked about ecological and social factors that lead to changes from dominant male parental care to dominant female parental care, if we asked about ecological and social circumstances leading to both within and between species variation in maternal care?

## B. FACULTATIVE VARIATION IN PATERNAL PROVISIONING

Facultative variation in parental care is seldom mentioned or studied except in relation to the probability of one sex or the other deserting a nesting attempt altogether (Maynard Smith, 1977; Beissinger and Snyder, 1987; Lazarus, 1990). There are few species in which males and females desert completely; these include frigate birds, currently under study (J. L. Osorno, personal communication) and Florida snail kites (*Rostrhamus sociabilis*) Besseinger and Snyder, 1987). The ideas in Maynard Smith (1977) were expanded by Lazarus (1990) to include consideration of facultatively expressed options for variation in parental care, depending on what the optimal response of partners would be if one partner deserted. Such models would seem to be useful in examination of species that fall short of full-scale desertion. Further on in Section IV, I discuss facultative variation in male and female parental care duties as a function of partner attractiveness (Burley, 1988).

Beletsky and Orians (1990, p. 608) discussed variation in male parental care in red-winged blackbirds, and noted that although males never incubate eggs, "nestling feeding by male red-winged blackbirds is remarkable for its geographic, local and intraindividual variability," and go on to note varying ecological conditions that seem to be associated with facultative expression of nestling provisioning by males (probably all males can feed nestlings). They suggest two explanations for facultative expression of male provisioning. These might be called the Trade-off Between Paternal Provisioning and Mate Attraction Hypothesis and the Variation in Female Competence Hypothesis.

The Trade-off Between Paternal Provisioning and Mate Attraction Hypothesis says that when ecological conditions are such that males' territories will support additional females, males may sacrifice paternal provisioning to attract additional mates. This hypothesis has also been discussed by Westneat *et al.* (1990) and Westneat and Sherman (1993). Smith (1995) experimentally tested this idea in starlings by increasing the availability of nesting boxes for males, which increased their opportunities to attract additional females to their territories; it also decreased the amount of time that males spent incubating during the early, but not the late, incubation period. Notably, females mated to experimental males compensated for the withdrawal of male incubation, indicating that they could do so.

The Variation in Female Competence Hypothesis says that when ecological conditions are such that it is difficult for females to adequately provision their nestlings, males increase their provisioning to young, a notion for which Whittingham (1989) provided experimental evidence, again in red-winged blackbirds.

The Trade-off Between Paternal Provisioning and Mate Attraction and the Variation in Female Competence hypotheses appear to be two aspects of the same phenomenon that can be expressed as a continuum of ecological opportunity for males in terms of some combined function of the intrinsic abilities of "their" females and the environments that these females are in that affect the abilities of females to provision nestlings alone (Fig. 1). Notice that I did not say that male provisioning depends on the abilities of females to provision alone; rather I want to focus on other ecological and social opportunities for males once male parental care has evolved.

Figure 1 is an expanded expression of a model (Gowaty, 1995) to explain intraspecific variation in extra-pair paternity in socially monogamous birds. It might also be used to explore interspecific variation; however, I stress that understanding my use of this model might most easily be achieved by thinking about intraspecific variation in females. I discuss it here because it provides a framework—part of the continuum of ecological and social

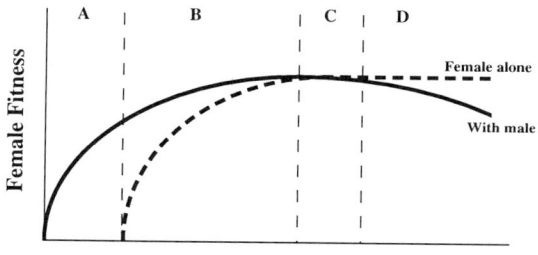

FIG. 1. An expanded version of the Constrained Female Hypothesis (CFH) to account for variation in extra-pair paternity within populations and species of socially monogamous birds. The original graphical representation of the CFH (Gowaty, 1995) held Panels B and C only. The hypothesis assumes that paternal care originated independently of its effects on female fitness or offspring fitness. This axis specifically refers to females' invulnerabilities to social coercion (manipulation or control) by males. The model says that female fitness is a function of interactions between variation in females and their environments. The solid line represents the fitness of the females when they are being "helped" by males; the dashed line represents female fitness if help by males is withdrawn from females. The original model (Panels B and C) described the vulnerability of females to social coercion by males and predicted that females in Panel C more frequently opt for extra-pair fertilization than females in Panel B, because of the threat of withdrawn paternal provisioning should males perceive their threat of lost paternity. Females in Panel A have no reproductive success without male help, that is, male parental care is essential to females. In Panel B, males have variable effects on females' fitnesses; while in Panel C, male parental care has no effect on females' fitnesses. Panel D demonstrates the theoretical possibility that male parental care is detrimental to females' fitnesses. See text for further discussion of how variation in females' need for male parental care theoretically affects females likelihoods of extra-pair fertilization.

opportunities for males—for examining intraspecific facultative variation in male parental care.

On the $x$ axis of Fig. 1 is an interaction variable representing females' intrinsic qualities and their environments relative to their abilities to escape or ignore males' efforts to constrain female behavior for the benefit of males; this axis represents variables that make females more or less vulnerable to social constraint by pair males. If males use paternal provisioning as a mechanism of social manipulation of females, the horizontal axis can be operationalized as females' abilities (intrinsic and environmental) to raise their offspring without male help. It predicts that when there exist advantages for females of multiple mates, females that are less vulnerable to social constraint by males will be most likely to engage in extra-pair paternity. There are two lines on the graph, the solid curve represents the fitnesses of females when they receive "help" from a male in parental duties; the dotted line represents the fitnesses of females if males fail to help or withdraw help from females. The dashed vertical lines indicate four areas of female–male interactions. In Panel A, at the far left-hand side of the graph, females are unable to achieve any reproductive success without the help of a male. In Panel B, at the left-middle of the graph, male help always benefits female fitness, but to varying degrees. In Panel C, at the right-middle of the graph, female fitness with and without males is the same, meaning that male help cannot increment female fitness; and at Panel D, at the far right of the graph, female fitness is decremented by male help.

Figure 1 is not meant to address the evolution of male parental care, rather it assumes that selection pressures exist that favor expression of male parental care. I want to stress this point. The model is not saying, as one previous reader thought, that male parental care depends on females abilities to raise their offspring alone. What it does say, is that once male parental care has evolved, the relationship of female fitness with and without male parental care as a function of variation among females provides a framework for potentially understanding the effect of male parental care on the likelihood that females can be socially constrained by males to avoid extra-pair fertilizations (perhaps by the threat of withdrawal of help in raising offspring). According to the model male parental care may be expressed at random with respect to female fitness; and it assumes that male intrinsic qualities are distributed at random with respect to female intrinsic qualities. The assumptions and predictions of the Constrained Female Hypothesis are discussed in Gowaty (1995). Among the main predictions is females in Panel C will take opportunities to be extra-pair fertilized by preferred gametic partners more frequently than will Panel B (or A) females. It also predicts that males socially paired to Panel C (or D) females will be able to take advantage of their social mates' quality and suffer less from supposed trade-offs between mating effort and parental effort than males socially paired to Panel B (or A) females, so that such males themselves may seek more extra-pair partners than males mated to Panel B (or A) females. Notice that this is a prediction about variation in extra-pair fertilization rates of males that is independent of male quality but not independent of the quality of his social mate.

Figure 1 suggests that facultative variation in male parental care including provisioning may be a correlate of variation in the abilities of females to raise their offspring alone. When intraspecific facultative variation in male parental care is observed, a potential explanation lies in the variable abilities of females to raise their offspring alone, so that the variation among females is a part of the continuum of ecological and social opportunities for males. For example, in socially monogamous species, males mated to Panel C or D females are more likely to engage in other activities because of trade-off costs would be lower than for males mated to Panel A or B females. This idea encourages attention to variation in females' foraging skills relative to availability of resources as a pivotal variable in analyses of distributions of varying male parental care. It suggests that intrinsic variations in females such as metabolic efficiency, as perhaps indicated by how fat they are, their skill as foragers, or the environmental potential of females and their young to avoid predation without help, or the quality and abundance of available food appropriate for offspring, may be important components of males' reproductive success and important sources of variation in male behavior towards young. The Constrained Female Hypothesis predicts that in a within-populational comparison of, say, red-winged blackbirds, that males mated to females on the right-hand side of the graph will be freed from the necessity of paternal provisioning to nestlings and should attempt to attract additional mates (the Trade-off Between Paternal Provisioning and Mate Attraction Hypothesis); while males with females and in environments represented on the left-hand side of the graph should help females provision nestlings (the Variation in Female Competence Hypothesis).

The Constrained Female Hypothesis is different from classical expressions of constraints on males (Orians, 1969; Emlen and Oring, 1977) because it situates variation in females' intrinsic qualities and females' options at the center of the analysis, and says that males' options, the continuum of ecological and social opportunities for males, are a function of variation in females and their environments. These ideas might be applied profitably in the future to studies of facultative variation in male parental care in other passerines, including those with higher frequencies of social monogamy and male parental care than are typical for red-winged blackbirds.

## C. Facultative Variation in Maternal Provisioning

Few studies look at variation in female parental care, much less faculative variation. Where are the studies of variation in ecological and social opportunities for females? However, data from a variety of sources indicate that variation in maternal care in birds exists (e.g., Gowaty, 1983; Bart and Tornes, 1989) and may be an important component in social behavior. I look forward to the day when we know far more about variation in females.

The differential allocation hypothesis (Burley, 1986, 1988, see as follows) was published almost 10 years ago. It says that parental allocations are a function of partner attractiveness. Also, it could partially account for within-sex variation in maternal provisioning, and certainly predicts such variation.

Facultative variation in maternal provisioning might arise via female willingness to invest more in the offspring of high phenotypic or genotypic quality males. This seems to be the case in peahens (*Pavo cristatus*), who lay larger eggs for more attractive males (Petrie, 1994). Such maternal effects could explain observations of higher viability in broods with young from extra-pair paternity.

In the sections that follow, I review many studies that hint at the idea that variation in female parental care exists and is important. I hope this review will help move variation in females to the center of our theoretical and empirical attention—if only for long enough for us to collectively learn more about the ecology and behavior of females.

## III. Male Parental Care and Female Fitness: The Adaptive Significance of Social Monogamy

### A. Why Social Monogamy?

Lack (1968) and others before him (e.g., Darwin 1871, Williams, 1966) explained the ubiquity of social monogamy in birds as an adaptive response to the requirement of male parental care (i.e., the notion that females could not raise their offspring without male help). This hypothesis was so strongly believed that it constituted an assumption of several of the most influential theories of the time (e.g., Orians, 1969; Emlen and Oring, 1977); these theories were attempts to explain the existence and/or evolution of nonmonogamous mating systems. Gordon Orians, Steve Emlen, and Lew Oring, all ornithologists, developed theories with what appeared to be special relevance to avian sociality. There is no doubt that social monogamy and male parental care in birds are statistically correlated; however, whether one causes the other and the direction of causality (male opportunities for parental care could depend on the existence of social monogamy or social monogamy might depend on male parental care) remain important questions. Researchers seldom tested this most prominent of the monogamy hypotheses until after 1980, indicating that most students of the evolution of mating systems experienced a strong intuitive commitment to the idea that females were handicapped without male help.

However, the combined weight of the experimental tests of the hypothesis that male parental care is essential to females' fitnesses presents a different picture of females' abilities to raise their offspring without male help. In most of the male removal tests on birds accomplished so far (reviewed by Bart and Tornes, 1989; Wolfe *et al.,* 1990; Fig. 2 in this paper), what might have been considered most interesting in an age in which facultative variation in *male* behavior was expected (Trivers, 1972), was that, both within and between species, *females vary too* in their abilities to raise their offspring without male help.

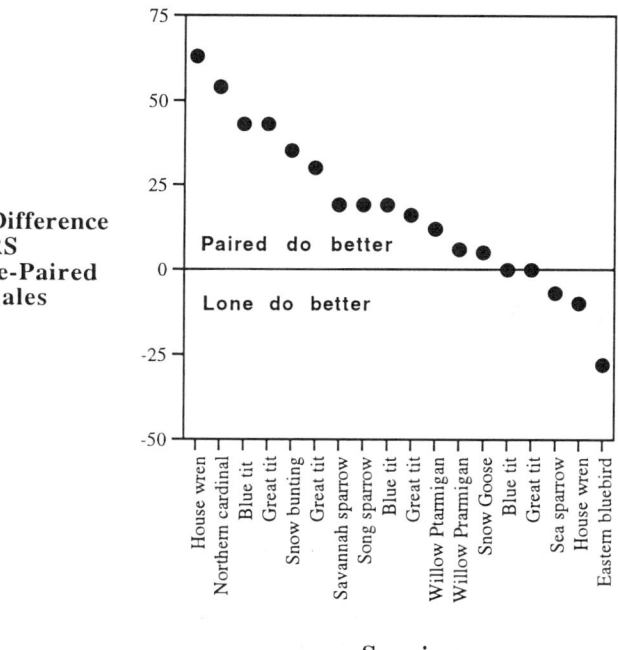

FIG. 2. Frequency distribution of results from male removal studies. Data are from tables in Bart and Tornes (1989). The $y$ axis represents the percentage differences in reproductive success for experimental females whose male partners were removed minus control females who remained paired.

Figure 2 is a graph of between-species variation in the effect of male parental care on the reproductive success of females. On the graph, I have expressed the percentage difference between the reproductive success of experimental (lone) females minus that for control (paired) females. Clearly, there is interesting variation between species in the extent to which male parental care affects females' reproductive success. Furthermore, for most of the species tested so far the percentage difference between females with and without males is relatively narrow, falling between −20 and +20%. To estimate within-species variation in the abilities of females to raise their offspring without males, I calculated the coefficient of variation (CV) in number of hatchlings fledged for lone females only for the six studies for which data were presented in such a way that I could calculate CV. For experimental female snow buntings (Lyon et al., 1987), the CV = 44%; for eastern bluebirds (Gowaty, 1983), 39%; for dark-eyed juncos (Wolfe et al., 1990), 38%; for great tits (Björklund and Westman, 1986), 37%; for song sparrows (Smith et al., 1982), 27%; and for savannah sparrows (Weatherhead, 1979), 14.5%. Again, clearly there seems to be intraspecific variation in the ability of females to raise their offspring without the help of males. The implications of this variation among females was overlooked by many researchers, but was the crucial observation facilitating the original forerunner (Gowaty, 1995) of Fig. 1.

Among the conclusions that have been drawn with these male removal studies is that there is little doubt that males have positive effects on some females' fitnesses in almost all species studied so far, so that for some females there exist important advantages of male care. That is, the requirement of male parental care for female fitness may explain the maintenance of social monogamy for some females or even for all females in some species. For many females male parental care has small or negligible effects on female reproductive success, suggesting that as a general explanation for social monogamy, the Male Care Is Essential Hypothesis is inadequate (Gowaty, 1995). Overlooked conclusions include the significance of variation in females abilities for males' options. Given that females vary in their need for male parental care, males socially paired with highly competent females or females in permissive environments would experience fewer negative consequences from the trade-offs between paternal care and seeking extra-pair partners or additional territorial partners.

B. FEMALE MANIPULATION OF MALES MAY INCREASE PATERNAL PROVISIONING

When male parental care is a resource essential to the fitness of females, females theoretically will be under selection to manipulate and control parental investments by males (Gowaty, 1996). Just as females' competencies to feed offspring probably vary, the tendency for females to attempt to manipulate and control male parental investments should also vary intraspecifically (as well, of course, as interspecifically).

Females can manipulate increases in male parental investments in several ways, including aggressive female territoriality. Slagsvold and Lifjeld (1994) concluded that aggression is often used by females to exclude other females from males' territories and attentions; they interpret female–female aggression as an evolved mechanism for reserving male parental investments for their own offspring. Gowaty (1981) and Gowaty and Wagner (1988) favored another adaptive explanation not inconsistent with a correlated effect of reserving paternal provisioning for a female's offspring. Female–female aggression may protect females' nests from the threat of conspecific nest parasitism. Despite the fact that selection operating from both these costs could operate simultaneously, the two ideas make nonoverlapping predictions about the intensity and frequency of female–female aggression. The hypothesis that female-female aggression functions to protect a female's nest from conspecific nest parasitism predicts that females' tendencies to be aggressive are most likely and most exaggerated when their nests are vulnerable to conspecific nest parasitism, during nest building and egg laying stages of nesting cycles. The hypothesis that female–female aggression functions to reserve paternal provisioning for pair females' young says that female–female aggression will be most pronounced whenever resident females come in contact with intruding females, no matter what the stage of the resident's nest. Of course, in species with short breeding seasons, the time when resident females may be most likely to come in contact with intruding females may be during nest building and egg laying, in which case temporal and spatial variation in female–female aggression may not lead to clear support of one or the other of these hypotheses. However, in species with longer breeding seasons such as eastern bluebirds that breed in South Carolina and Georgia, in which breeding synchrony of females may be relatively weak at times, especially later in nesting cycles, it is possible to discriminate between the selective force of threats to paternal provisioning and threats due to conspecific nest parasitism (Gowaty and Wagner, 1988). Our experimental studies suggest that female–female aggression functions to decrease the threat of conspecific nest parasitism. Specifically, our data indicate that, on average, females are most aggressive during nest building and egg laying, and female–female aggression is most unlikely during incubation and when nestlings are being fed. Our data also indicate that despite the mean tendencies that we observed, that the expression and intensity of aggression is highly variable. We did these tests before the significance of variation in intrinsic quality of females and their environments for mating and reproductive behavior occurred to me. Now we recognize that both the costs of conspecific nest parasitism and the cost of decrements of male parental care could simultaneously affect the likelihood that a female is aggressive *at all* to intruding females. Thus, I predict, based on the logic of Fig. 1, that intraspecifically, females on the left-hand side of the graph will be more likely to be aggressive and more likely to be aggressive at any time in breeding cycles that extraterritorial females intrude than females on the right-hand side of the graph. Specifically, Panel C and D females are less likely to engage in costly aggression than Panel A and B females because when the contested resource is male parental care, it is of lower value to Panel C and D females than Panel A and B females. The interaction of these two hypothetical costs to females of intruding females, rather than either of the hypotheses alone, may have much greater power to explain variation in female–female aggression.

Another way that females may attempt to manipulate and control male behavior for the benefit of females is to copulate with them more than once (Petrie, 1992). Because copulation is costly in several ways, including exposure to pathogens and parasites, why individuals of the same pair copulate more than once is problematic. Petrie (1992, p. 790) hypothesized that "repeated copulation may reduce the likelihood that a male partner mates with another female." She predicted that repeated copulation is most likely in species with high variance in male genetic quality or in access to resources that males provide to females and their young; she predicted that copulations between partners would occur outside of times when females are fertile if there is a risk of males copulating with other females at that time; she predicted that females should solicit copulations after mate absences and after extraterritorial intrusions by females; and Petrie (1992, p. 791) predicted that the highest copulation frequency should occur within pairs where males are highest quality or where females would "suffer the highest cost from losing a high quality mate." What may be even more important is a female's own intrinsic quality as it interacts with the environment that she is in. Obviously, environments for females include such things as variation in the abilities of males to help provision her offspring (e.g., his foraging skill). Specifically, Fig. 1 predicts that all else equal, females on the left-hand side of the graph (Panel A and B females) solicit copulations from their territorial partners more than females on the right-hand side of the graph. It predicts that in species represented only by Panel B females that females on the left-hand side of Panel B will solicit copulations from males more than females on the right-hand side of Panel B. It further predicts that between-female variation for females on the left-hand side of the graph only will be an interaction of "female quality" *and* male foraging skill, such that the more likely females are to be truly helped by male help, the more females will solicit copulations from their territorial partners. For example, in a species with only Panel B females this hypothesis predicts that if males' quality relative to their abilities to feed offspring or guard against predators varies, females' efforts to manipulate greater parental provisioning from such males will be a function of an interaction between her quality and his abilities. The greater the value of the helping resource offered by males to particular females, the more likely is a female to attempt to manipulate further parental investments from given males via copulation solicitation or other means. Evaluation of intraspecific variation in the value of male help interacting with female quality, skill, and females' environments to affect social behavior, such as within-pair copulation frequency, is an exciting frontier in avian studies.

Eens and Pinxten (1995) first demonstrated in songbirds such female manipulation of males through copulation solicitation. The female starlings they studied solicited copulations before and during egg laying, as well as long after egg laying. Primary females in polygynous consorts solicited copulations more than females in monogamous consorts, and "males refused to copulate in 26% of female solicitations" (Eens and Pinxten, 1995, p. 73). These authors interpreted this behavior as a mechanism females use to interrupt males' attempts to attract secondary females, because many of female copulation solicitations occurred when males were singing. Although their data clearly support the mate-guarding function of female copulation solicitation, the value of a monogamously mated male may not to be related to the frequency with which males feed nestlings, as there was no correlation between copulations and paternal provisioning, suggesting that some other explanation may have greater explanatory power. The previous discussion suggests that the missing covariate in their analysis may have been variation in females' abilities to feed offspring without male help, or the value of paternal provisioning to *particular* females.

C. MALE MANIPULATION OF FEMALES MAY INCREASE
   MATERNAL PROVISIONING

When theory focuses on the possibility of within-species and within-population variation in fitness and behavior of individuals, as Fig. 1 does, it becomes obvious when there is variation in female quality (by now I hope readers have realized that "female quality" is my shorthand for "intrinsic quality of females as it interacts with the quality of environments in which females find themselves") that there exists a range of possibilities of female and male contributions to offspring well-being. In some cases both males and females must provision at the limits of their abilities for offspring to reach independence (say, as represented by the far left-hand side of Fig. 1); in others the provisioning of just one parent can suffice to get offspring to independence (as represented by the right-hand side of Fig. 1). Between these extremes there exists the possibility for "negotiations" between the collaborating caregivers over who will contribute what and how much to offspring well-being (Fig. 1B). Whenever there are alternative opportunities for increasing survival or reproductive success of either of the collaborators, whenever the costs of parental investments for either of the collaborators varies, sexual conflict can arise. Thus, just as females may attempt to increase paternal provisioning, it is also likely that males attempt to manipulate females into increased maternal provisioning.

*Mate guarding*—behavior that decreases the likelihood that other males approach females or that decreases the likelihood that territorial females approach males (Gowaty and Bridges, 1991b)—is one frequently discussed paternity guard (e.g., Birkhead and Møller, 1992). Paternity guards are mechanisms males use to increase the probability that females share their gametes only with the guarding males. Thus, if females have options for alternative mating, paternity guards represent one set of ways that males may attempt to manipulate females into further maternal provisioning of his young. Male aggression against females (Smuts and Smuts, 1993) may function in the same way.

However, even in the case in which there is no likelihood that the total number of a males' young will decrease due to females' investments, males, like females, should be under selection to increase the relative parental investment of their partners—especially if the likelihood of female desertion from the partnership is likely after completion of a current nesting attempt. Mechanisms that males might use to increase maternal provisioning include affiliative conditioning of females' behavior, by continuing so-called "courtship" feeding as well as the use of "punishment" (Clutton-Brock and Parker, 1995). Bird females are almost always the limiting resource for males' reproductive success, and therefore males should experience especially strong selection to manipulate and control females' reproductive capacities and activities for their own benefit (Gowaty, 1992, 1996). Mechanisms that function within partnerships in these ways are not yet commonly studied in birds, but are presaged by studies of differential allocation (Burley, 1986, 1988, and as follows).

D. Parental Care and Polygamy

Just as social monogamy positively correlates with biparental care, social polygyny correlates with uniparental female care, and social polyandry with uniparental male care. The strong correlations among these patterns have suggested that the need for parental care and the outcome of contests for "emancipation from parental care" (cf., Emlen and Oring, 1977) determine mating systems.

Orians (1969) model was a female choice model that assumed that in a given environment female reproductive success was always enhanced by male help in parental care. The polygyny threshold model theoretically described conditions that would make it as advantageous for a female to settle with an already-mated as an unmated male.

Emlen and Oring (1977) explained variation in mating systems in terms of individuals' opportunities to be "emancipated from parental care." Such emancipation, along with the distribution in space and time of females, determined the environmental potential for polygyny for males.

Davies (1989) sexual conflict models describe the evolution of mating systems in terms of the conflicting interests of males and females over mating opportunities and apportionment of parental care duties. Specifically in his models females mate with more than one male in order to accrue help in parental care from each of her mates. And, indeed, male dunnocks, the species Davies made famous, do apportion their parental care as a positive function of their realized genetic paternity (Davies and Hatchwell, 1992).

Newer models (Gowaty, 1995, 1996) explaining the co-occurrence of traits that we associate with mating systems also are based on sexual conflict and the idea that fundamental sexual conflicts over the control of reproduction (see also Gowaty, 1992) drive mating systems evolution. These ideas, in contrast to the previous ones, suggest that the contested prize is not "emancipation from parental care" but the control of reproduction, including control of parental care and parental "fiddling" (see Section VIII,C that follows for a definition).

## IV. Sexual Conflict and Differential Parental Allocation

In an ideal, socially egalitarian world, biparental care would be split even-Steven between males and females. Also, in birds, in which males can do everything that females do except lay eggs, one might expect to see an equitable division of parental care duties between cooperating partners. Recently, sexual conflict theory and ideas about variation in partner attractiveness have led to some stunning data about allocation of parental investments by females and males cooperating in care of offspring, suggesting that the ideal—equality of investment—is seldom achieved.

Burley (1986, p. 415) first suggested the differential allocation hypothesis. It states "that mates of attractive individuals are 'willing' to contribute greater than average parental investment to obtain and/or maintain their attractive mates." Her experimental data are consistent with this idea and suggest that in species with biparental care, individuals will behave facultatively as a function, not only of their partners' attractiveness, but their own "desirability" as well. Burley's results have remarkable implications, not the least being that decisions about parental investments will involve trade-offs between quality and quantity of offspring, so that measures of sexual selection that depend only on variance in number of mates are sure to miss important insights. Burley's (1988) experimental tests of the differential allocation hypothesis showed that: (1) males mated to the most attractive females (experimentally manipulated by band color) had the highest rates of paternal expenditure (male contributions relative to total by males and females), and (2) the most attractive males had the lowest rates of paternal expenditure. Given the subtlety of the "trade negotiations" that these results indicate, there would appear to be opportunity for conflict between individuals mated to one another, as a function of differential attractiveness, and resultant differential opportunities for fitness gains outside the pair for attractive partners. I know of few studies in species with biparental care that have attempted to examine differential allocation patterns as a function of attractiveness in wild-living birds. Furthermore, I imagine that most attempts would fail to have the experimental elegance of Burley's studies, in which she has been able to control correlated traits, which may contribute to attractiveness in wild-living birds, by assigning individuals at random to experimental categories associated with attractiveness.

Slagsvold and Lifjeld (1989) hypothesized that significant sexual conflict over hatching asynchrony was likely in species of birds with biparental care. They hypothesized that in species in which only females incubate, hatching asynchrony would facilitate greater paternal investment than hatching synchrony. They reasoned that females benefit from hatching asynchrony because it would increase the duration of paternal feedings and minimize the time males have to attract additional mates. Slagsvold *et al.* (1994) tested this idea in blue tits, in which there exist different optimal patterns of hatching synchrony for males and females. In an experiment in which hatching asynchrony was manipulated, postbreeding survival of males and females was a function of hatching asynchrony, such that female parents survived better when young were even-aged; in contrast, male parents survived better when young were of variable ages. This seems to have been so because females preferentially feed the smallest chicks in broods, while males preferentially feed the largest chicks; largest chicks were larger from asynchronous broods that synchronous broods, so that male parents worked less to feed chicks from asynchronous broods and survived better after caring for chicks from asynchronous broods than synchronous broods; the opposite was the case for females. These differences in adult postbreeding survival as a function of hatching synchrony represent a conflict of interest between females and males that is "won" by females, because only blue tit females incubate and therefore females are in control of how synchronously eggs in a clutch hatch. Therefore, blue tit females are in a position to minimize their own investment at the expense of their partners. Hebert and Sealy (1993) were unable to provide support for the sexual conflict over hatching asynchrony hypothesis in yellow warblers (*Dendroica petechia*). Whether there is significant sexual conflict over hatching spread in other species of birds and whether hatching asynchrony is an option routinely used by female birds to manipulate male paternal investments is a question for the future.

## V. Extra-pair Paternity

### A. Data on Extra-pair Paternity

Before Trivers (1972), "faithfulness" between social partners was part of the definition of monogamy. After Trivers' (1972) prediction that males would cooperate with a primary female in raising offspring, but seek opportunities to copulate with additional females, many others suspected that social monogamy was not perfectly correlated with faithfulness (genetic monogamy of individuals). However, evaluation of that prediction and other questions about genetic parentage depended on the development of reliable genetic markers of parentage. The 10 years since the publication the classic paper of Jeffrey *et al.* (1985) describing "DNA fingerprinting" have proved revolutionary for our understanding of the genetical aspects of social monogamy in birds.

Table I is a list of studies in which the frequency of extra–pair paternity has been evaluated using a variety of genetical markers, most notably DNA fingerprinting.

In general, extra-pair paternity is more frequently reported than conspecific nest parasitism, a finding that may reflect an observational bias associated with DNA fingerprinting (Gowaty and Bridges MS), namely, that in order to infer paternal genotypes, maternal genotypes must be known. Many studies have assumed that the putative mother, the caregiving female, was the genetic mother and used the "maternal" genotype to infer paternal contributions to offspring DNA fingerprints. This assumption has probably led to some underestimation of the number of species in which conspecific nest parasitism occurs; however, in studies in which conspecific nest parasitism and extra-pair paternity both could be reliably estimated (all those based on single-locus markers in the table), conspecific nest parasitism occurs less frequently than extra-pair paternity (Table I). Thus, based on current data, the conclusion that even in birds maternity is more certain than paternity is warranted. Also, if selection acts so that caregivers adjust their provisioning to nestlings as a function of their genetic parentage, in species with biparental care at the same nest, females should be more reliable caregivers than males.

Table I demonstrates that the number of species with 0% extra-pair paternity is quite low and that the frequency within species with extra-pair paternity ranges from low (<10% of nestlings or chicks from extra-pair matings) to very high (>85% of nestlings). The most dramatic conclusion from Table I is that the number of socially monogamous species in which individual males and females are also genetically monogamous is remarkably small (Fig. 3), and it suggests that our definitions, expectations, and theories about social monogamy in birds must provide for the existence of extra-pair paternity and multiple mating by females (Gowaty, 1995).

Some studies have included evaluations of correlations between the likelihood of males providing care to nestlings that are not theirs ("cuckoldry"), and phenotypic or demographic characteristics of males and females and environmental variation. The density of neighbors has a positive effect on extra-pair paternity in some species (Gowaty and Bridges, 1991a) but not others (Dunn et al., 1994a). Nevertheless, the possibility that neighbor densities account for some of the variance in estimates of extra-pair paternity suggests that efforts to use comparative data (between species) (Møller and Birkhead, 1993) to discern adaptive patterns may suffer from the confounding bias that nests in different studies may have been at quite different densities (and even when neighbor densities are known, it is unlikely that they have constant effects across species on extra-pair paternity). However, the few studies in which the identity of genetic fathers of extra-pair offspring have been sought indicate that extra-pair males are usually territorial neighbors rather than floaters (e.g., Gibbs et al., 1990). Generally, younger males are more vulnerable to caring for offspring that are not theirs than older males (e.g., Gowaty and Bridges, 1991b; Sherman and Morton, 1988); or males with phenotypes that usually indicate younger, rather than older, age (Bereson et al.,) MS are more likely to have extra-pair offspring in their nests.

Because it was impossible to assign reliably paternity in most of the studies in Table I, what we are able to infer from these studies about male behavior and reproductive dynamics is still limited. For instance, measures

TABLE I
Molecular Studies of Genetic Parentage in Avian Species

| Species | Method[a] | EPO (%)[b] | N chicks | N broods | Broods with EPO (%) | CNP[c] | Social mating system[d] | Reference |
|---|---|---|---|---|---|---|---|---|
| Common loon (*Gavia immer*) | ml mini | 0 | 34 | 20 | 0 | No | m | Piper, personal communication |
| Northern fulmar (*Fulmarus glacialis*) | ml mini | 0[e] 4.6[f] | 85 | 0 | 85 | No | m | Hunter et al., 1992 |
| Shag | ml mini | 17.9 | 28 | 15 | 20 | No | m | Graves et al., 1991 |
| Mallard (*Anas platyrhynchos*) | ml mini | 3 | 298 | 46 | 17.4 | ? | m | Evarts and Williams, 1987 |
| Blue duck (*Hymenolaimus malacorhynchus*) | ml mini | 0 | 14 | 10 | 0 | ? | m | Triggs et al., 1991 |
| Black vulture (*Coragyps atratus*) | ml mini | 0 | 36 | 0 | 16 | No | m | Decker et al., 1993 |
| American kestrel (*Falco sparverius*) | ml mini | 8.3 | | | | | m | Villarroel, personal communication |
| Australian brush-turkey (*Alectura lathami*) | ml mini | 27.7 | 65 | ? | | | prom | Birks, MS |
| Pheasant | ml mini | 0 | 22 | 2 | 0 | ? | prom | von Schantz et al., 1989 |
| Pukeko (*Porphyrio porphyrio*) | ml mini | Many sires[g] | 73 | 12 | 0 | | pgyny/pland | Jamieson et al., 1994 |
| Tasmainian native hen (*Tribonyx mortierii*) | ml mini | 4.8 | 28 | 6 | 16 | No | pland | Gibbs et al., 1994 |
| Oystercatcher (*Haematopus ostralegus*) | ml mini | 1.5 | 65 | 26 | 1 | No | m | Heg et al., 1993 |
| Eurasian dotterel (*Charadrius morinellus*) | ml mini | 4.5 | 44 | 22 | 4.5 | No | m | Owens et al.,1995 |
| Spotted sandpiper | | | | | | | pland | Oring et al., 1992 |
| European bee-eater (*Merops apiaster*) | ml mini | 2 | 100 | 65 | | | c | Jones et al., 1991 |
| White-fronted bee-eater (*Merops bullockoides*) | alloz | 9; 12 | 97 | 65 | | Yes | c | Wrege and Emlen, 1987 |

|  | | | | | | | |
|---|---|---|---|---|---|---|---|
| Acorn woodpecker (*Melanerpes formicivorus*) | alloz | 2.2 | 186 | | | c | Mumme et al., 1985 |
| Acorn woodpecker (*Melanerpes formicivorus*) | alloz | 33.3 | 8 | | | c | Joste et al., 1985 |
| Red-cockaded woodpecker (*Picoides borealis*) | ml mini | 1.3 | 48 | 28 | 3.6 | c | Haig et al., 1994 |
| House martin (*Delichon urbica*) | ml mini | 19 | 73 | 20 | 35 | No | m | J. T. Lifjeld, personal communication |
| House martin (*Delichon urbica*) | ml mini | 15 | 62 | 19 | 32 | No | m | Riley et al., 1995 |
| Cliff swallow (*Hirundo pyrrhonota*) | alloz | Some | | 105 | 6 | Yes | m | Brown and Brown, 1988 |
| Barn swallow (*Hirundo rustica*) | ml mini | 22.2 | 45 | 11 | 45.5 | No | m | Smith et al., 1991 |
| Purple martin (*Progne subis*) | ml mini | 3.6[h] 70.8[i] | | 30 | | Yes | m | Morton et al., 1990 |
| Tree swallow (*Tachycineta bicolor*) | ml mini | 69 | 111 | 25 | 84 | Yes | m | Barber et al., MS |
| Tree swallow (*Tachycineta bicolor*) | ml mini | 51 | 181 | 34 | 68 | No | m | Dunn et al., 1994 |
| Tree swallow (*Tachycineta bicolor*) | ml mini | 53 | 119 | 23 | 87 | No | m | Dunn et al., 1994 |
| Tree swallow (*Tachycineta bicolor*) | ml mini | 38 | 86 | 16 | 42 | No | m | Lifjeld et al., 1992 |
| Scrub jay (*Aphelocoma coerulescens*) | ml mini | 0 | ~120–140 | 50 | ~0 | No | c | Quinn et al., 1990 |
| Blue tit (*Parus caeruleus*) | ml mini | 11 | 314 | 34 | 11 | No | m | Kempenaers et al., 1992 |
| Blue tit (*parus caeruleus*) | ml mini | 6 | | | | | m | Gullberg et al., 1992 |
| Great tit (*Parus major*) | ml mini | 15 | | | | | m | Gullberg et al., 1992 |
| Great tit (*Parus major*) | ml mini | 16.6 & 20.2 | | 32 | 47.1 & 53.3 | | m | Lubjuhn et al., 1993 |
| Stripe-backed wren (*Campylorhynchus nuchalis*) | ml mini | 10 | 69 | | | | c | Rabenold et al., 1990 |
| Stripe-backed wren (*Campylorhynchus nuchalis*) | ml mini | 4.6 | 43 | | | | c | Stevens, 1988 |
| House wren (*Troglodytes aedon*) | alloz | 0 | | | | Yes | m | Price et al., 1989 |
| Great reed warbler (*Acrocephalus arundinaceus*) | ml mini | 3.1 | 553 | 130 | 5.4 | No | pgyny | Hasselquist et al., 1995 |
| Superb fairy-wren (*Malurus cyaneus*) | ml mini | 76 | 181 | 40 | 38 | 95 | c | Mulder et al., 1994 |

(*continues*)

TABLE I (*Continued*)

| Species | Method[a] | EPO (%)[b] | N chicks | N broods | Broods with EPO (%) | CNP[c] | Social mating system[d] | Reference |
|---|---|---|---|---|---|---|---|---|
| Splendid fairy-wren (*Malurus splendens*) | alloz | 65 | 91 | | 100 | No | c | Brooker *et al.*, 1990 |
| Willow warbler (*Phylloscopus*) | ml mini | 0 | 120 | 19 | 0 | No | m | Gyllensten *et al.*, 1990 |
| Wood warbler (*Phylloscopus sibilatrix*) | ml mini | 0 | 56 | 13 | 0 | No | m | Gyllensten *et al.*, 1990 |
| Yellow warbler (*Dendroica petechia*) | ml mini & alloz & | 37 | 355 | 90 | 59 | No | m | Yezerinac *et al.*, in press Gelter and Tegelstrom, 1992 |
| Pied flycatcher (*Ficedula hypoleuca*) | ml mini | 13; 24 | 131 | 22 | 13.6 | Yes | m | Lifjeld *et al.*, 1991 |
| Pied flycatcher (*Ficedula hypoleuca*) | ml mini | 4 | 135 | 27 | 15 | No | m | J. T. Lifjeld, personal communication |
| Pied flycatcher (*Ficedula hypoleuca*) | ml mini | 7 | 98 | 17 | 18 | 0 | m | J. T. Lifjeld, personal communication |
| Bluethroat (*Luscinia svecica*) | ml mini | 21 | | 16 | 44 | No | m | J. T. Lifjeld, personal communication |
| Bluethroat (*Luscinia svecica*) | ml mini | 19 | 150 | 31 | 35 | 0 | m/pgyny | Gowaty *et al.*, MS |
| Eastern bluebird (*Sialia sialis*) | sl mini | 20 | 195 | 59 | 42.4 | Yes | m | Gowaty and Bridges, 1991a,b |
| Eastern bluebird (*Sialia sialis*) | alloz | 20 | 523 & 345 | | >16–25 | Yes | m | Gowaty and Karlin, 1984 |
| Eastern bluebird (*Sialia sialis*) | alloz | >5 | 210 | | | Yes | m | Meek *et al.*, 1994 |
| Eastern bluebird (*Sialia sialis*) | ml mini | 8.4 | 83 | 21 | 24 | Yes | m | Dickinson, personal communication |
| Western bluebird (*Sialia mexicana*) | ml mini | 16.5 20[j] | 115 | 29 44 | 34.5 | Yes | m | Hoi, MS. |
| Bearded tit (*Panurus biarmicus*) | ml mini | 0[k] | 27 col | clutches | 20 | Yes | m | Hartley *et al.*, 1995 |
| Alpine accentor (*Prunella collaris*) | ml mini | 0 | 110 | 38 | 0 | | c/pland | Burke *et al.*, 1989 |
| Dunnock (*Prunella modularis*) | ml mini | 0 | | | | No | pgyny/pland | |

500

| Species | Method | | | | | | | | Reference |
|---|---|---|---|---|---|---|---|---|---|
| Bull-headed shrike (*Lanius bucephalus*) | ml mini | 10.1 | | 99 | 24 | 16.7 | No | m | Yamagishi et al., 1992 |
| European starling (*Sturnus vulgaris*) | ml mini | 8.7 | | 92 | 22 | 31.8 | No | m | Smith and von Schantz, 1993 |
| European starling (*Sturnus vulgaris*) | ml mini | 9.7 | | 62 | 14 | 28.6 | Yes | m | Pinxten et al., 1993 |
| Eurpoean starling (*Sturnus vulgaris*) | alloz | 1.1 | | 365 | 95 | 2.1 | | m | Hoffenberg et al., 1988 |
| Hooded warbler (*Wilsonia citrina*) | ml mini | 29 | | 78 | 17 | 47 | No | m | Stutchbury et al., 1994 |
| Wilson's warbler (*Wilsonia pusilla*) | ml mini | 52 | | 59 | 17 | 53 | ? | m | Bereson et al., MS |
| Northern cardinal (*Cardinalis cardinalis*) | ml mini | 13.5 | | 37 | 16 | 19 | No | m | Ritchison et al., 1994 |
| Indigo bunting (*Passerina cyanea*) | ml mini | 35 | | 63 | 25 | 48 | No | m | Westneat, 1990 |
| Reed bunting (*Emberiza schoenicius*) | sl mini | 55 | | 216 | 58 | 86 | No | m | Dixon et al., 1994 |
| Corn bunting (*Miliaria calandra*) | ml mini | 4.5 | | 44 | 15 | 6.7 | No | pgyny | Hartley et al., 1993 |
| Field sparrow (*Spizella pusilla*) | ml mini | 14.5 | | 152 | 50 | | Yes, 1 chick | m | M. Carey, personal communication |
| Field sparrow (*Spizella pusilla*) | ml mini | 0 | | | | | | m | Petter et al., 1990 |
| White-throated sparrow (*Zonotrichia albicollis*) | ml mini | 11 | | 62 | 21 | 24 | Yes | m | E. Tuttle, personal communication |
| White-crowned sparrow (*Zonotrichia leucophrys*) | alloz | 34 | | 110 | 35 | 26 | No? | m | Sherman and Morton, 1988 |
| Red-winged blackbird (*Agelaius phoeniceus*) | ml mini | 35 | | 403 | | 55 | No | pgyny | Gray, 1994 |
| Red-winged blackbird (*Agelaius phoeniceus*) | sl mini | 24.5 | | 11 | 36 | 47.2 | No | pgyny | Gibbs et al., 1990 |
| Red-winged blackbird (*Agelaius phoeniceus*) | ml mini | 25 | | 235 | 68 | 41 | No | pgyny | Westneat, 1993 |

(*continues*)

TABLE I (Continued)

| Species | Method[a] | EPO (%)[b] | N chicks | N broods | Broods with EPO (%) | CNP[c] | Social mating system[d] | Reference |
|---|---|---|---|---|---|---|---|---|
| Bobolink (*Dolichonyx oryzivorous*) | alloz | 14.6 | 840 | 38 | 19[l] | No | m | Bollinger and Gavin, 1991 |
| Chaffinch (*Fringilla coelebs*) | ml mini | 17 | 47 | 23 | 13 | | m | Sheldon and Burke, 1994 |
| House sparrow (*Passer domesticus*) | ml mini & alloz | 13.2[l] 12.1[m] | 420 | 144 | | No | m | Wetton et al., 1992 |
| House sparrow (*Passer domesticus*) | ml mini | 13.6 | 536 | 183 | 26.1 | | m | Wetton and Parkin, 1991 |
| Zebra finch (*Taeniopygia guttata*) | ml mini | 2.4 | 82 | 25 | 8 | Yes | m | Birkhead et al., 1990 |
| Zebra finch (*Taeniopygia guttata*) | ml mini | 12.8 | 278 | 126 | | Yes | m | Burley et al., MS |

[a] ml mini = multilocus minisatellites; sl mini = single locus minisatellites; alloz = allozymes; sl micr = single locus microsatellites.
[b] Extra-pair offspring.
[c] Conspecific brood parasitism.
[d] m = monogamy; pgyny = polygyny; pland = polygandy; c = cooperative breeders; prom = promiscuous.
[e] Estimate from genetical markers.
[f] Estimate from observations of copulations.
[g] Multiple sires, all males had reproductive success.
[h] After second year.
[i] Second year.
[j] Colonial.
[k] Solitary.
[l] Estimate from allozymes.
[m] Estimate from DNA markers.

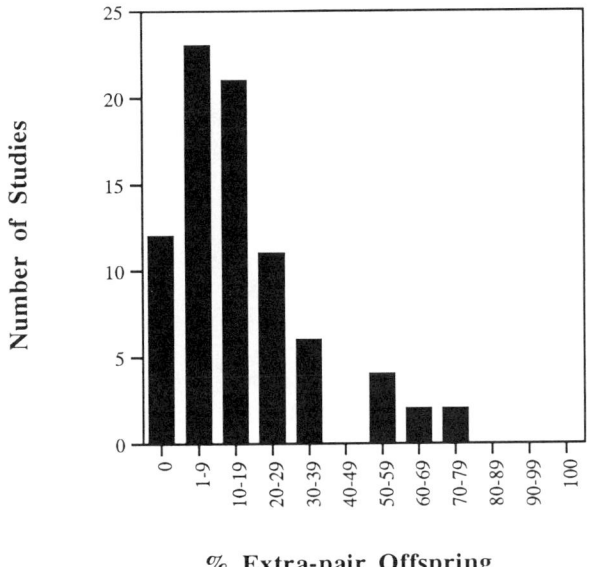

FIG. 3. Extra-pair paternity rates for passerines reported in studies in Table I.

of variance in male realized (genetic) paternity of wild-living birds are still rare (Gibbs *et al.*, 1990). Therefore, at this point, although it is an extremely good bet that many males are genetically polygynous as well as socially monogamous, we are on somewhat stronger footing when we conclude from these data that females are multiply mating and are genetically polyandrous while socially monogamous.

The data in Table I and Fig. 3 direct one's attention to the variation in the behavior of females. Multiple mating by females may mean either: (1) that offspring in a brood are genetically sired by more than one male (females are genetically polyandrous), or (2) that all the nestlings in a brood are genetically sired by an extra-pair male (females are socially monogamous with one male and simultaneously genetically monogamous with another!). In almost every species so far submitted to parentage analysis using DNA markers (Table I) there are some females who are socially monogamous with one male, but share their gametes with another male(s) entirely. This sort of variation in extra-pair paternity patterns is especially interesting in terms of the possibility of variation in females' abilites to control paternity of their offspring (Gowaty, 1995).

## B. Why Do Females Multiply Mate?

The most frequently asked question after a description of extra-pair paternity in a socially monogamous species is what are the advantages of mixed genetic paternity for females or why do females engage in extra-pair matings at all? The question from the male side is seldom asked, because since Trivers (1972) and before (Bateman, 1948), the reasonable adaptive explanation, that the number of offspring a male has is a function of the number of his mates, adequately explains multiple mating by males. Multiple mating by females has seemed more paradoxical because the quantity of offspring for a female is unlikely to increment as a function of additional mates. Also, unless there is variation in the quality of males, there is unlikely to be an advantage to multiple mating by females (Petrie and Lipsitch, 1994; Gowaty, 1996). Birkhead and Møller (1992) concluded the weight of current evidence favors the hypothesis that females prefer to share their gametes with males with good genes or good phenotypic qualities. Petrie's extraordinary study (1994) in peafowl (*Pavo cristatus*) is the first one I know of in birds to test experimentally the effect of female choice on the viability of offspring. Petrie specifically tested whether there are any viability advantages in offspring for females mated to males with elaborate trains containing the most eyespots, the character on which peahens base their choices of peacocks. Petrie assigned peahens at random to males and then raised the chicks under identical conditions. She found significant differences between fathers for offspring weight on Day 84 of chick life, a head start that she showed was associated with the likelihood that offspring subsequently survived.

A second explanation for multiple mating by females is that females thereby gain access to resources such as additional help in raising their offspring, something that partially explains the polygynandrous habits of dunnocks and alpine accentors (Davies, 1992; Hartley *et al.*, 1995). However, it might also explain some extra-pair paternity, if females "trade" fertilizations for some resource(s) that males broker (Gowaty, 1996). In keeping with this idea, Kempenaers (1993) reports that a female solicited extra-pair copulation with a male, who subsequently helped feed her nestlings, and Gray (1994) reports that female red-winged blackbirds gain material benefits including foraging privileges on the territories of males with which they copulated and in terms of predator mobbing by their extra-pair partners.

If female genetic monogamy is most likely to predominate in socially polygynous species and in socially promiscuous species (such as those with leks) as suggested by some considerations of the co-occurrence of genetical and social mating patterns of individuals (Gowaty, 1995), then the co-occurrence of multiple mating by females in socially monogamous species is another facet of the question. Put another way, why is multiple mating by females more common in socially monogamous than socially polygynous or promiscuous species? Gowaty (1995) suggests the hypothesis that females' decisions regarding social mates are constrained by ecological including social constraints in ways that particularly favor their extra-pair mating (possibly for good genes, variable progeny, increased offspring viabilities, or immediate benefits). The Constrained Female Hypothesis (Gowaty, 1995) is based on the assumption that because males' reproductive success is primarily limited by access to females, that selection favors males that attempt to manipulate females' mating decisions for their own benefit; likewise, whenever male manipulation of females decreases females' fitnesses, females should be under selection to resist males' efforts to manipulate their decisions. In many avian social systems, male territoriality may facilitate males' brokering females' access to resources crucial for female reproduction. Thus, females may often be manipulated into social situations in which they must trade copulations (and resultant fertilizations) for access to the resources they need; thus, social monogamy may represent females pairing for resources under social constraint by males. If this is the case, extra-pair mating by females may represent selection on females to resist males' efforts to control their reproductive capacities. This theory predicts that for females with both extra-pair and within-pair offspring in the same brood that the extra-pair offspring are of higher genetic and phenotypic quality than the within-pair offspring. More novel predictions include that higher-quality females or females in more permissive environments (Panel C and D females, Fig. 1) will be more likely to mate with extra-pair males than lower-quality females or females in less permissive environments (Panel A and B females, Fig. 1). It predicts that for females with young from extra-pair partners, genetical monogamy is more likely in Panel C and D females than Panel A and B females. It also predicts that variance in male reproductive success will be higher for males socially paired to Panel C and D females than to Panel A and B females.

A more formal statement of the aforementioned is graphed in the forerunner to Fig. 1 (Gowaty, 1995), which I formulated as an explanation for why females mate multiply. Assuming that females are favored who share their gametes with high-quality males and that mating is costly, females unconstrained by social or ecological constraints should choose simultaneously among available males, mate, and be on with it. In many socially monogamous species, females may be constrained by both ecological and social factors. Males may exploit variation in females' abilities to escape such constraints by brokering females' access to resources in trade for realized genetic paternity. Thus, genetic monogamy should occur when females are completely constrained by males or completely free of all social and ecological constraints, and multiple mating by females or genetic polyandry should occur when females make decisions about gametic partners under imperfect social constraint (Gowaty, 1995, 1996). The Constrained Female Hypothesis predicts that extra-pair paternity will be greater for Panel C than Panel B females.

### C. FEMALE CONTROL OF GENETIC PATERNITY

It is remarkable that avian researchers are testing whether females control genetic paternity (e.g., Wagner, 1991; Lifjeld and Robertson, 1992), because it has seemed obvious to some (Fitch and Shugart, 1984; McKinney *et al.*, 1983), since fertilization occurs inside of females' reproductive tracts, that fertilization is most likely a female-controlled function. It seems reasonable that once insemination has occurred, the outcome of any conflicts between females and males would be biased in females' favors, and furthermore, that females would have the option, at least, of biasing or managing outcomes of competitive interactions between inseminates of different males (see Waage, 1979 for a discussion in nonavian taxa). Thus, the new language of "female control" (e.g., Birkhead and Møller, 1993) suggests just how embedded the notion of "passive females" has been. Happily, the new "female perspectives" suggest that females are not just passive receptacles for sperm, but actively manipulate and manage the outcome of sperm wars and are the architects of sperm competition (Gowaty, 1994), so that sperm competition (Parker, 1970) questions can now be addressed in a more balanced way. For example, researchers are now asking: do winners of sperm contests win because of female choice mechanisms or male–male interactions (Lifjeld *et al.*, MS)?

Female control of paternity starts with behavioral control of copulations by females. Most birds lack intromittent organs, and in most bird species, females must evert their cloacae in order for sperm transfer to take place (Fitch and Shugart, 1984), meaning that females must be complicit with males in order for most copulations to be achieved. In fact, forced copulation is rare in birds (Gowaty and Buschhaus, MS), occurring mostly in Anseriformes (McKinney *et al.*, 1983), in which males have intromittent organs.

Female control of paternity also suggests that females may seek, not just accept, extra-pair copulations and fertilizations, something that was a surprising idea only a few years ago. Smith's (1988) report that female black-headed chickadees sought extra-pair copulations was greeted with surprise in some quarters, because the idea that benefits of multiple mating for females were low or nonexistent. Studies of females' actively seeking copulations include Sheldon (1994) on chaffinches, Eens and Pinxton (1995) on starlings, Gray (1994) on red-winged blackbirds, Wagner (1991) on razorbills, and more are probably on the way.

The use of female control of paternity (e.g., Lifjeld and Robertson, 1992) suggests that there are conflicts between males and females in this most cooperative of endeavors, sexual reproduction. Also, an area in which significant conflicts between males and females over mating is obvious, has to do with behaviors that lead to copulations. Mate guarding by males is one way that males use to attempt to keep females from copulating with other males, and if males are successful, they can control copulation indirectly (Kempanaers *et al.*, 1995). Other methods that males may use to manipulate and control female mating decisions include affiliative and aversive conditioning of females' behavior (Gowaty, 1996). Until very recently, mate guarding was considered behavior that males used to keep other males away from their females. Female control language suggests that this is another area of changed perspective, for now mate guarding is something males do to keep "their" females away from other males.

The best data so far on female control of paternity is from captive studies of zebra finches (Burley *et al.* in press, submitted). Despite the fact that 82% of extra-pair copulations were forced on females, there was no realized genetic paternity from these. Almost 30% of offspring in Burley's breeding colony were extra-pair young, but all these arose from unforced extra-pair copulations, that is, from a minority of extra-pair events.

The language of female control indicates the emerging idea that males and females compete over the control of reproduction (Gowaty, 1992, 1995, 1996); however, female control is currently used in two ways and I think it worth noting the (subtle) differences. In the first, *female control* indicates that females win out in intersexual contests over functions in which males and females have roughly equal interests. In the second, the prize is control of female reproduction, itself; in this case, females are resource providers, on which male reproduction depends, and females compete with males to remain in control of their own reproduction.

## VI. Genetic Paternity and Paternal Care

### A. Theory

Is paternal care costly to males? Is paternal care of the sharable or nonsharable variety? Can males assess their genetic paternity? Is the likelihood of extra-pair paternity predictable, regular in occurrence? What is the effect of withdrawal of male parental care on females' fitnesses? Is variation in parental care facultatively expressed or fixed by selection during evolutionary time? All of these questions but the first have been incorporated into models of male parental care and the risk of males' caring for offspring that are not theirs (Westneat and Sherman, 1993; Whittingham et al., 1992; Graffen, 1980; Ridley, 1978; Trivers, 1972). In this section I discuss how these disparate models might be reconciled by inclusion of notions about variation in females.

If males can assess their genetic paternity, intuitive ideas (Trivers, 1972) about natural selection suggest that males should reduce their parental effort, whenever there is a heightened probability that they would be expending their parental effort on other than their own genetic progeny, and withdrawal of paternal effort should be proportional to the risk of expending effort on nondescendants. Maynard Smith's (1977) and Grafen's (1980) models indicate that when withdrawal of sharable paternal effort reduces offspring survival, there is little benefit to males of withdrawing male parental care. Likewise, if males withdraw nonsharable parental efforts in proportion to their likelihood of misdirecting effort to nondescendants, the cost of risking survival of descendants reduces or wipes out the benefits of withdrawal, so that these modelers concluded that a parent should invest optimally to each offspring according to its needs regardless of its kinship status to the investing parent. Both of these models assumed that males could not assess their kinship status to particular offspring.

Whittingham et al. (1992) modeled the relationship between male parental care and genetic paternity as a function of offspring survival to breeding age. Their model took account of variation in offspring survival as a function of amount of male parental effort and pointed out that the relationship between paternal effort and genetic paternity would furthermore depend on fitness functions of nonparental activities. The results of their model are qualitatively similar to Westneat and Sherman (1993), whose models take into account the fitness losses from alternative activities when males are engaged in parental care. Their model assumes trade-offs between current and future reproductive effort, and if males can assess their likelihood of kinship to offspring, parents should provide less care to low-value clutches if future clutches promise to be of high value. Their model says that the effect of variation in genetic paternity on male parental care depends on three factors: trade-offs in current and future reproduction, variation in the risk of caring for nondescendants, and the ability of parental individuals to assess their risks. Owens (1993) points out a problem with the Westneat and Sherman (1993) model, namely, that the effect of unilateral reduction in parental care on the amount provided by the other parent is not taken into account; he suggests that game theoretic approach to the problem is essential for predictions and resolution of the puzzle of male parental care and genetic paternity. Yamamura and Tsuji (1993) have modeled parental care as a game and shown that, given the parameters of their model, paternal uncertainty at fertilization favors female care, a nonintuitive result that might profitably be considered by avian empiricists. Neither Whittingham *et al.* (1992) nor Westneat and Sherman (1993) take into account variation in the effects of male parental care on females' fitnesses, either as it might be expressed in between-species comparisons or, more interestingly, in within-species and within-population comparisons. It might be worth noting that all of the previous models, at least, implicitly assume that male parental care enhances the reproductive success of their social mates and that, therefore, the problem of the existence of male parental care is solved, a classical point of view that I hope this review will lead some readers to question.

Figure 1 can also be cast as a representation of the variable effects of withdrawal of male parental care on females' fitnesses. In Panel A, male parental care—either of the sharable or nonsharable form—is essential; any withdrawal of male care will have catastrophic effects on females' fitnesses or offspring survival. This portion of the graph corresponds to the assumptions of Maynard Smith (1977) and Graffen (1980). In Panel B, withdrawal of male parental care will have negative effects on females' fitnesses, so that withdrawal would have negative effects of different magnitudes, depending on the interactions of females' quality and their environments. In Panel C, male parental care has no effect on females' fitnesses, so withdrawal of male parental care will have no effects on females' fitnesses. In Panel D, male parental care has a negative effect on females' fitnesses, so withdrawal of male parental care should enhance females' fitnesses. The second and third panels represent the situation in many, perhaps most, passerine bird populations (Gowaty, 1995) in which the effects of male parental care on females' fitnesses vary such that some males have dramatic effects on females' fitnesses, some have little effect, and some have no positive effects. These portions of the graph seem to correspond to the assumptions of Whittingham *et al.* (1992). Thus, it would seem that a synthetic approach to variation in male parental care in relation to male genetic paternity would take into account how male parental care affects females' fitnesses (offspring recruitment) as a function of interactions between female quality and their environments.

If males can punish or threaten to punish females by the withdrawal of male parental care, variation in paternal care as a function of genetic paternity should occur when females of a species are mostly Panel B females, but not Panel A, C, or D females. The specific expectation for Panel B females is that paternal provisioning will be negatively correlated with realized genetic paternity. The lack of a relationship between paternal provisioning and genetic paternity for Panel A females occurs because selection for detection of extra-pair paternity is unlikely to be strong for Panel A females, because withdrawal of paternal care is likely to be catastrophic, not only punishing females but spiting the males themselves. This means that if there is only one offspring, that withdrawing a portion of care represented by the probability that the offspring is from an extra-pair fertilization will lead to reproductive failure. If there is more than one offspring, withdrawal of sharable components of parental care is likely to affect related and unrelated offspring, thus selecting against withdrawal. Furthermore, withdrawal of nonsharable components from nondescendants only is also likely to be costly, if withdrawal is associated with offspring death because of the cost of removal of dead offspring or the costs of failed reproduction on future reproduction with a female. Thus, it seems likely that for conditions in Panel A, extra-pair paternity will have no discernible relationship on paternal care.

The relationship between genetic paternity and paternal care for Panel B females is unlikely to be as uniform as for Panel A females. In Gowaty (1995), I hypothesized that females' tendencies to mate with extra-pair males is an expression of females making the best of a bad job under social constraint. I assumed that females are favored who share their gametes with males of high phenotypic (and/or genotypic) quality and that copulation is costly in terms of exposure to pathogens and parasites, so that if females are constrained or coerced in their choice of territorial partner or breeding location to settle with a less preferred partner, that females will engage in extra-pair fertilizations as a function of their abilities to raise their offspring without males' help, given that males withdraw help as a function of detection of nondescendants. This means that females' behavior is potentially manipulated by males' threats of withdrawn paternal provisioning. What the model says is that females best choices of genetic partners are constrained socially or ecologically, and that observations of extra-pair paternity represents females' intrinsic abilities (or their environments' permissiveness) to escape or ignore social or ecological constraint. This is best represented by Panel B of Fig. 1. For Panel B females, there should be a negative relationship between paternal provisioning and realized genetic paternity.

Females represented by Panel C should be invulnerable to social constraint through withdrawal of paternal provisioning. Thus, there should be no correlation, certainly no negative correlations, between paternal provisioning and genetic paternity for males mated to females represented by Panel C. The question immediately arises over why males would be under selection to help Panel C females at all. Female manipulation of male behavior comes to mind as one possible explanation for the maintenance of male provisioning in species with Panel C females, and in a within-species comparison one might ask why a male would stay with a Panel C female at all.

Panel D females theoretically experience fitness decrements from male parental care, thus it would seem unlikely that males would be able to punish females for multiple mating by withdrawal of paternal provisioning. Rather, males socially paired to Panel D females that mate multiply may not change their paternal provisioning patterns or they may even increase their provisionings or other interactions with offspring.

Overall, the model says that paternal care and extra-pair paternity are both functions of female's options rather than some simple function relating paternal care to extra-pair paternity. It predicts that negative correlations between provisioning and paterntiy will occur only when withdrawal of male parental care can punish females. It further predicts that withdrawal of paternal provisioning will be more common among species that produce more than one brood in a season or that are relatively long lived.

B. DATA

Some early attempts to evaluate the idea that parental care increases as a function of brood size (Dawkins and Carlisle, 1976) failed to support theory (e.g., Regelmann and Curio, 1983; Wiklund and Stigh, 1983; Bjerke et al., 1985). A reason for this may be that brood size does not represent genetic maternity and genetic paternity, because of conspecific nest parasitism and extra-pair paternity (Table I). More recent studies have asked questions about the relationship of genetic paternity to paternal provisioning.

*1. Paternal Provisioning Correlated with Extra-Pair Paternity*

Lubjuhn et al. (1993) used experimental presentations of a stuffed tawny owl (*Strix aluco*), along with playbacks of mixed-species mobbing chorus at nesting boxes of great tits (*Parus major*) to evaluate nest defense. Lubjuhn et al. (1993) estimated the level of extra-pair paternity at each nest using multilocus minisatellite DNA markers; they assumed that nestlings not excluded as genetic offspring were the genetic offspring of the caregiving male. The intensity of the caregiving male's nest defense was a significant, positive function of the number of his genetic offspring in a brood. The statistically significant regression coefficient, $r = .59$, suggests that genetic paternity is an important variable in the likelihood of paternal risk taking in defense of the brood; however, it also suggests that other variables affect paternal risk taking as well.

Møller (1988) reported that the percentage of offspring feeding provided by male caregivers was a positive function of the frequency of within-pair copulations, a negative function of the number of experimentally induced extra-pair copulations by females, as well as a negative function of the number of chases a male's partner was subjected to by other males. Møller (1991) reported that predator mobbing by male swallows of a stuffed little owl (*Athene noctua*) was a negative function of females' extra-pair copulation rates. Møller (1988, 1991) interpreted these correlations in terms of adaptive responses of males in the face of paternal uncertainty (which he assumed correlated with extra-pair copulation frequencies of females and the attention that other males paid to their mates). Møller (1988) estimated that the extra-pair fertilization rate was probably relatively low in swallows based on the observation that fewer than 18% of females engaged in extra-pair copulations. If the observed extra-pair copulation rate adequately represents the likelihood of extra-pair offspring for this species, *Hirundo rustica,* Møller's (1988, 1991) conclusions and his reasoning for them should hold. However, if swallows are like other members of their family, the nestlings from extra-pair paternity equal between 35 and 87% (Table I) of nestlings and many more than 18% of females probably have young from extra-pair fertilizations. These would be quite substantial even at the lower frequencies, and one might suspect that other explanations for these correlations might work better. For instance, I predict that in swallows, females solicit or submit to within-pair copulations as a mechanism for increasing paternal investments (sensu Petrie, 1992) as a function of their own quality, such that the females most likely to solicit or submit to frequent within-pair copulations are lower-quality females with extra-pair young in their nests. I predict that females most likely to engage in readily observed extra-pair copulations are higher-quality females for whom detection of extra-pair matings are unlikely to affect their current or future reproductive success. The negative correlation between paternal provisioning, paternal predator mobbing, and extra-pair copulations might reflect that higher-quality females are less constrained by withdrawal of paternal investments than lower-quality females who may engage in extra-pair copulations more surreptitiously than higher-quality females. Finally, I predict that there will be no correlation between realized genetic paternity (based on molecular genetic information) and paternal provisioning for this species. If subsequent observation of genetic paternity and variation in females agrees with these predictions, Møller's (1988, 1991) conclusion that males' certainty of paternity affects their provisioning may be wrong, which would demonstrate once again how important molecular genetic measures of parentage are for the understanding of social behavior.

Dixon *et al.* (1994) capitalized on naturally occurring variation in breeding behavior of reed buntings (*Emberiza schoenichus*) to investigate the relationship between genetic paternity and paternal provisioning. In their sample were 13 pairs with two broods within a single season; of these, 12 had proportions of extra-pair paternity that differed between the first and second broods. Their test of the effect of extra-pair paternity on paternal provisioning is an especially robust one because in all cases the individual males and females remained paired to the same individual, they nested on the same territories for the first and second broods, and even their neighbors did not change. The adult male provisioning rate was a significantly negative function of the proportion of extra-pair young in the nest.

In two studies my colleagues and I have investigated the relationship of extra-pair paternity to paternal provisioning rates in eastern bluebirds. In the first (Gowaty et al., submitted), based on protein polymorphisms and algorithms that facilitated correction of estimates of extra-pair paternity based on exclusions only, we found that males that fed offspring the most (those that provided feeds per chick a half standard deviation above the mean or higher) had a significantly lower estimated proportion of extra-pair paternity than males that fed offspring the least (those that provided feeds per chick a half standard deviation below the mean or lower). In the second (Gowaty et al., MS), based on single-locus minisatellite DNA markers, we found that the proportion of nestling provisioning provided by caregiving males was not significantly related to proportion of extra-pair young in a male's nest. Two-thirds of the males in our sample had no extra-pair offspring in their nests, yet these males sometimes did not feed their broods at all and others provided 100% of the feeds to their nestlings. In an analysis restricted only to those males with extra-pair offspring in their nests, we detected a significant negative relationship: males with more extra-pair young provision less than males with fewer extra-pair young. One conclusion from these observations is that there are many factors that contribute to paternal provisioning decisions. However, because males with the highest likelihoods of having extra-pair offspring in their nests are the most active and energetic mate guarders (Gowaty and Bridges, 1991b), we suspect that the explanation for the negative association between genetic paternity and provisioning is not that "cuckolded" males are lazy.

Wright and Cotton (1994) experimentally manipulated certainty of paternity in starlings by presentation of live decoy males to females in the presence of resident males. Their experiment had an effect on paternal provisioning rates, such that there was a significant sex difference in chick feeding rates within experimental pairs compared with controls. They interpret this to mean that when males withdrew provisionings, females compensated for the shortfall created by males.

Møller and Birkhead (1993) used an interspecific comparative study of paternal provisioning and molecular genetic paternity estimates, and concluded that paternal provisioning is significantly negatively related to genetic paternity. Dale (1995) discussed some of the limitations of pairwise comparisons, and criticized the conclusions of Møller and Birkhead on several grounds, some of which Møller and Birkhead (1995) addressed. Despite their conclusion that there is a negative relationship between genetic paternity and paternal provisioning, I think a general conclusion about cause and effect is premature. More descriptive and experimental studies of the relationship between genetic paternity and paternal provisioning against a backdrop of intraspecific variation in females and males are necessary before general conclusions about cause and effect are likely to be reliable.

2. *Paternal Provisioning Not Correlated with Genetic Paternity*

Westneat (1995) in a study of eastern red-winged blackbirds (*Agelaius phoeniceus*) based on multilocus DNA minisatellites, found no relationship between paternal provisioning and extra-pair paternity. Given the emerging focus on variation in females and conclusions of the current article, it is interesting that there was an effect of female age: males feed nestlings at the nests of young females significantly more than at the nests of older females.

Bereson *et al.* (MS) observed feeding rates of male Wilson's warblers but found no significant difference in feeding rates of males with and males without evidence of extra-pair paternity. Experimental attempts to vary confidence of paternity include Whittingham *et al.* (1993). They held tree swallow males in small boxes with one-way glass, so that the captive males could see out but others could not see in. They placed these cages so that the captive males could have clear views of their nesting boxes and females when females were at these nesting boxes, during nest building and egg laying (experimentals) and during incubation (controls). Females paired to experimental males copulated with other males, while females paired to controls did not. There was no effect of experimental treatment on the number of feeding visits by (released) males to their nestlings. The treatment clearly affected the number of times females copulated with other males. As Stamps (1995) points out though, from females' points of view these were not extra-pair, because females probably experienced themselves as widowed. If information about females' behavior was important for males' gauging their confidence of paternity, the experimental males should have been able to assess the likelihood that "their" females copulated with other males. The fact that the experiment had no effect on paternal provisioning could be because there has been no previous selection on males to assess female behavior and modify provisioning; it could be due to equal probabilities of paternal uncertainty for all broods; or it could be because withdrawal of male parental care would have no power to affect the likelihood of extra-pair mating by females, that is, withdrawal of male parental care would not be punishing to females.

C. FEMALE OPTIONS, GENETIC PATERNITY, AND PATERNAL CARE

Reference to Fig. 1 may shed some light on the observed variation in results among studies of genetic paternity and paternal provisioning. Empirical evidence of the effect of male parental care on females' fitnesses (Gowaty, 1983; Gowaty et al., MS) indicate that the fitness of eastern bluebird females falls into the right-hand side of Panels B and C of Fig. 1. Therefore, the observation of a negative correlation between male paternal provisioning and extra-pair paternity fits expectations derived from Fig. 1, and furthermore, provides a possible explanation for why the correlation between paternal provisioning and genetic paternity is not dramatically strong (Gowaty et al., MS). Namely, female eastern bluebirds vary in their vulnerabilities to social coercion by males and in the advantages that accrue via paternal provisioning, making it advantageous for some males to withhold paternal provisioning in the face of paternal uncertainty. Likewise, starling females probably fall into Panels B and C, so that there is likely to be an advantage for some males to withhold paternal provisioning.

Male removal studies of great tits (Sasvari, 1986; Bjorklund and Westman, 1986) indicate that paternal provisioning has variable effects on females' fitnesses, so that I expect that great tit females fall into Panel B as well. Therefore, the observed negative correlation (Lubjuhn et al., 1993) between the male nest defense and genetic paternity is the relationship Fig. 1 predicts between genetic paternity and paternal provisioning.

I know of no data on reed buntings indicating the effects of male parental care of females' fitnesses, but the negative correlations between paternal provisioning and genetic paternity allows one to predict that reed buntings females fall into panel B of Fig. 1, that is, that male parental care is not essential, but is helpful to females, and that some or all females to varying degrees are vulnerable to social coercion by males through the threat of withheld paternal provisioning.

In contrast, male parental care is probably not essential, and probably not even helpful, to many female red-winged blackbirds, thus, red-winged blackbird females probably frequently fall into Panel C of Fig. 1. If this is usually the case, the lack of correlation between genetic paternity and paternal provisioning is unsurprising. Observations of this sort beg the question of the adaptive significance of male parental care (that is, is what looks like parental care functioning in some other way than we typically expect in this and other species), and from this perspective are exceptionally important observations.

However, the lack of correlation in Wilson's warblers and tree swallows is more problematic in that these are fairly typical socially monogamous passerines, albeit, with relatively high rates of extra-pair paternity (Table

1). This result would make sense in terms of Fig. 1 if in these species variation in females is described entirely by fitness functions in Panel A or C. I know of no male removal studies in Wilson's warblers, so it is fair to predict that paternal provisioning is or almost always is essential (Panel A) or has no effect at all on females' fitnesses (Panel C); notice that these are predictions about variance among experimental females from whose territories males are removed. It is more likely that most tree swallow females fall into Panel C, given that male removal tests indicate that male parental care is nonessential to females and only rarely has a positive effect on females' fitnesses (Dunn and Robertson, 1992). It may be that the relatively high level of extra-pair paternity in tree swallows is possible because females are relatively invulnerable to social coercion by males (Gowaty, 1995), something that would explain the lack of correlation between genetic paternity and paternal provisioning.

These considerations allow some additional predictions about both the frequencies of extra-pair paternity and relationships between genetic paternity and paternal provisioning in species in which tests have not yet been accomplished. For example, based on variation in the effects of male removal on females' fitnesses (see, e.g., Wolfe et al., 1990), the Constrained Female Hypothesis (Gowaty, 1995, and expanded in Fig. 1) predicts: (1) negative correlations between genetic paternity and paternal provisioning in species in which most females fall into the left-hand portion of Panel B of Fig. 1, namely, blue tits (Sasvari, 1986) and pied flycatchers (Alatalo et al., 1982); (2) negative correlations between genetic paternity and paternal care only among those pairs with some extra-pair paternity in those species in which most females fall into the right-hand side of Panels B and C of Fig. 1, namely, Northern cardinals (Richmond, 1978), song sparrows (Smith et al., 1982), and dark-eyed juncos (*Junco hyemalis*) (Wolfe et al., 1990); and (3) no correlation between genetic paternity and paternal provisioning for species in which most females fall into Panel C of Fig. 1, perhaps including indigo buntings and seaside sparrows (*Ammodramus maritimus*) (Greenlaw and Post, 1985). The Constrained Female Hypothesis furthermore predicts that if all else is equal, which it seldom is, extra-pair paternity rates will vary among these three classes of species, such that extra-pair paternity will look like the following: Group 3 > Group 2 > Group 1. Given that all else is seldom equal, the most compelling tests of these ideas will be based on intraspecific variation in females abilities to raise offspring alone.

## VII. Maternal Care

Maternal care in birds is seldom a topic of interest. For example, there are almost no studies on variation in maternal care in species with uniparental female care, and the topic is seldom or never considered in species with biparental care. Only recently, have authors focused on any sources of variation in maternal care. Also, not unsurprisingly, the questions spring from notions about female choice of mates. When females mate multiply so that their broods are sired by more than one male, the possibility that these offspring vary in reproductive value to females arises. Females may seek extra-pair paternity in order to increase the quality (viability) of their offspring, which may arise solely as an effect of male genetic quality, something demonstrated by Petrie and Lipsitch (1994). Increases in quality of offspring also may be due to parental effects (such as increased access to resources brokered to females through extra-pair males (e.g., Kempenaers, 1993), or from differential provisioning of nestlings as a function of the reproductive value to the mother of offspring from one versus another genetic sire.

Differential provisioning to nestlings as a function of offspring sex has been searched for in several species of birds (reviewed in Gowaty and Droge, 1991; see also Clutton-Brock, 1991) and found in budgerigars (Stamps et al., 1987) and eastern bluebirds. However, in each of these cases differential provisioning was a paternal, not a maternal, trait. To my knowledge the only study of birds showing differential allcoation by mothers to nestlings of different sex was a study on pigeons (D. D. Droge, unpublished data). In fact, there have been relatively few studies of differential allocation of provisioning by sex, and I know of none that look at differential provisioning allocation by mothers as a function of the identity of the genetic sire of offspring. Part of the reason for this may be that little developed theory posits or discusses variation in maternal provisioning.

Focus on the value and function of female choice for female fitness will likely lead to changes in the empirical focus of studies in the future. I expect that in the next 10 years we will witness revolutions in our research agendas similar to the sweeping changes that Trivers (1972) instituted. The effects of variation in maternal care and in resultant fitness functions of females and males is one of these questions for the future.

## VIII. Adaptive Significance of Parental Care in Birds

### 2. Costs and Benefits of Parental Care

Clutton-Brock's (1991) review of parental care notes that the evolution of parental care has taken a decided backseat to questions about the evolution of mating and breeding systems, something that seems true to me even in birds where parental care is well developed in both sexes of parents. He notes that precious little work links into a unified theoretical framework that integrates studies of variation in parental care within species to knowledge about interspecific trends. I agree that lack of unified theory has inhibited progress in understanding existing variation in and maintenance of parental care, much less the origins of parental care. However, I think this situation is on the cusp of change for the following reasons: (1) the almost explosive growth of behavioral ecology in the last 10 years is resulting in massive increases in data, that themselves suggest further questions and new theories; (2) congruent with the increased participation of women in the study of behavioral ecology, theoretical angles concentrating on perspectives of females are increasing our fields of view (e.g., Lawton *et al.,* 1995; Stamps, 1995); and (3) increasing attention on intraspecific variation in parental care patterns will illuminate interspecific trends.

What we know now about the adaptive significance of parental care is about the costs and benefits to parents, especially males, of parental behavior, what I characterize as maintenance explanations for parental care. Clutton-Brock (1991) has an excellent review of studies of costs and benefits of parental care.

### B. If Uniparental Male Care Is Ancestral

Most theories about the adaptive significance of parental care begin with implicit or explicit assumptions that ancestral patterns were uniparental female or biparental, thus most theories (e.g., Lack, 1968) have focused on selection pressures favoring male parental care. However, if uniparental male care is ancestral in birds, explanations must focus on selective factors favoring maternal care. I see this as a challenge for the future, something that has not yet been adequately broached for birds.

### C. Parental "Fiddling": Any Behavior of Parents towards Offspring

*Parental care,* by its very definition, is behavior towards offspring that increases offspring fitness. The definition implies the most commonly noted adaptive advantage of parental care. Parental care is maintained in populations and is adaptive because it increases the likelihood of survival and reproduction of offspring. However, as Clutton-Brock (1991) notes for species in a wide variety of taxa and that is obvious for birds from the preceding review, there are very few studies of parental care that measure the costs of parental care to parents (think of Panel D females in Fig. 1)—and none that I know of that measure the costs of parental care to the offspring. Of course, given common definitions of parental care, it is unlikely that theorists or empiricists are going to jump at the idea that parental care may be costly to offspring survival and reproductive success. However, it seems that additional light on the adaptive function of parental bahavior to offspring will be shed, if we operationally define parental behavior as any behavior of parents to offspring, and only then ask the question of whether it is advantageous to the survival and reproductive successs of the offspring—or to the survival and reproductive success of the other parent. I am surprised that this angle on parental "care" has not been more prominently forefronted; after all, ideas about parent–offspring conflict (Trivers, 1974) were published over 20 years ago, and indeed are a thriving area of research in some quarters at least (e.g., Drummond, 1993; Mock and Parker, in press).

I expect that if the field of view is enlarged to include behavior of parents towards offspring independent of positive fitness consequences for offspring, that new questions will arise not only about behavior of parents that obviously do not enhance offspring fitness, but about costs to offspring of parental behavior (for example, selective or facultative abortion would fall under this category). I think such an expanded view will lead to new hypotheses about both origins and maintenance of parental care, something I now call parental "fiddling," an inclusive term that indicates things parents do to offspring that may have either positive (what we now call "care") or negative effects on offspring or alternative parent's fitness.

D. SEXUAL CONFLICT AND ORIGINS HYPOTHESES

Nick Davies (1989) was the first I know of to suggest that mating systems could arise via sexual conflict. Whenever the reproductive interests of the sexes conflict, powerful selection pressures should be set in motion (Gowaty, 1996), favoring not only patterns of mateship but of other social behavior traits, as well. It seems as likely that parental fiddling may have been an adaptive response to sexual conflict as much as to selection for increased survival of propagules. Three hypotheses for the origin of parental care that do not depend on the definition of parental care as what parents do to enhance offspring fitness are each aspects of a general idea about the force of sexual conflict to provide selection pressure for parental fiddling. Once parental fiddling is favored, the conditions are in place for parental fiddling to take the form of parental care; however, parental fiddling via sexual conflict need not take the form of enhanced fitness benefits for offspring.

*1. Sexual Conflict over the Sex Ratio*

The origin of parental fiddling with offspring may lie in sexual conflict over the sex ratio of offspring (Gowaty and Droge, 1991; Gowaty *et al.*, submitted; Gowaty and Plissner, submitted). Parent–parent or sexual conflict over sex ratio will arise whenever there are different optimal sex ratios for each sex of parent. Optimal sex ratios are a function of relative costs to the parent(s) of each sex of offspring (Fisher, 1958). Sources of differential Fisherian costs of offspring to parents include differential mortality of daughters and sons, local mate competition, and local resource competition (Charnov, 1982). Local resource competition may be an important source of progeny sex ratio variation in birds (Gowaty, 1993), given that passerines, in which son philopatry predominates, produce daughter-biased sex ratios, while Anseriformes, in which daughter-biased philopatry predominates, produce son-biased sex ratios. If local resource competition is such that philopatric sons extract costs from their mothers and fathers over food, nest sites, and roost sites, but only from their fathers over mates, the condition is met for different optimal sex ratios for mothers and fathers. In the case of typical passerines (Gowaty, 1993), this would mean that mothers and fathers should prefer daughter-biased sex ratios, but fathers' preferred sex ratios would be more daughter-biased than mothers'. In birds, females contribute the sex-associated chromosome. Also, in eastern bluebirds, the sex ratio is daughter biased (see references in Gowaty, 1993), something that mothers have complete control of until fathers begin to feed nestlings, because only mothers incubate. Perhaps, fathers' preference for a more daughter-biased sex ratio provided the original selection pressure for paternal feeding of nestlings in passerines (Gowaty and Droge, 1991), and explains why fathers feed their daughters more than they feed their sons in eastern bluebirds.

Sexual conflict over the sex ratio may have provided the original selection pressure for parental fiddling in birds and other taxa, as well. In species with sex-associated chromosomes, the sex of parent that contributes the sex chromosome automatically "wins" any sexual conflict over the sex ratio at fertilization; thus, the alternate parent should be under selection to modify subsequent sex ratios (at hatch; at fledging) more towards its optimal value. This idea predicts that the homogametic sex would experience initial selection pressures to fiddle with progeny after fertilization. Given that females are the heterogametic sex in birds, this idea postdicts that uniparental male fiddling was ancestral in birds.

2. *Sexual Conflict over Other Aspects of Reproduction*

In a similar vein to previous Section VIII,D,1, Slagsvold and Lifjeld's (1989) interesing idea that parent–parent conflict or sexual conflict over hatching asynchrony (discussed in previous Section IV) is won by female blue tits (Slagvold *et al.,* 1994), who alone incubate, provides a possible origins explanation for incubation by males. In species in which only females incubate, if sexual conflict over hatch spread results in significant fitness costs to males, incubation by males will be favored by selection. Thus, another origins hypothesis for parental fiddling that does not depend on immediate enhancement of offspring fitness is sexual conflict over other aspects of reproduction besides progeny sex ratios.

### 3. Sexual Conflict over Control of Female Reproductive Capacities

van Rhijn (1991) argues that uniparental male care is likely to have evolved from a system of no parental care through selection on males for postcopulatory mate guarding. His ideas are dependent on the benefits of guarding for males, which may include increased fertilization success or inhibition of females' mating with additional males so that sperm competition does not occur. Once postcopulatory mate guarding has evolved, the conditions are set for the evolution of further fiddling of eggs by males, something that may relieve females to invest in further eggs. This is an argument for the evolution of parental care via sexual conflict (parent–parent conflict) over the control of females' reproductive options insomuch as mate guarding keeps females from mating with and investing in the offspring of other males. It is a plausible scenario that differs from those developed under previous Section VIII,D,1. and VIII,D,2 in that it describes conditions preliminary to those that actually provide the selection pressure for paternal fiddling.

In Gowaty (1995) I argued that the evolution of social monogamy and associated biparental care occurs via selection on males to manipulate and control female reproduction. I argued (1996) that males could use both "nasty" and "nice" mechanisms for manipulation of females, including brokering females' access to essential resources. Once males are in situations in which they broker females' access to resources for reproduction, they are likely to be in close spatial and temporal proximity to propagules, something that would provide preadaptations necessary for the evolution of paternal fiddling, including care.

## IX. The Future

Because of the explosive growth of behavioral ecology during the last 20 years, we know more about the details of naturally occurring parental care in birds than ever before. It seems likely that the new perspectives-of-females approaches to social behavior will expose another mother lode of insight and information analogous to the revolutions brought about by molecular genetic markers of kinship. I expect that studies of the future will focus more on intraspecific and intrapopulational variation than ever before for a variety of reasons, including that the interpretation of selective pressures acting on individual variation within species is not handicapped with the interpretive difficulties of inferring the results of selection via interspecific comparisons. Among the nice things about sexual conflict theories are that conflict hypotheses almost inevitably characterize the conflicted parties as active, engaged individuals with the agency to affect evolution, and these theories force attention onto variation among individuals. Ironically, the sexual conflict theories will probably do more for centralizing female interests in the study of social behavior than any other forces that affect the trends in what we choose to study and think about.

## X. Summary

Biparental care is the norm in birds and is associated with social monogamy, the typical mating pattern in birds. Both uniparental female and uniparental male care occur, though as yet there is no generally agreed upon solution to the question of ancestral parental care patterns in birds. Paternal care has received more attention than maternal care in both biparental and uniparental care systems. Facultative variation in paternal care has been reported in a variety of species and multiple hypotheses have been offered to explain within- and between-species variation. Male removal studies indicate that there is almost always within-species variation in the abilities of females to raise their offspring without male help, something that has been generally overlooked. Since 1985, around 80 studies (Table I) of extra-pair paternity in birds have been published; these indicate that it is far more common to find extra-pair paternity as a regular feature of socially monogamous (Fig. 1), biparental care systems. So, in keeping with the traditional focus of avian biologists on male behavior, studies have asked questions about the relationship of genetic paternity to paternal care, even though in birds because of egg laying, maternity may be as uncertain as paternity. Results are mixed, indicating that in some species genetic paternity and paternal provisioning are significantly positively correlated; in others no such correlations have been detected. Studies about female manipulation of partner contributions to parenting outnumber studies about male manipulation of females, though both sorts of studies should compose an important component of future studies. There are few convincing origins stories for parental care in birds, something that might be partially remedied by a redefinition of the traits of interest to include any parental behavior directed towards their offspring, something I call parental fiddling. Several novel hypotheses for the origin of parental fiddling involve selection pressures arising via sexual conflict, sometimes called parent–

parent conflict. Three important sources of selection pressure favoring parental fiddling are parent–parent conflict over the sex ratio, parent–parent conflict over hatch spread and other aspects of reproductive timing or decisions, and parent–parent conflict over the control of maternal reproductive capacities. An elaboration (Fig. 1) of the Constrained Female Hypothesis to explain patterns of extra-pair paternity in socially monogamous birds suggests that an overlooked key to understanding not only facultative variation in male parental care, but a variety of other related social behavior features such as the relationship between genetic paternity and paternal care, female manipulation of male behavior, and male manipulation of female behavior, is intrinsic characteristics of females and their environments that affect females' abilities to escape or ignore social constraints on their reproduction imposed by males. The predictions of the Constrained Female Hypothesis as it applies to each of the reviewed aspects of avian parental care are listed in appropriate sections of this article. The Constrained Female Hypothesis suggests that a key feature to future understandings of the evolution of avian parental care systems lies in descriptions of within-species variation of females.

## Acknowledgments

I thank Colleen Barber, Raleigh Robertson, and Peter Boag for unpublished data on tree swallows; Sharon Birks for unpublished data on brush-turkeys; Rachel Barreson, Judith Rhymer, and Robert Fleisher for unpublished data on Wilson's warblers; Walter Piper for unpublished data on common loons; and Elaina Tuttle for unpublished data on white-throated sparrows. I thank J. H. Plissner for discussions and technical help during the preparation of the manuscript. I thank Jon Wright and Chuck Snowdon for comments on a previous version. During the preparation of this manuscript my research was supported by an ADAMHA RSDA, KO2-MH00706-04, and a grant from the NSF, IBN-9222005.

## References

Alatalo, R. V., Lundberg, A., and Ståhlbrandt, K. (1982). Why do pied flycatcher females mate with already mated males? *Anim. Behav.* **30,** 585–593.

Barber, C. A., Robertson, R. J., and Boag, P. T. MS. High rates of extra-pair paternity in tree swallows are not an artifact of nest boxes.

Bart, J., and Tornes, A. (1989). Importance of monogamous male birds in determining reproductive success: Evidence for house wrens and a review of male-removal experiments. *Behav. Ecol. Sociobiol.* **24,** 109–116.

Bateman, A. J. (1948). Intrasexual selection in Drosophila *Heredity* **2,** 349–368.

Beissinger, S., and Snyder, N. F. R. (1987). Mate desertion in the snail kite. *Anim. Behav.* **35,** 477–487.

Bereson, R. C., Reymer, J., and Fleisher, R. C. MS. Extra-pair fertilizations in Wilson's warblers and correlates of cuckoldry.

Birkhead, T. R., and Møller, A. P. (1992). "Sperm Competition in Birds: Evolutionary Causes and Consequences." Academic Press, San Diego.

Birkhead, T. R., and Møller, A. P. (1993). Female control of paternity. *TREE* **8**, 100–104.

Birkhead, T. R., Burke, T., Zann, R. A., Hunter, F. M., and Krupa, A. P. (1990). Extra-pair paternity and intraspecific brood parasitism in wild zebra finches, *Taeniopygia guttata*, revealed by DNA fingerprinting. *Behav. Ecol. Sociobiol.* **27**, 315–324.

Birks, S. M. MS Paternity in the Australian Brush-Turkey (*Alectura lathami*),: A promiscuous bird with uniparental male care.

Bjerke, T., Espmark, Y., and Fonstad, T. (1985). Nest defense and parental investment in the redwing, *Turdus iliacus*. *Ornis Scand.* **16**, 14–19.

Björklund, M., and Westman, B. (1986). Adaptive advantages of monogamy in the great tit (*Parus major*): An experimental test of the polygyny threshold model. *Anim. Behav.* **34**, 1436–1440.

Black, J. (ed.) (1995). "Partnerships in Birds." Oxford Univ. Press, Oxford.

Bollinger, E. K., and Gavin, T. A. (1991). Patterns of extra-pair fertilizations in bobolinks. *Behav. Ecol. Sociobiol.* **29**, 1–7.

Brooker, M. G., Rowley, I., Adams, M., and Baverstock, P. R. (1990). Promiscuity: An inbreeding avoidance mechanism in a socially monogamous species? *Behav. Ecol. Sociobiol.* **26**, 191–199.

Brown, C. R., and Brown, M. B. (1988). Genetic evidence of multiple parentage in broods of cliff swallows. *Behav. Ecol. Sociobiol.* **23**, 379–387.

Brown, J. (1987). "Helping and Communal Breeding in Birds." Princeton Univ. Press, Princeton, NJ.

Burke, T., Davies, N. B., Bruford, M. W., and Hatchwell, B. J. (1989). Parental care and mating behaviour of polyandrous dunnocks *Prunella modularid* to paternity by DNA fingerprinting. *Nature (London)* **338**, 249–251.

Burley, N. T. (1986). Sexual selection for aesthetic traits in species with biparental care. *Am. Nat.* **4**, 415–445.

Burley, N. T. (1988). The differential-allocation hypothesis: An experimental test. *Am. Nat.* **132**, 611–628.

Burley, N. T., Enstrom, D. A., and Chitwood, L. In press. Extra-pair relations in zebra finches: Differential male success results from female tactics. *Anim. Behav.*

Burley, N. T., Parker, P., and Lundy, K. 1996. Submitted. Sexual selection and extra-pair fertilization in a socially monogamous passerine, the zebra finch (*Taeniopygia guttata*). *Behav. Ecol.* **7**, 218–226.

Charnov, E. (1982). "Sex Allocation." Princeton Univ. Press, Princeton, NJ.

Clutton-Brock, T. H. (1991). "The Evolution of Parental Care." Princeton Univ. Press, Princeton, NJ.

Clutton-Brock, T. H., and Parker, G. (1995). Punishment in animal societies. *Nature (London)* **373**, 209–216.

Dale, J. (1995). Problems with pair-wise comparisons: Does certainty of paternity covary with paternal care? *Anim. Behav.* **49**, 519–521.

Darwin, D. (1871). "The Descent of Man, and Selection in Relation to Sex." Murray, London.

Davies, N. B. (1989). Sexual conflict and the polygamy threshold. *Anim. Behav.* **38**, 226–234.

Davies, N. B. (1992). "Dunnok Behavior and Social Evolution." Oxford Univ. Press, Oxford.

Davies, N. B., and Hatchwell, B. J. (1992). The value of male parental care and its influence on reproductive allocation by male and female dunnocks. *J. Anim. Ecol.* **61**, 259–272.

Dawkins, R., and Carlisle, T. R. (1976). Parental investment, mate desertion, and a fallacy. *Nature (London)* **262**, 131–133.

Decker, M. D., Parker, P. G., Minchella, D. J., and Rabenold, K. N. (1993). Monogamy in black vultures: Genetic evidence from DNA fingerprinting. *Behav. Ecol.* **4**, 29–35.

Dixon, A., Ross, D., O'Malley, C., and Burke, T. (1994). Paternal investment inversely related to degree of extra-pair paternity in the reed bunting. *Nature (London)* **371,** 698–700.

Drummond, H. (1993). Have avian parents lost control of offspring aggression? *Etologia* **3,** 187–198.

Dunn, P. O., and Robertson, R. J. (1992). Geographic variation in the importance of male parental care and mating systems in tree swallows. *Behav. Ecol.* **3,** 291–299.

Dunn, P. O., Robertson, R. J., Michaud-Freeman, D., and Boag, P. T. (1994a). Extra-pair paternity in tree swallows: Why do females mate with more than one male? *Behav. Ecol. Sociobiol.* **35,** 273–281.

Dunn, P. O., Whittingham, L. A., Lifjeld, J. T., Robertson, R. J., and Boag, P. T. (1994b). Effects of breeding density, synchrony, and expeience on extrapair paternity in tree swallows. *Behav. Ecol.* **5,** 123–129.

Eens, M., and Pinxten, R. (1995). Inter-sexual conflicts over copulations in the European starling: Evidence for the female mate-guarding hypothesis. *Behav. Ecol. Sociobiol.* **36,** 71–81.

Emlen, S. T., and Oring, L. W. (1977). Ecology, sexual selection, and the evolution of mating systems. *Science (Washington, D. C.)* **197,** 215–223.

Evarts, S., and Williams, C. J. (1987). Multiple paternity in a wild population of mallards. *Auk* **104,** 597–602.

Fitch, M. A., and Shugart G. W. (1984). Requirements for a mixed reproductive strategy in avian species. *Am. Nat.* **124,** 116–126.

Fisher, R. A. (1958). "The Genetical Theory of Natural Selection." Dover Publ., New York.

Fraga, R. M. (1992). Biparental care in bay-winged cowbirds *Molothrus badius*. *Ardea* **80,** 389–393.

Gelter, H. P., and Tegelstrom, H. (1992). High frequency o fextra-pair paternity in Swedish pied flycatchers revealed by allozyme electrophoresis and DNA fingerprinting. *Behav. Ecol. Sociobiol.* **31,** 1–8.

Gibbs, H. L., Weatherhead, P. J., Boag, P. T., White, B. N., Tabak, L. M., and Hoysak, D. J. (1990). Realized reproductive success of polygynous red-winged blackbirds revealed by DNA markers. *Science (Washington, D. C.)* **250,** 1394–1397.

Gibbs, H. L., Goldizen, A. W., Bullough, C., and Goldizen, A. R. (1994). Parentage analysis of multi-male social groups of tasmanian native hens (*Tribonyx mortierii*): Genetic evidence for monogamy and polyandry. *Behav. Ecol. Sociobiol.* **35,** 363–371.

Gowaty, P. A. (1981). The aggression of breeding eastern bluebirds *Sialia sialis* toward each other and intra- and inter-specific intruders. *Anim. Behav.* **29:**1013–1027.

Gowaty, P. A. (1983). Male parental care and apparent monogamy in eastern bluebirds (*Sialia sialis*. *Am. Nat.* **121,** 149–157.

Gowaty, P. A. (1992). Evolutionary biology and feminism. *Hum. Nat.* **3,** 217–249.

Gowaty, P. A. (1993). Differential dispersal, local resource competition, and sex ratio variation in birds. *Am. Nat.* **141,** 263–280.

Gowaty, P. A. (1994). Architects of sperm competition. *TREE* **9,** 160–161.

Gowaty, P. A. (1996). Battles of the sexes and origins of monogamy: In "Partnerships in Birds" (J. L. Black, ed.), Oxford Series in Ecology and Evolution. Oxford Univ. Press, Oxford, pp. 21–52.

Gowaty, P. A. (1996). Sexual dialectics, sexual selection, and variation in mating behavior. *In* "Feminism and Evolutionary Biology: Boundaries, Intersections, and Frontiers" (P. A. Gowatty, ed.), pp. Chapman Hall, New York.

Gowaty, P. A., and Bridges, W. C. (1991a). Nest box availability affects extra-pair fertilization and conspecific nest parasitism in eastern bluebirds, *Sialia sialis*. *Anim. Behav.* **41,** 661–676.

Gowaty, P. A., and Bridges, W. C. (1991b). Behavioral, demographic, and environmental correlates of extra-pair fertilizations in eastern bluebirds, *Sialia sialis. Behav. Ecol.* **2,** 339–350.

Gowaty, P. A., and Bridges, W. C. MS The logic of percentage studies using molecular genetic markers.

Gowaty, P. A., Robertson, R. J., Dufty, A., and Ball, G. MS. Geographic variation in the necessity of male parental care in eastern bluebirds: A multinational, experimental study.

Gowaty, P. A., and Buschhaus, N. (MS). Aggressive, resisted, and forced copulation in birds: Occurrances and the CODE hypothesis. MS.

Gowaty, P. A., and Karlin, A. A. (1984). Multiple maternity and paternity in single broods of apparently monogamous eastern bluebirds (*Sialia sialis*). *Behav. Ecol. Sociobiol.* **15,** 91–95.

Gowaty, P. A., and Droge, D. L. (1991). Sex ratio conflict and the evolution of sex-biased provisioning. "Acta XX Congressus Internationalis Ornithologici," Vol. II (B. D. Bell, R. O. Cossee, J. E. C. Flux, B. D. Heather, R. A. Hitchmough, C. J. R. Robertson, and M. J. Williams, eds.), pp. 932–945. Ornithological Congress Trust Board, New Zealand.

Gowaty, P. A., and J. H. Plissner. MS. Fathers provision daughter-biased broods more often and parents manipulate fledging time in eastern bluebirds, *Sialia sialis.*

Gowaty, P. A., and Wagner, S. J. (1988). Breeding season aggression of female and male eastern bluebirds (*Sialia sialis*) to models of potential conspecific and interspecific egg dumpers. *Ethology* **78,** 238–250.

Gowaty, P. A., Bridges, W. C., and D. L. Droge, MS. Correlates of provisioning in eastern bluebirds *Sialia sialis:* Sex bia of broods, age of nestlings and their genetic paternity.

Gowaty, P. A., Richardson, D., Bridges, W. C., and Burke, T. Paternal provisioning varies with genetic paternity in eastern bluebirds *Sialia sialis.* MS.

Graffen, A. (1980). Opportunity cost, benefit and degree of relatedness. *Anim. Behav.* **28,** 967–968.

Graves, J., Hay, R. T., Scallan, M., and Rowe, S. (1991). Extra-pair paternity in the shag *Phalocrocorax aristotelis* as determined by DNA fingerprinting. *J. Zool. (London)* **226,** 399–408.

Gray, Elizabeth Marie. (1994). "The Ecological and Evolutionary Significance of Extra-Pair Copulations in the Red-winged Blackbird (*Agelaius phoeniceus*)." A dissertation from the University of Washington, Seattle.

Greenlaw, J. S., and Post, W. (1985). Evolution of monogamy in seaside sparrows, *Ammodramus maritimus:* Tests of hypotheses. *Anim. Behav.* **33,** 373–383.

Gullberg, A., Tegelstrom, H., and Gelter, H. P. (1992). DNA fingerprinting reveals multiple paternity in great and blue tits (*Parus major* and *P. caeruleus*). *Hereditas* **117,** 103–108.

Gyllensten, U. B., Jakobsson, S., and Temrin, H. (1990). No evidence for illegitimate young in monogamous and polygynous warblers. *Nature (London)* **343,** 168–170.

Haig, S. M., Walters, J. R., and Plissner, J. H. (1994). Genetic evidence for monogamy in the cooperatively breeding red-cockaded woodpecker. *Behav. Ecol. Sociobiol.* **34,** 295–303.

Hartley, I. R., Shepherd, M., Robson, T., and Burke, T. (1993). Reproductive success of polygynous male corn buntings (*Miliaria calandra*) as confirmed by DNA fingerprinting. *Behav. Ecol.* **4,** 310–317.

Hartley, I. R., Davies, N. B., Hatchwell, B. J., Desrochers, A., Nebel, D., and Burke, T. (1995). The polygynandrous mating system of the alpine accentor, *Prunella collaris.* II. Multiple paternity and parental effort. *Anim. Behav.* **49,** in press.

Hasselquist, D., Bensch, S., and von Schantz, T. (1995). Low frequency of extra-apir paternity in the polygynous great reed warbler. *Acrocephalus arundinaceus. Behav. Ecol.* **6,** 27–38.

Hebert, P. N., and Sealey, S. G. (1993). Hatching asynchrony and feeding rates in yellow warblers: A test of the sexual conflict hypothesis. *Am. Nat.* **142,** 881–892.

Heg, D., Ens, B. J., Burke, T., Jenkins, L., and Kruijt, J. P. (1993). Why does the typically monogamous oystercatcher (*Haematopus ostralegus*) engage in extra-pair copulations? *Behaviour* **126,** 247–289.

Hoffenberg, A. S., Power, H. W., Romagnano, L. C., Lombardo, M. P., and McGuire, T. R. (1988). The frequency of cuckoldry in the European starling (*Sturnus vulgaris*). *Wilson Bull.* **100,** 60–69.

Hoi, H. MS. An alternative route to coloniality in the bearded tit: Females pursue extra-pair fertilizations.

Hunter, F. M., Burke, T., and Watts, S. E. (1992). Frequent copulation as a method of paternity assurance in the northern fulmar. *Anim. Behav.* **44,** 149–156.

Jamieson, I. G., Quinn, J. S., Rose, P. A., and White, B. N. (1994). Shared paternity among non-relatives is a result of an egalitarian mating system in the communially breeding bird, the pukeko. *Proc. R. Soc. London Ser. B* **257,** 271–277.

Jeffreys, A. J., Wilson, V., and Thein, S. L. (1985). Hypervariable "minisatellite" regions in human DNA. *Nature (London)* **314,** 67–73.

Jones, C. S., Lessells, C. M., and Krebs, J. R. (1991). Helpers-at-the-nest in European bee-eaters (*Merops apiaster*): A genetic analysis. *In* "DNA Fingerprinting: Approaches and Applications." (Burke, T., Dolf, G., Jeffreys, A. J., and Wolff, R., eds.), pp. 169–192. Birkhauser, Basel.

Joste, N., Ligon, J. D., and Stacey, P. B. (1985). Shared paternity in the acorn woodpecker (*Melanerpes formicivorus*). *Behav. Ecol. Sociobiol.* **17,** 39–41.

Kempenaers, B. (1993). A case of polyandry in the blue tit: Female extra-pair behavior results in extra male help. *Ornis Scand.* **24,** 246–249.

Kempenaers, B., Verheyen, G. R., VandenBroeck, M., Burke, T., Banbroeckhoven, C., and Dhondt, A. D. (1992). Extra-pair paternity results from female preference for high-quality males in the blue tit. *Nature (London)* **357,** 494–496.

Kempenaers, B., Verheyen, G. R., and Dhondt, A. A. (1995). Mate guarding and copulation behavior in monogamous and polygynous blue tits: Do males follow a best-of-a-bad-job strategy? *Behav. Ecol. Sociobiol.* **36,** 33–42.

Lack, D. (1968). "Ecological Adaptations for Breeding in Birds." Methuen and Co., London.

Lawton, M., Garstka, W. R., and Hanks, C. (1996). The mask of theory and the face of nature. *In* "Feminism and Evolutionary Biology: Boundaries, Intersections, and Frontiers" (P. A. Gowaty, ed.), Chapman Hall, New York.

Lazarus, J. (1990). The logic of mate desertion. *Anim. Behav.* **4,** 672–684.

Lifjeld, J. T., Dunn, P. O., and Westneat, D. F. in press. Sexual selection by sperm competition in birds: Male-male competition or female choice? *Behav. Ecol.*

Lifjeld, J. T., and Robertson, R. J. (1992). Female control of extra-pair fertilization in tree swallows. *Behav. Ecol. Sociobiol.* **31,** 89–96.

Lifjeld, J. T., Slagsvold, T., and Lampe, H. M. (1991). Low frequency of extra-pair paternity in pied flycatchers revealed by DNA fingerprinting. *Behav. Ecol. Sociobiol.* **29,** 94–101.

Lubjuhn, T., Curio, E., Muth, S. C., Brun, J., and Epplen, J. T. (1993). Influence of extra-pair paternity on parental care in great tits (*Parus major*). *In* "DNA Fingerprinting: State of the Science" (S. D. J. Pena, R. Chakraborty, J. T. Epplen, and A. J. Jeffreys, eds.), pp. 379–385. Birkhauser Verlag, Basel.

Lyons, B. E., R. D., Montgomerie, and Hamilton, L. D. (1987). Male parental care and monogamy in snow buntings. *Behav. Ecol. Sociobiol.* **20,** 377–382.

Maynard Smith, J. (1977). Parental investment: A prospective analysis. *Anim. Behav.* **25,** 1–9.

McKinney, F., Derrickson, S. R., and Mineau, P. (1983). Forced copulation in waterfowl. *Behavior* **86**, 250–288.
McKitrick, M. C. (1992). Phylogenetic analysis of avian parental care. *Auk* **109**, 828–846.
Meek, S. B., Robertson, R. J., and Boag, P. T. (1994). Extrapair paternity and intraspecific brood parasitism in eastern bluebirds revealed by DNA fingerprinting. *Auk* **111**, 739–744.
Mock, D., and Parker, G. in press. "The Evolution of Sibling Rivalry." Oxford Univ. Press, Oxford.
Møller, A. P. (1988). Paternity and paternal care in the swallow. *Anim. Behav.* **36**, 996–1005.
Møller, A. P. (1991). Defence of offspring by male swallows, *Hirundo rustica*, in relation to participation in extra-pair copulations by their mates. *Anim. Behav.* **42**, 261–267.
Møller, A. P., and Birkhead, T. R. (1993). Certainty of paternity covaries with paternal care in birds. *Behav. Ecol. Sociobiol.* **33**, 261–268.
Møller, A. P., and Birkhead, T. R. (1995). Certainty of paternity and paternity care in birds: A replay to Dale. *Anim. Behav.* **49**, 522–523.
Morton, E. S., Forman, L., and Braun, M. (1990). Extrapair fertilizations and the evolution of colonial breeding in purple martins. *Auk* **107**, 275–283.
Mulder, R. A., Dunn, P. O., Cocburn, A., Lazenby-Cohen, K. A., and Howell, M. J. (1994). Helpers liberate female fairy-wrens from constraints on extra-pair mate choice. *Proc. R. Soc. London Ser. B* **255**, 223–229.
Mumme, R. L., Koenig, W., Zink, M., and Marten, J. A. (1985). Genetic variation and parentage in a California population of acorn woodpeckers. *Auk* **102**, 305–312.
Orians, G. H. (1969). On the evolution of mating systems in birds and mammals. *Am. Nat.* **103**, 589–603.
Oring, L. W., Fleischer, R. C., Reed, J. M., and Marsden, K. E. (1992). Cuckoldry through stored sperm in the sequentially polyandrous spotted sandpiper. *Nature (London)* **359**, 631–633.
Owens, I. P. F. (1993). When kids just aren't worth it: Cuckoldry and parental care. *TREE* **6**, 269–271.
Owens, P. F., Dixon, A., Burke, T., and Thompson, Des. B. A. (1995). Strategic paternity assurance in the sex-role reversed Eurasian dotterel (*Charadrius morinellus*): Behavioral and genetic evidence. *Behav. Ecol.* **6**, 14–26.
Parker, G. (1979). Sperm competition and its evolutionary consequences in the insects. *Biol. Res.* **45**, 525–567.
Patrie, M. (1992). Copulation frequency in birds: Why do females copulate more than once with the same male? *Anim. Behav.* **44**, 790–792.
Petrie, M. (1994). Improved growth and survival of offspring of peacocks with more elaborate trains. *Nature (London)* **371**, 598–599.
Petrie, M., and Lipsitch, M. (1994). Avian polygyny is most likely in populations with high variability in heritable male fitness. *Proc. R. Soc. London Ser. B* **526**, 275–280.
Petter, S. C., Miles, D. B., and White, M. M. (1990). Genetic evidence of mixed reproductive strategy in a monogamous bird. *Condor* **92**, 702–708.
Pinxten, R., Hanotte, O., Eens, M., Verheyen, R. F. F., Dhondt, A. A., and Burke, T. (1993). Extra-pair paternity and intraspecific brood parasitism in the European starlings, *Sturnus vulgaris:* Evidence from DNA fingerprinting. *Anim. Behav.* **45**, 795–809.
Platt, J. R. (1964). Strong inference. *Science (Washington, D. C.)* **146**, 347–353.
Price, D. K., Collier, G. E., and Thompson, C. F. (1989). Multiple parentage in broods of house wrens: Genetic evidence. *J. Hered.* **80**, 1–5.
Quinn, J. S., Wolfenden, G. E., Fitzpatrick, J. W., and White, B. N. (1990). DNA fingerprinting analysis of Florida scrub jay parentage. Abstracts AOU/COS Joint Meeting.
Rabenold, P. P., Rabenold, K. N., Piper, W. H., Haydock, J., and Zack, S. W. (1990). Shared paternity revealed by genetic analysis in cooperatively breeding tropical wrens. *Nature (London)* **348**, 538–540.

Regelmann, K., and Curio, E. (1983). Determinants of brood defence in the great tit *Parus major* L. *Behav. Ecol. Sociobiol.* **13,** 131–135.

Richmond, A. (1978). An experimental study of advantages of monogamy in the cardinal. Ph.D. thesis, Indiana Univ., Bloomington, IN.

Ridley, M. (1978). Paternal care. *Anim. Behav.* **26,** 904–932.

Riley, H. T., Bryant, M., Carter, R. E., and Parkin, D. T. (1995). Extra-pair fertilizations and paternity defence in house martins, *Delichon urbica. Anim. Behav.* **49,** 495–509.

Ritchison, G., Klatt, P. H., and Westneat, D. F. (1994). Mate guarding and extra-pair paternity in northern cardinals. *Condor* **96,** 1055–1063.

Sasvari, L. (1986). Reproductive effort of widowed birds. *J. Anim. Ecol.* **55,** 553–564.

Sheldon, B. C. (1994). Sperm competition in the chaffinch: The role of the female. *Anim. Behav.* **47,** 163–173.

Sheldon, B. C., and Burke, T. (1994). Copulation behavior and paternity in the chaffinch. *Behav. Ecol. Sociobiol.* **4,** 149–156.

Sherman, P. W., and Morton, M. L. (1988). Extra-pair fertilizations in mountain white-crowned sparrows. *Behav. Ecol. Sociobiol.* **22,** 413–420.

Silver, R., Andrews, H., and Ball, G. F. (1985). Parental care in an ecological perspective: A quantitative analysis of avian subfamilies. *Am. Zool.* **25,** 823–840.

Slagsvold, T., and Lifjeld, J. T. (1989). Hatching asynchrony in birds: The hypothesis of sexual conflict over parental investments. *Am. Nat.* **134,** 239–253.

Slagsvold, T., and Lifjeld, J. T. (1994). Polygyny in birds: The role of competition between females for male parental care. *Am. Nat.* **143,** 59–94.

Slagsvold, T., Amundsen, T., and Dale, S. (1994). Selection by sexual conflict for evenly spaced offspring in blue tits. *Nature (London)* **370,** 136–138.

Smith, H. G. (1995). Experimental demonstration of a trade-off between mate attraction and paternal care. *Proc. R. Soc. London Ser. B.* in press.

Smith, H. G., and von Schantz, T. (1993). Extra-pair paternity in the european starling: The effect of polygyny. *Condor* **95,** 1006–1015.

Smith, H. G., Montgomerie, R., Poldmass, T., White, B. N., and Boag, P. T. (1991). DNA fingerprinting reveals relation between tail ornaments and cuckoldry in barn swallows, *Hirundo rustica. Behav. Ecol.* **2,** 90–98.

Smith, J. N. M., Yom Tov, Y., and Moses, R. (1982). Polygyny, male paternal care, and sex ratio in song sparrows: An experimental study. *Auk* **99,** 555–564.

Smith, S. M. (1988). Extra-pair copulations in black-capped chickadees: The role of the female. *Behavior* **197,** 15–23.

Stacey, P. B., and Koenig, W. D. (1990). "Cooperative Breeding in Birds: Long-term Studies of Ecology and Behavior." Cambridge Univ. Press, Cambridge.

Stamps, J. (1995). The role of females in extra-pair copulations in socially monogamous territorial animals. *In* "Feminism and Evolutionary Biology: Boundaries, Intersections, and Frontiers" (P. A. Gowaty, ed.), Chapman Hall, New York.

Stamps, J., Clark, A., Arrowood, P., and Kus, B. (1987). The effects of parent and offspring gender on food allocation in Budgerigars. *Behavior* **101,** 177–199.

Stevens, E. E. (1988). "Kin Selection and Cooperative Breeding in the Stripe-backed Wren (*Campylorynchus nuchalis*)." Ph.D. dissertation, Univ. of North Carolina, Chapel Hill, NC.

Stutchbury, B. J., Rhymer, J., and Morton, E. S. (1994). Extrapair paternity in hooded warblers. *Behav. Ecol.* **5,** 384–392.

Triggs, S., Williams, M., Marshall, S., and Chambers, G. (1991). Genetic relationships within a population of blue duck *Hymenolaimus malacorhynchos. Wildfowl* **42,** 87–93.

Trivers, R. L. (1972). Parental investment and sexual selection. *In* "Sexual Selection and the Descent of Man" (B. G. Campbell, ed.), pp. 136–179. Aldine, Chicago.

Trivers, R. L. (1974). Parent-offspring conflict. *Am. Zool.* **11**, 249–264.

van Rhijn, J. G. (1984). Phylogenetical constraints in the evolution of parental care strategies in birds. *Neth. J. Zool.* **34**, 103–122.

van Rhijn, J. G. (1985). A scenario for the evolution of social organization in Ruffs *Philomachus pugnax* and other charadriiform species. *Ardea* **73**, 25–37.

van Rhijn, J. G. (1990). Unidirectionality in the phylogeny of social organization, with special reference to birds. *Behavior* **115**, 153–174.

van Rhijn, J. G. (1991). Mate guarding as a key factor in the evolution of parental care in birds. *Anim. Behav.* **41**, 963–970.

Verner, J., and Willson, M. F. (1969). Mating systems, sexual dimorphism and the role of male North American passerine birds in the nesting cycle. *Ornithol. Monogr.* **9**, 1–76.

von Schantz, T., Goransson, G., Andersson, G., Froberg, I., Grahn, M., Helge, A., and Wittzell, H. (1989). Female choice selects for a viability-based male trait in pheasants. *Nature (London)* **337**, 166–169.

Waage, J. K. (1979). Adaptive significance of postcopulatory guarding of mates and nonmates by male *Calopteryx maculata* (Odonata). *Behav. Ecol. Sociobiol.* **6**, 147–154.

Wagner, R. H. (1991). Evidence that female razorbills control extra-pair copulations. *Behavior* **118**, 157–169.

Walters, J. (1984). The evolution of parental behavior and clutch size in shorebirds. *In* "Behavior of Marine Animals V. Shorebirds: Breeding Biology and Populations" (J. Burger and B. L. Olla, eds.), pp. 243–287. Plenum Press, New York.

Weatherhead, P. J. (1979). Ecological correlates of monogamy in tundra-breeding savannah sparrows. *Auk* **96**, 391–401.

Wesokowski, T. (1993). On the origin of parental care and the early evolution of male and female parental roles in birds. *Am. Nat.* **143**, 39–58.

Westneat, D. F. (1990). Genetic parentage in the indigo bunting: A study using DNA fingerprinting. *Behav. Ecol. Sociobiol.* **27**, 67–76.

Westneat, D. F. (1993). Polygyny and extrapair fertilizations in eastern red-winged blackbirds. *Behav. Ecol.* **4**, 49–60.

Westneat, D. F. (1995). Paternity and paternal behavior in the red-winged blackbird, *Agelaius phoeniceus. Anim. Behav.* **49**, 21–35.

Westneat, D. F., and Sherman, P. W. (1993). Parentage and the evolution of parental behavior. *Behav. Ecol.* **4**, 66–77.

Westneat, D. F., Sherman, P. W., and Morton, M. L. (1990). The ecology and evolution of extra-pair copulations in birds. *In* "Current Ornithology," Vol. 7 (D. M. Power, ed.), pp. 331–369. Plenum, New York.

Wetton, J. H., and Parkin, D. T. (1991). An association between fertility and cuckoldry in the house sparrow, *Passer domesticus. Proc. R. Soc. London Ser. B* **245**, 227–233.

Wetton, Jon H., Parkin, D. T., and Carter, R. E. (1992). The use of genetic markers for parentage analysis in *Passer domesticus* (house sparrows). *Hereditary* **69**, 243–254.

Whittingham, L. A. (1989). An experimental study of paternal behavior in red-winged blackbirds. *Behav. Ecol. Sociobiol.* **25**, 73–80.

Whittingham, L. A., Taylor, P. D., and Robertson, R. J. (1992). Confidence of paternity and male parental care. *Am. Nat.* **139**, 1115–1125.

Whittingham, L. A., Dunn, P. O., and Robertson, R. J. (1993). Confidence of paternity and male parental care: An experimental study in tree swallows. *Anim. Behav.* **46**, 139–147.

Wiklund, C. G., and Stigh, J. (1983). Nest defence and evolution of reversed sexual size dimorphism in snowy owls *Nyctea scandiaca. Ornis. Scand.* **4**, 58–62.

Williams, G. C. (1966). "Adaptation and Natural Selection." Princeton Univ. Press, Princeton, NJ.
Wolfe, L., Ketterson, E. D., and Nolan, V. Jr. (1990). Behavioral response of female dark-eyed juncos to the experimental removal of their mates: Implications for the evolution of male parental care. *Anim. Behav.* **39,** 125–234.
Wrege, P. H., and Emlen, S. T. (1987). Biochemical determination of parental uncertainty in white-fronted bee-eaters. *Behav. Ecol. Sociobiol.* **20,** 153–160.
Wright, J., and Cotton, P. A. (1994). Experimentally induced sex differences in parental care: An effect of certainty of paternity? *Anim. Behav.* **47,** 1311–1322.
Wright, J., and Cuthill, I. (1990). Biparental care: Short-term manipulation of partner contribution and brood size in the starling, *Sturnus vulgaris. Behav. Ecol.* **1,** 116–124.
Wright, J., and Cuthill, I. (1992). Monogamy in the European starling. *Behavior* **120,** 262–285.
Yamagishi, S., Nishiumi, I., and Shimoda, C. (1992). Extrapair fertilization in monogamous bull-headed shrikes revealed by DNA fingerprinting. *Auk* **109,** 711–721.
Yamamura, N., and Tsuji, N. (1993). Parental care as a game. *J. Evol. Biol.* **6,** 103–127.
Yezerinac, S. M., Weatherhead, P. J., and Boag, P. T. (1996). Extra-pair paternity and the opportunity for sexual selection in a socially monogamous bird (*Dendroica petechia*). *Behav. Ecol. Sociobiol.* in press.

# Parental Investment in Pinnipeds

FRITZ TRILLMICH

LEHRSTUHL FÜR VERHALTENSFORSCHUNG
UNIVERSITÄT BIELEFELD
D-33501 BIELEFELD, GERMANY

## I. INTRODUCTION

Pinnipeds, the true seals (Phocidae), eared seals (Otariidae), and walruses (Odobenidae) are excellent subjects for the study of parental care and parental investment. They combine high preparturient and postparturient energy expenditure for producing and rearing young (Kovacs and Lavigne, 1986a, 1992) with exclusive maternal care. Analysis of mother–young interactions is eased because maternal care consists mostly of milk transfer from mother to offspring and some degree of defense of the altricial newborn against other females and males (Bartholomew, 1970; see Section III that follows). These attributes combined with extreme sexual dimorphism and concomitant sexual differences in variance in lifetime reproductive success in some species (land breeding phocids, otariids, and walruses) that are larger than in most other mammalian groups (Alexander *et al.,* 1979; Boness, 1991; Clutton-Brock, 1988; Kovacs and Lavigne, 1992). Consequently, pinnipeds appear to offer an excellent opportunity to test parental investment theory.

Pinniped social structure is famous for the enormous degree of polygyny achieved by some species like the Northern elephant seal and the Northern fur seal (Bartholomew, 1952, 1970; Bartholomew and Hoel, 1953; Le Boeuf, 1974). This goes along with an enormous, up to sixfold, sexual mass dimorphism making the sexes in effect appear to belong to quite different species. Therefore, pinnipeds were considered ideal objects for the critical assessment of theories of differential investment in the sexes (Fisher, 1930; Trivers and Willard, 1973; Williams, 1979; Maynard Smith, 1980; reviewed in Frank, 1990). Nevertheless, despite a number of excellent studies of energetic costs of lactation, growth of pups, mass loss, and behavior of mothers in relation to sex, little is still known about the selective forces acting upon sex ratio and differential sex allocation in pinnipeds and the matching of these facts to the theoretical models is tenuous.

In addition to the study of mother–offspring interactions on land, the energetics of milk transfer and foraging strategies of pinnipeds have increasingly become amenable to investigation through the development of new methodology. Time depth recorders (Kooyman *et al.,* 1983) and the methodology of labeled water (Costa, 1987; Oftedal and Iverson, 1987) provided a wealth of new detailed information on the ecological conditions influencing the rearing strategies of seals, sea lions, and fur seals (Gentry and Kooyman, 1986; Trillmich, 1990; Bowen, 1991; Costa, 1993; Le Boeuf and Laws, 1994). The new methodologies opened up new approaches to the understanding of maternal strategies, particularly under energetic aspects (Huntley *et al.,* 1987). The broad spectrum of pinniped species of widely differing body masses and distributed worldwide offers rich material to study questions of adaptation by the comparative method. In addition, high environmental variance in food availability connected with the El Niño phenomenon in the Pacific Ocean provided new perspectives on the ecology and life history of pinnipeds (Trillmich and Ono, 1991). El Niño is a meteorological and oceanographic phenomenon that occurs at irregular intervals in the eastern tropical Pacific and causes enormous disturbances in the marine ecosystem. For pinnipeds the local reductions in primary and secondary production appear most influential. Similar disturbances of the marine ecosystem have also been documented in other pinniped-inhabited ocean systems (Croxall *et al.,* 1988). Study of the responses of individual species to such drastic changes in their environments have added a further dimension to comparative studies of pinniped maternal strategies.

However, while maternal *expenditure* has been measured in a number of species under a variety of circumstances as energetic cost, it has proven much more elusive to measure maternal *investment* in ultimate terms, namely, benefits to offspring at a cost in fitness (and not only energy) to the mother (Trivers, 1972). It was hoped that a thorough understanding of mothers as highly efficient mass and energy transfer devices would also lead to an understanding of the cost of reproduction. This seems logical if one assumes that energy expenditure is a limiting commodity (Drent and Daan, 1980), and that it should therefore closely correlate with fitness measures. However, while energy may still prove a proximate correlate of fitness, so far ultimate costs of reproduction are only poorly documented and the connection between energetics and fitness currency is far from clear.

I will first give information on the natural history of maternal strategies (to which I also refer as maternal rearing strategies or simply maternal care) in pinnipeds. However, because good reviews of maternal rearing strategies (Bonner, 1984; Gentry and Kooyman, 1986; Trillmich, 1990; Bowen, 1991; Le Boeuf and Laws, 1994), their energetics (Oftedal *et al.,*

1987a; Costa, 1991, 1993), and their reaction to environmental disturbance (Trillmich and Ono, 1991), as well as of fostering (Riedman, 1982; Boness, 1990; Bowen, 1991) and abuse of young (Le Boeuf and Campagna, 1994) are available, this article will concentrate on data about fitness effects of maternal care in pinnipeds.

I will, therefore, focus on the following topics:

• An abridge review of parental care patterns in pinnipeds summarizing reviews and adding a few pertinent data (see previous references for more details)

• The evidence for maternal investment and the relationship between energy expenditure and investment

• The evidence for and against differential investment in male and female offspring

• Briefly present the observation of postweaning care in pinnipeds for which it is not clear if investment is involved at all

For convenience, Table I gives an alphabetical list of common and scientific names of the pinniped species mentioned.

## II. Background Information on Pinniped Systematics and Phylogeny

Pinnipeds are traditionally classified as three families, the true seals (Phocidae, e.g., harbor seal, elephant seal), the eared seals (Otariidae, fur seals and sea lions), and the walruses (Odobenidae). Originally, paleontologists held that these families had arisen from different ancestors, the phocids from mustelid stock, the otariids and odobenids from ursid stock (Barnes *et al.*, 1985; Tedford, 1976; Repenning, 1980; see, however, Wiig, 1983). Molecular evidence tended to be more in favor for monophyly (Sarich, 1969; Arnason and Widegren, 1986), and modern cladistic morphological analyses also point to a monophyletic origin of pinnipeds (Wyss, 1987, 1988, 1989; Berta *et al.*, 1989). Presently the evidence is in favor of monophyly, and actually places the walruses closer to the phocids than to otariids (Wyss, 1987). In the context of this article, accepting monophyletic derivation of pinnipeds from arctoid (bearlike) ancestry means that differences in maternal strategies of the pinniped families must be interpreted as adaptations to environmental conditions in which the various species evolved and live today rather than be attributed to phylogenetic inertia.

TABLE I
ALPHABETICAL LIST OF COMMON AND SCIENTIFIC NAMES OF PINNIPED SPECIES MENTIONED IN THE TEXT

| Family | Common name | Scientific name |
|---|---|---|
| Phocidae, true seals | Baikal seal | *Phoca sibirica* |
| | Bearded seal | *Erignathus barbatus* |
| | Caspian seal | *Phoca caspica* |
| | Crabeater seal | *Lobodon carcinophagus* |
| | Grey seal | *Halichoerus grypus* |
| | Harbor seal | *Phoca vitulina* |
| | Harp seal | *Phoca groenlandica* |
| | Hawaiian monk seal | *Monachus schauinslandi* |
| | Leopard seal | *Hydrurga leonina* |
| | Hooded seal | *Cystophora cristata* |
| | Mediterranean monk seal | *Monachus monachus* |
| | Northern elephant seal | *Mirounga angustirostris* |
| | Ringed seal | *Phoca hispida* |
| | Southern elephant seal | *Mirounga leonina* |
| | Spotted seal | *Phoca largha* |
| | Weddell seal | *Leptonychotes weddelli* |
| Otariidae, eared seals | Antarctic fur seal | *Arctocephalus gazella* |
| | Australian sea lion | *Neophoca cinerea* |
| | California sea lion | *Zalophus californianus* |
| | Galápagos fur seal | *Arctocephalus galapagoensis* |
| | Galápagos seal lion | *Zalophus californianus wollebaeki* |
| | New Zealand fur seal | *Arctocephalus fosteri* |
| | Northern fur seal | *Callorhinus ursinus* |
| | South African fur seal | *Arctocephalus pusillus* |
| | South American fur seal | *Arctocephalus australis* |
| | South American sea lion | *Otaria byronia* |
| | Steller sea lion | *Eumetopias jubata* |
| | Subantarctic fur seal | *Arctocephalus tropicalis* |
| Odobenidae, walruses | Walrus | *Odobenus rosmarus* |

## III. PARENTAL CARE PATTERNS IN PINNIPEDS

### A. MATERNAL BEHAVIOR

With the exception of the walruses (Fay, 1982) and the Australian sea lion (*Neophoca cinerea;* Higgins, 1993), pinnipeds reproduce synchronized to the annual cycle. Phocid females copulate towards or at the end of lactation, otariids within 5–10 days after parturition. Pregnancy usually lasts 7–9 months and synchronization with the annual cycle is achieved through delayed implantation (King, 1983). Maternal behavior differs widely between phocids and otariids. Phocids are generally larger than otariid seals. Following the complications by Kovacs and Lavigne (1986a, 1992), most phocid females weigh more than 100 kg (5 species < 100 kg, 12 species > 100 kg; median mass 220 kg), whereas most otariid females weigh less than 100 kg (11 species < 100 kg, 3 species > 100 kg female body mass; median female mass between 40 and 50 kg). This size and mass difference between females of the two families relates to the major difference in patterns of maternal care between the groups. Maternal strategies have been characterized as the phocid *fasting* and the otariid *foraging cycle* or *attendance* strategy (Fig. 1; Bonner, 1984; Gentry and Kooyman, 1986; Trillmich, 1990; Bowen, 1991). The fasting strategy implies that phocid mothers come ashore, give birth to a pup, and lactate, usually without ever feeding during this period. They have stored all the materials for maintenance and lactation during previous foraging at sea. In contrast, the otariid foraging cycle strategy is shorthand for alternating attendances ashore nursing the pup and periods at sea foraging. One could, therefore, as well characterize the phocids as "capital" breeders and otariids as "income" breeders (Drent and Daan, 1980). Typically, phocid mothers take all the resources for lactation and their own as well as the pup's maintenance metabolism out of their stored reserves (capital), whereas obtained females forage during lactation to gain the energy for pup growth as well as their own and the pup's maintenance metabolism (i.e., they live from current income). Only the energy expended during the approximate 7-day perinatal fast (which varies between individuals and species from 5 to 12 days) is stored as additional fat reserves by otariid females (Costa and Trillmich, 1988). As a consequence of these differing breeding strategies, phocids can

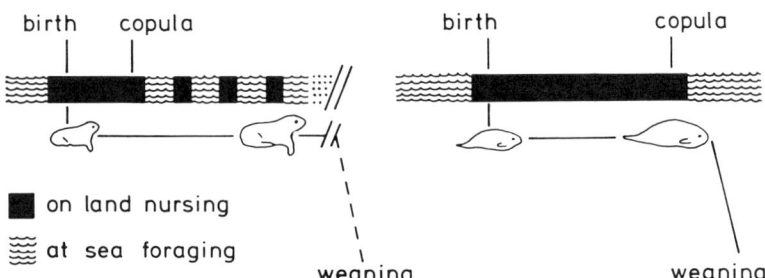

FIG. 1. Typical maternal care patterns in otariids (left) and phocids (right). The otariid rearing pattern, income breeding, consists of an initial perinatal period (about 7 days) when the fasting mother stays continuously ashore with her pup. Thereafter, she alternates foraging at sea (1–14 days) with short attendances ashore (1–2 days) where she suckles the pup. Phocid patterns are typically capital breeding, that is, the mother comes ashore or on the ice and stays with her pup (for 4 to about 60 days) fueling lactation from stored reserves.

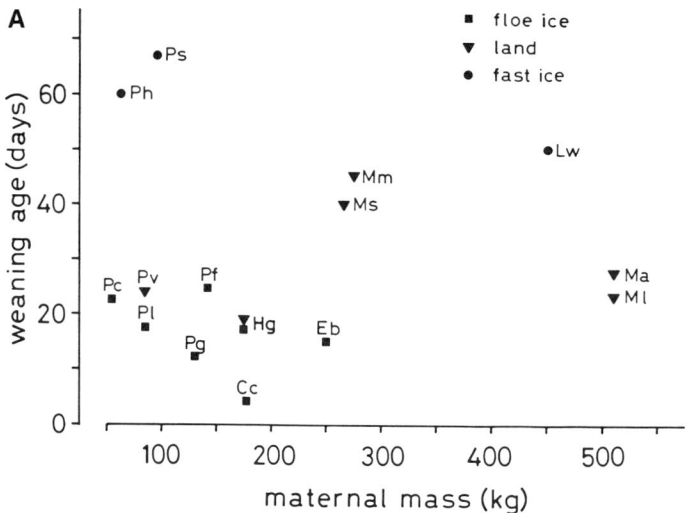

FIG. 2. (a) There is no relationship between duration of lactation and maternal mass in phocids and neither do the data show an orderly relationship between breeding habitat and lactation duration. Pc = *Phoca caspica*, Caspian seal; Ph = *Phoca hispida*, ringed seal; Pv = *Phoca vitulina*, harbor seal; Pl = *Phoca largha*, spotted seal; Ps = *Phoca sibirica*, Baikal seal; Pg = *Phoca groenlandica*, harp seal; Pf = *Phoca fasciata*, ribbon seal; Hg = *Halichoerus grypus*, grey seal, breeds on land and floe ice; Cc = *Cystophora cristata*, hooded seal; Eb = *Erignathus barbatus*, bearded seal; Ms = *Monachus schauinslandi*, Hawaiian monk seal; Mm = *Monachus monachus*, Mediterranean monk seal; Lw = *Leptonychotes weddelli*, Weddell seal; Ma = *Mirounga angustirostris*, Northern elephant seal; Ml = *Mirounga leonina*, Southern elephant seal. (Data from Bowen (1991) and Costa (1993). (b) Weaning age is not related to maternal mass in otariids. Arrows indicate the option of extending lactation for more than 18 months. Question marks indicate that earliest or latest age at weaning is not documented. Agal = *Arctocephalus galapagoensis*, Galápagos fur seal; Ag = *A. gazella*, Antarctic fur seal; At = *A. tropicalis*, Subantarctic fur seal; Cu = *Callorhinus ursinus*, Northern fur seal; Af = *A. forsteri*, New Zealand fur seal; Aa = *A. australis*, South American fur seal; Ap = *A. pusillus*, South African fur seal; Nc = *Neophoca cinerea*, Australian sea lion; Zc = *Zalophus californianus*, California sea lion. Data on *A. tropicalis* from Goldsworthy (1992), *A. australis* from Majluf (1987), other fur seals and California sea lion from Gentry *et al.* (1986), and *Neophoca cinerea* from Higgins (1990, 1993).

separate foraging and lactation in time, whereas otariids need access to rich feeding grounds during the breeding season. This may well explain why otariids usually breed close to cold, productive ocean regions; cold coastal currents; and upwelling zones. I will first sketch the typical phocid and otariid breeding pattern (mentioning the much less-known walruses only briefly) and then describe some new data that shake up old textbook truths about phocid maternal strategies.

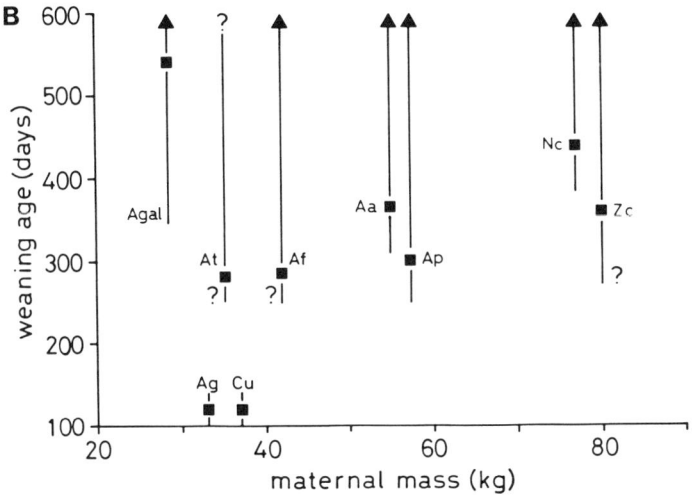

FIG. 2. (*Continued*)

Phocid mothers come onto the breeding substrate will all resources for maintenance and milk production during the period of lactation stored in their body. They fast during a short lactation period (4–60 days) (Fig. 1). Milk is generally extremely high in fat content (40–60%) during midlactation to late lactation (Oftedal *et al.*, 1987a) and pups fatten quickly. Weaning is abrupt at the end of the brief nursing period and pups often fast for considerable time after weaning, living off their accumulated fat reserves. The periods of sucking and independent foraging of these pups, therefore, are separated by a period of fasting after weaning. Duration of lactation, pup growth rates, and postweaning fast durations have been tabulated (Bowen, 1991; Muelbert and Bowen, 1993; Costa, 1993). Pup mass at weaning correlates well with maternal mass (Kovacs and Lavigne, 1986a, 1992), but there is no detectable orderly relationship between maternal mass and lactation duration or weaning age of young (Fig. 2a).

Within phocids it has often been claimed that lactation duration covaried with breeding habitat, both this claim seems at best weakly supported by the data. Species breeding on floe ice appeared to have the shortest lactation periods (Bonner, 1984; Oftedal *et al.*, 1987a; Bowen, 1991), down to a 4-day period in the hooded seal (Bowen *et al.*, 1985), an incredibly abbreviated lactation for a 180-kg mammal. The shortening of lactation was ascribed to the combined risks of habitat instability and predation (Stirling, 1977). Land breeding species were said to have longer lactation periods, and lactation was supposedly longest in species breeding on fast ice (Bowen, 1991). However, the numbers of species from which these means were derived are small (nine species breeding on floes, three on fast ice, and five on land), data points are not independent phylogenetically (i.e., the five land breeding species comprise two species of elephant seals and the two monk seals), and species with highly differing ecology are lumped into the same breeding habitat group (e.g., harbor seal, elephant seals, and monk seals all classified as land breeding, when the relation of the harbor seal to land differs widely from that of the other species). Comparison of antarctic ice breeding species (mean lactation of four species somewhere around 30 days; reported lactation durations vary enormously, and data for Leopard and Crabeater seal are guesses only; Bowen, 1991) and arctic ice breeding species (nine species: mean 26 days) suggests that the lack of land predators in the Antarctic may correlate with slightly longer lactation periods of antarctic phocids. On the other hand, the arctic Baikal and ringed seals, which are among the smaller species (female body mass 94 and 62 kg, respectively), have found a way to circumvent some problems of harsh climate and predation by constructing birth lairs where young appear to be relatively warm and safe (Smith and Stirling, 1975; Smith, 1980). These two species have the longest lactation periods of any phocid species, about 60–65 days.

Land breeding phocids generally breed on predator-free islands. If instability of the breeding substrate or predation were the main forces shaping the duration of lactation, these species should have the longest lactation periods, which is clearly not the case. Thus, intrinsic factors like body size and correlated fasting ability (implying energetic limits) as well as ecological factors like foraging options during lactation (see as follows) may be more influential than predation and perhaps breeding habitat stability in determining the duration of lactation. At this stage, all conclusions about a relationship between breeding habitat and lactation duration appear unwarranted. The importance of these factors is hard to test by comparative methods given the close relatedness and consequent nonindependence among a small number of sampling points (species) in a multivariate analysis. Thus, to test predictions of the existing hypotheses or to erect new ones, we need more detailed documentation of the factors related to intraspecific variability in lactation duration and of the intraspecific and interspecific importance of habitat stability, predation pressure, thermoregulation of pups, and maternal foraging options during lactation. In addition, hypotheses testing should correct explicitly for maternal size effects.

In contrast, the mostly smaller otariid seals breed invariably on land in often large colonies almost exclusively on predator-free islands. Mothers come ashore with moderate body reserves (although fat content as percentage of body mass can be the same as in phocids; Costa and Trillmich, 1988; Costa, 1991), stay on land fasting for about 1 week after parturition, and thereafter alternate between foraging trips to sea and returns to land where they suckle the pup for 1 or 2 days (Fig. 1; Bonner, 1984; Gentry et al., 1986; Trillmich, 1990; Bowen, 1991). The duration of foraging trips largely depends on distance of foraging areas and on food abundance in the foraging areas. This influences female foraging efficiency and is usually reflected in her body mass as well as the growth rate of her pup. Within a given species, the shorter the time a mother is away, the higher is the growth rate of her pup and the better her own body condition (Trillmich and Limberger, 1985; Croxall et al., 1988; Trillmich, 1990; Lunn and Boyd, 1993a,b). Lactation periods are longer (4 monnths to more than 2 years) than those of phocids and again show no relationship to maternal body mass (Fig. 2b). Subpolar species, the Northern fur seal (*Callorhinus ursinus*) and the Antarctic fur seal (*Arctocephalus gazella*), have the shortest lactation periods of about 120 days, temperate and tropical species have much longer ones. Milk transferred is generally less concentrated in fat than phocid milk (19–50% fat) (Oftedal et al., 1987a) but varies in composition with trip duration and time ashore (Trillmich and Lechner, 1986; Costa and Gentry, 1986; Arnould and Boyd, 1995). Pups grow more slowly than phocid pups (Bowen, 1991; Kovacs and Lavigne, 1992) and get less fat, but are weaned at a greater mass relative to maternal mass (roughly 40%) than phocid young (ca. 30%) (Bowen, 1991; Costa, 1991, 1993; Kovacs and Lavigne, 1992).

In many otariid species, the long period of nursing provides an opportunity for pups to practice swimming, diving, and independent foraging while still partly relying on maternal milk (Gentry et al., 1986; Trillmich, 1990; Bowen, 1991). Therefore, weaning can be a more gradual affair than in phocids. Indeed, Horning (1992) demonstrated for the Galápagos fur seal that young may forage independently of their mothers for over a year before weaning is completed.

However, this does not apply for two subpolar species, the Northern fur seal and Antarctic fur seal. In these species weaning is abrupt at about 4 month of age and time to weaning shows little variation (Fig. 2b; Gentry and Holt, 1986; Doidge et al., 1986). Interestingly, close observation of the weaning pattern suggests that in these species many (if not most) of the young wean themselves by leaving the colony while mothers are still returning. Macy (1982) observed that 77% of all Northern fur seal young wean before their mothers stop to return to the colony and the same was documented by Doidge et al. (1986) for the Antarctic fur seal. A calculation of energy demands of growing young (from data in Costa and Gentry, 1986; and Costa and Trillmich, 1988) and energy intake via maternal milk (Costa and Gentry, 1986) shows that young of these species leave the colonies, without weaning conflict, approximately when their energy balance (energy intake through milk per day minus energy metabolized per day) gets close to zero or becomes slightly negative. This is shown in Fig. 3 for Northern fur seal pups in which male pups get more milk per maternal attendance and are weaned at heavier weights than female pups (Costa and Gentry, 1986; see Section V, where the additional complication of potentially differing body compositions of male and female pups is also discussed).

Apparently, weaning at 4 months of age is a genetically fixed trait in the Northern and the Antarctic fur seals. Both species wean their pups at the same age at lower latitude breeding colonies irrespective of the pups' actual size. As productivity appears to be lower at these lower latitude sites, pups grow more slowly than in the subpolar areas. Thus, mothers weaning their pups at 120 days may risk pup survival (DeLong and Antonelis, 1991; Kerley, 1985). In these same lower-latitude sites, other sympatric species

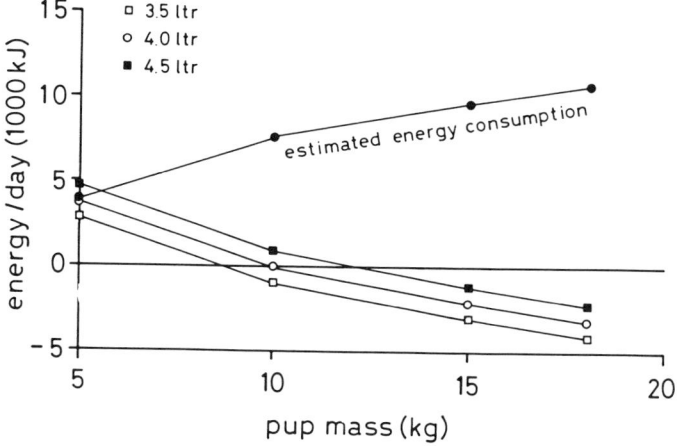

FIG. 3. Estimated energy consumption (upper curve) of a Northern fur seal young as a function of its mass. Lower curves: net energy available for growth and/or fat storage at milk yields of (from lowest curve) 3.5, 4.0, and 4.5 liters per female attendance. Male pups wean at a mean mass of 14.1 kg, females of 11.7 kg (Gentry et al., 1986). This corresponds to zero net energy income from a milk yield of 4.5 liters per attendance ashore for mothers of male pups and of about 3.75 liters per attendance for mothers of female pups, close to estimated milk yields. Estimates derived from Costa and Gentry (1986); energy expenditure for heavier young estimated from equation for subadult male Antarctic fur seals in Costa and Trillmich (1988). Given known milk composition (Costa and Gentry, 1986) and attendance as well as foraging (absence) times of mothers (Gentry and Holt, 1986), the curves for net energy available to young from given amounts of milk delivered per maternal attendance were derived as the difference between per day energy intake from milk minus daily metabolic energy expenditure. The lower curves show this net energy.

(the subantarctic fur seal, *Arctocephalus tropicalis,* in the case of the Antarctic fur seal and the California sea lion in the case of the Northern fur seal) wean their young much later and show greater variability in weaning age than the subtropical species (Trillmich, 1990). Great flexibility in the timing of weaning is typical for temperate and tropical fur seals and sea lions (compare Fig. 2b). In this way, mothers are able to buffer pups from unpredictable variation in food availability while pups are gradually increasing their foraging skills and become independent (Gentry *et al.,* 1986; Trillmich, 1990; Trillmich *et al.,* 1991).

Contrary to common belief, there is no substantial evidence indicating that pups leave the colonies to forage at sea together with their mothers and thus learn foraging skills or foraging locations directly from mothers. My own observations on individually marked 1- and 2-year-old Galápagos sea lions indicate instead that pups often leave the colony together with their mother after a period of suckling, but will return briefly thereafter without the mother. They usually leave for foraging later, independently of their mothers, after they have slept off their belly full of milk.

Odobenid maternal care is least documented. Fay (1982) has compiled the available data. Walruses are the most social pinnipeds, always living in large herds. Mothers separate briefly from the herd before parturition, give birth on an isolated patch of ice, and then join the herd again. They may fast for a while after birth of the pup, but accurate observations are lacking. Thereafter they are always accompanied by the pup during foraging and resting (Fay, 1982). Mother and pup stay together for 2, sometimes 3 years. Pups are usually suckled in the water (Miller and Boness, 1983). They are believed to depend largely on maternal milk for about 1 year and to be gradually weaned during the second year (Fay, 1982). Young can thus gain considerable foraging experience while still with the mother (Fay, 1982). Female calves are then integrated into the female herd, whereas male calves may stay for 2 to 3 years with the female herd, joining all-male herds thereafter.

This classification of maternal care patterns according to pinniped families is the classical textbook picture alluded to previously. It proves, however, too simplistic as research has produced clear evidence that some phocid species forage during the lactation period. Particularly ice breeding species should find food easily during the lactation period as they breed above foraging areas. Within the Phocidae, foraging during lactation was best demonstrated in Weddell seals (*Leptonychotes weddelli,* maternal mass ca. 450 kg) (Testa *et al.,* 1989), harp seals (*Pusa groenlandica,* maternal mass 130 kg; Lydersen and Kovacs, 1993), and harbor seals (*Phoca vitulina,* mass 85 kg) (Boness *et al.,* 1994; Thompson *et al.,* 1994) (see also Costa, 1993). The first is a large-sized species; the second a medium-sized species, both breeding on fast ice; and the third is a small, land breeding phocid. Lactation lasts about 40–50 days in Weddell seals, 12 days in the harp seal (Kovacs, 1987), and 24 days in harbor seals (Muelbert and Bowen, 1993).

Records of intensive diving activity of four female harp seals studied provide strong evidence that females were feeding during lactation (Lydersen and Kovacs, 1993). Similarly, Boness *et al.* (1994) showed that about 8–10 days after parturition female harbor seals on Sable Island began to dive quite intensively (100–250 dives per day on average). They found significant amounts of food in the stomachs of lactating females pointing to a major importance of the diving activities of females in supporting their energy demands during lactation. Measured mass loss rates of lactating harp (Lydersen and Kovacs, 1993) and harbor seal females (Bowen *et al.*, 1992a) could not have been sustained for the whole duration of the lactation period had females not foraged before the end of lactation. Females of these species live in circumstances that make it unnecessary to store all materials needed to complete lactation from body stores. The intercalation of foraging and nursing may enable the long lactation periods of the fast ice breeding species. If this hypothesis is correct, in the near future one would want to obtain more observations of foraging during lactation in phocids, particularly in the species that build lairs for their young and have unusually long lactation periods (Baikal and ringed seal), and in small species like the spotted and the Caspian seals.

Proof of the otariid-like patterns of foraging during lactation in typical phocids suggests that maternal strategies are more related to body reserves, distance to foraging areas, and other ecological variables (like perhaps predation) rather than to phylogenetic inertia as previously assumed to explain the dichotomy between otariid and phocid or income (foraging cycle) versus capital (fasting) maternal strategies.

Another new finding is that phocid pups may begin foraging with greater diving skills and perhaps also earlier than formerly thought. Studies of nursing ringed seal (*Phoca hispida*) pups showed surprising diving abilities. They spent 53% of their time in the water, and were able to dive to 89 m depth and stay submerged for 12 min, suggesting that they as well might gain considerable experience diving while still not being weaned (Lydersen and Hammill, 1993). Similar data exist for the bearded seal (*Erignathus barbatus*) pups that before weaning were shown to dive to 84 m and stay submerged for a maximum of 5.5 min (Lydersen *et al.*, 1994). Thompson *et al.* (1994) report that female harbor seals took their pups along when changing to another haul out site up to 30 km away, suggesting substantial swimming and perhaps diving abilities of these pups before weaning.

B. PATERNAL CARE?

Parental care in pinnipeds can be studied without paying any attention to the father's role. Usually there is no paternal care or it is incidental, for example, when a male protects pups from harassment by other males by keeping subadults and other competing males outside his territory and thus away from the pupping area. Sometimes (especially in elephant seals and South American sea lions) the males' role is outright negative because males (not necessarily the fathers) may trample, abduct, or violate pups (Le Boeuf and Campagna, 1994).

There is just one explicit claim for a paternal role of male pinnipeds, based on observations on the Galápagos sea lion (*Zalophus californianus wollebaeki*). Building on observations by Eibl-Eibesfeldt (1955), Eibl-Eibesfeldt and Hass (1959), and Nelson (1968), Barlow (1972, 1974) hypothesized that males of the Galápagos sea lion were kin selected to provide paternal care to pups venturing in the water and thereby exposing themselves to shark predation. The basic observation was one of (sometimes collective) mobbing of sharks by territorial males on Champion Island. Barlow (1972, 1974) interpreted this observation as paternal behavior of territorial bulls. Miller (1974) challenged his view and suggested the observed behavior was a by-product of behavior normally directed against conspecific territorial intruders, without necessarily having a selective advantage on its own. This interpretation seems unlikely as it assumes that sea lion males are unable to distinguish between sharks and male competitors (Barlow, 1974).

Our own observations of shark mobbing by sea lions gathered for more than 10 years of observations on Galápagos sea lions and information gained from interviews with divers document that shark mobbing behavior is also shown by females and juveniles around colonies, as well as by single sea lion females in open water. Such behavior apparently serves to move sharks away from the preferred resting places of sea lions (also observed for Galápagos fur seals, *Arctocephalus galapagoensis;* F. Trillmich, unpublished observations) or to chase them from feeding sites (e.g., when a sea lion is breaking up a big fish on the surface). Only in the case of shark mobbing by territorial males does this behavior coincidentally protect pups. In the other contexts the behavior clearly cannot be interpreted as paternal.

Territorial males rest and sleep much of the daytime in the water adjacent to their shoreline territories. They are, therefore, in the greatest danger from sharks, which usually attack unaware sea lions from below and behind. If mobbing keeps sharks away, males can benefit greatly from such behavior. The common risk of shark attacks would also explain why several males will converge in apparent cooperation when chasing sharks away, whereas they otherwise fight quite fiercely with each other over territorial boundaries. There is no reason to assume that through this behavior they would attempt to protect the pups of the other male in whose territory the shark appears.

However, no specific study on degrees of relatedness between territorial bulls and pups inside a male's territory and the influence of the male's behavior on pup survival has been made. If males were shown to be closely related to pups on their territory and to attack sharks more likely when pups are nearby, this would provide evidence for Barlow's (1972) hypothesis. Until such a study proves or disproves Barlow's hypothesis, a simple mobbing response by which males coincidentally protect pups appears to be the more parsimonious explanation of the sparse observational evidence.

## IV. Evidence for a Cost of Reproduction

### A. Maternal Investment

It is obvious that maternal care in pinnipeds is very costly in terms of energetic expenditure (Oftedal *et al.,* 1987a; Costa, 1991, 1993). In pinnipeds, it is particularly important to keep clear the conceptual difference between measures of expenditure, for example, energy transfer to a pup, and measures of investment, that is, fitness changes, because energy flow is so impressive. Even within a species mothers may differ widely in body mass and therefore in the ability to transfer resources to their young. The same absolute amount of transfer may mean very different expenditure, let alone investment, to females of different body size (Fig. 4). To give just one example of each of the two families: Northern elephant seal females start on their reproductive career at body weights as low (for an elephant seal) as 360 kg and may grow during their lifetime to 710 kg (Fig. 4a; Deutsch *et al.,* 1994). Similarly, Antarctic fur seal females begin to reproduce at a body mass of 22 kg and can grow to a size of 50 kg or more (Fig. 4b; Payne, 1979). Larger mothers transfer significantly larger amounts of resources to their pups, but it is very unlikely that they incur greater costs by doing so because relative to maternal body mass transfer diminishes as maternal mass increases (see Fig. 4a,b). To get at the cost of reproduction, fitness changes must be measured. Fitness costs could—among other possible trade-offs (Stearns, 1989)—consist of lowered future fertility or increased mortality of a breeding female. The fitness costs could also be passed onto the next cohort of young if females were less likely to provide adequate care after successfully weaning a pup in the preceding reproductive cycle. The evidence of such effects is still surprisingly limited.

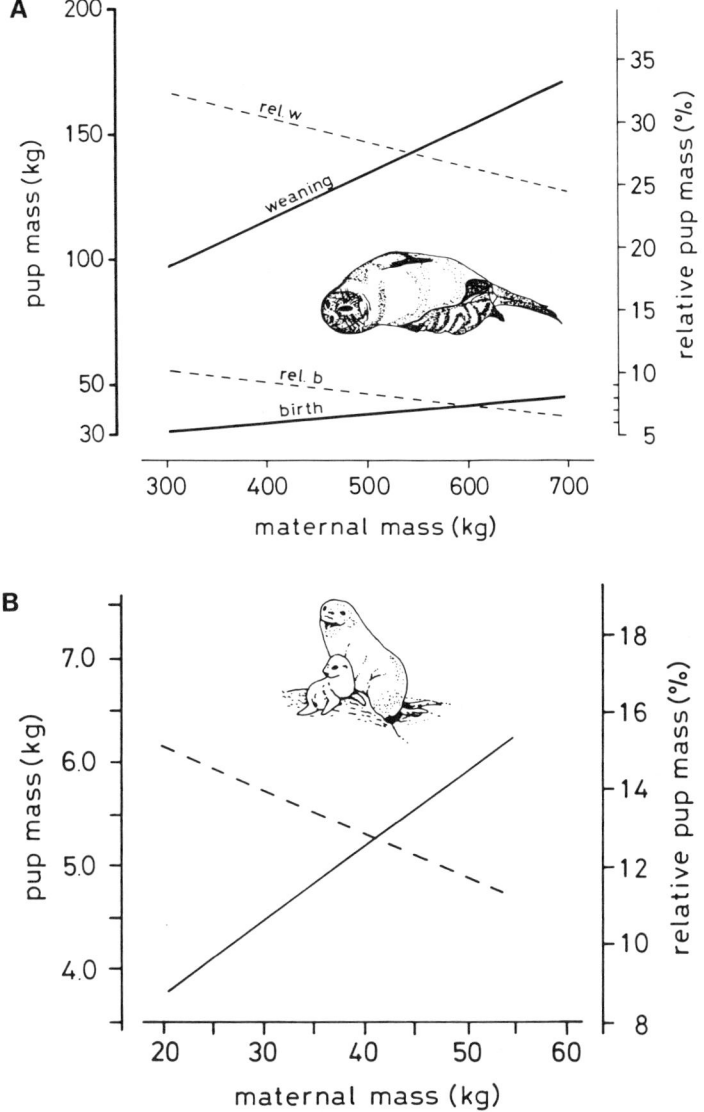

FIG. 4. Relative (rel.) and absolute mass in relation to maternal mass of (a) pup (at birth = b) and weaner (= w) Northern elephant seals (data from Deutsch *et al.*, 1994) and (b) pup Antarctic fur seals at birth (data from Costa *et al.*, 1988). No data exist for mother–pup pairs at weaning for Antarctic fur seals.

## 1. Phocids

Despite a number of highly successful long-term studies of seals, in particular on the Northern and Southern elephant seals (Le Boeuf and Laws, 1994) and the British grey seals (Fedak and Anderson, 1982; Anderson and Fedak, 1987), data on fitness costs of reproduction to females only exist for the Northern elephant seal (Reiter et al., 1981; Huber, 1987; Le Boeuf and Reiter, 1988; Reiter and Le Boeuf, 1991; Sydeman et al., 1991; Deutsch et al., 1994; Sydeman and Nur, 1994). Results are difficult to interpret due to great intraspecific variability. They serve here as an excellent example for the status of studies in phocids and of the kinds of problems encountered.

The best data exist on the effect of primiparity at different ages on fertility and survival of females. Northern elephant seals begin to reproduce when 2–6 years old. Females primiparous at the age of 3 and 4 years have much lower weaning success than older primiparous, and older experienced females (Reiter et al., 1981). This is, at least partly, explained by the small size of young females and their subdominant position in crowded female groups, making mother–pup separation and consequent pup loss more likely for young, inexperienced females than for bigger, older, and more experienced females. Small females can actually gain fitness (in terms of pup survival to weaning) by emigrating from dense colonies to less densely settled areas (Le Boeuf and Reiter, 1988). On the island of Año Nuevo, primiparous females were not less likely to pup in the subsequent year than multiparous females. Basically, once a female had begun to pup she continued to do so every year and natality rates were 97.5% (Le Boeuf and Reiter, 1988). Thus, on Año Nuevo, neither in primiparous nor in older, multiparous females was a fertility cost of reproduction observable.

In contrast, females on the Farallon Islands were less likely to pup in the following year if they had given birth for the first time (Huber, 1987). Huber followed a sample of 139 primiparous females and compared the subsequent fertility of females breeding for the first time at the ages of 3, 4, 5, and 6 years. The older these females began to reproduce, the greater their fertility in the subsequent 3 or 4 years. In particular, of the 3-year-old females, only 60% produced a pup in the year following the production of the first pup, while older (4 or 5 years old) and, therefore, bigger primiparous females had natalities of 0.87 and 0.90, respectively. This implies that reproduction at a small body size carries a greater fertility cost than later reproduction (as also found by Arnbom et al., 1994 for the Southern elephant seal). Nevertheless, early breeding is not maladaptive. Le Boeuf and Reiter (1988) showed that females that began breeding early and survived realized greater lifetime reproductive success than females that began breeding later. However, this effect is strongest at very low breeding densities and may disappear as density increases (Reiter and Le Boeuf, 1991).

The proximate mechanism mediating this effect most likely lies in the higher relative maternal expenditure of small Northern elephant seal females as compared with larger ones (Fig. 4a). Pups of larger mothers get more milk in the available time and, therefore, grow to a larger mean size than pups of small mothers, presumably at a higher cost to small mothers. After controlling for mass, age of the mothers did not significantly influence this relationship pointing to the great importance of mass in these animals. However, interpretation is complicated by the observation of Deutsch *et al.* (1994) that females of all sizes lost between 31 and 41% of body mass in rearing a pup with no correlation between maternal mass and percentage mass loss. The large variability in percentage mass loss and relatively small sample size ($n = 22$) make it likely that this statement will not remain the last word about the relative expenditure of mothers of differing sizes.

In any case, the data imply that fertility costs exist for Northern elephant seals that begin to breed at a young age on the Farallon Islands (Huber, 1987)—and for Southern elephant seals as well (Arnbom *et al.*, 1994)—but not for the Northern elephant seals on Año Nuevo. This is more surprising because apparently females immigrate from Año Nuevo to the Farallon Islands precluding genetic differentiation, unless females emigrating from the high-density site (Año Nuevo) differ systematically in their genetics from females choosing to breed there. It also seems unlikely that females forage in totally different areas of the Pacific Ocean. Unfortunately, there are slight differences between the two study sites in details of observation methods and man-hours spent surveying the breeding beaches. The problems in comparing the data sets are increased by relatively high rates of tag loss (6% per year) and insecurity about the exact probability of observing tagged females (33% per annum rate for failure to identify tagged females present, according to Le Boeuf and Reiter, 1988) as well as witnessing their pupping events every year (roughly 15%; see discussion in Le Boeuf and Reiter, 1988). As missing observations had to be partly supplemented by reasonable assumptions, some of the results reviewed previously may still be open to reinterpretation.

During the course of the long-term studies on both islands (Huber, 1987; Le Boeuf and Reiter, 1988), population density (measured as number of pup births) increased tremendously. On the Farallon Islands pup numbers increased tenfold (from 35 in 1975 to 367 in 1982) and age at primiparity increased with this increased population density. As data were lumped over the study period (1975–1983) while age at primiparity was increasing, it is not entirely clear how the change in population density influenced the estimates of subsequent natality for females primiparous at different ages given that primiparous females appear most sensitive to density effects at the breeding beaches (Reiter *et al.,* 1981). If most of the females primiparous at 3 years of age were observed during the first years of the study and most of the females primiparous at 5 years late in the study, this could have confounded density effects and estimates of subsequent fertility of females primiparous at different ages. To demonstrate a cost of reproduction, ideally one should compare females of the same cohort, some of which were primiparous at age 3 and some at age 4 or later. This would reduce the effects of variance among years in foraging conditions as well as breeding density. However, it is obviously difficult to get and follow a large enough sample of individually known pups of one cohort to perform such a test.

Sydeman *et al.* (1991) showed that age had a significant influence on weaning success. This contradicts similar data for the animals on Año Nuevo (Deutsch *et al.,* 1994). In addition, the authors also pointed to an influence on weaning success of experience, defined as the number of pups successfully weaned by a female up to the time. Experience defined in this way also increased weaning success of females up to about 6–8 years of age but later in life appeared to decrease it. The authors used multivariate statistical methods to prize out the effect while keeping the influence of age and year constant. They interpret the apparent decline in weaning success as a "burnout" of females that have reproduced many times, thus explaining "senescent" reduction in reproductive success. The effect does not show up as reduced fertility in the subsequent year of females having reproduced in the previous year, in comparison to females that were barren. As mass was not measured concurrently and taken into account in the analyses of Sydeman *et al.* (1991), the issue cannot be considered resolved because mass increases in elephant seal females in close correlation with age and experience, and is highly likely to influence female fertility. Data by Le Boeuf and his group (Deutsch *et al.,* 1994; B. J. Le Boeuf, personal communication) from Año Nuevo show neither such burnout effects nor reduced fertility of parturient mothers in subsequent years. The problems with the interpretation of these data sets highlight the difficulties encountered when trying to deduce life-history consequences of reproduction from population data gathered over a long period of fairly variable years at changing population density. Furthermore, in a field study it is extremely difficult to measure many potentially important variables concurrently (especially mass of large samples of mothers and pups) without undue disturbance or reducing sample size too much.

Mortality effects of early reproduction were reported by Reiter and Le Boeuf (1991). They showed that females first reproducing at an age of 3 years had lower survival in the 4 following years than females primiparous at 4 years. No such cost was documented for Northern elephant seals on the Farallon Islands (Huber, 1987; Sydeman *et al.*, 1991). Sydeman and Nur (1994) questioned the conclusions of Reiter and Le Boeuf (1991) on the basis of statistical problems inherent in a regression analysis of dependent data. They also mention the biological problem that female survival after the year of first reproduction was included in the analysis for females primiparous at 4 years of age (P4) but not for those primiparous at 3 years of age (P3). In the year after first reproduction P3 females survived better than P4 females in their first year after first reproduction, biasing the results in favor of higher survival of the P4 females. However, another analytical method (the Cox Proportional Hazards Model, Sydeman and Nur, 1994) suggested at least the same trend ($p = .09$) as the original analysis by Reiter and Le Boeuf (1991). It seems likely that a larger sample size could provide conclusive evidence that females beginning reproduction early incur greater mortality costs than females primiparous at a later age. At present, the issue cannot be decided unequivocally either way.

Thus, the evidence on fertility costs of reproduction is mixed, suggesting such costs exist on one island but not on the other. Mortality costs are documented on one island but not on the other, and costs of reproduction in terms of reduced size at birth, growth, or survival to adulthood of pups of mothers having reproduced in the previous year have not been found. In comparison with data sets on other mammals, for example, red deer (Clutton-Brock *et al.*, 1986) observations on the Northern elephant seal to date provide only limited evidence of a cost of reproduction, and some of that evidence is still controversial. This being the best-studied phocid species, the evidence for parental investment in the sense of Trivers (1972) is indeed surprisingly limited in phocids, despite our detailed knowledge about the energetics of maternal care.

*2. Otariids*

Data on otariid pinnipeds are only slightly more clear-cut and exist only for two species, the Galápagos fur seal and the Antarctic fur seal.

Galápagos fur seal females often extend lactation beyond one year (Fig. 2b; Trillmich, 1986a). Females that reproduced in the previous year and are still accompanied by a yearling in the next breeding season have much reduced natality (Trillmich, 1986b). In contrast to nonlactating females that have natalities of 90%, females that had reproduced in the previous year had natalities of only 45%. The total cost of reproduction became even higher because females that gave birth to a pup when still accompanied by the previous year's young were very likely to lose the newborn to competition for milk by the older sibling (Trillmich, 1990). The competition between yearling and newborn was the fiercer the smaller the previous year's young was at the time of the birth of its (half) sibling. When yearlings had reached about 15 kg (roughly 50% of the mass of a full-grown female), they did not compete any more with the next pup of their mother but rather weaned without aggressive conflict with the mother at the time of the next pup's birth. If older young were particularly small, they sometimes competed even as 2 year olds with their mothers' newborn and in some cases thus killed two subsequent pups of their mother by (sometimes aggressively) outcompeting them. In this way, fertility of females was greatly reduced by previous reproduction (F. Trillmich, unpublished data). The degree of competition between the older young and the newborn critically depends on the diving abilities of the older young. Horning (1992) showed that diving abilities develop slowly in Galápagos fur seal young and are mainly related to body mass in young older than 12 months. Young with insufficient diving abilities cannot forage efficiently for themselves in years of low marine productivity and, therefore, depend partly on maternal milk for maintenance. In the Galápagos fur seal, no mortality cost of reproduction was measurable in females, perhaps due to small sample sizes spread over a number of highly variable years.

Costs of reproduction were also studied in a long-term investigation of the Antarctic fur seal by the team from the British Antarctic Survey (Croxall *et al.*, 1988; Lunn *et al.*, 1994; Boyd *et al.*, 1995a). They analyzed the reproductive performance of female Antarctic fur seals in relation to age, reproductive experience, and environmental variation over a period of 10 years (from 1983 to 1992) at Bird Island, South Georgia. There was a (nonsignificant, $p = .075$) trend that females first pupping at 3 years of age survived less well (69%) than 4-year-old primiparae (86%). For females of all ages, fertility costs were marked; parturient females were significantly less likely to pup in the subsequent year than previously nonpupping females. Natality of primiparae (independent of age at primiparity) was lowered more by previous pupping than natality of multiparae. Whereas 40% of the former did not give birth after producing a pup in the previous year, only 22% of the latter did so. Data on mortality costs of reproduction were similarly clear-cut (Boyd *et al.*, 1995a). Females that had given birth in a previous year not only suffered reduced fertility but also reduced survival to the next reproductive season. No effect was demonstrable of a reduction in female fecundity or survival, depending on whether or not females that had pupped had also successfully weaned their pup or had lost it early on in the lactation period (Lunn *et al.*, 1994; Boyd *et al.*, 1995a).

There seemed to be a senescence effect in Antarctic fur seals (Lunn *et al.*, 1994) as also shown by Trites (1991) for Northern fur seals. Beyond an age of 12 or 13 years, fertility of females declined in Antarctic fur seals from a high of 80% natality down to around 60% natality. Trites (1991) showed on a large sample that female Northern fur seals older than about

13 years carried smaller fetuses than prime females. These findings are reminiscent of the burnout effects described for Northern elephant seals by Sydeman *et al.* (1991). However, if female Northern fur seals that become very old are more likely to be small animals, the finding of Trites (1991) could also simply be due to the allometric effect that smaller mothers produce smaller young. Boyd *et al.* (1995a) also suggest an alternative interpretation for the lowered fertility in old Antarctic fur seal females that may equally apply to the Northern fur seal data. According to this hypothesis, females may follow one of two alternative life-history strategies. They may either reproduce early, nearly continuously every year, and die relatively young, or else start reproduction later and reproduce only every other year, but survive longer. This would lead to an accumulation of low-fertility females in old age classes and explain the apparent decrease in pregnancy rate of old fur seal females without postulating senescence. This explanation is based on the fact that sustained high fertility has substantial mortality costs in the Antarctic fur seal.

## B. Relationship between Energy Expenditure and Investment

In these fur seal systems, high energy transfer from mother to young means, somewhat paradoxically, low cost to the mothers. Young Galápagos fur seals become big yearlings and young Antarctic fur seals large weanlings in years when mothers find much food at sea, consequently are able to return to their young at short intervals, and thus transfer much energy per unit time (in the form of milk). In contrast, in poor years energy transfer is low, and much of the energy is spent by the pup on its maintenance costs rather than growth (i.e., the pup has a high metabolic overhead). This comes about because conditions of food scarcity force mothers to spend much time at sea searching for prey and to return less frequently to the pup to suckle (Costa *et al.*, 1989; Trillmich, 1990; Trillmich *et al.*, 1991; Lunn *et al.*, 1994). Thus, little energy transfer in this system means high cost to the mother (or low efficiency) that is expressed as reduced fertility of females in both species, and increased sibling competition between young of different cohorts in the Galápagos fur seal. In addition, Boyd *et al.* (1995a) demonstrated a positive correlation between growth rates of pups in year $n$ and survival of mothers to year $n + 1$ in Antarctic fur seals, again contrary to an expected trade-off between expenditure on one reproductive episode and survival to the next. Thus, the proportion of resources allocated to reproduction and self-maintenance may be nearly the same in poor and rich years, but in poor years energy available is neither enough to support high growth rates of pups nor to ensure high survival and fertility of females.

Why are fertility (and perhaps mortality) effects more pronounced in otariids than in phocids (if we can take the two otariids and one (or two) phocids studied as representatives of their respective families)? This question may relate to the difference between income and capital breeders (Drent and Daan, 1980). Costa (1991, 1993) has reviewed the energetics of foraging and lactation in phocids and otariids. Basically, his argument is that phocids are adapted to low metabolic expenditures (i.e., have low basal metabolic rate, BMR) and apparently do not work as hard as otariids while foraging. For example, to gather all the energy necessary for staying on land for 28 days while lactating, Northern elephant seal mothers only have to increase their foraging energy expenditure (estimated as about 1.3 times BMR by Le Boeuf et al., 1988) by 15% per day (i.e., to 1.5 × BMR) while at sea, to gain and store all the energy necessary for the 28 days lactation ashore. During the stay ashore, lactating females metabolized (stored) energy at about 2.4 × BMR (Costa et al., 1986). Another large phocid, the Weddell seal, for which foraging energetics was actually measured (Kooyman et al., 1973) worked at about 2.0 × BMR. Thus, these large phocids may live a life in the slow lane. Their leisurely pace of energy expenditure is made possible by the complete separation of gestation and foraging from lactation (remember that implantation is delayed and thus gestation is completely separated from the lactation period in phocids). The animals can, therefore, spend most of the year gathering stores for the short period of intense lactation. These, admittedly few, data suggest low metabolic rates for phocids. In addition, available data are biased towards large phocids, which makes a direct comparison with the data from much smaller otariids problematic. One value of 6.0 × BMR for a male harbor seal measured during the reproductive season (Reilly and Fedak, 1991) stresses the need for more data on the free-ranging metabolism of small phocids.

In stark contrast, otariids were reported to operate at much higher metabolic rates. Mothers have to forage and lactate more or less simultaneously, and, in many species with extended lactation, must bear the cost of gestation at the same time. Data on the Antarctic fur seal suggest the following energetics for the duration of the 120-day lactation period: a female stays with its pup for about 1.5–2.0 days on land at 3.4 × BMR and then forages for about 4–5 days at sea at about 6.0 × BMR (Costa et al., 1989). Even though estimates of foraging energetics are based on data from apparently poor years with low food availability and thus may overestimate average female energy expenditure, the data suggest a mean daily energy expenditure (DEE) of about 5 × BMR, a much higher sustained metabolic rate correlated with the income breeding strategy of otariids. As diving in otariids is also more expensive than in phocids (Costa, 1993), otariids maintain high metabolic rates throughout the year even when not caring for young. It has been questioned whether estimates of DEE based on doubly labeled water methodology might not consistently overestimate energy expenditure by about 30% (Boyd *et al.*, 1995b). The issue is certainly not settled. However, assuming the metabolic measurements are reliable, such high continuous energy expenditure is likely to lead to "exhaustion" of the physiological machinery, which may result in reduced fertility and survival subsequent to successful pupping. If in years of poor prey availability otariids need to increase DEE during lactation even more than normal, decreases in survival to the subsequent year and natality become probable. That such effects are more pronounced in otariids with their perhaps more "hectic" lifestyle (Costa, 1993) than in phocids with their low-metabolism strategy seems logical. Clearly, more data are needed on the energetics of phocid as well as otariid species covering a wider range of body masses and comparing different measurement methods. Only such a test can decide whether the suggested difference between the lifestyles of these two groups is a general phenomenon.

## V. Evidence for Differential Investment in Male and Female Offspring?

### A. Theoretical Considerations

In the highly polygynous pinnipeds males are usually of much larger size and can achieve potentially much larger maximal lifetime reproductive success than females (Bartholomew, 1970). The only species for which this is well documented is the Northern elephant seal (Le Boeuf, 1974; Le Boeuf and Reiter, 1988), but the suggestion is highly convincing for other species of land breeding polygynous phocids like the grey seal and for most otariids as well, even if documentation is still limited (Boness, 1991; Le Boeuf, 1991). Given the fact that large size may be of greater importance for male reproductive success than for female success, it has been suggested, referring to theoretical arguments by Trivers and Willard (1973) and Maynard Smith (1980), that females may invest more in male offspring than in female offspring. Differential investment in the sexes was to be expected given that large males were likely to achieve much greater reproductive success than small ones in the fierce competition for access to females.

Initially, an attempt was made to measure energetic components of maternal expenditure as proximate equivalents, it was hoped, of ultimate costs. It seemed logical to do so because in phocids females store all the nutrients and energy in their body that are needed for the entire duration of lactation.

It was assumed that depletion of maternal body stores over the period of lactation provided a good measure of her fitness costs. This was again a logical assumption, because the loss of the blubber layer down to a bare minimum could eventually compromise a female's ability to thermoregulate when she returned to the sea to regain her energy after the stressful period of lactation. A similar argument was made for otariid mothers that have to expend energy at a high rate during their foraging sojourns to maintain themselves and their pups, and achieve high growth rates of pups. However, as shown previously, maternal energy drain does not simply correlate with selective processes that determine fitness.

Tests of the theoretical ideas of differential investment in the sexes have generally looked for evidence of the following: (1) uneven sex ratio at birth and weaning, (2) unequal mass of the sexes at birth and weaning as well as the growth and sometimes metabolic rate of pups from birth to weaning, and (3) shifts in sex ratio at birth with maternal size or condition. Again, energetics was the main yardstick for measuring allocation. Let us first see how good the evidence is for differential maternal investment in sons and daughters and thereafter ask how the evidence fits to theoretical models (Frank, 1990).

## B. Phocid Seals

Among polygynous phocid seals, data on the Northern and Southern elephant seals and the grey seal are most complete and will serve here to focus on the problems encountered in the investigation of differential investment and the present status of the evidence. Similar data are much more scanty for the other phocids mostly due to the enormous logistical problems in investigating these species that breed far more dispersed and on less stable habitats. Evidence to data indicates no differential investment in the sexes, but the interesting cases of reversed sexual dimorphism have not been investigated in as much detail as the elephant seals and the grey seal.

### 1. Elephant Seals

In the Northern elephant seal, males are born about 7–8% heavier than females (on a sample of 35 newborns, but the difference was not significant; see also Kretzmann et al., 1993), even though sex ratio at birth was slightly, but significantly, biased in favor of males (51.8% males) (Le Boeuf et al., 1989). The authors documented that there was no sex ratio bias according to season or maternal mass. Determining differences in birth weight between male and female pups from a sample of pups for which the size of mothers is unknown is problematic as shown by Arnbom et al. (1994). As mass of pups at birth increases significantly with maternal mass in a nonlinear manner, slight biases in the size distribution of mothers of male and female pups sampled can confound conclusions about the existence of differences in birth mass. For the Southern elephant seal, Arnbom et al. (1994) showed convincingly on a large sample that at all maternal sizes male pups were larger than female pups at birth. So, it seems safe to conclude that in elephant seals male pups are born heavier than female pups.

At weaning, sex ratio in the Northern elephant seal was not different from 0.5 and male pups were larger than female pups (Le Boeuf et al., 1989). This finding was later questioned by Kretzmann et al. (1993). They showed that the energetic cost of rearing a son and a daughter was identical. Milk transferred and pup metabolic rates were the same for male and female pups, and resulted in identical growth rates of the sexes. Kretzmann et al. (1993) suggested that perhaps the finding of different mass of sons and daughters at weaning (Le Boeuf et al., 1989) was due to the fact that male pups are more likely to get adopted than female pups (Reiter et al., 1978). If so, the difference in weaning mass between male and female pups might be due to pup behavior, namely the greater persistence of male pups at milk stealing, and not to an allocation decision of mothers. Le Boeuf et al. (1989) found no difference in weaning age, as reported in an earlier paper (Reiter et al., 1978). In the Southern elephant seal, no difference in mass at weaning between sons and daughters was found by McCann et al. (1989). Arnbom et al. (1993) documented a significant sex difference in mass at weaning that disappeared, however, when they controlled for maternal mass. Their result again points out the danger of lumping values of pups whose mothers differ in mass. Maternal mass explained 55% of variation in weaning mass, whereas sex of the pup explained only 7%.

Young Northern elephant seal mothers were significantly less likely to raise sons to weaning than daughters (Le Boeuf et al., 1989). Interestingly, mothers that reared a pup successfully to weaning were more likely to be resighted in later years than unsuccessful mothers. As this difference is unlikely to reflect differential dispersal to other colonies (because these were under observation and animals did not appear in any other colonies), this finding indicates quality differences among females. High-quality females were successful at weaning pups and survived well, whereas females that were unsuccessful at weaning pups were more likely to die in the subsequent year. In addition, females that reared a pup in 1 year were more likely to rear one the next year than unsuccessful females (Le Boeuf et al., 1989). No differences in female fertility or survival according to pup sex were found.

Maternal mass has the most pervasive influence on pup mass at birth as well as at weaning in Northern and Southern elephant seals. As females almost double their mass from the time of primiparity to old age, this is not surprising. Nevertheless, there is little evidence that maternal mass influences the sex ratio. Arnbom et al. (1994) found for Southern elephant seals that very small primiparous females were less likely to produce male pups, but in this species as well as in Northern elephant seals sex ratio for older females was always very close to 0.5. However, their findings also suggest that male fetuses may demand more resources from their mothers during early development than female fetuses because for all female sizes sons were born heavier than daughters. This fits in with another curious finding in Southern elephant seals, that is, birth mass of female offspring correlated more closely with maternal mass than that of male offspring (McCann et al., 1989; Arnbom et al., 1994). In addition, Arnbom et al. (1994) found that the relative difference in birth mass between male and female pups was larger in small mothers. This finding can be interpreted to mean that prenatally male pups draw resources at a maximal rate from the mother, thus ensuring largest possible size at birth, whereas female fetuses draw resources in a way that pays more attention to the mother's ability to expend energy. A similar phenomenon is also documented for the Antarctic fur seal (see as follows).

In conclusion, data on the Northern and Southern elephant seals suggest that scaled to the size of the mother sons are born heavier than daughters, but thereafter demand the same amount of energy during lactation. Fitness costs to the mother seem to be the same whether she has raised a son or a daughter.

What about benefits to the pup? The data on the Northern elephant seal bring another problem into focus. It is usually assumed that greater investment in male pups will increase their reproductive value more than that of female pups. However, Le Boeuf et al. (1994) found no difference in the survival of pups correlated with mass at weaning. Basically, survival to 1 year of age did not correlate with mass at weaning if body mass surpassed a threshold of about 85 kg; above the threshold pup survival to 1 or 2 years was independent of weaning mass. Pup standard length, however, correlated with survival to 1 year of age, but not to 2 years of age. In summary, there is no clear evidence that greater energy transfer to pups would necessarily increase their survival.

Furthermore, studies by Haley et al. (1994), Clinton (1994), and Le Boeuf et al. (1994) give an impression of the complications involved in assessing the often-made assumption of greater size at weaning being beneficial to later reproductive value. Within the size class of fully adult male elephant seals, actual body size mattered little for reproductive success (as estimated from the number of observed copulations). Large size was important to get into the ranks of dominant, reproductively active males, but within this class dominance in behavioral interactions with other males was more important than actual size (see also Tinker *et al.*, 1995 for similar data on grey seals).

Male elephant seals reach their final size by a tremendous growth spurt from the age of 3–5 years (Clinton, 1994). The growth spurt during the late juvenile and early subadult years contributes most to final size, and the growth during this period is likely to be influenced more by foraging conditions than by the maternal provisioning long ago. Consequently, size at weaning is certainly important as a threshold trait for survival during the first year. However, male pups may not benefit more from additional maternal expenditure than female pups (Le Boeuf *et al.*, 1989). This observation undermines the most frequently made assumption that additional investment in male pups is likely to produce greater fitness returns than the same additional investment in female pups.

## 2. Grey Seals

Seasonal changes in sex ratio had been documented for grey seals by Coulson and Hickling (1961; see also Anderson and Fedak, 1987; and for Weddell seals by Stirling, 1971). Generally, females arriving early are larger and more likely to give birth to male pups than late-arriving females (Anderson and Fedak, 1987). The date-related shift in sex ratio could arise if late-arriving females were predominantly young or in poor condition. They may then more likely have lost a male than a female fetus because of the male's greater energy demands during development.

What evidence is there for differential maternal energy expenditure on the sexes? Kovacs and Lavigne (1986b) suggested that males were born heavier than females, grew faster, and weaned heavier. This conclusion was based on regression analysis of data from pups whose age was estimated from external appearance. Their conclusions were criticized by Bowen *et al.* (1992b), who showed for a similar data set, that there was no difference in birth mass, growth rate, and weaning mass of grey seal young on Sable Island. Indeed, the growth rates measured by Kovacs and Lavigne (1986b) were not statistically different, only the intercepts of growth curves differed. However, Anderson and Fedak (1987) also reported higher growth rates of male pups. The comparisons are complicated by the fact that the studies of Kovacs and Lavigne (1986b) and Anderson and Fedak (1987) were made in Britain, whereas the study of Bowen *et al.* (1992b) was made in Canada. Nevertheless, Bowen *et al.* (1992b) argue convincingly that even if sex differences in birth mass are accepted, there is little evidence that the difference in growth rates are important for later reproductive value of these young as was argued by McCann *et al.* (1989) and Le Boeuf *et al.* (1989) for elephant seals (see also Campagna *et al.*, 1992).

Given the potential confounding effects of year (Kovacs and Lavigne worked in 1982 and 1983, Anderson and Fedak in 1979–1981, and Bowen et al. in 1987), populations (Britain versus Canada), and different methodologies, one can only follow Bowen et al. (1992b) and conclude that based on current data there is little evidence for differential maternal investment in grey seals. This is even more true because absolutely no data are available on fitness costs of maternal expenditure. Here, as in elephant seals, fitness benefits of higher-than-average maternal expenditure to a pup are also not documented. Given that surviving grey seal pups must almost quadruple mass before reproducing, mass at weaning may have little influence on later reproductive value of a pup beyond enabling its initial survival (Bowen et al., 1992b; following McCann et al., 1989; and Le Boeuf et al., 1989). The best that mothers may be able to do is "to wean a healthy pup regardless of sex" (Bowen et al., 1992b). However, this conclusion does not explain the fact that pups here as in elephant seals appear to be born heavier if their sex is male.

C. Fur Seals and Sea Lions

What about the evidence for differential investment in the sexes in otariids? Kretzmann et al. (1993) claim that in otariids the evidence for differential investment in the sexes is much better. This conclusion appears premature. In many otariid species it is well established that male pups are born heavier than females. The difference in birth mass amounts to about 10–15%. For example, in the Antarctic fur seal mean male mass at birth varied in the period from 1972 to 1991 between 4.9 and 5.9 kg, and female mass varied between 4.4 and 5.3 kg. The difference in mean mass of the sexes at birth amounted to between 500 and 700 g in these years (Lunn et al., 1993). Data for other fur seals and sea lions look generally quite similar, but data sets are not as extensive as this one (Gentry et al., 1986; Trillmich, 1986b; Croxall and Gentry, 1987; Ono et al., 1987). After birth, male pups tend to grow faster than female pups. This was shown in a few longitudinal studies (Trillmich, 1986b; Lunn et al., 1993; Goldsworthy, 1995 where individual pups were repeatedly weighed, but most growth rates are based on cross-sectional data in which groups of unknown pups were weighed at different mean ages. The latter method may overestimate differences in growth rates due to biased sampling as shown by Lunn et al. (1993) for the Antarctic fur seal.

A difference in growth rate between the sexes could logically come about by: (1) differences in milk intake on the income side, or (2) differences in metabolic rate, (3) activity, and (4) caloric value of tissue deposited on the side of expenditure. There is clear evidence for (1) from studies on Northern fur seals (Costa and Gentry, 1986), Galápagos fur seals (Trillmich, 1986a,b), and California sea lions (Oftedal et al., 1987b). Studies of milk intake using labeled water methodology or weighing of pups before and after sucking showed that male pups had greater intake than female pups. However, at equal mass male and female pups of Northern fur seals (Costa and Gentry, 1986) and California sea lions (Oftedal et al., 1987b) had the same milk intake, whereas in Antarctic fur seals (Goldsworthy, 1995) male pups had higher milk intake than female pups even when mass was the same. There was a hint for a contribution of the factor (2) that male pups may be more efficiently growing in Northern fur seals than female pups (Costa and Gentry, 1986), but the same was not obvious in California sea lions (Thompson et al., 1987). No clear differences in (3) overall activity of pups while ashore was found in any species even though male pups show more play fighting than female pups, for example, in the Steller sea lion (Gentry, 1974). At the risk of slightly overgeneralizing, one might say that male otariids are born heavier than females, grow slightly faster, but get about the same amount of milk at the same body size as female pups. Arnould et al. (1996) showed that differences in body mass (in Antarctic fur seal pups) do not translate into differences in energy content because (4) male pups had relatively more lean mass than female pups, whereas female pups had greater adipose stores.

Weaning is hard to document in otariids. Only in the subpolar fur seals is it clearly observable because they wean within a short period when about 4 months old (Fig. 2b). In all the other species weaning is a protracted process that can last for longer than a year during which young begin to forage independently (Trillmich, 1990; Bowen, 1991; Horning, 1992). It is, therefore, generally difficult to say if male and female young are weaned at different ages. Lunn et al. (1993) were able to show that male pups were weaned later in the Antarctic fur seal, but the difference is only about 5 days out of a 120-day lactation period. Francis and Heath (1991) suggested that in California sea lions females may actually be suckled longer than males due to the latter's habit of migrating away from the islands of birth. Thus, in the California sea lion, greater maternal expenditure early on in male pups may be balanced by longer maternal care for female offspring.

Growth and duration of lactation in many of the temperate and tropical otariid species appear to be strongly influenced by marine productivity in a given year (Trillmich, 1986b, 1990; Trillmich and Ono, 1991), but much less by differences between the sexes. Strong environmental influences on growth rate were also documented for the Antarctic fur seal (Lunn et al., 1993; 1994; Boyd et al., 1995a). They found that year-to-year variation explained 36% of the variation in growth rate of pups, and 45% of the variation in weaning mass, whereas sex of the pup explained only 6% of the latter, and maternal characteristics (duration of foraging trips, age, date of birth of pup, duration of nursing visits) explained only 11% of the variance. This finding highlights the great extent of the dependence of the otariid income breeding strategy on local food abundance.

Overall, these data give no evidence whatsoever of differential maternal investment in male and female pups, but suggest perhaps slightly greater energy expenditure on males. No influence on future maternal fertility or mortality depending on the sex of the pup has been found in any species and this is perhaps not very surprising given the overwhelming influence of maternal size and particularly of year-to-year variation on the cost of pup rearing as explained previously (Section IV,B). All of this can best be summarized in the words of Le Boeuf *et al.* (1989):

> In no pinniped has it been shown that higher energetic costs of producing sons translates to higher reproductive costs for the mother. This is important because selection acts on the reproductive costs of breeding, the effects on future reproductive performance. Maternal energetic costs associated with reproductive effort are only significant if they correlate with reproductive costs, and in elephant seals the association is tenuous.

### D. Comparison of the Data with Sex Ratio Theory

How do the empirical findings fit to theories of sex ratio evolution? I will follow Frank's (1990) review of sex ratio theory in highlighting a few of the problems that arise when one attempts to test predictions or assumptions of the theory with data from pinnipeds. My basic conclusion is that theories are presently not sufficiently rich in natural history detail to map in a useful way onto pinniped data, and looking at the problem the other way around, that data on pinnipeds are not yet substantial enough to test some of the most basic assumptions of the theory.

Most commonly, authors have compared their data to Fisher's theory of sex allocation. In Fisher's (1930) theory, mothers have a fixed amount of resource to spend on offspring and should allocate resources in such a way as to produce the same (fitness) return through male and female offspring. When expending resources on offspring, increases in return are modeled as linearly related to expenditure (Frank, 1990; see Fig. 5, b). None of these assumptions fits easily to the facts.

Firstly, research has amply shown that maternal resources vary tremendously from year to year and depend in a predictable way: (1) on marine productivity, and (2) on maternal size. Thus maternal resources cannot adequately be modeled as a finite, fixed "resource cake" that can be distributed in an optimal way over a lifetime of reproduction. Secondly, the problem of the nature of the relation of expenditure to returns, whether it is linear or nonlinear, cannot be addressed at present because the crucial

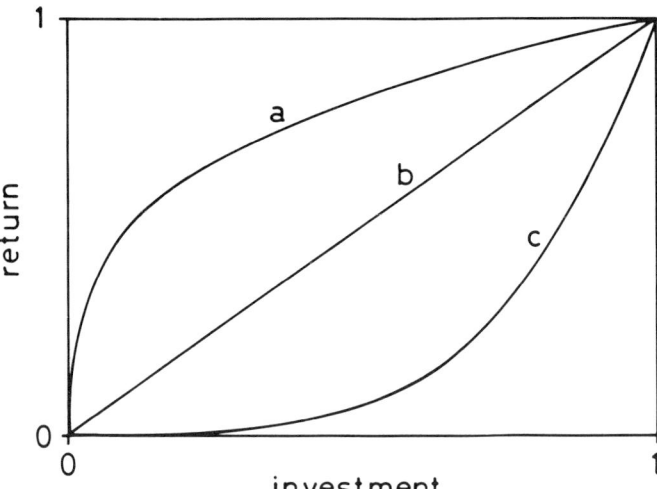

FIG. 5. Possible relationships between parental expenditure (or investment) and expected reproductive success of offspring (here called "returns," after Frank, 1990). Only in **b** is increase in returns linearly related to maternal expenditure, the condition necessary for Fisher's equal allocation argument to apply. If functions are nonlinear as in **a** or **c** (and of course many other shapes, e.g., sigmoidal, logistic, etc. could potentially be realized), sex allocation and sex ratio are expected to differ from 1:1.

data about the dependence of survival and later reproductive success on mass at weaning are missing or suggest a threshold rather than a linear relationship. Thirdly, if resources are expended nearly equally on the sexes, with perhaps slightly higher expenditure on males as the data appear to suggest, Fisher's model would predict biases in the sex ratio favoring females. The opposite is the case. Most pinniped species produce a slight preponderance of males (at birth). Part of this preponderance may derive from errors in sexing pups at birth. Usually there errors are not random, but more commonly a female is sexed as a male (my own observations on Galápagos fur seals; and I. L. Boyd, personal communication, on the Antarctic fur seal). However, sexing errors are unlikely to explain all of the bias in favor of males. Thus the contradiction is likely to remain.

Basically, these observations already show that it is presently impossible to map empirical data in an adequate manner onto the models. This also applies to the second theoretical idea that has repeatedly been quoted by authors. This is the idea of Trivers and Willard (1973) that mothers in good condition should produce the sex that needs more expenditure to produce high returns. Frank (1990) mentions a few corollaries of this hypothesis that have largely been ignored in the attempts to test the ideas on pinnipeds. The Trivers and Willard model assumes nonlinear returns with expenditure (at least for one sex, see Fig. 5, a or c). As stated previously, none of the species has been investigated in sufficient detail to address the question of the form of the return curve. Also, if the Trivers and Willard model applies, unequal sex ratios are expected populationwide, particularly in low-fecundity organisms like pinnipeds (Frank, 1986), which is obviously not the case. Also, despite the enormous variation in maternal mass within a given species, and the great influence this has on pup size at birth and at weaning, sex ratio did in no species covary with maternal mass.

This brings up the third theoretical model by Maynard Smith (1980). He assumes that sex ratio is fixed (by the mechanism of meiosis) and predicts that unequal investment in the sexes can be an evolutionary stable strategy (ESS) under those circumstances if for a given investment one sex survives better than the other (Frank (1990) questions whether this is a robust result). Maynard Smith also showed that increased investment in males was stable if male viability or success for a given level of investment decreases as the population-wide investment in males increases. Both his predictions may apply, but it is clearly impossible to test them with presently available data (even though I have claimed to be able to do that with data on the Galápagos fur seal; Trillmich, 1986b).

Thus, serious tests of sex ratio theories are problematic. This is no reason for despair, but it implies that it may be more useful at present to study in more detail, for example, the basic relationship between maternal expenditure and pup survival and later reproductive success. It would be extremely helpful to know if this relationship differs between the sexes. Such research might also provide access to the puzzling question of why the life history of pinniped mothers involves doubling of body mass during the reproductive career, a fact that greatly influences pup development, but that is hard to understand as a reproductive adaptation. Maybe it is a side effect of greater foraging efficiency of a larger diving mammal?

Lastly, I would suggest that it may be very fruitful to look at pups (or any young) not only as passive vessels into which maternal expenditure is being poured, but as active players in the game for maternal investment (Trivers, 1974). If male and female young, for whatever reason, have different needs and are able to communicate these to the mother, mothers will be selected to respond within the limits of their ability to the signals of their young (Godfray, 1991). This could explain why in so many pinniped species male and female fetuses extract different amounts of resources from their mothers during pregnancy without demonstrable differential fitness consequences to the mother. Again, this could be due to differences in body composition at birth (Arnould et al., 1996).

As suggested previously, data are already available that indicate that offspring pay attention to their mothers' condition and extract resources accordingly (for the corresponding theoretical argument, see Godfray, 1991). Alternatively, offspring of different sizes or sexes may be constrained to some degree by the varying condition of their mothers or else mothers may decide to hand out resources to offspring according to their own body condition. In the investigated phocids and otariids, males are apparently selected to extract more of the maternal resources than females (Reiter *et al.*, 1978; Costa *et al.*, 1988; Boyd and McCann, 1989; Arnbom *et al.*, 1993, 1994). This is evident from greater male birth mass, higher male than female growth rates in several species, and greater persistence in milk stealing of male elephant seal pups. It is also Goldsworthy's (1995) explanation for the influence of pup sex on maternal foraging pattern in Antarctic fur seals. Data by Costa *et al.* (1988) and Boyd and McCann (1989) give, at first sight, a confusing account of the relationship between sex of the pup and the relation of its mass at birth to maternal mass in the Antarctic fur seal. As shown in Fig. 4b, pup mass generally increases with maternal mass. However, when the authors analyzed the correlation between maternal mass and pup mass for the two sexes separately, they came to different conclusions about the details of this correlation. Whereas Costa *et al.* (1988) found a better correlation between maternal mass and female pup mass than male pup mass, Boyd and McCann (1989) found no correlation between female mass and maternal mass, and a clear correlation between male mass and maternal mass. However, this confusion can perhaps be resolved if we take into account that maternal mass differed greatly between these two data sets that come from the same site, but were taken in years of widely differing food abundance. The mothers in the data set of Costa *et al.* (1988) were observed when food was scarce and had a parturition mass of only 39.4 kg, whereas the mothers in the study of Boyd and McCann (1989) were in much better condition with a mean mass of 47.4 kg, about 20% heavier. If we assume that pup mass at birth increases in a nonlinear, perhaps sigmoidal way with maternal mass for both sexes, but that male fetuses extract more and, therefore, follow a curve that is shifted up (to higher birth mass) and to the right (to higher maternal mass), the results can be reconciled. At low maternal mass female fetuses are already in a region where pup mass increases steeply with maternal mass and, therefore, they acquire resources in relation to maternal mass, but at high maternal mass the relationship has reached the asymptote and female fetus mass is nearly independent of maternal mass. In contrast, male fetuses extract the maximum possible at low maternal mass, that is, their birth mass in relation to maternal mass is still at the flat initial part of the relationship in the Costa *et al.* (1988) data. When mothers are in better condition, as in Boyd and McCann's (1989) data, they are on the steep part of the relationship between maternal mass and male fetus mass and, therefore, their mass correlates well with maternal mass. If this hypothesis were correct, male pup mass should again become largely independent of maternal mass in mothers that are particularly large and/or in particularly good condition. Such a hypothesis could also explain that male fetuses are more likely to die or to be aborted when mothers are in very poor condition as after El Niño events (Trillmich and Limberger, 1985; Ono and Boness, 1991), resulting in a slightly female biased sex ratio at birth among the low number of pups born after these major environmental disturbances. The explanation offered here agrees with the finding of Arnbom *et al.* (1994) that small Southern elephant seal mothers are unlikely to produce male pups at all, that the relative difference in birth mass between sons and daughters is highest in small mothers, and that birth mass of male and female pups levels off at high maternal mass.

## VI. Postweaning Investment?

The strong bond between a mother and her offspring dissolves at weaning, quite abrupt in phocids and more gradual in the otariids. In most cases it is obvious that mother and pup separate widely around this time (see aforementioned). However, very little is known about the possibility that parent and offspring may meet later on in life. The great site fidelity of many seal species would make such a meeting between a mother and her female offspring likely, particularly in the land breeding species of otariids and phocids, but it seems possible in fast ice breeding species as well (e.g., the Weddell seal). When the bond between mother and juvenile is weakened very gradually, as in the odobenids and in many sea lions and fur seals that stay year round or for a major proportion of the year in their rookeries, mothers and daughters might well recognize each other years after weaning, making the establishment of a special relationship possible. Individual recognition has been shown in many pinniped species (Trillmich, 1981; Roux and Jouventin, 1987; Schusterman *et al.*, 1992; Gisiner and Schusterman, 1991) and could serve as a basis for reunions or preferential interactions when offspring meet their mothers years after weaning.

Hanggi and Schusterman (1990) provided the first evidence that long-term bonding based on imprinting like phenomena (Schusterman *et al.*, 1992) can indeed exists. In their study of a captive group of California sea lions, they observed differential interactions among females along kinship lines. Females interacted highly preferentially in an amicable way with relatives and much more aggressively with nonrelatives. Such relationships lasted over many years, for example, involving mature breeding daughters and their mothers. Even more interestingly, offspring of the same female born in different years seemed to acquire the same interaction biases, preferentially interacting with siblings even if they belonged to a different cohort. If such phenomena exist and are of equal importance under field conditions as under the captive conditions investigated in their study, interactions among relatives may be a major structuring factor in pinniped colonies. By being larger, mothers can potentially provide protection to their daughters against aggressive approaches by unrelated larger females. This may increase the reproductive success of daughters when these pup for the first time. Similarly, it seems conceivable that adoptive suckling is structured along matrilines. It is presently unknown if any of this represents maternal investment in the sense of Trivers (1972).

If these possibilities were realized, pinniped colonies may have a matrilineal structure where females in a particular site are more closely related than expected by chance. If so, this would introduce an important element of structuring into the seemingly chaotic, large pinniped colonies. Modern techniques of DNA analysis should allow us to approach these questions. In particular, they may show female–female interactions and the role of female choice in pinnipeds (Amos *et al.*, 1994) to be far more important in structuring sociality of pinnipeds than is evident at present.

## VII. Conclusion

As this article has shown, pinnipeds are a group of species for which excellent data on maternal strategies exist. Energetic expenditure of mothers for nursing their young is high and in phocids is also highly concentrated in time. Energetics of free-ranging animals have been measured in unusual detail in a number of sepcies. Moreover, sexual size dimorphism is extreme in the best-investigated species. Finally, for Northern elephant seals lifetime reproductive success and its correlation with size has been studied in detail for females as well as for males. Thus, at first sight, the group seemed to offer striking patterns of maternal care as well as strong evidence for extreme sexual selection on males and, therefore, promised relatively large and easier to analyze effects of parental investment than many other mammalian groups. What has the work on this group to contribute to our knowledge of parental investment strategies in mammals in general and to the theoretical questions about parental investment, parent–offspring conflict, and sex allocation?

One generally important point is the repeated finding that factors contributing to the explanation of variance in juvenile development to independence and maternal care patterns are ranked in the sequence of decreasing importance from ecological conditions, via maternal size, age, and experience, to sex of offspring as the explanatory variable that explains least of the variance. This does not detract from the importance of sex allocation questions, but it puts them into perspective. Environmental variance in resource abundance may be overwhelming in relation to quality differences among females and differences in allocation to the sexes and obscure underlying trade-offs influencing maternal reproductive strategy.

The question of benefits of maternal size is largely unresolved. In all pinnipeds, mothers continue to grow during their reproductive career. The importance of this growth might lie: (1) in its significance for the diving, that is, foraging, performance of a marine mammal; (2) in the role of size in social interactions and of size-related dominance on land during the breeding period; or (3) in the reduction of a physiological cost of reproduction by reducing relative maternal expenditure on reproduction. These explanations are not mutually exclusive and it is a challenge for future studies to determine the relative fitness contribution of the three factors. Size-related differences in maternal strategy are but one aspect of variability in maternal strategy. Interspecific and intraspecific variability, for example, in time to weaning, and the width of the reaction norms of components of maternal strategy (or the decision options open to mothers) need themselves to be considered adaptations of maternal strategy to more or less unpredictably variable environments. Similar considerations will certainly also influence the maternal strategies of other mammals and need to be addressed in the context of life-history evolution (Stearns, 1989).

The influence of maternal mass on offspring mass at birth makes direct comparison of birth mass of sons and daughters difficult unless samples are well balanced with respect to maternal size and age. This applies similarly to growth rate and weaning mass of young. More serious than these methodological problems is the failure to document clear benefits of additional maternal expenditure to young above a certain threshold size or mass. Presently, we cannot decide whether this finding on elephant seals is general or is perhaps an artifact of lumping animals from different cohorts and different-size mothers. However, it stresses that we cannot even take the simplest assumption about mammalian maternal investment for granted, namely that additional maternal expenditure will automatically be converted into increased benefit to young.

Perhaps even more unexpected is the finding that differences in mass cannot be equated with differences in energy content of young. In pinniped young, males and females may follow different growth strategies perhaps beginning prenatally, but certainly during the lactation period as shown for the Antarctic fur seal. Sexual differences in the growth of body components, like muscle mass and adipose stores, can greatly influence growth trajectories of juveniles and make the interpretation of body mass growth as well as comparison of sizes between the sexes much more complex than one would naively assume. Such complications may well apply for many other mammals and could potentially contribute substantially to, until now, unexplained variance. Taking such proximate, physiological facts into account might even change our perspective of the nature of the ultimate problem to be explained.

Similarly, the question of how weaning mass correlates with adult mass that arose in the discussion of maternal investment in grey seals and elephant seals, complicates the interpretation of fitness effects of maternal care. The tremendous growth spurt of males later in life around puberty is perhaps more important for their final adult size than size at weaning. This possibility calls for a reassessment of an often-made assumption about the correlation of weaning mass with adult mass. Size increase during the growth spurt will depend on food abundance at that time. If this is more closely correlated with final adult size than size at weaning, then selection on mothers to invest more in sons than in daughters might be much weaker than assumed so far. All of these findings highlight that energetic measures, however awe-inspiring, must be scaled carefully to ecological conditions and maternal characteristics before they can be considered useful fitness correlates. The hope that energy expenditure might provide a direct correlate of fitness currency was certainly not born out.

The limited data on the development of pinniped young to independence suggest that close communication is needed between mother and offspring to ensure an optimal weaning time that is neither too late for the mother nor too early for the offspring. Much of what is routinely interpreted as mother–offspring conflict about parental investment seems to be absent in a few pinniped species (like the Northern and Antarctic fur seals) and in others can be interpreted in terms of costly handicap communication (Grafen, 1990; Godfray, 1991). These observations reinforce the view repeatedly voiced lately (Mock and Forbes, 1992; Bateson, 1994) that we need to reevaluate the evidence for parent–offspring conflict and study underlying ultimate and proximate causes before jumping to conclusions. Detailed study of the development of offspring foraging options are likely to provide new insights into the mechanisms and the ultimate factors influencing the weaning process, and may enable better prediction of the costs and benefits for mother and offspring.

Lastly, this group seemed particularly well suited to test sex ratio theory and even theoreticians (Frank, 1990) often pointed to pinniped examples to illustrate theoretical assumptions or preditions. The careful study of the most spectacularly dimorphic species has shown that neither the assumptions nor the predictions of the theory can reliably be mapped onto data. Critical data (like return curves for parental investment) are still missing despite tremendous efforts expended to get the empirical data. On the other hand, simplifying assumptions of the theory about resources (e.g., the famous fixed resource cake) available for allocation are simply unrealistic in the face of complicating facts of life history (e.g., maternal growth) and ecology (i.e., variance in food abundance). This lesson from the study of pinnipeds will likely apply to many other mammals as well. More specific and detailed modeling of sex allocation decisions appears necessary to reach a better understanding of selection on mammalian maternal sex allocation strategies. Obviously, such models might greatly gain from better knowledge of underlying mechanisms (Krackow, 1995). First steps have been made in this direction by Frank (1986). Overall, testing assumptions of models is presently just as productive as testing predictions.

### Acknowledgments

I would like to thank R. L. Gentry, B. J. Le Boeuf, and I. L. Boyd for providing information on unpublished data. R. F. Benus, D. J. Boness, I. L. Boyd, and T. Meijer critically read the manuscript. Daryl Boness and Ian Boyd kindly shared their insights into pinniped maternal strategies and energetics with me and pointed out where my generalizations were not sufficiently supported by data. I am deeply grateful for their input. Remaining errors in overstretching few available data are clearly mine. A. York helped me in the interpretation of data on the Northern fur seal. E. Geißler drew the graphs, and P. Bergen helped with typing and making sure the references in the text and in the reference list tally. My sincere thanks to them all.

This paper is dedicated to Wolfgang Wickler on the occasion of his 65th birthday in deep gratitude for his wholehearted, longtime support of my Galápagos field work.

### References

Alexander, R. D., Hoogland, J. L., Howard, R. D., Noonan, K. M., and Sherman, P. W. (1979). Sexual dimorphism and breeding systems in pinnipeds, ungulates, primates, and humans. *In* "Evolutionary Biology and Human Social Behavior" (N. A. Chagnon and W. Irons, eds.), pp. 432–435. Duxbury Press, North Scituate, MA.

Amos, W., Twiss, S., Pomeroy, P. P., and Anderson, S. S. (1994). Male mating success and paternity in the grey seal, *Halichoerus grypus:* A study using DNA fingerprinting. *Proc. R. Soc. London, Ser. B* **252,** 199–207.

Anderson, S. S., and Fedak, M. A. (1987). Grey seal, *Halichoerus grypus,* energetics: Females invest more in male offspring. *J. Zool. (London)* **211,** 667–679.

Arnason, U., and Widegren, B. (1986). Pinniped phylogeny enlightened by molecular hybridization using highly repetitive DNA. *Mol. Biol. Evol.* **3,** 356–365.

Arnbom, T., Fedak, M. A., Boyd, I. L., and McConnell, B. J. (1993). Variation in weaning mass of pups in relation to maternal mass, postweaning fast duration, and weaned pup behaviour in Southern elephant seals (*Mirounga leonina*) at South Georgia. *Can. J. Zool.* **71,** 1772–1781.

Arnbom, T., Fedak, M. A., and Rothery, P. (1994). Offspring sex ratio in relation to female size in Southern elephant seals, *Mirounga leonina. Behav. Ecol. Sociobiol.* **35,** 373–378.

Arnould, J. P. Y., and Boyd, I. L. (1995). Temporal patterns of milk production in Antarctic fur seals (*Arctocephalus gazella*). *J. Zool. (London)* **237,** 1–12.

Arnould, J. P. Y., Boyd, I. L., and Socha, D. G. (1995). Milk consumption and growth efficiency in Antarctic fur seal (*Arctocephalus gazella*) pups. *Can. J. Zool.* **74,** 254–266.

Barlow, G. W. (1972). A paternal role for bulls of the Galapagos Islands sea lion. *Evolution* **26,** 307–308.

Barlow, G. W. (1974). Galapagos sea lions are paternal. *Evolution* **28,** 476–478.

Barnes L. G., Domning, D. P., and Ray, C. E. (1985). Status of studies on fossil marine mammals. *Mar. Mamm. Sci.* **1,** 15–53.

Bartholomew, G. A. (1952). Reproductive and social behaviour of the Northern elephant seal. *Univ. Calif. Publ. Zool.* **47,** 369–429.

Bartholomew, G. A. (1970). A model for the evolution of pinniped polygyny. *Evolution* **24,** 546–559.

Bartholomew, G. A., and Hoel, P. G. (1953). Reproductive behavior of the Alaska fur seal, *Callorhinus ursinus. J. Mammal.* **34,** 417–436.

Bateson, P. (1994). The dynamics of parent-offspring relationships in mammals. *Trends Ecol. Evol.* **9,** 399–403.

Berta, A., Ray, C. E., and Wyss, A. R. (1989). Skeleton of the oldest known pinniped, *Enaliarctos mealsi. Science (Washington, D. C.)* **244,** 60–62.

Boness, D. J. (1990). Fostering behavior in Hawaiian monk seals: Is there a reproductive cost? *Behav. Ecol. Sociobiol.* **27,** 113–122.

Boness, D. J. (1991). Determinants of mating systems in the otariidae (Pinnipedia). *In* "Behavior of Pinnipeds" (D. Renouf, ed.), pp. 1–44. Chapman & Hall, London.

Boness, D. J., Bowen, W. D., and Oftedal, O. T. (1994). Evidence of maternal foraging cycle resembling that of otariid seals in a small phocid, the habor seal. *Behav. Ecol. Sociobiol.* **34,** 95–104.

Bonner, W.N. (1984). Lactation strategies in pinnipeds: Problems for a marine mammalian group. *Symp. Zool. Soc London* **51,** 253–272.

Bowen, W. D. (1991). Behavioural ecology of pinniped neonates. *In* "Behaviour of Pinnipeds" (D. Renouf, ed.), pp. 66–127. Chapman & Hall, London.

Bowen, W. D., Oftedal, O. T., and Boness, D. J. (1985). Birth to weaning in 4 days: Remarkable growth in the hooded seal, *Cystophora cristata. Can. J. Zool.* **63,** 2841–2846.

Bowen, W. D., Oftedal, O. T., and Boness, D. J. (1992a). Mass and energy transfer during lactation in a small phocid, the habor seal (*Phoca vitulina*). *Physiol. Zool.* **65,** 844–866.

Bowen, W. D., Stobo, W. T., and Smith, S. J. (1992b). Mass changes of grey seal *Halichoerus grypus* pups on Sable Island: Differential maternal investment reconsidered. *J. Zool. (London)* **227,** 607–622.

Boyd, I. L., and McCann, T. S. (1989). Pre-natal investment in reproduction by female Antarctic fur seals. *Behav. Ecol. Sociobiol.* **24,** 377–385.

Boyd, I. L., Croxall, J. P., Lunn, N. J., and Reid, K. (1995a). Population demography of Antarctic fur seals: The costs of reproduction and implications for life-histories. *J. Anim. Ecol.* **64,** 505–518.

Boyd, I. L., Woakes, A. J., Butler, P. J., Davis, R. W., and Williams, T. M. (1995b). Validation of heart rate and doubly labelled water as measures of metabolic rate during swimming in California sea lions. *Funct. Ecol.* **9,** 151–160.

Campagna, C., Le Boeuf, B. J., Lewis, M., and Bisioli, C. (1992). Equal investment in male and female offspring in Southern elephant seals. *J. Zool. (London)* **226,** 551–561.

Clinton, W. L. (1994). Sexual selection and growth in male Northern elephant seals. *In* "Elephant Seals" (B. J. Le Boeuf and R. M. Laws, eds.), pp. 154–168. Univ. of California Press, Berkeley, CA.

Clutton-Brock, T. H. (ed.) (1988). "Reproductive Success." Univ. of Chicago Press, Chicago.

Clutton-Brock, T. H., Albon, S. D., and Guinness, F. E. (1986). Great expectations: Dominance, breeding success and offspring sex ratios in red deer. *Anim. Behav.* **34,** 460–471.

Costa, D. P. (1987). Isotopic methods for quantifying material and energy intake of free-ranging marine mammals. *In* "Marine Mammals Energetics" (A. C. Huntley, D. P. Costa, G. A. J. Worthy, and M. A. Castellini, eds.), pp. 43–66. Allen Press, Lawrence, KS.

Costa, D. P. (1991). Reproductive and foraging energetics of pinnipeds: Implications for life history patterns. *In* "Behaviour of Pinnipeds" (D. Renouf, ed.), pp. 300–344. Chapman & Hall, London.

Costa, D. P. (1993). The relationship between reproductive and foraging energetics and the evolution of the Pinnipedia. *Symp. Zool. Soc. London* **66,** 293–314.

Costa, D. P., and Gentry, R. L. (1986). Free-ranging energetics of Northern fur seals. *In* "Fur Seals: Maternal Strategies on Land at the Sea" (R. L. Gentry and G. L. Kooyman, eds.), pp. 79–101. Princeton Univ. Press, Princeton, NJ.

Costa, D. P., and Trillmich, F. (1988). Mass changes and metabolism during the perinatal fast: A comparison between Antarctic (*Arctocephalus gazella*) and Galápagos fur seals (*Arctocephalus galapagoensis*). *Physiol. Zool.* **61,** 160–169.

Costa, D. P., Le Boeuf, B. J., Ortiz, L., and Huntley, A. C. (1986). The energetics of lactation in the Northern elephant seal. *J. Zool. (London)* **209,** 21–33.

Costa, D. P., Trillmich, F., and Croxall, J. P. (1988). Intraspecific allometry of neonatal size in the Antarctic fur seal (*Arctocephalus gazella*). *Behav. Ecol. Sociobiol.* **22,** 361–364.

Costa, D. P., Croxall, J. P., and Duck, C. D. (1989). Foraging energetics of Antarctic fur seals in relation to changes in prey availability. *Ecology* **70,** 596–606.

Coulson, J. C., and Hickling, G. (1961). Variation in the secondary sex-ratio of the grey seal (*Halichoerus grypus* (Fabr.)) during the breeding season. *Nature (London)* **190,** 281.

Croxall, J. P., and Gentry, R. L. (eds.) (1987). "Status, Biology, and Ecology of Fur Seals." NOAA Technical Report NMFS 51. U.S. Dept. of Commerce, Springfield, VA.

Croxall, J. P., McCann, T. S., Prince, P. A., and Rothery, P. (1988). Reproductive performance of seabirds and seals at South Georgia and Signy Island, South Orkney Islands, 1976–1987: Implications for southern ocean monitoring studies. *In* "Antarctic Ocean and Resources Variability" (D. Sahrhage, ed.), pp. 261–285. Springer-Verlag, Heidelberg.

DeLong, R. L., and Antonelis, G. A. (1991). Impact of the 1982–1983 El Niño on the Northern fur seal population at San Miguel Island, California. *In* "Pinnipeds and El Niño" (F. Trillmich and K. A. Ono, eds.), pp. 75–83. Springer-Verlag, Heidelberg.

Deutsch, C. J., Crocker, D. E., Costa, D. P., and Boeuf, B. J. (1994). Sex- and age-related variation in reproductive effort of Northern elephant seals. *In* "Elephant Seals" (B. J. Le Boeuf and R. M. Laws, eds.), pp. 169–210. Univ. of California Press, Berkeley, CA.

Doidge, D. W., McCann, T. S., and Croxall, J. P. (1986). Attendance behavior of Antarctic fur seals. *In* "Fur Seals: Maternal Strategies on Land and at the Sea" (R. L. Gentry and G. L. Kooyman, eds.), pp. 102–114. Princeton Univ. Press, Princeton, NJ.

Drent, R. H., and Daan, S. (1980). The prudent parent: Energetic adjustments in avian breeding. *Ardea* **68,** 225–252.

Eibl-Eibesfeldt, I. (1955). Ethologische Studien am Galapagos Seelöwen, *Zalophus wollebaeki;* Sivertsen. *Z. Tierpsychol.* **12,** 286–303.

Eibl-Eibesfeldt, I., and Hass, H. (1959). Erfahrungen mit Haien. *Z. Tierpsychol.* **16,** 733–746.

Fay, F. H. (1982). Ecology and biology of the pacific walrus, *Odobenus rosmarus divergens*, Illiger., North American Fauna No.**74**, U. S. Department of the Interior, Washington, D. C.
Fedak, M. A., and Anderson, S. S. (1982). The energetics of lactation: Accurate measurements from a large wild mammal, the Grey seal (*Halichoerus grypus*). *J. Zool.* (*London*) **198**, 473–479.
Fisher, R. A. (1930). "The Genetical Theory of Natural Selection." Oxford Univ. Press, Oxford.
Francis, J. M., and Heath, C. B. (1991). The effects of El Niño on the frequency and sex ratio of suckling yearlings in the California sea lion. *In* "Pinnipeds and El Niño" (F. Trillmich and K. A. Ono, eds.), pp. 193–201. Springer-Verlag, Heidelberg.
Frank, S. A. (1986). The genetic value of sons and daughters. *Heredity* **56**, 351–354.
Frank, S. A. (1990). Sex allocation theory for birds and mammals. *Annu. Rev. Ecol. Syst.* **21**, 13–55.
Gentry, R. L. (1974). The development of social behavior through play in the Steller sea lion. *Am. Zool.* **14**, 391–404.
Gentry, R. L., and Holt, J. R. (1986). Attendance behavior of Northern fur seals. *In* "Fur Seals: Maternal Strategies on Land and at Sea" (R. L. Gentry and G. L. Kooyman, eds.), pp. 41–60. Princeton Univ. Press, Princeton, NJ.
Gentry, R. L., and Kooyman, G. L. (eds.) (1986). "Fur Seals: Maternal Strategies on Land and at Sea." Princeton Univ. Press, Princeton NJ.
Gentry, R. L., Costa, D. P., Croxall, J. P., David, J. H. M., Davis, R. W., Kooyman, G. L., Majluf, P., McCann, T. S., and Trillmich, F. (1986). Synthesis and conclusions. *In* "Fur Seals: Maternal Strategies on Land and at Sea" (R. L. Gentry and G. L. Kooyman, eds.), pp. 220–264. Princeton Univ. Press, Princeton, NJ.
Gisiner, R., and Schusterman, R. J. (1991). California sea lions pups play an active role in reunions with their mothers. *Anim. Behav.* **41**, 364–366.
Godfray, H. C. J. (1991). Signalling of need by offspring to their parents. *Nature* (*London*) **352**, 328–330.
Goldsworthy, S. D. (1992). Maternal care in three species of southern fur seal (*Arctocephalus* spp.) Ph.D. thesis, Monash Univ. Clayton, VIC, Australia.
Goldsworthy, S. D. (1995). Differential expenditure of maternal resources in Antarctic fur seals, *Arctocephalus gazella*, at Heard Island, southern Indian Ocean. *Behav. Ecol.* **6**, 218–228.
Grafen, A. (1990). Biological signals as handicaps. *J. Theor. Biol.* **144**, 517–546.
Haley, M. P., Deutsch, C. J., and Le Boeuf, B. J. (1994). Size, dominance and copulatory success in male Northern elephant seals. *Mirounga angustirostris. Anim. Behav.* **48**, 1249–1260.
Hanggi, E. B., and Schusterman, R. J. (1990). Kin recognition in captive California sea lions (*Zalophus californianus*). *J. Comp. Psychol.* **104**, 368–372.
Higgins, L. V. (1990). Reproductive behaviour and maternal investment of Australian sea lions. Ph.D. thesis, Univ. of California, Santa Cruz, CA.
Higgins, L. V. (1993). The non-annual, non-seasonal breeding cycle of the Australian sea lions. *J. Mammal.* **74**, 270–274.
Horning, M. (1992). Die Ontogenese des Tauchverhaltens beim Galapagos Seebären *Arctocephalus galapagoensis* (Heller 1904). Ph.D. Thesis, Universität Bielefeld, Germany.
Huber, H. R. (1987). Natality and weaning success in relation to age of first reproduction in Northern elephant seals. *Can. J. Zool.* **65**, 1311–1316.
Huntley, A. C., Costa, D. P., Worthy, G. A. J., and Castellini, M. A. (eds). (1987). "Marine Mammal Energetics." Allen Press, Lawrence, KS.
Kerley, G. I. H. (1985). Pup growth in the fur seals *Arctocephalus tropicalis* and *Arctocephalus gazella* on Marion Island. *J. Zool.* (*London*) **205**, 315–324.
King, J. E. (1983). "Seals of the World." Cornell Univ. Press, Ithaca, NY.

Kooyman, G. L., Kerem, D. H., Campbell, W. B., and Wright, J. J. (1973). Pulmonary gas exchange in freely diving weddell seal, *Leptonychotes weddelli. Respir. Physiol.* **17**, 190–283.

Kooyman, G. L., Billups, J. O., and Farwell, W. D. (1983). Two recently developed recorders for monitoring diving activity of marine birds and mammals. *In* "Experimental Biology at Sea" (A. G. Macdonald and I. G. Priede, eds.), pp. 197–214. Academic Press, New York.

Kovacs, K. M. (1987). Maternal behaviour and early behavioural ontogeny of harp seals, *Phoca groenlandica. Anim. Behav.* **35**, 844–855.

Kovacs, K. M., and Lavigne, D. M. (1986a). Maternal investment and neonatal growth in phocid seals. *J. Anim. Ecol.* **55**, 1035–1051.

Kovacs, K. M., and Lavigne, D. M. (1986b). Growth of grey seal (*Halichoerus grypus*) neonates: Differential maternal investment in the sexes. *Can. J. Zool.* **64**, 1937–1943.

Kovacs, K. M., and Lavigne, D. M. (1992). Maternal investment in otariid seals and walruses. *Can. J. Zool.* **70**, 1953–1964.

Krackow, S. (1994). Potential mechanisms for sex ratio adjustment in mammals and birds. *Biol. Rev.* **70**, 225–241.

Kretzmann, M. B., Costa, D. P., and Le Boeuf, B. J. (1993). Maternal energy investment in elephant seal pups: Evidence for sexual equality? *Am. Nat.* **141**, 466–480.

Le Boeuf, B. J. (1974). Male-male competition and reproductive success in elephant seals. *Am. Zool.* **14**, 163–176.

Le Boeuf, B. J. (1991). Pinniped mating systems on land, ice and in the water: Emphasis on the phocidae. *In* "Behaviour of Pinnipeds" (D. Renouf, ed.), pp. 45–65. Chapman & Hall, London.

Le Boeuf, B. J., and Campagna, C. (1994). Protection and abuse of young in pinnipeds. *In* "Infanticide and Parental Care" (S. Parmigiani and F. S. vom Saal, eds.), pp. 257–276. Harwood Academic Publisher, Chur, Switzerland.

Le Boeuf, B. J., and Laws, R. M. (eds.) (1994). "Elephant Seals—Population, Ecology, Behavior, and Physiology." Univ. of California Press, Berkeley, CA.

Le Boeuf, B. J., and Reiter, J. (1988). Lifetime reproductive success in Northern elephant seals. *In* "Reproductive Success. Studies of Individual Variation in Contrasting Breeding Systems" (T. H. Clutton-Brock, ed.), pp. 344–362. Univ. of Chicago Press, Chicago.

Le Boeuf, B. J., Costa, D. P., Huntley, A. C., and Feldkamp, S. D. (1988). Continuous, deep diving in female Northern elephant seals, *Mirounga angustirostris. J. Zool. (London)* **208**, 1–7.

Le Boeuf, B. J., Condit, R., and Reiter, J. (1989). Parental investment and the secondary sex ratio in Northern elephant seals. *Behav. Ecol. Sociobiol.* **25**, 109–117.

Le Boeuf, B. J., Morris, P., and Reiter, J. (1994). Juvenile survivorship of Northern elephant seals. *In* "Elephant Seals" (B. J. Le Boeuf, and R. Laws, eds.), pp. 121–136. Univ. of California Press, Berkeley, CA.

Lunn, N. J., and Boyd, I. L. (1993a). Influence of maternal characteristics and environmental variation on reproduction in Antarctic fur seals. *Symp. Zool. Soc. London* **66**, 115–129.

Lunn, N. J., and Boyd, I. L. (1993b). Effects of maternal age and condition in parturition and the perinatal period of Antarctic fur seals. *J. Zool. (London)* **229**, 55–67.

Lunn, N. J., Boyd, I. L., Barton, T., and Croxall, J. P. (1993). Factors affecting the growth rate and mass at weaning of Antarctic fur seals at bird island, South Georgia. *J. Mammal.* **74**, 908–919.

Lunn, N. J., Boyd, I. L., and Croxall, J. P. (1994). Reproductive performance of female Antarctic fur seals: The influence of age, breeding experience, environmental variation and individual quality. *J. Anim. Ecol.* **63**, 827–840.

Lydersen, C., and Hammill, M. O. (1993). Diving in ringed seal (*Phoca hispida*) pups during the nursing period. *Can. J. Zool.* **71,** 991–996.

Lydersen, C., and Kovacs, K. M. (1993). Diving behaviour of lactating harp seal, *Phoca groenlandica,* females from the Gulf of St. Lawrence, Canada. *Anim. Behav.* **46,** 1213–1221.

Lydersen, C., Hammill, M. O., and Kovacs, K. M. (1994). Diving activity in nursing bearded seal (*Erignathus barbatus*) pups. *Can. J. Zool.* **72,** 96–103.

Macy, S. K. (1982). Mother-pup interactions in the Northern fur seal. Ph.D. thesis, Univ. of Washington, Seattle, WA.

Majluf, M. P. J. (1987). Reproductive ecology of female South American fur seals at Punta San Juan, Peru. Ph.D. thesis, Univ. of Cambridge. United Kingdom.

Maynard Smith, J. (1980). A new theory of sexual investment. *Behav. Ecol. Sociobiol.* **7,** 247–251.

McCann, T. S., Fedak, M. A., and Harwood, J. (1989). Parental investment in Southern elephant seals, *Mirounga leonina. Behav. Ecol. Sociobiol.* **25,** 81–87.

Miller, E. H. (1974). A paternal role in Galapagos sea lions? *Evolution* **28,** 473–476.

Miller, E. H., and Boness, D. J. (1983). Summer behavior of Atlantic walruses *Odobenus rosmarus rosmarus* (L.) at Coats Island, N. W. T. (Canada). *Z. Säugetierk.* **48,** 298–313.

Mock, D. W., and Forbes, L. S. (1992). Parent-offspring conflict: A case of arrested development. *Trends Ecol. Evol.* **7,** 409–413.

Muelberg, M. M. C., and Bowen, W. D. (1993). Duration of lactation and postweaning changes in mass and body composition of harbour seal, *Phoca vitulina,* pups. *Can. J. Zool.* **71,** 1405–1414.

Nelson, B. (1968). "Galapagos. Islands of Birds". Longmans, Green and Co., London.

Oftedal, O. T., and Iverson, S. J. (1987). Hydrogen isotope methodology for measurement of milk intake and energetics of growth in suckling young. *In* "Marine Mammals Energetics" (A. C. Huntley, D. P. Costa, G. A. J. Worthy, and M. A. Castellini, eds.), pp. 67–96. Allen Press, Lawrence, KS.

Oftedal, O. T., Boness, D. J., and Tedman, R. A. (1987a). The behavior, physiology, and anatomy of lactation in the pinnipedia. *Curr. Mammal.* **1,** 175–245.

Oftedal, O. T., Iverson, S. J., and Boness, D. J. (1987b). Milk and energy intakes of suckling California sea lion *Zalophus californianus* pups in relation to sex, growth, and predicted maintenance requirements. *Physiol. Zool.* **60,** 560–575.

Oftedal, O. T., Bowen, W. D., and Boness, D. J. (1993). Energy transfer by lactating hooded seals and nutrient deposition in their pups during the four days from birth to weaning. *Physiol. Zool.* **66,** 412–436.

Ono, K. A., and Boness, D. J. (1991). The influence of El Niño on mother-pup behavior, pup ontogeny, and sex ratios in the California sea lion. *In* "Pinnipeds and El Niño" (F. Trillmich and K. A. Ono, eds.), pp. 185–192. Springer-Verlag, Heidelberg.

Ono, K. A., Boness, D. J., and Oftedal, O. T. (1987). The effect of a natural environmental disturbance on maternal investment and pup behavior in the California sea lion. *Behav. Ecol. Sociobiol.* **21,** 109–118.

Payne, M. R. (1979). Growth in the Antarctic fur seal *Arctocephalus gazella. J. Zool. (London)* **187,** 1–20.

Reilly, J. J., and Fedak, M. A. (1991). Rates of water turnover and energy expenditure of free-living male common seals (*Phoca vitulina*). *J. Zool. (London)* **223,** 461–468.

Reiter, J., and Le Boeuf, B. J. (1991). Life history consequences of variation in age at primiparity in Northern elephant seals. *Behav. Ecol. Sociobiol.* **28,** 153–160.

Reiter, J., Stinson, N. L., and Le Boeuf, B. J. (1978). Northern elephant seal development: The transition from weaning to nutritional independence. *Behav. Ecol. Sociobiol.* **3,** 337–367.

Reiter, J., Panken, K. J., and Le Boeuf, B. J. (1981). Female competition and reproductive success in Northern elephant seals. *Anim. Behav.* **29,** 670–687.

Repenning, C. A. (1980). Warm-blooded life in cold ocean currents. *Oceans* **13,** 18–24.

Riedman, M. L. (1982). The evolution of alloparental care and adoption in mammals and birds. *Quart. Rev. Biol.* **57,** 405–435.

Roux, J.-P., and Jouventin, P. (1987). Behavioural cues to individual recognition in the subantarctic fur seal, *Arctocephalus tropicalis*. In "Status, Biology, and Ecology of Fur Seals" (J. Croxall and R. L. Gentry, eds.), pp. 23–27. NOAA Technical Report NMFS 51. U.S. Dept. of Commerce, Springfield, VA.

Sarich, V. (1969). Pinniped phylogeny. *Syst. Zool.* **18,** 416–422.

Schusterman, R. J., Hanggi, E. B., and Gisiner, R. (1992). Acoustic signalling in mother-pup reunions, interspecies bonding, and affiliation by kinship in California sea lions (*Zalophus californianus*). In "Marine Mammal Sensory Systems" (J. Thomas *et al.*, eds.), pp. 533–551. Plenum Press, New York.

Smith, T. G. (1980). Polar bear predation of ringed and bearded seals in the land-fast sea ice habitat. *Can. J. Zool.* **58,** 2201–2209.

Smith, T. G., and Stirling, I. (1975). Variation in the density of ringed seal (*Phoca hispida*) birth lairs in the Amundsen Gulf, Northwest Territories. *Can. J. Zool.* **56,** 1066–1071.

Stearns, S. C. (1989). Trade-offs in life history evolution. *Funct. Ecol.* **3,** 259–268.

Stirling, I. (1971). Variation in sex ratio of newborn weddell seals during the pupping season. *J. Mammal.* **52,** 842–844.

Stirling, I. (1977). Adaptation of weddell and ringed seals to exploit the polar fast ice habitat in the absence or presence of surface predators. In "Adaptation within Antarctic Ecosystems" (G. A. Llano, ed.), pp. 741–748, Smithsonian Institution, Washington, D. C.

Sydeman, W. J., and Nur, N. (1994). Life history strategies of female Northern elephant seals. In "Elephant Seals: Population Ecology, Behavior, and Physiology" (B. J. Le Boeuf and R. M. Laws, eds.), pp. 137–153. Univ. of California Press, Berkeley, CA.

Sydeman, W. J., Hubert, H. R., Emslie, S. D., Ribic, C. A., and Nur, N. (1991). Age-specific weaning success of Northern elephant seals in relation to previous breeding experience. *Ecology* **72,** 2204–2217.

Tedford, R. H. (1976). Relationship of pinnipeds to other carnivores (Mammalia). *Syst. Zool.* **25,** 363–374.

Testa, J. W., Hill, S. E. B., and Siniff, D. B. (1989). Diving behavior and maternal investment in Weddell seals (*Leptonychotes weddellii*). *Mar. Mamm. Sci.* **5,** 399–405.

Thompson, P. M., Miller, D., Cooper, R., and Hammond, P. S. (1994). Changes in the distribution and activity of female harbour seals during the breeding season: Implications for their lactation strategy and mating patterns. *J. Anim. Ecol.* **63,** 24–30.

Thompson, S. D., Ono, K. A., Oftedal, O. T., and Boness, D. J. (1987). Thermoregulation and resting metabolic rate of California sea lion (*Zalophus californianus*) pups. *Physiol. Zool.* **60,** 730–736.

Tinker, M. T., Kovacs, K. M., and Hammill, M. O. (1995). The reproductive behavior and energetics of male gray seals (*Halichoerus grypus*) breeding on a land-fast ice substrate. *Behav. Ecol. Sociobiol.* **36,** 159–170.

Trillmich, F. (1981). Mutual mother-pup recognition in Galapagos fur seals and sea lions: Cues used and functional significance. *Behaviour* **78,** 21–42.

Trillmich, F. (1986a). Attendance behavior of Galapagos fur seals. In "Fur Seals: Maternal Strategies on Land and at Sea" (R. L. Gentry and G. L. Kooyman, eds.), pp. 168–185. Princeton Univ. Press, Princeton, NJ.

Trillmich, F. (1986b). Maternal investment and sex-allocation in the Galapagos fur seal, *Arctocephalus galapagoensis*. *Behav. Ecol. Sociobiol.* **19,** 157–164.

Trillmich, F. (1990). The behavioral ecology of maternal effort in fur seals and sea lions. *Behaviour* **114,** 3–20.

Trillmich, F., and Lechner, E. (1986). Milk of the Galapagos fur seal and sea lion, with a comparison of the milk of eared seals (Otariidae). *J. Zool. (London)* **209,** 271–277.

Trillmich, F., and Limberger, D. (1985). Drastic effects of El Niño on Galápagos pinnipeds. *Oecologia* **67,** 19–22.

Trillmich, F., and Ono, K. (eds.) (1991). "Pinnipeds and El Niño—Responses to Environmental Stress." Springer-Verlag, Heidelberg.

Trillmich, F., Ono, K. A., Costa, D. P., DeLong, R. L., Feldkamp, S. D., Francis, J. M., Gentry, R. L., Heath, C. B., Le Boeuf, B. J., Majluf, P., and York, A. E. (1991). The effects of El Niño on pinniped populations in the Eastern Pacific. *In* "Pinnipeds and El Niño" (F. Trillmich and K. A. Ono, eds.), pp. 247–288. Springer-Verlag, Heidelberg.

Trites, A. W. (1991). Fetal growth of Northern fur seals: Life history strategy and sources of variation. *Can. J. Zool.* **69,** 2608–2617.

Trivers, R. L. (1972). Parental investment and sexual selection. *In* "Sexual Selection and the Descent of Man" (B. Campbell, ed.), pp. 136–179. Heinemann Ltd., London.

Trivers, R. L. (1974). Parent-offspring conflict. *Amer. Zool.* **14,** 249–264.

Trivers, R. L., and Willard, D. E. (1973). Natural selection of parental ability to vary the sex ratio of offspring. *Science (Washington, D. C.)* **179,** 90–92.

Wiig, Ø. (1983). On the relationship of pinnipeds to other carnivores. *Zool. Scr.* **12,** 225–227.

Williams, G. C. (1979). The question of adaptive sex ratio in outcrossed vertebrates. *Proc. R. Soc. London, Ser. B* **205,** 567–580.

Wyss, A. R. (1987). The walrus auditory region and the monophyly of pinnipeds. *American Museum Natural History,* Novitates **2871,** 31 pp.

Wyss, A. R. (1988). Evidence from flipper structure for a single origin of pinnipeds. *Nature (London)* **334,** 427–428.

Wyss, A. R. (1989). Flippers and pinniped phylogeny: Has the problem of convergence been overrated? *Mar. Mamm. Sci.* **5,** 343–360.

# Individual Differences in Maternal Style
## Causes and Consequences for Mothers and Offspring

LYNN A. FAIRBANKS

DEPARTMENT OF PSYCHIATRY AND BIOBEHAVIORAL SCIENCES
UNIVERSITY OF CALIFORNIA
LOS ANGELES, CALIFORNIA 90024

## I. Introduction

Individual differences in the way primate mothers treat their infants have been noted ever since researchers began to collect detailed records of maternal behavior and mother–infant relationships. The field studies of the early 1960s produced rich, qualitative descriptions of maternal behavior and even with relatively small samples of mothers, wide variation was seen in maternal competence, ease in infant handling, and in the timing and severity of maternal rejection (Jay, 1963; DeVore, 1963). These differences were attributed partially to maternal experience, but also to individual differences in the temperament and personality of the mother. Quantitative studies of mother–infant interactions in the field and in captive social groups in the 1960s and 1970s confirmed that some mothers were consistently more restrictive of their infants' attempts to move out of contact, while others were relatively inattentive and rejecting (Rowell *et al.*, 1968; Hinde and Spencer-Booth, 1971; Struchsaker, 1971; Rosenblum, 1971; Nash, 1978; Altmann, 1980). The terms *protective, restrictive, laissez-faire,* and *rejecting* are commonly used to describe this variation.

There has been a growing interest in individual differences in temperament in recent years, including genetic, environmental and physiological mechanisms, and evolutionary implications (see Clark and Ehlinger, 1987; Wilson *et al.*, 1994; Clarke and Boinski, 1995, for reviews). Some of this literature focuses on interspecific variation and uses comparisons between closely related species to demonstrate links between social organization, temperament, and physiological reactivity (e.g., Clarke *et al.*, 1988). In an early description of species differences in maternal style, Kaufman and Rosenblum (1969) noted that bonnet macaque mothers were less restrictive than pigtail macaque mothers, and that these differences were related to the more relaxed and tolerant relationships among adults. Direct comparisons of other macaque species have also found significant differences among species in maternal style that are consistent with species differences in temperament and social organization (Thierry, 1985; Mason et al., 1993; Maestripieri, 1994).

This article will focus on spontaneously occurring individual differences in maternal behavior within primate species, and will argue that maternal style is both an expression of temperament-based differences among adult females, and a mechanism for transmitting individual differences across generations. The article begins with a description of the principal dimensions of maternal style that have been identified in studies of cercopithecine primates, and then examines the variables that have been proposed to explain this variation in maternal behavior. It will demonstrate that maternal style is responsive to social and ecological circumstances to a certain degree, but that there are also stable individual differences among females that persist in the face of changing conditions. The second half of the article will focus on research on the consequences of indivdual differences in maternal style for the mother and for the offspring. There is evidence that maternal style influences future fertility for the mother. Maternal style also predicts individual differences in temperament and response to novel and challenging situations for juvenile and adolescent offspring, and influences the maternal behavior of adult daughters. The article concludes with a brief discussion of some of the proximate mechanisms that mediate these effects. Particular emphasis is given to longitudinal research on the causes and consequences of individual differences in maternal behavior of vervet monkeys at the Sepulveda VA Medical Center Nonhuman Primate Research Facility.

## II. Individual Differences in Maternal Style

### A. Dimensions of Maternal Style: Protectiveness and Rejection

Research on individual differences in maternal style benefited greatly by the early work of Robert Hinde and his associates at Madingley, who established a standard set of behaviors and observational methods that have been widely adopted by others to describe mother–infant relationships. The Madingley system uses behavior categories that are objective and descriptive to measure mother–infant spatial relationships and interactions, including making and breaking contact by the mother and the infant, and rejection, restraint, and grooming by the mother. This common vocabulary has been used in studies of a variety of other species and has made cross-species comparisons feasible. The early studies of rhesus monkeys at Madingley, along with more recent research with other species and at other sites, have demonstrated a remarkably high degree of consistency in the basic dimensions of maternal behavior, even between studies using different species, environmental settings, and methods of analysis. This research supports the following facts about individual variation in maternal style in cercopithecine primates.

## 1. *Variation in Maternal Care Falls along Two Dimensions that Can be Labeled Protectiveness and Rejection*

Numerous studies have shown that variation in maternal style is not unidimensional. In an early quantitative study of maternal behavior, Hinde and Spencer-Booth (1971) demonstrated that differences between mothers in contact initiation and caregiving behavior, such as grooming, were correlated with one another, while rejection was positively correlated with the percentage of time the mother and infant were out of contact. Contrary to expectation, however, these two dimensions were only weakly negatively correlated with one another. Later work on the mother–infant relationships at the Madingley colony confirmed that restrictive and rejecting behaviors vary independently (Hinde and Simpson, 1975; Simpson and Howe, 1980).

Variation in maternal behavior along the two dimensions of protectiveness and rejection has also been demonstrated in our ongoing study of social behavior and development of vervet monkeys (*Cercopithecus aethiops sabaeus*) at the Sepulveda VA Medical Center Nonhuman Primate Research Facility. Vervet monkeys in the four north groups at the colony live in large outdoor enclosures that have been managed to approximate the natural social composition of free-ranging vervet monkey troops. Females remain in the natal group with their mothers and female kin, while males are removed when they reach adulthood and unrelated adult males are replaced at 3- to 4-year intervals. The behavior of all mother–infant dyads was documented from 1980 through 1989 using a consistent, reliability-tested observation system. After the first 5 years of data collection, complete behavioral records were available for 52 infants from birth to 6 months of age. A principal components analysis of maternal and infant behaviors at that time demonstrated that the behavior of the vervet monkey mothers was organized along similar dimensions to those found for rhesus monkeys (Fairbanks and McGuire, 1987). Table I shows a replication of the principal components analysis using the full sample of 127 surviving infants born between 1980 and 1989 at the colony and a subset of the original behaviors. The frequency per hour of eight behaviors initiated by the mother toward her infant, averaged over the first 6 months of the infant's life, was entered into the analysis. (See Fairbanks, 1988a, Fairbanks and McGuire, 1995,

TABLE I
PRINCIPLE COMPONENTS ANALYSIS OF MATERNAL
BEHAVIOR OF VERVET MONKEYS[a]

| Behavior initiated by mother to infant | Maternal protectiveness Factor 1 | Maternal rejection Factor 2 |
|---|---|---|
| Make ventral contact | 0.91 | 0 |
| Muzzle, inspect | 0.81 | 0 |
| Restrain | 0.80 | 0 |
| Approach | 0.60 | 0 |
| Groom | 0.52 | 0 |
| Break ventral contact | 0 | 0.82 |
| Reject | 0 | 0.80 |
| Leave | 0 | 0.72 |

[a] Factor loadings <.50 set to 0.

for description of behaviors and sampling methods.) Factor #1, which has been labeled maternal protectiveness, includes high loadings on contact-promoting and nurturant behaviors, including mother makes contact, restrains, approaches, grooms, and inspects her infant. Factor #2, labeled maternal rejection, has high loadings for mother rejects, breaks contact, and leaves infant. The fact that restraint and rejection fall on separate factors suggests that these two behaviors are not opposite ends of the same continuum but are separate dimensions of maternal behavior.

Independence of these same two dimensions of maternal style has been documented in free-ranging settings. In an early field study of baboon mothers, Altmann (1980) found a strong inverse relationship between the age that the mother stopped restraining and began rejecting her infant, but the sample only included 12 infants. More recent research has demonstrated that, with larger samples, maternal restriction and maternal rejection vary independently under free-ranging conditions. Principal components analysis of maternal behavior revealed separate factors for suckling rejection and maternal restriction among free-ranging Japanese monkeys at Jigokudani Monkey Park (Tanaka, 1989), and Berman (1990a) reported relatively low correlations between maternal restriction and restraint for free-ranging rhesus monkeys on Cayo Santiago. Results from these two free-ranging settings closely matched the studies of captive rhesus monkeys at Madingley, described previously, and a principal components analysis of maternal behavior of captive Japanese monkeys (Schino et al., 1995). The comparability of results from vervet monkeys with the pattern of correlations reported for rhesus monkeys at Madingley and on Cayo Santiago, and for free-ranging and confined Japanese monkeys, suggests that these two dimensions of maternal style generalize across species and circumstances.

2. *Individual Differences in Maternal Behavior Are Consistent over Time*

Even though the hourly rates of particular behaviors change dramatically with the infant's month of life, individual differences between mothers tend to be consistent. When mother–infant interactions were divided into 6-week blocks, individual differences in rhesus monkey maternal behavior were consistent from one time block to the next (Hinde and Spencer-Booth, 1971), with positive correlations for time off, relative rejections, the mother's role in contact initiations, and the proximity index. The absolute rejection rate was consistent until 5–6 months of age, when all mothers showed their highest rejection frequencies. Similar consistency over time blocks for individual mother–infant behaviors was also reported by Simpson and Simpson (1986) for rhesus monkeys from the Madingley colony and by Berman (1990a) for free-ranging rhesus monkeys on Cayo Santiago.

The mean frequency per hour that vervet monkey mothers restrain or reject their infants and the percentage time they spend in ventral contact for 127 mother–infant dyads at the Sepulveda colony is shown in Fig 1. In spite of dramatic effects of infant age on the rates of each of these maternal behaviors, the individual differences between mothers were remarkably stable (Table II). Mothers who had high rates of restraint, rejection, or contact in the first 2 months, continued to have relatively high rates of the same behavior at later ages. These results confirm the fact that individual differences in maternal style are stable over infant age periods.

3. *Individual Differences in Maternal Style Are Consistent across Infants of the Same Mother*

Longitudinal studies of maternal behavior have demonstrated that mothers tend to use a consistent maternal style across infants. Mother–infant contact has been shown to vary significantly between mothers and to be consistent across infants of the same mother at the Sepulveda vervet monkey colony (Fairbanks, 1989). Berman (1990a) demonstrated consistency in maternal style across infants for several maternal behaviors, including maternal rejection, contact time, and maternally initiated contact, for rhesus monkey mothers on Cayo Santiago.

Consistency in maternal style across infants of the same mother in the 10-year data set from the Sepulveda colony is represented in Fig. 2, which shows the maternal protectiveness scores, derived from the principal components analysis, for the 30 mothers who were observed with two or more

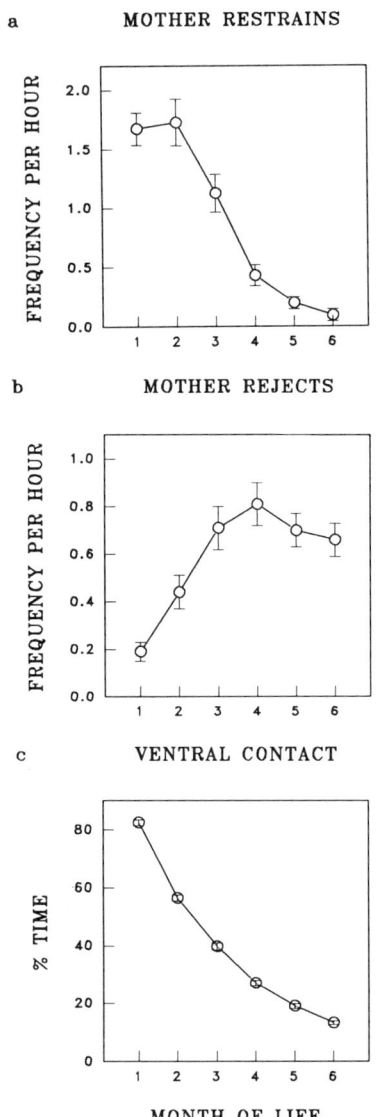

FIG. 1. Frequency per hour that (a) mother restrains infant, (b) mother rejects infant, and (c) percentage mother–infant ventral contact, by infant month of life for vervet monkey mother–infant dyads.

## TABLE II
### Pearson Correlations of Individual Differences in Maternal Behavior by Infant Age in Vervet Monkeys

|  | Months 1–2 with 3–4 | Months 3–4 with 5–6 |
|---|---|---|
| Reject | .56** | .52** |
| Restrain | .50** | .43** |
| Ventral contact | .48** | .58** |

** $p < .01$.

infants. (Factor scores have a mean of 0 and a standard deviation of 1, with a high score representing high rates of the behaviors that load positively on the factor.) Individual mothers are listed along the $x$ axis and each circle on the graph represents the maternal protectiveness factor score for one infant. For example, female I at the left side of the graph had six surviving infants and her maternal protectiveness score varied from a low of $-1.45$ to a high of 2.48 across infants. Female J, near the right side of the graph, had six surviving infants and her scores with these infants all fell within a narrow range from .26 to .74. Overall, Fig. 2 demonstrates a remarkable consistency within mothers and marked differences between mothers in the level of maternal protectiveness. A one-way analysis of variance using the 30 females with two or more infants in the data set produced a significant effect of mothers' identity on protectiveness score ($F = 6.55$, df = 29,91, $p < .001$). The level of maternal protectiveness toward each infant could

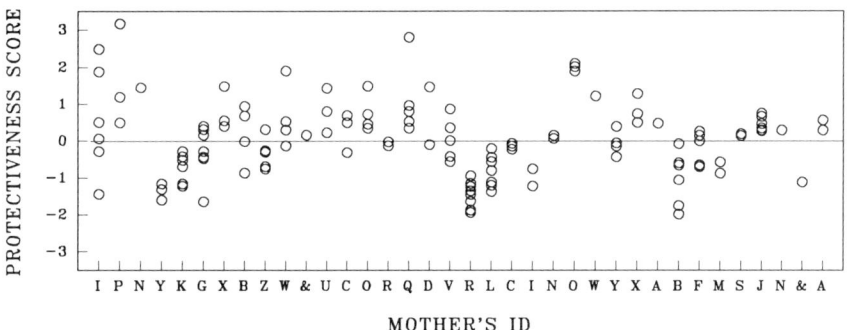

Fig. 2. Protectiveness factor scores for individual vervet monkey infants by identity of the mother.

be predicted with a high degree of accuracy by knowing how the mother had treated her other infants. The correlation between the protectiveness score toward one infant and the mother's average protectiveness score with her other infants for the 121 infants with at least one sibling in the sample was $r = .69$ ($p < .001$).

The same pattern of individual consistency across infants was found for maternal rejection. Maternal rejection factor scores differed significantly across mothers with more than one infant in the data set ($F = 4.15$, df = 29,91, $p < .001$), and the level of maternal rejection with one infant was significantly correlated with the mother's average maternal rejection with her other infants ($r = .59, p < .01$).

B. CONTEXTUAL AND LIFE-HISTORY DETERMINANTS OF MATERNAL STYLE

The first studies of differences in maternal style looked for social and demographic variables to explain the variation in maternal behavior. A variety of factors have been associated with individual differences in protectiveness, rejection, and mother–infant contact.

*1. Mothers Are More Protective When Their Infants Are at Greater Risk*

Many of the contextual variables that have been shown to be related to maternal style involve infant safety. In particular, mothers are more protective when their infants are at greater risk from social aggression. Early studies comparing individually housed mother–infant pairs with socially housed pairs demonstrated that mothers were more protective toward their infants in the presence of other group members (Wolfeim *et al.,* 1970). Social risk factors related to low dominance rank, the persistent attention of "aunts," and the presence of recently introduced adult males have all been related to higher levels of maternal protectiveness (White and Hinde, 1975; Rowell *et al.,* 1964; Altmann, 1980; Berman, 1984; Fairbanks and McGuire, 1987). The immediate past experience of losing an infant resulted in an increase in protectiveness toward the next infant for vervet monkey mothers at the Sepulveda colony (Fairbanks, 1988a). The social protection afforded by large matrilines and the presence of the infant's grandmother has been associated with a lower level of maternal protectiveness and an increase in rejection (Berman, 1980; Fairbanks, 1988b; Schino *et al.,* 1995).

Ecological variables related to mortality risk have also been shown to influence maternal responsiveness and mother–infant spatial relationships. Johnson and Southwick (1984) found that maternal behavior of free-ranging rhesus monkeys was relatively unaffected by structural differences in the natural habitat, but that mother–infant contact and proximity were greater in habitats that had the highest level of environmental risk for infant mortality. Trapping activities on Cayo Santiago were shown to increase maternal protectiveness and mother–infant contact for weeks after the trapping ended, even for mothers who were not directly involved (Berman, 1989). Variation in maternal responsiveness to infant alarm calls reflected the infant's risk of mortality for free-ranging vervet monkey mothers (Hauser, 1988). In a field study of long-tailed macaques, Karssemeijer *et al.* (1990) found that height in the canopy, which was related to risk of injury from falls, was a factor influencing mother–infant contact time. All of these studies suggest that mothers respond to increased risk to their infants by increasing maternal protectiveness and spatial proximity.

### 2. Maternal Style Changes with Age and Experience

A set of correlated variables that have been related to variation in maternal behavior includes the mother's age, parity, experience, and reproductive value. Logical predictions from parental investment theory would lead us to expect that, all else being equal, the youngest mothers would be more rejecting and the oldest mothers would be more protective or laissez-faire (Pianka and Parker, 1975). Very young mothers would have the most to lose in future reproductive potential by compromising their own health and welfare for their current infants; in contrast, the oldest mothers would have much less reason to terminate investment early because they would be unlikely to be able to produce another. These predictions, based on life-history changes in reproductive value, are opposite to the usual predictions based on the mother's experience in infant rearing. Young, inexperienced mothers are expected to put more effort into maternal care because they are less competent and less efficient than older, more experienced mothers.

Observations of one study of maternal behavior of provisioned Barbary macaques support the terminal investment hypothesis. Old mothers spent more time in contact with their infants and weaned them later than young mothers (Paul *et al.*, 1993). However, the results from most other studies are more in line with the experience predictions than with the reproductive value expectations. Primiparous mothers were found to be more protective and to suckle their infants more frequently than multiparous mothers in studies of rhesus and vervet monkeys (Hooley and Simpson, 1981; Fairbanks, 1988a; Gomendio, 1989a). Similarly, older mothers were more rejecting and less protective than younger mothers in rhesus and Japanese macaques (Hiraiwa, 1981; Berman, 1984; Schino, *et al.,* 1995).

### 3. Effects of Infant Sex on Maternal Style Are Inconsistent

The relationships between infant sex and maternal style has produced conflicting results. Many studies that focus on the first few months of life find no differences in the way mothers treat sons and daughters (Altmann,

1980; Tanaka, 1989; Silk, 1991). A matched comparison of maternal behavior toward male and female infants of the same mother in the vervet monkey colony produced no significant differences in an analysis of 15 different behaviors (L. A. Fairbanks, unpublished data). Sex differences in maternal restraint have been reported in studies of several macaque species, but the results have been inconsistent and often contradictory. For example, Eaton *et al.* (1985) reported that Japanese macaque mothers retrieved male infants more than female infants during the first 4 months of life, while Schino *et al.* (1995) found the opposite pattern of mothers restraining female infants and rejecting male infants more often. Primiparous rhesus macaque mothers restrained female infants more often, while multiparous mothers were more protective toward sons (Hooley, 1983). Results on sex differences are also complicated by complex interactions with infant sex ratio, mother's dominance rank, and maternal behavior (Simpson, 1983; Gomendio, 1990; Altmann and Samuels, 1992). At this point, it is difficult to make any generalizations about differences in maternal style toward male versus female infants.

## 4. Food Quality and Maternal Condition Influence Maternal Style

Data from other mammalian species suggest that when the food supply is severely restricted and lactating females are unable to maintain their body weight, mothers should be more rejecting and more likely to abandon or neglect their infants (Clutton-Brock, 1991; Lee *et al.,* 1991). Observations of vervet monkeys at Amboseli suggest that maternal neglect and infant mortality are related to maternal condition. Lee (1984) found that mothers living in the group with the lowest estimated food availability were more likely to wean their infants early compared with mothers in groups with a better quality food supply. In a later study of the same population, Hauser (1993) noted that mothers in the poorer quality habitats were more likely to stop responding to their infant's distress cries.

At the opposite extreme, in food-rich habitats, mothers who are in good condition should be able to provide food and care for their surviving offspring more efficiently than average mothers. Infants fed by well-nourished mothers would be likely to grow faster and achieve independence at an earlier age. Food-rich habitats would also provide better weaning foods for the infant. If early weaning by mothers in prime reproductive condition can increase the chance of producing an infant in the next year without increasing infant mortality, then mothers living in better ecological circumstances could increase their lifetime reproductive success by weaning their infants early (Lee *et al.,* 1991). Support for the hypothesis that better conditions lead to earlier weaning of surviving infants has come from free-ranging vervet monkeys and baboons at Amboseli. Vervet monkey mothers that lived in groups with a higher quality food supply were more rejecting toward their surviving infants and had shorter interbirth intervals than mothers in neighboring groups with poorer food quality (Hauser and Fairbanks, 1988). Similarly, high-ranking baboon mothers at Amboseli, who were likely to have better access to preferred foods, spent less time carrying their infants than lower-ranking mothers (Altmann and Samuels, 1992).

These results suggest that maternal style is responsive to ecological circumstances as a U-shaped function of maternal condition, with mothers at both extremes being more rejecting and average mothers being more indulgent and protective (Lee *et al.*, 1991). This hypothesis was supported by an analysis of maternal behavior at the Sepulveda vervet monkey colony (Fairbanks and McGuire, 1995). Mothers who were in marginal condition for reproduction, by virtue of their extremely young or old age and low body weight, were more rejecting than the average mother, and were more likely to lose their infants. Mothers in the best condition for reproduction were also more rejecting, and were able to convert their more efficient mothering style into higher fertility and shorter interbirth intervals without increasing infant mortality.

## C. Individual Consistency in Maternal Style across Conditions

While each of the variables discussed previously has been associated with differences in maternal behavior in one or more studies, consistency between studies and even between subsequent studies at the same site has not always been good. Several authors have noted that the effects of major variables, like the mother's rank, can be weak or inconsistent (Berman, 1984; Nicolson, 1987; Schino *et al.*, 1995). This variability can be partially explained by stable individual differences in maternal style that obscure the effects of demographic and contextual variables, particularly in relatively small samples.

Longitudinal research on maternal style in rhesus monkeys has demonstrated that there are stable individual differences in the way a mother treats her infants that persist when the influence of contextual variables that affect maternal behavior are controlled (Berman, 1990a). A similar pattern of consistency in maternal style across changing conditions has been found at the Sepulveda vervet monkey colony. Vervet monkey mothers respond to circumstances that predict greater risk for their infants by increasing their maternal protectiveness (Fairbanks, 1993a), but the magnitude of the changes are small compared with the magnitude of individual differences within conditions. This is illustrated in Fig. 3, which shows the maternal protectiveness scores for the same mothers who were observed with infants in years following the introduction of new adult males and in years with long-term resident males. Mothers who were observed under

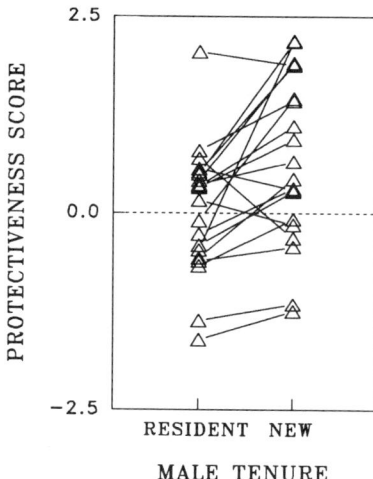

FIG. 3. Comparison of maternal protectiveness factor scores for individual mothers with long-term resident adult males and with new adult males (introduced in the past year) in the group.

both conditions were significantly more protective in the presence of new males than with resident males (matched $t = -3.49$, df $= 19$, $p < .01$). While almost all of the mothers adjusted their behavior in response to the new males, the majority stayed within one-half standard deviation of their baseline score (mean difference $= +.65$) and there was a strong positive correlation between a female's scores in the two conditions ($r = .64$, df $= 19$, $p < .01$). Mothers who were high on protectiveness in one condition were also high in the other, and a mother's score could be predicted more accurately by knowing her identity than by knowing her circumstances.

In the 10-year sample of mother–infant dyads at the Sepulveda colony, several variables are associated with maternal protectiveness that have been previously reported using smaller samples from the early years of the colony. As shown in Fig. 3, mothers are more protective in years with new adult males in the group compared with years with long-term resident males (Fairbanks and McGuire, 1987). Mothers who have lost infants in the preceding year are more protective toward their next infant compared with mothers who successfully raised their last infant (Fairbanks, 1988a), and young adult mothers who are living with the infant's grandmother are less protective than young adult females who have lost their mothers (Fairbanks, 1988b). Mother's rank also appears as a significant bivariate correlate of maternal protectiveness in this sample (Table III). When all

TABLE III
BIVARIATE CORRELATIONS AND SUMMARY TABLE OF STEPWISE REGRESSION ANALYSIS
PREDICTING MATERNAL PROTECTIVENESS SCORE TOWARD INDIVIDUAL INFANTS[a]

| Step | Variable entered | Bivariate $r$ | Entry level | Multiple R | RSQ | Change in RSQ | F to enter |
|---|---|---|---|---|---|---|---|
| 1 | Loss of grandmother | .43 | 1 | .43 | .19 | .19 | 27.21 |
| 2 | New versus resident male | .30 | 1 | .49 | .24 | .06 | 9.03 |
| 3 | Success of last pregnancy | −.27 | 1 | .55 | .31 | .06 | 10.44 |
| 4 | Rank | .30 | 1 | .59 | .35 | .04 | 7.48 |
| 5 | Mother's mean with other infants | .69 | 2 | .77 | .60 | .25 | 72.68 |

[a] $n = 121$.

of these variables are entered into a multiple regression analysis, each makes a significant independent contribution to predicting maternal protectiveness, but the best individual predictor of the mother's behavior toward her current infant is her average level of protectiveness with her other infants ($r = .69, p < .01$). The mother's behavior toward her other infants is still the best predictor of protectiveness to the current infant after all of the other variables are controlled by forcing them into the regression first (Table III).

The consistency of maternal rejection across infants of the same mother in this population is comparable to the results presented here for protectiveness (see Section II,A,3). Even though maternal behavior is responsive to circumstances to a certain degree, there are still strong stable components of maternal style that are consistent from infant to infant, over time, and across changing conditions.

### III. Consequences of Variation in Maternal Style for the Mother

Evolutionary biological theories of parental care predict a trade-off between effort expended in individual offspring and a mother's ability to produce and care for other offspring (Trivers, 1972; Pianka, 1976). Thus, we would expect that protective mothers who put more time and energy into infant care would suffer a loss of future fertility compared with mothers who were more rejecting. Consistent variation in maternal style should have consequences for the mother's lifetime reproductive success. Evidence is beginning to accumulate for several primate species that individual differences in maternal style do have consequnces for the mother's future reproduction.

A. MATERNAL REJECTION PROMOTES FUTURE FECUNDITY

Variation in maternal rejection has been shown to influence the mother's future fertility in macaques and in vervet monkeys. Rhesus monkey mothers from the Madingley colony who gave birth in the following season were more rejecting than mothers who skipped a year (Simpson et al., 1981; Gomendio, 1989b). Maternal rejection and mating activity peaked simultaneously for Japanese macaque females (Worlein et al., 1988; Collinge, 1991). Among provisioned free-ranging rhesus monkeys on Cayo Santiago and at Tughlaqabad, India, mothers who were more rejecting conceived sooner and gave birth earlier in the next season (Johnson et al., 1993; Berman et al., 1993). Rejection rate has also been associated with future fertility in free-ranging and captive vervet monkeys. Comparison of rates of maternal rejection between groups of free-ranging vervet monkeys at Amboseli demonstrated that groups with significantly higher rates of maternal rejection also had shorter average interbirth intervals (Hauser and Fairbanks, 1988), and captive mothers who rejected their infants more often had significantly shorter intervals to the birth of their next infant (Fairbanks and McGuire, 1987). An analysis with a larger sample of mothers from the captive colony indicated that rejecting mothers fell into two very different types (see previous Section II,B,4). Mothers in good condition for reproduction were able to increase their fertility by limiting suckling time through higher rejection rates, without increasing infant mortality (Fairbanks and McGuire, 1995). High rates of rejection were also found for adolescent mothers and mothers with low fat stores who were probably unable to sustain the full energetic costs of lactation. For these marginal mothers, more time was needed to recover from the effects of pregnancy and lactation, and their high rates of rejection were not associated with shorter-than-average interbirth intervals.

The mechanism by which maternal rejection shortens the interbirth interval is probably related to the fertility-suppressing effects of lactation. There is evidence in the human literature that lactation not only reduces fertility, but that the rate and timing of suckling may be as influential as the quantity of milk produced (Short, 1983). Frequent nursing bouts with short interbout intervals have been associated with delays in the onset of cycling in several primate studies (Lee, 1987; Stewart, 1988; Gomendio, 1989b).

B. MATERNAL PROTECTIVENESS AND DELAY OF THE NEXT CONCEPTION

Maternal protectiveness can interfere with a mother's future fertility in several ways. In the field, sustaining a high level of vigilance over infant safety might interfere with a mother's foraging efficiency, and longer periods of carrying a heavy infant would be likely to compromise the mother's ability to replace her fat stores. In captivity, protective mothers who spend more time in contact with their infants are likely to have more frequent suckling bouts. In addition to these direct effects of energy balance and lactation on fertility, higher levels of stress hormones associated with maternal protectiveness might also delay the resumption of cycling.

A few studies have provided evidence that a protective maternal style is associated with a reduction in fertility. Fairbanks (1988a) demonstrated that mothers who had lost an infant in the previous year were more protective than females who had successfully raised their last infant. The mothers with the past history of failure were more restrictive, they groomed and inspected their infants more often, and remained in proximity and in contact for a larger percentage of the time. Mothers in both groups successfully raised their current infants, but the more protective mothers had significantly longer intervals to the birth of their next infant.

In the 10-year sample of all surviving births at the vervet colony, there is a significant association of maternal protectiveness and interbirth interval. Interbirth intervals in the captive colony are typically 1 year and only a small percentage of females ever skip a year (8%). The high-quality food supply and low energy expenditure has allowed some females to produce two infants in the same year. Some multiparous females who give birth early in the season are able to produce a replacement infant if their first infant dies, while females who give birth late in the season do not. Because primiparous mothers only rarely have intervals that are less than a full year, regardless of maternal style, they have been excluded from this analysis). Approximately half (47%) of the multiparous mothers during the 10-year study period had interbirth intervals that were less than 12 months, and the other half had intervals that were 12 months or longer. The average level of maternal protectiveness for the mothers with short interbirth intervals was significantly lower than for mothers with longer interbirth intervals ($t = 2.12, p < .05$). Only 10% of the mothers who were highly protective (scores $>1.0$) gave birth in less than 12 months, while 65% of the mothers who were extremely low in protectiveness (scores $< -1.0$) were able to give birth again in less than 1 year (linear $\chi^2 = 6.85$, df $= 1, p < .01$).

### IV. Consequences of Variation in Maternal Style for the Offspring

#### A. Effects of Variation in Maternal Style for the Infant

The interactions of mothers and infants are complexly intertwined and the behavior of one necessarily affects the options and responses of the other. Infants play a large part in establishing and maintaining spatial proximity with their mothers, but most research has shown that individual differences in the mother–infant relationship in the first few months are largely controlled by the behavior of the mother (Hinde, 1974; Fairbanks and McGuire, 1987; Berman, 1990a).

### 1. Infant Mortality is Associated with Maternal Rejection and Neglect

The most serious consequence of variation in maternal style for the offspring is mortality. The young infant is completely reliant on its mother for survival and early neglect usually will lead to the infant's death. There is some evidence that a rejecting maternal style is associated with a higher level of infant mortality, while more attentive, protective mothering leads to a higher infant survival rate.

Infant mortality is common in the field, but the exact causes are rarely known. Altmann (1980) reported that ill health and infant death occurred at a higher rate among infants of laissez-faire mothers for baboons at Amboseli (5/7 versus 2/5 for infants of restrictive mothers). Studies of vervet monkeys at Amboseli have also indicated a relationship between maternal rejection and infant mortality. Rejection of suckling attempts and early weaning were associated with greater infant mortality (Lee, 1984), and infant mortality was high for infants whose mothers stopped responding to their distress cries (Hauser, 1993).

In the Sepulveda vervet monkey colony, maternal rejection is relatively high among high-ranking, prime-age mothers who are able to raise their infants successfully in the protected and food-rich captive environment with less maternal effort (Fairbanks and McGuire, 1995). No differences in infant mortality have been found between more rejecting high-ranking mothers and less rejecting middle- and low-ranking mothers in the vervet monkey colony. However, there is a significant increase in infant mortality related to maternal rejection and neglect by mothers in marginal condition for reproduction. Three-year-old mothers who are not fully mature, and mothers who are below a minimum body weight at the time of delivery, are more likely to reject and abandon their infants, leading to significantly higher rates of early infant mortality (Fairbanks and McGuire, 1995).

### 2. Rejecting Mothers May Have More Enterprising Infants

The general expectation of early researchers was that high levels of maternal rejection would result in behavioral problems for the infant (Hinde and Simpson, 1975; Rosenblum, 1971). In fact, most results have demonstrated that infants who survive relatively high levels of maternal rejection develop independence at an earlier age and may even be more enterprising and resourceful than their more protected peers (Sackett and Ruppenthal,

1973; Hinde, 1974; Nash and Wheeler, 1982; Simpson and Datta, 1991). Correlations of maternal rejection with infant behaviors typically demonstrate that infants of rejecting mothers play a greater part in promoting and maintaining proximity and contact with their mothers compared with infants of less rejecting mothers (Hinde, 1974; Simpson and Howe, 1980; Fairbanks and McGuire, 1987; Collinge, 1991). They also spend more time away from the mother exploring the environment and spend more time interacting with social companions (Hinde, 1974; Simpson and Simpson, 1986; Suomi, 1987; L. A. Fairbanks, unpublished data). Rhesus infants with higher rates of initiating social contact with group companions other than the mother were shown to be less fearful and more willing to explore a novel object (Simpson and Howe, 1986). In a later test situation, Simpson et al. (1989) compared infants who had experienced high levels of risk, either by having extremely rejecting mothers or by being exposed to social aggression from other group members, with infants of high-ranking mothers who were relatively immune from intragroup aggression. The infants in the high-risk group were more likely to enter a novel section of the cage and succeeded in getting a greater number of raisins than infants in the low-risk group.

While maternal rejection may promote infant independence, the experience for the infant is not completely benign. Infants of more rejecting mothers have more tantrums, give more distress cries, and appear to be more disturbed by forced separation from the mother (Hinde and Simpson, 1975; Fairbanks and McGuire, 1987; Suomi, 1987; Collinge, 1991). More research is needed to determine if the effects of maternal rejection differ between moderate and extreme levels, and if there are interactions between a rejecting maternal style and infant temperament (Suomi, 1987).

3. *Infants of Protective, Restrictive Mothers Are More Cautious in Response to Novelty*

A variety of studies have indicated that protective and restrictive early mothering produces infants who are relatively fearful and cautious when faced with novel and challenging situations. Infants with anxious or protective mothers are generally held back in their early explorations of the physical and social world. Vochteloo et al. (1993) demonstrated that when mothers are spatially restricted, their infants are developmentally delayed in moving beyond arm's reach and exploring other sections of the home cage. Growing infants use their mother as a secure base for increasingly larger and longer excursions away. There is a time period at about 4–5 months when the frequency of approaching and leaving by infants peaks as they make continued short trips to and from their mother's side. Vervet monkey infants with protective, restrictive mothers took a month longer to reach this point in the development of independence compared with infants with more laissez-faire, relaxed mothers (Fairbanks, 1988b). When the mother–infant relationship was disrupted by imposing a foraging regime that was unpredictable and variable in difficulty, mothers compensated by breaking contact with their infants more often but also by increasing the amount of time they groomed their infants and decreasing the length of time they spent out of contact (Andrews and Rosenblum, 1991, 1993). Infants raised under these circumstances did not show greater distress to separation from the mother, but they were significantly more cautious in a novel test cage. They stayed near their mothers more often, and spent less time playing and exploring compared with infants whose mothers were on a more consistent and easier foraging regime.

At this point we can only speculate about the consequences of being enterprising or cautious for infants living under free-ranging conditions. In an environment with predators and potentially lethal physical hazards, a certain degree of caution would seem to be advantageous. On the other hand, a bolder infant would have more opportunity to develop social relationships and gain experiences that might be valuable in later life. An infant that takes more initiative in meeting its needs might have a greater chance of surviving a brief separation, or even the death of its mother. Within a primate social group, there are individual differences in access to resources and relative risks from social companions. The response of the infant's mother to environmental and social cues is probably a very good predictor of the relative danger to the infant. Thus correlations between maternal behavior and infant cautiousness would serve to adapt the infant to the risks and opportunities faced by its matriline within the larger group.

### B. Maternal Style and Juvenile Response to Novelty

To what extent do the individual differences in the early mother–infant relationship influence behavior during the juvenile period? Two different challenge tests conducted at the Sepulveda vervet monkey colony suggest that the lessons learned in infancy continue to influence how juveniles respond to novel and potentially threatening situations.

In 1986, we conducted a test to evaluate individual differences in exploration and response to novelty. At that time, all but four of the animals in the original two groups were born in the colony and had never been outside of their home enclosure. The response of the monkeys to the opening of a tunnel connecting their home enclosure to one of two newly constructed enclosures was used as a measure of individual differences along a bold–cautious dimension in response to novel and potentially threatening circumstances (Fairbanks and McGuire, 1988). Data collected on the latency of each animal in the home group to enter the new enclosure showed some interesting patterns. Adult males were relatively conservative and waited until several animals had gone through the tunnel before they entered. The latency of entry by adult females could be predicted by the age of their youngest infant. Females with the smaller infants were more conservative and took longer to enter than females with older, more independent infants. On average, the juveniles had shorter latencies to enter the new enclosure than adults. Their order of entry was not related to their mother's current behavior, but it could be predicted by the quality of maternal care they had experienced as infants. Figure 4 shows the latency to enter the new enclosure for the 22 juveniles between 1 and 3 years of age by early maternal protectiveness. Juveniles who had more protective mothers were more cautious and took significantly longer to enter the new enclosure than juveniles who had less protective mothers ($r = .62, p < .01$).

The same pattern of results was observed 2 years later when novel food containers were placed in the four enclosures (Fairbanks and McGuire, 1993). The monkeys in all four groups responded with excitement and apprehension to the novel objects that were different in appearance and location from the old food bins. The order and latency to approach within 1 m demonstrated the same tendency for mothers to respond to the age and vulnerability of their youngest infants, and the same tendency for juvenile response to be predicted by the quality of early maternal care. This test also allowed us to determine that the similarity between maternal style and juvenile response to novelty could not be completely explained

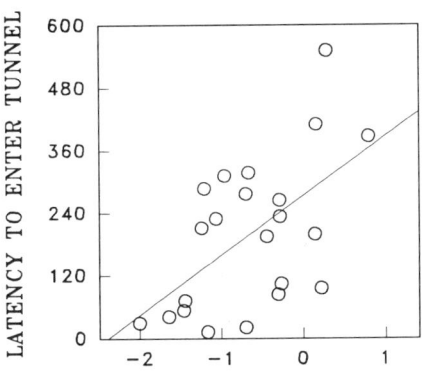

Fig. 4. Juvenile latency to enter a new enclosure by maternal protectiveness experienced in infancy.

by inherited temperamental similarities between mothers and offspring. The introduction of new males in the preceding year causes virtually all females to increase in maternal protectiveness (see Fig. 3), and new male introductions were performed in all groups at periodic intervals, independent of the style of the mothers. When juveniles who had been born in new male years were compared with juveniles born in resident male years, the new male juveniles were significantly more cautious and took longer to approach the novel object. This suggests that environmentally induced increases in maternal protectiveness can have enduring effects on juvenile response to novel and potentially threatening situations.

In the field, the juvenile period is a time of high mortality and in many primate populations the majority of juveniles do not reach adulthood (Dunbar, 1987). In spite of the risks, however, juveniles are generally more curious and quicker to approach in novel situations than are infants or adults (Fairbanks, 1993b). The value of juvenile curiosity has been the subject of much discussion, but the costs have not really been measured. More research in natural settings is needed to assess the effects of individual differences in juvenile response to novelty on survival to adulthood, and on adult competitive ability and reproductive success.

### C. Maternal Style and Adolescent Male Response to Strangers

Vervet monkey males emigrate from their natal group to breed between 3 and 6 years of age. This is a crucial time of life for young males as they must leave the predictable world they grew up in. Emigrating males face the challenges of adjusting to new home ranges, competing with adult males, and winning the support of adult females in a new group. Temperamental characteristics of an individual are likely to influence how they respond to these challenges.

At the Sepulveda vervet monkey colony, we performed a challenge test designed to assess individual differences along the bold–fearful dimension that would be associated with successful emigration and immigration. At the age when males would normally be emigrating from the natal troop, they were presented with an unfamiliar adult male at the edge of their home cage. Males were tested in their home cage in the company of two or three other adolescent males. The adult male stranger, confined to an individual cage, was placed in contact with the fence of the home enclosure and the response of each natal adolescent male was recorded for a 30-min period.

The presence of a strange adult male evoked a strong reaction from all of the adolescent males, but there were marked differences in the way individual males responded. The boldest males spent virtually the entire

30 min session in front of the intruder's cage, displaying and threatening to the stranger. They readily approached and challenged the strange adult males, and some even tried to make affiliative contact. The most fearful males, in contrast, spent most of the time in the far section of the enclosure and never approached within 1 m of the stranger. These males threatened and yawned at the strange male from a distance while avoiding direct confrontation.

The 36 adolescent males who were tested in the strange male challenge situation were given a boldness score based on the amount of time they spent in three sections of the enclosure at varying distances from the stimulus animal. A maximum score of 3.0 indicated the male spent the entire session in the section of the enclosure closest to the strange male, while a minimum score of 1.0 indicated that the male spent the entire session in the section farthest from the stranger. This score correlated positively with challenging and assertive behaviors, including display ($r = .56, p < .01$), penile erection ($r = .48, p = .01$), approaching within 1 m ($r = .68, p < .01$), sniffing ($r = .52, p < .01$), and touching ($r = .40, p < .05$) the stranger.

Individual differences in response of the adolescent males to the adult male stranger could be predicted by the maternal style they had experienced in infancy (Fig. 5). There was a strong, significant correlation between early maternal rejection and adolescent male boldness in the strange male challenge test ($r = .55, n = 36, p < .01$). There was also a significant inverse relationship of early mother–infant contact with the boldness score ($r = -.38, n = 36, p < .05$). Males who had relatively rejecting mothers and who spent less time in contact with their mothers as infants were more

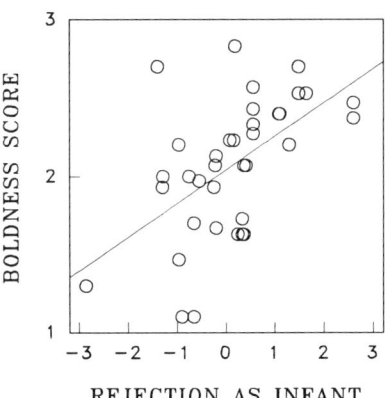

FIG. 5. Adolescent male boldness score in response to adult male stranger by maternal rejection experienced in infancy.

likely to be bold and challenge the adult male stranger as adolescents. Males who had had high levels of early mother–infant contact were more likely to be cautious and fearful in response to the male stranger.

Because maternal rejection is related to the mother's dominance rank in this sample, it is possible that family rank was the primary factor influencing individual differences in male boldness. The mother's current rank was not significantly correlated with the adolescent males' responses ($r = .10$, df = 35, NS), but the males' birth rank was ($r = .37$, df = 35, $p < .01$). When birth rank was controlled through partial correlation, maternal rejection was still a significant predictor of adolescent male boldness in response to the adult male stranger (partial $r = .44$, df = 34, $p < .01$). Infants who survived a moderately rejecting early experience became more independent and more socially assertive, and these characteristics prepared them for the social challenges of adolescence.

### D. Maternal Style and Maternal Behavior of Adult Daughters

Longitudinal research with rhesus monkeys on Cayo Santiago and vervet monkeys at the Sepulveda colony have both demonstrated similarity in maternal style between mothers and their adult daughters. In 1988, an analysis of continuity of mother–infant contact from infancy to adulthood was conducted with data from mothers who had been observed since infancy at the Sepulveda vervet monkey colony (Fairbanks, 1989). At that time, we had data for 15 adult daughters on the percentage time they had spent in contact with their mothers as infants and the percentage time they spent in contact with their infants as mothers. Data on all of the mothering behavior observed by these females when they were juveniles and young adults were also available, so the influence of early experience could be contrasted with observational learning of mothering style and with situational determinants of variation in mother–infant contact. The genetic contribution to similarity between mothers and daughters was estimated by calculating the average of the mothers' mother–infant contact across all infants, regardless of the daughters' experience. The correlations of these measures with mother–infant contact are shown in Table IV. Mother–infant contact for the adult daughters was positively correlated with the mother–infant contact they experienced as infants, with the average mother–infant contact of their mothers, and with the maternal behavior of their mothers that they observed as juveniles, but not with the mother–infant contact of other adult females in the group or with their birth rank. Partial correlations with the three interrelated variables that predicted adult mother–infant contact indicated that mother–infant contact experienced in infancy was the primary source of the individual differences.

TABLE IV
PEARSON CORRELATIONS OF EARLY EXPERIENCE OF MOTHER–INFANT (MI) CONTACT WITH ADULT MOTHER–INFANT CONTACT BEHAVIOR

|  | Mean MI contact of adult daughter | Partial correlation | |
|---|---|---|---|
| MI contact as infant (A) | .69** | Controlling B | .59* |
|  |  | Controlling C | .52* |
| Mean MI contact of mother (B) | .54* | Controlling A | .17 |
| Mean MI contact of mother observed as juvenile (C) | .44 | Controlling A | .12 |
| Mean MI contact of other adult females observed as juvenile (D) | −.07 |  |  |
| Birth rank (E) | .28 |  |  |

\* $p < .05$
\*\* $p < .01$

A similar set of analyses demonstrating cross-generational consistency in maternal style were performed by Berman (1990b) for rhesus monkeys on Cayo Santiago. The average rejection rate of rhesus mothers was significantly correlated with the rejection rate of their mothers. This similarity was partially due to shared dominance rank between mothers and adult daughters, but when rank was controlled, the mother–daughter similarity still held. The rejection rate for the rhesus mothers could not be predicted by the rejection rate they experienced as infants, but was correlated with their mothers' rejection of younger siblings, leading Berman to conclude that daughters may match their rejection behavior to that of their mothers by attending to their mothers' behavior toward their younger siblings.

In contrast to the results from Cayo Santiago, similarity of maternal rejection rates between mothers and daughters at the Sepulveda vervet monkey colony is mediated by similarity in rank. There have been several rank changes in the colony associated with the group fissions and with the death of older alpha females. When maternal rejection of mothers is compared with maternal rejection shown by their adult daughters, the correlation is positive and significant when mothers and daughters share the same rank ($r = .49$, df = 21, $p < .05$). When mothers and adult daughters differ in rank, the correlation disappears ($r = .19$, df = 23, NS).

Results at the Sepulveda colony and at Cayo Santiago both demonstrate continuity of maternal style across generations. In primate groups with a strong matrilineal structure, the social situation of a daughter will be more similar to that of her mother than to the average female in the group. Mothers and adult daughters will share a similar level of control over the behavior of other group members, and a similar risk from intragroup aggression. Because her mother was successful in raising her, we would expect that a young adult female would increase the probability of successfully raising her own infant by matching her maternal behavior to that of her mother. More research is needed to unravel the relative contributions of genetics, experience in infancy, social learning as a juvenile, and shared current circumstances in explaining these similarities.

## V. Proximate Mechanisms

When making functional arguments, we do not presume that the monkey mother is actually calculating the effect of her behavior on her future reproductive success or on her offspring's fitness. We expect that natural selection has produced proximate mechanisms, both emotional and physiological, that control maternal behavior. A mother with a negative energy balance may feel too tired to carry her infant, a low-ranking mother whose infant is at risk from other group members may feel anxious, and a mother with low levels of circulating prolactin may not want to nurse her infant. Research into proximate mechanisms controlling maternal style has focused on two different levels of analysis, psychological and physiological.

### A. Psychological Variables

At the psychological level, anxiety and the subjective experience of stress have been postulated as intervening proximate mechanisms influencing maternal protectiveness. In a field study of baboon mothers, Altmann (1980) measured glance rate as an indicator of general attentiveness or nervousness. A dichotomous classification into restrictive or laissez-faire maternal style could be made on the basis of glance rate for six of the seven mothers who were sampled. Restrictive mothers had higher glance rates than laissez-faire mothers. This result is consistent with the idea that maternal style is a reflection of the mothers' general anxiety level.

Further research has confirmed the relationship between objective measurements of maternal anxiety and maternal style. Maestripieri et al. (1992) demonstrated that the rate of scratching and other forms of autogrooming can be used as an indicator of anxiety level in primates because scratching increases in contexts that would be expected to produce anxiety, and decreases with administration of anxiolytic medications. Maestripieri (1993a) showed that both visual monitoring and scratching by rhesus monkey mothers increased in situations where the infant was in greater danger. Individual differences between mothers in level of maternal protectiveness could be predicted by individual differences in visual monitoring in the rhesus mothers (Maestripieri, 1993b), and maternal protectiveness was positively correlated with scratching rate in Japanese macaques (Troisi *et al.,* 1991). In both cases, mothers showing a higher level of anxiety were more protective toward their infants.

While maternal anxiety has been discussed as the emotional concomitant of protectiveness, there has been no comparable speculation about the feeling states that underlie maternal rejection. Feelings of fatigue or exhaustion, irritability, annoyance, or anger might all lead to increases in rejection of an infant's attempts to suckle or be carried. The lack of energy that goes with fatigue or ill health could probably be measured objectively, independent of maternal behavior. Irritation and annoyance would be more difficult to define operationally, but both would be likely candidates for the subjective emotions that accompany rejection of infants by high-ranking mothers. All of these psychological mechanisms await effective operational definitions before we can understand their roles in proximate control of maternal behavior.

B. Physiological Mechanisms

Given the major interest in reproductive physiology and the hormonal basis for fertility, lactation, and maternal responsiveness in mammals, it is surprising that there has been very little work on the physiology of individual differences in maternal style (for reviews, see Coe, 1990; Cortert and Fleming, 1990; Knobil and Neill, 1993). Most of the emphasis in research on physiological mechanisms of maternal behavior has been on the factors that influence the onset of maternal responsiveness, and on changes in hormonal and neurological variables through gestation, parturition, and lactation, and not on those that predict temperament-based variation in maternal style within a species.

If maternal protectiveness is related to individual differences in temperament along the shy–bold dimension, as we have suggested (Fairbanks, 1989; Fairbanks and McGuire, 1993), then we would expect the phsyiological concomitants of this dimension of temperament to correlate with the individual differences in maternal behavior described previously. Shy–bold differences in human children are believed to be caused by individual differences in physiological responsiveness to novel and stressful situations. Children who are fearful and inhibited in challenging situations have been shown to have higher and less variable heart rates and higher cortisol levels than children who are more relaxed and confident (Kagan *et al.,* 1988). In nonhuman primates, similar individual differences in physiological reactivity of heart rate and endocrine response to stress have been documented and related to behavioral differences in response to novel and stressful circumstances for infants, juveniles, and adult males (Suomi, 1983; Mendoza and Mason, 1989; Rasmussen and Suomi, 1989; Sapolsky, 1989; Clarke & Boinski, 1995). This would be a promising area for research into the physiological correlates of individual differences in maternal style, particularly for variation along the protectiveness dimension.

Studies of brain biochemistry have found a relationship between activity of the serotonin and dopamine neurotransmitter systems and behavioral differences in primates. Individual differences in these central nervous system metabolites, measured in cisternal cerebrospinal fluid (CSF), tend to be relatively stable within individuals, making them good candidates for proximal determinants of consistent individual differences in temperament (Higley et al., 1992a; Raleigh et al., 1992). Concentration of the serotonin metabolite, 5-hydroxyindole-acetic acid (5-HIAA), has been associated with behavioral differences in dominance, impulsivity, and aggressiveness in adult male vervet and rhesus monkeys (Raleigh et al., 1985; Higley et al., 1992b), and the timing of emigration from the natal troop has also been related to CSF serotonin activity for rhesus monkeys on Cayo Santiago (Kaplan et al., 1995). The dopamine system, with homovanillic acid (HVA) as its principal metabolite, has been associated with individual differences in fearfulness and novelty seeking in humans and dogs (Cloninger, 1987; Gurguis et al., 1990). Behavioral correlates of spontaneously occurring individual differences in CSF HVA have not yet been reported for nonhuman primates, but several studies have noted differences in HVA in groups that have been subjected to stressful developmental experiences such as peer rearing or separation from the mother (Bayart et al., 1990; Weiner et al., 1990; Higley et al., 1991). Preliminary results from the Sepulveda vervet monkey colony suggest that 5-HIAA and HVA are related to both maternal protectiveness and to offspring boldness (Raleigh and Fairbanks, in prep.). Protective mothers have offspring who are relatively cautious and fearful, and both the mothers and the offspring are relatively low in both 5-HIAA and HVA.

The physiological basis for maternal rejection, with its possibility of detrimental effects for the offspring, has received surprisingly little attention. None of the CSF or peripheral measures taken in the pilot study of vervet monkey mothers correlated well with individual differences in maternal rejection. This is an area that deserves attention in the future.

## VI. Conclusions

The research described here suggests that there are stable individual differences in the way adult females respond to the challenges of mothering.

Mothers will modify their maternal behavior according to the immediate demands of the current environment and according to their own personal history to a certain extent, but most of the variance in maternal behavior is due to stable individual differences in maternal style that persist across conditions. Research on the reproductive consequences of variation in maternal style suggest that there may be trade-offs between offspring production and infant mortality. Mothers who are more rejecting can produce more offspring, but mothers who are more protective may have increased infant survival. From the infant's point of view, a high degree of maternal rejection increases the risk of infant mortality, but a moderately high level of rejection in a generally safe environment appears to produce offspring who are more resourceful, more socially self-confident, and more enterprising. A high degree of maternal protectivness communicates fearfulness and produces offspring who are more cautious in unfamiliar situations.

Discussions of maternal style and temperament in the psychological literature often imply the value judgments that it is better to be a warm, nurturing mother than a rejecting one, and it is better for offspring to be bold than to be cautious. From an evolutionary biological point of view, however, these value judgments are irrelevant. If a mother can increase her lifetime reproductive success by being rejecting, she should do so (Clark and Ehlinger, 1987; Fairbanks, 1993a). If a juvenile can live longer by being cautious in novel circumstances, then cautiousness is a more desirable trait than boldness (Fairbanks, 1993b). We can speculate that it is adaptive for offspring to use their mothers as a source of information about environmental dangers and to match their thresholds of responsiveness to hers. Much more information about the life-history consequences of individual differences in temperament in natural populations is needed before we can begin to understand to what extent the variation we observe in maternal style in captive and provisioned settings reflects random versus adaptive variation, and how the environmental context influences the life-history consequences of individual differences (Wilson *et al.*, 1994).

Quantitative research on individual differences in maternal behavior is heavily weighted toward a few species specifically rhesus and Japanese macaques, baboons, and vervet monkeys. At this point, we do not know how far the basic facts about maternal style and the cross-generational continuity of individual differences described previously can be generalized to other primates. Whether the patterns observed within species can be extended to explain evolutionary trends and differences between species is also yet to be determined (Maestripieri, 1994; Clark and Boinski, 1995).

The research discussed here does not resolve the nature versus nurture dichotomy as a root cause of individual differences in maternal style (Roubertoux *et al.*, 1990). It does underscore the far-reaching influence of these individual differences in predicting offspring behavior at different stages of the life cycle. Primate infants inherit more than genes from their mother. In a cultural sense, they inherit her home range, her social companions, and her rank, and grow up in the developmental niche produced by her individual maternal style (West and King, 1987). For socially living primates, with a long period of dependency and overlapping generations, maternal behavior is the context of development and individual differences in maternal style can be used to predict offspring behavior. Genetic influences on individual differences in temperament can be identified when the contribution of maternal behavior to offspring development is eliminated (Schneider et al., 1991), but a true understanding of the contribution of individual differences in maternal style must involve control and manipulation of maternal behavior within the normal range (Suomi, 1987). Attempts were made in several of the studies described previously to estimate the contribution of maternal behavior independent of genetically inherited similarities between mothers and offspring, by manipulating a social variable that influenced maternal protectiveness or by using partial correlation analysis. More research is needed before we can understand how maternal environment and maternal genes combine to produce individual differences in offspring behavior.

### Acknowledgments

The research described in this article was supported by grants from the National Science Foundation (BNS 84-02292, BNS 87-09765, and BNS 91-108017) to the author, and from Veterans Administration Merit Review grants to Michael T. McGuire. I would like to thank Karin Blau, Michaela Heeb, Jill Kusnitz, Dan Diekmann, and Pess Morton for assistance with behavioral observations and animal care; and my colleagues, Michael McGuire and Michael Raleigh, for intellectual and logistical support throughout the years.

### References

Altmann, J. (1980). "Baboon Mothers and Infants." Harvard Univ. Press, Cambridge, MA.
Altmann, J., and Samuels, A. (1992). Costs of maternal care: Infant-carrying in baboons. *Behav. Ecol. Sociobiol.* **29,** 391–398.
Andrews, M. W., and Rosenblum, L. A. (1991). Attachment in monkey infants raised in variable- and low-demand environments. *Child Dev.* **62,** 686–693.
Andrews, M. W., and Rosenblum, L. A. (1993). Assessment of attachment in differentially reared infant monkeys (*Macaca radiata*): Response to separation and a novel environment. *J. Comp. Psychol.* **107,** 84–90.
Bayart, F., Hayashi, K. T., Faull, K. F., Barchas, J. D., and Levine, S. (1990). Influence of maternal proximity on behavioral and physiological responses to separation in infant rhesus monkeys (*Macaca mulatta*). *Behav. Neurosci.* **104,** 98–107.

Berman, C. M. (1980). Mother-infant relationship among free-ranging rhesus monkeys on Cayo Santiago: A comparison with captive pairs. *Anim. Behav.* **28,** 860–873.

Berman, C. M. (1984). Variation in mother-infant relationships: Traditional and nontraditional factors. *In* "Female Primates: Studies by Women Primatologists" (M. F. Small, ed.), pp. 17–36. Alan R. Liss, New York.

Berman, C. M. (1989). Trapping activities and mother-infant relationships on Cayo Santiago: A cautionary tale. *P. R. Health Sci. J.* **8,** 73–78.

Berman, C. M. (1990a). Consistency in maternal behavior within families of free–ranging rhesus monkeys: An extension of the concept of maternal style. *Am. J. Primatol.* **22,** 159–169.

Berman, C. M. (1990b). Intergenerational transmission of maternal rejection rates among free-ranging rhesus monkeys. *Anim. Behav.* **39,** 239–247.

Berman, C. M. (1992). Immature siblings and mother-infant relationships among free-ranging rhesus monkeys on Cayo Santiago. *Anim. Behav.* **44,** 247–258.

Berman, C. M., Rasmussen, K., and Suomi, S. J. (1993). Reproductive consequences of maternal care patterns during estrus among free-ranging rhesus monkeys. *Behav. Ecol. Sociobiol.* **32,** 391–399.

Clark, A. B., and Ehlinger, T. J. (1987). Pattern and adaptation in individual behavioral differences. *In* "Perspectives in Ethology," Vol. 7 (P. P. G. Bateson and P. H. Klopfer, eds.), pp. 1–47. Plenum, New York.

Clarke, A. S., and Boinski, S. (1995). Temperament in nonhuman primates. *Am. J. Primatol.,* **37,** 103–125.

Clarke, A. S., Mason, W. A., and Moberg, G. P. (1988). Differential behavioral and adrenocortical responses to stress among three macaque species. *Am. J. Primatol.* **14,** 37–52.

Cloninger, C. R. (1987). A systematic method for clinical description and classification of personality variants. *Arch. Gen. Psych.* **44,** 573–588.

Clutton-Brock, T. H. (1991). "The Evolution of Parental Care." Princeton Univ. Press, Princeton, NJ.

Coe, C. L. (1990). Psychobiology of maternal behavior in nonhuman primates. *In* "Mammalian Parenting: Biochemical, Neurobiological and Behavioral Determinants" (N. A. Krasnegor and R. S. Bridges, eds.), pp. 157–183. Oxford Univ. Press, New York.

Collinge, N. E. (1991). Variability in aspects of the mother-infant relationship in Japanese macaques during weaning. *In* "The Monkeys of Arashiyama" (L. M. Fedigan and P. J. Asquith, eds.), pp. 157–174. SUNY Press, New York.

Corter, C. M., and Fleming, A. S. (1990). Maternal responsiveness in humans: Emotional, cognitive, and biological factors. *Adv. Study Behav.* **19,** 83–135.

DeVore, I. (1963). Mother-infant relations in free-ranging baboons. *In* "Maternal Behavior in Mammals" (H. L. Rheingold, ed.), pp. 305–335. John Wiley & Sons, New York.

Dunbar, R. I. M. (1987). Demography and reproduction. *In* "Primate Societies" (B. B. Smuts, D. L. Cheney, R. M. Seyfarth, R. W. Wrangham, and T. T. Struhsaker, eds.), pp. 240–249. Univ of Chicago Press, Chicago.

Eaton, G. G., Johnson, D. F., Glick, B. B., and Worlein, J. M. (1985). Development of Japanese macaques (*Macaca fuscata*): Sexually dimorphic behavior during the first year of life. *Primates* **26,** 238–248.

Fairbanks, L. A. (1988a). Mother-infant behavior in vervet monkeys: Response to failure of last pregnancy. *Behav. Ecol. Sociobiol.* **23,** 157–165.

Fairbanks, L. A. (1988b). Vervet monkey grandmothers: Effects on mother-infant relationships. *Behaviour* **104,** 176–188.

Fairbanks, L. A. (1989). Early experience and cross-generational continuity of mother-infant contact in vervet monkeys. *Dev. Psychobiol.* **22,** 669–681.

Fairbanks, L. A. (1993a). What is a good mother? Adaptive variation in maternal behavior of primates. *Curr. Dir. Psychol. Sci.* **2,** 179–183.

Fairbanks, L. A. (1993b). Risk taking by juvenile vervet monkeys. *Behaviour* **124,** 57–72.

Fairbanks, L. A., and McGuire, M. T. (1987). Mother-infant relationships in vervet monkeys: Response to new adult males. *Int. J. Primatol.* **8,** 351–366.

Fairbanks, L. A., and McGuire, M. T. (1988). Long-term effects of early mothering behavior on responsiveness to the environment in vervet monkeys. *Dev. Psychobiol.* **21,** 711–724.

Fairbanks, L. A., and McGuire, M. T. (1993). Maternal protectiveness and response to the unfamiliar in vervet monkeys. *Am. J. Primatol.* **30,** 119–129.

Fairbanks, L. A., and McGuire, M. T. (1995). Maternal condition and the quality of maternal care in vervet monkeys. *Behaviour* **132,** 733–754.

Gomendio, M. (1989a). Differences in fertility and suckling patterns between primiparous and multiparous rhesus mothers (*Macaca mulatta*). *J. Reprod. Fertil.* **87,** 529–542.

Gomendio, M. (1989b). Suckling behaviour and fertility in rhesus macaques (*Macaca mulatta*). *J. Zool. (London)* **217,** 449–467.

Gomendio, M. (1990). The influence of maternal rank and infant sex on maternal investment trends in rhesus macaques: Birth sex ratio, inter-birth intervals and suckling patterns. *Behav. Ecol. Sociobiol.* **27,** 365–375.

Gurguis, G. N. M., Kelin, E., Mefford, I. N., and Uhde, T. W. (1990). Biogenic amines distribution in the brain of nervous and normal pointer dogs: A genetic animal model of anxiety. *Neuropsychopharmacology* **3,** 297–303.

Hauser, M. D. (1988). Variation in maternal responsiveness in free-ranging vervet monkeys: A response to infant mortality risk. *Am. Nat.* **131,** 573–587.

Hauser, M. D. (1993). Do vervet monkey infants cry wolf? *Anim. Behav.* **45,** 1242–1244.

Hauser, M. D., and Fairbanks, L. A. (1988). Mother-offspring conflict in vervet monkeys: Variation in response to ecological conditions. *Anim. Behav.* **36,** 802–813.

Higley, J. D., Suomi, S. J., and Linnoila, M. (1991). CSF monamine metabolite concentrations vary according to age, rearing, and sex, and are influenced by the stressor of social separation in rhesus monkeys. *Psychopharmacology* **103,** 551–556.

Higley, J. D., Suomi, S. J., and Linnoila, M. (1992a). A longitudinal assessment of CSF monamine metabolite and plasma cortisol concentrations in young rhesus monkeys. *Biol. Psych.* **32,** 127–145.

Higley, J. D., Mehlman, P. T., Taub, D. M., Higley, S. B., Suomi, S. J., Linnoila, M., and Vickers, J. H. (1992b). Cerebrospinal fluid monoamine and adrenal correlates of aggression in free-ranging rhesus monkeys. *Arch. Gen. Psych.* **49,** 436–441.

Hinde, R. A. (1974). "Biological Bases of Human Social Behaviour." McGraw-Hill, New York.

Hinde, R. A., and Spencer-Booth, Y. (1971). Towards understanding individual differences in rhesus mother-infant interaction. *Anim. Behav.* **19,** 165–173.

Hinde, R. A., and Simpson, M. J. A. (1975). Qualities of mother-infant relationships in monkeys. *In* (CIBA Symposium 33) "Parent-Infant Interaction" pp. 39–67. Elsevier, Amsterdam.

Hinde, R. A., and Spencer-Booth, Y. (1971). Towards understanding individual differences in rhesus mother-infant interaction. *Anim. Behav.* **19,** 165–173.

Hiraiwa, M. (1981). Maternal and alloparental care in a troop of free-ranging Japanese monkeys. *Primates* **22,** 309–329.

Hooley, J. M. (1983). Primiparous and multiparous mothers and their infants. *In* "Primate Social Relationships" (R. A. Hinde, ed.), pp. 142–145. Sinauer Assoc., Sunderland, MA.

Hooley, J. M., and Simpson, M. J. A. (1981). A comparison of primiparous and multiparous mother-infant dyads in *Macaca mulatta*. *Primates* **22,** 379–392.

Jay, P. (1963). Mother-infant relations in langurs. *In* "Maternal Behavior in Mammals" (H. L. Rheingold, ed.), pp. 282–304. John Wiley & Sons, New York.

Johnson, R. L., and Southwick, C. H. (1984). Structural diversity and mother-infant relations among rhesus monkeys in India and Nepal. *Folia Primatol.* **43,** 198–215.

Johnson, R. L., Berman, C. M., and Malik, I. (1993). An integrative model of the lactational and environmental control of mating in female rhesus monkeys. *Anim. Behav.* **46,** 63–78.

Kagan, J., Reznick, J. S., and Snidman, N. (1988). Biological bases of childhood shyness. *Science (Washington, D. C.)* **240,** 167–171.

Kaplan, J. R., Fontenot, M. B., Berard, J., Manuck, S. B., and Mann, J. J. (1995). Delayed dispersal and elevated monoaminergic activity in free-ranging rhesus monkeys. *Am. J. Primatol.* **35,** 229–234.

Karssemeijer, G. J., Vos, D. R., and van Hooff, J. A. R. A. M. (1990). The effect of some non-social factors on mother-infant contact in long-tailed macaques (*Macaca fascicularis*). *Behaviour* **113,** 273–291.

Kaufman, I. C., and Rosenblum, L. A. (1969). The waning of the mother-infant bond in two species of macaque. *In* "Determinants of Infant Behaviour," Vol. 4 (B. M. Foss, ed.), pp. 41–59. Methuen, London.

Knobil, E., and Neill, J. D. (1993). "The Physiology of Reproduction" 2nd ed. Raven Press, New York.

Lee, P. C. (1984). Ecological constraints on the social development of vervet monkeys. *Behavior* **91,** 245–262.

Lee, P. C. (1987). Nutrition, fertility and maternal investment in primates. *J. Zool. (London)* **213,** 409–422.

Lee, P. C., Majluf, P., and Gordon, I. J. (1991). Growth, weaning and maternal investment from a comparative perspective. *J. Zool. (London)* **225,** 99–114.

Maestripieri, D. (1993a). Maternal anxiety in rhesus macaques (*Macaca mulatta*) I. Measurement of anxiety and identification of anxiety-eliciting situations. *Ethology* **95,** 19–31.

Maestripieri, D. (1993b). Maternal anxiety in rhesus macaques (*Macaca mulatta*) II. Emotional bases of individual differences in mothering style. *Ethology* **95,** 32–42.

Maestripieri, D. (1994). Mother-infant relationships in three species of macaques (*Macaca mulatta, M. nemestrina, M. arctoides*). II. The social environment. *Behaviour* **131,** 97–113.

Maestripieri, D., Schino, G., Aureli, F., and Troisi, T. (1992). A modest proposal: Displacement activities as an indicator of emotions in primates. *Anim. Behav.* **44,** 967–979.

Mason, W. A., Long, D. D., and Mendoza, S. P. (1993). Temperament and mother-infant conflict in macaques: A transactional analysis. *In* "Primate Social Conflict" (W. A. Mason and S. P. Mendoza, eds.), pp. 205–227. SUNY Press, Albany, NY.

Mendoza, S. P., and Mason, W. A. (1989). Primate relationships: Social dispositions and physiological responses. *In* "Perspectives in Primate Biology," Vol. 2 (P. K. Seth and S. Seth, eds.), pp. 129–143. Today & Tomorrow's Printers and Publishers, New Delhi, India.

Nash, L. T. (1978). The development of the mother-infant relationship in wild baboons (*Papio anubis*). *Anim. Behav.* **26,** 746–759.

Nash, L. T., and Wheeler, R. L. (1982). Mother-infant relationships in non-human primates. *In* "Child Nurturance," Vol. 3 (H. E. Fitzgerald, J. A. Mullins, and P. Gage, eds.), pp. 27–61. Plenum, New York.

Nicolson, N. A. (1987). Infants, mothers, and other females. *In* "Primate Societies" (B. B. Smuts, D. L. Cheney, R. M. Seyfarth, R. W. Wrangham, and T. T. Struhsaker, eds.), pp. 330–342. Univ. of Chicago Press, Chicago.

Paul, A., Kuester, J., and Podzuweit, D. (19930. Reproductive senescence and terminal investment in female Barbary macaques (*Macaca sylvanus*) at Salem. *Int. J. Primatol.* **14,** 105–124.

Pianka, E. R. (1976). Natural selection of optimal reproductive tactics. *Am. Zool.* **16,** 775–784.
Pianka, E. R., and Parker, W. S. (1975). Age-specific reproductive tactics. *Am. Nat.* **109,** 453–464.
Raleigh, M. J., Brammer, G. L., McGuire, M. T., and Yuwiler, A. (1985). Dominant social status facilitates the behavioral effects of serotonergic agonists. *Brain Research* **348,** 274–282.
Raleigh, M. J., Brammer, G. L., McGuire, M. T., Pollack, D. B., and Yuwiler, A. (1992). Individual differences in basal cisternal cerebrospinal fluid 5-HIAA and HVA in monkeys: The effects of gender, age, physical characteristics, and matrilineal influences. *Neuropsychopharmacology* **7,** 295–304.
Rasmussen, K. L. R., and Suomi, S. J. (1989). Heart rate and endocrine responses to stress in adolescent male rhesus monkeys on Cayo Santiago. *P. R. Health Sci. J.* **8,** 65–71.
Rosenblum, L. A. (1971). The ontogeny of mother-infant relations in macaques. *In* "The Ontogeny of Vertebrate Behavior" (H. Moltz, ed.), pp. 315–368. Academic Press, New York.
Roubertoux, P. L., Nosten-Bertrand, M., and Carlier, M. (1990). Additive and interactive effects of genotype and maternal environment. *Adv. Study Behav.* **19,** 205–247.
Rowell, T. E., Din, N. A., and Omar, A. (1968). The social development of baboons in their first three months. *J. Zool. (London)* **155,** 461–483.
Rowell, T. E., Hinde, R. A., and Spencer-Booth, Y. (1964). "Aunt"-infant interactions in captive rhesus monkeys. *Anim. Behav.* **12,** 219–226.
Sackett, G. P., and Ruppenthal, G. C. (1973). Development of monkeys after varied experiences during infancy. *In* "Ethology and Development" (S. A. Barnett, ed.), pp. 52–87. Lippincott, Philadelphia.
Sapolsky, R. M. (1989). Styles of dominance and their endocrine correlates among wild olive baboons (*Papio anubis*) *Am. J. Primatol.* **18,** 1–13.
Schneider, M. L., Moore, C. F., Suomi, S. J., and Champoux, M. (1991). Laboratory assessment of temperament and environmental enrichment in rhesus monkey infants (*Macaca mulatta*). *Am. J. Primatol.* **25,** 137–155.
Schino, G., D'Amato, F., and Troisi, A. (1995). Mother-infant relationships in Japanese macaques: Sources of inter-individual variation. *Anim. Behav.* **49,** 151–158.
Short, R. V. (1983). The biological bases for the contraceptive effects of breast feeding. *In* "Advances in International Maternal and Child Health" (D. B. Jelliffe and E. F. B. Jelliffe, eds.), pp. 27–39. Oxford Univ. Press, Oxford.
Silk, J. B. (1991). Mother-infant relationships in bonnett macaques: Sources of variation in proximity. *Int. J. Primatol.* **12,** 21–38.
Simpson, M. J. A. (1983). Effect of the sex of an infant on the mother-infant relationship and the mother's subsequent reproduction. *In* "Primate Social Relationships" (R. A. Hinde, ed.), pp. 53–57. Sinauer Assoc., Sunderland, MA.
Simpson, M. J. A., and Datta, S. B. (1991). Predicting infant enterprise from early relationships in rhesus macaques. *Behaviour* **116,** 42–63.
Simpson, M. J. A., and Howe, S. H. (1980). The interpretation of individual differences in rhesus monkey infants. *Behaviour* **72,** 127–155.
Simpson, M. J. A., and Howe, S. (1986). Group and matriline differences in the behaviour of rhesus monkey infants. *Anim. Behav.* **34,** 444–459.
Simpson, M. J. A., and Simpson, A. E. (1986). The emergence and maintenance of interdyad differences in the mother-infant relationships of rhesus macaques: A correlational study. *Int. J. Primatol.* **7,** 379–399.
Simpson, M. J. A., Simpson, A. E., Hooley, J., and Zunz, M. (1981). Infant related influences on birth intervals in rhesus monkeys. *Nature (London)* **290,** 49–51.

Simpson, M. J. A., Gore, M. A., Janus, M., and Rayment, F. D. G. (1989). Prior experience of risk and individual differences in enterprise shown by rhesus monkey infant in the second half of their first year. *Primates* **30,** 493–509.

Stewart, K. (1988). Suckling and lactational anoestrus in wild gorillas. *J. Reprod. Fertil.* **83,** 627–634.

Struhsaker, T. T. (1971). Social behaviour of mother and infant vervet monkeys (*Cercopithecus aethiops*). *Anim. Behav.* **19,** 233–250.

Suomi, S. J. (1983). Social development in rhesus monkeys: Consideration of individual differences. *In* "The Behaviour of Human Infants" (A. Oliverio and M. Zappella, eds.), pp. 71–92. Plenum Press, New York.

Suomi, S. J. (1987). Genetic and maternal contributions to individual differences in rhesus monkey biobehavioral development. *In* "Perinatal Development: A Psychobiological Perspective" (N. A. Krasnegor, E. E. Blass, M. A. Hofer, and W. P. Smotherman, eds.), pp. 397–419. Academic Press, Orlando, FL.

Tanaka, I. (1989). Variability in the development of mother-infant relationships among free-ranging Japanese macaques. *Primates* **30,** 477–491.

Thierry, B. (1985). Social development in three species of macaques (*Macaca mulatta, M. fascicularis, M. tonkean*): A preliminary report on the first ten weeks of life. *Behav. Process.* **11,** 89–95.

Trivers, R. L. (1972). Parental investment and sexual selection. *In* "Sexual Selection and the Descent of Man" (B. Campbell, ed.), pp. 136–179. Aldine, Chicago.

Troisi, A., Schino, G., D'Antoni, M., Pandolfi, N., Aureli, F. and D'Amato, F. R. (1991). Scratching as a behavioral index of anxiety in macaque mothers. *Behav. Neural Biol.* **56,** 307–313.

Vochteloo, J. D., Timmermans, P. J. A., Duijghuisen, J. A. H., and Vossen, J. M. H. (1993). Effects of reducing the mother's radius of action on the development of mother-infant relationships in longtailed macaques. *Anim. Behav.* **45,** 603–612.

Weiner, S. G., Bayart, F., Faull, K. F., and Levine, S. (1990). Behavioral and physiological responses to maternal separation in squirrel monkeys (*Saimiri sciureus*). *Behav. Neurosci.* **104,** 108–115.

West, M. J., and King, A. P. (1987). Settling nature and nurture into an ontogenetic niche. *Dev. Psychobiol.* **20,** 549–562.

White, L. E., and Hinde, R. A. (1975). Some factors affecting mother-infant relations in rhesus monkeys. *Anim. Behav.* **23,** 527–542.

Wilson, D. S., Clark, A. B., Coleman, K., and Dearstyne, T. (1994). Shyness and boldness in humans and other animals. *Trends Ecol. Evol.* **9,** 442–446.

Wolfheim, J. H., Jensen, G. D., and Bobbitt, R. A. (1970). Effects of group environment on the mother-infant relationship in pigtailed monkeys (*Macaca nemestrina*). *Primates* **11,** 119–124.

Worlein, J. M., Eaton, G. G., Johnson, D. F., and Glick, B. B. (1988). Mating season effects on mother-infant conflict in Japanese macaques. *Macaca Fuscata. Anim. Behav.* **36,** 1472–1481.

# Mother–Infant Communication in Primates

DARIO MAESTRIPIERI AND JOSEP CALL

DEPARTMENT OF PSYCHOLOGY AND
YERKES REGIONAL PRIMATE RESEARCH CENTER
EMORY UNIVERSITY
ATLANTA, GEORGIA 30322

## I. Introduction

The mother–offspring relationship can be viewed as the product of a series of interactions between mother and offspring in the course of feeding, other maintenance activities, or just being together (Hinde, 1974). Reciprocal stimulation and communication between mother and offspring play a crucial role in the regulation of caregiving interactions and in the dynamic readjustment of their relationship. If we view communication as "the process by which the behavior of an individual affects the behavior of others" (Altmann, 1967, p. 326), most if not all behavioral interactions between mother and offspring would qualify as communication. The exchange of information between mother and offspring, however, often occurs through the use of species-specific signals through the visual, auditory, tactile, and olfactory modalities.

When offspring are significantly dependent on their caregivers for nutrition, thermoregulation, transport, and protection, their survival may be crucially dependent upon their ability to signal their needs to their mothers. The ability to raise offspring successfully, in turn, may depend significantly on the mother's ability to recognize and respond appropriately to her offspring's signals as well as to guide and coordinate, through the use of signals, the offspring's behavior. The exchange of signals between mother and offspring can also mediate the transmission of information about the physical and social environment, and the acquisition of skills on the part of the offspring.

In nonhuman primates, the exchange of signals between mother and infant has rarely been the subject of systematic examination. Over the last two decades, emphasis on developmental changes in mother–infant contact and proximity, and efforts to identify general adaptive trends in mother–infant relationships have resulted in neglecting the role played by vocal, visual, and tactile signals in regulating the behavioral interactions between mother and infant. Infant vocalizations have been the focus of studies of vocal development in primates (see Snowdon and Elowson, 1992; Symmes and Biben, 1992; Newman, 1995, for reviews). In these studies, however, the emphasis has primarily been on developmental changes in the production, comprehension, and usage of calls, and less attention has been paid to the immediate behavioral impact that infant vocalizations have on the receiver, usually the mother, who responds to the call as part of her parenting effort.

Understanding *how* primate mothers and infants communicate and *what* they communicate requires documenting the signal repertoire of mothers and infants in different species, and the context of use and behavioral responses to signals. In this article, we review the information available on the use of vocal, visual, and tactile signals by primate mothers and infants in a developmental–interactional perspective. Then we discuss patterns of mother–infant communication across primate species in a comparative and evolutionary perspective. Finally, we discuss the possible implications of some evolutionary trends across the primate order for the evolution of active information transfer and language in humans.

II. VOCAL COMMUNICATION

A. INFANT VOCALIZATIONS

A primate infant must be able to attract its mother's attention when necessary. Infant distress calls are found throughout the order Primates from the prosimians to the great apes, and weaning tantrums can almost be considered a universal phenomenon (see Andrew, 1963; Newman, 1985a). When physically and visually isolated from their mothers, infants emit loud and almost continuous vocalizations often referred to as isolation calls or lost calls (e.g., Newman and Symmes, 1982). Marked individual differences in the acoustic structure of infant distress vocalizations allow mothers to recognize their infants from auditory cues alone (galagos, *Galago* spp.: Klopfer, 1970; talapoins, *Miopithecus talapoin*: Gautier, 1974; squirrel monkeys, *Saimiri sciureus*: Kaplan *et al.*, 1978; Symmes and Biben, 1985; vervet monkeys, *Cercopithecus aethiops*: Cheney and Seyfarth, 1980; Japanese macaques, *Macaca fuscata*: Pereira, 1986; but see Simons *et al.*, 1968, and Simons and Bielert, 1973, for negative results in pigtail macaques, *Macaca nemestrina*).

From the first day of life, all prosimian infants are reported to utter distress vocalizations when alarmed or when contact with the mother is interrupted (e.g., Andrew, 1963; Doyle, 1979). In *Galago s. senegalensis*, infants younger than 12 weeks produced very short, broadband zek calls as well as pure ultrasounds, when isolated, in potential danger, soliciting nursing, or in response to adult contact calls (Zimmerman, 1989). Infants also used several different narrow-banded and frequency-modulated calls of low intensity (week, weeak, or wak) during nursing, or when interacting with the father or with other animals. Southern African lesser bush baby (*Galago s. moholi*) infants uttered three types of distress calls, the click, the crackle, and the chirp, as early as a few hours after birth and throughout infancy; these calls disappeared at weaning (Mascagni and Doyle, 1993). Mothers usually responded to distress calls by promptly approaching their infants and starting to groom and lick them. The acoustic structure of click calls, characterized by short, high-pitched pulses with a broadband frequency spectrum, uttered in fast repetitive sequences, could be particularly suited for accurate localization of the infant (Andrew, 1963; Mascagni and Doyle, 1993). Crackles and chirps could convey to the mother information other than cues for spatial localization, including the presence of danger and the fact that the infant is in physical pain (Mascagni and Doyle, 1993). The occurrence of zek or click calls in infants seems to be widespread among young *Galaginae* and *Lorisinae* (e.g., Daschbach *et al.,* 1981), but these calls are less common in the *Lemurinae* and *Cheirogaleinae* (Zimmerman, 1989). Ring-tailed lemur (*Lemur catta*) infants emitted contact calls when visual contact with the mother was lost and trills when contact was regained after separation (Macedonia, 1993). *Lemur catta* infants also emitted three different types of distress calls in contexts ranging from maternal rejection to aggression from an adult (Macedonia, 1993). A squawk call that acts as a distress signal leading to maternal retrieval is the only vocalization reported for sifaka (*Propithecus verreauxi coquereli*) infants (Richard, 1976).

New World monkeys are among the most vocal primates and infants are no exception. In squirrel monkeys, infant vocalizations acoustically distinct from adult vocalizations occur shortly after birth (Lieblich *et al.,* 1980; Symmes and Biben, 1985). In the first weeks of life, squirrel monkey infants vocalized spontaneously or in response to the calls of an adult with similar probability (Biben, 1992). Initially, most spontaneous vocalizations occurred in the context of nursing and moving around the mother's body. Infants seeking their mother's nipple emitted grumbles and tucks, and purrs during nursing (Winter, 1968). The full repertoire of neonatal vocalizations (isolation peeps, peeps, tucks, twitters, chucks, location trills, cackles, yaps, errs, purrs, shrieks, screams, oinks, keckers, and grumbles) was not expressed until infants started leaving their mothers, in the second month of life (Winter, 1968). In this period, the rate of peeps increased dramatically and peeps represented over 50% of the total amount of infant vocalizations, followed by grumbling (24%) and cackling (18%). The most likely function of the peeps seems to be maintenance of contact with mothers and other group members during exploration of the environment (Winter, 1968; Newman, 1985b; Biben, 1992). Grumbling was also associated with exploration and probably served a contact–maintenance function, whereas the context and function of cackling were less clear (Winter, 1968). The gradual decrease in spontaneous vocalizing in the third month may indicate increasing confidence when separated (Biben, 1992). Calls emitted in response to adult vocalizations were accompanied by orienting toward or touching the individual to whom the infant was responding. Adult vocalizations directed to infants were primarily caregiver calls (see as follows) uttered by mothers or other adult females (Biben, 1992). Infants emitted extended isolation peeps at greater separation distance from their mothers and the latter responded by similarly extending their own vocalizations (Masataka and Symmes, 1986).

In marmosets and tamarins (family *Callitrichidae*), infants are quite vocal from an early age and most vocalizations occur when infants are transferred from their mothers to another caregiver, are temporarily left on the ground, are hungry, or are in pain (e.g., Epple, 1968). Nursing calls such as chirps, rattles, and twitter-hooks were among the first vocalizations heard in cotton-top and saddle-backed tamarins (*Saguinus* spp.; Moody and Menzel, 1976; Snowdon, 1987). The rattle was hypothesized to have a pacifying effect on the mother. Twitter-hooks were also emitted by infants while exploring the environment, during rejection-related distress, and in association with squawks to solicit nursing, solid food, or grooming (Moody and Menzel, 1976). The phee calls of common marmoset (*Callithrix jacchus*) infants increased in frequency when infants began to investigate the environment and interact socially, and calls became longer, louder, and more frequent when distance from mothers increased (Epple, 1968). Distress was also associated with tsik calls, twitters, and squeals (Epple, 1968). In pygmy marmosets (*Cebuella pygmea*), infant vocalizations initially consisted of sequences of tsik and phee calls with randomly placed squeals. During subsequent stages of increasing infant independence, infant trills and J calls clearly indicated distress while presumably providing location information to the parent (Pola and Snowdon, 1975). Trills occurred in vocal bouts composed of several different calls mixed together and were uttered when parents left their infants on a branch or rope by themselves. Snowdon (1987) suggested that these babbling bouts may also give the infant extensive vocal practice. In the Goeldi's monkey (*Callimico goeldii*), infant vocal repertoire was composed of at least 15 different calls (Masataka, 1982). Some calls resembled the contact calls emitted by adults, and others occurred in such contexts as infant transfer, threats, and maternal rejection.

Three different calls were given depending on the distance separating the infant from its mother (Masataka, 1982).

Howler monkey (*Alouatta palliata*) infants purred loudly during nursing (Carpenter, 1965) and uttered quiet eh eh calls when exploring the environment away from their mothers (Baldwin and Baldwin, 1976). The same calls were sometimes uttered as an infant returned toward and sat next to its mother, perhaps increasing the probability that the mother would open her arms and allow the infant to nurse. Infants whimpered during travel and play or when they fell from a tree. Mothers responded quickly to the whimpers of fallen infants but not to those of juveniles. Mothers themselves whimpered in response to infant whimpers or caw J calls and retrieved the infant. Mothers rarely retrieved squeaking or barking infants from rough play (Baldwin and Baldwin, 1976).

Among Old World monkeys, information on vocal repertoires of colobine infants is limited when compared to that available for cercopithecine infants. Langur infants produced milk grumbles and isolation peeps, and their main distress calls were squeals and screams (Hanuman langur, *Presbytis entellus*, and Nilgiri langur, *Presbytis johnii*: Jay, 1963, 1965; Poirier, 1968; Hohmann, 1989a). The squeals of langur infants were similar to the squeals of red colobus infants (*Colobus badius*) and to the screams of the black and white colobus infants (*Colobus guereza*) in acoustic structure and context of occurrence (Marler, 1972; Struhsaker, 1975; Hohmann, 1989a).

Cercopithecine infants have a rich repertoire of distress calls. Talapoin monkey infants emitted at least four distinct vocalizations when separated from their mothers, depending on whether visual and auditory isolation from the mother was complete, incomplete, or separation prolonged (Gautier, 1974). All of these calls elicited maternal retrieval of the infant. In vervet monkeys, Struhsaker (1967) distinguished five types of calls given by infants when separated from their mothers (rrr, eee, squeal, scream, and rrah) and one call (eh eh) upon reunion with them. The significance of this latter call was not clear but it was hypothesized that it may enhance the mother–infant bond. Screams were used when infants were approached by certain individuals, for example, adult males, and usually evoked the rapid retrieval of the infant by the mother (Struhsaker, 1967). Infant rrrs (also referred to as whrrs or wrrs) and grunts were also heard when infants moved into a new area, followed another infant, or looked at another group (Seyfarth and Cheney, 1986; Hauser, 1989; Cheney and Seyfarth, 1990).

In the first weeks of infant life, vervet infant rrrs were highly effective in obtaining contact suckling or transport (Hauser, 1993). In Weeks 8–10, there was a dramatic increase in call rate and a sharp decrease in effectiveness. For infants who survived the first year, call rate decreased in Weeks 12–16, whereas effectiveness increased. In contrast, for infants who died within the first year, call rate continued to increase and effectiveness stayed low. Because the relationship between call rate and effectiveness was affected by habitat quality and presumably maternal condition, it was not clear whether infant calls were unreliable indicators of need or if they were reliable but some mothers were incapable of providing the care required (Hauser, 1993).

In macaques, infant distress calls may take different forms in early infancy (barks and shrieks in rhesus macaques, *Macaca mulatta;* Rowell and Hinde, 1962; a long cry in pigtail macaques, *Macaca nemestrina,* Grimm, 1967; trilled whistles in stumptail macaques, *Macaca arctoides;* Bertrand, 1969; Chevalier-Skolnikoff, 1974; screeches, hacks, and trills in Japanese macaques, *Macaca fuscata;* Green, 1975, 1981; whistles, twits, and yaps in liontail macaques, *Macaca silenus;* Green, 1981; Hohmann and Herzog, 1985; whimpers in longtail macaques, *Macaca fascicularis;* Palombit, 1992; hacks, clucks, babbles, and sneezes in bonnet macaques, *Macaca radiata;* Hohmann, 1989b) but most of these forms are lost with development leaving the coo (or whoo) as the main distress and isolation call (Newman, 1995).

Rhesus macaque infants emitted distress calls as early as the first day of life when mothers broke contact with them for the first time (Maestripieri, 1995a). Mothers restored contact almost immediately after infant vocalization. Mothers whose infants vocalized when contact was first broken continued to break contact with them in subsequent days at a lower rate than mothers whose infants did not vocalize. This suggests that mothers may use infant vocalization upon interruption of contact as an "honest" indicator of infant maturation (see as follows).

Studies of stumptail and rhesus macaques indicated that different distress calls may be used in different contexts (Lillehei and Snowdon, 1978; Levine *et al.,* 1987; Bayart *et al.,* 1990; Kalin *et al.,* 1992). In particular, Kalin *et al.* (1992) suggested that rhesus macaque infants use girns to seek contact comfort in close-range interactions, and coos when they are at a greater distance from the mother and retrieval or contact comfort is less immediately possible. Levine *et al.* (1987) and Bayart *et al.* (1990) found that two types of coo calls were emitted by infants when they were separated but adjacent to their mothers (*adjacent separation calls*), and a third type was emitted during total isolation (*isolation calls*). The short adjacent separation calls were not accompanied by a plasma cortisol response, whereas the long and intense isolation calls were associated with heightened adrenal activation. Levine *et al.* (1987) and Bayart *et al.* (1990) suggested that the adjacent separation call may indicate the infant's sudden need for contact and reassurance, whereas the isolation call may indicate that the infant is lost and facilitate localization and retrieval.

Maternal vocalizations in response to infant coos are uncommon but they have been reported in wild liontail macaques (Hohmann and Herzog, 1985). In general, macaque distress calls tend to elicit immediate retrieval if the caller is a young infant but they are often ignored when the vocalizer is older. In captive stumptail macaques, the number of infant distress calls increased and their effectiveness in eliciting maternal retrieval decreased in the first 8 weeks of life (Maestripieri, 1995b). In Weeks 10–12, infant calls decreased and effectiveness increased again. Individual differences in maternal responsiveness to infant calls were related to differences in infant tendency to leave the mother. Calls were less likely to be responded to if the mother was involved in allogrooming than if she was not.

Some distress calls probably function to solicit agonistic support from the mother rather than just re-establishment of contact. Newman (1995) reported that yearling rhesus females vocalized more than males when separated from their mothers, and suggested that this sex difference may be related to the relative importance of formation of alliances with their mothers and other matriline members for females and males. Gouzoules *et al.* (1984) and Gouzoules and Gouzoules (1989a) showed that rhesus and pigtail macaque juveniles facing threat or attack from adults gave different calls according to the intensity of the threat, and the dominance rank and kin relationship of the opponent. This variability was recognized and reacted to differentially by the mothers of the callers. Mothers gave weak responses to calls signaling the involvement of related individuals compared with those that occurred when opponents were unrelated. In pigtail macaques, juvenile females were found to be more proficient than males in both the production and proper contextual usage of the different recruitment screams (Gouzoules and Gouzoules, 1989b).

In the great apes, infants may utter soft vocalizations when searching for the nipple, to which the mother responds by repositioning the baby to allow suckling (gorilla, *Gorilla gorilla*: Dixson, 1981; chimpanzee, *Pan troglodytes*: van Lawick-Goodall, 1967; Rogers and Davenport, 1970; Nicolson, 1977; bonobo, *Pan paniscus*: de Waal, 1989). Despite the fact that repositioning helps the infant to reach the nipple, very rarely have mothers been observed placing the newborn directly on the nipple in response to its vocalizations (Rogers and Davenport, 1970; Nicolson, 1977).

Similar to monkey infants, great ape infants emit a variety of distress calls in contexts such as maternal rejection, brief separation during travel, and rough handling by other individuals (e.g., whines, wails, screeches, screams, hoot cries, howls, and pig-grunts in gorillas, Schaller, 1963; Hess, 1973; Fossey, 1979; Maple and Hoff, 1982; Harcourt *et al.*, 1993; whimpers, hoos, squeaks, screams, and grins in chimpanzees, Yerkes and Tomilin, 1935; Reynolds and Reynolds, 1965; van Lawick-Goodall, 1967, 1968b;

Rogers and Davenport, 1970; Clark, 1977; Nicolson, 1977; Yoshida et al., 1991; whimpers in bonobos, Kuroda, 1984, 1989; de Waal, 1989; whimpers, cries, screams, squeaks, and soft hoots in orang-utans, *Pongo pygmaeus,* Mackinnon, 1974a,b; Neimitz and Kok, 1976; Horr, 1977; Maple et al., 1978; Miller and Nadler, 1981; Chevalier-Skolnikoff, 1982). Detailed information concerning contextual usage of different distress calls is lacking. Also, maternal responses to different types of distress calls have not been investigated. Generally, mothers respond by retrieving and embracing the vocalizing infant (e.g., in gorillas: Schaller, 1963; Hess, 1973; Fossey, 1979; in chimpanzees: Yerkes and Tomilin, 1935; Reynolds and Reyonlds, 1965; van Lawick-Goodall, 1967; Clark, 1977; Nicolson, 1977; Miller and Nadler, 1981; Yoshida et al., 1991). Chimpanzee mothers of older infants, however, reacted by simply approaching to watch the infant or by watching it from a distance (Yerkes and Tomilin, 1935; Nicolson, 1977; Yoshida et al., 1991).

In addition to vocalizing when contact with the mother is interrupted, ape infants also vocalize during play, grooming, and food sharing with their mother. Chimpanzee and gorilla infants "laugh" when tickled by their mothers (van Lawick-Goodall, 1967; Plooij, 1979; Dixson, 1981) and chimpanzee, bonobo, and orang-utan infants whimper to request grooming or food from their mothers (van Lawick-Goodall, 1967, 1968a; Clark, 1977; Kuroda, 1984; Mackinnon, 1974a). Older infants may also display temper tantrums to obtain food from their mothers (Kuroda, 1984). According to Nishida (1990), in the midst of the tantrum infants sometimes glanced at their mother as if to monitor her response. Yerkes (1943) offered a similar interpretation of the behavior of a juvenile that in the midst of a tantrum glanced furtively at its mother as if to discover whether its action was attracting attention.

B. SUMMARY AND DISCUSSION

Primate infants possess a rich repertoire of vocalizations that are primarily addressed to their mothers. Infant distress calls most commonly occur when the infant is not in contact with its mother, and their most likely function is to signal the infant's need for nursing, transport, or protection, all activities that take place in contact with the maternal body. Infant distress calls are given with particular urgency when the infant is under threat or is physically prevented from making contact with its mother either by the mother herself or by other individuals, including humans.

Distress calls emitted by primate infants can be reasonably considered analogous or homologous to the cry of human infants (e.g., Newman, 1985a; Todt, 1988). From an evolutionary point of view, both primate distress calls and human crying can be regarded as part of a proximity-maintaining behavioral system that is important for survival (Bowlby, 1969). As crying has acoustic properties that elicit maximal attention in human adults (Lester and Boukydis, 1992), the distress calls of primate infants are in the auditory range where a mother's sensory threshold is the lowest (Struhsaker, 1967). It has been argued that although human crying promotes close contact with a distant caregiver, its communicative value may be limited (e.g., Papousek, 1992). When used in close-range interactions, prolonged crying becomes disruptive because of its aversive characteristics, and other presyllabic vocalizations are used by mother and child for communicative purposes (e.g., Papousek, 1992).

From the review of infant vocal repertoires in several species of primates, it is apparent that infants possess far more types of calls than would be necessary if these calls only served to signal a state of distress. Data from a wide range of species suggest that infant calls convey information that facilitates the re-establishment of contact and the occurrence of maternal support. This information may include the distance from the mother, the degree of accessibility, and the nature of the threat posed to the infant, including individual characteristics of opponents. The information content of infant calls thus seems to imply some knowledge on the part of the infant of information about the physical as well as the social environment.

Infant purrs emitted during contact and nursing perhaps convey some information to the mother regarding the infant's comfort state as well as contribute to reciprocal stimulation of arousal. Maternal responses to infant purrs, however, have not been investigated in detail nor is it clear why purrs have mostly been reported in New World monkeys and less frequently in Old World monkeys and apes. A possible explanation for this difference may lie in the relative roles played by vocal and visual communication in close-range interactions between mothers and infants in different species. Squirrel monkey mothers usually carry their infants on their backs, whereas continuous ventroventral contact is more common in Old World monkeys and apes. Ventroventral contact facilitates the exchange of information through eye contact and facial expressions, whereas dorsoventral contact may promote the use of vocalizations for subtle close-range interactions.

As is the case of adult vocalizations, species differences in the structural complexity and in the contextual usage of infant calls are likely to be related to characteristics of both the social and the ecological environment (e.g., Snowdon et al., 1982). Complex contact calls are most common among small-bodied arboreal New World monkeys, where dense foliage encourages spatial separation during foraging, and mothers and infants may often be out of each other's sight (Boinski and Mitchell, 1995). Infants in larger-bodied and more terrestrial Old World monkeys and apes use contact calls to a lesser extent, and parallels of the squirrel monkeys' peep calls may only be found in the most arboreal species (e.g., in liontail macaques; Green, 1981; Hohmann, 1991). Although most of the infant distress calls function to re-establish contact with the mother and solicit her support, perhaps it is not a coincidence that the highest degree of structural and semantic complexity of recruitment calls was found in species like rhesus and pigtail macaques in which dominance relationships rely critically on agonistic support from kin. The sex differences in the production and contextual usage of recruitment screams in immature macaques confirm that some characteristics of infant vocalizations may be related to patterns of agonistic support and alliance formation between adults.

Maternal responses to infant calls probably vary in relation to a number of factors, including infant age, call rate, characteristics of the mother and the infant, ongoing maternal activity, contextual cues, and perhaps memory of events prior to the occurrence of the call. Data on developmental changes in maternal responsiveness to infant calls in vervet monkeys (Hauser, 1993) and stumptail macaques (Maestripieri, 1995b) suggest that monkey mothers may use primarily infant age and call rate as criteria for responsiveness to distress signals. In fact, age and call rate might be reliable cues as to whether calls are "honest" signals of need.

Infant distress vocalizations are usually assumed to be costly to produce because of energy expenditure and increased probability to attract predators (Godfray, 1991). The costs of these signals, however, may be higher to younger infants than to older ones as, for example, the latter are more likely to escape a predator attack than the former. Therefore, distress calls by young infants are expected to be infrequent, limited to situations of actual need, and associated with high maternal responsiveness. As infants grow older and the costs of distress signals decrease, the latter are likely to become more frequent and increasingly used out of context, this being accompanied by reduced maternal responsiveness. When calls are poorly effective in eliciting maternal responses, infants, in turn, may reduce their call rate (see also Hauser, 1986).

C. PARENTAL VOCALIZATIONS

Primate parents address vocalizations to their infants spontaneously or in response to infant vocalizations. In black-and-white ruffed lemurs (*Varecia variegata variegata*) mothers emitted mews upon departure from their infants (Pereira *et al.,* 1988; Morland, 1990) and in *Galago crassicaudatus* and *G. senegalensis,* mothers responded to the distress calls of their separated infants with click, grunt, or groan vocalizations (Klopfer, 1970; Zimmermann, 1989). Although maternal vocal responses to infant calls have not been systematically investigated in lemurs, the absenteeism and infant parking that characterize several species of prosimians suggest that vocal interactions between mothers and infants may be important for recognition and behavioral coordination (Pereira *et al.*, 1988).

In some New World monkeys, mothers emit purrs during nursing, which are similar to infant purrs (e.g., titi monkeys, *Callicebus moloch*: Fragaszy *et al.*, 1982; squirrel monkeys: Newman, 1985b). Squirrel monkey mothers also utter specific vocalizations during infant carrying and retrieval (DuMond, 1968; Rosenblum, 1968; Baldwin, 1969; Kaplan *et al.*, 1978). These calls are also uttered by adult females while interacting with infants other than their own and were first termed *rrrt calls* (DuMond, 1968) and later *caregiver calls* (Biben *et al.*, 1989). The acoustic structure of the caregiver calls is markedly different from that of other adult calls such as chucks and peeps, and it includes low fundamental frequency, variable duration, and well-developed harmonic structure and tonality (Biben *et al.*, 1989; Biben, 1992; Symmes and Biben, 1992; Boinski and Mitchell, 1995).

Squirrel monkey caregiver calls are structurally different in relation to their contextual use. Biben *et al.* (1989) divided infant care patterns into the four categories of dorsal contact, nursing, retrieval, and inspection. The structure of calls associated with dorsal contact and nursing was essentially homogeneous. The calls associated with retrieval were hypothesized to carry a message of comfort in their flat initial segment, followed by an attention-getting note of urgency in the steep slopes and greater amplitude characteristic of the latter segment of the call (Biben *et al.*, 1989). Calls given while inspecting other females' infants were the most complex, with significantly greater duration and frequency modulation than that of calls in the other contexts. These features suggested an attention-getting function (Biben *et al.*, 1989; Biben, 1992). Mothers did not use these calls with their own infants (Biben *et al.*, 1989; Symmes and Biben, 1992).

In the first month of infant life, squirrel monkey mothers only emitted the short nursing calls (Biben, 1992). Interestingly, however, maternal vocalizations did not occur when the mother–infant dyads were isolated but only in the presence of other females (Symmes and Biben, 1992). The frequency of maternal calls increased sharply in the following months when infants started leaving their mothers for brief periods and mothers began using the retrieval calls. Retrieval calls were also emitted in response to the infants' peeps and were often repeated several times before infants restored contact with their mothers (Symmes and Biben, 1992). Infants also responded vocally to maternal calls as well as to retrieval and inspection calls made by other adult females (Biben, 1992).

Boinski and Mitchell (1995) analyzed maternal caregiver calls in wild squirrel monkeys in three contexts of interaction with infants, prior to nursing, during nursing, and at the end of a nursing bout. The acoustic structure of caregiver calls was different across these contexts. Infants responded to prenursing calls by approaching their mothers and initiating a nursing bout. Caregiver calls during nursing, in addition to purrs, possibly indicated the mother's continued willingness to nurse. End nurse calls signaled the mother's intention to end a nursing bout and occurred in conjunction with physical rejection of the infant. Boinski and Mitchell (1995) suggested that caregiver calls increase the coordination of activity between mothers and infants and reduce conflict over nursing.

In New World monkeys other than squirrel monkeys, maternal vocalizations occurring in caregiving interactions have been investigated only to a limited degree. In howler monkeys, mothers gave wrah-ha vocalizations while separated from their troop and waiting for their infant (Baldwin and Baldwin, 1973). Although the call could help the infant find its mother, the wrah-ha was not restricted to this circumstance. Adult females gave wrah-ha calls whenever they became separated from their troops (Baldwin and Baldwin, 1973). In *Leontopithecus rosalia,* parents or older siblings emitted vocalizations that notified infants of food items to be shared (Brown and Mack, 1978). In tamarins, persistent calling by infants during the weaning stage elicited a unique call type in the male parent, the loud croak, which was associated with the infant obtaining food (Moody and Menzel, 1976).

Old World monkey mothers appear to be far less vocal than squirrel monkey mothers. In langurs, the only reported examples of maternal vocalizations involve alarm calls (Jay, 1963; Poirier, 1968). In baboons, in the first weeks of infant life, "the mother makes almost no sound except that resulting from soft lipsmacking" (DeVore, 1963, p. 312). According to Cheney and Seyfarth (1990, pp. 119–120), vervet monkey mothers do not possess retrieval calls for their infants and "make no apparent attempt to let the infants know that they are moving off or where they are going." Macaque mothers, however, do emit loud coo calls when their infants are out of sight (pigtail macaques: Simons *et al.,* 1968; stumptail macaques: Bertrand, 1969; Japanese macaques: Green, 1975; rhesus macaques: Hansen, 1976). A severely handicapped and visually impaired Japanese macaque infant was often able to locate its mother after hearing her calls (Fedigan and Fedigan, 1977). The mother, however, did not appear to respond vocally to the lost calls of her infant. Individual differences in the structure of maternal coo calls could allow infants to recognize their mothers based on acoustic stimuli alone (rhesus macaques: Hansen, 1976; Japanese macaques: Masataka, 1985). Cercopithecine females have specific grunt vocalizations to express interest in other females' infants (rhesus macaques: Rowell and Hinde, 1962; de Waal, 1989; yellow baboons: *Papio cynocephalus,* Altmann, 1980; stumptail macaques: Bauers, 1993; Japanese macaques:

Green, 1976; Blount, 1985) but mothers usually do not grunt to their own infants (Bauers, 1993; D. Maestripieri, personal observation).

There is little evidence that great ape mothers frequently vocalize in response to their infants' vocalizations. Occasionally, chimpanzee mothers emitted a hoo vocalization in response to their infants' whimpers (van Lawick-Goodall, 1967) and gorilla mothers responded to their whimpering infants by looking and emitting a syllabled call (Harcourt *et al.*, 1993). More frequently, mothers vocalized to their infants spontaneously, especially in the context of infant retrieval and preparation for group travel (gorillas: Fossey, 1979; bonobos: Kano, 1992; chimpanzees: Yerkes and Tomilin, 1935; orang-utans: Mackinnon, 1974a). According to van Lawick-Goodall (1967, 1968b), the vocalizations used by chimpanzee mothers were similar to those of their infants, the most frequent calls being hoos and whimpers. Soft vocalizations were emitted by chimpanzee mothers while examining their infants (Nicolson, 1977), and a soft bark and a cough were used as threats toward begging infants or juveniles (van Lawick-Goodall, 1967; van de Rijt-Plooij and Plooij, 1987). Grunting and roaring threat vocalizations have also been reported for orang-utans, for example, when mothers attempted to prevent their older offspring from contacting their newborns or entering the nest (Horr, 1977).

### D. Summary and Discussion

Maternal vocalizations clearly distinguishable from other calls in the adult repertoire have been demonstrated in squirrel monkeys. Squirrel monkey mothers' long caregiver calls are given in response to external threat or infant distress and presumably serve a retrieval function. The coo calls of macaque mothers appear to serve as contact and retrieval signals as well. The fact, however, that these calls seem to be used far less frequently than facial and gestural signals (see as follows), that they are rarely uttered in response to infant distress calls, and that mothers utter them while carrying their dead babies (e.g., Green, 1981) makes the functional significance of these calls unclear. Ape mothers use vocalizations to retrieve their infants but it is not clear whether they are structurally different from other calls in the adult repertoire. Overall, although some maternal vocalizations are emitted in response to infant calls, there is no evidence suggesting that primate mothers and infants engage in turn taking or "conversations."

Biben *et al.* (1989) suggested that the prosodic variations within maternal caregiver calls in squirrel monkeys resemble the use of prosodic features of human motherese and that nursing vocalizations may be analogous to motherese in conveying information on states of pleasure or comfort, or in inducing such states. This functional interpretation of nursing vocalizations and the comparison with human motherese, however, must be viewed with caution for the following two reasons: (1) squirrel monkey maternal vocalizations did not occur when the mother–infant dyads were isolated but only in the presence of other females, and (2) it is not clear why these vocalizations have not been commonly reported in other primate species. Further microanalytic studies of maternal and infant nursing vocalizations and of close-range behavioral interactions associated with them are needed to clarify the functional significance of these early communicative interactions between mothers and infants.

Biben (1992) also hypothesized that squirrel monkey caregiver calls represent important sources of stimulation, promoting infant socialization and cognitive development. She ascribed this function, however, primarily to caregiver calls addressed to infants by adult females other than the mother. In her view, these females would take responsibility for infant's early socialization because the dorsal carrying position of infants does not allow direct interactions between mothers and infants. The hypothesis, however, that adult females actively participate in the socialization process of unrelated infants, seems to require evidence of reciprocal altruism phenomena, and this evidence, at present, is not available.

Mother–infant vocal interactions in squirrel monkeys undoubtedly represent a complex and interesting phenomenon and their study could be crucial for understanding the mother–infant relationship in these primates. However, in order to understand the functional importance of these vocal interactions in some species and explain their absence in others, it seems necessary that vocal communication be discussed in the light of other communicative modes available to primate mothers and infants.

### III. Visual and Tactile Communication

In many species of monkeys, the neonate spends long intervals staring into its mother's face and responding to her eye contact (pigtail macaques: Jensen, 1965; stumptail macaques: Chevalier-Skolnikoff, 1974; rhesus macaques: Mendelson, 1982; Japanese macaques: Ehardt and Blount, 1984). Conversely, the mother spends long periods looking at her infant's face, and she may at times lip smack, teeth-chatter, or pucker to restore eye contact (baboons: DeVore, 1963; stumptail macaques: Bertrand, 1969; Chevalier-Skolnikoff, 1974; pigtail macaques: Castell and Wilson, 1971; rhesus macaques: Mitchell and Brandt, 1975; Japanese macaques: Ehardt and Blount, 1984; see Redican, 1975, for a description of facial expressions). Similar and more complex communicative sequences are reported in the great apes, in which mother and infant often engage in prolonged interactions involving mutual gazing, hand movements, smiling, and looking away (e.g., chimpanzees: van Lawick-Goodall, 1967; Plooij, 1978, 1979; gorillas: Hoff et al., 1981).

It has been suggested that primates learn the correct shape and social use of facial expressions and gestures during their ontogeny, and there are some indications that the development and use of these signals follow a course parallel to that of responsiveness to social cues (Kenney et al., 1979; Mendelson, 1982; Zeller, 1992). Gestures may be built up from the basic biological structure and movement possibilities of the animals (Andrew, 1963). Anthoney (1968) remarked that neonatal sucking patterns in baboons are identical to those of lip smacking, and suggested that they subsequently generalize to involve sexual and greeting behaviors. In both monkeys and apes, some form of pouting appears early in life and functions as a request of contact comfort or the nipple (e.g., Chevalier-Skolnikoff, 1982). Pouting includes raised eyebrows and a slightly opened mouth with mouth corners brought forward and lips protruded. Pouting is initially spontaneous or elicited by tactile stimulation in the oral region, but after the first weeks of life in monkeys and the first months in apes, pouting and lip smacking are mainly eliciting by social stimuli, including maternal eye contact and lip smacking (Chevalier-Skolnikoff, 1974, 1982). Burton (1972) suggested that, in Barbary macaques (*Macaca sylvanus*), mother and father play an active role in their infant's performance of teeth chattering. According to Burton (1972, pp. 33–35):

> Each time the infant makes a sucking motion, both parents immediately chatter to it. By the third day at the earliest, and fifth generally, the infant is capable of returning weak chatter movements when it is chattered to, but invariably just moves closer to the animal chattering to it. That is, it returns sucking motions with an added component of pulling back the lips and opening and closing the mouth. Thus a preexisting action movement, already included in the repertoire by birth, is conditioned into a purely social gesture.

In Japanese macaques, lip smacking by mothers elicited orientation responses from their separated infants, and it may have facilitated the process of visual discrimination and recognition of the mother (Nakamichi and Yoshida, 1986).

Squirrel monkeys do not have the well-developed facial musculature of some Old World monkeys and apes, and lack many of the facial expressions seen in these primates (but see Marriott and Salzen, 1978). Genital display by young squirrel monkeys has been observed in captivity (Ploog, 1967; Rosenblum, 1968) but rarely in the wild (S. Boinski, personal communication). Initially, this signal was mainly directed to other group members while the infant was still on the mother. When the infant became more independent, the genital display was frequently addressed to the mother, often in association with rejection from the nipple (Ploog, 1967). The display probably has a general appeasement function and may signal desire for closer contact and nursing (Ploog, 1967; Newman, 1985b). Colobine infants do not appear to have a repertoire of facial expressions similar to that of cercopithecine infants, and, in general, show fewer expressions in response to familiar stimuli than macaque infants (e.g., Chevalier-Skolnikoff, 1982). According to Jay (1965, p. 224), from birth to 3 to 5 months, langur infants "displayed rudimentary and mostly unrecognizable motions that communicated only the most general states of discomfort or comfort."

Although macaque infants address facial expressions, such as lip smacking, the fear grin, and the play face, to other group members early in life (Redican, 1975; Kirkevold et al., 1982), the role of infant nonvocal signals in mediating early mother–infant interactions is not clear. Castell and Wilson (1971) reported that during the first 20 weeks of life, pigtail macaque infants puckered towards their mothers, on average, 12 times less often than mothers puckered towards infants. The display of the infant's pucker was not reliably associated with any behavioral response in the mother. Stumptail macaque infants lip smacked to their mothers infrequently and only during extremely stressful situations, such as when they were about to fall or were climbing over the dominant male and suddenly became afraid (Chevalier-Skolnikoff, 1974). No maternal responses to infant lip smacking were reported.

In the great apes, infants use facial expressions and other nonvocal signals, such as hand gestures and body postures, in a variety of interactions with their mothers. Facial expressions most commonly occur during play and feeding. Chimpanzee infants attempted to initiate play with their mothers by displaying the play face and a certain posture in front of the mother (van Lawick-Goodall, 1967; Plooij, 1978; van de Rijt-Plooij and Plooij, 1987), by pulling their mother's hand toward themselves, or by jumping up and down (van Lawick-Goodall, 1967; Plooij, 1979; van de Rijt-Plooij and Plooij, 1987). Orang-utan and gorilla infants as well displayed the play face in interaction with their mothers (Mackinnon, 1974a,b; Fossey, 1979; Chevalier-Skolnikoff, 1982). Pouting was initially used to request access to the nipple and later solid food from the mother (chimpanzees: van Lawick-Goodall, 1967; Tomasello et al., 1985; van de Rijt-Plooij and Plooij, 1987; bonobos: Kuroda, 1984; de Waal, 1989; orang-utans: Chevalier-Skolnikoff, 1982). In chimpanzees, pouting and teeth baring were also accompanied by throwing dirt or crouching and bobbing in front of the mother to request access to the nipple (Clark, 1977; Tomasello et al., 1994). In bonobos, pouting often occurred in conjunction with placing a hand in front of the mouth (Kuroda, 1984). Chimpanzee and bonobo infants requested food also by peering in the hand or mouth of their mother (Plooij, 1978; Silk, 1979; Kano, 1992), touching their mother's hand or mouth (van Lawick-Goodall, 1967; Plooij, 1978), and holding out a hand in a begging gesture (Kortlandt, 1962; Sugiyama, 1972; Silk, 1978; Tomasello et al., 1985). Infant orang-utans used a gesture consisting of placing an open hand underneath their mother's chin to request solid food from her (Chevalier-Skolnikoff, 1982; Bard, 1992).

Outside the contexts of play and feeding, great ape infants use facial expressions, gestures, and postures to signal distress and/or regain contact with their mother's body. For example, gorilla infants displayed distress facial expressions when they lost contact with their mothers or when there was some group disturbance (Fossey, 1979). Chimpanzee infants used body movements, such as stamping their feet, bouncing up and down, hitting, rocking or throwing objects, to attract their mother's attention (Goodall, 1965; Reynolds and Reynolds, 1965; Tomasello et al., 1994), and extended one or both arms toward their mother or pulled her to request transport during travel (van Lawick-Goodall, 1967; Clark, 1977; Tomasello et al., 1985, 1989, 1994; van de Rijt-Plooij and Plooij, 1987). Chimpanzee infants also tugged at their mothers' hand (Clark, 1977) or raised an arm to solicit grooming (van de Rijt-Plooij and Plooij, 1987).

In cercopithecine monkeys, mothers appear to use facial expressions and postures far more frequently than vocalizations to regulate contact and proximity with their infants. Harlow et al. (1963) and Hansen (1966) noted that rhesus macaque mothers retrieved their infants by using two types of visual signals, the "silly grin response" and the "affectional present." The silly grin involved lip smacking to the infant when the infant was at a distance, and the affectional present consisted of a posture similar to the sexual present. These signals were equally successful in determining infant retrieval and both practically disappeared after 90 days of infant life (Harlow et al., 1963; Hansen, 1966). Seay (1966) reported that primiparous rhesus mothers retrieved their infants by lip smacking more than did multiparae, while the reverse was true for the affectional present. Seay (1966) argued that lip smacking was used in situations where a threat was perceived, whereas the present was more likely to depend on maternal experience and learning. The use of lip smacking or pucker for infant retrieval has also been reported in baboons (DeVore, 1963) and pigtail macaques (Jensen and Gordon, 1970; Jensen et al., 1973; Castell and Wilson, 1971; Bolwig, 1980).

Infant retrieval in the context of group movement or foraging is also accomplished through gestures. For example, in howler monkeys and langurs, mothers have been observed to use both facial expressions and gestures to invite their infants to follow or mount (Jay, 1963; Carpenter, 1965; Poirier, 1968). Ape mothers, in particular, have often been reported to use an outstretched arm to recall infants in potentially dangerous situations or before travel (chimpanzees: Yerkes and Tomilin, 1935; Goodall, 1965; Reynolds and Reynolds, 1965; bonobos: Kano, 1992). Mothers also signaled their infants to climb on by bending their knees, presenting the hindquarters, looking back over their shoulder, or reaching back to gently tap their infants' backs (chimpanzees: Yerkes and Tomilin, 1935; Goodall, 1965; van Lawick-Goodall, 1967, 1968a; Clark, 1977; van de Rijt-Plooij and Plooij, 1987; gorillas: Schaller, 1963; Maple and Hoff, 1982). Hess (1973) observed that gorilla mothers sometimes used genital touching to induce their infants to crawl, climb, walk, and run. Goodall (1965) also reported a mother tapping on the trunk of a tree to encourage her infant to climb down. Mothers also solicit their infants to change their position during transport. For instance, mothers encouraged a shift from dorsal to ventral carrying by gently touching the infant on its back (van Lawick-Goodall, 1967, 1968a) or by a shrugging movement of one shoulder and slight flexion of the opposite elbow and one or both legs (van Lawick-Goodall, 1968b).

In both cercopithecine monkeys and apes, mother–infant communication through facial expressions, gestures, and postures has often been reported in interactions suggestive of tutoring or teaching. Hinde and Simpson (1975) described how some rhesus macaque mothers frequently left their young infants and engaged in behavioral sequences involving walking backwards, lip smacking, and presenting to the infant. Similarly, van de Rijt-Plooij and Plooij (1987) described how one chimpanzee mother repeatedly placed her infant on a rock or on the ground, walked a few steps away, and signaled to cling by flexing her knees and looking back at the infant. The sequence was repeated until the infant clung immediately whenever its mother displayed the signal. Flexing the knees and looking back were sometimes accompanied by extending one arm and grunting. Whenever the infant seemed not to be looking at its mother, she would wait, at times for several minutes, and ultimately gain its attention by touching it.

Interactions similar to those observed in rhesus macaques and chimpanzees have also been reported in Barbary macaques (Burton, 1972), patas monkeys (*Erythrocebus patas*: Chism, 1986), vervet monkeys (L. A. Fairbanks, personal communication), baboons (*Papio anubis*: Ransom and Rowell, 1972; *Papio cynocephalus*: Altmann, 1980; *Papio hamadryas*: Bolwig, 1980), and in other studies of chimpanzees (Yerkes and Tomilin, 1935; van Lawick-Goodall, 1967; Nicolson, 1977). In particular, according to Yerkes and Tomilin (1935) "sooner or later every mother encourages and variously aids her baby to learn to creep, stand erect, climb, and finally to walk and run . . ." (p. 333) and "each mother has her own preferred instructional or pedagogical methods" (p. 334). Although maternal encouragement in chimpanzees mainly occurs through the use of gestures and postures, some mothers were seen supporting their infants with a hand and walking backwards bipedally while the infants were taking their first steps (see also Bolwig, 1980, for a similar observation in pigtail macaques). Ransom and Rowell (1972) and Hinde and Simpson (1975) also noted that the infants that were encouraged by their mothers early in life showed some behavioral differences from infants that were not encouraged, for example, in terms of better coordination with the mother's activity (Hinde and Simpson, 1975) and more time spent following, rather than being carried by, their mothers (Ransom and Rowell, 1972).

Quantitative data on early communicative interactions between mothers and infants were gathered in two studies of rhesus and pigtail macaques (Maestripieri, 1995a,1996). In both rhesus and pigtail macaques, multiparous mothers broke contact with their infants early in life and engaged in backward walking and lip smacking (or puckering, in the case of pigtails) to their infants more than primiparous mothers. Rhesus infants that were left by their mothers in their first days of life broke and made contact with them for the first time earlier than infants that were not left by their mothers. In pigtail macaques, the infants' initiative in reducing distance after the pucker increased with their age and the latency of response gradually decreased in the first 6–7 weeks of life. Infants that received more puckers from their mothers spent a lower percentage of time in contact with them, and approached and left them more frequently than infants that received fewer puckers. In rhesus macaques, interruption of contact with infants early in their life had no apparent immediate benefits to mothers but increased the probability of infant kidnapping by other group members (see also Ransom and Rowell, 1972, for a similar observation in baboons). Rhesus mothers whose infants gave a distress vocalization after the first interruption of contact broke contact with them less frequently in subsequent days than mothers whose infants did not vocalize. Altogether, the findings of these two studies strongly suggest that experienced macaque mothers actively encouraged their infants' independent locomotion, that maternal encouragement was sensitive to infant competence, and that encouraged infants displayed some locomotor skills earlier in life than they would have without maternal encouragement.

A. Summary and Discussion

Primate infants, with the exception of the great apes, appear to display remarkably few visual and tactile signals compared with the richness of their vocal repertoire. In many species, several vocal patterns seem to be already present at birth. In contrast, some visual patterns are related primarily to functions such as nourishment and acquire their communicative function secondarily during development. Facial expressions, gestures, and postures could be more complex than vocalizations to perform because more parts of the body must be controlled before they can be given accurately (Zeller, 1992). In particular, the newborn's neuromuscular systems may not be well developed enough to allow for elaborate facial expressions (Redican, 1975). Studies of macaques have shown that visual signals may not complement vocal signals but may even replace them with increasing age (e.g., Green, 1981). This is in contrast with New World monkeys, where infant vocalizations can be replaced by adult vocalizations that serve the same function (e.g., contact calls).

In monkeys, the role of infant nonvocal signals in mediating early mother–infant interactions is not as clear as that of vocalizations. While monkey infants emit distress vocalizations to regain contact with their mothers, nonvocal signals are rarely observed in this context. In contrast, in the great apes, infant facial expressions, gestures, and postures are frequently used for soliciting contact, grooming, play, and food sharing from mothers. Thus, although both vocal and nonvocal signals may function to re-establish physical contact with the mother, nonvocal signals are also used to request maternal participation in specific activities that may or may not require contact with the maternal body.

Cercopithecine monkeys and great apes clearly differ from prosimians, New World monkeys, and colobines in the maternal repertoire of nonvocal signals. In cercopithecine monkeys and apes, maternal facial expressions, gestures, and postures occur frequently in the context of contact and proximity regulation, and, thus, they appear to serve a function similar to that of vocalizations in other monkeys. The "pedagogic" use of maternal signals observed in several species of cercopithecines and in chimpanzees, however, has not been observed in other monkeys.

The unique characteristic of some early communicative interactions between mothers and infants in cercopithecine monkeys and chimpanzees is that the mother appears to "intentionally" create the opportunity for the occurrence of the signal and its response in the infant. Usually, infants emit distress signals when they are unable to regain contact with their mothers, are hungry, or in pain. Similarly, mothers emit vocal or nonvocal signals to retrieve infants at times of danger or when ready for travel. In these cases, mothers and infants can be thought of as emitting a signal in response to a problem that requires the coordination of their behavior to be solved. When cercopithecine and ape mothers increase the distance from their infants while emitting retrieval signals, such as facial expressions, gestures, and postures, the problem (i.e., the momentary separation) seems to be created by the mother for no apparent goal other than learning or practicing its solution. In these circumstances, the mother's behavior and the signals she employs appear to be aimed at facilitating learning in the infant.

## IV. Parent–Infant Communication in Evolutionary Perspective

Parents and infants in all species of primates engage in communicative interactions involving species-specific vocal, visual, and tactile signals. Overall, the review of infant vocal repertoires across primate species highlights similarities rather than differences in contextual usage of infant calls and in the responses they elicit in the mother. Infant vocalizations, however, probably contain more information than previously suspected and the information content of these calls probably varies in relation to socioecological characteristics of different primate species.

It has been suggested that vocal signals are more directly related to emergency situations and have a more critical importance for infant survival than nonvocal signals (Redican, 1975). If this suggestion is correct, it could account for the great variability across the primate order in the use of facial expressions, gestures, and postures by mothers and infants. Infant nonvocal signals appear to occur with a certain frequency only in the great apes, where they are used to mediate complex transactions between mothers and infants. Also, whereas maternal facial expressions, gestures, and postures seem to play a negligible role in interactions with infants in prosimians and arboreal New World monkeys, these signals gain prominence in the maternal communicative repertoire of terrestrial cercopithecine monkeys and apes. In several species of cercopithecines and in chimpanzees, mothers display facial expressions, gestures, and postures in interactions suggestive of encouragement of infant behavior and teaching. Although at the moment this is only an intriguing possibility, these interactions may suggest the existence of another evolutionary trend in primate mother–infant communication: the tendency to use behavioral signals for pedagogic purposes in cercopithecine monkeys and apes.

The qualitative and anecdotal nature of a large part of observations concerning the exchange of behavioral signals between mothers and infants in nonhuman primates prevents any attempt to characterize evolutionary trends in mother–infant communication from being little more than speculative. Qualitative observations and speculations, however, can provide impetus for the production of more substantial sets of data and the formulation and testing of specific hypotheses. It is from this perspective that we now speculate about the possible implications of primate mother–infant communication patterns for the evolution of complex forms of information transfer in humans and ultimately language.

Traditional hypotheses for the reconstruction of hominid evolution have emphasized that activities such as hunting, gathering, or scavenging provided the opportunity for cooperation and exchange of information among hominids (e.g., Washburn, 1957; Isaac, 1978; Tanner, 1981). In most cases, the emphasis has been on interactions occurring between adult members of the same band or within the male–female pair (Lovejoy, 1981). The idea that innovative changes in the way information was acquired and transmitted may have occurred in the parent–offspring interaction has received less attention. King (1991), however, has emphasized that the evolution of the hominid lineage might have been characterized by an increased use of active information transfer from adults to immatures.

Active transfer of information between mother and offspring, in the form of outright teaching or molding the offspring's behavior through reinforcement and discouragement, can be favored not only by natural selection but also by maternal selection, a phenomenon whose role in behavioral evolution has only recently been appreciated (e.g., Kirkpatrick and Lande, 1989; Stamps, 1991; Avital and Jablonka, 1994). Theoretical studies of maternal effects indicate that the behavior of parents may have profound effects on the evolution of phenotypic traits in their offspring. Although maternal effects have been identified in a wide variety of taxa, maternal selection of behavioral traits in the offspring may be especially common in primates, in which infants are dependent on their mothers for a long period and mothers have a variety of opportunities to influence their sociobehavioral development. Examples of how maternal behavioral phenotypes can influence the phenotype of the offspring in primates include inheritance of dominance rank (e.g., Holekamp and Smale, 1991) and cross-generational consistency in mothering style (e.g., Fairbanks, 1989). Cases of outright teaching in primates are rare and, for the most part, are based on anecdotal observations, but perhaps it is not a coincidence that most, if not all, putative cases of teaching in primates, whether in the domain of early infant locomotion (see aforementioned), tool use, food selection and processing, use of alarm calls, and human sign language (see King, 1991; Caro and Hauser, 1992, for reviews), involve mothers and infants. While these observations do not necessarily suggest that the phenomenon of teaching in primates has the cognitive underpinnings characteristic of human teaching (e.g., Tomasello *et al.*, 1993), they do suggest that there may be a continuum in the ability to engage in active information transfer across the primate order and that this ability did not arise *de novo* after the pongid–hominid split (King, 1994). We emphasize that this continuity is particularly evident in some interactions between primate mothers and infants, and make the provocative suggestion that active information transfer perhaps evolved initially as a maternal character.

Because nonvocal signals, such as facial expressions, hand gestures, and postures, play an important role in some of these instances of primate teaching, the hypothesis that the language of instruction developed from the same repertoire of gestures as those entailed in the actions directed or described should also be reconsidered (see Hewes, 1973). In particular, it may be speculated that regardless of the nature of the ecological or social pressures that favored the prominence of nonvocal signals in the maternal communicative repertoire of cercopithecine monkeys and apes, perhaps at some point during the evolution of the primate order, these signals acquired a different functional significance and from serving a function of basic coordination of mother–infant activities they became instrumental in facilitating social learning during development and perhaps the cultural transmission of information from mother to offspring.

### Acknowledgments

This work was supported in part by NIH grant RR-00165 awarded by the National Center for Research Resources to the Yerkes Regional Primate Research Center. The Yerkes Regional Primate Research Center is fully accredited by the American Association for Accreditation of Laboratory Animal Care. We thank S. Boinski, K. Carroll, B. Donaghey, B. King, M. Tomasello, A. Weaver, and the editors for helpful comments on the manuscript.

### References

Altmann, S. A. (1967). The structure of primate social communication. In "Social Communication Among Primates" (S. A. Altmann, ed.), pp. 325–362. Univ. of Chicago Press, Chicago.
Altmann, J. (1980). "Baboon Mothers and Infants." Harvard Univ. Press, Cambridge, MA.
Andrew, R. J. (1963). The origin and evolution of the calls and facial expressions of the primates. *Behaviour* **20,** 1–109.
Anthoney, T. R. (1968). The ontogeny of greeting, grooming and sexual motor patterns in captive baboons (superspecies *Papio cynocephalus*). *Behaviour* **31,** 358–372.
Avital, E., and Jablonka, E. (1994). Social learning and the evolution of behaviour. *Anim. Behav.* **48,** 1195–1199.
Bayart, F., Hayashi, K. T., Faull, K. F., Barchas, J. D., and Levine, S. (1990). Influence of maternal proximity on behavioral and physiological responses to separation in infant rhesus monkeys. *Behav. Neurosci.* **104,** 98–107.
Baldwin, J. D. (1969). The ontogeny of social behaviour of squirrel monkeys *(Saimiri sciureus)* in a seminatural environment. *Folia Primatol.* **11,** 35–79.
Baldwin, J. D., and Baldwin, J. I. (1973). Interactions between adult female and infant howling monkeys *(Alouatta palliata)*. *Folia Primatol.* **20,** 27–71.
Baldwin, J. D., and Baldwin, J. I. (1976). Vocalizations of howler monkeys *(Alouatta palliata)* in southwestern Panama. *Folia Primatol.* **26,** 81–108.
Bard, K. A. (1992). Intentional behavior and intentional communication in young free-ranging orangutans. *Child Dev.* **63,** 1186–1197.

Bauers, K. A. (1993). A functional analysis of staccato grunt vocalizations in the stumptailed macaque *(Macaca arctoides)*. *Ethology* **94,** 147–161.
Bertrand, M. (1969). "The Behavioral Repertoire of the Stumptail Macaque." Karger, Basel.
Biben, M. (1992). Allomaternal vocal behavior in squirrel monkeys. *Dev. Psychobiol.* **25,** 79–92.
Biben, M., Symmes, D., and Bernhards, D. (1989). Contour variables in vocal communication between squirrel monkey mothers and infants. *Dev. Psychobiol.* **22,** 617–631.
Blount, B. G. (1985). "Girney" vocalizations among Japanese macaque females: Context and function. *Primates* **26,** 424–435.
Boinski, S., and Mitchell, C. M. (1995). Wild squirrel monkey *(Saimiri sciureus)* "caregiver" calls: Contexts and acoustic structure. *Am. J. Primatol.* **35,** 129–137.
Bolwig, N. (1980). Early social development and emancipation of *Macaca nemestrina* and species of *Papio*. *Primates* **21,** 357–375.
Bowlby, J. (1969). "Attachment and Loss." Vol. 1 Attachment. Basic Books, New York.
Brown, K., and Mack, D. S. (1978). Food sharing among captive *Leontopithecus rosalia*. *Folia Primatol.* **29,** 268–290.
Burton, F. D. (1972). The integration of biology and behavior in the socialization of *Macaca sylvana* of Gilbraltar. *In* "Primate Socialization" (F. E. Poirier, ed.), pp. 29–62. Random House, New York.
Caro, T. M., and Hauser, M. D. (1992). Is there teaching in nonhuman animals? *Q. Rev. Biol.* **67,** 151–174.
Carpenter, C. R. (1965). The howlers of Barro Colorado Island. *In* "Primate Behavior" (I. DeVore, ed.), pp. 250–291. Rinehart and Winston, New York.
Castell, R., and Wilson, C. (1971). Influence of spatial environment on development of mother-infant interaction in pigtail monkeys. *Behaviour* **39,** 202–211.
Cheney, D. L., and Seyfarth, R. M. (1980). Vocal recognition in free-ranging vervet monkeys. *Anim. Behav.* **28,** 362–367.
Cheney, D. L., and Seyfarth, R. M. (1990). "How Monkeys See the World." Univ. of Chicago Press, Chicago.
Chevalier-Skolnikoff, S. (1974). "The Ontogeny of Communication in the Stumptail Macaque *(Macaca arctoides).*" Karger, Basel.
Chevalier-Skolnikoff, S. (1982). A cognitive analysis of facial behavior in Old World monkeys, apes, and humans. *In* "Primate Communication" (C. T. Snowdon, C. H. Brown, and M. Petersen, eds.), pp. 303–368. Cambridge Univ. Press, Cambridge.
Chism, J. (1986). Development and mother-infant relations among captive patas monkeys. *Int. J. Primatol.* **7,** 49–81.
Clark, C. B. (1977). A preliminary report on weaning among chimpanzees of the Gombe National Park, Tanzania. *In* "Primate Bio-Social Development" (S. Chevalier-Skolnikoff and F. E. Poirier, eds.), pp. 235–260. Garland, New York.
Daschbach, N. J., Schein, M. W., and Haines, D. E. (1981). Vocalizations of the slow loris, *Nycticebus coucang (Primates, Lorisidae)*. *Int. J. Primatol.* **2,** 71–80.
DeVore, I. (1963). Mother-infant relations in free-ranging baboons. *In* "Maternal Behavior in Mammals" (H. L. Rheingold, ed.), pp. 305–335. John Wiley & Sons, New York.
de Waal, F. B. M. (1989). "Peacemaking Among Primates." Harvard Univ. Press, Cambridge, MA.
Dixson, A. F. (1981). "The Natural History of the Gorilla." Columbia Univ. Press, New York.
Doyle, G. A. (1979). Development of behaviour in prosimians with special reference to the lesser bushbaby, *Galago senegalensis moholi*. *In* "The Study of Prosimian Behavior" (G. A. Doyle and R. D. Martin, eds.), pp. 157–206. Academic Press, New York.
DuMond, F. V. (1968). The squirrel monkey in a seminatural environment. *In* "The Squirrel Monkey" (L. A. Rosenblum and R. W. Cooper, eds.), pp. 87–145. Academic Press, New York.

Ehardt, C. L., and Blount, B. G. (1984). Mother-infant visual interaction in Japanese macaques. *Dev. Psychobiol.* **17,** 391–405.

Epple, G. (1968). Comparative studies on vocalization in marmoset monkeys *(Hapalidae). Folia Primatol.* **8,** 1–40.

Fairbanks, L. A. (1989). Early experience and cross-generational continuity of mother-infant contact in vervet monkeys. *Dev. Psychobiol.* **22,** 669–681.

Fedigan, L. M., and Fedigan, L. (1977). The social development of a handicapped infant in a free-living troop of Japanese monkeys. *In* "Primate Bio-Social Development" (S. Chevalier-Skolnikoff and F. E. Poirier, eds.), pp. 205–222. Garland, New York.

Fossey, D. (1979). Development of the mountain gorilla *(Gorilla gorilla beringei)*: The first thirty-six months. *In* "The Great Apes" (D. A. Hamburg and E. R. McCrown, eds.), pp. 139–185. Benjamin Cummings, Menlo Park, CA.

Fragaszy, D. M., Schwartz, S., and Shimosaka, D. (1982). Longitudinal observations of care and development of infant titi monkeys *(Callicebus moloch). Am. J. Primatol.* **2,** 191–200.

Gautier, J. P. (1974). Field and laboratory studies of the vocalizations of tatapoin monkeys *(Miopithecus talapoin). Behaviour* **51,** 209–273.

Godfray, H. C. J. (1991). Signalling of need by offspring to their parents. *Nature (London)* **352,** 328–330.

Goodall, J. (1965). Chimpanzees of the Gombe Stream Reserve. *In* "Primate Behavior" (I. DeVore, ed.), pp. 425–481. Holt, Rinehart, and Winston, New York.

Gouzoules, S., Gouzoules, H., and Marler, P. (1984). Rhesus monkeys *(Macaca mulatta)* screams: Representational signalling in the recruitment of agonistic aid. *Anim. Behav.* **32,** 182–193.

Gouzoules, H., and Gouzoules, S. (1989a). a. Design features and developmental modification in pigtail macaque *(Macaca nemestrina)* agonistic screams. *Anim. Behav.* **37,** 383–401.

Gouzoules, H., and Gouzoules, S. (1989b). b. Sex differences in the acquisition of communicative competence by pigtail macaques *(Macaca nemestrina). Am. J. Primatol.* **19,** 163–174.

Green, S. (1975). Variation of vocal pattern with social situation in the Japanese monkey *(Macaca fuscata)*: A field study. *In* "Primate Behavior: Developments in Field and Laboratory Research," Vol. 4 (L. A. Rosenblum, ed.), pp. 1–102. Academic Press, New York.

Green, S. (1981). Sex differences and age gradations in vocalizations of Japanese and lion-tailed monkeys. *Am. Zool.* **21,** 165–183.

Grimm, R. J. (1967). Catalogue of sounds of the pigtailed macaque *(Macaca nemestrina). J. Zool. (London)* **152,** 361–373.

Hansen, E. W. (1966). The development of maternal and infant behavior in the rhesus monkey. *Behaviour* **27,** 109–149.

Hansen, E. W. (1976). Selective responding by recently separated juvenile rhesus monkeys to the calls of their mothers. *Dev. Psychobiol.* **9,** 83–88.

Harcourt, A. H., Stewart, K. J., and Hauser, M. D. (1993). Functions of wild gorilla 'close' calls. I. Repertoire, context, and interspecific comparison. *Behaviour* **124,** 89–122.

Harlow, H. F., Harlow, M. K., and Hansen, E. W. (1963). The maternal affectional system of rhesus monkeys. *In* "Maternal Behavior in Mammals" (H. L. Rheingold, ed.), pp. 254–281. John Wiley & Sons, New York.

Hauser, M. D. (1986). Parent-offspring conflict: Care elicitation behaviour and the "cry-wolf" syndrome. *In* "Primate Ontogeny, Cognition, and Social Behaviour" (J. G. Else and P. C. Lee, eds.), pp. 193–203. Cambridge Univ. Press, Cambridge.

Hauser, M. D. (1989). Ontogenetic changes in the comprehension and production of vervet monkey *(Cercopithecus aethiops)* vocalizations. *J. Comp. Psychol.* **103,** 149–158.

Hauser, M. D. (1993). Do vervet monkey infants cry wolf? *Anim. Behav.* **45,** 1242–1244.

Hess, J. P. (1973). Some observations on the sexual behaviour of captive lowland gorillas, *Gorilla g. gorilla* (Savage and Wyman). In "Comparative Ecology and Behaviour of Primates" (R. P. Micahel and J. H. Crook, eds.), pp. 507–581. Academic Press, London.

Hewes, G. W. (1973). Primate communication and the gestural origin of language. *Curr. Anthropol.* **14**, 5–24.

Hinde, R. A. (1974). "Biological Bases of Human Social Behaviour." Cambridge Univ. Press, Cambridge.

Hinde, R. A., and Simpson, M. J. A. (1975). Qualities of mother-infant relationships in monkeys. In (CIBA Foundation Symposium 33) "Parent-Infant Interaction," pp. 39–67.Elsevier, Amsterdam.

Hoff, M., Nadler, R., and Maple, T. (1981). Development of infant independence in a captive group of lowland gorillas. *Dev. Psychobiol.* **14**, 251–265.

Hohmann, G. (1989a). Comparative studies of vocal communication in two Asian leaf monkeys, *Presbytis johnii* and *Presbytis entellus*. *Folia Primatol.* **52**, 27–57.

Hohmann, G. (1989b). Vocal communication of wild bonnet macaques *(Macaca radiata)*. *Primates* **30**, 325–345.

Hohmann, G. (1991). Comparative analyses of age- and sex-specific patterns of vocal behaviour in four species of Old World monkeys. *Folia Primatol.* **56**, 133–156.

Hohmann, G., and Herzog, M. O. (1985). Vocal communication in lion-tailed macaques *(Macaca silenus)*. *Folia Primatol.* **45**, 148–178.

Holekamp, K. E., and Smale, L. (1991). Dominance acquisition during mammalian social development: The "inheritance" of maternal rank. *Am. Zool.* **31**, 306–317.

Horr, D. A. (1977). Orang-utan maturation: Growing up in a female world. In "Primate Bio-Social Development" (S. Chevalier-Skolnikoff and F. E. Poirier, eds.), pp. 289–321. Garland, New York.

Isaac, G. (1978). The food-sharing behavior of protohuman hominids. *Sci. Am.* **238**, 90–108.

Jay, P. C. (1963). Mother-infant relations in langurs. In "Maternal Behavior in Mammals" (H. L. Rheingold, ed.), pp. 282–304. John Wiley & Sons, New York.

Jay, P. (1965). The common langur of North India. In "Primate Behavior" (I. De Vore, ed.), pp. 197–249. Holt, Rinehart, and Winston, New York.

Jensen, G. D. (1965). Mother-infant relationship in the monkey *Macaca nemestrina*: Development of specificity of maternal response to own infant. *J. Comp. Physiol. Psychol.* **59**, 305–308.

Jensen, G. D., Bobbitt, R. A., and Gordon, B. N. (1973). Mothers' and infants' roles in the development of independence of *Macaca nemestrina*. *Primates* **14**, 79–88.

Jensen, G. D., and Gordon, B. N. (1970). Sequences of mother-infant behavior following a facial communicative gesture of pigtail monkeys. *Biol. Psychiatry* **2**, 267–272.

Kalin, N. H., Shelton, S. E., and Snowdon, C. T. (1992). Affiliative vocalizations in infant rhesus macaques *(Macaca mulatta)*. *J. Comp. Psychol.* **106**, 254–261.

Kano, T. (1992). "The Last Ape." Stanford Univ. Press, Stanford, CA.

Kaplan, J., Winship-Ball, A., and Sim, L. (1978). Maternal discrimination of infant vocalizations in squirrel monkeys. *Primates* **19**, 187–193.

Kenney, M. D., Mason, W. A., and Hill, S. D. (1979). Effects of age, objects, and visual experience on affective responses of rhesus monkeys to strangers. *Dev. Psychol.* **15**, 176–184.

King, B. J. (1991). Social information transfer in monkeys, apes, and hominids. *Yearb. Phys. Anthr.* **34**, 97–115.

King, B. J. (1994). "The Information Continuum: Evolution of Social Information Transfer in Monkeys, Apes, and Hominids." School of Am. Res., Santa FE, NM.

Kirkevold, B. C., Lockard, J. S., and Heestand, J. E. (1982). Developmental comparisons of grimace and play mouth in infant pigtail macaques *(Macaca nemestrina)*. *Am. J. Primatol.* **3,** 277–283.

Kirkpatrick, M., and Lande, R. (1989). The evolution of maternal characters. *Evolution* **43,** 485–499.

Klopfer, P. H. (1970). Discrimination of young in galagos. *Folia Primatol.* **13,** 137–143.

Kortlandt, A. (1962). Chimpanzees in the wild. *Sci. Am.* **206,** 128–138.

Kuroda, S. (1984). Interactions over food among pygmy chimpanzees. *In* "The Pygmy Chimpanzee" (R. Susman, ed.), pp. 301–324. Plenum Press, New York.

Kuroda, S. (1989). Developmental retardation and behavioral characteristics of pygmy chimpanzees. *In* "Understanding Chimpanzees" (P. G. Heltne and L. A. Marquardt, eds.), pp. 184–193. Harvard Univ. Press, Cambridge, MA.

Lester, B. M., and Boukydis, C. F. Z. (1992). No language but a cry. *In* "Nonverbal Vocal Communication: Comparative and Developmental Approaches" (H. Papousek, U. Jurgens, and M. Papousek, eds.), pp. 145–173. Cambridge Univ. Press, Cambridge.

Levine, S., Wiener, S. G., Coe, C. L., Bayart, F. E. S., and Hayashi, K. T. (1987). Primate vocalization: A psychobiological approach. *Child Dev.* **58,** 1408–1419.

Lieblich, A. K., Symmes, D., Newman, J. D., and Shapiro, M. (1980). Development of the isolation peep in laboratory-bred squirrel monkeys. *Anim. Behav.* **29,** 1–9.

Lillehei, R. A., and Snowdon, C. T. (1978). Individual and situational differences in the vocalizations of young stumptail macaques *(Macaca arctoides)*. *Behaviour* **65,** 270–281.

Lovejoy, C. O. (1981). The origin of man. *Science (Washington, D. C.)* **211,** 341–350.

Macedonia, J. M. (1993). The vocal repertoire of the ringtailed lemur *(Lemur catta)*. *Folia Primatol.* **61,** 186–217.

Mackinnon, J. (1974a). "In Search of the Red Ape." Holt, Rinehart, and Winston, New York.

Mackinnon, J. (1974b). The behaviour and ecology of wild orang-utans *(Pongo pygmaeus)*. *Anim. Behav.* **22,** 3–74.

Maestripieri, D. (1995a). First steps in the macaque world: Do rhesus mothers encourage their infants' independent locomotion? *Anim. Behav.* **49,** 1541–1549.

Maestripieri, D. (1995b). Maternal responsiveness to infant distress calls in stumptail macaques. *Folia Primatol.* **64,** 201–206.

Maestripieri, D. (1996). Maternal encouragement of infant locomotion in pigtail macaques *(Macaca nemestrina)*. *Anim. Behav.* **51,** 603–610.

Maple, T., and Hoff, M. (1982). "Gorilla Behavior." Van Nostrand Reinhold, New York.

Maple, T., Wilson, M. E., Zucker, E. L., and Wilson, S. F. (1978). Notes on the development of a mother-reared orangutan: The first six months. *Primates* **19,** 593–602.

Marler, P. (1972). Vocalizations of East African monkeys. II. Black and white colobus. *Behaviour* **42,** 175–197.

Marriott, B. M., and Salzen, E. A. (1978). Facial expressions in captive squirrel monkeys *(Saimiri sciureus)*. *Folia Primatol.* **29,** 1–18.

Masataka, N. (1982). A field study of the vocalizations of Goeldi's monkeys *(Callimico goeldii)*. *Primates* **23,** 206–219.

Masataka, N. (1985). Development of vocal recognition of mothers in infant Japanese macaques. *Dev. Psychobiol.* **18,** 107–114.

Masataka, N., and Symmes, D. (1986). Effect of separation distance on isolation call structure in squirrel monkeys *(Saimiri sciureus)*. *Am. J. Primatol.* **10,** 271–278.

Mascagni, O., and Doyle, G. A. (1993). Infant distress vocalizations in the southern African lesser bushbaby *(Galago moholi)*. *Int. J. Primatol.* **14,** 41–60.

Mendelson, M. (1982). Clinical examination of visual and social responses in infant rhesus monkeys. *Dev. Psychol.* **18,** 658–664.

Miller, L. C., and Nadler, R. D. (1981). Mother-infant relations and infant development in captive chimpanzees and orang-utans. *Int. J. Primatol.* **2,** 247–261.
Mitchell, G., and Brandt, E. M. (1975). Behavior of female rhesus monkey during birth. In "The Rhesus Monkey" (G. H. Bourne, ed.), pp. 231–244. Academic Press, New York.
Moody, M. I., and Menzel, E. W. (1976). Vocalizations and their behavioral contexts in the tamarin *Saguinus fuscicollis. Folia Primatol.* **25,** 73–94.
Morland, H. S. (1990). Parental behavior and infant development in ruffed lemurs *(Varecia variegata)* in a northeast Madagascar rain forest. *Am. J. Primatol.* **20,** 253–265.
Nakamichi, M., and Yoshida, A. (1986). Discrimination of mother by infant among Japanese macaques *(Macaca fuscata). Int. J. Primatol.* **7,** 481–489.
Neimitz, C., and Kok, D. (1976). Observations on the vocalizations of a captive infant orangutan *(Pongo pygmaeus). Sarawak Mus. J.* **24,** 237–250.
Newman, J. D. (1985a). The infant cry in primates: An evolutionary perspective. In "Infant Crying" (B. M. Lester, and C. F. Z. Boukydis, eds.), pp. 307–323. Plenum Press, New York.
Newman, J. D. (1985b). Squirrel monkey communicaiton. In "Handbook of Squirrel Monkey Research" (L. A. Rosenblum and C. L. Coe, eds.), pp. 99–126. Plenum Press, New York.
Newman, J. D. (1995). Vocal ontogeny in macaques and marmosets: Convergent and divergent lines of development. In "Current Topics in Primate Vocal Communication" (E. Zimmermann, J. D. Newman, and U. Jurgens, eds.), pp. 73–97. Plenum Press, New York, in press.
Newman, J. D., and Symmes, D. (1982). Inheritance and experience in the acquisition of primate acoustic behavior. In "Primate Communication" (C. T. Snowdon, C. H. Brown, and M. R. Petersen, eds.), pp. 259–278. Cambridge Univ. Press, Cambridge.
Nicolson, N. A. (1977). A comparison of early behavioral development in wild and captive chimpanzees. In "Primate Bio-Social Development" (S. Chevalier-Skolnikoff and F. E. Poirier, eds.), pp. 529–560. Garland, New York.
Nishida, T. (1990). Deceptive behavior in young chimpanzees: An essay. In "The Chimpanzees of the Mahale Mountains" (T. Nishida, ed.), pp. 285–290. Tokyo Univ. Press, Tokyo.
Palombit, R. (1992). A preliminary study of vocal communication in wild long-tailed macaques *(Macaca fascicularis).* I. Vocal repertoire and call emission. *Int. J. Primatol.* **13,** 143–181.
Papousek, M. (1992). Early ontogeny of vocal communication in parent-offspring interactions. In "Nonverbal Vocal Communication: Comparative and Developmental Approaches" (H. Papousek, U. Jurgens, and M. Papousek, eds.), pp. 230–261. Cambridge Univ. Press, Cambridge.
Pereira, M. E. (1986). Maternal recognition of juvenile offspring coo vocalizations in Japanese macaques. *Anim. Behav.* **34,** 935–937.
Pereira, M. E., Seeligson, M. L., and Macedonia, J. M. (1988). The behavioral repertoire of the black-and-white ruffed lemur, *Varecia variegata variegata (Primates: Lemuridae).* Folia Primatol. **51,** 1–32.
Ploog, D. (1967). The behavior of squirrel monkeys *(Saimiri sciureus)* as revealed by sociometry, bioacoustics, and brain stimulation. In "Social Communication Among Primates" (S. A. Altmann, ed.), pp. 149–184. Chicago Univ. Press, Chicago.
Plooij, F. X. (1978). Some basic traits of language in wild chimpanzees? In "Action, Gesture, and Symbol: The Emergence of Language" (A. Lock, ed.), pp. 111–131. Academic Press, London.
Plooij, F. X. (1979). How wild chimpanzee babies trigger the onset of mother-infant play and what the mother makes of it. In "Before Speech: The Beginning of Interpersonal Communication" (M. Bullowa, ed.), pp. 223–243. Cambridge Univ. Press, Cambridge.
Poirier, F. E. (1968). The Nilgiri langur *(Presbytis johnii)* mother-infant dyad. *Primates* **9,** 45–68.
Pola, Y. V., and Snowdon, C. T. (1975). The vocalizations of pygmy marmosets *(Cebuella pygmaea). Anim. Behav.* **23,** 826–842.

Ransom, T. W., and Rowell, T. E. (1972). Early social development of feral baboons. *In* "Primate Socialization" (F. E. Poirier, ed.), pp. 105–144. Random House, New York.

Redican, W. K. (1975). Facial expressions in nonhuman primates. *In* "Primate Behavior" (L. A. Rosenblum, ed.), pp. 103–194. Academic Press, New York.

Reynolds, V., and Reynolds, F. (1965). Chimpanzees of the Budongo Forest. *In* "Primate Behavior" (I. DeVore, ed.), pp. 368–424. Holt, Rinehart, and Winston, New York.

Richard, A. F. (1976). Preliminary observations on the birth and development of *Propithecus verreauxi* to the age of six months. *Primates* **17,** 357–366.

Rogers, C. M., and Davenport, R. K. (1970). Chimpanzee maternal behavior. *In* "The Chimpanzee" (G. H. Bourne, ed.), pp. 361–368. Karger, Basel.

Rosenblum, L. A. (1968). Mother-infant relations and early behavioral development in the squirrel monkey. *In* "The Squirrel Monkey" (L. A. Rosenblum and R. W. Cooper, eds.), pp. 207–233. Academic Press, New York.

Rowell, T. E., and Hinde, R. A. (1962). Vocal communication by the rhesus monkey (*Macaca mulatta*). *Proc. Zool. Soc. London* **138,** 279–294.

Schaller, G. B. (1963). "The Mountain Gorilla: Ecology and Behavior." Univ. of Chicago Press, Chicago.

Seay, B. (1966). Maternal behavior in primiparous and multiparous rhesus monkeys. *Folia Primatol.* **4,** 146–168.

Seyfarth, R. M., and Cheney, D. L. (1986). Vocal development in vervet monkeys. *Anim. Behav.* **34,** 1640–1658.

Silk, J. B. (1978). Patterns of food sharing among mother and infant chimpanzees at Gombe National Park, Tanzania. *Folia Primatol.* **29,** 129–141.

Silk, J. B. (1979). Feeding, foraging, and food sharing behavior of immature chimpanzees. *Folia Primatol.* **31,** 123–142.

Simons, R. C., Bobbitt, R. A., and Jensen, G. D. (1968). Mother monkeys' (*Macaca nemestrina*) responses to infant vocalizations. *Percept. Motor Skills* **27,** 3–10.

Simons, R. C., and Bielert, C. F. (1973). An experimental study of vocal communication between mother and infant monkeys (*Macaca nemestrina*). *Am. J. Phys. Anthropol.* **38,** 455–461.

Snowdon, C. T. (1987). A naturalistic view of categorical perception. *In* "Categorical Perception" (S. Harnard, ed.), pp. 332–354. Cambridge Univ. Press, Cambridge.

Snowdon, C. T., and Elowson, A. M. (1992). Ontogeny of primate vocal communication. *In* "Topics in Primatology," Vol. I, T. Nishida, W. C. McGrew, P. Marler, M. Pickford, and F. B. M. de Waal, eds.), pp. 279–290, Univ. of Tokyo Press, Tokyo.

Snowdon, C. T., Brown, C. H., and Petersen, M. (1982). Social and environmental determinants of primate vocalizations. *In* "Primate Communication" (C. T. Snowdon, C. H. Brown, and M. Petersen, eds.), pp. 63–66. Cambridge Univ. Press, Cambridge.

Stamps, J. A. (1991). Why evolutionary issues are reviving interest in proximate behavioral mechanisms. *Am. Zool.* **31,** 338–348.

Struhsaker, T. T. (1967). Auditory communication among vervet monkeys. *In* "Social Communication Among Primates" (S. A. Altmann, ed.), pp. 281–324. Univ. of Chicago Press, Chicago.

Struhsaker, T. T. (1975). "The Red Colobus Monkey." Univ. of Chicago Press, Chicago.

Sugiyama, Y. (1972). Social characteristics and socialization of wild chimpanzees. *In* "Primate Socialization" (F. E. Poirier, ed.), pp. 145–163. Random House, New York.

Symmes, D., and Biben, M. (1985). Maternal recognition of individual infant squirrel monkeys from isolation call playback. *Am. J. Primatol.* **9,** 39–46.

Symmes, D., and Biben, M. (1992). Vocal development in nonhuman primates. *In* "Nonverbal Vocal Communication: Comparative and Developmental Approaches" (H. Papousek, U. Jurgens, and M. Papousek, eds.), pp. 123–140. Cambridge Univ. Press, Cambridge.

Tanner, N. M. (1981). "On Becoming Human." Cambridge Univ. Press, Cambridge.
Todt, D. (1988). Serial calling as a mediator of interaction processes: Crying in primates. *In* "Primate Vocal Communication" (P. Goedeking, D. Symons, and D. Todt, eds.), pp. 88–107. Springer-Verlag, Berlin.
Tomasello, M., George, B. L., Kruger, A. C., Farrar, M. J., and Evans, A. (1985). The development of gestural communication in young chimpanzees. *J. Hum. Evol.* **14,** 175–186.
Tomasello, M., Gust, D., and Frost, G. T. (1989). A longitudinal investigation of gestural communication in young chimpanzees. *Primates* **30,** 35–50.
Tomasello, M., Kruger, A. C., and Ratner, H. H. (1993). Cultural learning. *Behav. Brain Sci.* **16,** 495–552.
Tomasello, M., Call, J., Nagell, K., Olguin, R., and Carpenter, M. (1994). The learning and use of gestural signals by young chimpanzees: A transgenerational study. *Primates* **35,** 137–154.
van de Rijt-Plooij, H. H. C., and Plooij, F. X. (1987). Growing independence, conflict and learning in mother-infant relations in free-ranging chimpanzees. *Behaviour* **101,** 1–86.
van Lawick-Goodall, J. (1967). Mother-offspring relationships in free-ranging chimpanzees. *In* "Primate Ethology" (D. Morris, ed.), pp. 287–346. Aldine, Chicago.
van Lawick-Goodall, J. (1968a). The behaviour of free-living chimpanzees in the Gombe Stream Reserve. *Anim. Behav. Monogr.* **1,** 161–311.
van Lawick-Goodall, J. (1968b). A preliminary report on expressive movements and communication in the Gombe stream chimpanzees. *In* "Primates. Studies in Adaptation and Variability" (P. C. Jay, ed.), pp. 313–374. Holt, Rinehart, and Winston, New York.
Washburn, S. L. (1957). Australopithecines: The hunters or the hunted? *Am. Anthropol.* **59,** 612–614.
Winter, P. (1968). Social communication in the squirrel monkey. *In* "The Squirrel Monkey" (L. A. Rosenblum and R. W. Cooper, eds.). pp. 235–253. Academic Press, New York.
Yerkes, R. M. (1943). "Chimpanzees, a laboratory colony." Yale Univ. Press, New Haven, CT.
Yerkes, R. M., and Tomilin, M. I. (1935). Mother-infant relations in chimpanzees. *J. Comp. Psychol.* **20,** 321–348.
Yoshida, H., Norikoshi, K., Kitahara, T., and Yoshihara, K. (1991). A study of the mother-infant relationship of chimpanzees (*Pan troglodytes*) during the first four years of infancy in Tama Zoological Park. *Jpn. J. Anim. Psychol.* **41,** 88–100.
Zeller, A. (1992). Communication in the social unit. *In* "Social Processes and Mental Abilities in Nonhuman Primates" (F. D. Burton, ed.), pp. 61–89. Edwin Mellen, New York.
Zimmermann, E. (1989). Aspects of reproduction and behavioral and vocal development in Senegal bushbabies (*Galago senegalensis*). *Int. J. Primatol.* **10,** 1–16.

# Infant Care in Cooperatively Breeding Species

CHARLES T. SNOWDON

DEPARTMENT OF PSYCHOLOGY
UNIVERSITY OF WISCONSIN
MADISON, WISCONSIN 53706

## I. Introduction

There are several avian and mammalian species where parental care is neither restricted to the mother nor shared between father and mother, but where many other group members serve as helpers, assisting with infant care while deferring their own reproduction. These species have varying degrees of reproductive division of labor. Sherman et al. (1995) have argued that the avian and mammalian cooperative breeders share many characteristics with eusocial insects. These species create a variety of problems for traditional adaptive explanations of parental care because nonreproductive helpers take care of infants rather than reproducing themselves. Moreover, in some cases the nonreproductive helpers assist with unrelated infants, making the behavior even more difficult to explain in adaptive terms. This article reviews results on several species of cooperatively breeding birds and mammals. Most of the data that we have on the likely function of and the ecological variables affecting cooperative breeding come from field studies on birds, while most of our information on the behavioral and physiological mechanisms that underlie cooperative parental care come from studies of mammals, particularly the dwarf mongoose, and marmosets and tamarins, small nonhuman primates from South America. Direct comparisons between birds and mammals will, therefore, be difficult, because only in a few cases do we have comparable types of data on both.

Several issues that will be addressed:

1. What type of help do helpers provide and what are the effects of these caretakers on the infants and on mothers and fathers? Do helpers really help? What are the effects of helpers on the survival of the young?

2. What are the conditions that lead helpers to help? Are there particular ecological or social conditions that are more conducive to the occurrence

of helping? Do parents always accept helpers? What is known about the physiological mechanisms that underlie infant care? Because fathers' direct involvement in infant care is rare in most mammals, except in cooperatively breeding species, do similar behavioral and physiological mechanisms that affect maternal care also affect paternal care?

3. What is the effect of helping on the reproductive success of the helpers? Do they benefit from being helpers? What other reproductive options are available to them?

4. What mechanisms prevent helpers from becoming reproductively active themselves? Is there any physiological inhibition of fertility or do behavioral mechanisms suffice to inhibit their reproduction? Who is responsible for maintaining their inhibition of reproduction?

5. Given our current knowledge, what are the research areas that should be pursued in the future? How can we better integrate information on birds and mammals?

## II. How Do Helpers Help?

### A. Patterns of Helping in Callitrichid Primates

The family Callitrichidae contains the marmosets and tamarins, small-bodied primates of the Neotropics. The body weights range from the 90 to 110 g of pygmy marmosets (*Cebuella pygmaea*) to the lion tamarins (*Leontopithecus rosalia*) that weigh about 750 g as adults. There are four genera: the marmosets (*Callithrix sp.*), the pygmy marmoset (*Cebuella pygmaea*), the tamarins (*Saguinus sp.*), and the lion tamarins (*Leontopithecus sp.*). All species are characterized by twin births with the total infant to maternal weight ratio being greater than 15% (Leutenegger, 1980; Goldizen, 1987a). Gestation length ranges from 130 days in lion tamarins to 184 days in the cotton-top tamarin (*Saguinus oedipus*). Many of the species living in the Amazon basin and in northeastern Brazil have two births a year (Snowdon and Soini, 1988; Stevenson and Rylands, 1988; de la Torre *et al.*, 1995). Thus, the mother can become pregnant while she is still nursing her current infants. (However, the number of surviving infants (singletons versus twins) and the pattern of nursing twins (nursing both twins simultaneously versus nursing each twin successively) does affect the timing of the subsequent pregnancy in cotton-top tamarins (Ziegler *et al.*, 1990b)). The high birth weight of infants coupled with the high fecundity of females and the fact that marmosets and tamarins do not leave their young in nests, but must carry them throughout the day, have led to a clear reproductive division of labor. In all species fathers and other nonreproductive group members assist in infant care that includes carrying one or both twins, sharing food with infants during the weaning process, being vigilant and defending the family group from predators, and providing thermoregulatory assistance at night nests. As small-bodied primates with a high metabolic rate and a 13 h per night inactive period, marmosets and tamarins face severe metabolic challenges each night. Even captive tamarins have reduced heart rate and body temperature and huddle as a group each night. Additional helpers might help infants maintain body temperature overnight, especially in species living in more temperate habitats (Snowdon and Soini, 1988). Between transporting young, providing them food, protecting them from predators, and assisting them with thermoregulation, the costs to helpers are thought to be high but the benefits to infant survival are thought to be equally high.

*1. Does Helping Improve Infant Survival*

Although the previous presentation suggests that helpers should significantly increase the survival of infants, the null hypothesis that helpers provide no benefits to infants must be evaluated. Garber *et al.* (1984), who studied the moustached tamarin (*Saguinus mystax*) in the field in Peru, found that the number of surviving infants in a group was directly related to the number of adult males in the group. Although the number of infants born is not easily observed, it is assumed that there is only one breeding female giving birth to twins in the typical group in moustached tamarins and in the examples that follow. In groups with one male, the mean number of infants that survived was 0.8; in groups with two males, 1.2 infants survived, and in groups with three males, 2.0 infants survived. There was no effect on infant survival of the number of adult females in a group.

Koenig (1995) analyzed data from published field studies on the common marmoset (*Callithrix jacchus*) and found that groups with one or two adult males had a mean of 1.1 infants surviving, whereas groups with more than two adult males had 2.0 infants surviving. As with the moustached tamarins, there was no effect of the number of adult females on infant survival, but no surviving infants were found in groups with fewer than four adults.

Similar data have been obtained on wild cotton-top tamarins by Savage *et al.* (in press). There was a linear relationship between the number of helpers in a group and the percentage of infants surviving to 1 year ($r = .62, p = .06$). Only with group sizes of five or more did infant survival reach 100%. Interestingly, these field results parallel data from captive colonies of cotton-top tamarins. Johnson *et al.* (1991) reported a 24.8% infant survival to 1 year in captive pairs, but an 80.3% infant survival in groups with parents and older siblings. We have analyzed 18 years of data from our cotton-top tamarin colony with similar results on survival (Greene and Snowdon, in preparation). We found only 37% infant survival with two caretakers (both parents), 70% infant survival with three to four caretakers (parents plus one or two helpers), and 100% infant survival only with five or more caretakers in a group. Thus, in cotton-top tamarins in both captive and field conditions, the presence of helpers significantly affects infant survival (see Fig. 1).

2. *Comparison of Field and Captive Data*

Most of our information about infant caretaking in cooperatively breeding monkeys has come from studies of captive animals, thus many of the phenomena and also the mechanisms leading to cooperative infant care could be artifacts of captive studies. Fortunately, comparative studies with both common marmosets (*Callithrix jacchus*) and cotton-top tamarins (*Saguinus oedipus*) provide data comparing caretaking in field and captive settings. Yamamoto *et al.* (1996) compared infant caretaking of common marmosets in a wild population and in two captive populations, one an indoor facility in the Northern Hemisphere and the other an outdoor facility in Natal, Brazil, in the natural range of common marmosets. There were no differences among the three conditions in the total amount of infant carrying at different ages. Infants were carried 100% of the time for the first 4 weeks, this decreased in subsequent weeks as infants started to become independent. Wild groups carried infants more often in subsequent

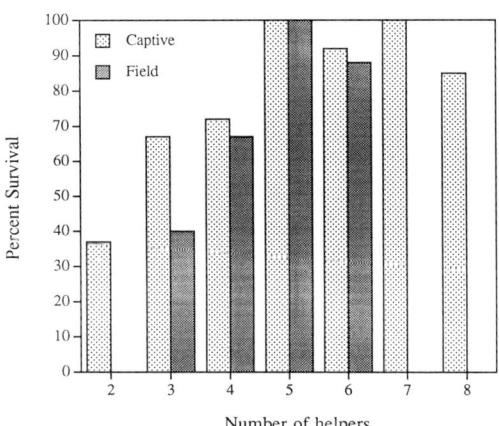

FIG. 1. Percentage survival of cotton-top tamarin infants as a function of the number of helpers available. (Captive data from Greene and Snowdon, in preparation. Field data plotted from Savage *et al.*, in press.

weeks, but the difference was not significantly greater. In all three conditions fathers and mothers carried infants approximately equally at about 20% of the time. Helpers in the wild population carried infants significantly more than helpers in the captive populations, suggesting a greater importance for helpers in field conditions. In all three conditions adult helpers carried infants more often than subadult and juvenile helpers.

We have completed a study of infant care in wild cotton-top tamarins in Colombia (Savage et al. in press). Twelve litters were observed in the field between 1988 and 1993 for a total of 391 h of observation during periods of infant caretaking (minimum of 32 h per litter). Groups were observed 3–4 days each week after the birth of infants and all animals were individually marked so that the age, sex, and reproductive status of each caretaker was known. Infants were carried on 90% or more of the scan samples through Week 7 and on 35% of the samples by Week 10, with complete independence of the infants not observed until Week 13. These data contrast with the field data on common marmosets of Yamamoto et al. (1996) that found infants were carried only 60% of the time in Week 7 and 20% of the time in Week 10, but parallel data from Tardif et al. (1990) comparing captive cotton-top tamarins and common marmosets.

In the wild cotton-top tamarins, adult males and females carried infants equally during the first week, but female carrying decreased to about 10% by Week 5, whereas adult male caretaking was greater than 60% of the time by Week 6. As with studies on captive cotton-top tamarins (McGrew, 1988) and wild moustached tamarins (Garber et al., 1984), with increased group size the amount of carrying by any individual adult male decreased, yielding a correlation of $r = -.76$ between group size and duration of carrying by adult males.

In our captive colony of cotton-top tamarins we have found that first-time parents have very low reproductive success (<10% infant survival, Greene and Snowdon, in preparation). In the field, we observed two litters born to primiparous mothers (though with other experienced adults in the group). In both cases these primiparous mothers carried infants 90% of the time in Weeks 1 and 2, whereas multiparous females carried infants 50% of the time in the first week and 25% of the time in the second week. This suggests that first-time mothers might have to learn to allow other group members to assist her with infant caretaking.

Animals in the field that migrated into a group after the conception date of the infants were significantly less likely to carry infants than were longer-term residents, although both immigrants and longer-term residents were equally active in performing vigilance behavior. Males, in general, were more vigilant than females, and in larger groups there was male role division with one male showing vigilance behavior while another male carried the infants, whereas in smaller groups a male more often simultaneously was infant caretaker and showed vigilance behavior. This suggests that one reason why there is greater reproductive success in larger goups is the division of labor among males as infant carriers or showing vigilance. Males within a group exchange these roles frequently, so the role division is apparent only over a short term. An animal that has been vigilant for a period of time will subsequently carry the infants while another male engages in vigilance. Because it is extremely difficult, if not impossible, to integrate immigrants into established captive groups of cotton-top tamarins (Price and McGrew, 1991), we cannot compare the role of immigrants and residents in captive studies. Caine (1993) has argued that vigilance and defense against predation should be considered a major influence on the evolution of cooperative infant care, and our field data support her arguments that vigilance behavior serves an important function in infant survival.

In summary, for the two species where we have parallel studies in captivity and in the field, the general patterns of infant carrying and the division of labor are very similar. In the field there is a delay of about 2 weeks before infants are spending 50% of the time off a caretaker (see Table I). In cotton-top tamarins, at least, vigilance behavior occurs separately from infant care, especially in larger groups, suggesting that additional helpers

TABLE I
Duration of Infant Care in Callitrichid Species as a Function of Infant to Maternal Body Weight and Mean Distance Traveled

| Species | Time to 50% time off (weeks) | | Infant:maternal weight ratio (%)[a] | Mean distance traveled (m)[b] |
|---|---|---|---|---|
| | Captive | Field | | |
| Common marmoset | 5[c] | 7–8[d] | 22.8 | 700 |
| Pygmy marmoset | 5.5[e] | | | |
| Cotton-top tamarin | 7[f] | 9–10[g] | 20.4 | 1750 |
| Saddle-back tamarin | 7–10[h] | | 19.8 | 2325 |
| Golden lion tamarin | 7[i] | | 17.4 | 1500 |

[a] Garber (1994).
[b] Tardif (1994).
[c] Ingram (1977), Arruda et al. (1986).
[d] Albuquerque (1994).
[e] Wamboldt et al. (1988).
[f] Cleveland and Showdon (1984).
[g] Savage et al. (in press).
[h] Vogt et al. (1978).
[i] Hoage (1978).

allow a greater division of labor between carrying and vigilance than is possible in smaller groups.

## 3. Comparison of Marmosets and Tamarins

Although many view the Callitrichidae as a unitary family with common infant care patterns (e.g., Goldizen, 1987a; Dunbar, 1995), there are important differences between species and between genera that suggest somewhat similar mechanisms and different functional significance for cooperative infant care in different species. In this section I will first review some of the different patterns of infant care that have been described and then review the hypothesis that have been suggested to account for these differences.

Wamboldt et al. (1988) described infant care in captive pygmy marmosets, performed separation experiments (15–30 min), and observed responses of caretakers to both related and unrelated infants. Adult females did most of the infant carrying in the first 3 weeks, with adult males and subadults doing much of the subsequent carrying. Adult females were most responsive to the separated infant in the first 3 weeks, while adult males and subadults retrieved the infant most after 3 weeks. Adult females responded only to their own offspring, while adult males retrieved all infants whether related to them or not.

In a field study of the buffy-headed marmoset (*Callithrix flaviceps*), Ferrari (1987, 1992) reported that adult females carried infants more of the time in the first week after birth and adult males carried more after the first week. There was no difference in carrying time between the presumed reproductive male and other adult males. Food begging by infants and food sharing by adults were frequently observed, with the donor giving high-pitched chatter calls that the infant responded to both by vocalizing and approaching. In contrast to cotton-top tamarins (see Feistner and Chamove, 1986), both sexes appeared to be equally involved in food sharing.

In wild common marmosets Albuquerque (1994) studied three litters in two large groups, and found that infants were carried 100% of the time through Week 3 but never after Week 11; the 50% time-off point was between 7 and 8 weeks. The mother, putative father, and other adult males were the most frequent infant carriers, and juveniles carried infants infrequently. In the same population, Digby and Ferrari (1994) reported that two females frequently bred in the same group. Digby (1995b) found that all group members more than 5 months old assisted the dominant female even from the day of birth. The subordinate female was more protective of her infants and did not let other group members have access to them. By Days 11–20 the dominant female carried her infants only 25% of the time, while by Day 30 the subordinate female still carried her infants

75–100% of the time. The infant survival rate was 62% for offspring of dominant females, but only 33% for subordinate females. Infanticide occurred when both females gave birth at the same time, suggesting that competition for helpers is severe.

In captive common marmosets, Ingram (1977) reported equal caretaking by males and females, with adults doing more than subadults, who, in turn, did more than juveniles. There were considerable individual differences between families and one parent would compensate for a lower level of infant care effort by the other. Arruda et al. (1986) separated fathers from captive pairs with infants for 3-day periods at infant age 15, 30, and 45 days, and found that mothers compensated for the absence of fathers and that infants showed no signs of "depression" found in macaques during maternal separation. Infants were off 50% of the time by Week 5, 2 weeks sooner than in Albuquerque's (1994) field study. Yamamoto (1993), summarizing several studies of captive common marmosets in Brazil, reported that infants have the strongest relationship with fathers, second with mothers. Older juveniles were rarely observed to carry infants but they had a high tolerance of infants. Infants showed more interest in making contact with parents than they did with older siblings. Box (1977) found both parents carried infants in the first week, with the adult male carrying less with successive litters as more helpers from previous births were available to carry infants. With more caretakers, there was more frequent carrying of only one twin at a time, and younger animals often carried only one infant. Rothe et al. (1993) and Koenig and Rothe (1991) studied larger groups of common marmosets (8–10 animals) and found mothers to be the major caretakers in the early weeks, with helpers doing relatively little until later. This research group reported high variability in the distribution of infant care between parents and older siblings not only between groups but also within groups from one litter to the next.

In a field study of one group of moustached tamarins, Heymann (1990) reported that young monkeys begged or stole food mainly from the adult male. Adult males carried infants about 50% of the time, adult females about 10% of the time, and two subadult females carried infants 20% and 10% of the time, respectively. Goldizen (Terborgh and Goldizen, 1985; Goldizen, 1987a, 1988) reported in wild, saddle-back tamarins (*Saguinus fuscicollis*) that adult males carried infants nearly twice as often as the adult females, with 3-year-old females carrying as much as the reproductive female, 2 year olds carrying about half as much (12%), and 1 year olds carrying only 6% of the time. Two year olds were not allowed access to infants very often until later weeks. Adult males groomed infants more than adult females and were more often in contact with the infants (Goldizen, 1989). Juveniles were most often nearest neighbors to adult males, suggesting a possible interest in the infants. The primary male was most often the food sharer, while the secondary male spent the most time in vigilance and the reproductive female the least. A similar pattern of adult males carrying more than adult females was reported by Vogt et al. (1978), who studied captive saddle-back tamarins living in a large greenhouse. Fifty percent time off occurred in Week 10 for the first litter, but 50% time off occurred at Week 7 in the third litter when nonreproductive helpers were involved in caretaking, suggesting that helpers might promote earlier independence.

In lion tamarins mothers are the main caretakers through Week 3, with fathers the main caretakers from Weeks 4 to 12 (Hoage, 1978, 1982). Younger females and juvenile males carried infants from Weeks 3 to 8. Infants were carried only 50% of the time by Week 7, parallel to studies on other tamarin species, but 2 weeks later than found in studies of captive marmosets. In contrast to studies on other species, Hoage found that there was preferential carrying of one's own sex.

The cotton-top tamarin is the probably the most-studied callitrichid species. Cleveland and Snowdon (1984) found that males carried infants more than females and that adults carried infants more than did subadults, who carried infants more than did juveniles. Infants were carried 50% of the time in the first 4 weeks but were off 50% of the time by Week 7. Although males carried infants more than females did, once the infants were independent, they more often initiated contact with females than with males. The main factor influencing the amount of maternal carrying was the presence or absence of siblings. Later, Ziegler et al. (1990b) found a negative correlation ($r = -.88$, $p < 0.5$) beween the number of helpers present and the amount of time that mothers carried infants. However, McGrew (1988) reported no effect of helpers on the amount of time that mothers spent in maternal care, but instead found that the amount of infant carrying by adult males decreased with increasing numbers of helpers. This finding is supported by Tardif et al. (1990), who found a correlation of $r = -.60$ between number of helpers and time fathers carried infants. There were no similar effects for mothers, although they did find that mothers with no helpers carried twins much less than they did singletons. Our field data on cotton-top tamarins (Savage et al., in press) support these latter findings. These results are counterintuitive because the greater energetic demands (that would require multiple helpers) are on the mother, yet increasing the number of helpers reduces paternal effort more than maternal effort. The findings of Cleveland and Snowdon (1984) and Ziegler et al. (1990b) might explain the paradox. A single additional helper reduces maternal effort to a low value and from then on male effort is reduced linearly by increasing the number of helpers.

Tardif *et al.* (1990) reported that among cotton-top tamarins, both parents were equally likely to retrieve a nonharassed infant, but fathers were more likely to intervene with harassed infants. Older, nonreproductive helpers carried more often than younger helpers in the early weeks, but these effects decreased in Weeks 5–8 (Tardif *et al.*, 1992). Subadults with no prior infant care experience carried as much as those with experience, but there was great interindividual variability. Nonreproductive helpers more often carried one twin infant than they carried both twins.

Price (1990a,b, 1991, 1992a,b) has also carried out extensive studies of infant care in cotton-top tamarins. At the time of parturition fathers and older siblings shared in eating of the placenta with the mother, and fathers were observed to carry infants as soon as 1 h after birth. In contrast to previous reports, fathers did not assist the mothers with the delivery. Primiparous females were more likely than multiparous females to allow males to carry infants after birth (Price, 1990b), and males carrying infants were more likely to mate with the female than if they were not carrying infants, suggesting that infant care may also be a courtship strategy (Price, 1992a, see also Dunbar, 1995).

Price (1992b) also studied the relationship between parents before and after the birth of their infants. Mothers spent more time in proximity of and grooming males in the 4 weeks prior to birth, and males spent more time maintaining proximity to and grooming mothers after the birth of infants. Price interprets this behavior as the mother securing male assistance with infant care by showing positive behavior prior to parturition, while males attempt to secure mating opportunities with the female after parturition.

Price (1991) reported high levels of competition among captive cotton-top tamarins to carry infants. Individuals in large groups were less likely to reject infants, and were more likely to take over infant care and resist transfer of the infants to other caregivers. Moreover, singletons were rejected less often than twins. Younger helpers and mothers rejected infants more often than older siblings and fathers. Price interpreted these results as indicating high levels of competition for infant care, but the results are equally explained with reference to the energetic effort of carrying. Mothers have the highest energetic demands, including the costs of lactation and the initiation of subsequent pregnancy. Younger animals are smaller and less able to carry infants than older animals. In large groups caretaking effort is shared among more animals and thus rejections are less common. Although I believe there is competition to take care of infants, it is difficult to distinguish the effects of competition from those of energetic costs. Tardif *et al.* (1993) have provided additional data that support the energetic interpretation: cotton-top tamarin mothers relinquish care of twins to other group members earlier than they do to singletons. They transfer twins at Day 3 but carry singletons extensively to Day 6 or 7. Lion tamarins relinquish twins between Days 9 and 11, while they do not relinquish singletons until Days 14–17.

We are studying pygmy marmosets and cotton-top tamarins to test predictions both within and between species based on the amount of effort involved in infant care (G. G. Achenbach and R. L. Addington, in progress). As group size increases, we predict that the amount of time helpers carry both infants simultaneously should decrease, because the energetic load of carrying can be divided among more helpers. Preliminary data from 22 litters of cotton-top tamarins and 13 litters of pygmy marmosets showed that in cotton-top tamarins, increased numbers of helpers reduced the time that both mothers and fathers carried infants; whereas in pygmy marmosets, an increased number of helpers reduced infant carrying by fathers but not mothers. In both pygmy marmosets and cotton-top tamarins there was a significant positive relationship between the size of the family and the time that infants were carried separately. (Cotton-top tamarins, $r = .93$, $p < .001$, pygmy marmosets, $r = .94$, $p < .006$.)

Tardif et al. (1986b) also compared infant care of cotton-top tamarins and common marmosets, and found that common marmoset infants reached independence much sooner than cotton-top tamarin infants. Whereas mothers of both species carried equally often in the first and second months, fathers of both species carried equally often in the first month, but cotton-top males carried infants much more during the second month. This study illustrates a general finding that is summarized in Table I. Data from captive colonies indicate that common marmoset and pygmy marmoset infants reach the 50% time-off stage about 2 weeks sooner than tamarin or lion tamarin infants. Field data from cotton-top tamarins and common marmosets indicate the same relationship, but the 50% time-off point is reached about 2 weeks later in field compared with captive populations. This contradicts the proposal of Tardif et al. (1993) that wild animals should become independent sooner than captive animals.

Examination of infant to maternal weight ratio data suggests that the species differences in time to independence of carriers cannot be accounted for by differences in infant size. Tardif (1994) suggests that the energetics of infant care and growth alone cannot explain the differences in caretaking patterns between marmosets and tamarins. Mean daily weight gains are comparable between marmosets and tamarins. However, daily path length is quite different, with tamarins traveling about 50–100% more per day than marmosets. All marmosets excavate and make use of tree exudates that are highly localized sources of high energy. Thus, marmosets have smaller home ranges and travel less during the day. Tamarins are more frugivorous than marmosets and do not have stationary sources of exudates. Thus, they travel more. Greater locomotor endurance is needed not only when carrying infants but also for the infants when they become independent. On the basis of these ecological considerations, infant tamarins should be carried for a longer period than infant marmosets, and because helpers probably work harder in tamarins than in marmosets, group size effects on carrying should be more pronounced in tamarins.

Garber (1994) has argued that despite the common features of twinning and cooperative care of infants, there are significant differences among the four genera. *Saguinus* and *Leontophithecus* have more female migration, and among males patrilineal kinship and long-term social bonds, while *Callithrix* has stronger matrilineal relationships where daughters may inherit breeding positions from mothers. *Cebuella* appears to expel maturing offspring and thereby to maintain small groups. *Leontopithecus* is characterized by exclusive maternal care for nearly 3 weeks, by a lack of reproductive inhibition of females (see Section VI later) and a much shorter gestation period, suggesting an early evolutionary divergence. *Saguinus* share other traits with *Leontopithecus* and are thought to be closer to this genus than to *Cebuella* or *Callithrix,* which are quite similar to each other. While the ecological and phylogenetic variables proposed can explain some of the species and genera differences, there is much yet to be explained.

## B. Patterns of Helping in Other Mammals

There are other Neotropical primates with the potential to show cooperative breeding. Goeldi's monkey (*Callimico goeldii*) is closely related to marmosets and tamarins, but females give birth to a single infant rather than twins. Fathers do not start carrying the infant until the third week (Jurke and Pryce, 1994) and there is little involvement by other group members. Female infants are transferred to fathers significantly earlier than male infants, and, in contrast to marmosets and tamarins, fathers are more involved in maintaining proximity to infants than vice versa.

The titi monkey, *Callicebus,* is a monogamous primate where fathers are the primary caretakers. Fragaszy *et al.* (1982) found that older juveniles appeared startled and annoyed when infants were in physical contact with them. At 6 months of age, young were observed with family members on 85% of samples. Of these, 79% of the observations were with fathers, 14% with mothers, and only 7% with siblings. When infants were independent, most of their interactions were with fathers, suggesting that a strong social attachment exists between them. Mendoza and Mason (1986) found that experienced fathers carried infants 50% of the time in the first week and this increased to 80–90% of observations in Weeks 5–8. By the third month males were the primary attachment figure for infants. Males were more tolerant of infants at this age than mothers, who rejected infants frequently. In a series of separation studies, Mendoza and Mason found that the male–female pair had a greater attachment for each other (as measured by behavioral and cortisol responses during separation and reunion) than either parent did for the infant. In choice tests the infants preferred obtaining proximity to the father to the mother.

Wright (1984) studied titi monkeys in the field and found that fathers carried infants 92% of the time by Week 3. Fathers and juveniles were mainly involved in play initiation with the infants. Fathers shared food with infants, while mothers did not. Wright (1984) also studied night monkeys (*Aotus sp.*) and found fathers carried 51% of the time, mothers 33% of the time, and juveniles 15% of the time in the first week. After the first week, maternal contact was limited to suckling, and the father and juvenile shared infant carrying.

Several field studies have demonstrated the importance of cooperative breeding in some carnivore species. Moehlman (1979) studied jackals (*Canis mesomelas*) and found a significant correlation ($r = .967$) between the number of helpers and the number of pups that survived. Each helper led to an average of 1.5 more pups surviving. Jackal helpers regurgitate food to pups and to mothers during lactation. They guard pups when parents are foraging. The more helpers there were, the less time the pups were unguarded. Helpers are generally related to pups and so increase their inclusive fitness by helping. They may also benefit by having greater survivorship as a chance to inherit their natal territory or a part of their natal territory after serving as helpers.

Rasa (1987) studied dwarf mongooses (*Helogale parvula*) and found that subordinate females have infrequent cycles, but if they do become pregnant, their offspring are usually killed. Adult males physically prevent subordinate males from copulating. These nonreproductive helpers baby-sit for infants while the parents forage. Subordinate females often carry pups as the group moves from one termite mound to another. Helpers provide food to pups. In groups larger than five adults there was no predation of young, but groups smaller than five adults had infant predation rates of 70%. Large groups experienced 25% of the observed raptor attacks, whereas small groups experienced 75% of observed raptor attacks. Thus, helpers play an important role in protecting pups from predation. Daughters succeed the breeding females from within the breeding groups, so in addition to the inclusive fitness benefits from helping relatives, dwarf mongoose females gain direct fitness benefits from inheriting a breeding position.

Creel and Macdonald (1995) summarized studies on a variety of social carnivores, noting that tolerance of other animals provides benefits by allowing cooperative hunting and defense, and promoting a reduced vulnerability to predation. Large groups meet the costs of reproduction by providing alloparental care that leads to a greater number of litters, larger litters, and increased survival of offspring. Litter mass and litter growth are higher in species with communal or biparental care. Because there is high within-group relatedness in social carnivores, nonreproductive or minimally reproductive helpers are able to benefit from increased inclusive fitness.

The role of helpers has also been studied in gerbils (*Meriones unguicucatus*) and voles (*Microtus ochrogaster*). In captive prairie voles, Solomon (1991) removed the previous litter or left two juveniles behind. There was no effect of helpers on litter size or survival, but pups with helpers present opened their eyes sooner and were 13% heavier at weaning. Pups were left alone less often and the thermoregulatory effects of helpers being in contact with the infants may have led to the increased body weight. The presence of helpers had no effect on the mother's time budget, but fathers spent less time with infants, a result similar to that found with marmosets and tamarins.

In a rare study that demonstrates that helpers benefit from providing infant care, Salo and French (1989) formed pairs of gerbils (*Meriones unguicucatus*) in which either both partners had prior infant care experience, one partner had prior experience and one was naive, or both partners were naive. Litters were born sooner (<50 days) after pairing to pairs in which at least one partner had prior experience compared with naive pairs (150 days). Pups born to pairs with experienced males had a faster time for eye opening and a more rapid weight gain compared with groups with an experienced female only or to naive groups. The naive pairs also formed poor-quality nests. Thus, prior infant care experience can affect several parameters of reproductive success in gerbils.

The best example of a eusocial mammalian species is the naked mole rat (*Heterocephalus glaber*) (Sherman et al., 1991). These colonial animals live in complex underground burrow systems in seasonally dry regions in East and South Africa. One breeding female, or queen, is found in each colony with dozens of other nonreproductive females. In the field, the queen produces a litter averaging 4.3 animals, and she and the reproductive male do most of the direct infant caretaking (including nursing, handling, and grooming). The nonreproductive helpers contribute indirectly to infant care by digging new tunnels (up to 3.5 km per colony per year), keeping existing tunnels free of obstructions, locating food (mainly underground tubers) and bringing food to the reproductive animals, and defending against snakes. There appears to be some evidence for possible age-specific division of labor, with the largest nonbreeding animals doing most of the defense (Lacey and Sherman, 1991).

Jarvis *et al.* (1994) compared two species of colonial mole rats, the naked mole rat and the hairy Damaraland mole rat (*Crytomus damarensis*). Both species live in highly arid environments. DNA fingerprinting of naked mole rats indicates an extremely high coefficient of relatedness (.81) within colonies and only slightly lower relatedness between adjacent colonies. In contrast, the Damaraland mole rats show a high degree of incest avoidance and outbreeding. Thus, the eusocial lifestyle cannot be explained by the benefits of inclusive fitness alone. Jarvis *et al.* (1994) suggest that eusociality is primarily an ecological adaptation to an environment with unpredictable rainfall and food resources. Only with the high degree of task specialization involving tunnel construction, food transport, predator defense, and reproduction can the mole rats survive.

## C. Patterns of Helping in Birds

There is extensive literature on cooperative breeding birds (see Stacey and Koenig (1990) for a summary of several long-term studies). As noted in Section I, there is much more long-term field data on demography and survival in birds than in mammals, but far less detailed descriptions of activities and timing of activities for birds than for mammals. The mechanisms underlying cooperative breeding in birds are also less well understood.

The first question is: Do helpers really help? Table II summarizes the results for the species described by Stacey and Koenig (1990) where there are nonbreeding helpers. I have excluded the species where the other helpers are also breeders. The evidence that helpers contribute to the reproductive success of the breeders' offspring is quite variable. In several species the helpers appear to add significantly to offspring survival (see green woodhoopoe, *Campylorhynchus* wrens, white-fronted bee-eaters, red-cockaded and acorn woodpeckers), in other species there is little or no apparent beneficial effect on infant survival by helpers (see pinyon jays, groove-billed anis, and pukekos). In most species helpers are male; in the species with both sexes helping, females often disperse after 1 year of helping so males predominate as helpers after the first year. Where data are available on relatedness, unrelated helpers are rarely more than 10–20% of the helpers.

Most of the studies base the effect of helpers on survival on correlational measures of group size and number of chicks fledged, but, as noted by Brown *et al.* (1982), these correlations do not demonstrate a causal relationship. Large groups may be attracted to better territories with more abundant

## TABLE II
### Role of Helpers in Several Cooperatively Breeding Birds

| Species | Do helpers affect survival of young?[b] | Does helping improve direct fitness of helper? | Sex of Helper | Unrelated helpers? |
|---|---|---|---|---|
| Splendid Fairy wren | Moderate (>4) | Yes | M, F | ? |
| Green woodhoopoe | Small (>7) | ? | F | 20–30% |
| Red-cockaded woodpecker | Strong | ? | M | 5–11% |
| Arabian babbler | Moderate (>5) | Yes | ? | ? |
| Hoatzin | Moderate ($r = .43$) | No | M, F | ? |
| *Campylorhynchus* wrens | Strong (>4) | No | M, F | 13%, Year 1; 44%, Year 2 |
| Pinyon Jays | None | No | M | ? |
| Florida scrub jay | Moderate (>2) | Yes | M, F | 12% |
| Mexican jay | ? | ? | M, F | Extensive |
| Galápagos mockingbird | Moderate (in drought) | No | M, F | 40% |
| Groove-billed anis | No | No | ? | High |
| Pukeko | No | ? | M, F | Rare |
| Acorn woodpeckers | Strong (50% per adult) | No | M, F | Low |
| White-fronted bee-eaters | Strong, linear | Some | M, F | 11.5% |
| Pied kingfishers | Strong: Lake Victoria; weak: Lake Naivasha | Primary helpers: moderate; secondary helpers: high | M | Primary: rare; Secondary: 100% |
| Noisy miners | Moderate | ? | M | Rare? |

[a] Only species with nonbreeding helpers included. Derived from Stacey and Koenig (1990).

[b] ($>n$) indicates a threshold rather than a linear effect of group size $n$ helping with group size $=$ or $> n$ increasing survival of young.

resources, or the parents in large groups might be generally more successful. Brown et al. (1982) experimentally manipulated group size in gray-crowned babblers (*Turdiodes squamiceps*). Twenty groups with prior fledgling success were located and in 9 groups the size was reduced from six to eight animals to the parents plus one juvenile. Eleven other groups served as unmanipulated controls. The original group size, territory quality, and age and experience of parents were comparable between conditions. Control groups fledged a mean of 2.4 offspring, while the reduced-size groups fledged a mean of 0.8 chicks, which is comparable to the success rate of natural units with only three members.

The Florida scrub jay (*Aphelocoma coerulescons*) has been extensively studied by Woolfenden and Fitzpatrick (1984). Helpers are found in about 50% of breeding groups with equal sex ratio of helpers in the first year, but primarily males in subsequent years. Most helpers are found with at least one parent, but if a parent is replaced, helpers will help the breeders on that territory. Helpers do everything that parents do except reproduce: they are similar to breeders in foraging, resting, and territorial and predator defense. They carry nesting material, but do not help to build nests or to incubate eggs. About 0.9 more offspring are produced in groups with helpers than in groups without helpers, and pairs with helpers are successful even in years with poor resources. However, although there are more fledglings in groups with helpers than in groups without helpers, the survival of offspring to the next breeding season does not differ. The main efffect of helpers is that the breeding adults with helpers have a higher survival rate and therefore produce more offspring than breeding adults without helpers.

Emlen et al. (1995) have carried out long-term studies on the white-fronted bee-eater (*Merops bullockoides*), a colonial species found in Africa. Helpers assist by digging the nest chamber, bringing food to the female while she is producing eggs, assisting with incubation, defending the young as nestlings and fledglings, and bringing food to the young after hatching. As many as 50% of the young in unassisted nests die of starvation, but the presence of a helper markedly increases survival. Kinship is a significant predictor of helping (Emlen and Wrege, 1988). Among all helpers who made choices between breeders to assist, 94% chose the pair that was most closely related to them. However, once the choice to help was made, the intensity of helping did not differ as a function of relatedness. Helpers had no effect on the survival of breeders, but each additional helper in a group increased the mean number of fledglings by 0.44 (Emlen and Wrege, 1989). There is evidence that fathers recruit sons as helpers by actively disrupting their son's breeding. Fully 91% of all mating harassments were between males with father–son harassments leading to 84% recruitment of sons as helpers (Emlen and Wrege, 1992).

Koenig and Mumme (1987) studied acorn woodpeckers (*Melanerpes formicivorus*) and found the number of helpers had no influence on any measure of breeding success. However, they did find a higher rate of fledgling success in poor and fair years in groups with helpers versus groups with no helpers. In contrast to the Florida scrub jay, there were also more surviving young the following spring in groups with at least one helper present. Helpers do much less than breeders in most all aspects of behavior related to care of the young.

Walters (1990), studying red-cockaded woodpeckers (*Picoides borealis*), found that helpers increased survival of young in groups of up to four adults, primarily by reducing predation. Groups with helpers produced on average 0.62 more fledglings than groups without helpers. Rabenold (1990) found in *Campylorhynchus* wrens in Venezuela that helpers were extremely important. Unassisted pairs were unlikely to produce young even on territories that in previous years had supported successful breeding. The failure rate was 79% in pairs and trios but only 32% in groups with at least four adults. There was a mean of 2.4 surviving infants in pairs with two helpers, but only 0.4 survivors in groups with either one or no helpers. The main effect of helpers appeared to be predator defense. Groups of four were more vigorous in mobbing and attracting members of other species to mob too. In an experiment where wrens without helpers were protected from predation, 1.67 juveniles per nest survived compared with only 0.4 surviving juveniles in unprotected pairs. Thus, for both red-cockaded woodpeckers and wrens, the main role of helpers is protection from predation.

Reyer (1990) described helping in the pied kingfisher (*Ceryle rudis*), studied at two different habitats in Kenya. At Lake Victoria suitable nest areas are relatively far from the lake and strong winds make fishing energetically difficult. At Lake Naivasha nest sites are closer to the lake and fishing is relatively easy. Most breeding pairs at Lake Naivasha were unassisted and most breeding pairs at Lake Victoria had helpers. At Lake Naivasha adult breeders actively rejected potential helpers. Reyer experimentally manipulated nests by increasing clutch size at Lake Naivasha and found that 8 of 10 pairs accepted helpers. In a parallel experiment he reduced clutch size at Lake Victoria and found that 7 of 8 pairs rejected helpers. Thus, the acceptance or rejection of helpers depended upon ecological conditions and brood size. He also found two types of helpers. Primary helpers were related to the breeding pair and were present from the beginning of the breeding season. Secondary helpers arrived only after chicks had hatched and were generally unrelated to the breeding pair.

In summary, there is great diversity in the role of helpers in cooperatively breeding birds. In some species reproductive success of young is improved by helpers, but in other species helpers may benefit the parents or have little effect except in poor breeding conditions. In some species the main role of helpers is predator defense; in other species helpers appear to be important in assisting parents in provisioning the young; and in still other species helpers appear to be involved in all of the same activities as the parents.

### III. What Are the Factors That Influence Helping Behavior?

#### A. Behavior Mechanisms

The birth of new infants is a major event in a cotton-top tamarin group and we have found there are a variety of behavioral and physiological changes in response to the birth of infants. Pryce (this volume) discusses some of the hormonal influences on maternal behavior and maternal motivation in marmosets and tamarins. Here I will discuss the influences of new infants on juveniles, eldest daughters, and fathers.

The birth of new infants has been hypothesized to be a traumatic event for the next oldest siblings because parental resources are diverted from the older infants to the newest offspring. We have completed a study on 12 juveniles ranging from 6.5 to 15 months of age at the time of the birth of the next set of twins to a group (G. G. Achenbach, in preparation). Data were gathered from 8 weeks prior to birth until 12 weeks following the birth. The birth of infants was highly disruptive, leading to increased conflict between juveniles and parents, decreased play among juveniles, and increased proximity of juveniles to both caretakers and infants. However, the effects of infants on juveniles were highly transient, with most of the effects occurring on the day of birth but abating within 2 weeks after birth. The disturbance does not seem related to conflict over resources provided by parents, however, because the increased rate of contact and aggression were observed only when the parents were carrying the new infants. There was no change in behavior between parents and juveniles when parents were not carrying infants. The likelihood of juveniles gaining contact with infants was negatively correlated with the number of other group members ($r = .93$, $p < .05$), but the age or sex of the juvenile had no relation to their access to infants. By Weeks 4–6 the juveniles were playing an active role in carrying the infants. Thus, the increased conflict observed at birth between juveniles and parents is not the traditional parent–offspring conflict over parental and other resources, but appears to be a competition for access to the infants.

Eldest daughters in a family group often show detectable levels of both luteinizing hormone and estrogen, though these are not organized into ovarian cycles. However, we have found that with the birth of new infants, these low and essentially randomly fluctuating levels of hormones are reduced further (Snowdon, et al., 1993). These females also decreased their scent marking and increased their time in contact with other group members, especially those carrying infants. Like the juveniles, the eldest daughters were also increasingly involved in aggressive interactions soon after birth. The eldest daughters did not often have access to infants, because they carried infants significantly less than other group members. It appears that eldest daughters and juveniles were similar both in being attracted to infants and in being prevented by other group members from providing much caretaking of infants. These data, together with data from macaque and baboon species, where paternal care is rarely shown (Snowdon, 1990), suggests that the birth of infants attracts the interest and attention of many group members, but that only when the mother provides "permission" do adult males, subadults, or juveniles get the chance to carry and take care of infants. Thus, motivation to interact with infants may be nearly universal, but the expression of infant care behavior is controlled by mothers. In cooperatively breeding species, mothers relinquish control more often to fathers and nonreproductive helpers than do mothers in other mammalian species.

B. HORMONAL MECHANISMS: PROLACTIN

Several studies in both birds and mammals have demonstrated that males taking care of infants have elevated prolactin levels. Thus, in ring doves (*Streptopecia rosalia*) (Goldsmith et al., 1981), spotted sandpipers (*Actitis macularia*) (Oring et al., 1986), and Bengalese finches (*Lonchura striata*) (Seiler et al., 1992) males have high prolactin levels during incubation or early nestling stages. Gubernick and Nelson (1989) found elevated levels of prolactin soon after taking care of young in male California mice (*Peromyscus californicus*), a monogamous species where males exhibit parental care. In contrast, virgin males and expectant fathers showed no elevation of prolactin. Dixson and George (1982) showed that male common marmosets that were carrying infants had higher prolactin levels than did males in families without dependent infants or males paired with a cycling female without any offspring present.

However, prolactin has multiple functions and it has also been implicated as an indicator of stress, and is involved in the suppression of sexual behavior of subordinate talapoin monkeys (*Miopithecus talapoin*) (Bowman et al., 1978; Eberhart et al., 1983; Hansen et al., 1980). Because infant caretaking may be stressful to a caretaker, it is important to separate cause and effect. We have developed and validated a urinary prolactin assay for the cotton-top tamarin that allows us to take daily samples without handling the animals (Ziegler *et al.*, 1996). Because most primates tend to urinate soon after waking, we can enter a cage early in the morning and simply hold a container under the animal of interest until a urine sample is obtained. Cotton-top tamarins appear to be in a state of torpor throughout the night and we have never observed nocturnal urination. Thus, this first morning void of urine provides a sample of the previous 12 h of hormonal activity. We can obtain samples throughout the prepartum and postpartum period without interfering with male or female parental behavior, and we can collect a large number of samples without reducing the donor's red blood cell count.

We initially proposed two hypotheses about the role of prolactin in male caretaking.

1. Prolactin is necessary in order for parental care to be expressed. Thus, prolactin levels should be elevated before a male starts to take care of infants, and there should be a direct relationship between prolactin levels and the amount of caregiving displayed above a threshold level.

2. Prolactin is elevated as a consequence of male caretaking because infant caretaking is stressful. Thus, prolactin increases should be apparent only after males start to take care of infants, and less experienced males, such as first-time fathers and older siblings, should have higher prolactin levels than experienced fathers who would experience the least stress.

To test these hypotheses, we gathered daily urinary prolactin data from first-time and more experienced fathers starting 2 weeks prior to the birth of infants through 6 weeks after infant birth. We also collected samples at 12–14 weeks before or after the birth of infants, when no infant caretaking was occurring. We also collected samples from the eldest sons in family groups; from males paired with females without any dependent offspring present, and from males housed alone, but adjacent to, a female.

As part of our validation procedure during 1 week we captured three males and three females. We injected them with a saline vehicle and a week later with 0.4 mg metaclopramide (a dopamine antagonist expected to increase prolactin levels) dissolved in 0.5 ml of sterile saline. Urinary samples were collected in the morning prior to injection and then the following morning 16 h after injection. All six tamarins showed an increase in prolactin levels following either the saline or metaclopramide injections. Five of six tamarins increased cortisol levels after the saline injections and all six increased cortisol levels following the metaclopramide injections. Thus, for tamarins the stress of capture and injection was enough to increase urinary levels of both cortisol and prolactin. That both cortisol and prolactin increased in response to capture and injection suggests that our urinary assays are sensitive enough to determine stress-induced prolactin release.

A comparison of prolactin levels of lactating versus cycling females indicated that the assay was sensitive to changes in prolactin associated with nursing and maternal care.

Analyses of urinary prolactin data collected from males as described previously showed there were significant differences across male conditions. Fathers having contact with infants had significantly higher prolactin levels than paired males without infants and males living alone adjacent to a female. The prolactin levels of eldest sons were intermediate between those of fathers and other males, suggesting a confirmation of our first hypothesis that prolactin elevation was associated with infant care. However, when we compared the prolactin levels of fathers sampled from the 2 weeks prior to parturition with the levels sampled from the 2 weeks following the birth of infants, there were no differences, and both levels were higher than the levels in the same fathers when there were no dependent infants. Even in the absence of dependent infants, fathers had significantly higher prolactin levels than did males without infant care experience (see Fig. 2a).

When we examined the cortisol levels of both fathers and eldest sons who were involved in infant care, there was no difference from the baseline levels, but the cortisol levels were elevated in newly paired males and males housed alone, adjacent to females (Fig. 2b). Thus, prolactin elevations in males taking care of infants cannot be explained as a response to the stress of infant care. In contrast to the capture and injection study reported previously where both prolactin and cortisol were elevated, the dissociation of cortisol and prolactin here suggests that we can measure the nonstress release of prolactin.

There was a great individual variation in mean daily prolactin levels among the fathers. We found no correlation between prolactin levels and either the amount of time a male was observed to carry infants ($r = -.14$) or the number of helpers ($r = .35$). However, there was significant correlations between prolactin levels and male age ($r = .92$), the number of previous births the father had been exposed to ($r = .91$), and the number of births with surviving infants the male had experienced ($r = .83$). Multiple regression analyses indicated that the number of previous births and the number of successful births still correlated with prolactin levels when age was held constant (see Fig. 3).

The data did not support either of our two hypotheses. The stress hypothesis was clearly wrong. While prolactin levels did increase together with cortisol levels in response to capture and injection, prolactin was elevated, while cortisol levels remained low during infant caretaking, suggesting that prolactin release was not in response to the stress of infant care. Furthermore, prolactin levels were highest in the males with the greatest amount of infant care experience. There was no relationship, however, between

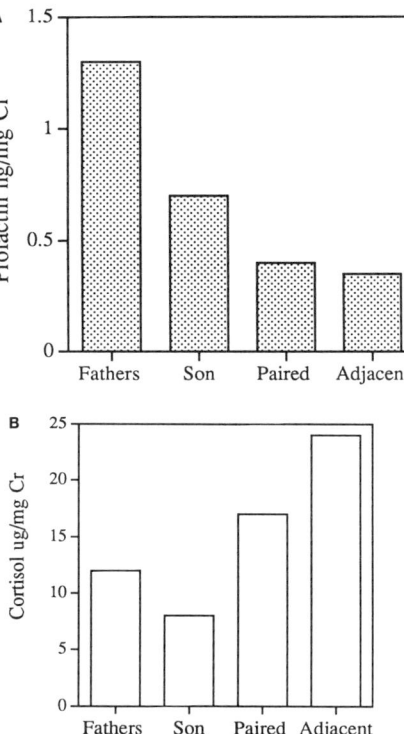

Fig. 2. Urinary hormone levels in fathers ($n = 6$) and eldest sons ($n = 3$) during periods of infant care and by males newly paired ($n = 7$) or housed singly, but adjacent ($n = 4$) to a female. (a) Urinary prolactin, (b) urinary cortisol. Cr = creatinine. (Adapted from Ziegler et al., 1996.)

the actual amount of caretaking shown and prolactin levels, suggesting that prolactin may not directly control infant caretaking either nor be released directly in response to current contact with the infant.

However, the correlation of prolactin with prior experience rather than current infant care behavior is intriguing. The amount of caretaking a male displays is negatively correlated with the number of helpers available (McGrew, 1988; Tardif, 1994), thus first-time fathers without other helpers will be doing more carrying than fathers in larger groups. Infant mortality is highest in first-time fathers, but survival increases greatly with the second litter. The increased prolactin levels with increased parental experience suggests a third hypothesis, that is, prolactin levels may increase during and after contact with infants as a result of stimuli that elicit prolactin release such as skin stimulation as seen in rats (*Rattus norvegicus*) (Fleming

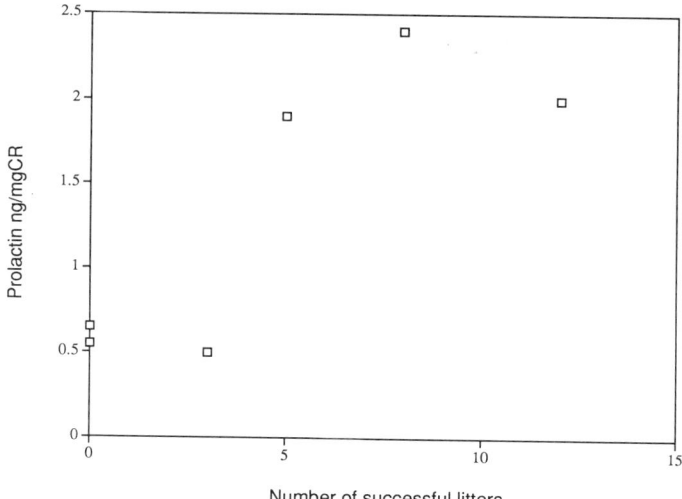

FIG. 3. Relationship between mean urinary prolactin levels and number of prior successful litters in fathers during the first 2 weeks after the birth of infants. Cr = creatinine. (Adapted from Ziegler *et al.*, 1996.)

*et al.*, 1995; Fleming *et al.*, this volume). Warren and Shortle (1990) have reported that prolactin levels are higher in women after the birth of the second child than after the first. With exposure to infants a priming effect on male prolactin release occurs that remains at a tonic level until the next litter arrives, leading to another increase in tonic levels. Males that have greater cumulative success taking care of infants will have increasing levels of prolactin due to greater total stimulation from infants received. This cumulative increase in prolactin may lead to more successful infant care through longer-term experiential effects, even though levels of infant carrying expressed with a given litter may not be related to prolactin levels. Thus, there is a common pattern of increased prolactin with increased experience in female rats, male cotton-top tamarins, and women.

There are some interesting parallels between these data on prolactin in tamarins and the data on voles reported by Wang and Insel (this volume) for changes in arginine vasopressin. In male voles, the experience of affiliation or pair-bonding leads to elevated vasopressin levels, and males displaying parental care also have increased vasopressin levels. In tamarins, our study included vasectomized, female-paired males who had lived for several years with their mates and displayed a strong pair-bond (Porter, 1994) in the absence of any infants, yet there was no evidence of prolactin elevation in these males over that of unpaired males. Our study also included eldest sons that were involved in infant caretaking but that had a slight, but not significant, rise in prolactin level. Thus, neither pair formation nor infant care by themselves led to increased prolactin levels. Only males that were pair-bonded and actively involved in infant care displayed high levels of prolactin, and these levels were still elevated even when no dependent infants were present.

C. ECOLOGICAL VARIABLES

Emlen (1994) notes that two explanations for cooperative breeding have emerged: one arguing the benefits of philopatry, the other arguing that ecological constraints affect cooperative breeding. In reality, Emlen argues, the real goal should be determining for each population or species which factors affect the delayed reproduction and helping behavior found in cooperatively breeding species. Reyer (1990) has shown experimentally that habitats with differing ease of obtaining food can lead to acceptance or rejection of helpers, and by manipulating clutch size to increase or decrease feeding demands, pied kingfishers will change their willingness to accept helpers. Rabenold (1990) has shown that wrens that are protected from predation can produce almost as many surviving young unassisted as unprotected wrens with two helpers. Therefore, both predation pressures and ease of obtaining food can be identified as important ecological variables. Both energetic variables and predator defense are the most common explanations for cooperative breeding.

Other ecological variables promoting cooperative breeding that have been suggested are limited breeding opportunities, such as found in Florida scrub jays or red-cockaded woodpeckers, so that delaying breeding is a strategy for increasing the likelihood of obtaining a breeding position. Helpers may be essential for providing food and transportation when the energetic demands on the breeding female are high, as in naked mole rats and marmosets and tamarins. Thus, when a mother is unable to successfully rear offspring alone, then she should tolerate or recruit other helpers, not only her mate, but other nonbreeding animals to provide assistance.

Finally, it is important to consider the null hypothesis of no direct benefit from helping. The number of species in Table II where helpers have little or no effect on survival of offspring suggests either that the appropriate ecological variables have not yet been found or that helpers are making the best of a bad situation from their perspective, but add little or nothing to the survival of the infants they help.

IV. WHAT BENEFITS DO HELPERS RECEIVE?

There is an important conceptual distinction between the benefits that helpers provide to those they help and the benefits that the helpers gain by being helpers rather than attempting their own reproduction. The previous sections have described the activities that helpers do and how these activities might benefit the parents and infants. Because, in most cases, helpers are related to the parents and offspring they help, an inclusive fitness benefit can often be assumed. In this section I discuss the costs of helping and examine potential direct fitness benefits to helpers.

### A. Costs of Taking Care of Infants

There are clearly costs of infant care, beyond deferring reproduction. Price (1992a) showed that cotton-top tamarin caretakers have less time to spend in foraging for themselves, have decreased mobility while carrying heavy infants, and may experience some decrease in affiliative behavior. Tamarins carrying infants spent less time in vigilance behavior and often were found in concealed parts of the cages. Tamarins carrying infants spent less time grooming, but because the infants are attractive to other group members, carriers of infants often spent more time in social contact with others, though high levels of aggression occur. Yamamoto (1993) reports that mothers in families without helpers show a decreased number of feeding bouts.

Dietz *et al.* (1994) reported changes in weights of wild lion tamarins over the reproductive cycle. Adult males did not gain weight during the period of infant care, but gained weight after infant independence. In contrast, lactating females did not lose weight during lcatation, suggesting that there is an energetic cost to caretaking in males that may alleviate energetic stress to females during lactation. Unfortunately, field data on maternal feeding rates are not available so we do not know if infant carrying by other group members allows mothers to obtain more food.

Although there are no solid field data, it is possible that helpers carrying infants are at a greater risk of predation for themselves and the infants, especially if they move more slowly during group movements. The energy expenditures of infant care can, especially during droughts or other poor conditions, place the helper at risk for surviving.

### B. Benefits of Being a Helper

Several potential benefits have been hypothesized to explain the behavior of helpers in all species where helpers have been found (Jennions and Macdonald, 1994). These include:

1. Increased chances of survival by remaining as a subordinate in a group rather than by dispersing. Remaining in a natal group leads to benefits of

increased foraging efficiency and predator detection while avoiding the high mortality often associated with dispersal.

2. Increased chance of filling a reproductive niche in the future. If suitable breeding sites or territories are extremely limited, an individual can wait until a breeding vacancy occurs in the natal group, or may be able to bud a territory from the natal territory (Woolfenden and Fitzpatrick, 1984).

3. Increased reproductive success when the helpers can reproduce. Helpers may acquire skills in infant caretaking, foraging ability, predator detection, and defense that lead to greater reproductive success as a result of deferring reproduction instead of reproducing immediately. A related means of increasing reproductive success is that demonstrating infant care ability may be a useful courtship strategy that increases the likelihood of mating successfully.

4. Increased benefits through inclusive fitness. Helpers may have a low probability of obtaining breeding positions on their own, but by taking care of related individuals, they will increase their inclusive fitness.

5. The null hypothesis of no benefits must be considered. Helpers may simply be making the most of a bad deal.

## 1. Marmosets and Tamarins

The benefits of infant care can be seen in increased fitness either directly or through kin selection; increased parental experience also seems to be critical for subsequent parental behavior. Infant caretaking might be the payment that immigrant animals make for being able to stay in the group. Price (1990a) has demonstrated that captive males carrying infants are more likely to mate with the reproductive female and this finding suggests that unrelated animals in the field might gain breeding opportunities by carrying infants that are not their own. Finally, infant carrying may serve to establish cooperative relationships between group members, both those related to and unrelated to the infants that could serve to increase group vigilance and defense against predators as well as leading to cooperation in locating and utilizing food resources.

There is relatively little direct evidence supporting most of these points. The role of infant care experience in future reproductive success for marmosets and tamarins was initially accepted as a major benefit for helpers, but in recent years, the evidence has become less clear. Epple (1978) provided data on captive saddle-back tamarins as a function of the prior infant care experience of males and females with infant care. Figure 4 shows that infant survival is low when both parents are inexperienced but high when both parents have prior infant care experience.

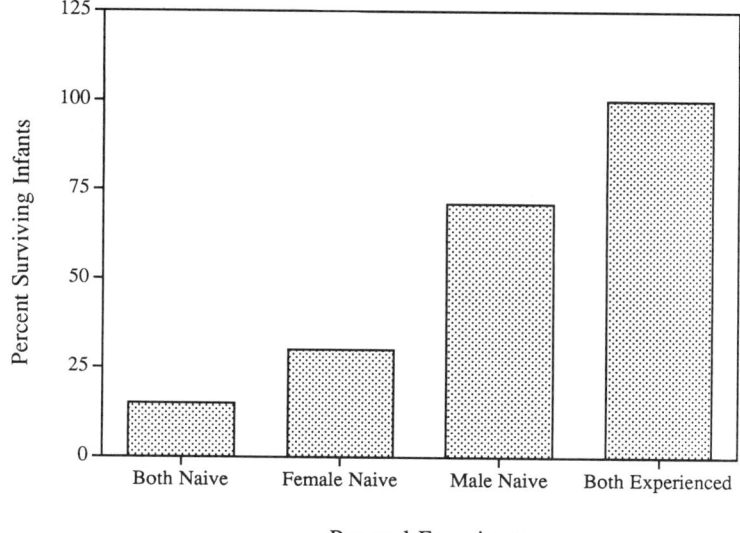

FIG. 4. Percentage of infant survival as a function of prior sibling care experience in saddleback tamarin parents. (Plotted from Epple, 1978.)

Johnson *et al.* (1991) reported an interaction of parity and prior infant care experience in cotton-top tamarins (see Fig. 5). In this study the effects of experience appeared to be greatest for fathers. Tardif *et al.* (1984) found that sibling rearing experience was more important for female cotton-top tamarins than for common marmosets (see Fig. 6). The survival rate for common marmosets was consistently higher both as a function of prior infant care experience and of parity.

Parity effects have been reported for several species. Hoage (1982) reported a 42% infant survival of primiparous lion tamarins, but an 80% infant survival for multiparous golden lion tamarins. Greene and Snowdon (in preparation) report 6.7% infant survival for primiparous cotton-top tamarins, with a 44% survival rate with the second birth, and survival above 50% thereafter. Variation in infant survival beyond the second parity was due to the number of helpers present. However, Tardif *et al.* (1986a) reported no parity effect in their colony of cotton-top tamarins, but in their colony, they typically form pairs with one multiparous and one nulliparous animal, whereas managers of other colonies have formed pairs with both members having sibling experience but neither partner having had direct experience with its own infants.

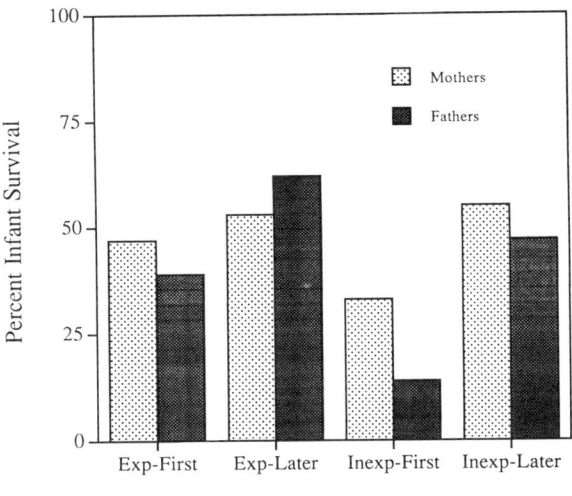

FIG. 5. Percentage of infant survival as a function of prior sibling care experience and parity in cotton-top tamarin parents. Exp, experienced; Inexp, inexperienced; First, first litter; Later, subsequent litters. (Plotted from Johnson et al., 1991.)

Rothe and Darms (1993) have challenged the role of experience as an adaptive explanation for helping behavior, especially in common marmosets. They find no evidence that helpers are critical to infant survival (in contrast to the data presented previously on tamarins). In the first few days when the need for helpers should be greatest, mothers often do not allow others access to the infants. The breeding adults, rather than nonbreeding helpers, do most of the infant care and show compensation for deficiencies in caretaking by others. Parents receive little relief in caretaking until their first infants are 2 years of age, which can be four litters later. Rothe and Darms conclude that young marmosets do not need direct exposure to infants, but that parental care skills depend on general social competence. There have been no studies that have demonstrated that the absolute amount of prior sibling care experience has a direct effect on parental competence. Most marmosets and tamarins only migrate to the adjacent territory, so that there is a good likelihood that neighboring animals are highly related. Rothe and Darms argue that it is probably rare for wild marmosets or tamarins to be involved in caring for a completely unrelated infant, even after dispersal. Thus, helping is best explained as a contribution

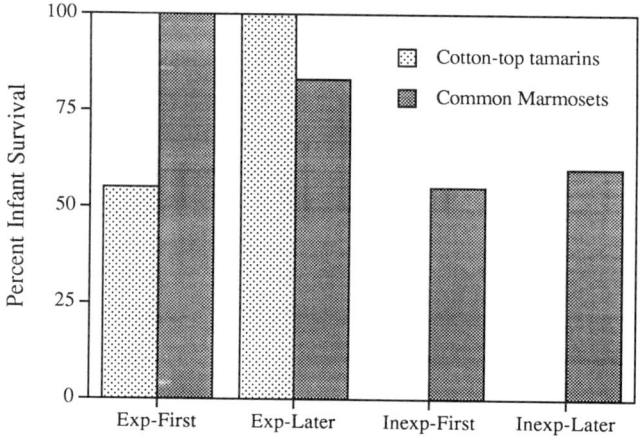

FIG. 6. Percentage of infant survival as a function of prior sibling care experience and parity in cotton-top tamarin and common marmoset mothers. See Fig. 5 for abbreviation explanations. (Plotted from Tardif et al., 1984.)

to inclusive fitness, to gaining skills in general social experience, and to infant care as a courtship strategy.

Dunbar (1995) has proposed an evolutionary stable strategy to account for helping. He proposes the following three types of helpers: (1) immatures remaining on the natal territory to acquire breeding territory from parents; (2) adult males acting as nonbreeding satellites on other males' territories to inherit a female; and (3) coalitions of males defending a female and mating polyandrously. Dunbar's analysis supports the second point and suggests that roving male polygyny is the most adaptive strategy for callitrichid males. However, this analysis seems overly simplistic given the variety of infant care patterns found in different species as well as the variation in home range size, defense of territories, and distribution of resources found in different species. In a common marmoset population where intrusions by individuals from other groups occur nearly daily, aggression toward intruders is high (C. M. Lazaro-Perea, personal communication), suggesting that there can be costs to entering a new group. On the other hand, Savage et al. (1996) report that migrations between groups of cotton-top tamarins occur suddenly and without any obvious aggression aimed at expelling the migrant. Terborgh and Goldizen (1985) have demonstrated group formation in saddle-back tamarins with two males breeding polyandrously with a female.

In different species and even within the same individuals at different stages of life, several adaptive strategies may operate. In tamarins, in contrast to marmosets, the evidence for the need to acquire infant care skills seems very strong, so any experience, whether with related or unrelated infants, can improve direct reproductive success if an individual later breeds. Because much of an individual's helping is with related infants, there will be an increase in inclusive fitness in many cases of helping. Infant care can be a courtship strategy within one's natal group that may serve also to gain acceptance in a new group. Savage *et al.* (in press) have found that new migrants do less infant care than longer-term residents, but they still engage in infant care. Finally, whether one can gain a breeding position in one's natal territory, by migrating to a neighboring territory or by forming a polygamous group, depends on local demographic conditions. The best strategy for a nonreproductive marmoset or tamarin is to continuously monitor both the natal group and adjacent groups and join the group that provides the best possibility of acquiring a breeding position. There seems to be no one option that is best for all individuals, other than to be flexible about all possible options (Sussman and Garber, 1987).

## 2. Birds

As Table II shows, there is relatively little evidence that helpers benefit from increasing direct reproductive success through helping. The number of species where helpers contribute to survival of young is greater than the number of species where helpers increase their own direct fitness.

In Florida scrub jays, Woolfenden and Fitzpatrick (1984) showed that more experienced helpers will produce more fledglings when they reproduce on their own. However, there is a greater mortality for helpers that help more than 1 year, so the apparent increase in reproductive success with increased helping may be negated by the high mortality. A young jay that can obtain a breeding territory after 1 year of helping will have greater total inclusive fitness than one that continues to help. Thus, the main advantage to Florida scrub jay helpers appears to be the possibility of gaining territory for their own breeding.

In white-fronted bee-eaters, Emlen and Wrege (1989) found that breeding success did not differ between birds with prior experience as helpers and those that had not been helpers. Again there is little direct fitness benefit to helping behavior. Koenig and Mumme (1987) found female acorn woodpeckers that did not help had a significantly longer time as breeders. Thus for females, helping can have a negative effect on breeding success. There is no evidence that acorn woodpeckers have increased likelihood of inheriting a territory or breeding position by being a helper. Birds that reproduce in their first year without having been helpers are as successful as birds that delay breeding. There is an increase in inclusive fitness by helping, but not from aiding siblings. Instead, the presence of helpers increases the survival of male breeders that are helped.

Walters (1990) reports that red-cockaded woodpeckers who stay and help are more likely to inherit the natal territory, and thus increase direct as well as indirect fitness. Rabenold (1990) found extreme costs to helping in wrens. Those that helped extensively had a 47% survival rate, while those birds helping less had an 83% survival rate. Thus, extensive helping leads to greatly reduced survival. Reyer (1990) found two types of helpers in pied kingfishers. Primary helpers were highly related to the breeder and increased inclusive fitness by helping. Secondary helpers were unrelated to the breeders they assisted and therefore had a very low increase in indirect fitness. However, secondary helpers had a high probability of inheriting the territory when a breeder died, whereas primary helpers had a very low probability of territory inheritance. Thus, secondary helpers do gain direct fitness benefits from helping, while primary helpers gain only indirect fitness benefits.

## C. Summary

Direct fitness benefits from helping are relatively infrequent among helpers in both cooperatively breeding birds and mammals. Tamarins appear to increase direct fitness from the experience of caring for other infants, and may benefit from using infant care as a courtship strategy or to gain acceptance to a new group, but these benefits are less clear in marmosets. In gerbils, Salo and French (1989) demonstrated that prior infant care experience by at least one parent decreases the time to first reproduction. In some species of cooperatively breeding birds, extensive helping actually leads to greater mortality. We simply do not have enough field data to see if helpers are at increased risk in mammals. In most endangered species where suitable habitat for breeding is at a premium (Florida scrub jay, red-cockaded woodpecker), direct fitness costs are most evident because helpers have an increased chance of acquiring a territory. However, if we assume that these species once did have adequate habitat and have not always been endangered, then these direct fitness benefits of helping must be related to the endangered status of the habitats. Because the great majority of helpers are related to the breeders they assist, most of the value of helping must be measured through indirect fitness. Even here the source of benefits is not always obvious. Helping may have a greater effect on the survival and future reproductive success of the breeders than on the survival of the offspring that are helpers. There is still much to understand about the benefits helpers receive.

## V. What Mechanisms Inhibit Reproduction by Helpers?

If helpers are to provide assistance to the breeders, it is important that they not compete by breeding on their own. If helping is valuable, then multiple breeding animals within a group are likely to be detrimental to the reproductive success of all individuals. We have extensive information concerning the mechanisms of reproductive inhibition in cooperatively breeding mammals, especially females, but little direct data on this topic in cooperatively breeding birds.

There have been several studies on a number of captive species of marmosets and tamarins that provide evidence of reproductive suppression by dominant females. Abbott *et al.* (1993) have reviewed the data for all of the species that have been studied. In common marmosets, one finds only one female reproducing in a group, and most subordinate females are reproductively suppressed, not even ovulating. In groups constructed of three males and three females, extensive aggression occurs in the first hours of group formation, resulting in only one female that continues to ovulate. If a subordinate female receives an olfactory block, she will ovulate, but still retain her subordinate status, suggesting that cues for reproductive suppression are mediated by the olfactory system. Abbott and his colleagues have hypothesized that stress resulting from the initial aggressive encounters is responsible for the inhibition of ovulation.

In cotton-top tamarins, French *et al.* (1984) used urinary estrogen assays to show that daughters living in family groups were reproductively inhibited, although Tardif (1984) reported evidence of serum progesterone elevations in older females living in family groups. Subsequent studies by Ziegler *et al.* (1987) and Savage *et al.* (1988) showed that ovulation as measured by urinary luteinizing hormone was suppressed, but that reproductively inhibited females, when moved from a family group to living with a novel male, would ovulate and conceive in as little as 8 days. When scent marks from the mother were collected and transferred to the daughter's cage each day, there was a significant delay in the onset of ovulation and conception compared with females receiving transfers of water only. A similar study on saddle-back tamarins by Epple and Katz (1984) also found that transfer of olfactory stimuli from the reproductive female inhibited ovulation in the recipient female. Snowdon *et al.* (1993) described an additional reproductive inhibition with the birth of new infants in eldest daughters in cotton-top tamarins. These females showed evidence of hormonal activity from 18 months of age but the levels of both luteinizing hormone and estrogen were quite low and were not organized into ovarian cycles. With the birth of new infants, these low levels become still further reduced.

Female cotton-top tamarins have the capacity to reproduce as early as 18 months when puberty occurs, and we have histological evidence that antral follicle cells develop in the ovary, but become atretic prior to being ovulated (Ziegler *et al.*, 1990a). However, we have studied more than 25 reproductively inhibited females prior to being moved to breeding positions with novel males at ages ranging from 18 to 45 months. We have never found evidence of ovulation prior to the change in social environment, regardless of the female's age.

In golden lion tamarins, French and Stribley (1987) found no evidence of reproductive inhibition. Instead, mothers and daughters ovulated synchronously. However, because no mating by subordinate females has been observed within captive lion tamarin groups (French *et al.*, 1989) and because reproductive synchrony allows the dominant female to predict and control reproduction by daughters, these results still lead to suppression of reproduction by daughters. However, field data (Dietz and Baker, 1993) indicates that multiple pregnancies are relatively common in wild golden lion tamarins, although subordinate females experience high infant mortality.

All of the data presented so far are consistent with the Dominant Control Model (see Table III). The model assumes that the reproductive female benefits from recruiting nonreproductive animals to assist with infant care and must prevent reproductive competition from these helpers. The mechanisms that appear to produce the reproductive suppression are aggression and related stress effects; pheromonal cues that block ovulation; active interference with copulation; and in a field study (Digby, 1995a,b), infanticide. The Dominant Control Model makes the following two major predictions: (1) in the presence of a dominant female (or her olfactory cues), no other female will attempt to breed; and (2) in the absence of a dominant female or her cues, a previously subordinate female should attempt to breed immediately.

TABLE III
DOMINANT CONTROL MODEL

| | |
|---|---|
| Functions | Recruit helpers to care for young |
| | Prevent competition for resources (food, shelter, helpers, etc.) |
| Mechanisms | Aggressive behavior leading to social stress |
| | Active intervention |
| | Chemical signals |
| | Infanticide |
| Predictions | In presence of dominant female no other female will attempt to breed |
| | In absence of dominant or cues associated with her, subordinate female will breed |

A variety of studies fail to support the Dominant Control Model. First, there is evidence of breeding in the presence of dominant females. Much of the data come from field studies where mechanisms of reproductive suppression might be imperfect, but studies of common marmosets (Digby and Ferrari, 1994), cotton-top tamarins (Savage et al., 1996), and lion tamarins (Dietz and Baker, 1993; Baker et al., 1993) have reported multiple pregnancies or multiple births within a group. In dwarf mongooses, Creel and Waser (1991) have reported that multiple births are found both in large groups and by older subordinate females.

There are also studies reporting that in certain conditions females fail either to ovulate or reproduce in the absence of the dominant female. Female cotton-top tamarins living alone do not ovulate (Ziegler et al., 1987) and 6 of 10 common marmosets living along stopped ovulating (Tardif et al., 1994). Cotton-top tamarin females living with related males in the absence of a dominant female did not ovulate (Widowski et al., 1990) but all of the females ovulated when paired with a novel male or even when housed in cages adjacent to a novel male where direct physical contact was not possible (Widowski et al., 1992). Finally, two studies report low, basal levels of cortisol in subordinate common marmosets (Saltzman et al., 1994) and cotton-top tamarins (Ziegler et al., 1995). If cortisol reflects chronic social stress in these subordinate females, then stress appears unlikely to be involved in inhibiting reproduction (see Fig. 7).

These departures from the Dominant Control Model have suggested to Teresa Porter and me that a Self-Restraint Model is more appropriate for explaining the reproductive inhibition in marmosets and tamarins (see Table IV). The functions of self-restraint are to minimize inbreeding by restricting reproduction until an unrelated male is present, to restrict reproduction until sufficient helpers are available, to restrict reproduction to when other resources are available, and to delay reproduction until adequate parental care skills are acquired (see also Stephens, 1989). Many of the deviations from the Dominant Control Model can be explained by this model. In golden lion tamarins, Dietz and Baker (1993) reported that many groups had multiple females pregnant, but the infants survived only in groups where a new male, unrelated to the daughter, had arrived prior to the estimated conception date. In cotton-top tamarins, Savage et al. (1996) found two examples of groups with two pregnant females, although only one female gave birth to live young. In both cases a new male joined the group prior to probable conception. Thus, the presence of an unrelated male can lead to multiple breeding females. Digby (1995b) reports in common marmosets that infanticide of the subordinate female's infants occurred if both females gave birth at the same time, but if females gave birth out of phase with each other and could presumably time-share the helpers, there

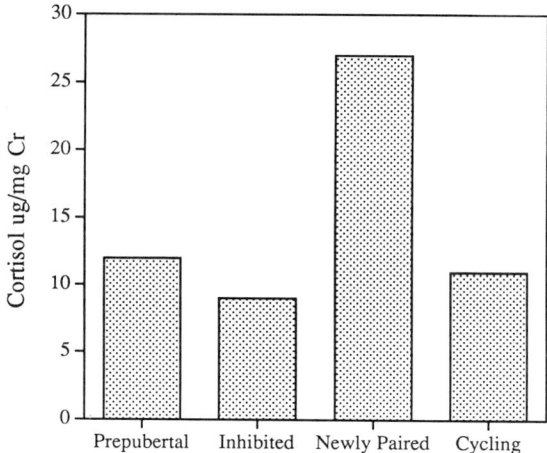

FIG. 7. Mean urinary cortisol levels for female cotton-top tamarins as a function of social and reproductive status. Prepubertal, females younger than 18 months; Inhibited, females older than 18 months but not ovulating; Newly paired, females during the first 6 weeks after pairing with a novel male. All females began to ovulate in this time. Cycling, nonpregnant, nonlactating females paired with a male for more than 6 months (median = 60 months). Cr = creatinine. (Adapted from Ziegler et al., 1995.)

TABLE IV
SELF-RESTRAINT MODEL

| | |
|---|---|
| Functions | Restrict reproduction until unrelated mate is available |
| | Restrict reproduction until sufficient helpers are available |
| | Restrict reproduction until other resources are adequate |
| | Defer reproduction until adequate parental care skills are learned |
| Mechanisms | Chemical signals provide information about the presence of a dominant female, but they do not act as pheromones suppressing reproduction |
| | Subordinate female regulates the timing and extent of exposure to olfactory signals |
| | Recognition and response to cues of unfamiliar animal of opposite sex |
| | Assessment of helpers and availability of resources |
| | Breeding out of phase with dominant female to time-share helpers |
| Predictions | Multiple females will attempt reproduction when a novel male joins the group |
| | Dominant female will not breed if helper pool is small or other resources are scarce |
| | Multiple females will breed if food resources are nonseasonal and therefore to allow helpers to be time-shared by multiple females |
| | Older subordinates are more likely to breed at the same time as the dominant female than younger subordinates |

was a greater chance of both sets of infants surviving. Savage *et al.* (1996) found strong seasonal breeding in wild cotton-top tamarins, but not in captive animals living out of doors at the same site. In a drought year no wild females gave birth at the normal season, but two females did give birth 6 months later, out of phase with the annual cycle. These results suggest that females adjust breeding to the availability of food resources because the captive tamarins had a constant and steady supply of food throughout the year.

The mechanisms of the Self-Restraint Model are unlikely to involve aggression and stress, because a subordinate female can control her interactions with the dominant female. While olfactory cues may be important for a subordinate female to identify the presence of a dominant female, the olfactory cues are unlikely to function as a pheromone that automatically suppresses reproduction. Subordinate females would be expected to recognize and respond differently to familiar versus unfamiliar males, and would also assess the availability of helpers and other resources.

We think that the Self-Restraint Model can account for much of the variability in reproduction in captive and wild marmosets and tamarins better than the Dominant Control Model, and the formulation of the Self-Restraint Model leads us to look for new mechanisms, especially mechanisms that could be effective in field conditions. It is interesting that while the Dominant Control Model has dominated the thinking of those who study cooperatively breeding mammals, almost all of the authors writing about cooperatively breeding birds look at reproductive inhibition or delay from the perspective of the helper rather than the breeder. Thus, those studying birds have implicitly accepted the Self-Restraint Model.

Unfortunately, little work has been done on the hormonal status of breeders and helpers in cooperative breeding birds, but Reyer *et al.* (1986) provide an illuminating study on pied kingfishers. In this species there are two types of helpers: primary helpers, who are related to the breeding animals and who rarely inherit territories; and secondary helpers, who are generally unrelated to the breeders and have a high probability of inheriting the territory. Interestingly, Reyer *et al.* found that primary, but not secondary, helpers had reduced testosterone levels coupled with reduced testes size and low spermatogenesis. These low levels were not related to age, degree of stress, or amount of helping. In contrast, there were no differences in testosterone levels or testes size between secondary helpers and breeders. These investigators explain these results by noting that secondary helpers must engage in high levels of fighting in order to acquire a territory and that almost all of their fitness depends upon being able to gain a territory. On the other hand, primary helpers gain most of their fitness via helping their kin, and have little chance of gaining a territory, therefore the low levels of testosterone and testes size may be adaptive. It seems unlikely that suppression in primary helpers and lack of suppression in secondary helpers is due to behavior of the breeding male, and suggests that, instead, primary helpers are limiting their own hormonal function.

## VI. Future Research Directions

There are several areas where future research is desirable. First, it would help to have more long-term field data on marmosets and tamarins in different habitats and facing different ecological pressures to evaluate different ecological factors that affect the expression of helping behavior as well as to evaluate the degree to which helpers are actually necessary and whether helpers themselves receive any direct benefits from helping. For example, data from captive animals suggests that helpers might be more important for tamarins than marmosets, and Ferrari and Lopes Ferrari (1989), Goldizen (1990), and Harrison and Tardif (1994) have suggested important ecological differences between the habitats and feeding ecology of marmosets and tamarins. Second, it would be desirable to have more detailed data on the behavioral and physiological mechanisms leading to helping and to reproductive inhibition in birds to match the nature of data available on cooperatively breeding mammals. Although there are abundant data on the behavioral aspects (reviewed by Gowaty, this volume) and physiological controls (reviewed by Buntin, this volume) of parental care in birds, we have a paucity of data on behavioral and physiological mechanisms in helpers.

As Fairbanks (this volume) has shown, individual differences in maternal style can have different outcomes for infant monkeys. In the species she reviews, mothers are nearly the exclusive caretakers of infants, so it would be interesting to know whether helpers in cooperatively breeding species can compensate for differences in maternal style to produce more uniform outcomes. Individual differences in infant behavior have rarely been examined. Do different types of infants elicit different amounts or different types of caretaking from others?

In captivity, tamarins, in particular, have considerable variance in reproductive success. Group size and parental parity are both important variables, but we know little about the variables that affect whether parents accept or reject infants other than the prepartum hormonal data reported for tamarins by Pryce (this volume). We need more research on the physiological and behavioral factors that lead to competent infant care by both parents and helpers.

We can make predictions about the energetic costs of carrying infants and how these costs predict the carrying by older versus younger helpers and when one infant or both infants will be carried, but the variability of caretaking behavior reported suggests other variables than energetic costs must be found to explain the variance. Because there is strong competition to take care of infants, how is the competition resolved? Can we identify which individuals control who has access to infants? What are the outcomes of the competition? Although learning infant care skills has been invoked frequently as an advantage to helpers, there are no data from field or captive studies that relate the quality or amount of infant care experience to subsequent reproductive success.

In light of the extensive studies involving the responses of infant macaques to separation and reunion from caregivers, and Mendoza and Mason's (1986) studies using choice tests with titi monkeys, only Arruda *et al.* (1986) have studied the effects of separation from the father on infant behavior in marmosets, and there have been no studies of infant choice between parents or helpers. It may be that separation effects in marmosets and tamarins are buffered by multiple caregivers, but we do not know.

Both Rothe and Darms (1993) and Price and McGrew (1991) have noted that marmoset and tamarin groups are destabilized by the loss of one parent and the introduction of a stepparent. The remaining parent does not intervene in the conflict between other group members. Often several group members have to be removed to end conflicts. Yet migration does occur in the wild and the reproductive animal of either sex can change groups from year to year without reports of conflict. What are the mechanisms that lead wild populations to remain stable with changes in breeders when captive groups are so unstable?

Finally, many of the mechanisms that have been thought to support reproductive inhibition in subordinates in captivity do not always work in the field. More research is needed to determine the mechanisms that do maintain reproductive inhibition in subordinates in field conditions. Moreover, we need to know for a greater variety of species the conditions under which the reproductive inhibition fails and multiple breeders occur.

## VII. Summary

This article has reviewed studies on infant care by nonreproductive helpers in both mammals and birds to describe the type of help that helpers provide and to evaluate whether helpers improve the survival of the infants they helped. The behavioral and physiological mechanisms that lead to helping were examined and the effect of helping on the reproductive success of helpers was evaluated. Finally, the mechanisms leading to the reproductive inhibition of helpers were examined.

Unfortunately, variation and flexibility emerged as major themes, rather than a uniform theoretical explanation of communal infant care. Marmosets and tamarins appear to be similar in many ways with twin births, a high infant to maternal weight ratio, a postpartum conception, and extensive caretaking by other group members. Yet helpers appear to play a more important role in the survival of infant tamarins where both field and captive data illustrate that infant survival is correlated with group size. The experience of helping infants appears to be more important in subsequent parental care success for tamarins than for marmosets. The differences between marmosets and tamarins in the need for helpers cannot be explained by differences in maternal energetics, but the smaller home range sizes and shorter travel paths of marmosets due to their ability to extract gum from trees can explain some differences.

We presented some preliminary data on urinary prolactin measurements in fathers and eldest sons involved in infant care that suggest repeated exposure to infants leads to progressively increased levels of prolactin, a result with interesting parallels to work on rats and humans, but many questions, both behavioral and physiological, concerning the mechanisms of nonmaternal infant care are needed in both mammals and birds.

In cooperatively breeding birds the significance of helpers for infant survival ranges greatly from no benefit to extreme importance. In some species the costs of providing care appear to be greater than any benefits received either directly or through inclusive fitness. The best data for the benefits to helpers are found in threatened or endangered species with highly limited breeding sites. In these species helpers increase their own reproductive success, improve the longevity of breeding relatives, and may result in territory inheritance. The work of Reyer (1990) on pied kingfishers shows considerable flexibility of helping behavior within a species. Parents will tolerate helpers only under conditions where providing food is difficult. Reyer has shown that related helpers differ from unrelated helpers in the timing and nature of helping, in their hormonal status, and in the relative amount of direct versus indirect fitness benefits received. More studies on within-species variation in helping and on the physiological mechanisms underlying this variation are greatly needed.

The presence of nonreproductive helpers implies some mechanism of reproductive inhibition, whether behavioral or physiological, and several studies have demonstrated various mechanisms not only in marmosets and tamarins, but also in dwarf mongooses. However, little information is available on mechanisms of reproductive inhibition in birds. We suggest that a

Self-Restraint Model, which focuses attention on the decisions of both inhibited and breeding animals, is more parsimonious than the prevailing Dominant Control Model.

Despite the great amount of information on cooperatively breeding birds and mammals, we are still in the infancy of what we know about the mechanisms and function of cooperative breeding.

### Acknowledgments

This research has been supported by USPHS Research Grant MH 35,215 and a Research Scientist Award MH 00,177. I am grateful to my research colleagues, Toni E. Ziegler, Tina M. Widowski, Anne Savage, Teresa Porter, Jennifer Greene, Rebecca Addington, and Gretchen Achenbach for their contributions to the ideas and research results reported here, and to Jay S. Rosenblatt for a careful critique of an earlier version of this article.

### References

Abbott, D. H., Barrett, J., and George, L. M. (1993). Comparative aspects of the social suppression of reproduction in female marmosets and tamarins. In "Marmosets and Tamarins: Systematics, Ecology and Behavior" (A. B. Rylands, ed.), pp. 152–163. Oxford Univ. Press, Oxford.

Albuquerque, F. (1994). "Distribucao do cuidado a prole em grupos de *Callithrix jacchus* no ambiente natural." Unpublished Dissertation, Universidade Federal do Rio Grande do Norte, Natal, Brazil.

Arruda, M. F., Yamamoto, M. E., and Bueno, O. F. A. (1986). Interactions between parents and infants and infants-father separation in the common marmoset. *Primates* 27, 215–228.

Baker, A. J., Dietz, J. M., and Kleiman, D. G. (1993). Behavioural evidence for monopolization of paternity in multi-male groups of golden lion tamarins. *Anim. Behav.* 46, 1091–1103.

Bowman, L. A., Dilley, S. R., and Keverne, E. B. (1978). Suppression of estrogen induced LH-surges by social subordination in talapoin monkeys. *Nature (London)* 275, 56–58.

Box, H. O. (1977). Quantitative data on the carrying of young captive monkeys (*Callithrix jacchus*) by other members of their family groups. *Primates* 18, 475–484.

Brown, J. L., Brown, E. R., Brown, S. D., and Dow, D. P. (1982). Helpers: Effect of experimental removal on reproductive success. *Science (Washington, D. C.)* 215, 421–422.

Caine, N. G. (1993). Flexibility and co-operation as unifying themes in *Saguinus* social organization and behaviour: The role of predation pressures. In "Marmosets and Tamarins: Systematics, Behaviour and Ecology" (A. B. Rylands, ed.), pp. 200–219. Oxford Univ. Press, Oxford.

Cleveland, J., and Snowdon, C. T. (1984). Social development during the first twenty weeks in the cotton-top tamarin (*Saguinus oedipus*). *Anim. Behav.* 32, 432–444.

Creel, S. R., and Creel, N. M. (1991). Energetics, reproductive suppression and obligate communal breeding in carnivores. *Behav. Ecol. Sociobiol.* 28, 263–270.

Creel, S., and Macdonald, D. (1995). Sociality, group size and reproductive suppression among carnivores. In "Advances in the Study of Behavior," Vol. 24 (P. J. B. Slater, J. S. Rosenblatt, C. T. Snowdon, and M. Milinski, eds.), pp. 203–257. Academic Press, San Diego.

Creel, S. R., and Waser, P. M. (1991). Failures of reproductive suppression in dwarf mongooses (*Helogale parvula*): Accident or adaptation. *Behav. Ecol.* **2,** 7–15.
de la Torre, S., Campos, F., and de Vries, T. (1995). Home range and birth seasonality of *Saguinus nibricollis graellsi* in Ecuadoran Amazonia. *Am. J. Primatol.* **37,** 39–56.
Dietz, J. M., and Baker, A. J. (1993). Polygyny and female reproductive success in golden lion tamarins, *Leontopithecus rosalia. Anim. Behav.* **46,** 1067–1078.
Dietz, J. M., Baker, A. J., and Miglioretti, D. (1994). Seasonal variation in reproduction, juvenile growth and adult body mass in golden lion tamarins (*Leontopithecus rosalia*). *Am. J. Primatol.* **34,** 115–132.
Digby, L. J. (1995a). Social organization in a wild population of *Callithrix jacchus:* II. Intragroup social behavior. *Primates* **36,** 361–375.
Digby, L. J. (1995b). Infant care, infanticide and female reproductive strategies in polygynous groups of common marmosets (*Callithrix jacchus*). *Behav. Ecol. Sociobiol.* **37,** 51–61.
Digby, L. J., and Ferrari, S. F. (1994). Multiple breeding females in free-ranging groups of *Callithrix jacchus. Int. J. Primatol.* **15,** 389–397.
Dixson, A. F., and George, L. (1982). Prolactin and parental behavior in a male New World primate. *Nature (London)* **299,** 551–553.
Dunbar, R. I. M. (1995). The mating system of callitrichid primates: II. The impact of helpers. *Anim. Behav.* **50,** 1071–1089.
Eberhart, J. A., Keverne, E. B., and Meller, R. E. (1983). Social influences on circulating levels of cortisol and prolactin in male talapoin monkeys. *Physiol. Behav.* **30,** 361–369.
Emlen, S. T. (1994). Benefits, constraints and the evolution of the family. *Trends. Ecol. Evol.* **9,** 282–285.
Emlen, S. T., and Wrege, P. H. (1988). The role of kinship in helping decisions among white-fronted bee eaters. *Behav. Ecol. Sociobiol.* **23,** 305–315.
Emlen, S. T., and Wrege, P. H. (1989). A test of alternative hypotheses for helping behavior in white-fronted bee-eaters. *Behav. Ecol. Sociobiol.* **25,** 303–319.
Emlen, S. T., and Wrege, P. H. (1991). Breeding biology of white-fronted bee-eaters at Nakuru: The influence of helpers on breeding success. *J. Anim. Ecol.* **60,** 309–326.
Emlen, S. T., and Wrege, P. H. (1992). Parent-offspring conflict and the recruitment of helpers among bee-eaters. *Nature (London)* **356,** 331–333.
Emlen, S. T., Wrege, P. H., and Demong, N. J. (1995). Making decisions in the family: An evolutionary perspective. *Am. Sci.* **83,** 148–157.
Epple, G. (1978). Reproductive and social behaviors of marmosets with special reference to captive breeding. *Primate Med.* **10,** 50–62.
Epple, G., and Katz, Y. (1984). Social influences on estrogen excretion, and ovarian cyclicity in the saddle-back tamarin (*Saguinus fuscicollis*). *Am. J. Primatol.* **6,** 215–227.
Feistner, A. T. C., and Chamove, A. S. (1986). High motivation toward food increases food sharing in cotton-top tamarins. *Dev. Psychobiol.* **19,** 439–452.
Ferrari, S. F. (1987). Food transfer in a wild marmoset group. *Folia Primatol.* **48,** 203–206.
Ferrari, S. F. (1992). The care of infants in a wild marmoset (*Callithrix flaviceps*) group. *Am. J. Primatol.* **26,** 109–118.
Ferrari, S. F., and Lopes Ferrari, M. A. (1989). A re-evaluation of the social organization of the Callithrichidae with reference to the ecological differences between genera. *Folia Primatol.* **52,** 132–147.
Fleming, A. S., Corter, C., and Steiner, M. (1995). Sensory and hormonal control of maternal behavior in rat and human mothers. *In* "Motherhood in Human and Nonhuman Primates" (C. R. Pryce and R. D. Martin, eds.), pp. 106–114. Karger, Basel.
Fragaszy, D. M., Schwartz, S., and Shimosaka, D. (1982). Longitudinal observations of care and development of infant titi monkeys (*Callicebus moloch*). *Am. J. Primatol.* **2,** 191–200.

French, J. A., and Stribley, J. A. (1987). Synchronization of ovarian cycles within and between social groups in golden lion tamarins (*Leontopithecus rosalia*). *Am. J. Primatol.* **12,** 469–478.

French, J. A., Abbott, D. H., and Snowdon, C. T. (1984). The effect of social environment on estrogen excretion, scent marking, and sociosexual behavior in tamarins (*Saguinus oedipus*). *Am. J. Primatol.* **6,** 155–167.

French, J. A., Inglett, B. J., and Detlefs, T. M. (1989). The reproductive status of nonbreeding group members in captive golden lion tamarin social groups. *Am. J. Primatol.* **18,** 73–86.

Garber, P. A. (1994). Phylogenetic approaches to the study of tamarin and marmoset social systems. *Am. J. Primatol.* **34,** 199–219.

Garber, P. A., Moya, L., and Malaga, C. (1984). A preliminary field study of the moustached tamarin monkey (*Saguinus mystax*) in Northeastern Peru: Questions concerned with the evolution of a communal breeding system. *Folia Primatol.* **42,** 17–32.

Goldizen, A. W. (1987a). Tamarins and marmosets: Communal care of offspring. *In* "Primate Societies" (B. B. Smits, D. L. Cheney, R. M. Seyfarth, T. T. Struhsaker, and R. W. Wrangham, eds.), pp. 34–43. Univ. of Chicago Press, Chicago.

Goldizen, A. W. (1987b). Facultative polyandry and the role of infant carrying in wild saddle-back tamarins (*Saguinus fuscicollis*). *Behav. Ecol. Sociobiol.* **20,** 99–109.

Goldizen, A. W. (1988). Tamarin and marmoset mating systems: Unusual flexibility. *Trends. Ecol. Evol.* **3,** 36–40.

Goldizen, A. W. (1989). Social relationships in a cooperatively polyandrous group of tamarins (*Saguinus oedipus*). *Behav. Ecol. Sociobiol.* **24,** 79–89.

Goldizen, A. W. (1990). A comparative perspective on the evolution of tamarin and marmoset social systems. *Int. J. Primatol.* **11,** 63–83.

Goldsmith, A. R., Edwards, C., Koprucu, M., and Silver, R. (1981). Concentrations of prolactin and luteinizing hormone in doves in relation to incubation and development of the crop gland. *J. Endocrinol.* **90,** 437–443.

Gubernick, D. G., and Nelson, R. (1989). Prolactin and paternal behavior in the biparental California mouse, *Peromyscus californicus*. *Horm. Behav.* **23,** 203–210.

Hansen, S., Keverne, E. B., Martensz, M. D., and Herbert, J. (1980). Behavioural and neuro-endocrine factors regulating prolactin and LH discharges in monkeys. *In* "Non-Human Primate Models for the Study of Human Reproduction," pp. 148–158. Karger, Basel.

Harrison, M. L., and Tardif, S. D. (1994). Social implications of gummivory in marmosets. *Am. J. Phys. Anthropol* **95,** 399–408.

Heymann, E. W. (1990). Social behavior and infant carrying in a group of moustached tamarins, *Saguinus mystax*, on Padre Isla, Peruvian Amazonia. *Primates* **31,** 183–196.

Hoage, R. J. (1978). Parental care in *Leontopithecus rosalia rosalia*: Age and sex differences in carrying behavior and the role of prior experience. *In* "Biology and Conservation of the Callitrichidae" (D. G. Kleiman, ed.), pp. 293–305. Smithsonian Institution Press, Washington, D. C.

Hoage, R. J. (1982). "Social and Physical Maturation of Captive Lion Tamarins, *Leontopithecus rosalia rosalia*." Smithsonian Institution Press, Washington, D. C.

Ingram, J. C. (1977). Interactions between parents and infants, and the development of independence in the common marmoset (*Callithrix jacchus*). *Anim. Behav.* **25,** 811–827.

Jarvis, J. U. M., O'Riain, M. J., Bennett, N. C., and Sherman, P. W. (1994). Mammalian eusociality: A family affair. *Trends Ecol. Evol.* **9,** 47–51.

Jennions, M. D., and Macdonald, D. W. (1994). Cooperative breeding in mammals. *Trends Ecol. Evol.* **9,** 89–93.

Johnson, L. D., Petto, A. J., and Sehgal, P. K. (1991). Factors in the rejection and survival of captive cotton-top tamarins (*Saguinus oedipus*). *Am. J. Primatol.* **25,** 91–102.

Jurke, M. H., and Pryce, C. R. (1994). Parental and infant behaviour during early periods of infant care in Goeldi's monkey, *Callimico goeldii. Anim. Behav.* **48,** 1095–1112.

Koenig, A. (1995). Group size, composition and reproductive success in wild common marmosets (*Callithrix jacchus*). *Am. J. Primatol.* **35,** 311–317.

Koenig, A., and Rothe, H. (1991). Social relationships and individual contribution to cooperative behaviour in captive common marmosets (*Callithrix jacchus*). *Primates* **32,** 183–195.

Koenig, W. D., and Mumme, R. L. (1987). "Population Ecology of the Cooperatively Breeding Acorn Woodpecker." Princeton Univ. Press, Princeton, NJ.

Lacey, E. A., and Sherman, P. W. (1991). Social organization of naked mole-rat colonies: Evidence for divisions of labor. *In* "The Biology of the Naked Mole-Rat" (P. W. Sherman, J. U. M. Jarvis, and R. D. Alexander, eds.), pp. 275–336. Princeton Univ. Press, Princeton, NJ.

Leutenegger, W. (1980). Monogamy in callithrichids: Consequences of phyletic dwarfism? *Int. J. Primatol.* **1,** 95–98.

McGrew, W. C. (1988). Parental division of caretaking versus family composition in cotton-top tamarins. *Anim. Behav.* **36,** 285–286.

Mendoza, S. P., and Mason, W. A. (1986). Parental division of labour and differentiation of attachments in a monogamous primate (*Callicebus moloch*). *Anim. Behav.* **34,** 1336–1347.

Moehlman, P. D. (1979). Jackal helpers and pup survival. *Nature (London)* **277,** 382–383.

Oring, L. W., Fivizzani, A. J., el-Halawani, M. E., and Goldsmith, A. (1986). Seasonal changes in prolactin and luteinizing hormone in the polyandrous spotted sandpiper. *Actitis mecularia. Gen. Comp. Endocrinol.* **62,** 394–403.

Porter, T. A. (1994). "The development and maintenance of heterosexual pair associations in cotton-top tamarins (*Saguinus oedipus oedipus*)." Unpublished Ph.D. dissertation, University of Wisconsin, Madison.

Price, E. (1990a). Infant carrying as a courtship strategy of breeding male cotton-top tamarins. *Anim. Behav.* **40,** 784–786.

Price, E. (1990b). Parturition and perinatal behavior in captive cotton-top tamarins (*Saguinus oedipus*). *Primates* **31,** 523–535.

Price E. (1991). Competition to carry infants in captive families of cotton-top tamarins (*Saguinus oedipus*). *Behavior* **118,** 66–88.

Price, E. (1992a). The costs of infant carrying in captive cotton-top tamarins. *Am. J. Primatol.* **26,** 23–33.

Price, E. (1992b). Sex and helping: Reproductive strategies of breeding male and female cotton-top tamarins, *Saguinus oedipus. Anim. Behav.* **43,** 717–728.

Price, E. C., and McGrew, W. C. (1991). Departures from monogamy in colonies of captive cotton-top tamarins. *Folia Primatol.* **57,** 16–27.

Rabenold, K. N. (1990). *Campylorhynchus* wrens: The ecology of delayed dispersal and cooperation in the Venezuelan savannah. *In* "Cooperative Breeding in Birds: Long-Term Studies of Ecology and Behavior" (P. B. Stacey and W. D. Koenig, eds.), pp. 159–196. Cambridge Univ. Press, Cambridge.

Rasa, O. A. E. (1987). The dwarf mongoose: A study of behavior and social structure in relation to ecology in a small, social carnivore. *Adv. Study Behav.* **17,** 121–163.

Reyer, H.-U. (1990). Pied Kingfishers: Ecological causes and reproductive consequences of cooperative breeding. *In* "Cooperative Breeding in Birds: Long-Term Studies of Ecology and Behavior" (P. B. Stacey and W. D. Koenig, eds.), pp. 527–557. Cambridge Univ. Press, Cambridge.

Reyer, H. U., Dittami, J., and Hall, M. R. (1986). Avian helpers at the nest: Are they psychologically castrated? *Ethology* **71,** 216–228.

Rothe, H., and Darms, K. (1993). The social organization of marmosets: A critical evaluation of recent concepts. *In* "Marmosets and Tamarins: Systematics, Behavior and Ecology" (A. B. Rylands, ed.), pp. 176–199. Oxford Univ. Press, Oxford.

Rothe, H., Darms, K., Koenig, A., Radespiel, U., and Juenemann, B. (1993). Long-term study of infant care behavior in captive common marmosets (*Callithrix jacchus*): Effect of nonreproductive helpers on the parents' carrying performance. *Int. J. Primatol.* **14,** 79–93.

Salo, A. L., and French, J. A. (1989). Early experience, reproductive success, and the development of parental behaviour in Mongolian gerbils. *Anim. Behav.* **38,** 693–702.

Saltzman, W., Schultz-Darken, N. J., Scheffler, G., Wegner, F. H., and Abbott, D. H. (1994). Social and reproductive influences on plasma cortisol in female marmoset monkeys. *Physiol. Behav.* **56,** 801–810.

Savage, A., Ziegler, T. E., and Snowdon, C. T. (1988). Sociosexual development, pair-bond formation and mechanisms of fertility suppression in female cotton-top tamarins (*Saguinus o. oedipus*). *Am. J. Primatol.* **14,** 345–359.

Savage, A., Giraldo, L. H., Soto, L. H., and Snowdon, C. T. (1996). Demography, group composition and dispersal in wild cotton-top tamarin (*Saguinus oedipus oedipus*) groups. *Am. J. Primatol.* **38,** 85–100.

Savage, A., Snowdon, C. T., Giraldo, L. H., and Soto, L. H. (in press). Parental care patterns and vigilance in wild cotton-top tamarins (*Saguinus oedipus*). *In* "Adaptive Radiation of Neotropical Primates" (M. A. Norconk, A. L. Rosenberger, and P. A. Garber, eds.), Aldine de Gruyter, Hawthorne, New York.

Seiler, W. K., Gahr, M., Goldsmith, M. A. R., and Guttinger, H. R. (1992). Prolactin and gonadal steroids during the reproductive cycle of the Bengalese finch (*Lonchura striata var domesticus*) a non-seasonal breeder with biparental care. *Gen. Comp. Endocrinol.* **88,** 83–90.

Sherman, P. W., Jarvis, J. U. M., and Alexander, R. D. (eds) (1991). "The Biology of the Naked Mole-Rat." Princeton Univ. Press, Princeton, NJ.

Sherman, P. W., Lacey, E. A., Reeve, H. K., and Keller, L. (1995). The eusociality continuum. *Behav. Ecol.* **6,** 102–108.

Snowdon, C. T. (1990). Mechanisms maintaining monogamy in monkeys. *In* "Contemporary Issues in Comparative Psychology" (D. A. Dewsbury, ed.), pp. 225–251. Sinauer Assoc., Sunderland, MA.

Snowdon, C. T., and Soini, P. (1988). The tamarins, genus *Saguinus*. *In* "Ecology and Behavior of Neotropical Primates," Vol. 2 (R. A. Mittermeier, A. B. Rylands, G. A. B. Fonseca, and A. F. Coimbra-Filho, eds.), pp. 223–298. World Wildlife Fund, Washington, D. C.

Snowdon, C. T., Ziegler, T. E., and Widowski, T. M. (1993). Increased hormonal suppression of eldest daughter cotton-top tamarins following birth of infants. *Am. J. Primatol.* **31,** 11–21.

Solomon, N. G. (1991). Current indirect fitness benefits associated with philopatry in juvenile prairie voles. *Behav. Ecol. Sociobiol.* **29,** 277–282.

Stacey, P. B., and Koenig, W. D. (1990). "Cooperative Breeding in Birds: Long-Term Studies of Ecology and Behavior." Cambridge Univ. Press, Cambridge.

Stephens, M. E. (1989). Tamarins: Alpha female, queen bee or white rat? *Int. J. Primatol.* **10,** 115–121.

Stevenson, M. F., and Rylands, A. B. (1988). The marmosets, genus *Callithrix*. *In* "Ecology and Behavior of Neotropical Primates," Vol. 2 (R. A. Mittermeier, A. B. Rylands, G. A. B. Fonseca, and A. F. Coimbra-Filho, eds.), pp. 131–222. World Wildlife Fund, Washington, D. C.

Sussman, R. W., and Garber, P. A. (1987). A new interpretation of the social organization and mating system of the Callitrichidae. *Int. J. Primatol.* **8,** 73–90.

Tardif, S. D. (1984). Social influences on sexual maturation of female *Saguinus oedipus*. *Am. J. Primatol.* **6,** 199–209.

Tardif, S. D. (1994). Relative energetic cost of infant care in small-bodied neotropical primates and its relation to infant-care patterns. *Am. J. Primatol.* **34,** 133–143.

Tardif, S. D., Richter, C. B., and Carson, R. L. (1984). Effects of sibling experience on future reproductive success in two species of Callitrichidae. *Am. J. Primatol.* **6,** 377–380.

Tardif, S. D., Carson, R. L., and Clapp, N. K. (1986a). Breeding performance of captive-born cotton-top tamarin (*Saguinus oedipus*) females: Proposed explanations for colony differences. *Am. J. Primatol.* **11,** 271–275.

Tardif, S. D., Carson, R. L., and Gangaware, B. L. (1986b). Comparison of infant care in family groups of the common marmoset (*Callithrix jacchus*) and the cotton-top tamarin (*Saguinus oedipus*). *Am. J. Primatol.* **11,** 103–110.

Tardif, S. D., Carson, R. L., and Gangaware, B. L. (1990). Infant care behavior of mothers and fathers in a communal care primate, the cotton-top tamarin (*Saguinus oedipus*). *Am. J. Primatol.* **22,** 73–85.

Tardif, S. D., Carson, R. L., and Gangaware, B. L. (1992). Infant care behavior of non-reproductive helpers in a communal care primate, the cotton-top tamarin (*Saguinus oedipus*). *Ethology* **92,** 155–167.

Tardif, S. D., Harrison, M. L., and Simek, M. A. (1993). Communal infant care in marmosets and tamarins: Relation to energetics, ecology and social organization. In "Marmosets and Tamarins: Systematics, Behaviour and Ecology" (A. B. Rylands, ed.), pp. 220–234. Oxford Univ. Press, Oxford.

Tardif, S. D., Hyde, K., and Digby, L. J. (1994). Evidence for suppression of ovulation in singly–housed female common marmosets (*Callithrix jacchus*). *Lab. Primate Newsltr.* **33**(2), 1–4.

Terborgh, J., and Goldizen, A. W. (1985). On the mating system of the cooperatively breeding saddle–back tamarin (*Saguinus fuscicollis*). *Behav. Ecol. Sociobiol.* **16,** 293–299.

Vogt, J. L., Carlson, H., and Menzel, E. (1978). Social behavior of a marmoset (*Saguinus fuscicollis*) group I. Parental care and infant development. *Primates* **19,** 715–726.

Walters, J. R. (1990). Red-cockaded woodpeckers: A 'primitive' cooperative breeder. In "Cooperative Breeding in Birds: Long Term Studies of Ecology and Behavior" (P. B. Stacey and W. D. Koenig, eds.), pp. 69–101. Cambridge Univ. Press, Cambridge.

Wamboldt, M. Z., Gelhard, R. E., and Insel, T. R. (1988). Gender differences in caring for infant *Cebuella pygmaea:* The role of infant age and relatedness. *Dev. Psychobiol.* **21,** 187–202.

Warren, M. P., and Shortle, B. (1990). Endocrine correlates of human parenting: A clinical perspective. In "Mammalian Parenting: Biochemical, Neurobiological and Behavioral Determinants" (N. A. Krasnegor and R. S. Bridges, eds.), pp. 209–226. Oxford Univ. Press, New York.

Widowski, T. M., Ziegler, T. E., Elowson, A. M., and Snowdon, C. T. (1990). The role of males in stimulation of reproductive function in female cotton-top tamarins, *Saguinus o. oedipus*. *Anim. Behav.* **40,** 731–741.

Widowski, T. M., Porter, T. A., Ziegler, T. E., and Snowdon, C. T. (1992). The stimulatory effect of males on the initiation, but not the maintenance, of ovarian cycling in cotton-top tamarins (*Saguinus oedipus*). *Am. J. Primatol.* **26,** 97–108.

Woolfenden, G. E., and Fitzpatrick, J. W. (1984). "The Florida Scrub Jay: Demography of a Cooperatively Breeding Bird." Princeton Univ. Press, Princeton, NJ.

Wright, P. C. (1984). Biparental care in *Aotus trivirgatus* and *Callicebus moloch*. In "Female Primates: Studies by Women Primatologists" (M. E. Small, ed.), pp. 59–75. Alan R. Liss, New York.

Yamamoto, M. E. (1993). From dependence to sexual maturity: The behavioural ontogeny of Callitrichidae. *In* "Marmosets and Tamarins: Systematics, Behaviour, and Ecology" (A. B. Rylands, ed.), pp. 235–254. Oxford Univ. Press, Oxford.

Yamamoto, M. E., Box, H. O., Albuquerque, F. S., and Miranda, M. F. A. (1996). Carrying behaviour in captive and wild marmosets (*Callithrix jacchus*): A comparison between two colonies and a field site. *Primates* **37,** in press.

Ziegler, T. E., Bridson, W. E., Snowdon, C. T., and Eman, S. (1987). Urinary gonadotropin and estrogen excretion during the post-partum estrus, conception and pregnancy in the cotton-top tamarin (*Saguinus oedipus oedipus*). *Am. J. Primatol.* **12,** 127–140.

Ziegler, T. E., Snowdon, C. T., and Uno, H. (1990a). Social interactions and determinants of ovulation in tamarins (*Saguinus*). *In* "Socioendocrinology of Primate Reproduction" (T. E. Ziegler and F. B. Bercovitch, eds.), pp. 113–133. Wiley-Liss, New York.

Ziegler, T. E., Widowski, T. M., Larson, M. L., and Snowdon, C. T. (1990b). Nursing does not affect the duration of the postpartum to ovulation interval in cotton-top tamarins (*Saguinus oedipus*). *J. Reprod. Fertil.* **90,** 563–570.

Ziegler, T. E., Scheffler, G., and Snowdon, C. T. (1995). The relationship of cortisol levels to social environment and reproductive functioning in female cotton-top tamarins, *Saguinus oedipus*. *Horm. Behav.* **29,** 407–424.

Ziegler, T. E., Wegner, F. H., and Snowdon, C. T. (1996). A hormonal role for male parental care in a New World primate, the cotton-top tamarin (*Saguinus oedipus*). *Horm. Behav.* (in press).

# Index

## A

Acceptance
  alien young, 398–399
  maternal, effects of lamb odor manipulation, 403–405
Acetylcholine, release at parturition, 410–411
Activation
  genomic, maternal memory, 234
  neural sites during interaction with pups, 282–283
  noradrenergic system, at parturition, 389–390, 394
  olfactory bulb, effect on amygdala, 322
Adaptation
  effect on fitness of organism, 145–146
  in explaining helping behavior, 671
Adaptive significance
  avian social monogamy, 487–494
  parental care in birds, 517–521
Adaptive strategy, socialization of maternal behavior as, 440–441
Age
  infant, effect on maternal behavior, 583
  mother, and maternal style changes, 587
  offspring, and parental care flexibility, 137–138
  primate weaning, 431–432
Aggression
  female–female, 490–491
  juvenile and daughter, effect of infant births, 661–662
  maternal
    elicitation and maintenance, 275–277, 286
    postpartum, 216–217
    reduction in sheep, 446
Aggressiveness
  differential, 35
  during nesting and posthatching, 180–182

Alien young
  acceptance and rejection, 398–399
  versus own, discrimination, 402–406, 412–413
Allocation
  differential parental, and sexual conflict, 494–496
  resource, and life history components, 30–31
  sex, Fisher's theory, 562–563
Alloparenting, voles, 365–366
Altricial state
  parental responses to, hormonal effects, 185–196
  and precocial state, in ontogeny, 96–98
Amboseli, free-ranging monkeys, 588–589
Amniotic condition, reptilian, 146–148
Amniotic fluid, covering neonate, maternal reactions, 396–397, 409, 438
Amphibia, parental care, 109–139
Amygdala
  effect of olfactory bulb activation, 322
  in expression of maternal behavior, 311–316
  Fos-labeled cells, 319
Anaptia
  acute, retrieval during, 276
  lidocaine-induced, 253, 255
Androgen, effect on parent responses to altricial young, 185–186
Anesthesia
  and denervation, effects on licking, 256–259
  during early lactation, 349–351
  mystacial pads, 260–261
Año Nuevo, phocid pupping, 548–550
Anurans
  parental care
    evolution, 131–132
    geographic distribution, 123–127
    postoviposition, 113–115

691

Anxiety, primate, as proximate mechanism, 602–603
Approach–withdrawal behavior
  at parturition, 444
  replaced by approach–maternal behavior, 446
Arginine vasopressin, changes during sheep parturition, 389
Attendance
  constraints, 133–134
  egg and tadpole, 116–119
Attraction
  mate, and paternal provisioning, trade-offs, 483–484
  processing of infant stimuli, 460–462
Attractiveness
  differential, 495
  neonatal lamb, 396–397

# B

Balon, E.K, concept of reproductive guilds for fishes, 60–63
Bar pressing, for infant sensory reinforcement, 455–460
Basal metabolic rate, phocids and otariids, 554–555
Basolateral nucleus, Fos-lir levels, 314–316, 320
Bed nucleus of stria terminalis
  oxytocin receptor levels, 225–227
  in vole parental behavior, 371–372
Behavior, *see also* Maternal behavior; Parental behavior
  affiliative
    in female prairie vole, 379–380
    opioid effects, 228–231
  alloparental, 365–366
  assessment, by burying beetle, 14–15
  avian
    defensive–aggressive, 179–182
    incubation, 162–179
  challenging, by adolescent male monkeys, 598–600
  piscine, associated with reproduction, 75–76, 82–85
  play-mothering, 433–436
  recovery of function after neural damage, 262–266
  related types of parental investment, 152–153
  species-specific, and somatosensory reflexes, 249–250
Behavior mechanisms, in expression of helping, 661–662
Benefits
  differential investment, to pups, 558–559
  to helpers, 656, 667–674
  parental care
    amphibians, 132–138
    birds, 517–518
  reptilian reproductive solutions, 154
Bias
  daughter-biased sex ratios, 520
  phylogenetic, among anurans, 123, 127–128
  in pinniped sex ratio, 563
Biochemistry
  brain, 604
  regulation of rat parental care, 235–237
Biogenic amines, and parental behavior, 231–233
Biology, evolutionary, 90–92
Biparental care
  amphibian, 135–136
  in burying beetle, 14–16
  as norm in birds, 478
  versus uniparental male care, 33–39
Birds
  helping
    benefits, 673–674
    patterns, 657–661
  parental behavior, neural and hormonal control, 161–202
  variation among females, 477–523
Biting, primate infants by parents, 453–455
Bleats, by maternal female sheep, 387–388
Blood–brain barrier
  penetration by steroid hormones, 443
  and prolactin effect on CNS, 177–179
Blumer, L.S., on parental care
  definition, 79–80
  ethological forms, 63
Body size
  female parent bug, 20
  pinniped
    male and female pups, 558–559
    mothers, 539–541
  relationship to reproductive strategy, 18

Body stocking, to disguise lamb odor, 403–405
Boldness score, in response to adult male stranger, 599–600
Bonding, selective, 398, 401, 403–406
Brain
  avian, prolactin binding sites, 196–197
  biochemistry, 604
  Fos-lir pattern, 314–320
  periaqueductal gray, 283–288
  rat, nursing behavior changes in, 278–282
Breder and Rosen, compendium of data on fishes, 58–60
Breeding, see also Reproduction
  cooperative, and infant care, 643–683
  patterns, phocid and otariid, 537–544
  terrestrial, amphibian, evolutionary trends, 130–136
Breeding pair, prolactin levels during incubation, 163–164
Bromocriptine, effect on maternal behaviors, 346–347
Brood
  asynchronous and synchronous, 495–496
  eggs, by aquatic frogs, 118
  failure, in burying beetle, 16
  number, regulation in invertebrates, 19–23
Brood patch, sensory input from, 176–177
Brood sac, milk production, 8–9
Burying beetle
  biparental care, 14–16
  clutch replacement, 21–22
  differential aggressiveness, 35
  exploitation of food resource, 5

## C

Caecilians
  parental care, geographic distribution, 122
  postoviposition, 111–112
California sea lion, differential investment, 560–562
Callitrichids, patterns of helping, 644–654
Calls
  monkey caregiver, 623–626
  primate infant, 614–622

Cannibalism, lagomorph, as deficient behavior, 337
Captive population studies, infant survival, helping effects, 646–654
Caregiving
  duration, regulation, 28–30
  invertebrate parental, associated decision making, 17–18
  marmosets and tamarins, 649–654
  relationship to fertilization mode, 127–128, 139
Care-helping, by females, preadult experience, 439–440, 450–453
Carnivores, cooperative breeding, 655–656
Carrion, exploited by burying beetle, 5
Carrying
  adolescent–infant, 435
  infants by wild and captive groups, 646–655
  maternal-like, 449–451
  mother–infant, 431, 436–440
Cayo Santiago, free-ranging rhesus monkeys, 582–583, 587, 592, 600–601
CB-154, interaction with prolactin, 221–222
Central nervous system, prolactin
  activity, 177–179
  binding sites, 196–197
Cercopithecine monkeys
  adult females, vocalizations, 624–625
  infant distress calls, 617
  use of facial expressions, 629–630, 632–633
Cerebrospinal fluid
  conduit for prolactin, 177–179
  lactogenic hormone activity, 223–224
  oxytocin levels at parturition, 392–393
  serotonin activity, 604
Cesarean delivery, macaques, and maternal behavior, 428
c-fos
  as cerebral marker, 296
  expression
    in medial preoptic area, 234
    in neurons during activation, 371–372
Chemical information, from mother to young via milk, 335–336
Cholecystokinin, role in rat parental behavior, 227
Cichlid fishes, oral-brooding, 93–95

Circadian periodicity, lagomorph nursing, 335
Classification
  effects associated with energy expenditures, 78
  evolution of reproductive patterns in fishes, 60–63
Clinging contact, by simian infants, 429–431, 450, 460–462
Clutch
  replacement, by burying beetles, 16, 21–22
  size
    amphibian, 129, 137
    invertebrate, 32–33
    reptilian, 149
Cockroach, viviparous, pregnancy, 8–10
Columbiform species
  feeding by parental regurgitation, 202
  incubation, prolactin and steroid hormone role, 168–172
  parental care expression, 161–162
Communication
  brood to caregivers, 28
  parent–infant, in evolutionary perspective, 633–635
  between pinniped mother and young, 569
  visual and tactile, 626–633
  vocal, 614–626
Conflict
  parent–offspring, and termination of care, 27–30
  parent–parent, in birds, 520–523
Consistency
  individual differences in maternal behavior, 583–586
  maternal style across conditions, 589–591
Consolidation, maternal experiences
  blocking, 312
  role of protein synthesis, 323–324
Constrained Female Hypothesis, 485–486, 505–506, 516, 523
Constraints, imposed by parental information gathering, 19–20
Consummatory behavior, hormonal effects, 446–447
Contact
  clinging, by simian infants, 429–431, 450, 460–462
  mother–young
    and bonding, 398
    as function of infant risk, 586–587
    thermal limitation, 272–273
Contact calls, primate infant, 615–622
Coo calls
  as infant isolation call, 618–619
  maternal, 624
Copulation
  behavioral control by females, 506–507
  ovarian hormonal effects, 447
  patterns in biparental invertebrates, 34
  solicitation by female birds, 491–492
Corpora allata
  during pregnancy in viviparous cockroach, 8–10
  removal, effect on earwig parental period, 11–14
Cortex
  norepinephrine content, 232–233
  plasticity, and nursing behavior changes, 278–282
Corticosterone
  effect on parental feeding activity, 186–187
  synergy with prolactin in promoting feeding, 194
Corticotropin-releasing factor
  in oxytocin release, 393
  in rat parental behavior, 227–228
Cortisol levels
  males involved in infant care, 664
  subordinate females, 677
Costs
  copulation, to female birds, 491
  energetic, lactation, 533–534
  high parental investment, 27
  infant caretaking, 668
  parental care
    amphibians, 109–110, 132–136, 138
    birds, 517–518
    invertebrates, 30–33
  reproduction, pinnipeds, 546–555
  reptilian reproductive solutions, 154
Courtship
  and invertebrate paternal care, 33–36
  strategy, infant care as, 652, 673
Cowbird
  cerebral prolactin binding sites, 197

prolactin levels during breeding season, 174
Crocodylians
  egg chamber, 150–151
  parental investment, 152–153
Crop milk, prolactin-induced production, 168
Crop sac, prolactin-induced growth, 189–193
Cross-fostering, meadow voles, 369, 380
Crying, simian infant
  from biting by parents, 453–455
  eliciting maternal behavior, 428–429
  in socialization–neuroendocrine model, 460–462
Cues
  in assessment of brood number, 22
  environmental, effect on developmental events, 96
  facilitating parental regurgitation feeding, 191–193
  informational, 17–18
  internal and external, for caregiving duration, 28–30
  olfactory, 386–387, 395–399
  photoperiodic, 164
  pup
    distal, 316–318
    experiential aspects, 296–299
    odors, 303–305
  sensory, to stimulate care, 39–40
  situational, role of hormonal stimulation, 170
Cycloheximide, blocking of consolidation, 312

## D

Daughters
  adult monkey, maternal behavior, 600–602
  effect of infant births, 661–662
  sex ratio biased by, 520
Decision making, in parental caregiving by invertebrates, 17–18
Defense
  and ecological risks, 26
  nest and young, hormonal and neural mechanisms, 179–182
  against predators, role of helpers, 660

Definitions, parental care, 76–81, 479
  exhibited by teleost fishes, 66–67
Denervation
  in basolateral nucleus, 314
  effects on rat maternal behavior, 248–251
  infraorbital nerves, 253–266
  vomeronasal or chemosensory, 304
Density
  breeding, effect on primiparous females, 550
  neighbor, effect on extra-pair paternity, 497
  vasopressin receptors, in lateral septum, 374–376
  yolk, determination of fish life history, 70
2-Deoxyglucose, enhanced activation, 321
Desertion, parental, timing, 28–30
Development
  arrest, egg diapause as, 151
  fishes, ontogenetic intervals, 68–71
  offspring, monitoring by parents, 28–29
  rat, perspective on parental behavior, 215–217
  tadpole, 119–121
  vole litter, species differences, 366–369
Differential allocation, and sexual conflict, 494–496
Digging
  burrow, aspect of maternal behavior, 337–338
  hormone effects, 342–343, 346–348
5$\alpha$-Dihydrotestosterone, action on hair follicle, 352
Discrimination
  own versus alien young, 402–406, 412–413
  between twin lambs, 400–402
Distress calls, primate infant, 614–622
Distribution, parental care in amphibians
  geographic, 122–127
  phylogenetic, 111–113
  between sexes, 127–128
Diversity
  parental care in fishes, 54–65
  reptilian, 146–148
Diving skills
  fur seal young, 552
  harp seal pups, 544
Division of labor
  under biparental care, 37–38

Division of labor (*continued*)
  male behaviors related to infant care, 647–649
DNA fingerprinting
  data, on extra-pair paternity, 496–503
  naked mole rat, 657
Dominant Control Model, 676–679
Dopamine
  blockade, and induction of nursing posture, 269–270
  experience-related levels, 410
  role
    lagomorph nest building, 347
    maternal experience, 309–310
    rat parental behavior, 231–232
    sensorimotor responsiveness dependent on, 265–266
Dyads, mother–infant
  interaction with preadult females, 433–436, 461–462
  at Sepulveda colony, 590–591

## E

Earwig, nest building, 10–14
Ecdysteroids, effect on earwig parental cycle, 12–13
Ecological groups, fishes, constructed by Kryzhanovsky, 55–58
Ecological opportunity, for avian males, continuum, 484–486
Ecology
  associated variables of cooperative breeding, 667
  paternal care, future studies, 38–39
  pressures
    and expression of helping behavior, 680
    as prime mover of care, 4–6
  related circumstances, maternal style responsive to, 588–589
Eggs
  attendance, 123, 127
  attendance and transport, amphibian, 116–118
  dumping, 31–34
  protection, among invertebrates, 25–26
  reptilian
    amniotic, 146–148
    chamber, 150–151
    pattern of formation, 148–150
Egg size, and parental care
  amphibians, 128–129
  fishes, 86–89, 99
Elephant seal, differential investment, 556–559
*Eleutherodactylus,* egg attendance, 123, 127
El Niño, effect on pinnipeds, 534
Embryo
  amphibian, gestation in oviduct, 121–122
  fish
    development, 56–58
    period of ontogeny, 68
Encouragement, primate infants by mothers, 630–631
$\beta$-Endorphin, role
  maternal care at parturition, 236–237
  rat parental behavior, 228–231
Energetic load, of carrying, 653
Energy
  consumption, as function of pup mass, 542
  expenditures
    correlation with egg size, 87–89
    parental, 77–78
    relationship to maternal investment, 553–555
    reproductive, 82–85
    requirements for reproduction in fishes, 72–76
Environment
  conditions relating to parental care
    amphibians, 134, 136–137
    fishes, 85
    voles, 369–371
  explorational vocalizations, 616
  harsh and stable, as prime movers of care, 4–6
  quality, and maternal condition, 588–589
  related reproductive styles in fishes, 61
  role in lamb individual odor signature, 401–402
Epigenesis
  versus neo-Darwinism, 90–92
  versus preformation, 91
Epigenetic perspective, parental care in fishes, 53–99
Ergocornine, effect on lagomorph nest building, 343

INDEX 697

Estradiol, role
  avian incubation behavior, 166, 168–169, 174
  initiation of lagomorph maternal behavior, 341–348
  maternal responsiveness, 390–391
  rat parental behavior, 218–220
Estradiol benzoate, plus progesterone, 343–348, 355
Estradiol-17$\beta$
  regulation of maternal behavior, 441–460
  urinary levels, 450–454, 464
Estrogen, role in avian incubation behavior, 166, 176
Ethological forms, parental care, by L.S. Blumer, 63
Eusociality
  insects
    physiological regulation, 16–17
    termites, 32
  mole rat, 656–657
Evolution
  parental care
    avian female, 480–481
    invertebrate, 5–6, 20
    parent–infant communication, 633–635
    propagule, 87–88
    reproductive styles in fishes, 64, 92–95
    terrestrial breeding in amphibians, 130–136
    uniparental male care, 34
Experience
  functional adaptiveness, 297–299
  maternal
    effect on lamb recognition, 409–411
    and oxytocin receptor upregulation, 394
  play-mothering, 436–440
  prior
    correlation with prolactin, 665–666
    and infant survival, 669–671
  simian preadult, 423–425, 462–464
Experiential factors
  parental care, neurochemical elements, 233–235
  in postpartum regulation of maternal care, 295–325
Experimental strategies, steroid hormones, 342–348

Extra-pair paternity
  correlation with paternal provisioning, 511–514
  DNA fingerprinting data, 496–503
  intraspecific variation, 484–486

F

Face
  expressions, as primate communication, 628–630, 632–633
  trigeminal sensory innervation, 247–251
Fanning, fish eggs, during guarding, 92–93
Farallon Islands, phocid pupping, 548–551
Fasting strategy, phocid, 537–540
Fear, of newborn, by poor mothers, 450–451, 465
Fecundity, future, promoted by maternal rejection, 592
Feeding, see also Provisioning
  avian
    rates, 514
    by regurgitation, 190–193
    steroid hormone effects, 186–187
  invertebrate offspring, facilitation, 23–25
  mouth-opening phase, 250
  reptilian, switch-in-feeding preference, 153
  tadpole, 121
Females
  amphibian
    costs of caregiving, 135
    tadpole attendance and transport, 118–120
  control
    genetic paternity, 506–507
    reproductive capacities, 521
  fitness, and male parental care, 487–494, 509, 511
  higher and lower quality, 485–486, 505
  and males, differential investment in phocids, 556–560
  nonlaying hens, role of gonadal steroids, 184
  options, and genetic paternity, 515–516
  pinniped, interaction along kinship lines, 566–567
  preadult
    infant-directed behavior, 432–436
    species-typical experience, 437–440

Fertility
  otariid, costs, 552–554
  phocid, effect of primiparity, 548–551
Fertilization
  extra-pair, 485–486, 512
  internal, in fishes, 93–95
  mode, relationship to caregiving, 127–128, 139
  traded by females for resources, 504–505
Field studies
  bird parental care, 477–523
  helping effects on infant survival, 645–654
  voles, 362
Fishes
  life history, general model, 65–72
  parental care
    diversity, 54–65
    epigenetics, 90–98
    review of recent work, 81–90
  reproduction, energy requirements, 72–76
Fitness
  benefits, of helping, 674
  costs, of maternal investment, 546–548, 556
  female, and male parental care, 487–494, 509, 511
Flexibility
  amphibian parental care, 136–138
  timing of weaning, 543
Food
  abundance, and maternal mass, 565
  appropriate amount per offspring, 20–21
  begging, 628–629, 649
  decreased intake
    as cost of parental care, 133
    transitory and reversible, 339–340, 355
  lagomorph fecal pellets, 336
Food resources
  availability, and maternal condition, 588–589, 679
  and paternal care, 36–38
  for salamander hatchlings, 131
  scarce and specialized, as prime mover of care, 4–6
Foraging
  energetics, 554
  reduced
    associated costs, 36
    as cost of care, 31
  strategy, otariid seals, 537, 540–544

Fos-lir, pattern in brain, 314–320
Free-ranging settings
  and enterprising infants, 596
  monkey studies, 582–583, 586–587
Frogs
  parental care, geographic distribution, 123–127
  poison dart tadpole
    feeding, 121
    transport, 134
  postoviposition, 113–115
  tropical tree, egg-brooding, 117–118
Fur seal, differential investment, 560–562

## G

Galliform species
  incubation, prolactin and hormone role, 165–168
  parental care expression, 161–162
Gamma-aminobutyric acid, increase after birth, 408–410, 415
Gastric brooding, as tadpole transport, 120–121
Generational continuity, maternal style, 601–602
Genetic parentage, avian species, molecular studies, 498–502
Genital display, by young squirrel monkeys, 627–628
Genotype, lamb, in olfactory signature, 400–401
Geographic distribution, parental care in amphibians, 122–127
Gestation, internal, in amphibian oviduct, 121–122
Glance rate, in restrictive primate mothers, 602
Glutamate, increase after birth, 408–409, 415
Gonadal steroids
  disruption of broody care, 184–185
  effect on sitting behavior in doves, 172
Gonadotropins, levels during egg laying, 167
Grey seal, differential investment, 559–560
Growth rate, otariid pups, 541–542, 560–561
Growth strategies, pinniped young, 568–569
Guarding
  nest, by invertebrate males, 37–38

transition from nonguarding in fishes,
        92–93

## H

Hair pulling, lagomorph maternal behavior,
        338–339
    hormonal effects, 342–343, 347, 352–354
Haloperidol, cataleptic dosage, 269–270,
        274
Hatching, *see also* Posthatching
    asynchrony and synchrony, 495–496
    timing, 151
Hatchlings
    crocodylian, vocalization, 153
    salamander, food supply, 131
Helpers
    mechanisms of helping, 644–661
    nonbreeding adult male, hormonal
        correlates, 188
Hippocampus
    in expression of maternal behavior,
        311–312
    norepinephrine content, 232–233
Historical perspective, maternal behavior in
    rabbits, 333–356
Hormonal mechanisms
    avian incubation behavior, 162–176
    defense of nest and young, 179–181
    in expression of helping, 662–667
    parental responses toward young,
        182–196
    rat parental behavior, 217–225
Hormones
    lactogenic, role in rat parental behavior,
        221–225
    parturitional, 299–302, 316–317
    and primate maternal behavior, 441–460
5-Hydroxyindole-acetic acid, relation to
    maternal protectiveness, 604
Hydroxypregnanolone, prepartum urinary
    levels, 453–454
Hyperphagia, by ring dove parents,
    193–199
Hypothalamic oxytocinergic system, in
    lagomorphs, 353
Hypothalamo–pituitary–adrenal axis,
    prolactin effect, 194
Hypothalamus, ventromedial, oxytocin
    receptor levels, 226–227

## I

Incubation, avian
    hormonal and neural mechanisms,
        162–179
    by males of monogamous species, 481
Independence
    developed by rejected infants, 594–595
    marmoset infants, 653
Individual differences, in maternal style,
    579–606
Infant, *see also* Offspring; Young
    care, in cooperatively breeding species,
        643–683
    direction of
        maternal behavior, 448–460
        preadult female behavior, 432–436
    effects of variation in maternal style,
        593–596
    at great risk, and protectiveness of
        mother, 586–587
    primate
        characteristics, 428–429
        vocalizations, 614–622
Information
    chemical, from mother to young via milk,
        335–336
    conveyed by infant calls, 621
    gathering, parental, constraints imposed
        by, 19–20
Inhibitory mechanisms, reproduction by
    helpers, 675–683
Intromittent organs, avian, and uniparental
    male care, 479–480
Invertebrates
    eusocial, 16–17
    mechanistic studies, 17–19
    parental care
        physiology, 6–19
        prime movers, 4–6
        theory, 19–33
    paternal care, uniparental versus
        biparental, 33–39
Investment
    differential, in male and female offspring,
        555–566
    equality, and sexual conflict, 494–496
    high parental costs, 27
    parental
        pinniped, 533–570
        reptilian, 148–153
        R.L. Trivers on, 77–79

Investment (*continued*)
  paternal
    in biparental invertebrates, 35
    manipulation by female, 490–491
    postweaning, 566–567
Isolation, physical, effect on maternal behavior, 434–436

**J**

Jamaican land crab, multiple functions of parental care, 4–5
Jaw opening
  following perioral contact, 250
  infraorbital nerve denervation effect, 253
  prolonged contact prior to, 255–256
Juvenile
  fish, period of ontogeny, 69–71
  lagomorph, female aggression toward, 341
  monkey
    carrying of infants, 650–651, 655
    gaining contact with infants, 661–662
    response to novelty, 596–598
  pinniped, development, and maternal care patterns, 567–568
  rat, parental behavior, 223
  vole, alloparental behavior, 365–366
Juvenile hormone
  high levels in burying beetle, 14–16
  during pregnancy in viviparous cockroach, 8–10
  topical application to earwigs, 10–14

**K**

Kinship
  lines, female pinniped interaction along, 566–567
  predictor of helping, 659
Kryzhanovsky, S.G., ecological grouping of fishes, 55–58
Kyphosis
  characterization, 245
  spinal pathway, 270–271
  state changes during, 286

**L**

Laboratory studies
  rabbit maternal behavior, 336–341
  vole social organization, 361–363
Lactation
  early, anesthesia during, 349–350
  foraging during, 543–544
  phocid, duration, 539–540
  separation from foraging, 554
Lamb
  amniotic fluid-covered, 396–397
  individual odor signature, 399–402
  recognition, olfactory mediation, 397–399, 406–411
Language, female control, 506–507
Larva, fish, period of ontogeny, 68–69
Latency
  to become maternal, 309, 313
  to display parental behavior, 216, 221–224
  to enter new enclosure, 597
  response, to infant crying, 454–455
  of sensitization, 252
Lateral septum
  c-fos expression, 372
  vasopressin receptor density, 374–377
Learning, social
  about infant stimuli, 462–463
  imitative and nonimitative, 434–436
Licking
  anesthesia and denervation effects, 256–259
  lamb anogenital region, 387–388
  during parturition, 284
  role of pup odors, 304–305
Lidocaine
  effect on licking of pups, 257
  induced anaptia, 253, 255
Life history
  components, and resource allocation, 30–31
  determinant of maternal style, 586–589
  fishes, general model, 65–72, 95–96
  nonparental, 26–27
  strategy
    pinniped females, 553
    size-dependent switches, 18
    and vole behavioral development, 367–368

Lifestyles, primate, in maternal behavior regulation, 427–428
Lip smacking, in monkeys and apes, 627–631
Lithophilous fishes, embryo development, 56
Litter
　alteration, effect on nursing behavior, 268–269
　development rate, vole species differences, 366–369
　interaction with mother, prior to nursing bout, 261
　separation from mother, 300
Lordosis, cutaneous stimuli, 250

## M

Madingley colony, primate mother–infant relationships, 580–583
Males
　amphibian, tadpole attendance and transport, 118–120
　avian
　　helpers, hormonal correlates, 188
　　nonincubating, plasma prolactin levels, 164, 178
　　parental care, 487–494
　caregivers, urinary prolactin levels, 663–667
　and females, differential investment in phocids, 556–560
　infant carrying and vigilance behaviors, 647–651
　invertebrate, uniparental versus biparental care, 33–39
　lagomorph, protective of juveniles, 341
　monkey adolescent, response to strangers, 598–600
　vole
　　fathers, presence in natal nest, 368
　　paternal behavior, vasopressin effect, 376–377
Mammals
　nonprimate, maternal behavior, endocrine regulation, 441–447
　patterns of helping, 654–657
Manipulation
　female birds by males, 492–493
　lamb odor, effects on maternal acceptance, 403–405
　male birds by females, 489–492
Mapping, neurophysiological, somatosensory cortex, 278–281
Marmosets
　benefits of helping, 669–673
　infant care, 649–654
Mate
　compensation, in burying beetle, 38
　copying, in invertebrates, 34
　guarding, by male birds, 493, 507
Maternal behavior
　dependence on perioral stimuli, 252–262
　endocrine regulation, 441–447
　neurobiological consequences, 277–286
　physiology, 299–322
　pinniped, phocid and otariid, 536–544
　primate
　　individual differences, 583
　　infant-directed, 448–460
　　and preadult socialization, 432–441
　　regulation, 426–432
　　species-typical, 464
　rabbit
　　initiation, hormonal role, 341–348
　　laboratory studies, 336–341
　　maintenance, 348–351
　rat, and somatosensation, 244–252
　recovery of function, 262–266
　vole
　　direct and indirect, 363
　　oxytocin role, 377–380
Maternal care
　avian, variation, 516–517
　odobenids, 543
　at parturition, $\beta$-endorphin role, 236–237
　postpartum, phases, 349
　regulation
　　postpartum, experiential factors, 295–325
　　in primates, 426–432
　and somatosensation, Norway rat, 243–288
Maternal investment
　phocid and otariid pinnipeds, 546–553
　relation to energy expenditure, 553–555
Maternal mass
　correlation with pup mass, 564–566, 569

Maternal mass (*continued*)
  effect on pup mass at birth, 557–558
  and reproductive costs, 546–547
Maternal style
  consequences of variation, 591–602
  individual differences, 580–591
Mating
  harrassments, 659
  multiple, by female birds, 503–506
  reduction of opportunities, 133
  relationship to parental care in fishes, 85–86
Medial preoptic area, *see also* Preoptic area
  c-fos expression, 234
  estradiol effect, 218–220
  estradiol-17$\beta$ receptors, 444
  in expression of maternal behavior, 314–320
  in induction of maternal responsiveness, 392–394, 413
  interaction with trigeminal afferents, 285–286
  prolactin infusion, 221–225
  in vole parental behavior, 371–372
Memory, maternal, genomic activation, 234–235
Microtine rodents, parental behavior, 361–380
Milk, *see also* Crop milk
  chemical information from, 335–336
  ejection
    during nursing bout, 273
    and oxytocin release, 311–312
  intake
    by otariid pups, 561
    and rabbit pup weight gain, 340
  secreted by cockroach brood sac, 8–9
Minisatellites, single- and multilocus, 513–514
Mitral cells
  complex with granule cells, 411
  transmission of olfactory signals, 407–408
Mobbing
  predator
    by groups of avian helpers, 660
    by male birds, 512
  shark, by sea lions, 545–546
Models
  based on sexual conflict, 494
  dominant control, 676–679

  hierarchical, for energy expenditure, 74
  life history of fishes, 65–72, 95–96
  rat
    biochemical regulation of parental care, 235–237
    hysterectomy–ovariectomy, 218–220
    nonhypophysectomized, 221–222
    self-restraint, 678–679, 683
    socialization–neuroendocrine, 460–462
Mole rat, eusociality, 656–657
Mongoose, dwarf, patterns of helping, 655–656
Monkeys
  maternal style, individual differences, 579–606
  mother–infant communication, 613–635
  Old World and New World, 427–465
  patterns of helping, 644–655
Monogamy, social
  adaptive significance, 487–494
  with genetic monogamy, 497–502
Mortality
  fish offspring, 88–89
  helpers, 674
  infant, association with maternal neglect, 594
  juvenile invertebrate, 21–22
  phocid females, effects of early reproduction, 550–551
Mothering
  absentee system in lagomorphs, 333–335
  appropriate, and functional adaptiveness of experience, 297–299
  play
    experience, 436–440
    preadult female behavior, 433–436
  style, juvenile monkeys, 600
Mother–pup interaction
  effect on Fos-lir distribution, 316–320
  neural site activation during, 282–283
  physical, 251–252
  after trigeminal denervation, 263–265
Mother–young interaction
  contact
    and bonding, 398
    as function of infant risk, 586–587
    thermal limitation, 272–273
  individual recognition, 320–322
  role of vocalizations, 632
  subsequent to parturition, 348–351

Motivation
  to look at infant replica, 457
  maternal, effect of prepartum state, 449
Motivational states, including attraction and neophobia, 462–463
Mouthbrooding, evolution, 93–95
Muzzling, effect on pseudo-licking, 259–261
Mystacial pads
  anesthesia, 260–261
  saline-injected, 253–255

## N

Naloxone, injections concurrent with morphine, 229
Natality, otariid primiparae and multiparae, 551–552
Natural selection
  assumed in safe harbor hypothesis, 88–89
  production of proximate mechanisms, 602–604
Neighbor density, effect on extra-pair paternity, 497
Neo-Darwinism, versus epigenesis, 90–92
Neonate, sheep, maternal responsiveness, 387–397
Nerves
  infraorbital, denervation, 253–266
  trigeminal sensory, innervation of trunk and face, 247–251
Nest
  avian
    conspecific parasitism, 490–491
    defense, 179–182, 511
  deprivation, and sitting interest, 170–171
  lagomorph, maternal, 333–341, 353–354
Nest building
  earwig, 10–14
  and provisional behavior, 24–25
  reptilian, and parental investment, 152
Nestlings, provisioning
  paternal decisions, 513–514
  patterns, 481–482
Neural mechanisms
  avian incubation behavior, 176–179
  defense of nest and young, 181–182
  parental responses toward young, 196–199

Neural sites, activation during interaction with pups, 282–283
Neuroanatomy, maternal experience, 310–322
Neurobiology
  lamb odor recognition, 406–411
  maternal behavior, 277–286
  parental behavior, 371–380
  rat somatosensory system, 247–249
Neurochemistry
  basis of vole parental behavior, 372–380
  and experience, in rat parental care, 233–235
  maternal experience, 307–310, 323–324
Neuroendocrine regulation
  maternal behavior
    nonprimate mammals, 441–447
    simian, 425–426
  sheep maternal responsiveness, 390–394
Neurosecretory products, in earwig parental care, 12–13
Nipples
  alteration, effect on nursing behavior, 268–269
  pheromone, 335, 344–345, 348, 355–356
  searching-related vocalizations, 619
Noradrenaline, release at parturition, 410–412
Noradrenergic system
  activation at parturition, 389–390, 394
  in consolidation of maternal experience, 323–324
  role in learning and memory, 308–309
Norepinephrine, role in rat parental behavior, 232–233
Novelty, responses of primate infants and juveniles, 595–598
Nursing
  behavior changes, and cortical plasticity, 278–282
  dam-initiated, 259–261, 285–286
  deficit, 350–351
  and lagomorph absentee mothering, 334–335
  primate caregiver calls during, 623–626
  related infant vocalizations, 615–617
  somatosensory alteration effects, 257–258
  stages, 245–246
  suckling- and nonsuckling-induced, 267–270

Nursing bout
  duration in lagomorphs, 340
  sequence of mother-litter interactions prior to, 261
  termination, 271-274
Nurturance, maternal, and trigeminal somatosensation, 252-267

## O

Odors
  lamb
    individual signature, 398-406
    recognition, neurobiology, 406-411
  maternal, within nest, 321-322
  as pup cues, 303-305
Odor-taste, as simian infant stimulus, 460-462
Offspring, see also Infant; Young
  amphibian, and parental care flexibility, 136-138
  behavior molding, 634
  consequences of variation in maternal style, 593-602
  fishes
    guarded and nonguarded, 61
    survivorship, 88-89
  invertebrate
    development monitoring by parents, 28-29
    feeding facilitation, 23-25
    parental behavior toward, 518-519
  pinniped male and female, differential investment, 555-566
  reptilian, limit on mass, 149-150
  survival, genetic paternity as function of, 508
Olfactory bulbs
  noradrenergic innervation, 308
  oxytocin increase during parturition, 389-390
  specialized changes in, 321-322
  synaptic organization, 407-409
Olfactory mediation, lamb recognition, 397-399, 406-411
Ontogeny
  saltatory, theory, 67-68, 95-96
  as spiral trajectory, 71-72

Operant behavior, relation to blood hormone levels, 455-457
Opioids
  effects on affiliative behavior, 229-230
  neural sensitivity, shifts as function of parity, 234-235
  promotion of maternal responsiveness, 393
Origins hypotheses, and sexual conflict, 519-523
Orosensations, trigeminal, role in maternal behavior, 265-266
Ostracophilous fishes, embryo development, 58
Otariids
  breeding patterns, 540-544
  costs of maternal investment, 551-553
Oviduct, reptilian, modifications, 148-149
Oviposition
  burying beetle, 14-15
  earwig, 11-13
  relation to egg attendance, 133-134
Ovulation, inhibition, 676-677
Oxytocin
  cell bodies immunoreactive to, 353
  cerebrospinal fluid levels at parturition, 392-393
  increase in olfactory bulbs during parturition, 389-390
  induced suckling, 311-312
  regulation of maternal behavior, 441-446
  role
    rat parental behavior, 225-227
    vole maternal behavior, 377-380

## P

Parasitism
  brood, among invertebrates, 22-23
  conspecific, avian nest, 490-491
Paraventricular nucleus
  dopamine release during parturition, 389
  in induction of maternal responsiveness, 392-394, 413
  site of oxytocin activity, 225-226
Parental behavior
  avian
    fiddling, 518-521
    neural and hormonal control, 161-202

piscine, relating to parental care, 85
rat, biochemical basis, 215–237
vole
   environmental factors, 369–371
   neurobiological basis, 371–380
   patterns, 363–366
   social organization, 361–363
Parental investment
   in pinnipeds, 533–570
   reptilian
      behavioral types, 152–153
      physiological types, 150–152
      structural types, 148–150
   R.L. Trivers on, 77–79
Parent–offspring conflict, and termination of care, 27–30
Parent–parent conflict, in birds, 520–523
Parity effect, and infant survival, 670
Pars intercerebralis, neurosecretory products, 12–13
Partner preference, among voles, 362–363
Parturition
   endocrine changes during, 445–448
   and establishment of maternal behavior, 261–262
   hormones, 299–302
   lagomorph maternal behavior
      prior to, 336–340
      subsequent to, 348–351, 354
   licking behavior during, 284
   physiological changes during, 388–390
   process in sheep, 386
Paternal care
   avian, and genetic paternity, 507–517
   invertebrate
      courtship, 33–36
      future ecological studies, 38–39
      resources, 36–38
   pinniped, 545–546
   vole
      association with monogamy, 363–365
      vasopressin effect, 376–377
Paternity
   extra-pair
      correlation with paternal provisioning, 511–514
      DNA fingerprinting data, 496–503
      intraspecific variation, 484–486
   genetic
      female control, 506–507
      and female options, 515–516
   and paternal care, 507–517
Patterns
   copulation, in biparental invertebrates, 34
   defensive, specific to avian sex, 181
   Fos-lir in brain, 314–320
   helping
      birds, 657–661
      callitrichid primates, 644–654
      mammals, 654–657
   oxytocin receptors in vole brain, 377–380
   phylogenetic, bi- or uniparental care, 480–482
   pinniped parental care, 536–546
   prolactin secretion, posthatching, 187–196
   reproduction in fishes, classification, 60–63
   reptilian egg formation, 148–150
   simian maternal behavior, 429–431
   vole parental behavior, 363–366
Pelagophilous fishes, positive buoyancy, 56–58
Phenotype
   maternal behavioral, 634
   related aspects of parents, adaptiveness, 145–146
Pheromones
   brood, 29
   nipple, 335, 344–345, 348, 355–356
   trail and alarm, 26
Phocids
   breeding patterns, 537–540
   costs of maternal investment, 548–551
   differential investment, 556–560
Phylogenetic tree, living primates, 424
Phylogeny
   and parental care distribution in amphibians, 111–113
   pattern of parental care, 480–481
   pinniped, 535
   related bias, among anurans, 127–128
   and study of invertebrate parental care, 39–40
Physiology
   associated parental investment in reptiles, 150–152
   basis of rat parental behavior, 217–233
   invertebrate care, 6–19
   maternal behavior, 299–322
   mechanisms of maternal style, 603–604

Physiology (*continued*)
  regulation of galliform nest defense, 180–181
  related changes during parturition, 388–390
Phytophilous fishes, embryo development, 56
Pinnipeds, parental investment, 533–570
Plasticity
  behavioral, physiology, 6–7
  cortical, and nursing behavior changes, 278–282
  sensory factor role, 295
Play
  associated facial expressions in primates, 628–629
  mothering
    experience, 436–440
    preadult female behavior, 433–436
Polygamy, avian, and parental care, 493–494
Polygyny
  in pinnipeds, 533
  potential, under paternal care, 34–36
  threshold model, 494
Posthatching, *see also* Hatching
  prolactin
    decline, 183–185
    secretion patterns, 187–196
Postoviposition, *see also* Oviposition
  in amphibians, 111–113, 138–139
  anurans, 124–125
  voluntary behaviors as, 110
Postpartum state
  maternal aggression, 216–217
  maternal care
    phases, 349
    regulation, experiential factors, 295–325
  patterns of simian maternal behavior, 430–431
Pouting, as communication in young primates, 627–628
Precocial state
  and altricial state, in ontogeny, 96–98
  avian parental responses to, hormonal effects, 182–185
Predation
  defense by avian helpers, 659–660
  increased vulnerability to, 133
  as prime mover of care, 4–6
  risk
    as cost of care, 31
    for phocid seals, 539–540
    shark, 545–546
Preformation, versus epigenesis, 91
Pregnancy
  endocrinology, 447–448
  related estradiol-17$\beta$ levels, 451
  viviparous cockroach, 8–10
Preoptic area, *see also* Medial preoptic area
  lesions, effect on incubation behavior, 177–178
  prolactin receptors, 196–198
Prepartum state
  effect on maternal motivation, 449
  estradiol-17$\beta$ levels, 451–454
Primates
  callitrichids, patterns of helping, 644–654
  maternal style
    consequences of variation, 591–602
    individual differences, 580–591
  mother–infant communication, 613–635
  nonhuman simian, maternal behavior regulation, 423–465
Prime movers of care, 4–6
Primiparity, effect on phocid female fertility, 548–551
Principal components analysis, maternal behavior of monkeys, 581–582
Progesterone
  induced maternal responsiveness, 391
  regulation of maternal behavior, 441–460
  role
    avian incubation behavior, 166, 168–170, 174
    initiation of lagomorph maternal behavior, 341–348, 355–356
    rat parental behavior, 220–221
Prolactin
  cerebral binding sites, 178–179, 196–197
  in initiation of lagomorph maternal behavior, 341–348
  plasma
    during avian incubation, 162–165
    role in incubation behavior, 165–176
  posthatching
    decline, 183–185
    secretion patterns, 187–196
  regulation of maternal behavior, 441–448

role in parental behavior
  birds, 200–202
  rats, 221–225
urinary, in caregivers, 662–667, 682
Protection
  as function of egg attendance, 116–117
  young, among invertebrates, 25–27
Protectiveness
  maternal
    and delay of next conception, 592–593
    scores, 589–591
    and rejection, in primate mothers, 580–586
Provisioning, *see also* Feeding
  adjustment, among wasps, 20–21
  behaviors, in invertebrate nest builders, 24–25
  nestlings by parents, 482
  offspring by males, 37
  paternal
    correlation with extra-pair paternity, 511–514
    and genetic paternity, 509–511
  paternal and maternal
    facultative variation, 483–487
    increase, 489–493
  young and mates by passerine species, 185–186
Psammophilous fishes, embryo development, 56
Psychological variables, as proximate mechanisms, 602–603
Pucker, as primate communication, 628–629, 631
Punishment, female birds, by withdrawal of male care, 509–511
Pups
  cues
    eliciting maternal behavior, 297–299
    odors, 303–305
  killing, corticotropin-releasing factor effect, 227–228
  otariid, growth rate, 541–542, 560–561
  phocid
    differential investment, 556–560
    mass at weaning, 539
  reinforcement, 305–306
  retrieval and licking, 252–259
  role in maternal investment, 564–565

stimulation of maternal ventrum, 267–268
Purr, during nursing
  by mother, 623
  by primate infant, 615, 617, 621

**Q**

Quality
  female birds
    higher and lower, 485–486
    variation, 492–493
  female phocids, and weaning success, 557
  habitat, and maternal condition, 588–589

**R**

Rabbit, maternal behavior, 333–356
Rat
  maternal behavior patterns, 443–447
  Norway, somatosensation and maternal care, 243–288
  parental behavior, biochemical basis, 215–237
Recognition
  individual
    and chemosensory cues, 305
    and maternal responsiveness, 307–310
    between mother and young, 320–322
  lamb
    disruption, 403
    olfactory mediation, 397–399, 406–411
  selective, sheep and goats, 298
Recovery of function, after neural damage, 262–266
Recruitment calls, by primate infants, 622
Reflexes, somatosensory, and species-specific behavior, 249–250
Regulation
  biochemical, rat parental care, 235–237
  brood number in invertebrates, 19–23
  caregiving duration, 28–30
  maternal behavior
    socialization component, 463
    steroid hormone role, 342–348
    testosterone role, 352

Regulation (*continued*)
  maternal care
    postpartum, experiential factors, 295–325
    in primates, 426–432
  physiological
    eusocial insects, 16–17
    galliform nest defense, 180–181
  thermal, and rapid litter development, 368
Regurgitation, feeding of ring dove young, 190–193, 195–199
Reinforcement
  dopamine role, 310
  pup, 305–306
Rejection
  alien young, 398–399
  infants, 652
  maternal
    promotion of future fecundity, 592
    rates, 601
    and protectiveness, in primate mothers, 580–586
Remodeling, fish larvae, 68–69
Removal studies, male birds, 488–489
Reproduction, *see also* Breeding
  amphibians, terrestrial modes, 129
  experience, relation to parental care, 233–235
  fishes
    energy requirements, 72–76
    evolution of styles, 64, 92–95
    presented by Breder and Rosen, 58–60
    by helpers, inhibitory mechanisms, 675–680
  insects, physiology, 7–19
  pinnipeds, costs, 546–555
  related hormones, and infant-directed behavior, 449–460
  reptilians
    and amniotic condition, 147–148
    reversals, 146
Reproductive guilds, E.K. Balon's concept, 60–63
Reproductive state, relationship to infant-directed behavior, 448–449
Reproductive success
  amphibians, 134–135
  avian females, 489
  helpers, 669, 681–682

Reptilia, parental care, 145–155
Research
  cooperative breeding, future directions, 680–681
  on parental care in fishes, 83–84
Resources
  allocation, and life history components, 30–31
  availability, assessment by invertebrate parents, 19–21
  brokered by male birds, 504–505
  local, competition, 520
  maternal, yearly variation, 562–563
Respiratory organs, embryonic, in fishes, 56–58
Responsiveness
  maternal
    and individual recognition, 307–310
    to infant calls, 619–620
    neuroanatomy, 310–320
    postpartum, 300
    role of sensory input, 302–305
    sheep, 387–397
  parental
    invertebrates, 29
    toward avian young, 182–199
    voles, 366
  sensorimotor, dopamine-dependent, 265–266
Restriction, maternal
  correlation with rejection, 581–583
  and glance rate, 602
Retention testing, dam interaction with pups, 300–301
Retrieval
  associated calls by monkey caregivers, 623, 625
  dopamine role, 231–232
  effect of long separation, 264–265
  elicitation by perioral stimuli, 252–256
  emergence from mouthing, 284
  infant, via gestures, 629–630
  involving motor mechanisms, 287–288
  maternal
    in response to infant calls, 619
    species-typical, 437–440
  rat pups, 245
Retrieve-to-bite interval, in primates, 454–455

Ring dove
  incubation, prolactin and steroid
    hormone role, 168–172
  parental hyperphagia, 193–199
  prolactin effect, 195–196
  young, feeding by regurgitation, 190–193
Risk
  ecological, and defense mechanisms, 26
  infants exposed to
    and protectiveness of mother, 586–587
    relation to independence, 595
  predation
    as cost of care, 31
    for phocid seals, 539–540
RNA, messenger
  prolactin receptor, 224
  vasopressin, 374–376

S

Safe harbor hypothesis, 88–89
Salamanders
  parental care, geographic distribution, 123
  postoviposition, 111–112
  terrestrial breeding, 130–131
Salt-marsh beetle, coping with harsh environment, 4
Sargent, R.C. et al., definition of parental care, 80
Scattering, lagomorph, as deficient behavior, 337
Scent, foreign, masking olfactory signature, 402–403
Science Citation Index, 81
Seals, parental investment, 533–570
Selectivity, maternal, to neonate, 397–411
Self-Restraint Model, 678–679, 683
Semelparous species, and parental care theory, 30
Sensitization, pup-induced, 252
Sensory control, maternal responsiveness, 395–397
Sensory input
  from brood patch, 176–177
  role in gaining maternal experience, 302–306
Sensory processing, olfactory signals, 406–409

Separation
  calls elicited by, 617–619
  infant responses to, 681
  mother–litter, 300
  from pups, effect on retrieval, 264–265
Sepulveda colony, vervet monkeys, social behavior studies, 581–583, 586–591, 593–602
Serotonin
  related to monkey aggressiveness, 604
  role in rat parental behavior, 233
Sex
  amphibian, and parental care distribution, 127–128
  avian, defensive patterns specific to, 181
  avian helpers, 658
  pinniped offspring, differential investment, 555–566
  primate infant, effects on maternal style, 587–588
  reptilian, temperature-dependent determination, 151–152
Sex ratio
  seasonal changes, 559
  theory, 562–566
Sexual conflict
  and differential parental allocation, 494–496
  and origins hypotheses, 519–521
Sheep
  and goats, selective recognition, 298
  individual recognition, 321
  maternal behavior, endocrine regulation, 445–446
  parental care, 385–416
Siblings, effect on amount of maternal carrying, 651
Signature, odor, lamb, 399–406
Sitting activity
  correlation with prolactin fluctuations, 162–165
  progesterone and prolactin effects, 169–171
Size
  body
    female parent bug, 20
    pinniped mother, 539–541
    relationship to reproductive strategy, 18
  clutch
    adjustment, 21

Size, clutch (*continued*)
    amphibian, 129, 137
    reptilian, 149
    and subsequent reproduction, 32–33
    egg, and parental care
        amphibian, 128–129
        fishes, 86–89, 99
Skin, somatosensory cortex representation, 279–281
Sniffing, intruder, by rat mother, 275–277
Socialization, preadult, and maternal behavior, 432–441
Socialization–neuroendocrine model, 460–462
Social opportunity, for avian males, continuum, 484–486
Social organization, voles, 361–363
Somatosensation
    and maternal aggression, 275–277
    and maternal behavior, 244–252
    trigeminal, and maternal nurturance, 252–267
    ventral trunk, and nursing behavior, 267–275
Spawning, behaviors specific to, 75–76
Species differences, vole, in litter development rate, 366–369
Species richness, anurans, 123, 127
Spinal pathway, kyphosis, 270–271
Squab, prolactin release induced by, 189
State changes, in relation to maternal behavior, 283–286
Steroid hormones
    effect on parental activity expression, 200
    in regulation of maternal behavior, 342–348
    role in incubation behavior, 165–176
Stimuli
    artificial infant, 455–457
    cutaneous, lordosis, 250
    eliciting prolactin, 665–666
    infant, major groups, 460–462
    perioral
        maternal behavior dependent on, 252–262
        and ventral trunk, role in maternal aggression, 275–277
    pup-associated, 322–323
    somatosensory, 287–288, 354–355
    suckling, alterations, 273–274

vaginocervical, 391–394
visual, from incubating partner, 171
Strangers, adolescent male monkey response, 598–600
Strategies
    adaptive, socialization of maternal behavior as, 440–441
    courtship, infant care as, 652, 673
    experimental, steroid hormones, 342–348
    fasting and foraging, 537–544
    growth, pinniped young, 568–569
    life history
        pinniped females, 553
        and vole behavioral development, 367–368
    maternal, energetic aspects, 534–535
    reproductive, relationship to body size, 18
Stress, infant caretaking, 663–664
Substitution, behavioral, 263, 287
Suckling
    hormone release induced by, 298–299
    and nonsuckling, nursing induced by, 267–270
    oxytocin-induced, 311–312
    stimuli, alterations, 273–274
Survival
    effect of avian helpers, 657–661
    infant
        effects of helping, 645–646
        role of care-helping experience, 439–440
    offspring, genetic paternity as function of, 508
    phocid female, effect of primiparity, 548–551
Survivorship, offspring
    amphibian, 139
    fishes, 88–89
Synchronization
    hatching, 495–496
    ontogenetic changes, 96
Systematics, pinniped, 535

**T**

Tadpole
    attendance and transport, 118–121
    feeding, 121

Tamarins
  infant care, 649–654
  costs and benefits, 668–673
  prolactin data, 666–667
Tantrums, by primate infants, 620
Teleost fishes, definitions of parental care, 66–67, 79–80
Temperament, individual differences, 579–580, 603–605
Temperature, sex determination dependent on, 151–152
Tending behavior, in fishes
  associated with reproduction, 75–76
  research topics, 86
Termination
  care, and parent–offspring conflict, 27–30
  nursing bout, 271–274
Territory, inheritance by helpers, 679–680
Testosterone
  effect on sitting behavior in doves, 172, 176
  levels throughout pregnancy, 341–342
  in regulation of maternal behavior, 352
Theory
  epigenetic, 90–92
  genetic paternity and paternal care, 507–511
  parental care, and invertebrates, 19–33
  saltatory ontogeny, 67–68, 95–96
  sex ratio, 562–566
Thermal limitation, mother–young contact, 272–273
Thermal regulation, and rapid litter development, 368
Timing
  avian sitting period, 171
  hatching, 151
  parental desertion, 28–30
  parental energy expenditures, 77–78
  postpregnancy maternal behavior, progesterone role, 220–221
  weaning, flexibility, 543
Toads
  parental care, geographic distribution, 123–127
  postoviposition, 113–115
Touching, as primate communication, 630
Trade-offs
  in current and future reproduction, 508–509
  between offspring production and infant mortality, 605
  between paternal provisioning and mate attraction, 483–484
Transport, egg and tadpole, 117–121
Trivers, R.L., definition of parental investment, 77–79
Trunk, trigeminal sensory innervation, 247–251
Turkey, nesting, steroid effects, 173–174
Twins, lamb, and phenotypic characteristics, 400–401

U

Uniparental care
  amphibian, costs and benefits, 133–135
  avian, female and male, 479–481
  male
    ancestral, 518
    versus biparental care, 33–39
Urination, marine turtle, over egg chamber, 151

V

Vaginocervical stimulation, ewes, 391–394
Value judgments, about maternal style, 605
Variation
  extra-pair paternity patterns, 503
  facultative, in paternal and maternal provisioning, 483–487
  intraspecific and intrapopulational, 521–522
  maternal style, consequences for mother and infant, 591–602
  in primate maternal care, 581–583
Vasoactive intestinal polypeptide, immunization of birds, 167
Vasopressin
  levels in male voles, 666
  in vole parental behavior, 374–377
Ventral trunk, somatosensation, and nursing behavior, 267–275
Ventromedial hypothalamic nucleus
  estradiol receptors, 220
  lesions, effect on prolactin, 177
  prolactin injections, 198–199

Vigilance, and infant survival, 647–649
Visual appearance, as infant stimulus, 460–462
Visual nervous system, primate, 429
Visual stimuli, from incubating partner, 171
Vocalizations
  crocodylian hatchlings, 153
  primate infant, 614–622
  primate parent, 622–626
Voles
  male, vasopressin levels, 666
  parental behavior, 361–380
Vomeronasal organ, lesions, 395–396

## W

Walruses, parental investment, 533–570
Washing, lamb, disruption of ewe recognition, 403–404
Wasps, adjustment of provisioning, 20–21
Weaning
  in otariid species, 541–542, 561–562
  phocid pup mass at, 539, 557–559
  primate, ages, 431–432
Whimper
  by infant and mother, 617–620
  mother response to, 625
Withdrawal, paternal care, 508–511

## Y

Yolk density, determination of fish life history, 70
Young, *see also* Infant; Offspring
  avian
    defense mechanisms, 179–182
    parental responses, 182–199
  invertebrate, protection, 25–27

# Contents of Previous Volumes

**Volume 15**

Sex Differences in Social Play: The Socialization of Sex Roles
MICHAEL J. MEANEY, JANE STEWART, AND WILLIAM W. BEATTY

On the Functions of Play and Its Role in Behavioral Development
PAUL MARTIN AND T. M. CARO

Sensory Factors in the Behavioral Ontogeny of Altricial Birds
S. N. KHAYUTIN

Food Storage by Birds and Mammals
DAVID F. SHERRY

Vocal Affect Signaling: A Comparative Approach
KLAUS R. SCHERER

A Response-Competition Model Designed to Account for the Aversion to Feed on Conspecific Flesh
W. J. CARR AND DARLENE F. KENNEDY

**Volume 16**

Sensory Organization of Alimentary Behavior in the Kitten
K. V. SHULEIKINA-TURPAEVA

Individual Odors among Mammals: Origins and Functions
ZULEYMA TANG HALPIN

The Physiology and Ecology of Puberty Modulation by Primer Pheromones
JOHN G. VANDENBERGH AND DAVID M. COPPOLA

Relationships between Social Organization and Behavioral Endocrinology in a Monogamous Mammal
C. SUE CARTER, LOWELL L. GETZ, AND MARTHA COHEN-PARSONS

Lateralization of Learning Chicks
L. J. ROGERS

Circannual Rhythms in the Control of Avian Migrations
EBERHARD GWINNER

The Economics of Fleeing from Predators
R. C. YDENBERG AND L. M. DILL

Social Ecology and Behavior of Coyotes
MARC BEKOFF AND MICHAEL C. WELLS

**Volume 17**

Receptive Competencies of Language-Trained Animals
LOUIS M. HERMAN

Self-Generated Experience and the Development of Lateralized Neurobehavioral Organization in Infants
GEORGE F. MICHEL

Behavioral Ecology: Theory into Practice
NEIL B. METCALFE AND PAT MONAGHAN

The Dwarf Mongoose: A Study of Behavior and Social Structure in Relation to Ecology in a Small, Social Carnivore
O. ANNE E. RASA

Ontogenetic Development of Behavior: The Cricket Visual World
RAYJOND CAMPAN, GUY BEUGNON, AND MICHEL LAMBIN

**Volume 18**

Song Learning in Zebra Finches (Taeniopygia guttata): Progress and Prospects
PETER J. B. SLATER, LUCY A. EALES, AND N. S. CLAYTON

Behavioral Aspects of Sperm Competition in Birds
T. R. BIRDHEAD

Neural Mechanisms of Perception and Motor Control in a Weakly Electric Fish
WALTER HEILIGENBERG

Behavioral Adaptations of Aquatic Life in Insects: An Example
ANN CLOAREC

The Cicadian Organization of Behavior: Timekeeping in the Tsetse Fly, A Model System
JOHN BRADY

**Volume 19**

Polyterritorial Polygyny in the Pied Flycatcher
P. V. ALATATO AND A. LUNDBERG

Kin Recognition: Problems, Prospects, and the Evolution of Discrimination Systems
C. J. BARNARD

Maternal Responsiveness in Humans: Emotional, Cognitive, and Biological Factors
CARL M. CORTER AND ALISON S. FLEMING

The Evolution of Courtship Behavior in Newts and Salamanders
T. R. HALLIDAY

Ethopharmacology: A Biological Approach to the Study of Drug-Induced Changes in Behavior
A. K. DIXON, H. U. FISCH, AND K. H. MCALLISTER

Additive and Interactive Effects of Genotype and Maternal Environment
PIERRE L. ROUBERTOUX, MARIKA NOSTEN-BERTRAND, AND MICHELE CARLIER

Mode Selection and Mode Switching in Foraging Animals
GENE S. HELFMAN

Cricket Neuroethology: Neuronal Basis of Intraspecific Acoustic Communication
FRANZ HUBER

Some Cognitive Capacities of an African Grey Parrot (*Psittacus erithacus*)
IRENE MAXINE PEPPERBERG

**Volume 20**

Social Behavior and Organization in the Macropodoidea
PETER J. JARMAN

The $t$ Complex: A Story of Genes, Behavior, and Population
SARAH LENINGTON

The Ergonomics of Worker Behavior in Social Hymenoptera
PAUL SCHMID-HEMPEL

"Microsmatic Humans" Revisited: The Generation and Perception of Chemical Signals
BENOIST SCHAAL AND RICHARD H. PORTER

Lekking in Birds and Mammals: Behavioral and Evolutionary Issues
R. HAVEN WILEY

**Volume 21**

Primate Social Relationships: Their Determinants and Consequences
ERIC B. KEVERNE

The Role of Parasites in Sexual Selection: Current Evidence and Future Directions
MARLENE ZUK

Conceptual Issues in Cognitive Ethology
COLIN BEER

Responses in Warning Coloration in Avian Predators
W. SCHULER AND T. J. ROPER

Analysis and Interpretation of Orb Spider Exploration and Web-Building Behavior
FRITZ VOLLRATH

Motor Aspects of Masculine Sexual Behavior in Rats and Rabbits
GABRIELA MORALÍ AND CARLOS BEYER

On the Nature and Evolution of Imitation in the Animal Kingdom: Reappraisal of a Century of Research
A. WHITEN AND R. HAM

**Volume 22**

Male Aggression and Sexual Coercion of Females in Nonhuman Primates and Other Mammals: Evidence and Theoretical Implications
BARBARA B. SMUTS AND ROBERT W. SMUTS

Parasites and the Evolution of Host Social Behavior
ANDERS PAPE MØLLER, REIJA DUFVA, AND KLAS ALLANDER

The Evolution of Behavioral Phenotypes: Lessons Learned from Divergent Spider Populations
SUSAN E. RIECHERT

Proximate and Developmental Aspects of Antipredator Behavior
E. CURIO

Newborn Lambs and Their Dams: The Interaction That Leads to Sucking
MARGARET A. VINCE

The Ontogeny of Social Displays: Form Development, Form Fixation, and Change in Context
T. G. GROOTHUIS

**Volume 23**

Sneakers, Satellites, and Helpers: Parasitic and Cooperative Behavior in Fish Reproduction
MICHAEL TABORSKY

Behavioral Ecology and Levels of Selection: Dissolving the Group Selection Controversy
LEE ALAN DUGATKIN AND HUDSON KERN REEVE

Genetic Correlations and the Control of Behavior, Exemplified by Aggressiveness in Sticklebacks
THEO C. M. BAKKER

Territorial Behavior: Testing the Assumptions
JUDY STAMPS

Communication Behavior and Sensory Mechanisms in Weakly Electric Fishes
BERND KRAMER

**Volume 24**

Is the Information Center Hypothesis a Flop?
HEINZ RICHNER AND PHILIPP HEEB

Maternal Contributions to Mammalian Reproductive Development and the Divergence of Males and Females
CELIA L. MOORE

Cultural Transmission in the Black Rat: Pine Cone Feeding
JOSEPH TERKEL

The Behavioral Diversity and Evolution of Guppy, *Poecilia reticulata,* Populations in Trinidad
A. E. MAGURRAN, B. H. SEGHERS, P. W. SHAW, AND G. R. CARVALHO

Sociality, Group Size, and Reproductive Suppression among Carnivores
SCOTT CREEL AND DAVID MACDONALD

Development and Relationships: A Dynamic Model of Communication
ALAN FOGEL

Why Do Females Mate with Multiple Males? The Sexually Selected Sperm Hypothesis
LAURENT KELLER AND HUDSON K. REEVE

Cognition in Cephalopods
JENNIFER A. MATHER

ISBN 0-12-004525-7